Bio-inorganic Hybrid Nanomaterials

Edited by
Eduardo Ruiz-Hitzky,
Katsuhiko Ariga and Yuri Lvov

Further Reading

Vollath, D.

Nanomaterials

An Introdution to Synthesis, Properties and Applications

2008

ISBN: 978-3-527-31531-4

Willner, I., Katz, E. (Eds.)

Bionanomaterials

Synthesis and Applications for Sensors, Electronics and Medicine

2008

ISBN: 978-3-527-31454-6

Ajayan, P. M., Schadler, L. S., Braun, P. V., Keblinski, P.

Nanocomposite Science and Technology

Second Completely Revised Edition

2008

ISBN: 978-3-527-31248-1

Rao, C. N. R., Müller, A., Cheetham, A. K. (Eds.)

Nanomaterials Chemistry

Recent Developments and New Directions

2007

ISBN: 978-3-527-31664-9

Kickelbick, G. (Ed.)

Hybrid Materials

Synthesis, Characterization, and Applications

2007

ISBN: 978-3-527-31299-3

Kumar, Challa S. S. R. (Ed.)

Nanotechnologies for the Life Sciences

10 Volume Set

2007

ISBN: 978-3-527-31301-3

Bio-inorganic Hybrid Nanomaterials

Strategies, Syntheses, Characterization and Applications

Edited by
Eduardo Ruiz-Hitzky, Katsuhiko Ariga and Yuri Lvov

WILEY-VCH
WILEY-VCH Verlag GmbH & Co. KGaA

The Editors

Prof. Dr. Eduardo Ruiz-Hitzky
Instituto de Ciencia de Materiales
de Madrid
Consejo Superior de Investigaciones Cientificas
Cantoblanco
28049 Madrid
Spain

Dr. Katsuhiko Ariga
National Institute for Material Science
1-1 Namiki
305-0044 Tsukuba, Ibaraki
Japan

Prof. Yuri Lvov
Institute for Micromanufacturing
Louisiana Technical University
911 Hergot Avenue
Ruston, LA 71272
USA

Library of Congress Card No.: applied for

British Library Cataloguing-in-Publication Data
A catalogue record for this book is available from the British Library.

Bibliographic information published by the Deutsche Nationalbibliothek
Die Deutsche Nationalbibliothek lists this publication in the Deutsche Nationalbibliografie; detailed bibliographic data are available on the Internet at <http://dnb.d-nb.de>.

Printed in the Federal Republic of Germany
Printed on acid-free paper

Cover Design WMX-Design, Heidelberg
Typesetting Thomson Digital, India
Printing Strauss GmbH, Mörlenbach
Binding Litges & Dopf GmbH, Heppenheim

ISBN: 978-3-527-31718-9

Contents

Bio-inorganic Hybrid Nanomaterials. Edited by E. Ruiz-Hitzky, K. Ariga, Y. M. Lvov

ISBN: 978-3-527-31718-9

Preface

Materials from living matter and inorganic materials are apparently on opposite sides in the materials' world. In this context, biological materials including polysaccharides, proteins, nucleic acids, and lipids, have soft and flexible natures, and often show incredible functions with high specificity and efficiency, which cannot be easily re-generated or replicated by combination of man-made materials. Therefore, direct use of such a class of biological materials sounds like a rational way to construct highly sophisticated functional systems.

The best way to cope with the high functionality and stability of bio-related materials for practical applications is to create hybrids consisting of materials of biological origin and inorganic materials. However, simply mixing these materials together into a messy slurry is not a wise strategy. In biologically derived materials both components have their own well-organized, meaningful nanostructures, and therefore, hybridization of inorganic and biological elements in controlled structures with nanometer-scale precision is the most desirable strategy. The obtained materials can be called "bio-inorganic nanohybrids." They are well-blessed children from both worlds, and should succeed in providing essences and advantages of both biological and inorganic materials. Research in bio-inorganic nanohybrids can be defined as an interdisciplinary field resulting from the interfaces between biotechnology, materials science, and nanotechnology. Such a new field is closely related to significant topics such as biomineralization processes, bioinspired materials and biomimetic systems. The incoming development of novel bio-nanocomposites introducing multifunctionality and taking profit from the characteristics of both types of constituents is nowadays an amazing research line taking advantage from the synergistic assembling of biopolymers with inorganic nanosized solids.

Mother of pearl and marine shells, corals, teeth, bones, and microbe inclusions (such as sulfur or iron nanocrystals) are examples of bio-inorganic nanocomposites. In many cases, these composites have biopolymers and inorganic parts organized on the nanoscale, such as the regular alternation of proteins and calcium carbonate nanolayers in nacre. When struck, the layers glide over one another absorbing the shock. If cracks develop, plates simply grow back together. Such natural nanocomposite materials often combine unique mechanical properties based on the nanoscale organization of hard and soft materials and have the ability for regeneration

Bio-inorganic Hybrid Nanomaterials. Edited by E. Ruiz-Hitzky, K. Ariga, Y. M. Lvov

ISBN: 978-3-527-31718-9

and self-reproduction. However, these nanocomposites have a fatal drawback in their application. In most cases, their functions are optimized only at ambient conditions and their structural stability is maintained in a limited environment. In contrast, inorganic materials usually have incredible stability and stiffness, even in extreme conditions. In addition, they sometimes offer us the opportunity to prepare precise structures by both top-down fabrications and bottom-up assembly. Of course, superior aspects in electronic, photonic, magnetic, and mechanical properties can be expected in many inorganic materials. Nevertheless, no one can believe that the highly sophisticated functions seen in living systems may be constructed by assembly and fabrication of materials of exclusively inorganic nature.

The focus of this "Bio-Inorganic Hybrid Nanomaterials" book is to cover the wide spectrum of recent developments in natural and artificial bio-hybrid materials, which is the result of the successful assembly of 15 chapters by world-wide experts in their corresponding fields. Fundamental aspects on the preparation of bio-inorganic nanohybrids using various nanostructures including mesoporous materials, nanoparticles, gels, organoclays, membranes, and nanotubules with advanced methodologies such as the sol–gel process, self-assembly, intercalation, template synthesis and layer-by-layer adsorption upon the concept of supramolecular chemistry, biomimetics, and biomineralization are thoroughly described. Not limited to basic sciences, several chapters introduce practical applications of bio-inorganic nanohybrids, as exemplified in biodegradation, bone tissue re-generation, controlled delivery, and enzymatic activity. Readers can enjoy independent chapters and also feel the good harmony of the balanced assembly of the chapters. We hope that every reader can find potential possibilities of bio-inorganic nanohybrids with certain wonder, surprise, and impression, just after closing these pages.

Eduardo Ruiz-Hitzky
Katsuhiko Ariga
Yuri Lvov

Contributors

Mitsuru Akashi
Osaka University
Department of Applied Chemistry
Graduate School of Engineering
2-1, Yamada-oka,
and
The 21st Century COE Program
for "Center for Integrated Cell and
Tissue Regulation"
2-2, Yamada-oka, Suita
Osaka 565-0871
Japan

Pilar Aranda
Instituto de Ciencia de Materiales
de Madrid
Consejo Superior de Investigaciones
Científicas (CSIC)
Cantoblanco
28049-Madrid
Spain

Daniel Arcos
Universidad Complutense
Departamento de Quimica Inorgánica
y Bioinorganica
Facultad de Farmacia
Plaza Ramón y Cajal s/n
28040 Madrid
Spain

Katsuhiko Ariga
National Institute for Materials
Science (NIMS)
Supermolecules Group
1-1 Namiki
Tsukuba 305-0044
Japan

Mihail Barboiu
Institut Européen des Membranes
Adaptative Supramolecular
Nanosystems Group
UMR CNRS 5635
Place Eugène Bataillon CC 047
34095 Montpellier
France

Roberta Brayner
Université Pierre et Marie Curie
Université Denis Diderot
Interfaces, Traitements, Organisation
et Dynamique des Systèmes (ITODYS)
UMR-CNRS 7086
2 Place Jussieu
75252 Paris Cedex 05
France

Frank Caruso
The University of Melbourne
Centre for Nanoscience and Nanotechnology
Department of Chemical and Biomolecular Engineering
Victoria 3010
Australia

Soo-Jin Choi
Ewha Womans University
Center for Intelligent Nano-Bio Materials
Division of Nanoscience and Department of Chemistry
Seoul 120-750
Korea

Jin-Ho Choy
Ewha Womans University
Center for Intelligent Nano-Bio Materials
Division of Nanoscience and Department of Chemistry
Seoul 120-750
Korea

Thibaud Coradin
Université Pierre et Marie Curie
Chimie de la Matière Coudensée de Paris (CMCP)
UMR-CNRS 7574
4 Place Jussieu
75252 Paris Cedex 05
France

Margarita Darder
Instituto de Ciencia de Materiales de Madrid
Consejo Superior de Investigaciones Científicas (CSIC)
Cantoblanco
28049-Madrid
Spain

Fernand Fiévet
Université Denis Diderot
Interfaces, Traitements, Organisation et Dynamique des Systemès (ITODYS)
UMR-CNRS 7086
2 Place Jussieu
75251 Paris Cedex 05
France

Claude Forano
Université Blaise Pascal
Laboratoire des Matériaux Inorganiques
CNRS UMR 6002
63177 Aubiere Cedex
France

Kazuko Fujii
National Institute for Materials Science (NIMS)
Special Subjects Group
1-1 Namiki
Tsukuba 305-0044
Japan

Narasimhan Gokulakrishnan
National Institute for Materials Science (NIMS)
Supermolecules Group
1-1 Namiki
Tsukuba 305-0044
Japan

Jonathan P. Hill
National Institute for Materials Science (NIMS)
Supermolecules Group
1-1 Namiki
Tsukuba 305-0044
Japan

Masanobu Kamitakahara
Tohoku University
Graduate School of Environmental Studies
6-6-20, Aoba
Aramaki, Aoba-ku
Sendai 980-8579
Japan

Jacques Livage
Université Pierre et Marie Curie
Chimie de la Matière Condensée de Paris (CMCP)
UMR-CNRS 7574
4 Place Jussieu
75252 Paris Cedex 05
France

Yuri M. Lvov
Louisiana Tech University
Institute for Micromanufacturing
911 Hergot Ave.
Ruston, LA 71272
USA

Stephen Mann
University of Bristol
Centre for Organized Matter Chemistry
School of Chemistry
Bristol, BS8 1TS
United Kingdom

Toshiki Miyazaki
Kyushu Institute of Technology
Graduate School of Life Science and Systems Engineering
2-4, Hibikino
Wakamatsu-ku, Kitakyushu-shi
Fukuoka 808-0196
Japan

Toshiyuki Mori
National Institute for Materials Science (NIMS)
Nano Ionics Materials Group
Fuel Cell Materials Center
1-1 Namiki
Tsukuba 305-0044
Japan

Jae-Min Oh
Ewha Womans University
Center for Intelligent Nano-Bio Materials
Division of Nanoscience and Department of Chemistry
Seoul 120-750
Korea

Chikara Ohtsuki
Nagoya University
Department of Crystalline Materials Science
Graduate School of Engineering
Furo-cho, Chikusa-ku
Nagoya 464-8603
Japan

Masami Okamoto
Toyota Technological Institute
Advanced Polymeric Nanostructured Materials Engineering
Graduate School of Engineering
2-12-1 Hisakata, Tempaku
Nagoya 468 8511
Japan

Avinash J. Patil
University of Bristol
Centre for Organized Matter
Chemistry
School of Chemistry
Bristol, BS8 1TS
United Kingdom

Vanessa Prévot
Université Blaise Pascal
Laboratoire des Matériaux
Inorganiques
CNRS UMR 6002
63177 Aubiere Cedex
France

Ronald R. Price
Atlas Mining Corporation
Nanoclay and Technology Division
1200 Silver City Road
Eureka, UT 84628
USA

Eduardo Ruiz-Hitzky
Instituto de Ciencia de Materiales
de Madrid
Consejo Superior de Investigaciones
Científicas (CSIC)
Cantoblanco
28049-Madrid
Spain

Yury A. Shchipunov
Russian Academy of Sciences
Institute of Chemistry
Far East Department
690022 Vladivostok
Russia

María Vallet-Regí
Universidad Complutense
Facultad de Farmacia
Departamento de Quimica Inorgánica
y Bioinorganica
Plaza Ramón y Cajal s/n
28040 Madrid
Spain

Ajayan Vinu
National Institute for Materials Science
(NIMS)
Nano Ionics Materials Group
Fuel Cell Materials Center
1-1 Namiki
Tsukuba 305-0044
Japan

Yajun Wang
The University of Melbourne
Centre for Nanoscience and
Nanotechnology
Department of Chemical and
Biomolecular Engineering
Victoria 3010
Australia

Junji Watanabe
Osaka University
Department of Applied Chemistry
Graduate School of Engineering
2-1, Yamada-oka,
and
The 21st Century COE Program
for "Center for Integrated Cell
and Tissue Regulation"
2-2, Yamada-oka, Suita
Osaka 565-0871
Japan

1
An Introduction to Bio-nanohybrid Materials

Eduardo Ruiz-Hitzky, Margarita Darder, Pilar Aranda

1.1
Introduction: The Assembly of Biological Species to Inorganic Solids

The assembly of molecular or polymeric species of biological origin and inorganic substrates through interactions on the nanometric scale constitutes the basis for the preparation of bio-nanohybrid materials (Figure 1.1). The development of these materials represents an emerging and interdisciplinary topic at the border of Life Sciences, Material Sciences and Nanotechnology. They are of great interest due to their versatile applications in important areas as diverse as regenerative medicine and new materials with improved functional and structural properties [1–5]. It must be remarked that the development of bio-nanocomposites also represents an ecological alternative to conventional polymer nanocomposites, as the properties of the biodegradable polymers used ensure that the materials produced are environmentally friendly and renewable. Typical examples of this type of bio-nanocomposites result from the combination of polysaccharides such as starch, cellulose or polylactic acid (PLA) with microparticulated solids, which are usually called *green nanocomposites* or *bioplastics* [6,7].

Recently, special attention has been paid to strategies for synthetic approaches to bio-nanohybrids. One of these approaches is related to the preparation of bioinspired or biomimetic materials following the examples found in Nature, as for instance, bone [8], ivory [9] and nacre [10–13]. These materials show excellent structural properties due to the special arrangement at the nanometric level of their assembled components, that is biopolymers and inorganic counterparts. For instance, nacre represents a good example of a natural bio-nanocomposite, also known as *native biomineral,* formed by the stacking of highly oriented calcium carbonate (aragonite) platelets cemented by a fibrous protein (lustrin A). The resulting supra-architectures show exceptional mechanical properties compared to monolithic calcium carbonate [11,12].

Nowadays, bio-nanocomposites mimicking these natural materials have been prepared with the aim being to develop new biohybrids with improved mechanical properties together with biocompatibility and, in some cases, other interesting

Bio-inorganic Hybrid Nanomaterials. Edited by E. Ruiz-Hitzky, K. Ariga, Y. M. Lvov

ISBN: 978-3-527-31718-9

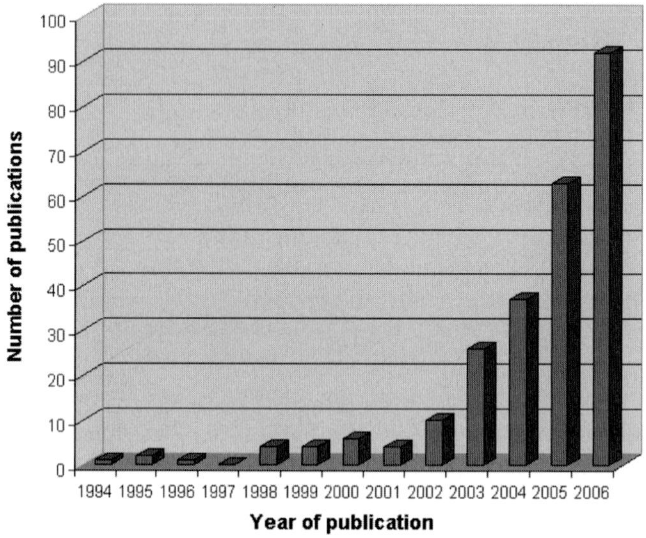

Fig. 1.1 Number of publications per year related to bio-nanohybrid materials. Data collected from the ISI Web of Knowledge [v3.0]-Web of Science. Keywords for search: (biopolymer* AND nanocomposite*) OR (natural polymer* AND nanocomposite*) OR (bio-nanohybrid*) OR (biohybrid* AND nano*).

features such as functional behavior [14–18]. In this context, the development of biohybrid systems based on biomimetic building [18,19], that is following biomineralization processes similar to those that take place in the cell wall of diatoms, where nanostructured silica nanospheres are assembled by the participation of cationic polypeptides called silaffins (SILica AFFINity) [20,21] appears to be of great significance. In relation to these natural systems, the mechanisms of interaction between colloidal silica and peptides have been particularly studied with the aim being to understand the biomineralization processes and consequently to develop, in a controlled manner, new improved synthetic bio-nanocomposites [22–24].

Inorganic solids assembled with biological species are of diverse nature with different chemical compositions, structures and textures, which determine the properties of the resulting bio-nanohybrids. In this way, single elements such as transition metals and carbon particles, metal oxides and hydroxides, silica, silicates, carbonates and phosphates, are typical inorganic components of bio-nanohybrids (Table 1.1). The affinity between the inorganic and the bio-organic counterparts, which determines the stability of the resulting bio-composites, depends on the interaction mechanisms governing the assembly processes.

As indicated above, the development of bio-nanohybrids by mimicking biomineralization represents an extraordinarily useful approach. This is, for instance, the case for those bio-nanocomposites based on bone biomimetic approaches, which show excellent structural properties and biocompatibility. They are prepared by

Tab. 1.1 Selected Examples of Bio-Nanohybrid Materials Involving Different Types of Inorganic Solids.

Inorganic moiety	Biological species	Bio-nanohybrid features	Authors/References
silica nanoparticles	poly-L-lysine (PLL)	biomimetic nanocomposites with controlled morphology	Patwardhan *et al.* [39]
siloxane networks	living bacteria	encapsulation by sol-gel	Fennouh *et al.* [62]
calcium carbonate	chitosan and poly(aspartate)	biomimetic preparation towards artificial nacre	Sugawara and Kato [88]
hydroxyapatite (HAP)	collagen	biomimetic porous scaffolds for bone regeneration	Yokoyama *et al.* [94]
layered clay minerals (montmorillonite)	chitosan	functional bio-nanocomposite for ion-sensing applications	Darder *et al.* [129]
fibrous clay minerals (sepiolite)	caramel	bio-nanocomposite as precursor of multifunctional carbon–clay nanostructured materials	Gómez-Avilés *et al.* [153]
organoclays	PLA	*green nanocomposites* as biodegradable bioplastics	Paul *et al.* [144]
layered double hydroxides (LDHs)	deoxyribonucleic acid (DNA)	bio-nanocomposite as non-viral vector for gene transfection	Choy *et al.* [159]
gold nanoparticles	chitosan	bio-nanohybrid processable as self-supporting films for biosensor applications	dos Santos *et al.* [164]
magnetite nanoparticles	phosphatidylcholine	*magnetocerasomes* for targeted drug delivery	Burgos-Asperilla *et al.* [73]
carbon nanotubes (CNTs)	galactose	modified CNTs able to capture pathogens by protein binding	Gu *et al.* [194]
layered perovskites ($CsCa_2Nb_3O_{10}$)	gelatin	bio-nanocomposite thin films with dielectric properties	Ruiz *et al.* [220]

assembling hydroxyapatite (HAP), which is the main mineral constituent of bones and teeth, with biopolymers, for example collagen [25–27]. The coating of the micro- or nano-particulated solids with biopolymers often occurs through hydrogen-bonding or metal-complexing mechanisms. In this way, the assembly of magnetic iron oxide nanoparticles (e.g., magnetite) with biopolymers (e.g., dextran) allows the preparation of magnetic bio-nanocomposites applied in NMR imaging, hyperthermia treatments or bio-carriers as drug delivery systems (DDS) [28,29].

The assembly of biopolymers with inorganic layered solids can lead to bio-nanocomposites in which the biopolymer becomes intercalated between the layers of the inorganic hosts [3]. The intercalation is a complex process that may simultaneously involve several mechanisms. Thus, in addition to hydrogen bonding, it has been invoked that certain biopolymers interact with the inorganic layers through ionic bonds. This is the case for polysaccharides, proteins and nucleic acids that can act as polyelectrolytes intercalating, via ion-exchange reaction, solids provided with positively or negatively charged layers, such as layered double hydroxides (LDHs) or smectite clay minerals (see below).

Microfibrous crystalline silicates such as sepiolite, similarly to amorphous silica, contain silanol groups (Si—OH) covering the external mineral surface. These groups can be effectively involved in hydrogen bonding by their association to OH, NH and other polar groups belonging to the biopolymers used. Silica generated by the sol–gel method from tetraethyl orthosilicate (TEOS) in the presence of chitosan, gives biopolymer-silica nanocomposites whose morphology can be determined by the experimental conditions adopted for the preparation [30]. Chitosan and collagen can also be assembled with sepiolite to give the corresponding biopolymer-sepiolite nanocomposites, which exhibit good mechanical properties resulting from the combination of the fibrous inorganic substrate with the biopolymer [31–34]. The interaction mechanisms governing the formation of sepiolite-based bio-nanocomposites are mainly ascribed to hydrogen bonding, but it must be taken into account that sepiolite exhibits cationic exchange capacity (CEC ~15 meq/100 g). Thus this silicate could also interact with positively charged polymers, such as chitosan, through electrostatic bonds.

Although to only a minor extent, other mechanisms can be invoked, as for instance covalent bonding (grafting) between hydroxy groups on the surface of the inorganic substrates and functional groups of the biopolymers [35].

The aim of this chapter is to provide a general overview of the preparation and main characteristics of bio-nanohybrids, with emphasis on the different types of inorganic solids that can be involved in the formation of this class of materials. Special attention will be devoted to the diverse mechanisms that govern the interaction between the components of biohybrids, illustrating them with selected examples. Relevant features and potential or actual applications of recently developed bio-nanocomposites will be discussed on the basis of their structure–property relationships.

1.2 Bio-nanohybrids Based on Silica Particles and Siloxane Networks

Biominerals are produced by living organisms following a set of processes known as biomineralization, which results in a wide variety of biological materials including shells, bones, teeth, ivory and magnetic nanoparticles in magnetotactic bacteria. Biomolecules secreted by living organisms control the nucleation and growth of inorganic minerals (carbonates, phosphates, silica and iron oxide) leading to such a diversity of biological-inorganic hybrid materials, which usually exhibit a hierarchical arrangement of their components from the nanoscale to the macroscopic scale. The skeletons of diatoms and radiolarians are astonishing examples of biosilicification giving rise to amorphous hydrated SiO_2 (biosilica), also formed in sponges and many higher plants [24]. As mentioned in Section 1, polycationic peptides, called silaffins, are involved in this process, controlling the assembly of silica nanoparticles to form

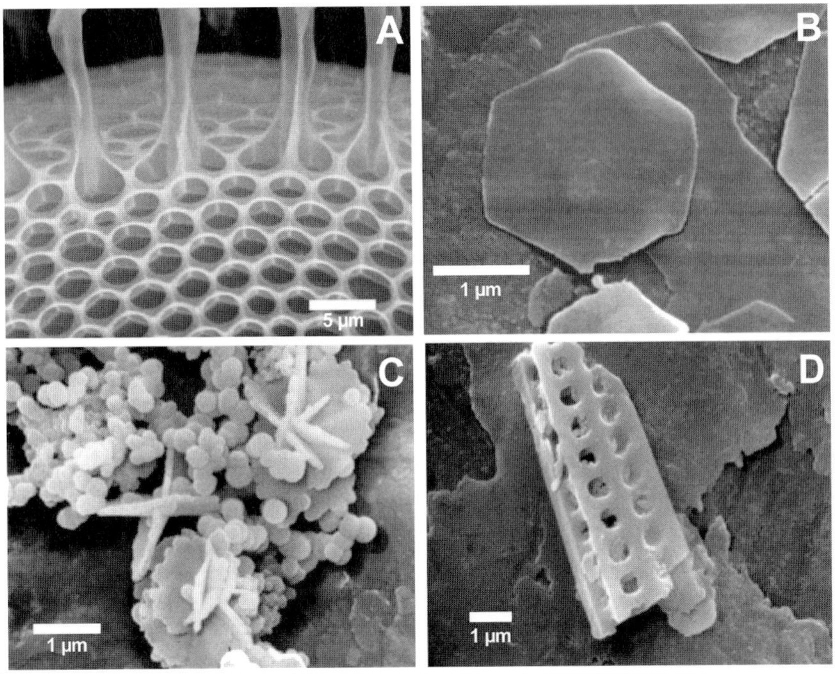

Fig. 1.2 Scanning electron micrographs of (A) the silica wall of the diatom *Stephanopyxis turris* (reproduced from [21] by permission of Wiley-VCH) and (B–D) singular morphologies of silica synthesized using poly-L-lysine and pre-hydrolyzed tetramethyl orthosilicate (TMOS) under different experimental conditions: (B) unperturbed solution, (C) flowed through a 1/8″ I.D. tube and (D) stirred for 25 min. Reproduced from [39] by permission of The Royal Society of Chemistry.

these siliceous structures [20,21,36]. Similarly, silica needles in the skeleton of marine sponges involve a central filament containing silicatein, an enzyme that catalyses the synthesis of biosilica [37]. Materials scientists try to understand and reproduce these biosilicification processes taking place in nature, with the aim being to develop bioinspired or biomimetic hybrid nanostructured materials with controlled morphologies and structural properties similar to those of biosilica [24,38–42], as shown in Figure 1.2. As recently reviewed by Coradin *et al.* [43], proteins (collagen, gelatin, and silk) and polysaccharides (alginate, carrageenans, chitosan, as well as cellulose and its derivatives) are the main biomacromolecules involved in the synthesis of biopolymer/silica nanocomposites, while silicic acid, sodium silicate and different silicon alkoxides are employed as precursors of the silica or the polysiloxane networks assembled with the biopolymer chains. Following biomimetic processes, lysozyme and bovine serum albumin (BSA) promote the precipitation of silica particles from sodium silicate solutions, leading to entrapment of the protein [44]. Similarly, polysaccharides such as cationic and hydrophobic derivatives of cellulose also promote silica precipitation, acting as efficient templates to develop organic–inorganic hybrid nanocomposites in combination with tetrakis(2-hydroxyethyl)orthosilicate (THEOS) [45]. Chitosan is another natural polysaccharide involved in this type of silica-based hybrid material prepared by the sol–gel method. For instance, it has been assembled with siloxane networks derived from aminopropylsiloxane (APS) [46] or TEOS [30]. A similar chitosan–polysiloxane biohybrid material has been recently prepared from chitosan and 3-isocyanatopropyltriethoxysilane, where chitosan is bound to the polysiloxane network by covalent bridges. This new functional material offers photoluminiscent features and bioactive behavior, since it promotes apatite formation in simulated body fluid [47]. Chitosan–silica hybrids present as microparticulate materials showing different shapes have been prepared by the sol–gel method using TEOS or polyethoxysiloxane oligomers in the presence of the biopolymer. These materials can be used as a stationary phase in HPLC [48].

In these examples as well as in analogous materials, the interaction of the biological and the inorganic components has synergetic effects leading to hybrid materials with improved mechanical resistance, higher thermal and chemical stability and biocompatibility, and, in some cases, with functional properties. Biopolymer/silica nanocomposites are suitable for the design of membranes and coatings, drug delivery systems and also for the encapsulation of bioactive molecules such as enzymes, antibodies, yeast and plant cells or even bacteria, resulting in functional biomaterials for different biotechnological applications, including biosensors and bioreactors [43,49].

Silica-based bio-nanocomposites for drug delivery purposes have been processed as nanospheres by means of spray-drying or CO_2 supercritical drying techniques. Hybrid nanoparticles based on algal polysaccharides such as alginate and carrageenan are potential carriers for the targeted delivery of drugs due to their ability to go into the intracellular space of cells and to their lack of cytotoxicity [50,51]. In other cases, silica nanoparticles serve as a support of biocide molecules and their dispersion in hydroxypropylcellulose allows the preparation of coatings and films with fungicide and pesticide activity [52]. Following a similar approach, Zhang and Dong [53] have developed functional materials based on the dispersion of $Ru(bpy)_3^{2+}$-doped silica nanoparticles in the biopolymer chitosan. The resulting hybrid material can be easily spread onto the surface of electrodes as a stable electroactive coating, allowing the development of chemiluminiscence sensors.

Similarly to the above-mentioned entrapment of proteins by biomimetic routes, the sol–gel procedure is a useful method for the encapsulation of enzymes and other biological material due to the mild conditions required for the preparation of the silica networks [54,55]. The confinement of the enzyme in the pores of the silica matrix preserves its catalytic activity, since it prevents irreversible structural deformations in the biomolecule. The silica matrix may exert a protective effect against enzyme denaturation even under harsh conditions, as recently reported by Frenkel-Mullerad and Avnir [56] for physically trapped phosphatase enzymes within silica matrices (Figure 1.3). A wide number of organoalkoxy- and alkoxy-silanes have been employed for this purpose, as extensively reviewed by Gill and Ballesteros [57], and the resulting materials have been applied in the construction of optical and electrochemical biosensor devices. Optimization of the sol–gel process is required to prevent denaturation of encapsulated enzymes. Alcohol released during the

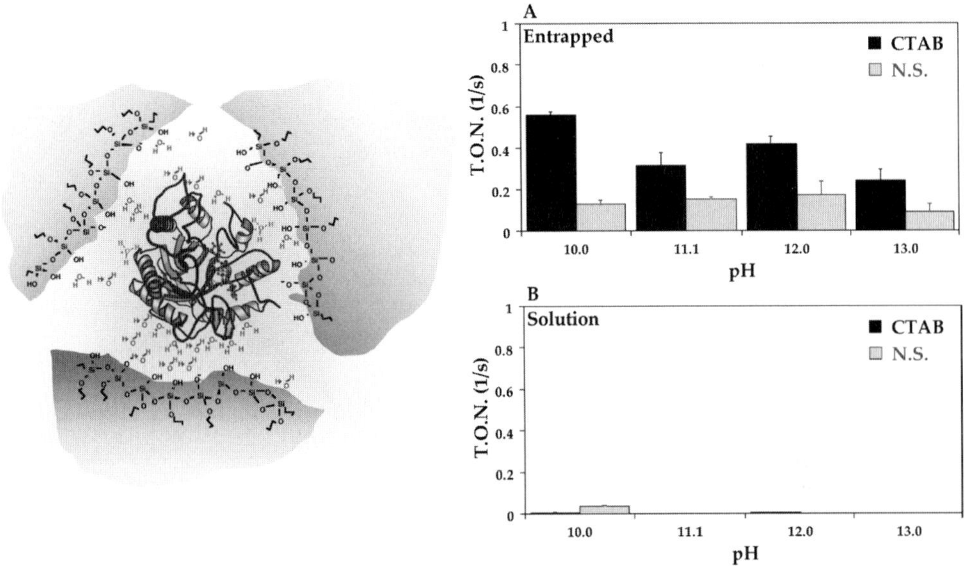

Fig. 1.3 Schematic representation of the entrapped enzyme in a silica matrix (left side). Enzymatic activity, under extreme alkaline conditions, of acid phosphatase (A) immobilized in silica sol–gel matrices with or without CTAB, or (B) in solution. Reprinted with permission from [56]. Copyright 2005, American Chemical Society.

hydrolysis process can be harmful for the entrapped biologicals and, thus, several methods propose its removal by evaporation under vacuum [58] or the use of polyol-based silanes that generate biocompatible alcohols [59]. Catalytic activity is also preserved when silica-polysaccharide bio-nanocomposites are used as immobilization hosts. This is the case for three-dimensional hybrid matrices resulting from the combination of THEOS with xanthan, locust bean gum or a cationic derivative of hydroxyethylcellulose, which have been reported as excellent networks for the long-term immobilization of $1 \rightarrow 3$-β-D-glucanase and α-D-galactosidase [60].

In addition to enzymes and antibodies, silica-based hybrid nanocomposites with a suitable porosity can successfully entrap more complex systems including yeasts, algae, lichens, plant cells and bacteria [49]. The huge volume of biological tissues, in comparison to enzymes, may hinder the polymerization processes resulting in fractures in the silica matrices. To overcome this drawback, lichen particles were embedded in a flexible network, derived from 3-(trimethoxysilyl)propyl methacrylate (MAPTS) and tetramethoxysilane (TMOS), that offers improved mechanical features (Figure 1.4A). This lichen-modified material was used to develop electrochemical sensors for the determination of heavy metal ions by anodic stripping voltammetry [61]. Similarly, algal tissue can be immobilized in sol–gel derived matrices based on TMOS and methyltrimethoxysilane (MTMOS) (Figure 1.4B).

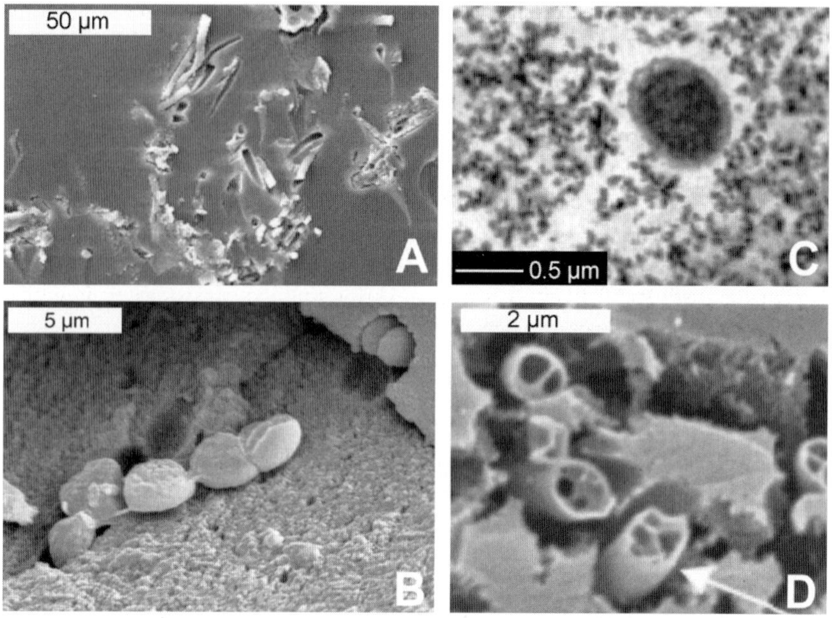

Fig. 1.4 Scanning electron micrographs of (A) the lichen *Pseudocyphellaria hirsuta*, (B) the alga *Anabaena*, and (D) the bacteria *E. coli* entrapped in sol–gel generated organopolysiloxane matrices (reprinted with permission from [65]. Copyright 2006, American Chemical Society). (C) Transmission electron micrograph of the same bacteria, *E. coli*, embedded in a silica matrix containing 10% glycerol (reproduced from [63] by permission of The Royal Society of Chemistry).

One of the first works reporting the entrapment of *E. coli* proposed its incorporation in a TMOS-derived silica network in which the water content was kept at about 70 wt% in order to guarantee the cells' viability, but when silica gel was dried the bacterial activity decreased [62]. In order to overcome this drawback, the authors explored other possibilities such as the incorporation of glycerol in the silica matrix to increase bacteria viability (Figure 1.4C), leading to almost 50% of viable bacteria after one month of ageing [63], or the addition of "quorum sensing" molecules involved in intercellular communication, which increase the cells' viability to 100% after one month [64]. Similar results have been achieved recently by Ferrer *et al.* [65], who showed that gluconolactone-bearing organopolysiloxane matrices are more efficient than pure silica in extending *E. coli* the cells' viability due to their increased biocompatibility (Figure 1.4D).

New materials that mimic liposomes have been recently reported as a new family of organic–inorganic hybrid compounds generated by a coupled process of sol–gel and self-assembly of long-chain containing organoalkoxysilanes [18,66–71]. These nanohybrids essentially refer to biomimetic materials derived from the assembly of a surfactant covalently bonded to a silica-based network. The name "cerasomes" was introduced by Ariga and coworkers [66] combining the terms "liposome" and "ceramic," this last making reference to the silica network. As "ceramic" is derived from the Greek word κεραμικόζ (*keramikos*) making reference to "inorganic non-metallic materials whose formation is due to the action of heat" [72], the term cerasome can be confusing as they are usually formed in soft conditions. Ruiz-Hitzky suggests the use of HOILs (Hybrid Organic–Inorganic Liposomes) for this class of compounds [18]. Anyway, "cerasome" is actually the most popular term for these hybrid materials. Interestingly, the bilayers formed by the surfactant tails are able to incorporate different organophyllic species [71] making these materials potentially applicable as Drug Delivery Systems (DDS). More recently, these types of bilayers have been grafted onto magnetic nanoparticles giving rise to the so-called "magnetocerasomes" [73], which are nanohybrids simultaneously having lipophilic character and magnetic properties (see below, Section 5).

1.3 Calcium Phosphates and Carbonates in Bioinspired and Biomimetic Materials

As pointed out in Section 2, a wide number of biominerals are synthesized in nature by living organisms using organic templates. Some well-known examples include bone and ivory, where the collagen matrix controls the growth of hydroxyapatite (HAP) mineral [8,9,74], or nacre in pearls and shells, showing a brick-like structure of aragonite layers cemented by proteins [11,12]. This assortment of biological–inorganic hybrid materials, showing a hierarchical arrangement from the nano- to the macroscale, serves as a model for the development of new biomimetic and bioinspired materials. *In vitro* studies have demonstrated the controlled nucleation and growth of carbonates and phosphates by soluble proteins and peptides combined with insoluble polysaccharide matrices (cellulose, chitin, collagen), leading to biomimetic materials that reproduce the exceptional features of native biominerals [1,75].

Besides nacre in pearls and shells, calcium carbonate is also present in sea urchin spines, coral skeleton, eggshell, and the exoskeleton of arthropods, forming organic–inorganic hybrid structures by assembly with biomacromolecules (soluble proteins and insoluble matrices) [76–78]. Calcite and aragonite are the calcium carbonate polymorphs that constitute the biominerals found in nature, since they show a higher stability than vaterite. However, *in vitro* studies have confirmed that the presence of functionalised macromolecules as soluble proteins and insoluble matrices have a considerable effect on calcium carbonate crystallization, allowing the formation of the less stable polymorphs and even of amorphous $CaCO_3$ [79,80]. Regarding soluble matrices, living organisms secrete biomacromolecules with a high content of glutamic and aspartic acids, bearing carboxyl groups that can interact with calcium ions. A similar effect has been found using polymers provided with sulfonic, hydroxy and even ether groups. Many of these studies have been carried out using the same biopolymers that act as insoluble matrices for $CaCO_3$ crystallization in nature, such as collagen and chitin [79,81]. Calcium carbonate polymorphs are also formed on other natural and synthetic polymers including elastin that controls the formation of calcite [82], poly(ethylene glycol) that forces the selective formation of aragonite [83], poly(α-L-aspartate) that promotes vaterite formation with a helical morphology [84], or

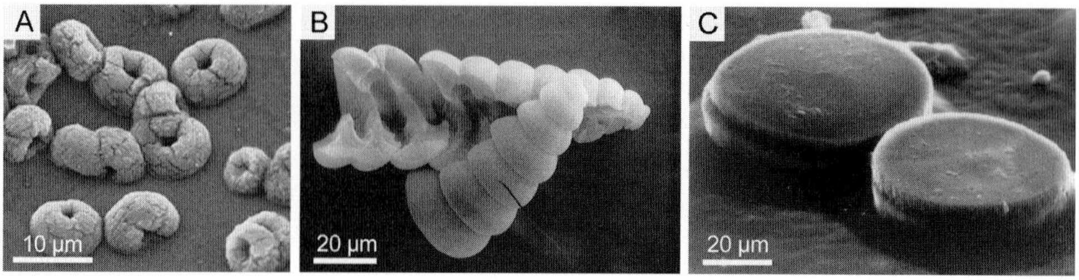

Fig. 1.5 SEM micrographs of (A) donut-shaped $CaCO_3$ crystals grown on polyacrylic acid grafted chitosan, (B) $CaCO_3$ hollow helix, fractured by micro-manipulation, formed on poly(α-L-aspartate), and (C) double layered aragonite thin films grown on a chitosan matrix in the presence of poly(aspartate) and $MgCl_2$ by alternate spin coating and crystallization. (A) and (B) adapted from [84] and [85] with permission from Elsevier, and (C) from [88] (reproduced by permission of The Royal Society of Chemistry).

poly(acrylate)-grafted chitosan giving rise to $CaCO_3$ crystals of unusual morphology [85] similar to those created using poly(*N*-isopropylacrylamide-*co*-(4-vinylpyridine)) as the platform for mineralization [86]. As confirmed by these reports, materials scientists are able to produce calcium carbonate organic hybrid materials with defined morphologies (Figure 1.5) by tuning the polymers and biomacromolecules that control the nucleation and growth of calcium carbonate crystals. Many studies are currently centered on the crystallization of $CaCO_3$ as thin films, as illustrated in Figure 1.5C, trying to mimic the nacre of shells [87–89]. The reason is that mollusk shells, where the organic matrix constitutes only 1 % of the total weight, present a fracture toughness about 3000 times higher than that of pristine calcium carbonate and offer a good example of ultra-lightweight hybrid materials with exceptional mechanical strength and interesting optical properties.

Calcium phosphate minerals are present in living organisms as the most important constituents of biological hard tissues (bones, teeth, tendons, and cartilage) to provide them with stability, hardness and function [74,90]. Among the different calcium phosphates, hydroxyapatite (HAP), with a chemical composition $Ca_{10}(PO_4)_6(OH)_2$ and Ca/P ratio of 1.67, is the most widely studied due to its huge incidence in the field of regenerative medicine. Bone can be regarded as a natural nanocomposite containing HAP nanocrystallites in a collagen-rich matrix also enclosing non-collagenous proteins. Besides providing structural support, bone serves as a reservoir of calcium and phosphate ions involved in numerous metabolic functions. Due to the significant role of bone in humans, most of the synthetic hybrid materials that mimic its structure and composition are currently devoted to biomedical applications for regeneration of injured bone and this fact has led to a wide number of scientific publications on this topic in the recent years.

Among synthetic materials for bone grafting, nanocomposites are replacing metals, alloys, ceramics, polymers, and composites, due to their advantageous properties: large surface area, high surface reactivity, relatively strong interfacial bonding, flexibility, and enhanced mechanical consistency. It has been proved that nanocrystalline HAP offers better results than microcrystalline HAP with respect to osteoblast cells adhesion, differentiation and proliferation, as well as biomineralization [74,91]. In addition to suitable mechanical properties, the requirements for these implants employed as a bioresorbable scaffold for bone regeneration are biocompatibility, suitable porosity with interconnected pores, as those observed in natural bone, to allow for the transport of nutrients and metabolic wastes, and controlled biodegradability, serving as a temporary scaffold for generation of new tissue [92,93].

Most of the bio-nanocomposites tested as implants for bone regeneration are based on the assembly of HAP nanoparticles with collagen, trying to reproduce the composition, biocompatibility and suitable mechanical properties of natural bone. The aim is to develop implants that mimic the nanostructuration, porosity and surface roughness of bone in order to facilitate the spreading of osteoblasts required for bone regeneration (Figure 1.6). Nanocrystalline HAP–collagen nanocomposites synthesized by biomimetic routes show a structure and composition very close to those of natural bone, providing better results in bone regeneration than nanocomposites prepared by conventional methods such as blending or mixing [94,95]. Murugan and Ramakrishna [74] have also reviewed the use of tissue-engineered nanocomposites, consisting in the isolation of specific cells by biopsy and their growth on a suitable nanocomposite scaffold, which is subsequently transplanted into the damaged bone site to promote its regeneration.

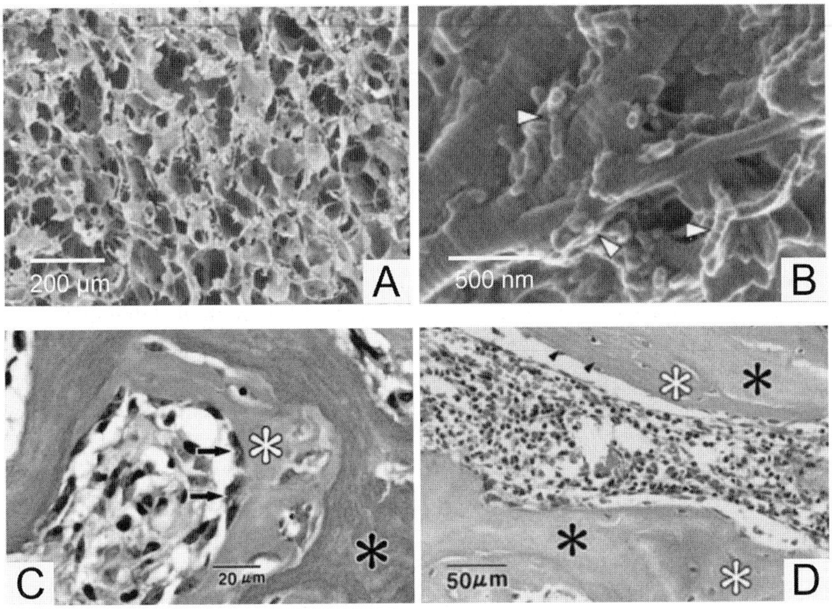

Fig. 1.6 (A and B) Scanning electron micrographs of the porous hydroxyapatite–collagen nanocomposite scaffolds at different magnifications. Arrowheads in B indicate the hydroxyapatite nanocrystals on the collagen fibrils. Histology at (C) 1 week and (D) 4 weeks after implantation in the bone marrow showing formation of new bone (white asterisk) attached directly to the nanocomposite (asterisk). Arrows indicate cuboidal osteoblasts on the surface of new bone. Adapted from [94], reproduced by permission of Wiley-VCH.

Besides the previously mentioned collagen, a wide variety of natural polymers have been involved in the synthesis of bio-nanohybrid materials with potential application in bone repair and dental prostheses. For instance, some recent examples refer to bio-nanocomposites based on the combination of HAP with alginate [96,97], chitosan [98,99], bovine serum albumin (BSA) [100], sodium caseinate [101], hyaluronic acid [102], silk fibroin [103,104], silk sericin [105], or polylactic acid (PLA) [106,107]. These examples illustrate the increasing interest in the subject of HAP-based biohybrid materials, which has led to almost 400 articles appeared in scientific journals in 2006 alone.

Given that macroporosity is an important requirement for materials used as implants in regenerative medicine, many of the developed HAP-based materials try to mimic not only the composition but also the porous structure of true bone. For this reason, different techniques such as fiber bonding, phase separation, solvent casting/particle leaching, gas foaming, and emulsification/freeze-drying, are being tested to create superporous matrices with interconnected pores [92,108]. A bioinspired procedure has been followed by Deville *et al.* [109] consisting in the directional freezing of an HAP suspension. This technique produces an arrangement of HAP in well defined layers due to the growth of ice, resulting in a multilayered structure that resembles nacre.

Bio-nanocomposites based on calcium phosphates can perform other innovative functions such as acting as a reservoir for the controlled release of bioactive compounds once the material is implanted in the bone defect. For instance, the incorporation of a morphogenetic protein that promotes bone regeneration in an HAP–alginate–collagen system [110] or a vitamin in a Ca-deficient HAP–chitosan nanocomposite [111] are recent examples of this kind of application.

In addition to hydroxyapatite, other calcium phosphate minerals such as tricalcium phosphate (TCP), substituted apatites, as well as cements and biphase mixtures with calcium phosphate content have also been studied for clinical applications, as extensively reviewed by Vallet-Regí and González-Calbet [112]. Although to a lesser extent than hydroxyapatite-based analogues, several biohybrid materials based on tricalcium phosphate (TCP, $Ca_3(PO_4)_2$ with $Ca/P = 1.50$) have been reported in the last few years. Some representative examples refer to microcomposites involving β-TCP in combination with structural proteins, polysaccharides or biodegradable polyesters such as collagen [113], chitosan [114] and PLA [115]. Calcium phosphate cements are also used as scaffolds for bone regeneration due to their biocompatibility, being gradually replaced by new bone after implantation. Their assembly with biopolymers results in highly stable biocomposites in which adhesion, proliferation and viability of osteoblasts are enhanced. This is the case of biocomposites based on chitosan combined with tetracalcium phosphate (TTCP, $Ca_4(PO_4)_2O$) and dicalcium phosphate anhydrous (DCPA, $CaHPO_4$) applied in periodontal and bone repair [116,117]. Among substituted apatites, fluorapatite is being studied as a component of hierarchically grown gelatin-based bio-nanocomposites showing a fractal character (Figure 1.7). The material synthesized by biomimetic routes has a composition very close to that of mature tooth enamel, with the formula $Ca_{5-x/2}(PO_4)_{3-x}(HPO_4)_x(F_{1-y}(OH)_y)\cdot 2.3$ wt% gelatin ($x = 0.82$; $0 < y \leq 0.1$), showing potential application in dentino- and osteogenesis [118,119].

Fig. 1.7 Scanning electron micrographs showing fractal pattern formation by hierarchical growth of fluorapatite–gelatin nanocomposites: (A) half of a dumbbell aggregate viewed along the central seed axis, (B) dumbbell aggregate at an intermediate growth state, and (C) central seed exhibiting tendencies of splitting at both ends ("small" dumbbell). Adapted from [119], reproduced by permission of Wiley-VCH.

1.4 Clay Minerals and Organoclay Bio-nanocomposites

In Earth, besides phosphates and carbonates, one of the most abundant groups of inorganic solids in interaction with the Biosphere is represented by the family of the so-called clay minerals [120]. They are silicates of aluminum and/or magnesium structured in tetrahedral and octahedral environments arranged as sheets that share oxygen atoms. Clay minerals showing two types of morphology have been used for the preparation of bio-nanocomposites: layered and microfibrous silicates. The first group includes montmorillonite, which is an aluminosilicate belonging to the smectite minerals group that show cation-exchange and expandability properties. The second group is represented by sepiolite, a microfibrous hydrated magnesium silicate whose main characteristics are determined by its textural behavior and surface reactivity.

Smectites are 2 : 1 charged layered silicates from natural (montmorillonite, hectorite, beidellite, saponite etc.) or synthetic (synthetic fluorohectorites, such as

Laponite) origin, built up by two tetrahedral silica sheets sandwiching one Al/Mg octahedral central sheet. In montmorillonite, the octahedral sheet is built up from Al^{3+} ions partially replaced by other cations such as Fe^{3+} and Mg^{2+}. The presence of divalent cations in that sheet and, to a lesser extent, Al^{3+} in the tetrahedral sheets, generates a negative charge compensated by extra-framework cations that are usually located as hydrated species in the interlayer region of the silicate [120]. Natural smectites are hydrophilic materials in which the accessibility of water to the interlayer region determines their colloidal properties, leading to gelatinous and viscous fluids and favoring its combination with water-soluble biomolecules. This is particularly significant for smectites containing alkaline ions in the interlayer region that may induce the swelling of the clay in water, promoting delaminated systems that allow their combination with molecular and macromolecular species, resulting in nanostructured system formation, including biopolymer-clay nanocomposites (bio-nanocomposites) [121–123].

Saccharides such as sucrose easily intercalate montmorillonite from aqueous solutions [124] giving basal spacing values of 1.64 and 1.83 nm [125], respectively, ascribed to mono- or bi-layer molecular arrangements in the interlayer space of the clay mineral. Recently, the intercalation of sucrose and its further *in situ* polymerization activated by microwave irradiation appears as a new procedure for the preparation of the bio-nanocomposites called caramel-clay nanostructured materials [126]. In this case, the bio-nanocomposite shows a value of about 1.9 nm corresponding to the intercalated caramel, which is characterized by chemical analyses, FTIR and ^{13}C-NMR spectroscopies [126]. These bio-nanocomposites act as precursors for the carbon–clay nanocomposite formation giving rise to electrical conductor solids. The starting caramel-montmorillonite is an insulating material that increases in conductivity (σ) from $\sigma < 10^{-12}$ S/cm to $\sigma \approx 0.1$ S/cm after the carbonization process. The most salient feature reported for these carbonaceous nanostructured materials is the combination of their electrical and textural behaviors suitable for potential applications as electrodes for electroanalysis, electrocatalysis, and in energy storage (supercapacitors) devices [127]. In this context, it must be taken into account that these materials are prepared from natural resources, therefore being considered as low cost environmentally friendly functional materials. This type of carbonaceous material is also formed by the direct pyrolysis of the carbohydrate–clay system containing small amounts of sulfuric acid [128].

Positively charged polysaccharides act as polyelectrolytes, being able to intercalate montmorillonite and other smectites by ion-exchange reactions from solutions [3]. By controlling the equilibrium concentration of chitosan in diluted acetic acid, it is possible to control the access of the biopolymer to the interlayer space of montmorillonite, leading to bio-nanocomposites with one, two or even more layers of intercalated polymer [129,130]. The incorporation of chitosan as a monolayer is governed by the cation-exchange capacity (CEC) of montmorillonite (about 90 mEq/100 g), whereas larger amounts of the polysaccharide, exceeding the CEC, lead to multilayer configurations. This last arrangement takes place by ion-exchange incorporation of chitosan aggregates formed in the concentrated solutions, which are self-assembled through hydrogen bonding, the charge excess being compensated with acetate anions. In this manner, these counter-ions are entrapped in the interlayer space and, therefore, the resulting bio-nanocomposites show anionic exchange properties, as confirmed by various techniques including ^{13}C NMR solid-state studies [130]. Intercalation of chitosan in Na-montmorillonite using solutions containing a high biopolymer/clay ratio may produce exfoliation of the layer silicate, in agreement with XRD and TEM results [131]. By solvent casting, these exfoliated materials have good film-forming ability, showing enhancement of the tensile strength and a decrease in the elongation-at-break compared to chitosan without montmorillonite. This behavior opens the way to new applications towards development of the so-called *green nanocomposites* (bioplastics) with potential applications in tissue engineering and food packaging.

Chitosan–clay bio-nanocomposites are very stable materials without significant desorption of the biopolymer when they are treated with aqueous salt solutions for long periods of time. In this way, they act as active phases of electrochemical sensors for detection of ions (Figure 1.8). The particular nanostructuration of the biopolymer in the interlayer region drives the selective uptake of monovalent versus polyvalent anions, which has been applied in electrode arrays of electronic tongues [132].

Chitosan–clay bio-nanocomposites showing the ability to incorporate anionic species can be used to prepare functionalized biohybrids. An example is the uptake of anionic dyes such as fast green and naphthol yellow S which are low-toxicity

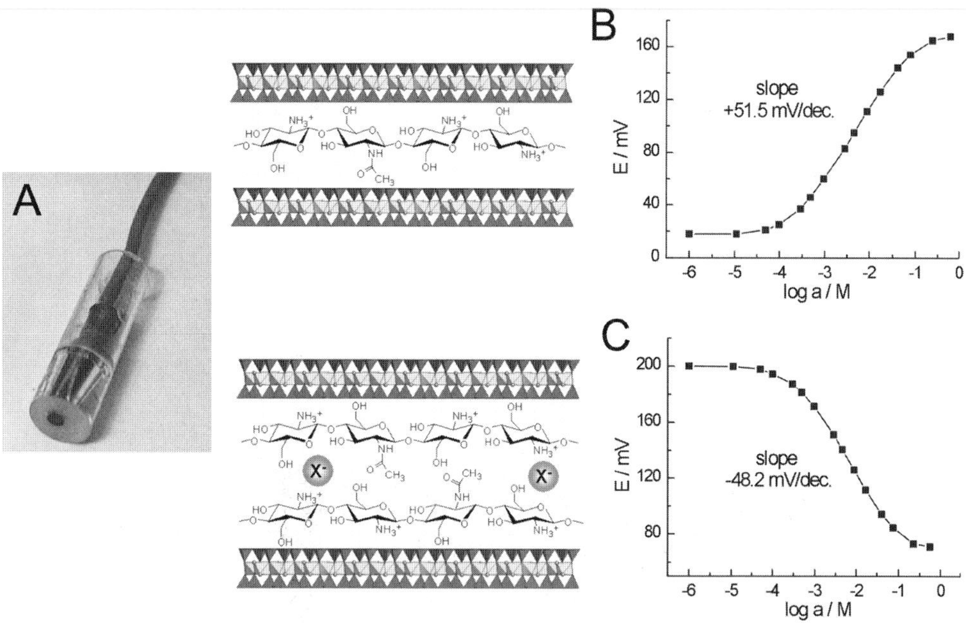

Fig. 1.8 (A) Design of a chitosan-montmorillonite sensor. Arrangement of chitosan chains in the clay interlayer space as a monolayer (B) or a bilayer (C) resulting in opposite behavior when measuring NaCl solutions of increasing concentration by direct potentiometry.

compounds even used as additives in food and cosmetics [133]. Interestingly, these dye/chitosan–clay hybrids are being used in applications as non-cytotoxic photoprotector systems providing considerable photostabilization of entomopathogenic fungus such as the conidia of *Aschersoni.* spp., which is used as a model biocontrol agent. In this context, assuming that the chitosan–clay bio-nanocomposites are biocompatible materials and also show capacity to absorb ionic species, they could, potentially, be used as vectors for controlled drug delivery of anionic drugs.

As chitosan is an excellent adsorbent for organic compounds its assembly with clay minerals could be interesting with regard to enhancing its mechanical properties for practical uses. Chang and Juang [134] have prepared composite beads by combination of activated clay and chitosan, developing adsorbents for dyes and organic acids (tannic and humic acids). Chitosan–clay biohybrids can be useful as intermediates for nanocomposites as they incorporate chemical functionality into the inorganic substrates. Qiu *et al.* [135] reported that monomers, such as acrylic acid, can react with these biohybrids producing their immobilization. Further polymerization gives rise to improved photostable polyacrylic acid/chitosan–clay nanocomposites, which exhibit water-superabsorbent properties.

The ion-exchanging ability of smectite clay minerals allows the development of intercalation compounds by treatment with salt solutions of organic cations, such as long-chain alkylammonium ions giving the so-called organoclay materials [121,122,136]. Organoclays show a better affinity towards non-polar species, which in this way may be assembled to the internal and the external surfaces of the modified clay mineral. Various authors have reported the preparation of bio-nanohybrid materials based on organoclays [137–144]. As this topic is specifically treated in Chapter 8, we will refer to it briefly here with some examples. Probably the most popular biohybrids are the poly L-lactic acid (PLA)–organoclay composites, usually prepared by a melt intercalation process [137,142,143,145]. These bio-nanocomposites (Figure 1.9A) show improved characteristics with respect to neat PLA, such as thermomechanical and gas barrier properties. This improvement, together with the PLA biodegradability behavior, makes these reinforced PLA bioplastics excellent

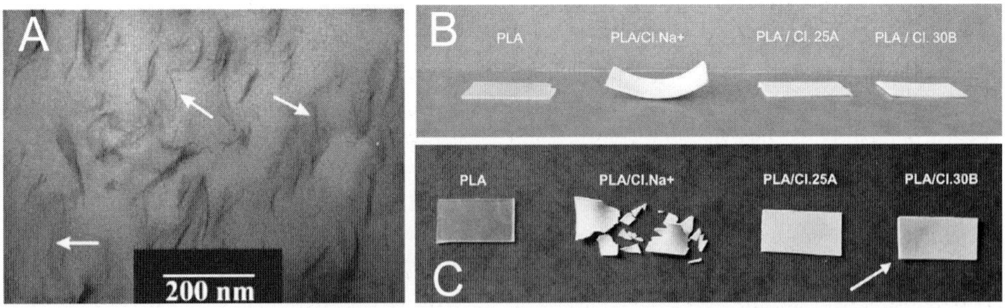

Fig. 1.9 (A) Exfoliation of clay platelets (white arrows) in a commercial polylactide matrix using a "masterbatch" process. (B, C) Visual aspect of unfilled PLA, microcomposite based on CloisiteNa^{+}, and nanocomposites based on Cloisite25A and Cloisite30B after (B) two and a half months hydrolysis and (C) after five and a half months hydrolysis. (A) adapted from [144] reproduced by permission of Wiley-VCH, and (B, C) from [147] with permission from Elsevier.

candidates for applications in the food industry and for biomedical purposes. It is important to signal that the biodegradability of these bio-nanocomposites strongly depends on the nature of the organoclay used as nanofiller [4,146,147]. In fact, the use of natural unmodified montmorillonite results in the formation of microcomposites that undergo fast degradation in comparison to nanocomposites formed from organomontmorillonites (Figure 1.9B, C). However, the incorporation of organoclays in bio-nanocomposites by assembly with biopolymers is not always advantageous for improving the final properties of the resulting biohybrids. This is, for instance, the case with chitosan-Cloisite 30B nanocomposites in which mechanical and thermal properties get worse in comparison with analogous bio-nanocomposites prepared from Na^+-exchanged montmorillonite [131]. This apparently contradictory behavior points out the importance of both the relative hydrophilicity of the clay layers and the nature of the biopolymer in the characteristics of the resulting nanocomposite and its final properties.

On the other hand, we have signaled (see above) the role of microfibrous clay minerals (sepiolite and palygorskite) in developing new nanocomposites. Sepiolite is structurally arranged by blocks formed by an octahedral sheet of magnesium oxide/hydroxide packed between two tetrahedral silica layers [148,149]. The periodic inversion of the SiO_4 tetrahedra determines a regular discontinuity of the sheets, being the origin of the structural cavities (*tunnels*) extended along the *c* axis, that is the axis of the microfibers [148] and the presence of silanol groups (Si—OH) [150,151]. Sepiolite shows a lower CEC than smectites (~15 compared to ~100 mEq/100 g) but, in addition to acting as an ion-exchanger able to uptake charged biopolymers, the presence of silanol groups on its external surface favors its association through hydrogen bonds. This last mechanism of interaction has been invoked to explain the formation of bio-composites with structural proteins (e.g., collagen) and polysaccharides (e.g., chitosan) via the amino and hydroxy groups of the respective biopolymers [152]. In the case of the fibrous proteins such as collagen, the interaction results in an exceptional arrangement of the biopolymer oriented in the same direction as the sepiolite fibers [34]. Further incorporation of low amounts (<1 %) of glutaraldehyde confers to the collagen–sepiolite biohybrids a strong improvement in mechanical properties, with excellent persistence after several months of implantation for bone repair applied to *in vivo* assays [33].

Chitosan shows high affinity for sepiolite assembly as, in addition to the hydrogen bonding between Si—OH surface groups and the biopolymer hydroxy groups, the negatively charged surface of the silicate could be balanced by the protonated amino groups of the biopolymer [3,31]. Multilayer coverage of chitosan can take place at high equilibrium concentrations of chitosan, which is adsorbed as polymer aggregates, similarly to chitosan–smectite interactions (see above). Sepiolite becomes strongly integrated within the biopolymer structure (Figure 1.10A), in a similar way to the chitosan–montmorillonite material (Figure 1.10B), providing nanohybrids that show good mechanical properties. In this way, the elasticity modulus of these bio-nanocomposites is superior to those of the components measured separately, which is the typical synergetic effect reported for structural polymer–clay nanocomposites [31]. Alternatively, sepiolite bio-nanocomposites have been used as precursors

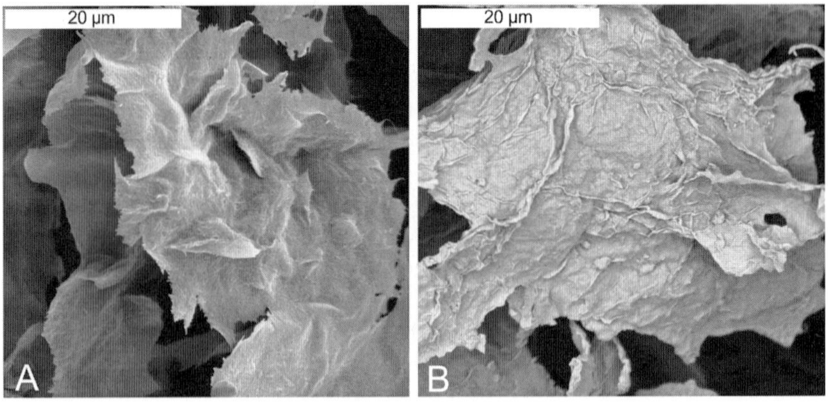

Fig. 1.10 SEM micrographs of chitosan-based bio-nanocomposites involving (A) sepiolite and (B) montmorillonite as the inorganic moiety.

of multifunctional materials, as reported by Gómez-Avilés *et al.* [153]. In this case, a precursor caramel-sepiolite bio-nanocomposite, prepared by application of MW irradiation to sucrose–sepiolite mixtures, can be turned into the corresponding carbonaceous derivative after pyrolysis at 800 °C in a N_2 atmosphere. These materials show significant electronic conductivity and, interestingly, they can be further functionalized by reaction with organosilanes. Grafting of aminopropyltrimethoxysilane and subsequent protonation of the amino groups, results in a material that can act simultaneously as an electronic collector and a sensing phase when used as an electrochemical sensor for anion detection. It is able to discriminate anions by their size and charge due to its textural behavior [153].

The structure and morphology of palygorskite are strongly related to those of sepiolite, although it shows higher Al content and narrower tunnels [120]. This microfibrous silicate acts similarly to sepiolite, assembling biopolymers via hydrogen bonding. In this way, hydroxyethyl and hydroxypropyl cellulose associate to palygorskite modifiying the rheological behavior of the clay dispersions [154]. Palygorskite can also be combined with modified polysaccharides, such as polyacrylamide (PAA) grafted-starch, giving rise to bio-nanohybrids with extraordinary superabsorbent properties as they may absorb up to 500 g of water per g of bio-nanocomposite [155].

Layered Double Hydroxides (LDHs), also called *hydrotalcite-like materials* or even *anionic clays*, are mixed hydroxides that can be described by the general formula $[M^{2+}{}_{1-x}M^{3+}{}_{x}(OH)_2]\ [A^{n-}{}_{x/n}{\cdot}zH_2O]$ in which M^{2+} and M^{3+} are metal ions and A^{n-} is the anion that compensates the deficit of negative charge in the layers. In contrast to the above-mentioned smectite clay minerals, the main characteristics of LDHs are the high charge density in the layers and the anion exchange ability. In this way, LDHs can be assembled with negatively charged biopolymers leading to biohybrids equivalent to those prepared from smectites, often showing intercalated or partially delaminated phases. An example of this type of bio-nanocomposite is the Zn-Al LDH assembly with negatively charged polysaccharides (alginate, pectin, κ-carrageenan and ι-carrageenan) [156,157]. These biohybrids are prepared by the so-called template synthesis of the LDH inorganic host in the presence of the biopolymer. The self-assembly of the polysaccharide in the medium in which the LDH is synthesized determines the incorporation of biopolymer aggregates, in which a part of the electrical charge is compensated by the LDH host and the other moiety by extra-framework cations. This situation is the inverse to that reported for chitosan–smectite bio-nanocomposites and is the basis for their application as active phases of potentiometric sensors. In this way, potentiometric sensors for calcium ions have been developed, profiting from the ability of the biopolymers alginate, ι-carrageenan, pectin and κ-carrageenan to complex calcium ions [157]. The comparative study demonstrates that sensors based on the two last bio-nanocomposites show a lower sensitivity in the determination of Ca^{2+} ions, probably due to the complex structure of pectin and to the low content of negatively charged groups in the κ-carrageenan chains with respect to the other polysaccharides.

The ability of LDHs to interact with negatively charged biopolymers has been applied to the preparation of biohybrid materials incorporating negatively charged DNA in Mg-Al LDH, as firstly reported by Choy and coworkers [158–160]. These bio-nanohybrids exhibit the unique property of acting as non-viral vectors in gene therapy to transfer nucleic acids to the cell interior via an endocytosis mechanism (Figure 1.11A). LDH-based biohybrids can also be used as carriers for controlled drug delivery, as diverse bioactive compounds are negatively charged [3]. LDHs can also be assembled with enzymes producing their effective immobilization between the inorganic layers, whilst at the same time allowing the diffusion of substrates and products (Figure 1.11B). An illustrative example is the entrapping of urease within Zn-Al LDH giving rise to bio-nanohybrids able to act as active phases in capacitance biosensors [161]. This topic will be more deeply considered in Chapter 15.

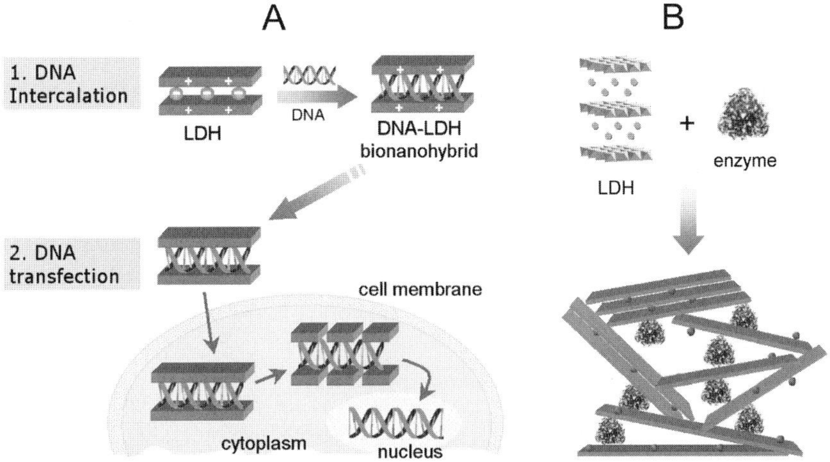

Fig. 1.11 Applications of LDHs as (A) non-viral vector in gene therapy for transfection of DNA to the cell nucleus, and (B) as matrix for enzymes immobilization in the development of biosensors.

1.5 Bio-Nanohybrids Based on Metal and Metal Oxide Nanoparticles

The dispersion of metallic, semiconductor or magnetic nanoparticles in polymeric matrices results in easy-handling nanocomposite materials that can be used in a wide number of applications. Nowadays, the use of biopolymers for this purpose is increasing since they are an eco-friendly alternative to the usually employed polymers, mainly for applications in the health area. For instance, bio-nanocomposites based on silver oxide nanoparticles have been prepared by embedding them in a gelatin matrix. This new bio-nanohybrid material shows luminescent intensity higher than non-modified nanoparticles after activation with a femtosecond-laser. This feature, together with the appropriate mechanical properties of the gelatin–Ag_2O nanoparticles nanocomposite, makes possible its application in light-emitting materials for all-optical logic devices or data-storage media [162].

The branched polysaccharide dextran is assembled with alkanethiol-modified gold nanoparticles and the resulting nanocomposite is then functionalized to facilitate the specific binding of target biomolecules. This biorecognition process can be easily detected by particle plasmon resonance (PPR), based on the optical properties of gold nanoparticles [163].

Charged polysaccharides can also serve as templates for the growth of metallic, semiconductor and magnetic nanoparticles. For instance, chitosan has been reported as a catalyst and stabilizing agent in the production of gold nanoparticles by the reduction of tetrachloroauric (III) acid by acetic acid. The biopolymer controls the size and the distribution of the synthesized Au nanoparticles and allows the preparation

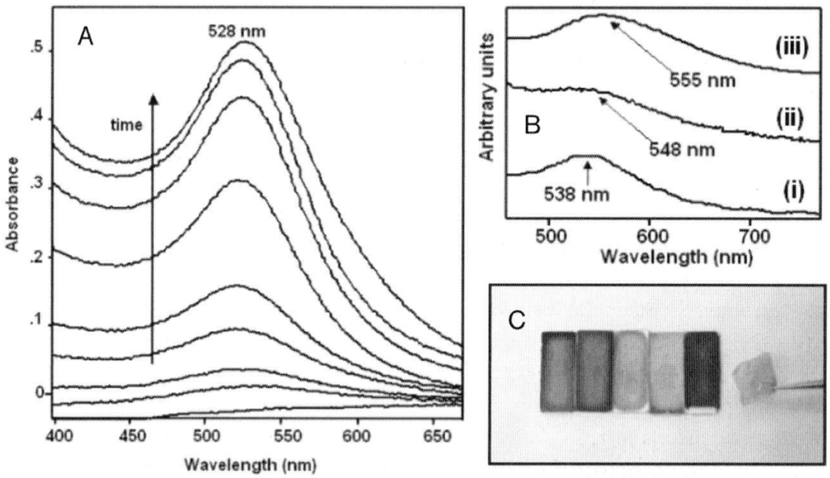

Fig. 1.12 (A) Increase in surface plasmon absorption as Au nanoparticles are produced from a reaction mixture containing 1% chitosan, 1% acetic acid and 0.01% tetrachloroauric (III) acid ($HAuCl_4$); (B) shift of surface plasmon absorption for films prepared from the previous mixture (i), and from mixtures with lower chitosan concentration (ii) or lower $HAuCl_4$ amount (iii); six different self-sustained nanocomposite films showing the control over the optical properties. Reprinted with permission from [164], 2004, American Chemical Society.

of self-supported gold–chitosan bio-nanocomposite films (Figure 1.12). This bio-nanohybrid material shows potential application in the development of biosensors and in trace chemical analysis [164]. The same polysaccharide, chitosan, is involved in a new CdS quantum dots–chitosan bio-nanocomposite. In this case, the ability of chitosan to complex transition metal ions is used to entrap Cd^{2+}, followed by immersion of the resulting Cd^{2+}–chitosan films in Na_2S solution to yield the CdS semiconductor nanoparticles. The size of the formed quantum dots is limited by the polysaccharide, which also avoids agglomeration of the nanoparticles. The resulting bio-nanocomposites exhibit increased thermal stability and improved aqueous solubility [165]. A similar procedure was followed with the anionic polysaccharide alginate, used as template for the controlled growth of Ni, Co or NiCo magnetic nanoparticles. In this case, the nanocomposites are processed as microcapsules resulting from gelation of alginate in the presence of Ni^{2+}, Co^{2+} or a mixture of both ions, which are further reduced under H_2/N_2 flow [166]. A recent contribution by Srivastava *et al.* [167] reports the formation of magnetic bio-nanocomposites by assembly of cationic FePt nanoparticles and DNA.

On the other hand, metal oxide nanoparticles such as Al_2O_3–Zr_2O have been used recently to reinforce biological matrices such as collagen, enhancing their mechanical and thermal properties and leading to hybrid materials with potential use in biomedical and bionic applications [168]. Of particular interest for such types of applications, including NMR imaging, hyperthermia treatment, and targeted drug delivery, is the use of nanoparticles based on metal oxides showing magnetic properties [169,170]. Ferrites (magnetite or maghemite) prepared as nano- or micro-particles [29,171] become biocompatible after assembly with biopolymers (poly-DL-lactate, starch, pectin, alginate) or biocompatible polymers (poly-acrylates, polyvinyl alcohol) [172–174]. Magnetic bio-nanocomposites from the assembly of magnetic nanoparticles and fibers of biopolymers such as poly(hydroxyethylmethacrylate) (PHEMA) or PLA have been prepared recently by the electrospinning technique [175]. This type of material was developed for accumulation of delivered drugs in a precise target area, due to its superparamagnetic properties and its ability to release the transported drug. Bio-nanocomposites based on dextran and magnetic nanoparticles such as magnetite or maghemite have been described as physiologically well-tolerated materials [28], being studied for use in different biomedical applications [29].

Alternatively, magnetic nanoparticles can be incorporated within a bilayer of phospholipids arranged as a liposome (Figure 1.13A). These systems, called magnetic liposomes or magnetoliposomes, were developed with the aim of entrapping and transport molecular species or genic matter acting in DDS under the guidance of a magnetic field [176,177]. We have indicated above the interest in functionalizing magnetic oxide nanoparticles in the preparation of magnetocerasomes (see above, Section 2) by grafting long-chain cationic surfactants onto magnetite through silane or phosphonic moities, which are further co-assembled with phosphatidylcholine (PC) (Figure 1.13B). In this context, magnetocerasomes show the ability to incorporate lipophilic target species, transporting them through artificial membranes being useful for the design of new carriers for DDS [73].

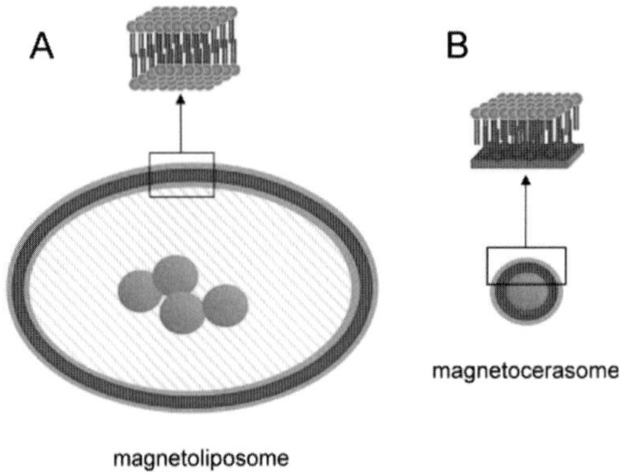

Fig. 1.13 Schematic representations of (A) magnetoliposome and (B) magnetocerasome configurations involving magnetic nanoparticles and phosphatidylcholine.

Nanoparticles of TiO_2 can be combined with biomolecules of different nature giving functional bio-nanocomposites. For instance, lysozyme has been encapsulated within titania nanoparticles resulting in nanohybrids showing antimicrobial activity. Here again, it is postulated that the inorganic moiety protects the enzyme against denaturation [178]. Moreover, bacterial immobilization on a nanocrystalline TiO_2 matrix allows the development of bio-nanocomposites acting as the active phase of photoelectrodes [179]. This is an excellent example of how nanoparticles associated with biomolecules act not only as a mechanical support but also may confer functional properties to the bio-nanocomposite materials. In relation to this type of material, organically modified layered titanates were assembled with PLA to prepare *green nanocomposites* with improved mechanical properties. The photodegradation of PLA in sunshine is enhanced by the presence of the inorganic component, as the titanate shows photocatalytic activity similar to that of TiO_2 [180].

Another semiconductor oxide such as ZnO is particularly promising in nanodevice applications. A pioneering work refers to the preparation of biofunctional ZnO nanorod arrays grown on a thermoplastic polyurethane flexible substrate and further assembled with proteins [181]. This innovative approach aims to develop new biological sensing systems showing flexible properties and biocompatibility, which are useful for the detection of complementary biomolecules on the acceptor side, such as antibody–antigen bioconjugation by photoluminescence spectroscopy [181].

1.6
Carbon-based Bio-nanohybrids

Carbon nanotubes (CNTs) were discovered in 1991 by Iijima [182] and since then they have attracted much attention in many research fields. CNTs can be described as tubular structures rolled up from a graphene sheet. Depending on the number of tubular walls CNTs can be classified as single-walled carbon nanotubes (SWCNTs) that show diameters in the range 0.7–1.5 nm, and multi-walled carbon nanotubes (MWCNTs), which are formed by 2–30 concentric tubes with diameters in the 2–10 nm for the inner tubular layer and additional thickness of about 0.7 nm for every additional layer [183]. Differences in the chiral vectors that describe how graphene sheets are rolled up, as well as topological defects present on the tube surface, determine the distinct electronic structures and therefore either the semiconducting or metallic behavior of CNTs. Besides their electrical properties, CNTs show exceptional mechanical properties due to their very high aspect ratio, which has made them very attractive for applications as nanofillers of different polymers [184–187].

Due to the poor solubility of CNTs in water, they are very frequently submitted to treatments with mixtures of concentrated sulfuric and nitric acids, which result in the formation of carboxyl and hydroxyl groups on their surface. Besides solubility in polar solvents, these activated CNTs are able to react with other functional groups favoring coupling to different compounds. In this way, chitosan–MWCNTs bio-nanocomposites have been prepared by a simple solution–evaporation method from activated CNTs, in order to favor their interaction with the amino groups of chitosan. The homogeneous distribution of MWCNTs in the biopolymer determines a significant improvement in the mechanical properties with 93 and 98 % enhancement in the tensile modulus and strength, respectively, with only a 0.8 %wt MWCNTs content [188]. The use of CNTs previously grafted to chitosan that are further blended to pure chitosan allows the incorporation of larger amounts of nanotubes, up to 50 %wt CNTs [189]. As compared with ungrafted CNTs, chitosan-grafted-CNTs show improved dispersion ability in the chitosan matrix, which results in significantly improved storage modulus and water stability. Spinks and coworkers [190] prepared SWCNTs–chitosan fibers confirming that improved dispersion of CNTs in chitosan results in enhanced mechanical properties without compromising swelling behavior. These materials also show pH actuation according to the variable pH swelling response, which makes them attractive for applications as bioactuators and artificial muscle materials.

The combination of CNTs and carbohydrates is also a research area of interest with regard to preparing bio-nanocomposites for biological applications. Thus, the preparation of multivalent carbohydrate-CNTs conjugates based on lactose-attached schizophyllan [191], lipid-terminated sugar polymers [192], and covalently attached sugar polymers [193,194] has been reported. In this way, *p*-*N*-acryloamidophenyl α-D-glucopyranoside has been assembled with SWCNTs (Figure 1.14A) with the aim being to create materials able to show specific interaction with carbohydrate recognition proteins, which can be of interest for the recognition of bacterial toxins and viral proteins, or quantitative estimation of carbohydrate–carbohydrate interactions [195]. Functionalization of SWCNTs via atom transfer radical polymerization (Figure 1.14B) allows the direct grafting of sugar- and phosphorylcholine-based biopolymers, making the modified CNTs water soluble and biocompatible [35]. Interestingly, the resulting bio-nanocomposites based on the anchorage of the sugar moiety may have interest in carbohydrate–protein recognition, whereas the phosphorylcholine derivative could prevent nonspecific protein adsorption and cellular adhesion.

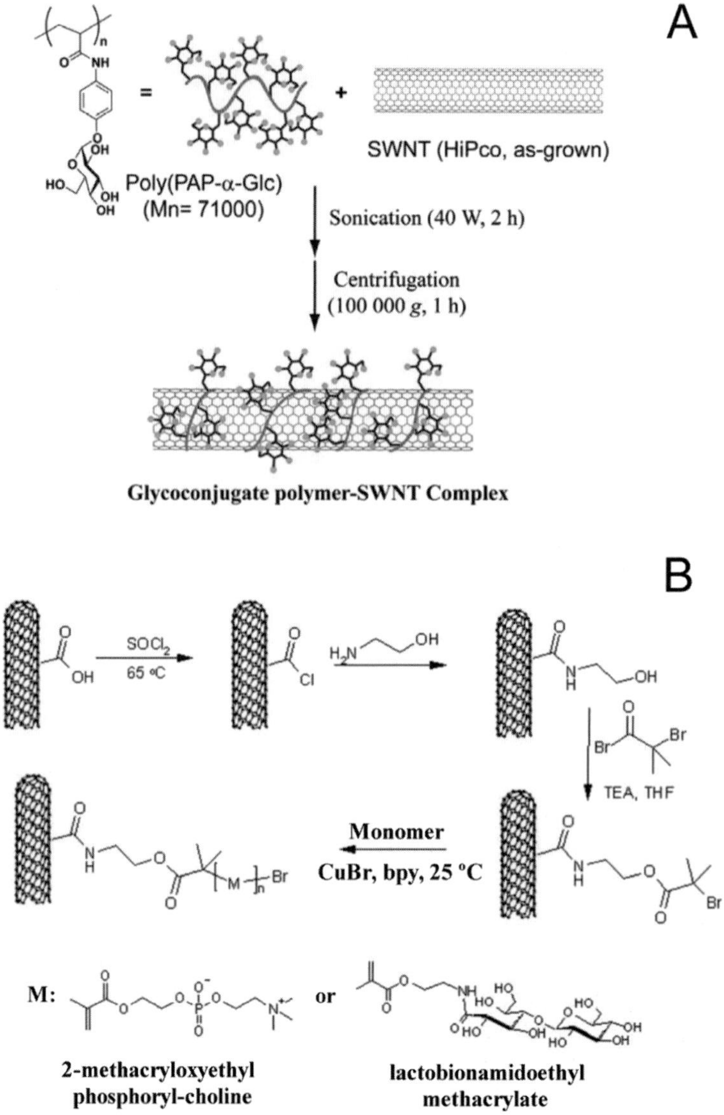

Fig. 1.14 (A) Single-wall carbon nanotubes wrapped by glycoconjugate polymer with bioactive sugars. (B) Modification of carboxyl-functionalized single-walled carbon nanotubes with biocompatible, water-soluble phosphorylcholine and sugar-based polymers. (A) adapted from [195] with permission from Elsevier, and (B) from [35] reproduced by permission of Wiley-VCH.

An illustrative example of carbohydrate–protein recognition has been reported by Chen and coworkers [192] using a C18-mucin mimic polymer that interacts hydrophobically with SWCNTs. This biohybrid is able to specifically recognize the lectin *Helix pomatia* agglutinin (Figure 1.15A). Galactose-modified CNTs are able to capture

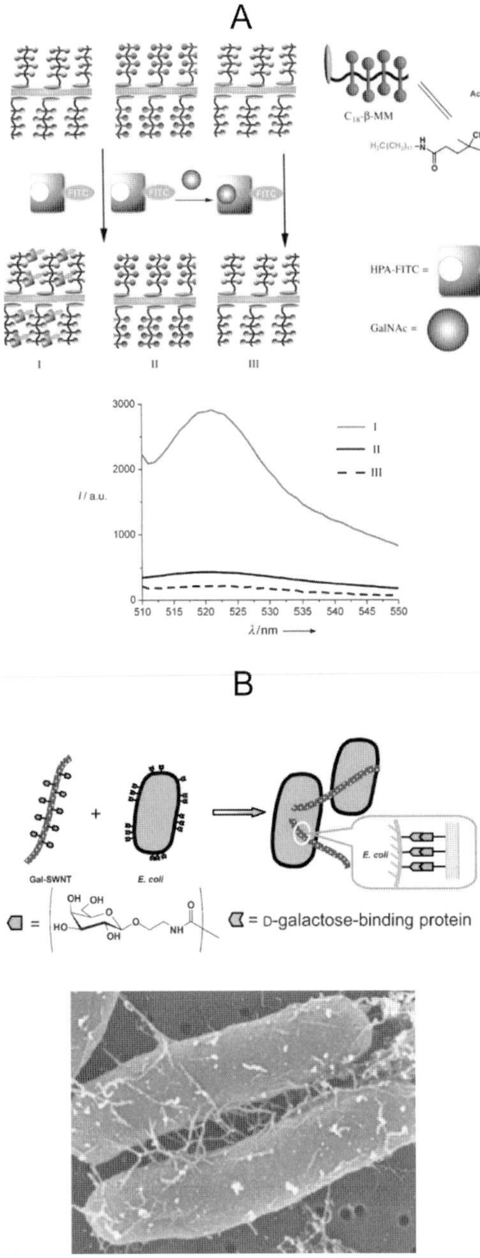

Fig. 1.15 (A) Scheme for (I) specific binding of the lectin *Helix pomatia* agglutinin (HPA) to C18-α-MM-coated SWNTs, (II) lack of binding of HPA to C18-β-MM-SWNTs, and (III) inhibition of HPA binding by soluble GalNAc, as well as the corresponding fluorescence spectra (510–550 nm, excitation wavelength 492 nm). (B) Schematic representation and SEM image showing the specific binding of galactose modified-SWNTs to galactose-binding proteins in *E. coli* cells.
(A) Adapted from [192] (reproduced by permission of Wiley-VCH) and (B) from [194] (reproduced by permission of The Royal Society of Chemistry).

pathogens, for example *E. coli*, following a similar type of specific binding reaction (Figure 1.15B) [194]. Other biopolymers, such as DNA, have also been combined with CNTs to create bio-nanocomposites that can serve for instance as biological transporters and near-infrared agents for selective cancer cell destruction [196]. In relation to this application, it is important to signal that recent studies demonstrate that CNTs can cause DNA destabilization and conformational transition that are sequence-dependent [197].

Taking into account their electrical and mechanical properties, CNTs have been extensively investigated, not only as additives in the preparation of conventional bio-nanocomposites, but also in the development of biohybrids for various biomedical applications, including tissue regeneration, biosensors and other biomedical devices such as microcatheters [198]. For these types of application, a key point is their biocompatibility [199], as the presence of metallic compounds from the catalysts used in their synthesis introduces significant toxicity [200]. In this way, Koyama and coworkers have performed systematic studies on the biological response of CNTs used as microcatheters by measuring $CD4^+$ and $CD8^+$ T-cells in *in vivo* experiments, confirming their almost inert behavior in high-purity microcatheters [198]. These authors also indicate that several obstacles besides the use of defect-free and high-purity CNTs must be solved before their use in biomedicine is completely safe. These requirements include the preparation of biohybrids with a homogeneous dispersion of CNTs in polymers, control of the CNTs chirality, and long-term systematic biological studies of the materials. In this line of research, bio-nanocomposites from MWCNTs and poly(L-lactide), a polymer susceptible to enzymatic and hydrolytic degradation to L-lactic acid, have been prepared with the aim being to obtain a new generation of implant materials. Studies on related polylactic-*co*-glycolic acid combined with CNTs and poly(vinyl alcohol) suggest an important influence of the distribution of components on the surface of the material on its biocompatibility properties [201]. Other CNT-biocompatible polymer systems that have been explored with the aim of developing scaffolds for tissue regeneration applications are SWCNTs–poly(propylene fumarate) biohybrids [202]. These bio-nanocomposite systems show improved mechanical properties and good electrical properties that could be of interest in stimulating cell growth and tissue regeneration by facilitating the physioelectrical signal transfer. However, cell culture tests suggest that the presence of MWCNTs in the nanocomposites inhibits the growth of fibroblast cells [203].

The interesting electrical conductivity properties of CNTs have been used to develop different types of biosensors. Strategies to prepare these devices include for instance the use of layer-by-layer techniques to build the active phase on an electrode by entrapping an enzyme, such as glucose oxidase (GOD), within the CNTs layer [204,205]. The entrapping by electrostatic interactions seems to be preferable to the formation of true covalent bonding of CNTs, not only to preserve the activity of the protein but also to avoid variation in the conductive response of the active phase. In this way, biosensors with relatively good electrochemical response have been built by a one-step electrodeposition method that grows the active phase from a chitosan–CNT–GOD solution [206]. In this approach the biopolymer acts as a continuous phase in which CNTs and the enzyme are entrapped. In other approaches the enzyme and the CNTs are first precipitated onto the surface of the electrode, and then a polymer is deposited to wrap the active phase, thus avoiding its dispersion in the solution. One example of this method is the preparation of hybrid films of MWCNTs and hemoglobin incorporating poly(sodium-*p*-styrene-sulfonate) and cetyltrimethylammonium bromide [207]. Mioglobin has also been entrapped together with MWCNTs on a glassy carbon electrode with Nafion acting as agglutinant [208]. In both cases, the presence of CNTs in the film promotes electron transfer properties between the entrapped protein and the electrode, which could be of interest for the development of biosensors, catalytic bioreactors and other biomedical devices.

Although the electrical properties of CNTs are altered by modification of their surface, in certain cases the direct anchorage of the enzyme to the nanotubes has also been explored. These synthetic routes imply a previous activation of CNTs with acid to create carboxyl and hydroxy groups on their surface to which the positively charged species can be directly attached. Thus, poly(acrylonitrile-*co*-acrylic acid) (PANCAA) has been coupled to modified MWCNTs with the aim being to prepare nanofibrous membranes in which it is possible to further immobilize enzymes [209]. For instance, catalase has been successfully bonded to the activated membrane, resulting in higher enzyme loading and activity, possibly due to the high specific surface area and good electrical conductivity afforded by the CNTs (Figure 1.16). Alternatively, enzymes can be grafted to a biopolymer, such as chitosan, previously combined with CNTs and

$$2H_2O_2 \rightarrow 2H_2O + O_2 \quad (1)$$

$$CAT\ (Por\text{-}Fe^{III}) + H_2O_2 \rightarrow Cpd\ I\ (Por^{+\cdot}\text{-}Fe^{IV}{=}O) + H_2O \quad (2)$$

$$Cpd\ I\ (Por^{+\cdot}\text{-}Fe^{IV}{=}O) + H_2O_2 \rightarrow CAT\ (Por\text{-}Fe^{III}) + H_2O + O_2 \quad (3)$$

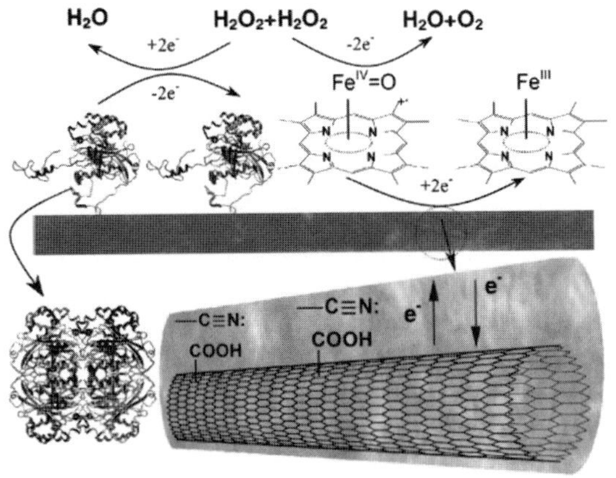

Fig. 1.16 Schematic representation of the nanofibrous poly(acrylonitrile-*co*-acrylic acid) membrane containing MWCNTs, as well as the promoted electron transfer from hydrogen peroxide to the immobilized catalase through the PANCAA/MWCNTs nanofiber. Reprinted from [209] (reproduced by permission of Wiley-VCH).

processed as a film onto a glassy carbon electrode [210]. In this example, glucose dehydrogenase was bonded to chitosan using glutaric dialdehyde as a bridge between the amino groups of chitosan and the enzyme.

1.7
Bio-nanohybrids Based on Layered Transition Metal Solids

Vanadium pentoxide xerogel ($V_2O_5{\cdot}1.6H_2O$) consists of V_2O_5 layers arranged as stacked ribbon-like particles about 20 nm wide, 2 nm thick and 1 μm long [211], containing water molecules and protons that allow the intercalation of a large variety of ions, molecules and polymers [212]. Several dipeptides such as Ala-Gln, Val-Gln and Ala-Gly (Figure 1.17A) can be intercalated in this host matrix by *in situ*

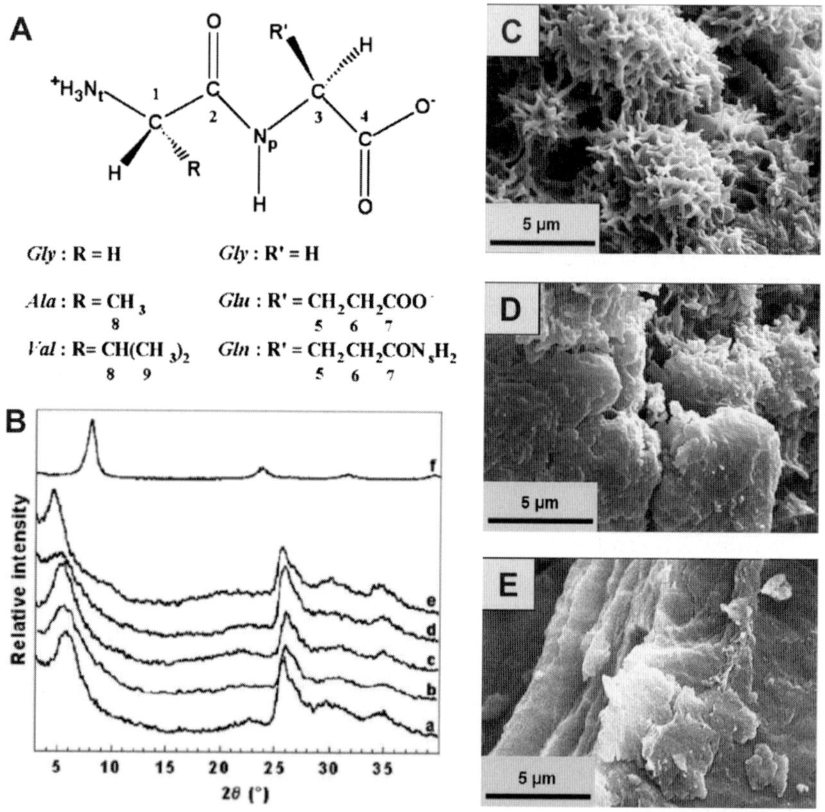

Fig. 1.17 (A) Structure of the dipeptides Val-Gln, Ala-Gln, Ala-Gly, Gly-Gln and Gly-Glu. (B) X-ray diffraction patterns of the bio-nanohybrids including V_2O_5 xerogel and (a) Ala-Gly, (b) Gly-Gln, (c) Ala-Gln, (d) Gly-Glu and (e) Val-Gln. (f) X-ray diffraction pattern of a $V_2O_5{\cdot}1.8H_2O$ xerogel film. SEM micrographs of the $(Ala\text{-}Gly)_{0.27}V_2O_5{\cdot}1.0H_2O$ hybrid synthesized at (C) pH = 1.5, and (D) pH = 1.0. (E) SEM micrograph of $V_2O_5{\cdot}1.8H_2O$ xerogel. Adapted from [213] with permission from Elsevier.

synthesis of the hydrated vanadium pentoxide, as confirmed by X-ray diffraction (Figure 1.17B), that is by mixing the dipeptides with an acidified solution of sodium metavanadate. The resulting materials present a sponge-like morphology, as shown in Figure 1.17C, D, in which the peptide bonds between the amino acids are preserved despite the low pH of the system [213]. Melanin is a biomacromolecule mainly derived from cystine and tyrosine amino acids, usually associated with proteins in aggregates denoted as melanosomes. Two recent studies report the intercalation of a melanin-like compound (3,4-dihydroxy-phenylalanine) between the layers of vanadium pentoxide [214,215]. Both studies open the way to the preparation of melanin-based nanocomposites, which could profit from the interesting optical (UV absorption) and solid-state electrical (photo- and semi-conductivity) properties of this biopolymer [123]. Other biomolecules with different functionality have been assembled with vanadium pentoxide xerogel, the aim being to develop new bio-nanocomposites with synergic properties afforded by both types of components. Chitosan intercalates V_2O_5 xerogel leading to an increase in the interlayer distance of about 0.4 nm, which indicates the arrangement of the polysaccharide in a monolayer [216]. The potential interest of the resulting bio-nanocomposites is their use as the active phase in electrochemical sensors, in analogy to chitosan–clay nanocomposites [129,130]. The electrical conductivity provided by the inorganic counterpart can be expected to facilitate the electrochemical response. However, the oxidant character of V_2O_5 xerogel may lead to biopolymer degradation to the detriment of the bio-nanocomposite stability and activity. It has been found that adsorption taking place only on the external surface of the layered solid results in microcomposite assemblies, increasing the stability of the systems in comparison to bio-nanocomposites formed by intercalation. This is the case for hybrid materials based on the immobilization of glucose oxidase on vanadium pentoxide [217].

Layered tetralkylammonium-modified manganese oxide shows the ability to intercalate myoglobin and hemoglobin that can retain their peroxidase activity [218]. Intercalation of methylcellulose into lithiated phases of manganese oxides was carried out through a mechanism of delamination–restacking giving rise to methyl cellulose–Li_xMoO_3 nanocomposites [219]. The presence of the insulating biopolymer between the inorganic layers decreases the inner conductivity of the pristine solid from 10^{-2} to 10^{-6} S/cm, preventing their potential use as electrodes for electrochemical devices. Further research to modulate the electrical conductivity of these bio-nanocomposites is required in order to develop new cathode materials for rechargeable Li-batteries.

Mixed oxides with a layered structure, such as calcium niobate perovskites exchanged with quaternary ammonium ions, can produce stable colloidal dispersion of the perovskite layers able to be assembled with biopolymers in aqueous solutions. For instance, assembly with gelatin leads to restacking of the perovskite nanosheets, which are homogeneously distributed in the biopolymer and highly oriented with the (*a*, *b*) plane parallel to the resulting films (Figure 1.18) [220]. These biohybrid films show an increase in dielectric permittivity that can be useful for application in the microwave industry or in high frequency devices. Similarly, hemoglobin has been intercalated as a mono- or bi-layer in tetrabutylammonium-modified layered niobate

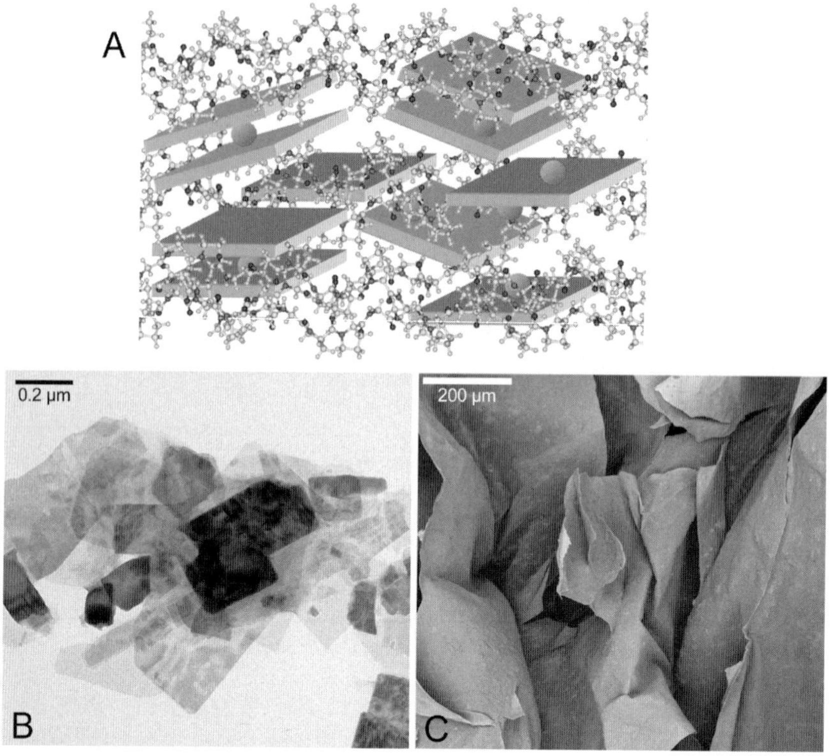

Fig. 1.18 (A) Schematic representation of gelatin-perovskite bio-nanocomposite. (B) TEM image of TBA-modified perovskite and (C) SEM image of this TBA-perovskite after assembly with gelatin.

through a mechanism of delamination–restacking. The immobilization of this protein improves its thermal stability and enhances its activity in organic media [221].

Besides smectites, LDHs or vanadium pentoxide xerogel mentioned above, there are other layered inorganic hosts able to assemble biomolecules to give bio-nanocomposite materials. In this way certain transition metal phosphates and chalcogenides have been used to prepare hybrids involving structural and functional proteins [123]. For instance, gelatin can be intercalated into layered α-zirconium phosphate ($Zr(HPO_4)_2{\cdot}nH_2O$) by treatment at acidic pH [222]. The process gives rise to a bio-nanocomposite that exhibits a basal spacing of 2.7 nm, due to the separation of the α-zirconium phosphate layers by the protein chains. The same phosphate was used to intercalate lysozyme and protamine, which represent a class of globular proteins of medium and low molecular weight, respectively [222]. Similarly, Zr-phosphate nanocomposites including hemoglobin, myoglobin, lysozyme, chymotrypsin and glucose oxidase have been synthesized at pH 7.2, giving well defined intercalated compounds (Figure 1.19) [223]. The basal spacing values deduced from the XRD patterns are compatible with the molecular dimensions of the different intercalated proteins.

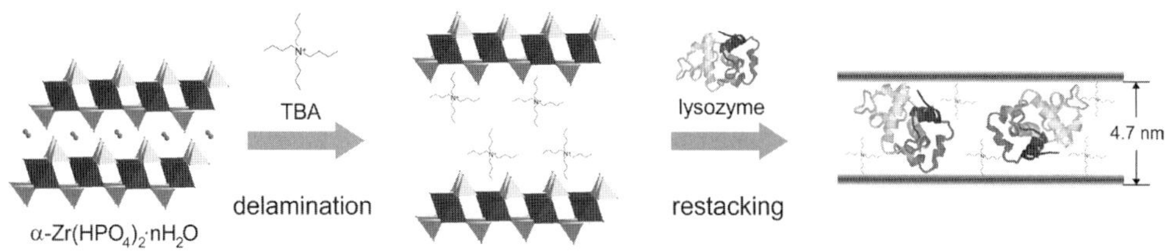

Fig. 1.19 Intercalation of enzymes at the galleries of layered α–zirconium phosphate following a delamination-restacking mechanism.

Transition-metal phosphorus trichalcogenides such as $MnPS_3$ are able to intercalate amino acids and peptides by ion exchange. In this way, increases in the basal spacing of 0.7 and 3–4 nm are observed for the intercalation of poly-L-lysine and lysozyme, respectively [224]. Interestingly, the enzymatic activity of the immobilized protein has been detected, suggesting that the enzyme is protected against denaturation.

1.8 Trends and Perspectives

Bio-nanohybrids represent an emerging group of advanced nanostructured materials that are receiving increasing interest in view of their versatility towards potential applications. The synthetic approaches that can be employed in the preparation of bio-nanocomposites may be principally based on detailed knowledge of the interfacial interactions between both components, that is biopolymers and inorganic solids. The mechanisms governing the assembly between the components, which can be the determining factor in the structural arrangement of the resulting materials, require further research. Within this context, understanding the biomineralization processes that take place in nature can also assist in the development of new bio-nanohybrid materials. In this way, bio-nanocomposites, and their resulting properties, can ideally be tunable towards suitable characteristics that are necessary for specific applications.

Above we have shown the attractiveness of the so-called green nanocomposites, although the research on these materials can still be considered to be in an embryonic phase. It can be expected that diverse nano- or micro-particles of silica, silicates, LDHs and carbonates could be used as ecological and low cost "nanofillers" that can be assembled with polysaccharides and other biopolymers. The controlled modification of natural polymers can alter the nature of the interactions between components, affording new formulations that could lead to bioplastics with improved mechanical and barrier properties.

Another important group of bio-nanohybrids are the new materials needed for biomedical purposes, such as the development of artificial biological tissues and particularly those related to bone implants. Future progress within this field will require investigation of the use of nanoparticulate inorganic solids based on diverse ceramics, and even metal–ceramic composites, as an alternative to HAP and the related compounds currently employed. In this context, the use of multicomponent polymeric and inorganic nanoparticulated systems could also be envisaged, employing techniques based on cryogenic synthesis to form assemblies with macroporosity useful for new materials for bone reparation. The use of bio-nanocomposites in DDS, including DNA non-viral vectors in gene therapy, also represents interesting biomedical applications of biohybrids. Alternative charged biopolymers coupled to diverse nanofillers (silica, phosphates, silicates, carbonates, etc.) appear attractive for the development of new biocompatible materials with the ability to uptake, carry and deliver molecular drugs in living organisms. By introducing magnetic properties, these bio-nanocomposites can be useful for drug transport to specific sites in the organs through guidance with external magnetic fields.

Among the new potential applications of bio-nanocomposites, we have highlighted that diverse inorganic components such as clay-based materials, metal nanoparticles and conductive carbon nanotubes can act as active phases in different type of devices, such as electrochemical sensors and biosensors. As certain inorganic solids have been employed as a protective matrix for the entrapment of living cells and enzymes, we can contemplate the extension of their use for the preparation of different bioactive nanocomposites that can be integrated in biosensors and bioreactors with a view to developing a new generation of advanced devices.

Finally, it can be envisaged that the future development of novel bio-nanohybrids will lead to new improved properties and multifunctionality derived from the synergistic combination of nanosized inorganic solids, with different structural and textural features, with molecular or even highly organized species of biological origin that are extraordinarily abundant in Nature.

Acknowledgements

The authors acknowledge financial support from the CICYT, Spain (Project MAT2006-03356), and the Comunidad de Madrid, Spain (Project S-0505/MAT/000227).

References

1 Dujardin, E. and Mann, S. (2002) *Advanced Materials*, **14**, 775–788.
2 Ruiz-Hitzky, E. and Darder, M. (eds) (2006) Special Issue on Trends in Biohybrid Nanostructured Materials, Current Nanoscience, **2**, 153–294.
3 Ruiz-Hitzky, E., Darder, M. and Aranda, P. (2005) *Journal of Materials Chemistry*, **15**, 3650–3662.
4 Darder, M., Aranda, P. and Ruiz-Hitzky, E. (2007) *Advanced Materials*, **19**, 1309–1319.
5 Darder, M., Aranda, P. and Ruiz-Hitzky, E. (2007) *Anales de Química*, **103**, 21–29.
6 Ray, S.S. and Bousmina, M. (2005) *Progress in Materials Science*, **50**, 962–1079.
7 Pandey, J.K., Kumar, A.P., Misra, M., Mohanty, A.K., Drzal, L.T. and Singh, R.P. (2005) *Journal of Nanoscience and Nanotechnology*, **5**, 497–526.

8 Fratzl, P., Gupta, H.S., Paschalis, E.P. and Roschger, P. (2004) *Journal of Materials Chemistry*, **14**, 2115–2123.

9 Su, X.W. and Cui, F.Z. (1999) *Materials Science and Engineering C*, **7**, 19–29.

10 Schaffer, T.E., Ionescu-Zanetti, C., Proksch, R., Fritz, M., Walters, D.A., Almqvist, N., Zaremba, C.M., Belcher, A.M., Smith, B.L., Stucky, G.D., Morse, D.E. and Hansma, P.K. (1997) *Chemistry of Materials*, **9**, 1731–1740.

11 Zaremba, C.M., Morse, D.E., Mann, S., Hansma, P.K. and Stucky, G.D. (1998) *Chemistry of Materials*, **10**, 3813–3824.

12 Smith, B.L., Schaffer, T.E., Viani, N., Thompson, J.B., Frederick, N.A., Kindt, J., Belcher, A., Stucky, G.D., Morse, D.E. and Hansma, P.K. (1999) *Nature*, **399**, 761–763.

13 Katti, D.R., Pradhan, S.M. and Katti, K.S. (2004) *Reviews on Advanced Materials Science*, **6**, 162–168.

14 Sellinger, A., Weiss, P.M., Nguyen, A., Lu, Y., Assink, R.A., Gong, W. and Brinker, C.J. (1998) *Nature*, **394**, 256–260.

15 Almqvist, N., Thomson, N.H., Smith, B.L., Stucky, G.D., Morse, D.E. and Hansma, P.K. (1999) *Materials Science and Engineering C*, **7**, 37–43.

16 Pezzotti, G., Asmus, S.M.F., Ferroni, L. P. and Miki, S. (2002) *Journal of Materials Science Materials in Medicine*, **13**, 783–787.

17 Tang, Z., Kotov, N.A., Magonov, S. and Ozturk, B. (2003) *Nature Materials*, **2**, 413–418.

18 Ruiz-Hitzky, E. (2003) *The Chemical Record*, **3**, 88–100.

19 Sanchez, C. (2001) Biomimetisme et Matériaux. Observatoire Françaisе des Techniques Avancées, Série ARAGO 25, OFTA, Paris.

20 Wetherbee, R. (2002) *Science*, **298**, 547–547.

21 Sumper, M. and Brunner, E. (2006) *Advanced Functional Materials*, **16**, 17–26.

22 Coradin, T. and Livage, J. (2001) *Colloids and Surfaces B: Biointerfaces*, **21**, 329–336.

23 Coradin, T., Durupthy, O. and Livage, J. (2002) *Langmuir*, **18**, 2331–2336.

24 López, P.J., Gautier, C., Livage, J. and Coradin, T. (2005) *Current Nanoscience*, **1**, 73–83.

25 Ikoma, T., Muneta, T. and Tanaka, J. (2000) *Key Engineering Materials*, **192-1**, 487–490.

26 Kikuchi, M., Itoh, S., Ichinose, S., Shinomiya, K. and Tanaka, J. (2001) *Biomaterials*, **22**, 1705–1711.

27 Itoh, S., Kikuchi, M., Koyama, Y., Takakuda, K., Shinomiya, K. and Tanaka, J. (2004) *Cell Transplantation*, **13**, 451–461.

28 Berry, C. (2005) *Journal of Materials Chemistry*, **15**, 543–547.

29 Tartaj, P., Morales, M.P., Veintemillas-Verdaguer, S., González-Carreño, T. and Serna, C.J. (2003) *Journal of Physics D-Applied Physics*, **36**, R182–R197.

30 Retuert, J., Quijada, R., Arias, V. and Yazdani-Pedram,M. (2003) *Journal of Materials Research*, **18**, 487–494.

31 Darder, M., López-Blanco, M., Aranda, P., Aznar, A.J., Bravo, J. and Ruiz-Hitzky, E. (2006) *Chemistry of Materials*, **18**, 1602–1610.

32 Olmo, N., Lizarbe, M.A. and Gavilanes, J.G. (1987) *Biomaterials*, **8**, 67–69.

33 Olmo, N., Turnay, J., Herrera, J.I., Gavilanes, J.G. and Lizarbe, M.A. (1996) *Journal of Biomedical Materials Research*, **30**, 77–84.

34 Pérez-Castells, R., Alvarez, A., Gavilanes, J., Lizarbe, M.A., Martínez del Pozo, A., Olmo, N. and Santarén, J. (1987) (eds L.G. Schultz H. vanOlphen , F.A. Mumpton), Proceedings of the International Clay Conference, Denver, 1985, The Clay Mineral Society, Bloomington, 359–362.

35 Narain, R., Housni, A. and Lane, L. (2006) *Journal of Polymer Science Part A: Polymer Chemistry*, **44**, 6558–6568.

36 Poulsen, N., Sumper, M. and Kröger, N. (2003) *Proceedings of the National Academy of Sciences of the United States of America*, **100**, 12075–12080.

37 Schröder, H.C., Boreiko, A., Korzhev, M., Tahir, M.N., Tremel, W., Eckert, C., Ushijima, H., Müller, I.M. and Müller,W.E.G. (2006) *Journal of Biological Chemistry*, **281**, (17), 12001–12009.

38 Cha, J.N., Stucky, G.D., Morse, D.E. and Deming, T.J. (2000) *Nature*, **403**, 289–292.

39 Patwardhan, S.V., Mukherjee, N., Steinitz-Kannan, M. and Clarson, S.J. (2003) *Chemical Communications*, 1122–1123.

40 Belton, D., Paine, G., Patwardhan, S.V. and Perry, C.C. (2004) *Journal of Materials Chemistry*, **14**, 2231–2241.

41 Sumper, M. and Kröger, N. (2004) *Journal of Materials Chemistry*, **14**, 2059–2065.

42 Gautier, C., Lopez, P.J., Hemadi, M., Livage, J. and Coradin, T. (2006) *Langmuir*, **22**, 9092–9095.

43 Coradin, T., Allouche, J., Boissière, M. and Livage, J. (2006) *Current Nanoscience*, **2**, 219–230.

44 Coradin, T., Coupé, A. and Livage, J. (2003) *Colloids and Surfaces B Biointerfaces*, **29**, 189–196.

45 Shchipunov, Y.A., Kojima, A. and Imae, T. (2005) *Journal of Colloid and Interface Science*, **285**, 574–580.

46 Fuentes, S., Retuert, P.J., Gonzalez, G. and Ruiz-Hitzky, E. (1997) *International Journal of Polymeric Materials*, **35**, 61–70.

47 Silva, S.S., Ferreira, R.A.S., Fu, L.S., Carlos, L.D., Mano, J.F., Reis, R.L. and Rocha, J. (2005) *Journal of Materials Chemistry*, **15**, 3952–3961.

48 Rashidova, S.Sh., Shakarova, D.Sh., Ruzimuradov, O.N., Satubaldicva, D. T., Zalyalieva, S.V., Shpigun, O.A., Varlamov, V.P. and Kabulov, B.D. (2004) *Journal of Chromatography B*, **800**, 49–53.

49 Livage, J., Coradin, T. and Roux, C. (2004) in Functional Hybrid Materials (P. Gómez-Romero andC. Sanchez) Wiley-VCH,Weinheim, Ch. 11.

50 Boissière, M., Meadows, P.J., Brayner, R., Hélary, C., Livage, J. and Coradin, T. (2006) *Journal of Materials Chemistry*, **16**, 1178–1182.

51 Boissière, M., Tourrette, A., Devoisselle, J.M., Di Renzo, F. and Quignard, F. (2006) *Journal of Colloid and Interface Science*, **294**, 109–116.

52 Böttcher, H., Jagota, C., Trepte, J., Kallies, K.-H. and Haufe, H. (1999) *Journal of Controlled Release*, **60**, 57–65.

53 Zhang, L. and Dong, S. (2006) *Analytical Chemistry*, **78**, 5119–5123.

54 Jin, W. and Brennan, D. (2002) *Analytica Chimica Acta*, **461**, 1–36.

55 Avnir, D., Coradin, T., Lev, O. and Livage, J. (2006) *Journal of Materials Chemistry*, **16**, 1013–1030.

56 Frenkel-Mullerad, H. and Avnir, D. (2005) *Journal of the American Chemical Society*, **127**, 8077–8081.

57 Gill, I. and Ballesteros, A. (2000) *Trends in Biotechnology*, **18**, 282–296.

58 Ferrer, M.L., del Monte, F. and Levy, D. (2002) *Chemistry of Materials*, **14**, 3619–3621.

59 Gill, I. and Ballesteros, A. (1998) *Journal of the American Chemical Society*, **120**, 8587–8598.

60 Shchipunov, Y.A., Karpenko, T.Y., Bakunina, I.Y., Burtseva, Y.V. and Zvyagintseva, T.N. (2004) *Journal of Biochemical and Biophysical Methods*, **58**, 25–38.

61 Darder, M., Colilla, M., Lara, N. and Ruiz-Hitzky, E. (2002) *Journal of Materials Chemistry*, **12**, 3660–3664.

62 Fennouh, S., Guyon, S., Livage, J. and Roux, C. (2000) *Journal of Sol-Gel Science and Technology*, **19**, 647–649.

63 Nassif, N., Roux, C., Coradin, T., Rager, M.N., Bouvet, O.M.M. and Livage, J. (2003) *Journal of Materials Chemistry*, **13**, 203–208.

64 Nassif, N., Roux, C., Coradin, T., Bouvet, O.M.M. and Livage, J. (2004) *Journal of Materials Chemistry*, **14**, 2264–2268.

65 Ferrer, M.L., Garcia-Carvajal, Z.Y., Yuste, L., Rojo, F. and del Monte, F. (2006) *Chemistry of Materials*, **18**, 1458–1463.

66 Katagiri, K., Ariga, K. and Kikuchi, J.-I. (1999) *Chemistry Letters*, 661–662.

67 Katagiri, K., Hamasaki, R., Ariga, K. and Kikuchi, J.-I. (2002) *Journal of the American Chemical Society*, **124**, 7892–7893.

68 Katagiri, K., Hamasaki, R., Ariga, K. and Kikuchi, J.-I. (2003) *Journal of Sol-Gel Science and Technology*, **26**, 393–396.

69 Ariga, K. (2004) *The Chemical Record*, **3**, 297–307.

70 Ariga, K., Vinu, A. and Miyahara, M. (2006) *Current Nanoscience*, **2**, 197–210.

71 Ruiz-Hitzky, E., Letaief, S. and Prévot, V. (2002) *Advanced Materials*, **14**, 439–443.

72 http://en.wikipedia.org/wiki/Ceramic.

73 Burgos-Asperilla, L., Darder, M., Aranda, P., Vázquez, L., Vázquez, M. and Ruiz-Hitzky, E. (2007) *Journal of Materials Chemistry*, DOI: 10.1039/b706011d.

74 Murugan, R. and Ramakrishna, S. (2005) *Composites Science and Technology*, **65**, 2385–2406.

75 Xu, A.-W., Ma, Y. and Cölfen, H. (2007) *Journal of Materials Chemistry*, **17**, 415–449.

76 Ameye, L., DeBecker, G., Killian, C., Wilt, F., Kemps, R., Kuypers, S. and Dubois, P. (2001) *Journal of Structural Biology*, **134**, 56–66.

77 Borelli, G., Mayer-Gostan, N., Merle, P. L., DePontual, H., Boeuf, G., Allemand, D. and Payan, P. (2003) *Calcified Tissue International*, **72**, 717–725.

78 Raabe, D., Romano, P., Sachs, C., Al-Sawalmih, A., Brokmeier, H.-G., Yi, S.-B., Servos, G. and Hartwig, H.G. (2005) *Journal of Crystal Growth*, **283**, 1–7.

79 Falini, G. (2000) *International Journal of Inorganic Materials*, **2**, 455–461.

80 Kato, T., Sugawara, A. and Hosoda, N. (2002) *Advanced Materials*, **14**, 869–877.

81 Manoli, F., Koutsopoulos, S. and Dalas, E. (1997) *Journal of Crystal Growth*, **182**, 116–124.

82 Manoli, F. and Dalas, E. (1999) *Journal of Crystal Growth*, **204**, 369–375.

83 Wang, W., Wang, G., Liu, Y., Zheng, C. and Zhan, Y. (2001) *Journal of Materials Chemistry*, **11**, 1752–1754.

84 Gower, L.A. and Tirrell, D.A. (1998) *Journal of Crystal Growth*, **191**, 153–160.

85 Neira-Carrillo, A., Yazdani-Pedram, M., Retuert, J., Diaz-Dosque, M., Gallois, S. and Arias, J.L. (2005) *Journal of Colloid and Interface Science*, **286**, 134–141.

86 Kuang, M., Wang, D., Gao, M., Hartmann, J. and Möhwald, H. (2005) *Chemistry of Materials*, **17**, 656–660.

87 Kato, T. (2000) *Advanced Materials*, **12**, 1543–1546.

88 Sugawara, A. and Kato, T. (2000) *Chemical Communications*, 487–488.

89 Wakayama, H., Hall, S.R. and Mann, S. (2005) *Journal of Materials Chemistry*, **15**, 1134–1136.

90 Dorozhkin, S.V. and Epple, M. (2002) *Angewandte Chemie-International Edition*, **41**, 3130–3146.

91 Palin, E., Liu, H. and Webster, T.J. (2005) *Nanotechnology*, **16**, 1828–1835.

92 Thomas, V., Dean, D.R. and Vohra, Y.K. (2006) *Current Nanoscience*, **2**, 155–177.

93 Widmer, M.S. and Mikos, A.G. (1998) in Frontiers in Tissue Engineering (eds C.W. Jr Patrick , A.G. Mikos , L.V. McIntire), Elsevier Science Ltd, Oxford, Ch. II.5.

94 Yokoyama, A., Gelinsky, M., Kawasaki, T., Kohgo, T., König, U., Pompe, W. and Watari, F. (2005) *Journal of Biomedical*

Materials Research Part B, **75B**, 464–472.

95 Wang, Y.J., Yang, C.R., Chen, X.F. and Zhao, N.R. (2006) *Advanced Engineering Materials*, **8**, 97–100.

96 Tampieri, A., Sandri, M., Landi, E., Celotti, G., Roveri, N., Mattioli-Belmonte, M., Virgili, L., Gabbanelli, F. and Biagini, G. (2005) *Acta Biomaterialia*, **1**, 343–351.

97 Parhi, P., Ramanan, A. and Ray, A.R. (2006) *Journal of Applied Polymer Science*, **102**, 5162–5165.

98 Rusu, V.M., Ng, C.-H., Wilke, M., Tiersch, B., Fratzl, P. and Peter, M.G. (2005) *Biomaterials*, **26**, 5414–5426.

99 Kong, L., Gao, Y., Lu, G., Gong, Y., Zhao, N. and Zhang, X. (2006) *European Polymer Journal*, **42**, 3171–3179.

100 Nayar, S., Sinha, M.K., Basu, D. and Sinha, A. (2006) *Journal of Materials Science, Materials in Medicine*, **17**, 1063–1068.

101 Ritzoulis, C., Scoutaris, N., Demetriou, E., Papademetriou, K., Kokkou, S., Stavroulias, S. and Panayiotou, C. (2004) *Journal of Biomedical Materials Research*, **71A**, 675–684.

102 Ishikawa, Y., Komotori, J. and Senna, M. (2006) *Current Nanoscience*, **2**, 191–196.

103 Senna, M. (2005) *Materials Science and Engineering A*, **412**, 37–42.

104 Kino, R., Ikoma, T., Monkawa, A., Yunoki, S., Munekata, M., Tanaka, J. and Asakura, T. (2006) *Applied Polymer Science*, **99**, 2822–2830.

105 Takeuchi, A., Ohtsuki, C., Miyazaki, T., Kamitakahara, M., Ogata, S.-I., Yamazaki, M., Furutani, Y., Kinoshita, H. and Tanihara, M. (2005) *Journal of the Royal Society Interface*, **2**, 373–378.

106 Kothapalli, C.R., Shaw, M.T. and Wei, M. (2005) *Acta Biomaterialia*, **1**, 653–662.

107 Mathieu, L.M., Mueller, T.L., Bourban, P.-E., Pioletti, D.P., Müller, R. and Månson, J.-A.E. (2006) *Biomaterials*, **27**, 905–916.

108 Rezwan, K., Chen, Q.Z., Blaker, J.J. and Boccaccini, A.R. (2006) *Biomaterials*, **27**, 3413–3431.

109 Deville, S., Saiz, E., Nalla, R.K. and Tomsia, A.P. (2006) *Science*, **311**, 515–518.

110 Sotome, S., Uemura, T., Kikuchi, M., Chen, J., Itoh, S., Tanaka, J., Tateishi, T. and Shinomiya, K. (2004) *Materials Science and Engineering C*, **24**, 341–347.

111 Liu, T.-Y., Chen, S.-Y., Li, J.-H. and Liu, D.-M. (2006) *Journal of Controlled Release*, **112**, 88–95.

112 Vallet-Regí, M. and González-Calbet, J.M. (2004) *Progress in Solid State Chemistry*, **32**, 1–31.

113 Zou, C., Weng, W.J., Deng, X.L., Cheng, K., Liu, X.G., Du, P.Y., Shen, G. and Han, G.R. (2005) *Biomaterials*, **26**, 5276–5284.

114 Liu, H., Li, H., Cheng, W.J., Yang, Y., Zhu, M.Y. and Zhou, C.R. (2006) *Acta Biomaterialia*, **2**, 557–565.

115 Aunoble, S., Clement, D., Frayssinet, P., Harmand, M.F. and LeHuec, J.C. (2006) *Journal of Biomedical Materials Research Part A*, **78A**, 416–422.

116 Takagi, S., Chow, L.C., Hirayama, S. and Eichmiller, F.C. (2003) *Dental Materials*, **19**, 797–804.

117 Xu, H.H.K. and Simon, C.G. (2005) *Biomaterials*, **26**, 1337–1348.

118 Simon, P., Schwarz, U. and Kniep, R. (2005) *Journal of Materials Chemistry*, **15**, 4992–4996.

119 Tlatlik, H., Simon, P., Kawska, A., Zahn, D. and Kniep, R. (2006) *Angewandte Chemie-International Edition*, **45**, 1905–1910.

120 Bergaya, F. Theng, B.K.G. and Lagaly, G. (eds) (2006) Handbook of Clay Science,Elsevier, Amsterdam.

121 Ruiz-Hitzky, E., Aranda, P. and Serratosa, J.M. (2004) in Handbook of Layered Materials (eds S.M. Auerbach, K.A. Carrado,

P.K.Dutta), Marcel Dekker, New York, pp. 91–154.

122 Ruiz-Hitzky, E. and Van Meerbeeck, A. (2006) in Handbook of Clay Science (eds F. Bergaya B.K.G. Theng, G. Lagaly), Elsevier, Amsterdam, Ch. 10.3.

123 Ruiz-Hitzky, E., Aranda, P. and Darder, M. (2007) in Bottom-Up Nanofabrication: Supramolecules, Self-Assemblies, and Organized Films (eds. Ariga, K. and Nalwa, H.S.) American Scientific Publishers 9, Ch. 73.

124 Bradley, W.F. (1945) *Journal of the American Chemical Society*, **67**, 975–981.

125 Greenland, D.J. (1956) *Journal of Soil Science*, **7**, 319–328.

126 Darder, M. and Ruiz-Hitzky, E. (2005) *Journal of Materials Chemistry*, **15**, 3913–3918.

127 Aranda, P., Darder, M., Fernández-Saavedra, R., López-Blanco, M. and Ruiz-Hitzky, E. (2006) *Thin Solid Films*, **495**, 104–112.

128 Bakandritsos, A., Steriotis, Th. and Petridis,D. (2004) *Chemistry of Materials*, **16**, 1551–1559.

129 Darder, M., Colilla, M. and Ruiz-Hitzky, E. (2003) *Chemistry of Materials*, **15**, 3774–3780.

130 Darder, M., Colilla, M. and Ruiz-Hitzky, E. (2005) *Applied Clay Science*, **28**, 199–208.

131 Xu, Y. and Hanna, M.A.Chitosan/clay nanocomposite films preparation and characterization Abstract in 2005 IFT Annual Meeting July 15–20, New Orleans, Louisiana.

132 Colilla, M. (2004) PhD Dissertation, Autonomous University of Madrid, Madrid.

133 Cohen, E., Joseph, T., Kahana, F. and Magdassi, S. (2003) *Photochemistry and Photobiology*, **77**, 180–185.

134 Chang, M.Y. and Juang, R.S. (2004) *Journal of Colloid and Interface Science*, **278**, 18–25.

135 Qiu, H.X., Yu, J.G. and Zhu, J.L. (2005) *Polymers and Polymer Composites*, **13**, 167–172.

136 Ogawa, M. and Kuroda, K. (1997) *Bulletin of the Chemical Society of Japan*, **70**, 2593–2618.

137 Ray, S.S. and Okamoto, M. (2003) *Macromolecular Rapid Communications*, **24**, 815–840.

138 Park, H.M., Misra, M., Drzal, L.T. and Mohanty, A.K. (2004) *Biomacromolecules*, **5**, 2281–2288.

139 Wibowo, A.C., Misra, M., Park, H.-M., Drzal, L.T., Schalek, R. and Mohanty, A. K. (2006) *Composites Part A: Applied Science and Manufacturing*, **37**, 1428–1433.

140 Kalambur, S. and Rizvi, S.S.H. (2005) *Journal of Applied Polymer Science*, **96**, 1072–1082.

141 Park, H.M., Mohanty, A.K., Drzal, L.T., Lee, E., Mielewski, D.F. and Misra, M. (2006) *Journal of Polymers and the Environment*, **14**, 27–35.

142 Maiti, P., Yamada, K., Okamoto, M., Ueda, K. and Okamoto, K. (2002) *Chemistry of Materials*, **14**, 4654–4661.

143 Paul, M.-A., Alexandre, M., Degeé, P., Henrist, C., Rulmont, A. and Dubois, P. (2003) *Polymer*, **44**, 443–450.

144 Paul, M.-A., Delcourt, C., Alexandre, M., Degeé, P., Monteverde, F., Rulmont, A. and Dubois, P. (2005) *Macromolecular Chemistry and Physics*, **206**, 484–498.

145 Pluta, M., Galeski, A., Alexandre, A., Paul, M.-A. and Dubois, P. (2002) *Journal of Applied Polymer Science*, **86**, 1497–1506.

146 Ray, S.S., Yamada, K., Okamoto, M. and Ueda, K. (2003) *Macromolecular Materials and Engineering*, **288**, 203–208.

147 Paul, M.-A., Delcourt, C., Alexandre, M., Degeé, P., Monteverde, F. and Dubois, P. (2005) *Polymer Degradation and Stability*, **87**, 535–542.

148 Brauner, K. and Preisinger, A. (1956) *Mineralogy and Petrology*, **6**, 120–140.

149 Santarén, J., Sanz, J. and Ruiz-Hitzky, E. (1990) *Clays and Clay Minerals*, **38**, 63–68.

150 Ahlrichs, J.L., Serna, C. and Serratosa, J.M. (1975) *Clays and Clay Minerals*, **23**, 119–124.

151 Ruiz-Hitzky, E. (2001) *Journal of Materials Chemistry*, **11**, 86–91.

152 Darder, M. and Ruiz-Hitzky, E. (2007) in Clay-Based Polymer Nanocomposite, Clay Minerals Society, Vol. 14 (eds. Carrado, K.A. and Bergaya, F.), The Clay Minerals Society, Chantilly (Virginia, EEUU), Ch. 8.

153 Gómez-Avilés, A., Darder, M., Aranda, P. and Ruiz-Hitzky,E. (2007) *Angewandte Chemie-International Edition*, **46**, 923–925.

154 Chang, S.H., Ryan, M.E. and Gupta, R. K. (1991) *Journal of Applied Polymer Science*, **43**, 1293–1299.

155 Li, A., Liu, R. and Wang, A. (2005) *Journal of Applied Polymer Science*, **98**, 1351–1357.

156 Leroux, F., Gachon, J. and Besse, J.-P. (2004) *Journal of Solid State Chemistry*, **177**, 245–250.

157 Darder, M., López-Blanco, M., Aranda, P., Leroux, F. and Ruiz-Hitzky, E. (2005) *Chemistry of Materials*, **17**, 1969–1977.

158 Choy, J.H., Kwak, S.Y., Park, J.S., Jeong, Y.J. and Portier, J. (1999) *Journal of the American Chemical Society*, **121**, 1399–1400.

159 Choy, J.H., Kwak, S.Y., Jeong, Y.J. and Park, J.S. (2000) *Angewandte Chemie-International Edition*, **39**, 4042–4045.

160 Kwak, S.Y., Jeong, Y.J., Park, J.S. and Choy, J.H. (2002) *Solid State Ionics*, **151**, 229–234.

161 Forano, C., Vial, S. and Mousty, C. (2006) *Current Nanoscience*, **2**, 283–294.

162 Gleitsmann, T., Bernhardt, T.M. and Wöste, L. (2006) *Applied Physics A*, **82**, 125–130.

163 Lee, S. and Pérez-Luna, V.H. (2005) *Analytical Chemistry*, **77**, 7204–7211.

164 dosSantos, D.S., Jr, Goulet, P.J.G., Pieczonka, N.P.W., Oliveira, O.N. Jr, and Aroca, R.F. (2004) *Langmuir*, **20**, 10273–10277.

165 Li, Z., Du, Y., Zhang, Z. and Pang, D. (2003) *Reactive and Functional Polymers*, **55**, 35–43.

166 Brayner, R., Coradin, T., Fiévet-Vincent, F., Livage, J. and Fiévet, F. (2005) *New Journal of Chemistry*, **29**, 681–685.

167 Srivastava, S., Samanta, B., Arumugam, P., Han, G. and Rotello, V. M. (2007) *Journal of Materials Chemistry*, **17**, 52–55.

168 Cao, Y., Zhou, Y., Shan, Y., Ju, H. and Xue, X. (2006) *Advanced Materials*, **18**, 1838–1841.

169 Katz, E., Shipway, A.N. and Willner, I. (2004) in Nanoparticles. From Theory to Applications (ed. G. Schmid),Wiley VCH,Weinheim, Ch. 6.

170 Rawat, M., Singh, D. and Saraf, S. (2006) *Biological and Pharmaceutical Bulletin*, **29**, 1790–1798.

171 Ha?feli, U. Schütt, W. and Teller, J. Zborowski, M. (eds) (1997) Scientific and Clinical Applications of Magnetic Carriers, Plenum Press, New York.

172 Bergemann, C., Mullerschulte, D., Oster, J., Abrassard, L. and Lubbe, A.S. (1999) *Journal of Magnetism and Magnetic Materials*, **194**, 45–52.

173 Gómez Lopera, S.A., Plaza, R.C. and Delgado, A.V. (2001) *Journal of Colloid and Interface Science*, **240**, 40–47.

174 Arias, J.L., Gallardo, V., Gómez Lopera, S.A., Plaza, R.C. and Delgado, A.V. (2001) *Journal of Controlled Release*, **77**, 309–321.

175 Tan, S.T., Wendorff, J.H., Pietzonka, C., Jia, Z.H. and Wang, G.Q. (2005) *ChemPhysChem*, **6**, 1461–1465.

176 Menager, C. and Cabuil, V. (1995) *Journal of Colloid and Interface Science*, **69**, 251–253.

177 Kuznetsov, A.A., Filippov, V.I., Alyautdin, R.N., Torshina, N.L. and Kuznetsov, O.A. (2001) *Journal of Magnetism and Magnetic Materials*, **225**, 95–100.

178 Luckarift, H.R., Dickerson, M.B., Sandhage, K.H. and Spain, J.C. (2006) *Small*, **2**, 640–643.

179 Lu, Y., Liu, Y., Xu, J., Xu, C., Liu, B. and Kong, J. (2005) *Sensors*, **5**, 258–265.

180 Hiroi, R., Ray, S.S., Okamoto, M. and Shiroi, T. (2004) *Macromolecular Rapid Communications*, **25**, 1359–1364.

181 Liu, T.-Y., Liao, H.-C., Lin, C.-C., Hu, S.-H. and Chen, S.-Y. (2006) *Langmuir*, **22**, 5804–5809.

182 Iijima, S. (1991) *Nature*, **354**, 56–58.

183 Shaito, R., Dresselhaus, G. and Dresselhaus, M.S. (1999) Physical Properties of Carbon Nanotubes, Imperial College Press, London.

184 Tjong, S.C. (2006) *Materials Science and Engineering: R: Reports*, **53**, 73–197.

185 Moniruzzaman, M. and Winey, K.I. (2006) *Macromolecules*, **39**, 5194–5205.

186 Coleman, J.N., Khan, U., Blau, W.J. and Gunko,Y.K. (2006) *Carbon*, **44**, 1624–1652.

187 Coleman, J.N., Khan, U. and Gunko,Y.K. (2006) *Advanced Materials*, **18**, 689–706.

188 Wang, S.-F., Shen, L., Zhang, W.-D. and Tong, Y.-L. (2005) *Biomacromolecules*, **6**, 3067–3072.

189 Shieh, Y.-T. and Yang, Y.-F. (2006) *European Polymer Journal*, **42**, 3162–3170.

190 Spinks, G.M., Shin, S.R., Wallace, G. G., Whitten, P.G., Kim, S.I. and Kim, S. J. (2006) *Sensors and Actuators B: Chemical*, **115**, 678–684.

191 Hasegawa, H., Fujisaka, T., Numata, M., Umeda, M., Matsumoto, T., Kimura, T., Okumura, S., Sakurai, K. and Shinkai, S. (2004) *Chemical Communications*, 2150–2151.

192 Chen, X., Lee, G.S., Zettl, A. and Bertozzi, C.R. (2004) *Angewandte Chemie-International Edition*, **43**, 6112–6116.

193 Matsura, K., Hayashi, K. and Kinmizuka, N. (2003) *Chemistry Letters*, **32**, 212–213.

194 Gu, L., Elkin, T., Jiang, X., Li, H., Qu, L., Tzeng, T.-Z.J., Joseph, R. and Sun, Y.-P. (2005) *Chemical Communications*, 874–876.

195 Dohi, H., Kikuchi, S., Kuwahara, S., Sugai, T. and Shinohara, H. (2006) *Chemical Physics Letters*, **428**, 98–101.

196 Kam, N.W., OConnell, M., Wisdom, J. A. and Dai, H. (2005) *Proceedings of the National Academy of Sciences of the United States of America*, **102**, 11600–11605.

197 Li, X., Peng, Y. and Qu, X. (2006) *Nucleic Acids Research*, **34**, 3670–3676.

198 Koyama, S., Haniu, H., Osaka, K., Koyama, H., Kuroiwa, N., Endo, M., Kim, Y.A. and Hayashi, T. (2006) *Small*, **2**, 1406–1411.

199 Smart, S.K., Cassady, A.I., Lu, G.Q. and Martin, D.J. (2006) *Carbon*, **44**, 1034–1047.

200 Tao, F., Gonzalez-Flecha, B. and Kobzik, L. (2003) *Free Radical Biology and Medicine*, **35**, 327–340.

201 Polizu, S., Maugey, M., Poulin, S., Polulin, P. and Yahia, L.H. (2006) *Applied Surface Science*, **252**, 6750–6753.

202 Shi, X.F., Hudson, J.L., Spicer, P.P., Tour, J.M., Krishnamoorti, R. and Mikos, A.G. (2006) *Biomacromolecules*, **7**, 2237–2242.

203 Zhang, D., Kandadai, M.A., Cech, J.H., Roth, S. and Curran, S.A. (2006) *Journal of Physical Chemistry B*, **110**, 12910–12915.

204 Wang, Y.D., Joshi, P.P., Hobbs, K.L., Johnson, M.B. and Schmidtke, D.W. (2006) *Langmuir*, **22**, 9776–9783.

205 Liu, G.D. and Lin, Y.H. (2006) *Journal of Nanoscience and Nanotechnology*, **6**, 948–953.

206 Luo, X.-L., Xu, J.-J., Wang, J.-L. and Chen, H.-Y. (2005) *Chemical Communications*, 2169–2171.

207 Chen, L. and Lu, G. (2006) *Journal of Electroanalytical Chemistry*, **597**, 51–59.

208 Li, Y., Lin, X. and Jiang, C. (2006) *Electroanalysis*, **18**, 2085–2091.

209 Wang, Z.-G., Xu, Z.-K., Wan, L.-S., Wu, J., Innocent, C. and Seta, P. (2006) *Macromolecular Rapid Communications*, **27**, 516–521.

210 Zhang, M., Smith, A. and Gorki, W. (2004) *Analytical Chemistry*, **76**, 5045–5050.

211 Livage, J. (1991) *Chemistry of Materials*, **3**, 578–593.

212 Rojas-Cervantes, M.L., Casal, B., Aranda, P., Savirón, M., Galván, J.C. and Ruiz-Hitzky, E. (2001) *Colloid and Polymer Science*, **279**, 990–1004.

213 Durupthy, O., Steunou, N., Coradin, T. and Livage, J. (2006) *Journal of the Physics and Chemistry of Solids*, **67**, 944–949.

214 Oliveira, H.P., Graeff, C.F.O., Zanta, C.L.P.S., Galina, A.C. and Goncalves, P.J. (2000) *Journal of Materials Chemistry*, **10**, 371–375.

215 Arashiro, E., Zampronio, E.C., Brunello, C.A., Lassali, T.A.F., Oliveira, H.P. and Graeff, C.F.O. (2001) *International Journal of Inorganic Materials*, **3**, 727–731.

216 Darder, M., Aranda, P. and Ruiz-Hitzky, E., unpublished results.

217 Gleezer, V. and Lev, O. (1993) *Journal of the American Chemical Society*, **115**, 2533–2534.

218 Gao, Q., Suib, S.L. and Rusling, J.F. (2002) *Chemical Communications*, 2254–2255.

219 Wang, L., Schindler, J., Kannewurf, C.R. and Kanatzidis, M.G. (1997) *Journal of Materials Chemistry*, **7**, 1277–1283.

220 Ruiz, A.I., Darder, M., Aranda, P., Jiménez, R., Van Damme, H. and Ruiz-Hitzky, E. (2006) *Journal of Nanoscience and Nanotechnology*, **6**, 1602–1610.

221 Gao, L., Gao, Q., Wang, Q., Peng, S. and Shi, J. (2005) *Biomaterials*, **25**, 5267–5275.

222 Ding, Y., Jones, D.J., Maireles-Torres, P. and Rozière, J. (1995) *Chemistry of Materials*, **7**, 562–571.

223 Kumar, C.V. and Chaudari, A. (2000) *Journal of the American Chemical Society*, **122**, 830–837.

224 Coradin, T., Coupé, A. and Livage, J. (2003) *Journal of Materials Chemistry*, **13**, 705–707.

2
Biomimetic Nanohybrids Based on Organosiloxane Units

Kazuko Fujii, Jonathan P. Hill, Katsuhiko Ariga

2.1
Introduction

Some kinds of biomolecules such as proteins have highly sophisticated functions compared to those of man-made artificial systems. These functions originate from an appropriate molecular structure, which is formed through the harmonized actions of peptide segments in a rational sequence. Proteins often form organized clusters where functional relays result in highly specific and efficient conversion of material, energy, and information. Therefore, biological systems can be regarded as the ultimate goals of current science and technologies including supramolecular chemistry and nanotechnology [1–4]. The field of chemistry that focuses on mimicking biological functions and creating materials obtained from studying biological supermolecules is called biomimetic chemistry [5–8]. Biomimetic chemistry attempts to create functions comparable with true biological systems but using purely artificial systems, and is an approach which is rather difficult to implement successfully. Alternatively, organization of biological components within artificial structures and/or hybridization of biomolecules with artificial materials are two methods often used as practicable approaches. Biomolecules such as proteins have been assembled into artificial structures using various supramolecular techniques such as the Langmuir-Blodgett (LB) method [9–14] and layer-by-layer (LbL) assembly [15–23].

Two examples are shown in Figures 2.1 and 2.2. Figure 2.1 depicts an example of a hybrid electrode for glucose sensing where glucose oxidase is immobilized at the working electrode through the LB process [24,25]. Immobilization of proteins using the LB technique may cause their denaturation due to the high surface tension on a water surface. In order to avoid this drawback, glucose oxidase was spread on water in the form of a lipid-coated enzyme that could be obtained as a precipitate simply by mixing aqueous solutions of the enzyme. This lipid-coated glucose oxidase is soluble in organic solvents and forms a stable monolayer on a water surface with minimal surface denaturation. The monolayer containing glucose oxidase was subsequently transferred onto a Pt electrode and employed for amperometric glucose sensing. The LB transfer process usually requires efficient packing of lipid components, which

Bio-inorganic Hybrid Nanomaterials. Edited by E. Ruiz-Hitzky, K. Ariga, Y. M. Lvov

ISBN: 978-3-527-31718-9

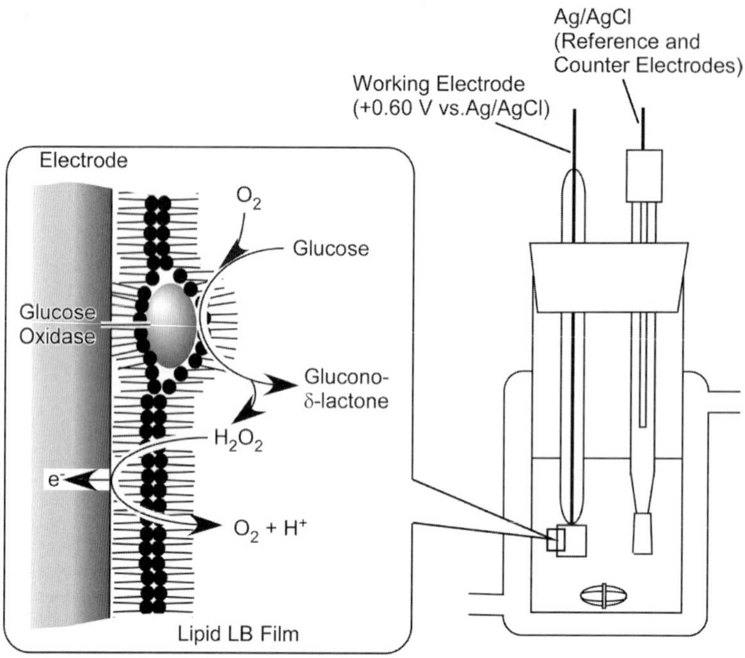

Fig. 2.1 Hybrid electrode for glucose sensing where glucose oxidase is immobilized on the working electrode through the LB process. Adapted from [25], Y. Okahata *et al.*, *Thin Solid Films* **1989**, *180*, 65.

may be disadvantageous for substrate diffusion in the film. In contrast, LbL adsorption provides multilayer structures which are permeable to small molecules. Immobilization of enzymes on solid supports is exemplified in Figure 2.2. Glucose oxidase, which is negatively charged under ambient pH conditions, can be assembled alternately with cationic polyelectrolytes as thin films on a quartz plate (Figure 2.2A) [26]. With the aid of peroxidase and a redox dye in solution, the film obtained on transparent quartz can work as a color-indicator of glucose in an analyte solution. The great freedom in film construction of the LbL technique allows us to prepare a multi-enzyme reactor, as illustrated in Figure 2.2B, where glucoamylase and glucose oxidase are sequentially assembled on a porous filter plate [27,28]. This nanoreactor can convert starch into glucose, in the first process, that is subsequently oxidized into gluconolactone at the second enzyme (glucose oxidase) layer.

These examples are still at a primitive stage when compared with real bio-organisms, but certainly suggest attractive directions for accomplishment of certain biomimetic functions. If introduction of the covalent hybridization concept into such biomimetic systems is successful, unusual mechanical and thermal stabilities can be added to these functions. For example, hybridization of inorganic structures such as glasses and mesoporous silicates with highly functionalized biomaterials such as DNA strands, proteins, and cell membranes should create novel materials that possess both biological functions and high stability (Figure 2.3). Such hybrid

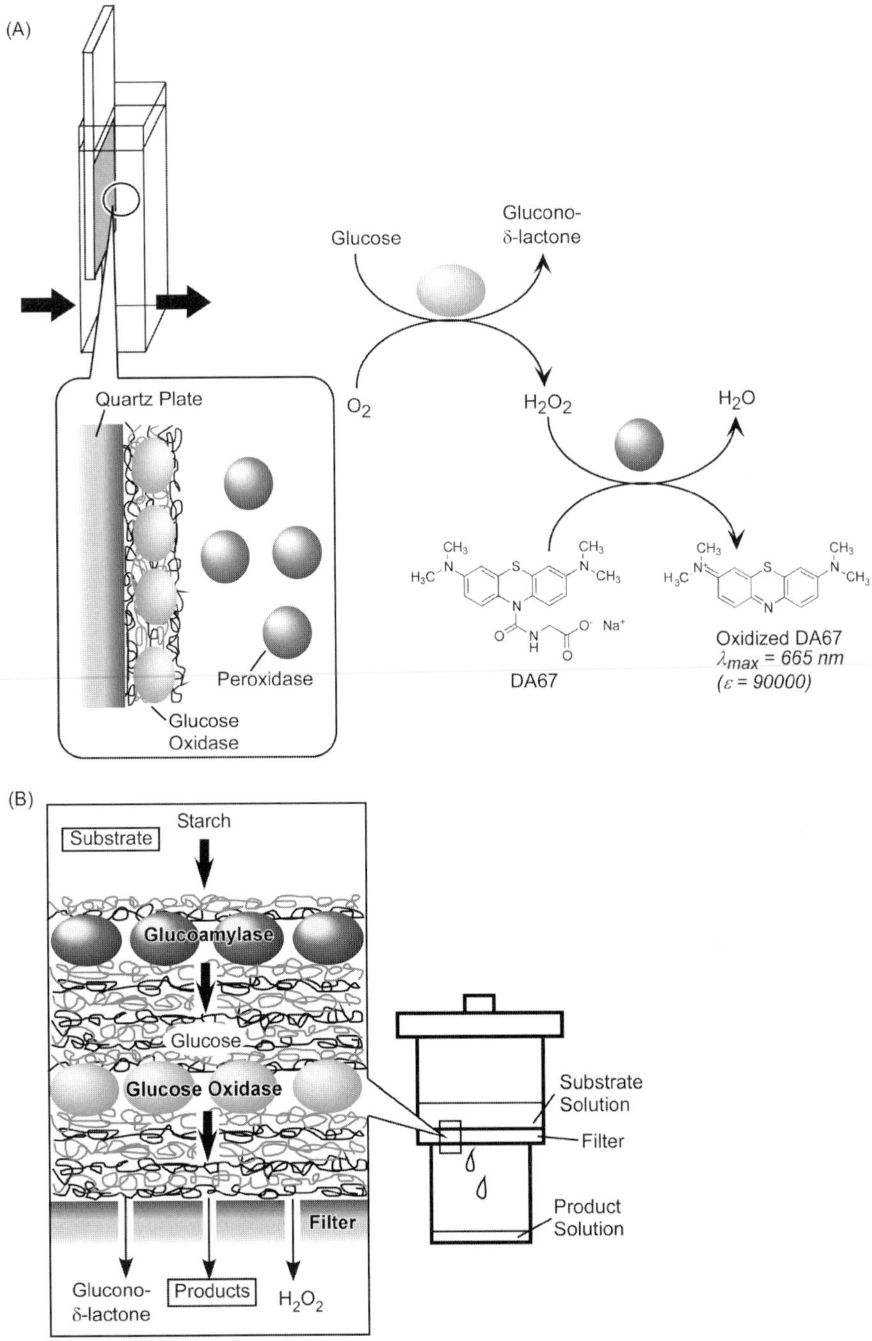

Fig. 2.2 Enzyme reactors prepared by LbL assembly: (A) reactor on quartz plate for color-indication of glucose; (B) multi-enzyme reactor for starch digestion on ultrafilter. Adapted from [26], M. Onda *et al.*, *Biotechnol. Bioeng.* **1996**, *51*, 163 and [27], M. Onda *et al.*, *J. Ferment. Bioeng.* **1996**, *82*, 502.

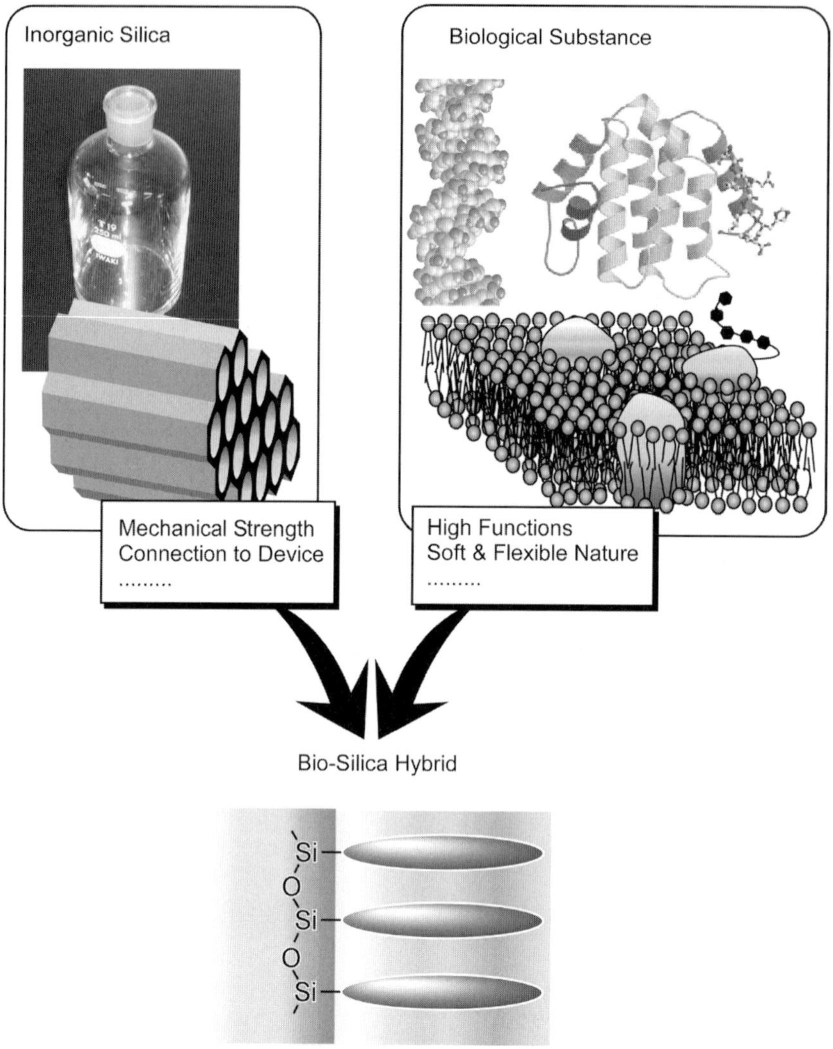

Fig. 2.3 Concept of hybridization of inorganic structures with biomaterials to provide bio-silica hybrids.

materials are useful for attaining biological functions under harsh conditions, which cannot be achieved by simple combinations of bio-components and artificial materials. In addition, higher functions seen in biological systems can be connected to artificial devices such as sensors through appropriate hybridization. One practical method for covalent hybridization of bio-structures into inorganic media is by the use of organosilane compounds. Condensation with appropriate organosilane compounds should result in bio-silica hybrids [29], as shown in Figure 2.3. In particular, synthesis of biomembrane mimics with a siloxane backbone is a good initial target. In this chapter approaches for the preparation of biomimetic nanohybrids, mainly as membrane mimics and based on organosiloxane units, are introduced. Monolayer, multilayer, vesicular, and mesostructured hybrids with silica-like frameworks will be described.

2.2 Monolayer on Solid Support

Lipid monolayer and bilayer structures can be regarded as the fundamental unit of biomembrane mimics. Organosilane compounds are often used for the preparation of self-assembled monolayers (SAM) on a surface of an oxide such as glass (Figure 2.4) [30,31]. When alkylsilane compounds such as octadecyl trichlosilane are used, strong covalent linkages between silanol groups in the organosilane component and the oxide surface are formed. The preparation method is quite simple – immersion of the solid support in a solution containing the corresponding organosilane compound induces self-assembly of the organosilane components as a monolayer on the support surface. In addition, immobilization of dyes, proteins and nucleic acids is possible using this self-assembled monolayer technique [32–35]. Incorporation of silane-immobilized monolayers into multi-layer structures is also possible [36] using a trichlorosilane amphiphile with a long chain terminated in a double bond, which can then be oxidized to hydroxy groups, after SAM immobilization. A second monolayer can then be covalently immobilized to the freshly prepared hydroxy surface. The

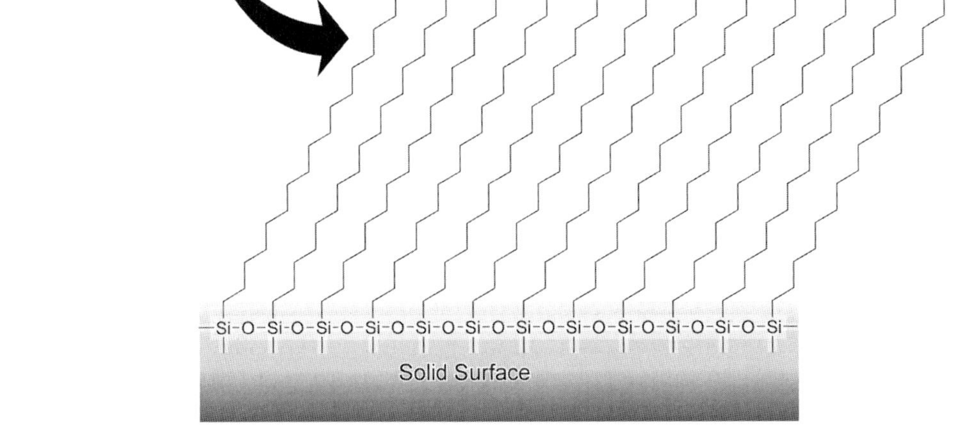

Fig. 2.4 Self-assembled monolayer (SAM) structures from organosilane compounds.

formation of self-assembled monolayers is a powerful tool for surface modification, permitting control of wettability and biocompatibility [37–39]. SAM-modified surfaces have often been used for material sensing, as reported by various researchers [40–42]. Ichimura and coworkers demonstrated a very unique use of silane-immobilized SAM for the orientation control of a liquid crystalline phase [43]. In their method, a photoisomerizable monolayer composed of an azobenzene derivative was immobilized on a solid surface and liquid crystalline layers were then deposited on the monolayer. Photoisomerization of the monolayer then results in a change in the orientation of the thick liquid crystalline layer. Because molecular information is therefore amplified into bulk structural change, the surface monolayer was called the command surface.

Okahata and coworkers prepared a monolayer of dialkylorganosilane on a porous glass plate using a self-adsorption process, and successfully regulated material permeation through the glass [44]. The organosilane amphiphiles penetrated deeply into the pores so that the monolayer structures filled the pores. These condensed structures of lipid membranes led to well-controllable material permeation. In a more advanced system, the same research group demonstrated permeation control through "the thinnest lipid membrane" by combining two techniques – the SAM method and the LB technique [45,46]. In their method, an organosilane monolayer was first prepared and polymerized at the air–water interface and then transferred onto a solid substrate for immobilization (Figure 2.5). A detailed description of this research is given below.

Prior to LB transfer, the surface pressure – molecular area (π–A) isotherms of dialkylsilane under various pH and temperature conditions were investigated. The pH condition of the subphase (water phase under the monolayer) is a crucial factor for the monolayer state. The condensed phase was formed directly without formation

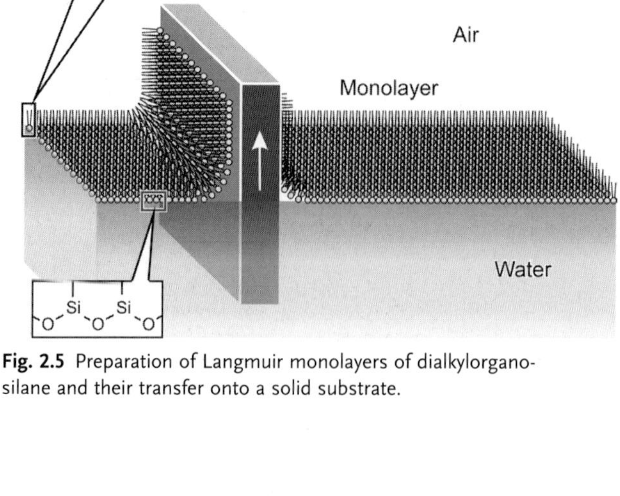

Fig. 2.5 Preparation of Langmuir monolayers of dialkylorganosilane and their transfer onto a solid substrate.

of an expanded state after monolayer spreading at pH 2, while the monolayer on pure water (surface pH 5.8) showed a transition from an expanded phase to a condensed phase. The limiting areas were 0.54 and 0.46 nm^2 at pH 2 and 5.8, respectively, rather close to the corresponding values obtained from the conventional monolayers of dialkyl compounds such as phospholipids [47]. In order to investigate structural changes of organosilane monolayers at pH 2 and pH 5.8, these monolayers (30 layers) were transferred onto a CaF_2 plate and subjected to FTIR analysis. The LB films prepared at pH 2 showed peaks assignable to Si–O–Si at 1026 and 1095 cm^{-1}, while Si–O–C stretching bands at 1080 and 1103 cm^{-1} were observed for the LB films transferred from pure water. This spectroscopic evidence indicates that the organosilane monolayers are easily polymerized to form Si–O–Si linkages under acidic conditions but not at neutral pH. Detailed investigation of the isotherms at low surface pressures, according to the reported method [48], revealed that the degree of polymerization at pH 2 was about 300. Formation of Si–O–Si linkages in the organosilane Langmuir monolayer was also confirmed by Blaudez and coworkers using polarization-modulated infrared reflection absorption spectroscopy [49]. Additionally, Ulman calculated bond lengths and suggested that the distance between the aliphatic chains becomes 0.43 nm upon formation of the Si–O–Si linkage, which is 10% less than the distance corresponding to the minimum of the van der Waals potential between hydrocarbon chains [50]. Fontaine and coworkers estimated that the energy gained by formation of the siloxane bond could compensate for the unfavorable chain compression [51].

In order to investigate the phase transition in the monolayer state, the temperature dependence of the π–A isotherm was measured at pH 2. The molecular area at 20 mN m^{-1}, which is the pressure for the LB transfer of the polymerized monolayer, is plotted as a function of temperature (Figure 2.6). Thermal expansion obviously changes at around 45 °C, indicating that the polymerized monolayer forms a disordered phase above this temperature. The observed temperature (45 °C) can be regarded as the phase transition point from the crystalline phase to the liquid crystalline phase of the polymerized organosilane monolayer.

The monolayer transfer to porous glass plates, whose average pore diameters were 5 or 20 nm, was then investigated. The transfer ratio of the polymerized monolayer on the porous glass plate with an average pore size of 5 nm was almost unity, while a somewhat smaller ratio was observed for the unpolymerized monolayer (0.85). The transfer ratio of the polymerized monolayer also depended on the pore size of the glass plate. The ratio decreased significantly if the glass plate had larger pores (average, 20 nm). This behavior can be explained by considering the coverage area of the polymerized species. From the unit area and degree of polymerization, the area occupied by a single polymer molecule can be estimated to be about 150 nm^2. This value is obviously larger than the pore area (20 nm^2) of the 5 nm glass but smaller than that (300 nm^2) of the 20 nm glass. Therefore, full coverage of a pore by a single organosilane polymer led to successful monolayer transfer only in the case of the glass plate with 5 nm pores. After LB transfer of the monolayer onto the porous glass plate (5 nm pore), heat treatment ensured immobilization of the monolayer covalently onto the glass surface. The glass plate with the immobilized monolayer was

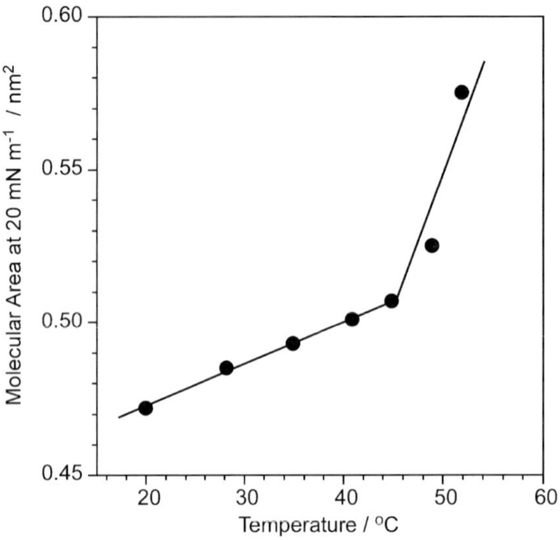

Fig. 2.6 Molecular area of dialkylorganosilane monolayer at 20 mN m^{-1} as a function of temperature. Reprinted with permission from [46], K. Ariga and Y. Okahata, *J. Am. Chem. Soc.*, **1989**, *111*, 5618. ©1989, American Chemical Society.

next attached to the bottom of a polyethylene tube and soaked in a quartz cell (Figure 2.7). A water-soluble fluorescent probe (naphthalene derivative) was dissolved in the upper aqueous solution (1 mM), and permeation through the monolayer immobilized on the porous glass plate was evaluated spectroscopically by monitoring intensity changes of the fluorescence at 340 nm (excited at 280 nm) in the lower solution phase. Permeation coefficients at different temperatures are summarized in

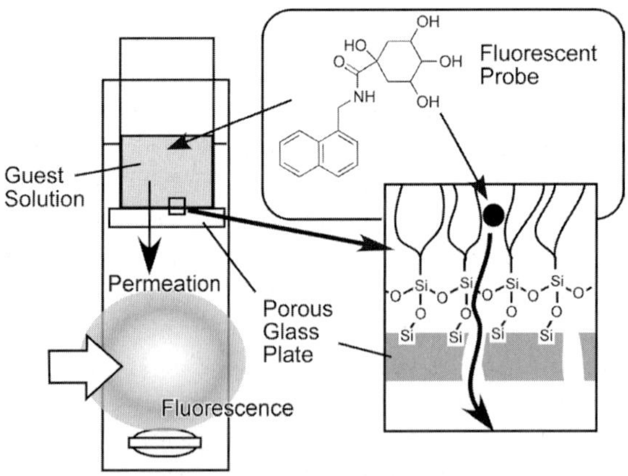

Fig. 2.7 Set-up of permeation experiment through organosilane monolayer immobilized on a porous glass plate. Reprinted with permission from [46], K. Ariga and Y. Okahata, *J. Am. Chem. Soc.*, **1989**, *111*, 5618. ©1989, American Chemical Society.

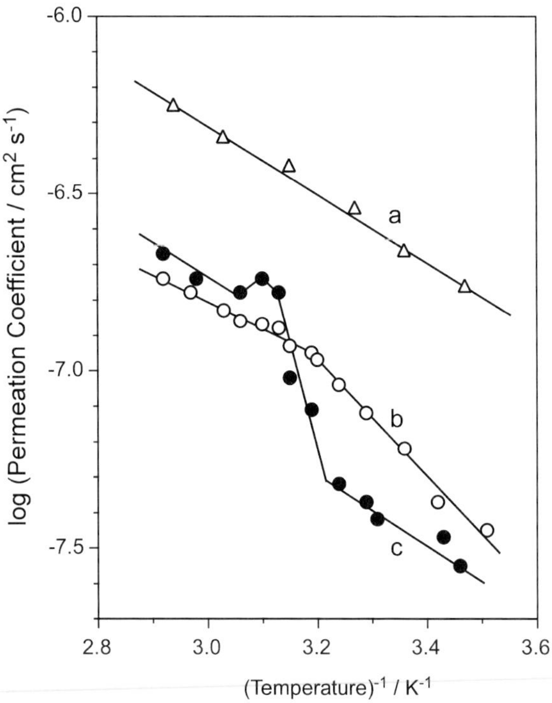

Fig. 2.8 Arrhenius plot of permeation coefficients: (a) bare glass plate; (b) glass plate with unpolymerized monolayer; (c) glass plate with polymerized monolayer. Reprinted with permission from [46], K. Ariga and Y. Okahata, *J. Am. Chem. Soc.*, **1989**, *111*, 5618. ©1989, American Chemical Society.

the form of an Arrhenius plot (Figure 2.8). The organosilane monolayers immobilized on the porous glass plate (plot (b) for unpolymerized monolayer and plot (c) for polymerized monolayer) clearly suppressed the permeability of the fluorescent probe as compared with a bare glass plate (plot (a)). Interestingly, permeation through the polymerized monolayer exhibited a discontinuity at around 45 °C (plot (c)), which is very close to the phase transition temperature (crystal state to liquid crystalline state) of the polymerized monolayer, which was estimated by the π–A isotherm measurement. The results obtained indicate strikingly that the abrupt permeation change at about 45 °C is caused by the phase transition of the polymerized monolayer from a crystalline state to a liquid crystalline state. This was the first example of permeation control using only 2 nm thick monolayers, which corresponds to the thinnest biomembrane-mimicking film operating as a permeation valve. Inconclusive changes at similar temperatures were also observed for the unpolymerized monolayer (plot (b)). Poor coverage of the pore surface resulted in incomplete regulation of the permeation. Therefore, pre-polymerization on water is an indispensable process for achieving well-controlled permeation by a single monolayer.

Although this is a very early example of the SAM/LB combination and was performed during the 1980s, this concept is still used by other researchers. In particular, this technique has been used for preparing two-dimensional patterned structures. For example, Kajiyama and coworkers studied extensively controlled phase separation based on detailed structural analyses of mixed monolayers of hydrocarbon-type and fluorocarbon-type organosilane compounds [52–54]. Matsumoto and coworkers studied similar phase separation structure and component exchange [55]. Formation of a non-equilibrium pattern of ring-in-ring morphology in a liquid expanded phase of an alkylsilane in a Langmuir monolayer was demonstrated by Wang and coworkers [56]. Knobler and coworkers reported phase-separated, two-component self-assembled organosilane monolayers and their use in the selective adsorption of a protein [57]. Adsorption experiments with bovine serum albumin showed that the protein is preferentially adsorbed at the CH_3-terminated regions of the patterned monolayers.

Because the siloxane backbone of these monolayers provides covalent linkage to some kinds of electrodes, functions of electrodes modified by organosilane monolayers have also been investigated. Okahata *et al.* immobilized monolayers of dialkyl organosilane compounds onto SnO_2 electrodes by forming covalent linkages [58]. Permeability of an electrochemical probe ($Fe(CN)_6^{4-}/Fe(CN)_6^{3-}$) through the monolayer was investigated by monitoring changes in the oxidation peak current of the cyclic voltammograms (CV) (Figure 2.9). Covering the SnO_2 electrode with the organosilane monolayer suppressed drastically the redox response of the $Fe(CN)_6^{4-}/Fe(CN)_6^{3-}$ couple, which was hardly affected by repeated measurements. In contrast, the electrode covered with monolayers of conventional fatty acids could not efficiently block permeation of the redox probe and the redox responses of the water-soluble probes gradually increased with repeated measurement. Comparison between these two electrodes indicates the crucial importance of covalent bonding between the monolayer and the electrode surface for stable control of material permeation through the monolayer structure. The effect of temperature on the permeability of the $Fe(CN)_6^{4-}/Fe(CN)_6^{3-}$ couple through the organosilane monolayers was also investigated. Although the permeation of the redox probe was suppressed below the phase transition temperature of the monolayer, the redox response of the probe drastically increased near the phase transition temperature. Above the phase transition temperature, a slight decrease in the probe permeation was observed. These results suggest that the immobilized monolayer has a maximum permeability to the probe molecule near the phase transition temperature, probably because of alkyl chain disorder during the unstable coexistence of crystalline and liquid crystalline states of the monolayer phase. Interestingly, the permeability of the aqueous probe molecules was also suppressed by addition of some kinds of alcohols to the water phase. Alcohol molecules with long straight alkyl chains efficiently blocked the probe permeability through the organosilane monolayer, while an alcohol with bulky moieties had less effect. The alcohol molecules were probably adsorbed in the organosilane monolayer, filling the pinholes formed on the monolayer. Because the monolayer was composed of the organosilane compound with long alkyl chains, a fatty alcohol was inserted preferentially into the monolayer. These results correspond to shape discrimination of alcohol guests by a monolayer-level structure.

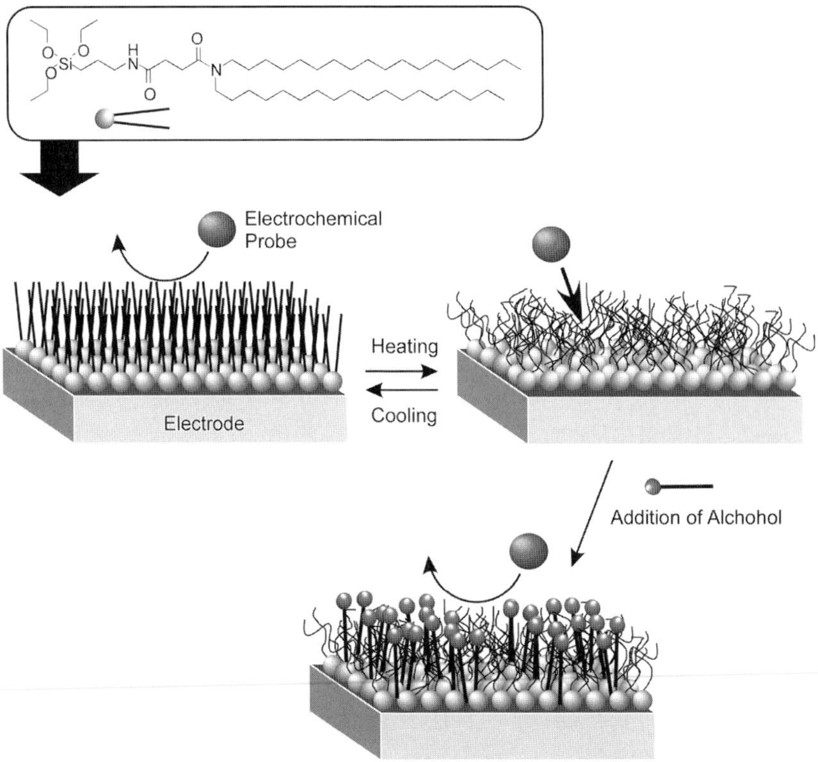

Fig. 2.9 Permeation of electrochemical probe ($Fe(CN)_6^{4-}$/ $Fe(CN)_6^{3-}$) through the dialkylsiloxane monolayer with the effect of temperature and the effect of addition of alcohol.

As more advanced bio-silica hybrid structures, Ariga, Hisaeda, and coworkers developed electrodes modified with a vitamin B_{12} derivative with the aid of organosilane monolayers [59]. The compounds used were heptapropyl and heptaoctyl esters of vitamin B_{12} derivatives with a Co(II) or Co(III) center and are shown in Figure 2.10 (A; chemical formula, B; models of VB_{12}-4). The effect of the ester chain length on the stability and orientation of the vitamin B_{12} in a lipid monolayer is discussed on the basis of their π–A isotherms [60]. The isotherm of the heptaoctyl derivatives has a relatively steep shape with a molecular area of about 3 nm^2, which is fairly close to the molecular area estimated by molecular modeling for face-on orientation. In contrast, the heptapropyl derivatives showed unstable characteristics with collapse even at low pressures. The results obtained in this study indicate that stable accommodation of the vitamin B_{12} functionality in the lipid assembly can be achieved by the introduction of a long chain. In the next step, the vitamin B_{12} derivatives were mixed with dialkylsilane at the air–water interface [59] (Figure 2.11). The vitamin B_{12} derivatives having seven octyl chains were stably incorporated into the matrix organosilane monolayer, while the short chain derivatives were easily expelled from the matrix monolayer upon compression. Analyses of the molecular areas revealed again that

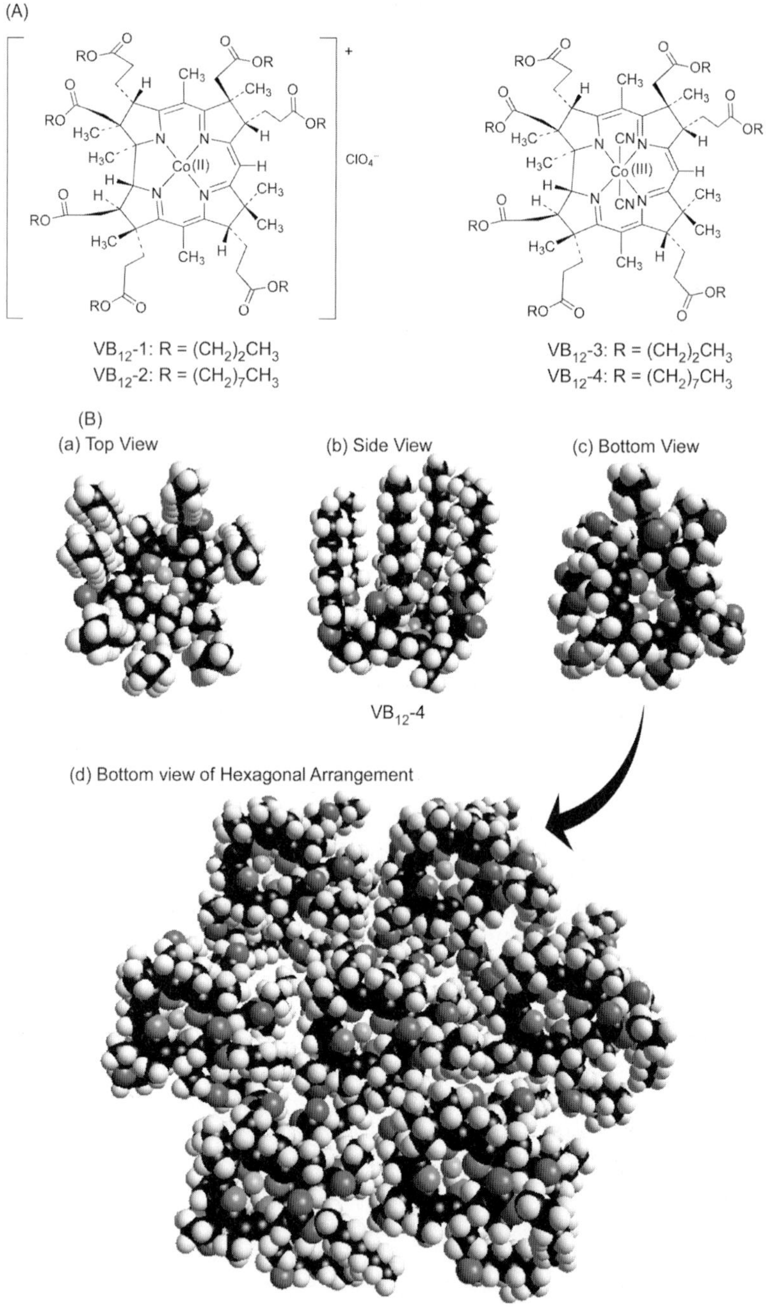

Fig. 2.10 (A) Vitamin B_{12} derivative. (B) Structural models of VB_{12}-4: (a) top view; (b) side view; (c) bottom view; bottom view of hexagonal arrangement. Adapted from [60], K. Ariga *et al.*, *Colloid Surf. A*, **2000**, *169*, 47.

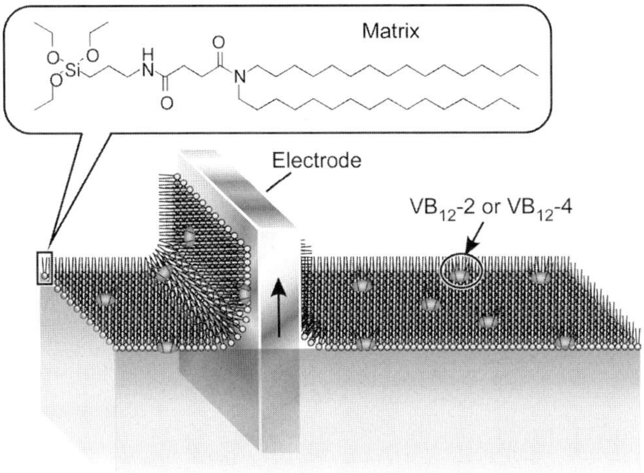

Fig. 2.11 Mixed Langmuir monolayer of dialkylsiloxane and vitamin B_{12} derivative, and its transfer onto a solid substrate.

the long chain derivatives had a face-on orientation in the matrix monolayer. The mixed monolayers of the vitamin B_{12} with long chains and the organosilane amphiphile were transferred as an X-type LB film onto a quartz plate that showed reasonable UV absorbance. The monolayer containing the vitamin derivative was similarly immobilized B_{12} on an indium-tin-oxide (ITO) electrode. A cyclic voltammetric measurement of this electrode in aqueous solution showed a Co(II)/Co(I) redox couple at −0.65 V vs. Ag/AgCl, which is in good agreement with the corresponding value of the vitamin B_{12} derivative dissolved in methanol. This indicates that the modified electrode can be used as a reactive electrode with a vitamin B_{12} function. Actual material conversion by an electrode modified with a vitamin B_{12} derivative was demonstrated using thicker films obtained by the sol–gel process (Figure 2.12) [61], rather than from the monolayer-modified electrode. The vitamin B_{12} derivative, heptapropyl cobyrinate perchlorate (BV_{12}-1 in Figure 2.10A), was readily trapped onto an ITO electrode by a sol–gel reaction. The modified electrode complex exhibited the Co(II)/Co(I) redox couple at −0.42 V vs. Ag–AgCl. This electroactive complex showed a high reactivity towards organic halides, and the controlled-potential electrolysis of benzyl bromide using the sol–gel modified electrode at −1.20 V vs. Ag–AgCl in aqueous solution containing 0.1 M KCl afforded dehalogenated products, bibenzyl and toluene, with a total turnover number of >1000 for 1 h.

2.3
Layered Alkylsiloxane

Organosilane monolayers are interesting objects for scientific research and can be used in sensitive detectors. However, bulk structures are sometimes required for a

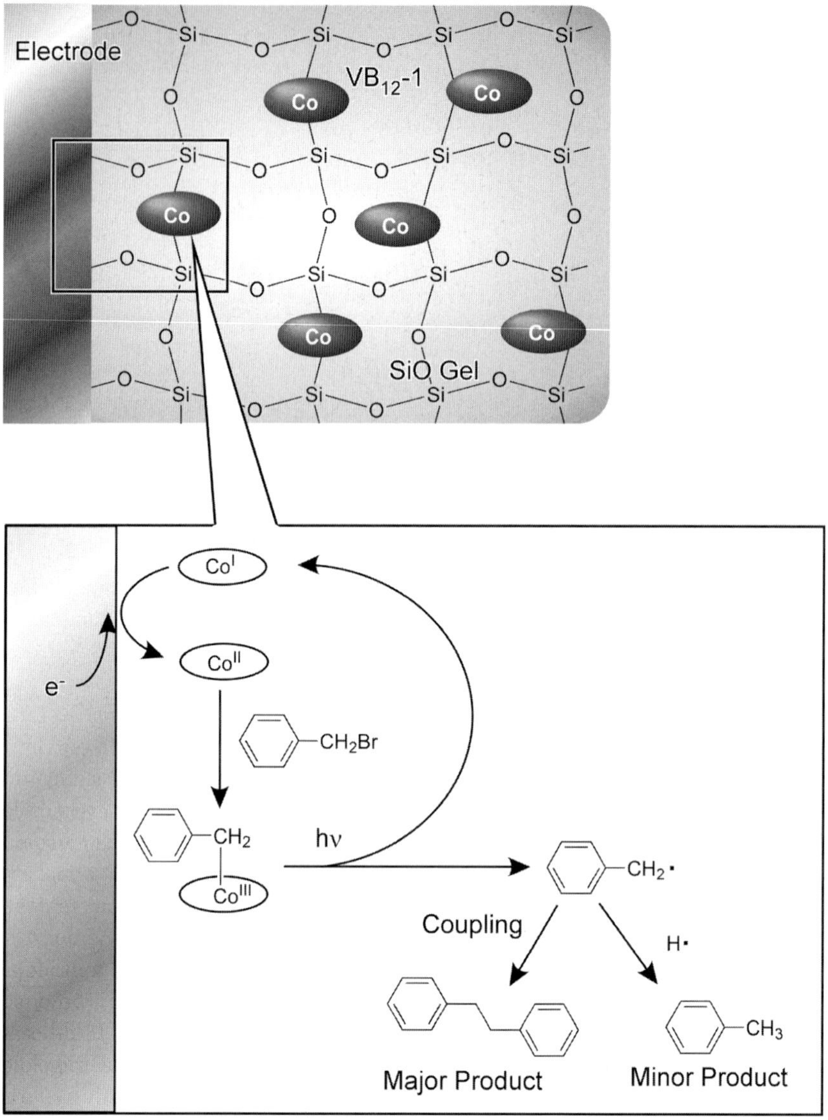

Fig. 2.12 Electrode modified with vitamin B_{12} by the sol–gel process and electrolysis of benzyl bromide at the electrode surface. Adapted from [61], H. Shimakoshi *et al.*, *Dalton Trans*, **2003**, 2308.

wide range of applications. In the following sections, three-dimensionally extended organosilane structures such as bio-silica hybrids are introduced sequentially.

The first category is layered organosilanes. Layered alkylsilanes are synthesized from akyltrialkoxysilane [$CH_3(CH_2)_{n-1}Si(OR)_3$, R is often CH_3 or C_2H_5] and tetraethyl orthosilicate (TEOS) [62–66]. Alkoxy groups are hydrolyzed and condensed into

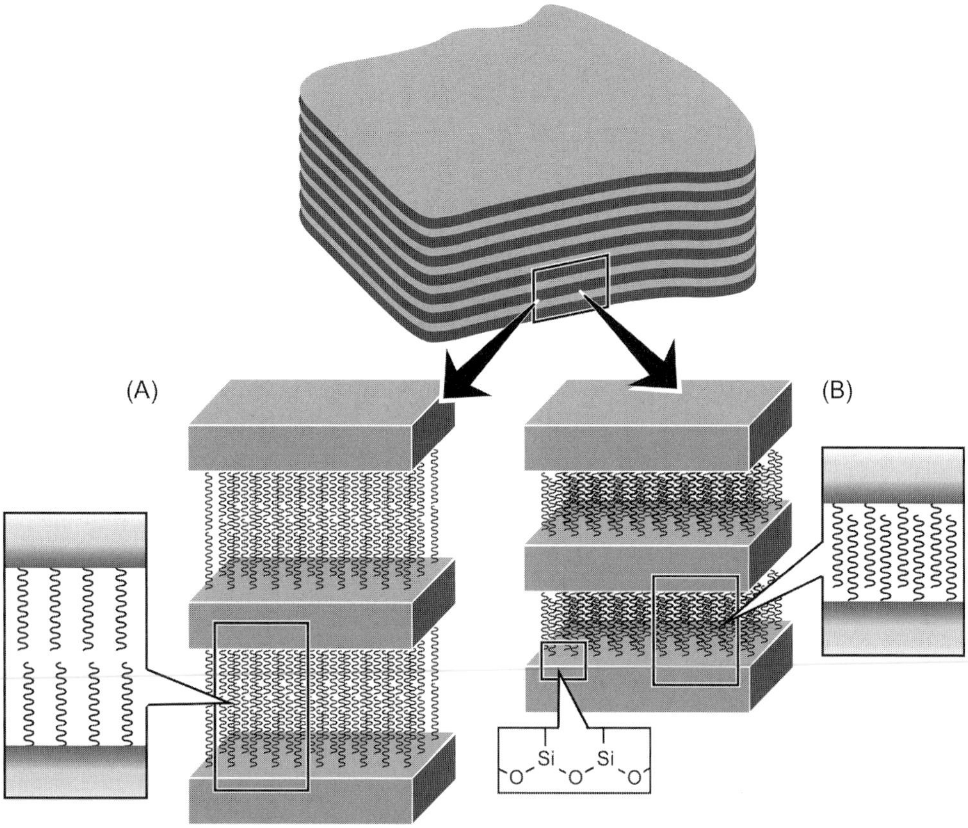

Fig. 2.13 Two types of layered alkylsiloxanes: (A) bilayer-type; (B) monolayer-type.

extended Si–O–Si networks, resulting in the formation of layered alkylsilanes. The layered alkylsilanes can be classified into two broad categories, a bilayer structure (Figure 2.13A) and a monolayer structure (Figure 2.13B). Alkyl chains in the bilayer structures form a separate bilayer motif, resulting in bilayer thickness as a unit structure. The monolayer motifs are formed through interdigitation of alkyl chains, which dictates the monolayer thickness of the repeat unit. In both structures, organic moieties and siloxane layers are covalently linked through Si–C bonds. Alkyl chains mostly adopt a trans-conformation, especially in the bilayer structures. IR investigation of the bilayer structure revealed the antisymmetric vibration band of CH_2 groups at 2917 cm^{-1}, indicating that the alkyl chains possess an all-trans conformation [67]. Analyses on the monolayer structure with interdigitated alkyl chains by X-ray diffraction (XRD) and nuclear magnetic resonance (NMR) also confirmed that most of the alkyl chains of the monolayer motifs are in a trans-conformation [68].

These structures can be differentiated by variation of the synthetic conditions such as organic/inorganic component ratios, synthesis temperature and structure of the components, as suggested in the detailed consideration by Fujii and coworkers [69].

In particular, synthesis temperature is considered to be the most influential factor for differentiation between the monolayer and bilayer structures. In a typical synthetic procedure, both monolayer and bilayer structures can be synthesized at room temperature and at higher temperature. At lower temperature, such as room temperature, extended organization in alkyl chain aggregates is promoted to form an ordered structure like alkane crystals. This situation is favorable for formation of the bilayer structure. In contrast, synthesis at higher temperature (>110 °C) also induces formation of a siloxane sheet based on an Si–O–Si linkage and aggregation of the alkyl chains, but further organization cannot be triggered because of the possibility of side-chain mobility. Post-synthesis cooling processes should promote interdigitation of alkyl chains for better ordering, resulting in the formation of monolayer structures. Catalyst and solvent identity can also be influential factors. Acidic catalysts promote hydrolysis but cause moderate condensation. Use of acids such as HCl allows the alkyl chains to organize into layered structures. Conversely, basic catalysts accelerate the extension of the Si–O–Si network that is not favorable for formation of ordered layered structures.

Some monolayer-type layered alkylsiloxanes showed melting behavior [68,69]. Reversible melting was confirmed by differential scanning calorimetry (DSC), where weight loss was not detected by thermogravimetry or differential thermal analysis (TG-DTA), as reported by Fujii and coworkers [69]. They observed directly changes of the layered alkylsiloxane into transparent fluid substances. This is a very rare example of a meltable layered alkylsiloxane. A rough phase diagram of the layered alkylsiloxane is summarized in Figure 2.14 where synthesis temperature, ratio of organic and inorganic components, and alkyl chain length are used as influencing parameters. Layered alkylsiloxanes with reversible melting behavior can be obtained with rather high organic contents, where the molar ratio of TEOS/alkyltrialkoxysilane should be not more than 1, and with high synthesis temperatures (>110 °C) [68,69]. Increase in the TEOS content (TEOS/alkyltrialkoxysilane >4) results in a non-meltable structure because of excessive extension of the siloxane network. Layered alkylsiloxane synthesized at room temperature without using TEOS provided meltable layered alkylsiloxane, which showed irreversible melting behavior [70]. In this case, formation of the Si–O–Si network would be insufficient and the remaining unreacted Si–OH group would extend the network irreversibly upon melting, finally resulting in destruction of the layered structures.

Other organic components such as lipids [71,72] and peptides [73] have also been incorporated into layered organosiloxane structures. In addition, formation of more complicated structures has been proposed, as depicted in Figure 2.15, where is shown an organosiloxane-based layered structure containing highly-hyperpolarizable chromophore arrays with and without an inserted layer of refractive index modifying Ga oxide or In oxide layer [74]. Inclusion of liquid crystalline materials into layered structures has also been reported [75]. Among these modified analogues, layered organophyllosilicates have also been scrutinized. In their structures, a metal hydroxide (such as $Mg(OH)_2$) sheet is inserted between two siloxane layers (Figure 2.16). They are categorized into two classes, structures without [76–83] and with [84,85] tetrahedral Si in their siloxane layer. The former structures can be

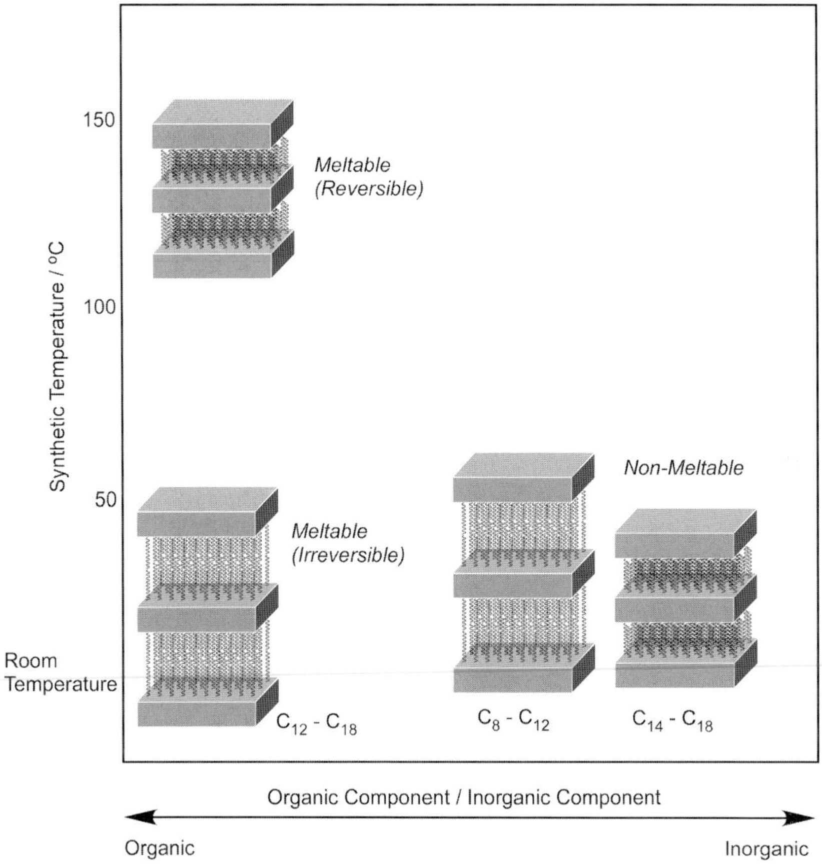

Fig. 2.14 Phase diagram of layered alkylsiloxane structures.

obtained from aqueous dispersion of organotrialkoxysilanes and metal oxide at room temperature, while reactions between organotrialkoxysilanes, metal oxide, and silica sol at about 100 °C provide the latter structure. Formation of layered structures should be most influenced by the formation processes of the metal hydroxide sheets. Various organic components such as a methacryloxy group (see Figure 2.16) [86], an amino group [87], and an alkylammonium group [85,88] can be incorporated. These hybrid structures were used for removal of heavy metal ions [87,89] and immobilization of proteins such as myoglobin [90].

Fujii and coworkers reported the synthesis and detailed structural analyses of alkylammonium/magnesium phyllosilicate hybrids [88], which were prepared by hydrothermal reaction from a mixture of octadecyldimethyl(3-trimethoxysilylpropyl)-ammonium chloride, silica sol, and magnesium hydroxide $Mg(OH)_2$. The structure of the hybrid compound was studied by XRD, TEM, electron diffraction, high-resolution solid-state NMR, TG-DTA/MS, and elemental analysis. The resulting analytical information confirmed the unit structure, which consists of a 2 : 1

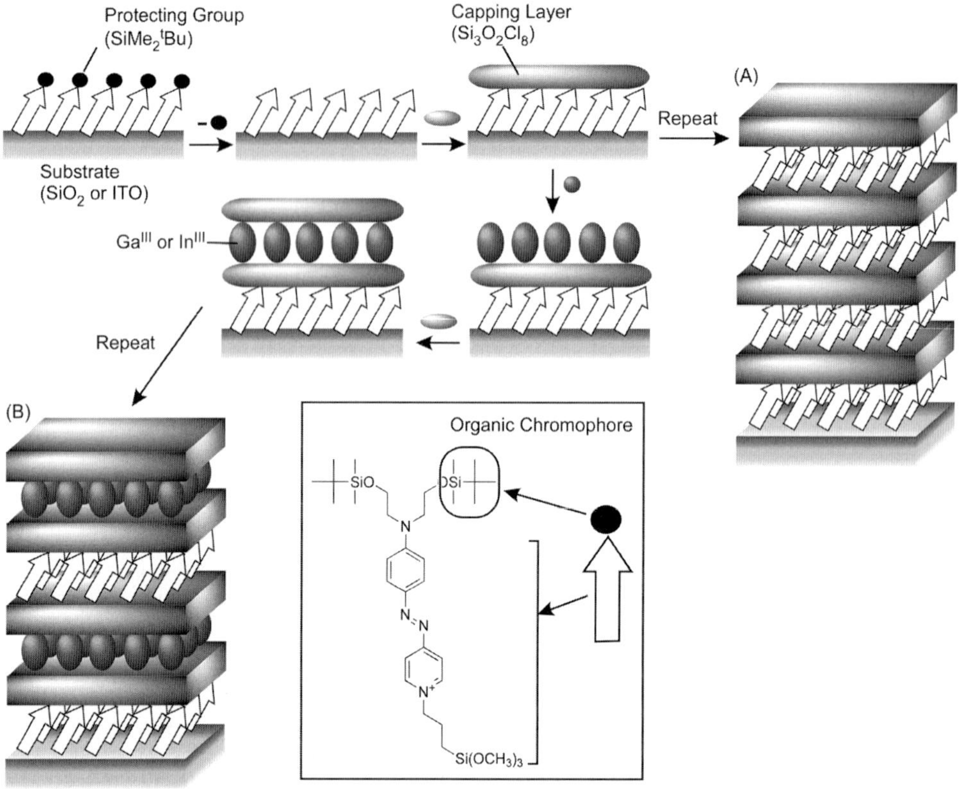

Fig. 2.15 Organosiloxane-based layered structure containing high-hyperpolarizable chromophore arrays: (A) without inserted layer of refractive index modifying Ga and In oxide layers; (B) with inserted layer of refractive index modifying Ga and In oxide layers.

trioctahedral phyllosilicate part and an alkylammonium part grafted to the magnesium phyllosilicate part through the Si−C bond. Three possible models of the layered structure were first proposed. One model shown in Figure 2.17A consists only of smectite-like layers that sandwich alkyl chains. However, on the basis of certain experimental evidence this model can be rejected. For instance, the basal spacing expected for this model with a bilayer structure of the alkyl chains is about 7 nm, which is inconsistent with the observed *d* value of 4 nm. A model of the monolayer structure with interdigitated alkyl chains (Figure 2.17B) may satisfy the observed basal spacing, but it cannot explain the lack of long-range periodicity in the experimental data. Finally, the authors proposed the model structure shown in Figure 2.17C, in which the alkyl chains of both parts are inserted alternately. This structure is comprised of three layers; the central one is the inorganic layer containing two silicic tetrahedral sheets and a magnesium octahedral sheet, and the outermost two layers are the organosiloxane sheets containing Si−O−Si networks. The alkyl chains can take a chain–chain spacing of about 0.41 nm with an all-trans

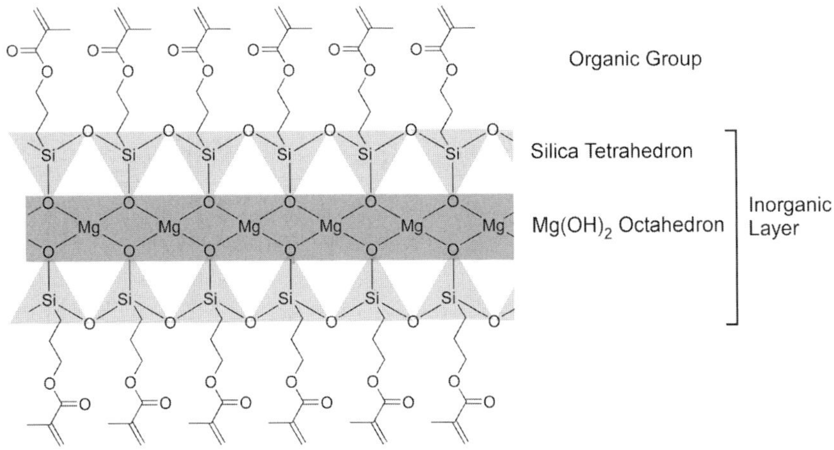

Fig. 2.16 Layered magnesium organophyllosilicates. Adapted from [76], Y. Fukushima and M. Tani, *Bull. Chem. Soc. Jpn.*, **1996**, *69*, 3667.

conformation. The basal spacing for this structure model is estimated to be 3.93 nm. These values agree well with the values obtained experimentally. The proposed structure includes amphiphilic organic layers between well-defined organic sheets. Therefore, these materials may have great potential for biological applications such as controlled and/or sustained drug delivery.

2.4 Organic–Inorganic Hybrid Vesicle "Cerasome"

All of the above-mentioned examples describe organosiloxane hybrid sheet-like structures. However, cell-mimicry requires spherical structures that can form an inner space as a container. Liposomes and lipid bilayer vesicles are known as models of a spherical cell membrane, which is a direct mimic of a unicellular membrane. However, the limited mechanical stability of conventional lipid vesicles is often disadvantageous for some kinds of practical application.

Cell-like vesicle structures have also been investigated using silica-related materials. For example, Pinnavaia and coworkers reported vesicle-type mesostructured silica (Figure 2.18A) [91] using gemini-type $C_nH_{2n+1}NH(CH_2)_2NH_2$ surfactants as templates with silica precursors derived from TEOS. The vesicular silica has a shell structure consisting of one or more undulated silica sheets of about 3 nm thickness with mesopores (average diameters from 2.7 to 4.0 nm) running both parallel and orthogonal to the silica sheets. These structural characteristics make the framework structure bi-continuous and highly accessible. Silica cross-linking structures also provide unprecedented thermal (1000 °C) and hydrothermal stabilities (more than 150 h at 100 °C). This vesicle-type silica is expected to be used in catalysis and

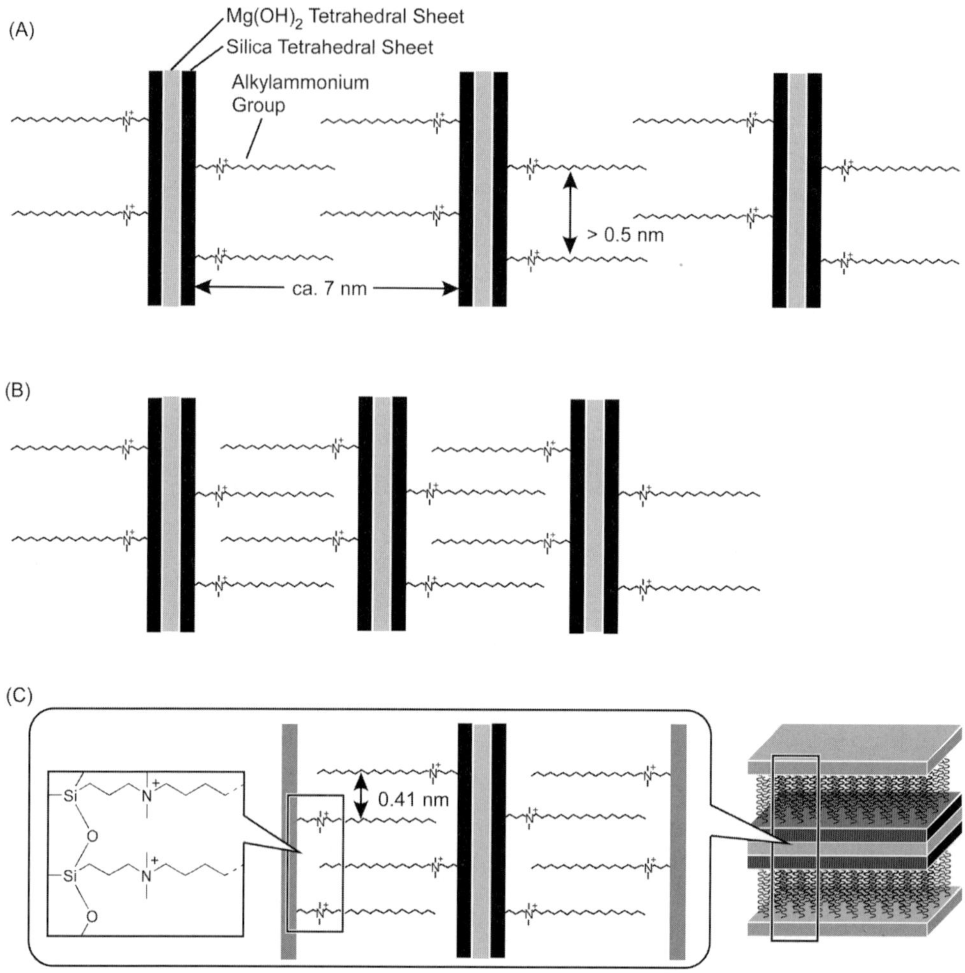

Fig. 2.17 Three structure candidates (A, B, and C) of alkylammonium/magnesium phyllosilicate hybrids. Reprinted with permission from [88], K. Fujii *et al.*, *Chem. Mater.*, **2003**, *15*,1189. ©2003, American Chemical Society.

molecular separations. Caruso, Möhwald and coworkers reported the formation of hollow silica vesicles through LBL assembly on colloidal nanoparticle templates (Figure 2.18B) [92]. Polyelectrolytes and smaller silica particles were initially formed on a colloidal core, which was subsequently destroyed, resulting in a hollow silica/polymer hybrid vesicle. Calcination of the hybrid vesicles left a hollow vesicle composed of silica. Frey and coworkers synthesized a novel type of amphiphilic spherosilsesquioxane derivative [93]. Aggregation of the uncondensed amphiphile led to micellar and vesicular structures that can be cross-linked to liposome-like silica particles at elevated pH. Sommerdijk and colleagues prepared an amphiphilic block copolymer consisting of hydrophilic poly(ethyleneoxide) blocks and hydrophobic

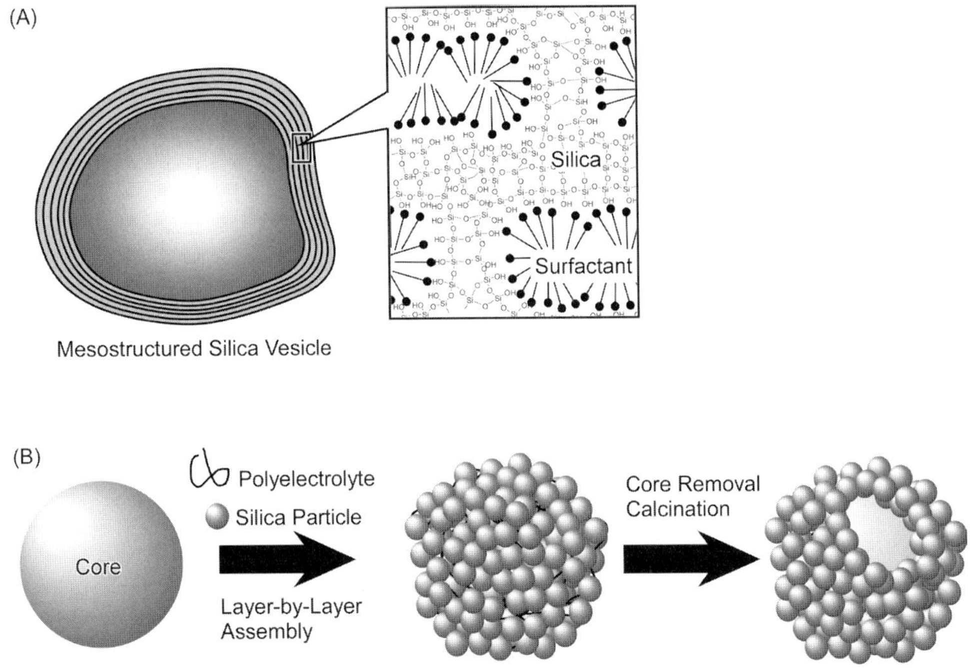

Fig. 2.18 Vesicle structures with silica wall; (A) mesostructured silica vesicle; (B) hollow capsule composed of silica particles prepared by LbL assembly.

poly(methylphenylsilane) segments [94]. It was also demonstrated that the vesicle aggregates could be broken up using UV irradiation, indicating that these vesicles were potentially interesting as controlled release systems.

Unlike the approaches mentioned above, Katagiri *et al.* pursued the covalent linkage of a siloxane framework to a lipid bilayer vesicle [95,96], and sophisticated organization of multi-cellular systems [97,98]. The resulting vesicles have a siloxane network covalently attached to the bilayer membrane surface and are called cerasome (ceramics + soma) (Figure 2.19). Alkoxysilane-bearing amphiphiles (Si-Lipid 1) were first dispersed in an acidic aqueous solution using a vortex mixer. The dispersion behavior depended significantly on the solution pH. Upon mixing 10 mmol of Si-Lipid 1 with aqueous HCl (pH 1, 5.0 ml), precipitation immediately occurred, probably because of an excessive rate of hydrolysis and condensation. In contrast, basic conditions resulted in an uneven dispersion of aqueous Si-Lipid 1. Medium acidic conditions (pH 3) were found to be most appropriate for stable formation of cerasomes. Under the optimized conditions, a stable dispersion was obtained after vortex mixing for 15 min at room temperature. This translucent sample was morphologically stable for long periods of time. The hydrolysis process was monitored quantitatively by ^1H-NMR in acidic D_2O. A broad proton signal due to $SiOCH_2CH_3$ ($\delta = 3.65$ ppm) gradually decreased in intensity being replaced by the sharp quartet signal of $HOCH_2CH_3$ ($\delta = 3.50$ ppm). The formation of the siloxane

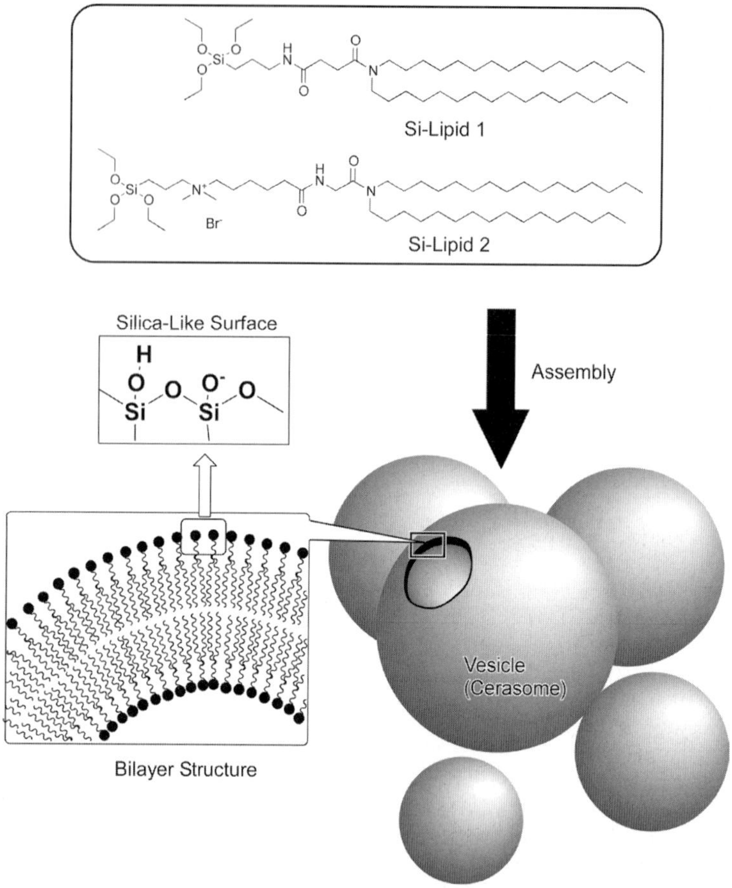

Fig. 2.19 Preparation of cerasome from alkoxysilane-bearing amphiphiles. Adapted from [29], K. Ariga, *Chem. Rec.* **2004**, *3*, 297.

bonds was confirmed by FT-IR spectra, which contained the peaks assigned to the Si–O–Si and Si–OH groups.

Formation of the vesicular structures was confirmed by transmission electron microscopy (TEM) with the aid of hexaammonium heptamolybdate tetrahydrate. The image of the multi-lamellar cerasomes with a bilayer thickness of about 4 nm and vesicular diameter of 150 nm was clearly visible (Figure 2.20). Interestingly, the TEM image of the vesicular aggregates was observed in the same specimen (Figure 2.21). That some of the aggregates maintained the original spherical structure implies suppression of the collapse and fusion of the cerasome, probably through the formation of the intra- and intermembrane siloxane network. This indicates that a multi-cellular model can be obtained by association with the stable cerasome.

Formation of cerasome assemblies is not limited to unintentional aggregation as described above. Subjecting the cerasome structure to LbL techniques could result in

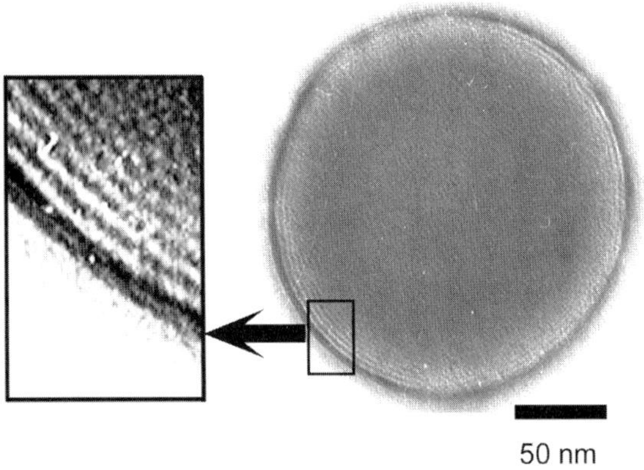

Fig. 2.20 A TEM image of the multi-lamellar cerasome. Adapted from [29], K. Ariga, *Chem. Rec.* **2004**, *3*, 297.

multi-cellular mimics in a pre-designed way. Figure 2.22 illustrated two types of cerasome LbL assemblies reported by Katagiri *et al.* [97,98]. LbL assemblies between the cationic polyelectrolyte (poly(diallyldimethylammonium chloride), PDDA) and anionic vesicles have been investigated using a quartz crystal microbalance (QCM) that from its frequency shiftsprovides sensitive mass detection of materials deposited on the surface.

When conventional anionic vesicles composed of dihexadecyl phosphate were alternately assembled with cationic PDDA on a QCM plate, the QCM frequency showed only small shifts (plot (a) in Figure 2.23). The observed frequency shifts corresponded to lipid bilayer formation on the surface, indicating that the vesicles of dihexadecyl phosphate were collapsed during the LbL assembling process, probably due to insufficient structural stability of the conventional vesicle. In contrast, LbL assembly between the anionic cerasome prepared from Si-Lipid 1 and cationic PDDA (see Figure 2.22A) exhibited significantly large frequency shifts of the QCM response (plot (b) in Figure 2.23) [98]. The large steps in the frequency shifts were in good agreement with the cerasome deposition in accordance with their spherical structure.

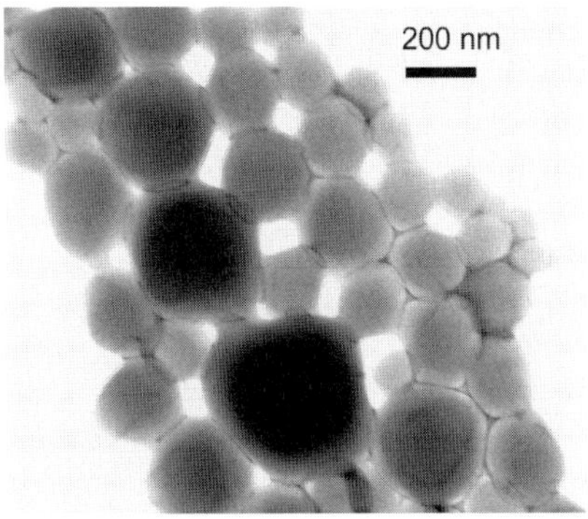

Fig. 2.21 A TEM image of the cerasome aggregates. Adapted from [29], K. Ariga, *Chem. Rec.* **2004**, *3*, 297.

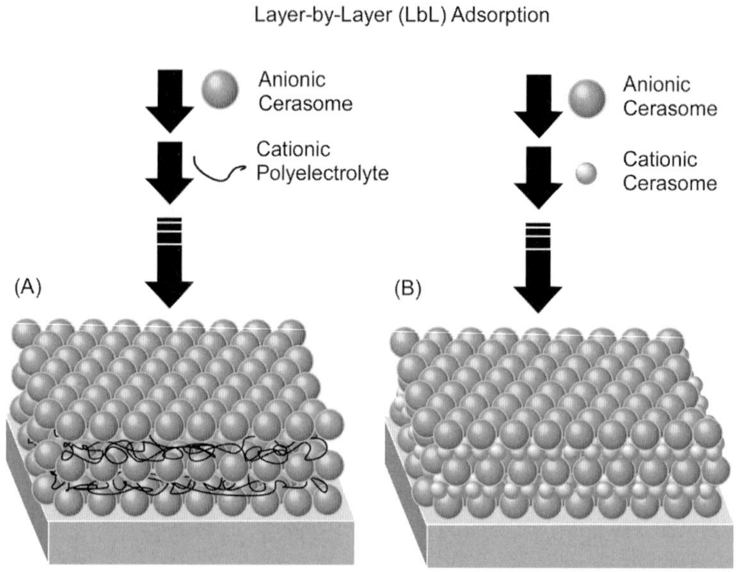

Fig. 2.22 LbL assembly of cerasome; (A) LbL assembly between anionic cerasome and cationic polyelectrolyte; (B) LbL assembly between anionic cerasome and cationic cerasome.

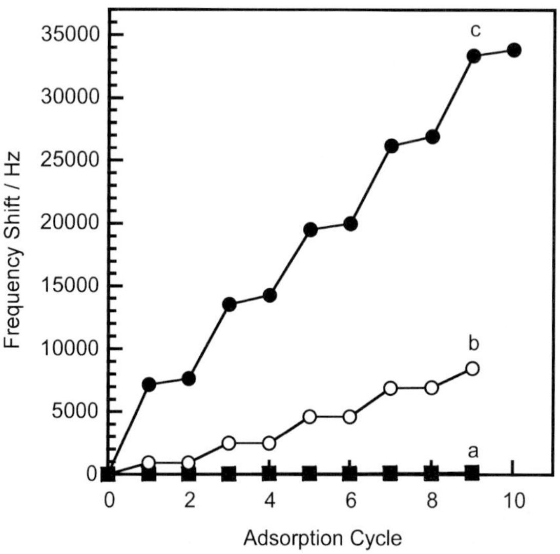

Fig. 2.23 QCM responses on LbL assemblies: (a) LbL assembly between dihexadecyl phosphate vesicle and PDDA; (b) LbL assembly between anionic cerasome (larger step) and PDDA (smaller step); (c) LbL assembly between anionic cerasome (larger step) and cationic cerasome (smaller step). Reprinted with permission from [97], K. Katagiri *et al.*, *J. Am. Chem. Soc.* **2002**, 124, 7892, ©2002 American Chemical Society, and from [98], K. Katagiri *et al.*, *Lamgmuir* **2002**, 18, 6709, ©2002 American Chemical Society.

The morphology of the assembled structure was also confirmed by atomic force microscopy (AFM).

Cerasomes with cationic surface charge can be prepared using Si-Lipid 2 and are shown in Figure 2.19. Using both the anionic and cationic cerasomes, direct LbL assembly of cerasome structures in the absence of polyelectrolyte counterions became possible (Figure 2.22B) [97]. TEM observation of the cerasomes revealed that cationic cerasomes are smaller in diameter (20–100 nm) than the anionic cerasomes (70–300 nm). Large steps and small steps of the frequency changes observed in the odd- and even-number steps (plot (c) in Figure 2.23) corresponded to adsorption of the anionic and cationic cerasome, respectively. The presence of closely packed cerasome particles like a stone pavement in both layers was clearly confirmed by AFM observations of the surface of the assembled structures. This kind of assembly can be regarded as a multi-cellular mimic and subsequently could be used as a bioreactors or a biosensor. Further functionalization of the cerasome surface using various biomolecules, such as enzymes and antibodies, through covalent linkage indicates a great potential for creating various kinds of biomimetic silica nanohybrids.

2.5
Mesoporous Silica Prepared by the Lizard Template Method

Immobilization of bio-functions within nano-sized spaces with well-designed structures could provide unique systems where orientation, conformation, and coordination of the entrapped biomolecules have certain limitations, possibly resulting in unusual selectivity and efficiency in their functions. Mesoporous materials such as mesoporous silica and mesoporous carbon could satisfy such demands, because these materials possess well-defined pore structures in regular geometry. Their pore diameters are defined in the range 2–50 nm [99–105]. These materials have attracted much attention due to commercial interest in their use as catalysts, catalyst supports and in the adsorption of biomolecules owing to their large specific surface area, large specific pore volume and large pore diameter. Covalent linkage of biomolecules to mesoporous silica structures is certainly an interesting target for the development of biomimetic nanohybrids.

Recently, Ariga, Aida and coworkers developed a new synthetic method for mesoporous silica that enables both dense modification of the pore interior with bio-functions and good accessibility by external guests [106]. The concept of this method is illustrated in Figure 2.24. Organosilane compounds used as a structure-directing reagent possess a condensable head and a cleavable alkyl tail. The organosilane was covalently attached to the silica framework upon sol–gel reaction with TEOS, resulting in mesoporous silica channels which are filled with the organic group of the template. Cleavage and removal of the alkyl tail by selective hydrolysis of the ester at the C-terminal leaves open pores with a surface covalently functionalized by the alanine residue. As one can easily imagine, the template behaves like a "lizard," whose head bites the silica wall and whose tail can be cleaved off. Therefore,

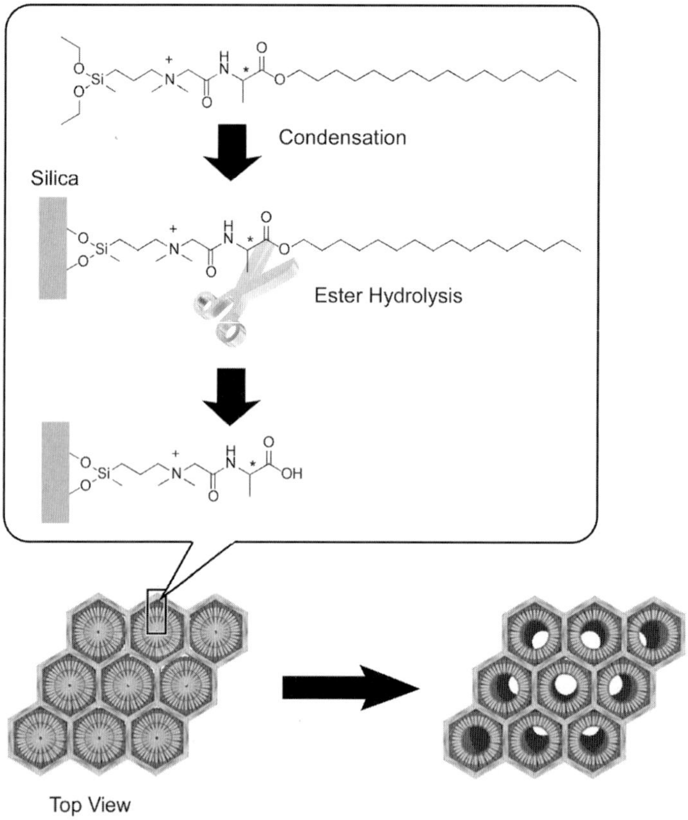

Fig. 2.24 Concept of lizard template method for the preparation of mesoporous silica. Adapted from [106], Q. Zhang *et al.*, *J. Am. Chem. Soc.* **2004**, *126*, 988.

the "lizard template" term was applied to describe this method. XRD analysis of the as-synthesized (unhydrolyzed) silica (Figure 2.25a) showed a set of peaks characteristic of a hexagonal silicate structure with a *d*-spacing for the (100) diffraction peak of 4.2 nm. The corresponding patterns remained after hydrolytic removal of the hexadecyl tail (Figure 2.25b) or after calcination at 550 °C (Figure 2.25c), although a decrease in *d*-spacing was observed in the latter case. Selective hydrolysis of the template ester was clearly demonstrated by FT-IR measurement of the silica materials (Figure 2.26A for the alkyl tail region and Figure 2.26B for the amino acid region). The spectrum of the as-synthesized silica (spectrum (a)) clearly shows peaks due to alkyl chains at 2924 cm^{-1} ($\nu_{as}(CH_2)$) and 2854 cm^{-1} ($\nu_s(CH_2)$) together with peaks characteristic of the alanine residue at 1742 cm^{-1} (ν(ester C=O)), at 1685 cm^{-1} (amide I), and at 1555 cm^{-1} (amide II). The former two peaks for the alkyl chain disappeared completely after hydrolysis (see spectrum (b)), although amide I at 1685 cm^{-1} and amide II at 1555 cm^{-1} remained intact, and the ester peak was converted to the peak of free carboxylic acid at 1733 cm^{-1}. All the peaks disappeared

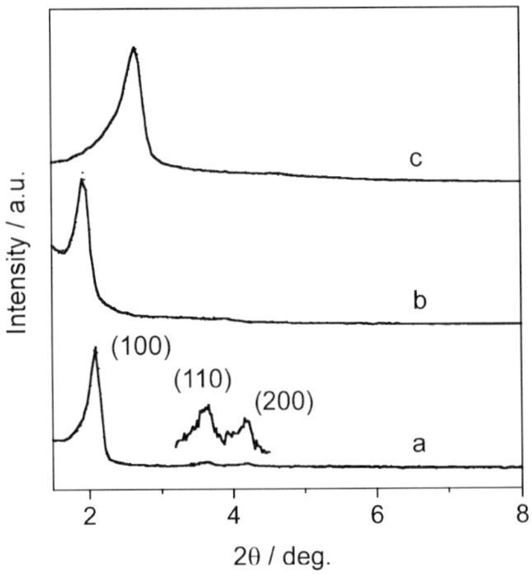

Fig. 2.25 XRD patterns: (a) as-synthesized; (b) hydrolyzed; (c) calcined materials. Reprinted with permission from [106], Q. Zhang *et al.*, *J. Am. Chem. Soc.* **2004**, *126*, 988. ©2004 American Chemical Society.

after calcination (spectrum (c)). In order to quantify the remaining organic content, thermogravimetric analyses (TGA) of the silica materials were conducted. Upon decomposition of the organic moiety between 180 and 420 °C, the as-synthesized and hydrolyzed silica materials showed weight losses of 46 and 27 %, respectively. The latter value is close to the theoretical value (24 %) calculated from the molecular weight change upon hydrolysis of the ester linkage and the organic content of the unhydrolyzed one.

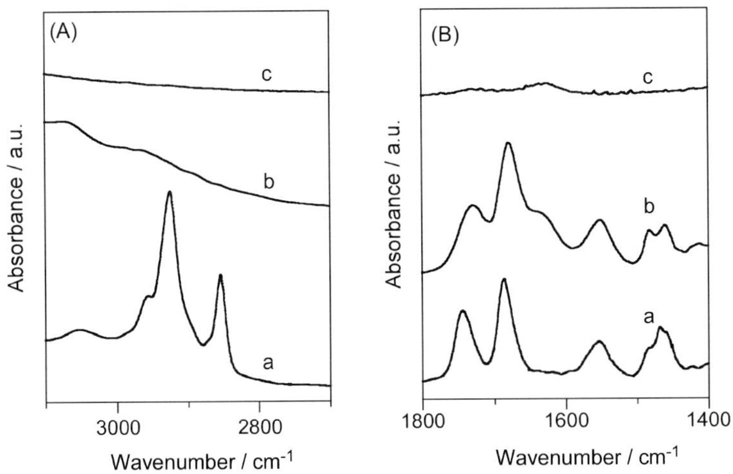

Fig. 2.26 FT-IR spectra of (a) as-synthesized, (b) hydrolyzed, and (c) calcined materials: (A) 3100–2700 cm^{-1}; (B) 1800–1400 cm^{-1}. Reprinted with permission from [106], Q. Zhang *et al.*, *J. Am. Chem. Soc.* **2004**, *126*, 988. ©2004 American Chemical Society.

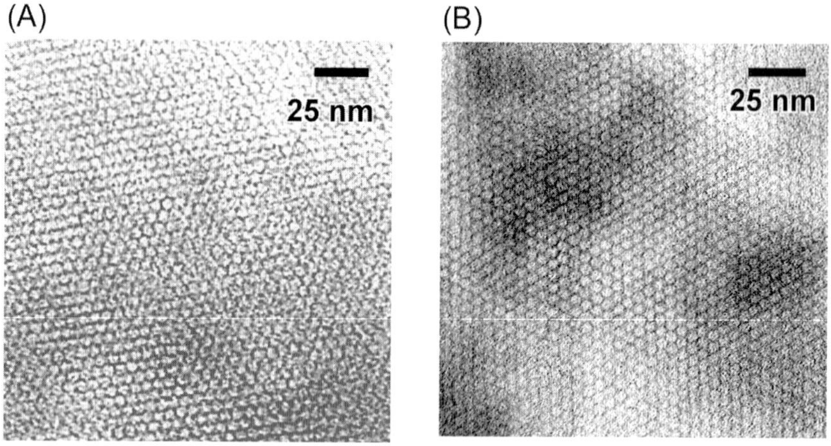

Fig. 2.27 HR-TEM images of mesoporous silica prepared by the lizard template method: (A) before hydrolysis (as-synthesized); (B) after hydrolysis. Reprinted with permission from [106], Q. Zhang *et al.*, *J. Am. Chem. Soc.* **2004**, *126*, 988. ©2004 American Chemical Society.

A high resolution transmission electron micrographic (HR-TEM) image of as-synthesized silica clearly showed the hexagonally aligned silicate channels (Figure 2.27A). A similar image was observed for the hydrolyzed material, ensuring preservation of the hexagonal porous structure after hydrolytic treatment (Figure 2.27B). Pore structure formation on hydrolysis was also demonstrated by nitrogen adsorption/desorption isotherms. Although as-synthesized silica was inert for nitrogen adsorption (profile (a) in Figure 2.28), the hydrolyzed material displayed a clear nitrogen adsorption/desorption isotherm (type IV) with a distinct capillary condensation step and a narrow pore size distribution (profile (b) in Figure 2.28). Temperature-programmed desorption (TPD) analysis with NH_3 as a basic guest confirmed the exposure of the alanine C-terminal in the silicate channel. From the desorption profile, it was found that the amount of adsorbed ammonia in the hydrolyzed porous silica was comparable to the amount of immobilized alanine C-terminal.

Aida and coworkers used the hybrid structure for reactor applications and illustrated the catalytic capability of unhydrolyzed materials on the acetalization of a ketone, such as cyclohexanone, in ethanol (EtOH) under mild conditions [107]. The proposed mechanism of catalysis is illustrated in Figure 2.29. Cyclohexanone and EtOH are simultaneously incorporated into the silica channels at the hydrophobic inner domain and hydrophilic (ionic) outer shell, respectively, and then activated through hydrogen-bonding interactions with the amide NH and carbonyl groups of the peptide functionalities, located at the core–shell interface ((a) and (c) in Figure 2.29). The acetalization should involve transient carbocationic intermediates ((c) in Figure 2.29), whose generation is favored by a highly polar environment containing concentrated ammonium salt functionalities.

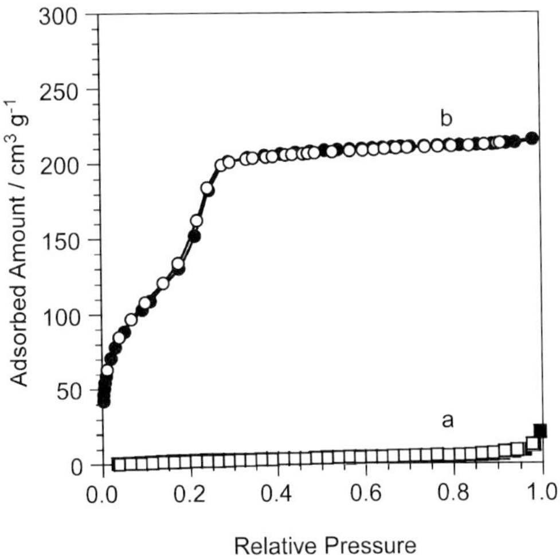

Fig. 2.28 Nitrogen adsorption (open symbol)/desorption (closed symbol) isotherms: (A) as-synthesized, (B) hydrolyzed materials. Reprinted with permission from [106], Q. Zhang *et al.*, *J. Am. Chem. Soc.* **2004**, *126*, 988. ©2004 American Chemical Society.

As shown above, in the amphiphilic core–shell architecture of the immobilized rod-like micelle, hydrophobic and hydrophilic reactants can be incorporated simultaneously into the silica channels and activated through hydrogen-bonding interactions with the peptidic functionalities located at the core–shell interface. This design strategy made use of templated hydrothermal synthesis of mesostructured silica with engineered amphiphiles, which allowed three-dimensional design of catalytic and substrate-binding sites. Therefore, the present work provides a novel and general strategy for rational molecular design of bio-inspired solid catalysts as a novel type of artificial enzyme [108–112]. The lizard templating method is expected to be used for the fabrication of mesoporous silica with a variety of bio-functional groups. It will provide artificial active sites for a wide range of biological functions.

2.6
Future Perspectives

In this chapter, various examples of biomimetic nanohybrids based on organosiloxane units, including siloxane-based monolayers, layered organosiloxanes, organic–inorganic hybrid vesicle cerasomes, and mesoporous silica hybrids from the functionalised lizard template method, have been introduced. Appropriate design of organosilane compounds and their assemblies led to the formation of various types of nano and micro objects with biomolecular components. Biomimetic functions such as permeation control and enzyme-like catalysis were demonstrated in some cases. The examples presented indicate the great potential of siloxane-based nanohybrids for biomimetic function and applications. For future development of the

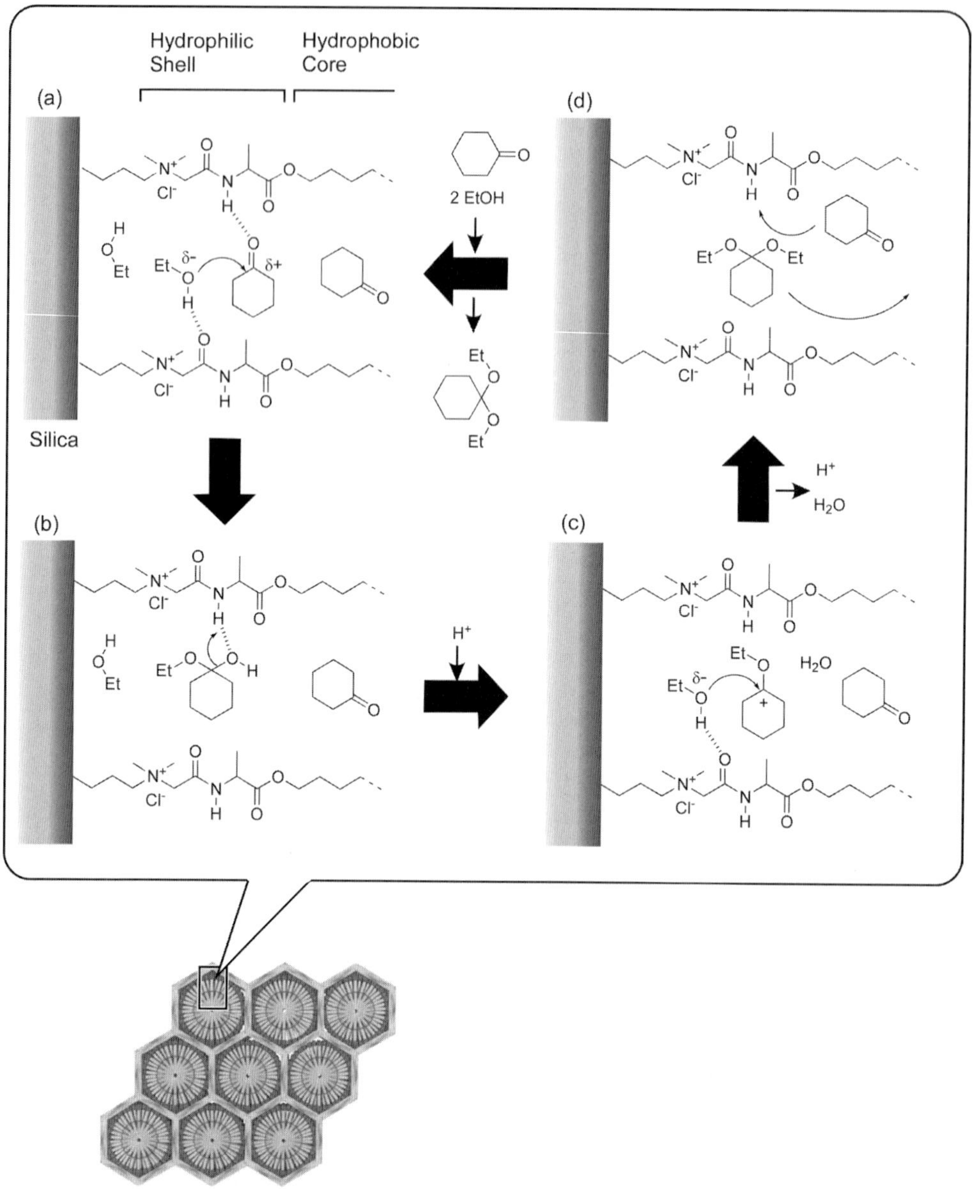

Fig. 2.29 Reaction mechanism on acetalization of cyclohexanone within unhydrolyzed mesoporous silica materials. Adapted from [107], W. Otani *et al.*, *Chem. Eur. J.* **2007**, *13*, 1731–1736.

corresponding fields, combination of these nanohybrids made by bottom-up approaches with finely fabricated artificial structures prepared by top-down approaches is of great importance. Hybridization of biomimetic hybrids and artificial structures should result in novel devices with biological activities. The possibilities of biomimetic nanohybrids are not limited to such practical applications. These nanohybrids offer well-defined nanospaces where desired biomolecules can be entrapped. Functions and properties of biomaterials in the controlled nanospaces, as well as their fundamental physics and chemistry, are not well known. Many basic sciences still remain unexplored for biomolecules within nanohybrid structures.

Acknowledgment

The research described in this chapter was partially supported by the "Ground-based Research Program for Space Utilization" promoted by the Japan Space Forum and Grant-in Aid for Scientific Research on Priority Area (No. 18033059 "Chemistry of Coordination Space") from the Ministry of Education, Science, Sports, and Culture, Japan.

References

1 Goodsell, D.S. (2000) *American Scientist*, **88**, 230.
2 Whitesides, G.M. (2001) *Scientific American*, **285**, 78.
3 Liu, D.R. and Schultz, P.G. (1999) *Angewandte Chemie-International Edition*, **38**, 36.
4 Niemeyer, C.M. (2001) *Angewandte Chemie-International Edition*, **40**, 4128.
5 Mann, S. (1993) *Nature*, **365**, 499.
6 Breslow, R. (1995) *Accounts of Chemical Research*, **28**, 146.
7 Haupt, K. and Mosbach, K. (2000) *Chemical Reviews*, **100**, 2495.
8 Hecht, S. and Frechet, J.M.J. (2001) *Angewandte Chemie-International Edition*, **40**, 74.
9 Taneva, S., Ariga, K., Okahata, Y. and Tagaki, W. (1989) *Langmuir*, **5**, 111.
10 Taneva, S., Ariga, K., Tagaki, W. and Okahata, Y. (1989) *Journal of Colloid and Interface Science*, **131**, 561.
11 Ariga, K. and Okahata, Y. (1994) *Langmuir*, **10**, 2272.
12 Ebara, Y., Ebato, H., Ariga, K. and Okahata, Y. (1994) *Langmuir*, **10**, 2267.
13 Ariga, K., Isoyama, K., Hayashida, O., Aoyama, Y. and Okahata, Y. (1998) *Chemistry Letters*, 1007.
14 Ariga, K., Nakanishi, T. and Michinobu, T. (2006) *Journal of Nanoscience and Nanotechnology*, **6**, 2278.
15 Lvov, Y., Ariga, K. and Kunitake, T. (1994) *Chemistry Letters*, 2323.
16 Lvov, Y., Ariga, K., Ichinose, I. and Kunitake, T. (1995) *Journal of the American Chemical Society*, **117**, 6117.
17 Lvov, Y., Ariga, K., Ichinose, I. and Kunitake, T. (1995) *Journal of the Chemical Society-Chemical Communications*, 2313.
18 Lvov, Y., Ariga, K., Ichinose, I. and Kunitake, T. (1996) *Thin Solid Films*, **285**, 797.
19 Ariga, K., Onda, M., Lvov, Y. and Kunitake, T. (1997) *Chemistry Letters*, 25.
20 Onda, M., Lvov, Y., Ariga, K. and Kunitake, T. (1997) *Japanese Journal of Applied Physics*, **36**, L1608.
21 Lvov, Y., Onda, M., Ariga, K. and Kunitake, T. (1998) *Journal of Biomaterials Science-Polymer Edition*, **9**, 345.
22 Caruso, F., Furlong, D.N., Ariga, K., Ichinose, I. and Kunitake, T. (1998) *Langmuir*, **14**, 4559.

23 Ariga, K., Hill, J.P. and Ji, Q. (2007) *Physical Chemistry Chemical Physics*, **9**, 2319.
24 Okahata, Y., Tsuruta, T., Ijiro, K. and Ariga, K. (1988) *Langmuir*, **4**, 1373.
25 Okahata, Y., Tsuruta, T., Ijiro, K. and Ariga, K. (1989) *Thin Solid Films*, **180**, 65.
26 Onda, M., Lvov, Y., Ariga, K. and Kunitake, T. (1996) *Biotechnology and Bioengineering*, **51**, 163.
27 Onda, M., Lvov, Y., Ariga, K. and Kunitake, T. (1996) *Journal of Fermentation and Bioengineering*, **82**, 502.
28 Onda, M., Ariga, K. and Kunitake, T. (1999) *Journal of Bioscience and Bioengineering*, **87**, 69.
29 Ariga, K. (2004) *Chemical Record*, **3**, 297.
30 Sagiv, J. (1980) *Journal of the American Chemical Society*, **102**, 92.
31 Schreiber, F. (2000) *Progress in Surface Science*, **65**, 151.
32 Fragneto, G., Thomas, R.K., Rennie, A. R. and Penfold, J. (1995) *Science*, **267**, 657.
33 Mooney, J.F., Hunt, A.J., McIntosh, J. R., Liberko, C.A., Walba, D.M. and Rogers, C.T. (1996) *Proceedings of the National Academy of Sciences, USA*, **93**, 12287.
34 Chtisey, L.A., Lee, G.U. and O'Ferrall, C.E. (1996) *Nucleic Acids Research*, **24**, 3031.
35 Tanii, T., Hosaka, T., Miyake, T., Zhang, G.J., Zako, T., Funatsu, T. and Ohdomari, I. (2004) *Applied Surface Science*, **234**, 102.
36 Nftzer, L. and and Sagiv, J. (1983) *Journal of the American Chemical Society*, **105**, 674.
37 Angst, D.L. and and Simmons, G.W. (1991) *Langmuir*, **7**, 2236.
38 Yang, Z., Galloway, A. and Yu, H. (1999) *Langmuir*, **15**, 8405.
39 Ruckenstein, E. and Li, Z.F. (2005) *Advances in Colloid and Interface Science*, **113**, 43.
40 Moore, L.W., Sprjnger, K.N., Shi, J.-X., Yang, X., Swanson, B.I. and Li, D. (1995) *Advanced Materials*, **7**, 729.
41 Chaki, N.K. and Vijayamohanan, K. (2002) *Biosensors and Bioelectronics*, **17**, 1.
42 Shirahata, N., Masuda, Y., Yonezawa, T. and Koumoto, K. (2002) *Langmuir*, **18**, 10379.
43 Ichimura, K., Suzuki, Y., Seki, T., Hosoki, A. and Aoki, K. (1988) *Langmuir*, **4**, 1214.
44 Okahata, Y., Ariga, K. and Shimizu, O. (1986) *Langmuir*, **2**, 538.
45 Okahata, Y., Ariga, K., Nakahara, H. and Fukuda, K. (1986) *Journal of the Chemical Society-Chemical Communications*, 1069.
46 Ariga, K. and Okahata, Y. (1989) *Journal of the American Chemical Society*, **111**, 5618.
47 Philips, M.C. and Chapman, D. (1968) *Biochimica et Biophysica Acta*, **163**, 301.
48 Bull, H.B. (1945) *Journal of the American Chemical Society*, **67**, 8.
49 Blaudez, D., Bonnier, M., Desbat, B. and Rondelez, F. (2002) *Langmuir*, **18**, 9158.
50 Ulman, A. (1992) *Advanced Materials*, **2**, 573.
51 Fontaine, P., Goldmann, M. and Rondelez, F. (1999) *Langmuir*, **15**, 1348.
52 Ge, S.R., Takahara, A. and Kajiyama, T. (1994) *Journal of Vacuum Science and Technology*, **A12**, 2530.
53 Ge, S., Takahara, A. and Kajiyama, T. (1995) *Langmuir*, **11**, 1341.
54 Kojio, K., Ge, S., Takahara, A. and Kajiyama, T. (1998) *Langmuir*, **14**, 971.
55 Matsumoto, M., Tanaka, M., Azumi, R., Manda, E., Tachibana, H., Kondo, Y. and Yoshino, N. (1997) *Molecular Crystals and Liquid Crystals*, **297**, 31.
56 Wang, R., Parikh, A.M., Beers, J.D., Shreve, A.P. and Swanson, B. (1999) *Journal of Physical Chemistry B*, **103**, 10149.

57 Fang, J. and Knober, C.M. (1996) *Langmuir*, **12**, 1368.

58 Okahata, Y., Yokobori, M., Ebara, Y., Ebato, H. and Ariga, K. (1990) *Langmuir*, **6**, 1148.

59 Ariga, K., Tanaka, K., Katagiri, K., Kikuchi, J., Shimakoshi, H., Ohshima, E. and Hisaeda, Y. (2001) *Physical Chemistry Chemical Physics*, **3**, 3442.

60 Ariga, K., Tanaka, K., Katagiri, K., Kikuchi, J., Ohshima, E. and Hisaeda, Y. (2000) *Colloids and Surfaces A-Physicochemical and Engineering Aspects*, **169**, 47.

61 Shimakoshi, H., Nakazato, A., Tokunaga, M., Katagiri, K., Ariga, K., Kikuchi, J. and Hisaeda, Y. (2003) *Dalton Transactions*, 2308.

62 Shimojima, A., Sugawara, Y. and Kuroda, K. (1998) *Journal of the American Chemical Society*, **120**, 4528.

63 Shimojima, A., Umeda, N. and Kuroda, K. (2001) *Chemistry of Materials*, **13**, 3610.

64 Shimojima, A. and Kuroda, K. (2002) *Langmuir*, **18**, 1144.

65 Shimojima, A. and Kuroda, K. (2006) *Chemical Record*, **6**, 53.

66 Fujimoto, Y., Shimojima, A. and Kuroda, K. (2006) *Journal of Materials Chemistry*, **16**, 986.

67 Parikh, A.N., Schivley, M.A., Koo, E., Seshadri, K., Aurentz, D., Mueller, K. and Allara, D.L. (1997) *Journal of the American Chemical Society*, **119**, 3135.

68 Bourlinos, A.B., Chowdhury, S.R., Jiang, D.D., An, Y., Zhang, Q., Archer, L.A. and Giannelis, E.P. (2005) *Small*, **1**, 80.

69 Fujii, K., Fujita, T., Iyi, N., Kodama, H., Kitamura, K. and Yamagishi, A. (2003) *Journal of Materials Science Letters*, **22**, 1459.

70 Shimojima, A., Sugawara, Y. and Kuroda, K. (1997) *Bulletin of the Chemical Society of Japan*, **70**, 2847.

71 Parikh, A.N., Beers, J.D., Shreve, A.P. and Swanson, B.I. (1999) *Langmuir*, **15**, 5369.

72 Sapuri-Butti, A.R., Li, Q., Groves, J.T. and Parikh, A.N. (2006) *Langmuir*, **22**, 5374.

73 Song, S., Ren, S., Wang, J., Yang, S. and Zhang, J. (2006) *Langmuir*, **22**, 6010.

74 Evmeneko, G., van der Boom, M.E., Kmetko, J., Dugan, S.W., Marks, T.J. and Dutta, P. (2001) *Journal of Chemical Physics*, **115**, 6722.

75 Achten, R., Koudijs, A., Giesbers, M., Reddy, R.A., Verhulst, T., Tschierske, C., Marcelis, A.T.M. and Sudhölter, E.J.R. (2006) *Liquid Crystals*, **33**, 681.

76 Fukushima, Y. and Tani, M. (1996) *Bulletin of the Chemical Society of Japan*, **69**, 3667.

77 Brukett, S.L., Press, A. and Mann, S. (1997) *Chemistry of Materials*, **9**, 1071.

78 Whilton, N.T., Burkett, S.L. and Mann, S. (1998) *Journal of Materials Chemistry*, **8**, 1927.

79 da Fonseca, M.G., Barone, J.S. and Airoldi, C. (2000) *Clays and Clay Minerals*, **48**, 638.

80 da Fonseca, M.G., Silva, C.R., Barone, J.S. and Airoldi, C. (2000) *Journal of Materials Chemistry*, **10**, 789.

81 Silva, C.R., Fonseca, M.G., Barone, J.S. and Airoldi, C. (2002) *Chemistry of Materials*, **14**, 175.

82 Jabber, M., Miechè-Brendlè, J., Roux, M., Dentzer, J., Dred, R.L. and Guth, J. (2002) *New Journal of Chemistry*, **26**, 1597.

83 Minet, J., Abramson, S., Bresson, B., Sanchez, C., Montouillout, V. and Lequeux, N. (2004) *Chemistry of Materials*, **16**, 3955.

84 Carrado, K.A., Xu, L., Csencsits, R. and Muntean, J.V. (2001) *Chemistry of Materials*, **13**, 3766.

85 Fujii, K. and Hayashi, S. (2005) *Applied Clay Science*, **29**, 235.

86 Fukushima, Y. and Tani, M. (1995) *Journal of the Chemical Society-Chemical Communications*, 241.

87 da Fonseca, M.G., Silva, C.R. and Airoldi, C. (1999) *Langmuir*, **15**, 5048.

88 Fujii, K., Hayashi, S. and Kodama, H. (2003) *Chemistry of Materials*, **15**, 1189.

89 Jaber, M., Miehe-Brendle, J., Michelin, L. and Delmotte, L. (2005) *Chemistry of Materials*, **17**, 5275.

90 Patil, A.J., Muthusamy, E. and Mann, S. (2005) *Journal of Materials Chemistry*, **15**, 3838.

91 Kim, S.S., Zhang, W. and Pinnavaia, T.J. (1998) *Science*, **282**, 1302.

92 Caruso, F., Caruso, R.A. and Möhwald, H. (1998) *Science*, **282**, 1111.

93 Knischka, R., Dietsche, F., Hanselmann, R., Frey, H. and Mülhaupt, R. (1999) *Langmuir*, **15**, 4752.

94 Kros, A., Jansen, J.A., Holder, S.J., Nolte, R.J.M. and Sommerdijk, N.A.J. M. (2002) *Journal of Adhesion Science and Technology*, **16**, 143.

95 Katagiri, K., Ariga, K. and Kikuchi, J. (1999) *Chemistry Letters*, 661.

96 Katagiri, K., Hamasaki, R., Ariga, K. and Kikuchi, J. (2003) *Journal of Sol-Gel Science and Technology*, **26**, 393.

97 Katagiri, K., Hamasaki, R., Ariga, K. and Kikuchi, J. (2002) *Journal of the American Chemical Society*, **124**, 7892.

98 Katagiri, K., Hamasaki, R., Ariga, K. and Kikuchi, J. (2002) *Lamgmuir*, **18**, 6709.

99 Yanagisawa, T., Shimizu, T., Kuroda, K. and Kato, C. (1990) *Bulletin of the Chemical Society of Japan*, **63**, 988.

100 Kresge, C.T., Leonowicz, M.E., Roth, W.J., Vartuli, J.C. and Beck, J.S. (1992) *Nature*, **359**, 710.

101 Beck, J.S., Vartuli, J.C., Roth, W.J., Leonowicz, M.E., Kresge, C.T., Schmitt, K.D., Chu, C.T.W., Olson, D. H., Sheppard, E.W., McCullen, S.B., Higgins, J.B. and Schlenker, J.L. (1992) *Journal of the American Chemical Society*, **114**, 10834.

102 Inagaki, S., Fukushima, Y. and Kuroda, K. (1993) *Journal of the Chemical Society-Chemical Communications*, 680.

103 Vinu, A., Hossain, K.Z. and Ariga, K. (2005) *Journal of Nanoscience and Nanotechnology*, **5**, 347.

104 Vinu, A., Miyahara, M. and Ariga, K. (2006) *Journal of Nanoscience and Nanotechnology*, **6**, 1510.

105 Ariga, K., Vinu, A. and Miyahara, M. (2006) *Current Nanoscience*, **2**, 197.

106 Zhang, Q., Ariga, K., Okabe, A. and Aida, T. (2004) *Journal of the American Chemical Society*, **126**, 988.

107 Otani, W., Kinbara, K., Zhang, Q., Ariga, K. and Aida, T. (2007) *Chemistry - A European Journal*, **13**, 1731–1736.

108 Breslow, R. (1991) *Accounts of Chemical Research*, **24**, 159.

109 Ariga, K. and Anslyn, E.V. (1992) *Journal of Organic Chemistry*, **57**, 417.

110 Smith, J., Ariga, K. and Anslyn, E.V. (1993) *Journal of the American Chemical Society*, **115**, 362.

111 Kneeland, D.M., Ariga, K., Lynch, V. M., Huang, C.Y. and Anslyn, E.V. (1993) *Journal of the American Chemical Society*, **115**, 10042.

112 Murakami, Y., Kikuchi, J., Hisaeda, Y. and Hayashida, O. (1996) *Chemical Reviews*, **96**, 721.

3
Entrapment of Biopolymers into Sol–Gel-derived Silica Nanocomposites

Yury A. Shchipunov

3.1
Introduction

Sol–gel processing forms the basis for various routes employed for the fabrication of a wide diversity of functional materials. To impart a structural organization at various length scales, the syntheses are performed using templates. Most consist of a self-organized ensemble of surfactants and co-polymers [1–10]. They have been successfully applied to control the geometry and dimensions of pores that are periodically arranged as in the initial structures. Mesoporous silica materials of the MCM family, which were first synthesized by a team from the Mobil oil company [11,12], are a well-known example.

Sol–gel processes are also widespread in the nature. The processes are called biomineralization and biosilica formation is found over a wide range from single-celled organisms to higher plants and animals [13,14]. Well-known examples of siliceous tissues are presented by diatoms and sponges. It has been demonstrated that the biomineralization or biosilicification processes in their case are completely controlled by biopolymers that regulate the size, structure, shape, spatial orientation and organization of hierarchically ordered biosilica [13–25]. The catalyzing and structure-driving role of biopolymers was substantiated *in vitro* when proteins called silicatein and silaffin, extracted, respectively, from the silica specula of sponges and the cell walls of diatoms, induced the precipitation of silica in solutions containing silicic acid or silanes at around neutral pH. Sol–gel processes did not take place in the absence of these proteins [26,27].

The strong interest in biomineralization is due to its high efficiency and the superiority of the properties of biosilica over those of silica fabricated in geological processes and industrially. It proceeds at mild, ambient conditions in

Bio-inorganic Hybrid Nanomaterials. Edited by E. Ruiz-Hitzky, K. Ariga, Y. M. Lvov

ISBN: 978-3-527-31718-9

aqueous media at circum-neutral pH, causing silica nucleation from media in which the concentration of silicic acid is significantly lower than the critical value needed for the precipitation, whereas the anthropogenic and geological processes are performed at high temperature and pressure using acids or alkalis. Even though the biosilicification occurs from diluted solutions, the rate of biosilica formation is 10^6 times higher, as estimated in Ref. [28], than that of silica formation by abiotic processes. The produced materials are also notably different. The silicious tissues in the Nature are highly structured at all levels, from macromolecular (nanoscopic) to macroscopic. Their fracture toughness can be three or four orders of magnitude higher than that of silica from the inorganic world [14–16].

The decided advantages of biomineralization processes and biosilica over industrial processes and products have inspired numerous attempts at using biopolymers for manipulation of the sol–gel processes *in vitro*. Successful structure-driven syntheses were first performed only when proteins separated from sponges and diatoms were applied by the groups of Stucky and Kroger, respectively [26,27]. The next stage began with the elucidation of the principal amino acids in their composition. This gave an impetus to the use of synthetic polypeptides and block co-polypeptides. Their regulating effects on the sol–gel processes, resulting in the formation of nanosized silica particles of various shape, was demonstrated in numerous publications (see, e.g., Refs. [29–40]). These successful manipulations on the morphology and structure of the sol–gel derived silica gave confidence that biopolymers could be applied as structure-driving templates *in vitro*.

Nevertheless, natural biopolymers cannot usually be involved in the common sol–gel processing to fabricate homogeneous hybrid materials. The main restriction is the poor biocompatibility of common silica precursors. The problem is of major importance for the immobilization of enzymes which has great potential for the development of biocatalysts for biotechnology and biosensors [41–51]. There have been numerous attempts to solve the problem. This chapter is devoted to a comprehensive consideration of the approaches developed for the entrapment of biopolymers in a silica matrix by means of sol–gel processing. Their distinctive features are discussed to give an insight into the decisive advantages and disadvantages. The currently available evidence suggests, in my opinion, that the use of a novel silica precursor containing ethylene glycol residues instead of alcohol in tetramethoxy- and tetraethoxysilanes (TEOS) represents the acme among the developed approaches. This has been applied successfully for the synthesis of a variety of hybrid silica nanomaterials containing both polysaccharides and proteins [52–54]. They were combined in any ratio without precipitation and phase separation over a wide pH range from 2 to 10. Furthermore, highly labile enzymes were immobilized at optimal conditions (pH and low temperature) that reduced their denaturation to a minimum in the course of processing, resulting in retention of the activity and stabilization [55,56]. The experimental data illustrating the opportunities of the approach are given at the end of this chapter.

3.2 Sol–Gel Processes

3.2.1 Chemistry

The cross-linking of sol particles resulting in the transition of the initial solution into the jellified state is called sol–gel processing. The phenomenon occurs widely in nature. It is also basic to a variety of technological processes known as sol–gel technology. Its use has increased sharply since the 1980s for the production of adsorbents, catalysts, ceramics, glasses, membranes, fibers, fine powders, microparticles, nanocomposites, insulating materials, protective coatings, and so on, owing to the flexibility of the sol–gel processes and the low temperature regimes [8,57–59].

The sol–gel route for silica fabrication is based on the strong tendency of silicic acid in solution to take part in condensation reactions [8,18,58]. It can be presented in the general form as:

$$(OH)_3Si{-}OH + HO{-}Si(OH)_3 \rightarrow (OH)_3Si{-}O{-}Si(OH)_3 + H_2O \qquad (1)$$

The reaction results in the formation of a Si–O–Si siloxane bond in which two silicon atoms are bridged by oxygen. The remaining three Si–OH sylanol groups can also be involved in condensation reactions. This will lead to the polymerization of silicic acid with the formation of branched inorganic polymers.

The use of alkoxides for fabrication of silica materials has some advantages that will be discussed later. They serve as a precursor; a silicic acid is generated in the course of their hydrolysis that thereafter enters into condensation reactions (1). The main processes may be presented by Scheme 3.1.

$$\text{Alkoxide} + H_2O \xrightarrow{\text{Hydrolysis}} \begin{matrix}\text{Silanol}\\ \text{groups}\end{matrix} \xrightarrow{\text{Condensation}} \text{Sol} \xrightarrow{\text{Cross-linking}} \text{Gel}$$

Scheme 3.1

The process can be separated into three basic stages [8,57,58]: (1) Partial hydrolysis of the alkoxide after mixing with water, leading to the formation of sylonol group(s) in the hydrolyzing molecule; (2) polycondensation through these reactive groups, resulting in the formation of oligomeric products appearing as sol nanoparticles; (3) cross-linking of the sol nanoparticles, accompanied by the transition to a gel state.

3.2.1.1 Hydrolysis

The reaction is initiated by mixing a precursor with water. Solvent molecules attack the alkoxide leading to a nucleophilic substitution of the alkoxy group (–OR) by a hydroxyl group (–OH). This can be presented in the general form by:

$$Si(-OR)_4 + nH_2O \rightarrow (HO-)_nSi(-OR)_{4-n} + nHO{-}R \qquad (2)$$

where R is a hydrocarbon radical and $n \leq 4$. The result of substitution is the formation of silanol group(s).

The hydrolysis of silica alkoxides is a very slow reaction, a sharp acceleration is observed when the pH is shifted to acidic or alkaline side. This is caused by catalysis by protons or hydroxy groups acting as strong electrophilic or nucleophilic agents, respectively [8,18,58].

3.2.1.2 **Condensation**

The condensation reactions occur only through the mediation of sylanol groups generated by hydrolysis. There are two main opportunities. The condensation can be carried out by oxolation or olation mechanisms. The oxolation occurs with the participation of two sylonol groups. In the course of the reaction a hydroxo ligand is exchanged for an "oxo" one [8]:

$$\begin{aligned}&(\mathrm{RO}-)_{4-n}\mathrm{Si}(-\mathrm{OH})_n + (\mathrm{HO}-)_n\mathrm{Si}(-\mathrm{OR})_{4-n}\rightarrow\\&(\mathrm{OH})_{n-1}(\mathrm{RO}-)_{4-n}\mathrm{Si{-}O{-}Si}(-\mathrm{OR})_{4-n}(\mathrm{OH})_{n-1} + \mathrm{H_2O}\end{aligned} \tag{3}$$

The result is a dimer in which the two silicon atoms are linked by the Si–O–Si siloxane bond. The olation is realized through involvement of both sylanol and alkoxy groups. It can be represented by the following general reaction:

$$\begin{aligned}&\mathrm{Si}(-\mathrm{OR})_4 + (\mathrm{HO}-)_n\mathrm{Si}(-\mathrm{OR})_{4-n}\rightarrow\\&(\mathrm{RO}-)_3\mathrm{Si{-}O{-}Si}(-\mathrm{OR})_{4-n}(\mathrm{OH})_{n-1} + \mathrm{HO{-}R}\end{aligned} \tag{4}$$

The olation leads also to the formation of a dimer with siloxane bond. If the other alkoxy groups in its molecule have been also substituted to hydroxyls, the dimer keeps on going further condensation reactions. This results in the formation of a polymeric chain that can be linear, branched or cyclic. As general, it includes all these segments.

It is worth mentioning that the condensation is a reversible reaction. A silixane bond can be disrupted under attack by alcohol and hydroxyl ion that serves as a catalyst [8,58]:

$$\begin{aligned}&(\mathrm{RO}-)_3\mathrm{Si{-}O{-}Si}(-\mathrm{OR})_{4-n}(\mathrm{OH})_{n-1} + \mathrm{HO{-}R}\rightarrow\\&\mathrm{Si}(-\mathrm{OR})_4 + (\mathrm{HO}-)_n\mathrm{Si}(-\mathrm{OR})_{4-n}\end{aligned} \tag{5}$$

The reversibility provides the restructuring of the silica formed, but sometimes it can cause its depolymerization [8,18].

3.2.1.3 **Sol–Gel Transition**

The rates of the reactions represented by Equations (2)–(5) vary with the pH in a different manner [8,18,58]. The hydrolysis proceeds too slowly in the neutral region, whereas the condensation kinetics is maximal. As soon as the pH is shifted to the acidic or alkaline side, the situation is reversed. By changing the pH, one may regulate the processes and thus the composition of the fabricated silica [8,58,59]. When a precursor is introduced into an acidic or alkaline solution, the hydrolysis can be significantly faster than the condensation and there is a predominant generation of oligomeric products which form colloidal particles, usually called a sol. When the condensation reactions (3) and (4) take place between sylanol groups locating on the surface of various sol nanoparticles in the course of their collision, the silixane bond thus formed brings about their cross-linking. This leads initially to nanoparticle

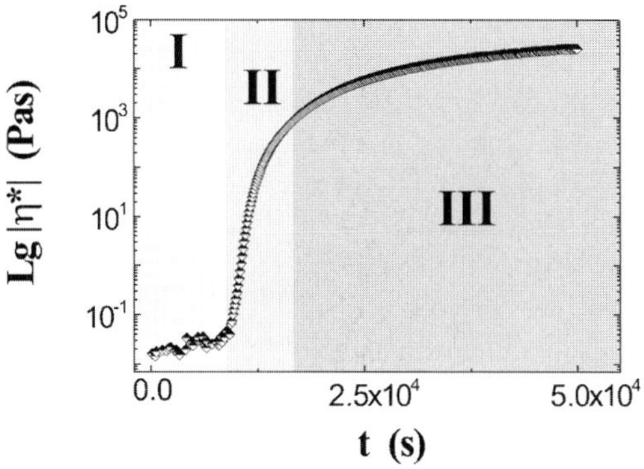

Fig. 3.1 Time dependences of the complex viscosity measured at a frequency of 1 Hz. An aqueous solution was prepared by dissolving 30 wt.% of precursor and 0.2 wt.% of *N*-[(trimethoxysilyl)propyl]-*N,N,N*-trimethylammonium chloride in 0.01 M sulfuric acid. The stages of the sol–gel processes discussed in the text are: I – condensation and sol formation, II – sol–gel transition, III – gel maturation. (Unpublished results).

aggregation into larger clusters, then aggregation of clusters and, finally, formation of a continuous three-dimensional network. By following, for example, the complex viscosity of a solution, one can note, as seen in Figure 3.1, its abrupt increase. The sudden change in rheological properties characterizes a transition from a sol solution to a gel state.

There are three regions, shown by gray gradation, in a temporal dependence of the complex viscosity (Figure 3.1). They are not completely consistent with the stages of sol–gel processing considered above. The first region, in which the complex viscosity does not experience a significant change, corresponds to the precursor hydrolysis and sol formation. As usual, it is impossible to separate these two stages. This is most easily understood when the sol–gel transition has been started. The third stage is obvious from an abrupt increase in the complex viscosity. The transition is shown as region II in Figure 3.1. Region III is also shown; here the complex viscosity is also increasing, but not so sharply as in the preceding region. Furthermore, its rate of increase slows with time. This is an additional stage that has not been previously discussed. The change in rheological characteristics in region III is caused by gel maturation. Although the reactions (2)–(5) slow down after the sol–gel transition they are not completely terminated. There is condensation as well as disruption of siloxane bonds that result in restructuring and strengthening of the gel [8,18,58].

3.2.2
Silica Precursors

The sol–gel processes are based on the polycondensation reactions that proceed with participation of silicic acid. There are three main opportunities to perform them. They are considered below.

3.2.2.1 Orthosilicic Acid

Orthosilicic acid, $Si(OH)_4$, could be an appropriate precursor, but it cannot be stored in monomeric form at reasonable concentrations in aqueous solutions. It enters easily into polycondensation reactions (1) that result in its polymerization. As a monomer, the orthosilicic acid exists in aqueous solutions at a concentration of less than 100 ppm [18]. This is too small to fabricate sol–gel derived materials, although diatoms and sponges have the property of concentrating silica from seawater containing only a few mg per liter [60]. Sol–gel processing in the laboratory can be performed with a rather concentrated solution of orthosilicic acid. This requires freshly prepared acid; the procedure is time consuming which restricts its widespread use.

3.2.2.2 Sodium Metasilicate

The salt Na_2SiO_3, known as water-glass, is used instead of orthosilicic acid. It is stored in basic media. To perform sol–gel processes, hydrochloric acid is added. When the pH drops below 7, a transition into a gel state is observed. The chemical reaction can be represented as follows:

$$Na_2SiO_3 + 2HCl + (x-1)H_2O \rightarrow SiO_2 \cdot xH_2O + 2NaCl \tag{6}$$

It is obvious from Equation (6) that the processes are accompanied by sodium chloride production. The salt content reaches a significant amount that sometimes poses a problem. The NaCl is withdrawn by dialysis that makes the silica fabrication inconvenient owing to this additional time consuming procedure. As a further disadvantage of sodium metasilicate, it is believed to be not very flexible in regulation of the silica morphology [61].

3.2.2.3 Alkoxides

They are silicon esters with the general formula $Si(OR)_4$ in which the silicon atom is linked to organic groups through an oxygen bridge. It is also possible to represent them as derivatives of orthosilicic acid the hydrogens of which are substituted by organic groups. The most commercially important alkoxide is tetraethoxysilane or tetraethyl orthosilicate containing ethanol residues [57]. It is also known under its abbreviation as TEOS. Its structural formula is represented in Figure 3.2. When brought into contact with water or aqueous vapor, the TEOS is hydrolyzed and its products are involved in polycondensation reactions that yield hydrated silicon dioxide (silica) in accordance with the following general equation:

$$Si(OC_2H_5)_4 + (x+2)H_2O \xrightarrow{\text{acid or alkali}} SiO_2 \cdot xH_2O + 4C_2H_5OH \tag{7}$$

This is the basis for commercial sol–gel technologies.

TEOS is of considerable current use on an industrial scale owing to some advantages over sodium metasilicate. Limitations on the sol–gel processing with TEOS arise in the entrapment of biopolymers in silica. They are caused primarily by the alcohol that is separated in the course of precursor hydrolysis Equation (7). Most

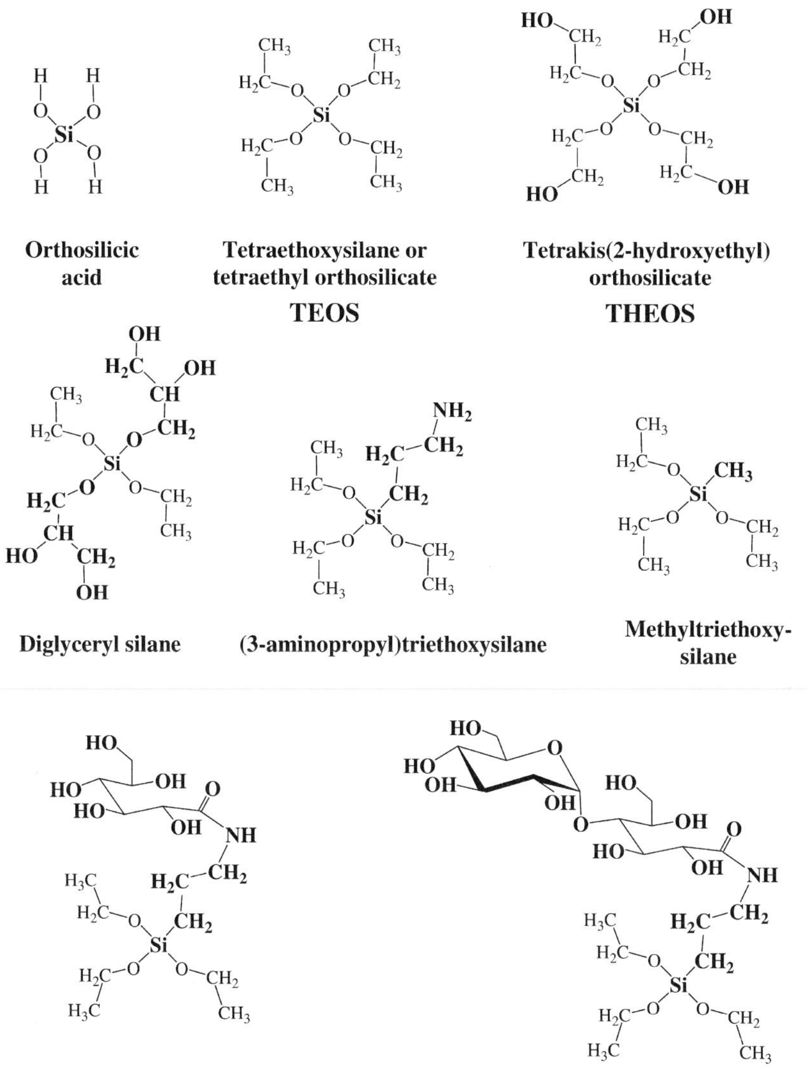

Fig. 3.2 Structural formulas of the silica precursors that are discussed in the chapter. Special features in the structures are shown in bold.

polysaccharides precipitate when the ethanol is added to their aqueous solutions. Proteins are also sensitive to the presence of alcohol because of unfolding of their macromolecules that can cause a denaturating effect [62–64]. For this reason there have been rather numerous attempts at improving the biocompatibility of sol–gel processing with entrapped biopolymers. The developed approaches are discussed below.

3.2.3
Two-Stage Approach to Biopolymer Entrapment

The first belief in the possibility of enzyme stabilization on a silica matrix was stated by Dickey in 1955, but he did not give experimental evidence, only mentioning that his experiments were unsuccessful [65]. A sol–gel procedure for enzyme immobilization in silica was first developed by Johnson and Whateley in 1971 [66]. The entrapped trypsin retained about 34 % of its tryptic activity observed in solution before the encapsulation. Furthermore, the enzyme was not released from the silica matrix by washing, demonstrating the increased stability and working pH range. Unfortunately, the article did not attract attention, although their method contained all the details that may be found in the present-day common approach. This was probably due to its publication in a colloid journal that was not read by biochemists.

The next publication appeared in 1984. Venton *et al.* [67] entrapped antiprogesterone antiserum into a sol–gel derived silica. It retained about 56 % of progesterone binding capacity. The authors mentioned the considerable promise of the sol–gel technique for biopolymer immobilization, but their study was not further continued.

Interest in sol–gel processing was awakened by the work of Avnir *et al.* in 1990 who performed successful experiments with such enzymes as β-glucosidase, alkaline phosphatase, chitinase and aspartase [68]. This gave impetus to their own systematic study of the entrapment of biopolymers in a silica matrix as well as those of other teams [69–79]. The results have been summarized and discussed in numerous review articles (see, e.g., Refs. [41–43,45–49,51,80–85]).

The present-day commonly used approach for the immobilization of biomaterial is a two-stage procedure shown schematically in Figure 3.3. The first stage serves to produce a sol solution. It is performed by admixing a silica precursor, TEOS as a rule, into water acidified by hydrochloric acid. The HCl-catalyzed hydrolysis and succeeding condensation reactions result in the formation of a stable silica sol in accordance with Equations (2)–(4). Biopolymers are entrapped at the second stage. Before its admixing into the sol solution, a buffer is added to shift the pH into the neutral region. The increase in pH accelerates the condensation reactions [58] that make the sol particles cross-linked. As a result of the sol–gel transition, the added biopolymers fall inside pores of the three-dimensional silica network.

The immobilization procedure performed in two stages allows one to exclude the detrimental effect of acid on the entrapped proteins [44,71,86]. It was demonstrated [87] by the example of a set of oxidases that their activity was retained if the entrapment was carried out at a pH as close to their isoelectric point (pI) as possible. Because the pI value of most enzymes is in the neutral region, the two-stage procedure favors the retention of their functionality. Therefore, the pH shift to the optimal region provides a means of extending the sol–gel entrapment to a wide range of enzymes [45,71].

The two-stage procedure eliminates the detrimental effect of acid, but an alcohol, which is produced in the course of precursor hydrolysis Equation (2), remains in solution. This can cause unfolding of the biomacromolecule and denaturation of

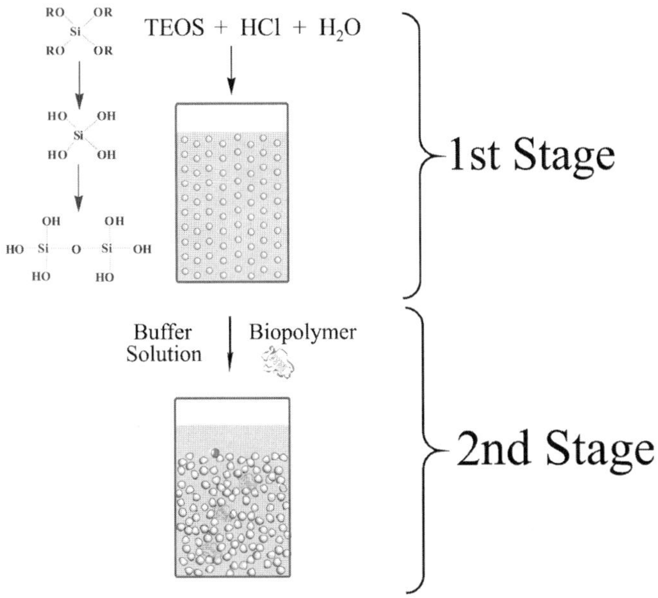

Fig. 3.3 The common two-stage sol–gel process used to entrap biopolymers in a silica matrix (see Scheme 3.1). The first stage serves to hydrolyze alkoxide Equation (2) in the acidic or alkaline media. This is also attended with condensation reactions Equations (3) and (4) resulting in the formation of oligomeric silica that self-organizes in the form of sol nano-particles. Biopolymers are entrapped in the second stage by shifting the pH into the neutral region. A buffer solution is initially added and then biopolymer or a buffer solution of biopolymer is added to the sol-containing solution. The pH shift triggers the cross-linking of sol nanoparticles that leads to formation of a silica three-dimensional network in the pores of which the biopolymers are encapsulated.

enzymes [62–64,88,89]. This presents a severe problem that does not permit the two-stage approach to be considered as a convenient method for biopolymer encapsulation.

To improve the biocompatibility of this procedure, the alcohol was removed from the sol solution produced in the first stage [90–92]. This was done under vacuum by rotary evaporation. As demonstrated in cited works, a labile enzyme, horseradish peroxidase, and bacteria, genetically engineered *E. coli* strain, held their full activity and viability after entrapment in the alcohol-free silica matrix. These experiments favored the view that the efficacious remedy for improvement of biocompatibility of sol–gel processing was removal of toxic alcohol. However, the suggested method is hardly appropriate for widespread use. The treatment under vacuum is time consuming and in addition, there is jellification of sol solution during the process of alcohol evaporation. Water is introduced to avoid this but the dilution complicates the gel fabrication in the second stage (Figure 3.3).

In a procedure developed by Carturan *et al.* [79,93] the alcohol is removed in a gas flow. This is also used to deliver the silica precursor. An alkoxide is initially evaporated to saturate a heated inert gas which serves as a carrier. When the gas is passed over biomaterial adhering to a solid support, the silica precursor nucleates on the wet surface. Silica formed in the course of sol–gel processing results in a shell formation, whereas alcohol produced in accordance with reactions (2) and (4) is swept away by the gas flow. The method was successfully applied for the encapsulation of plant and animal cells, their aggregates, pancreatic islets and alginate microcapsules loaded with cells [94–97]. Its application to biomacromolecules seems highly dubious.

The alcohol removal in the method by Mann *et al.* [98–100] is achieved at the stage of fabrication of the sol. The latter present hybrid nanoparticles of about 50 nm in diameter consisting of an organic core and a silica shell. Their fabrication was begun by mixing TEOS and tetrapropylammonium hydroxide, followed by prolonged refluxing (96 h) and particle separation by repeated centrifugation and washing. The infiltration of nanoparticles thus prepared into an organized bacterial matrix or starch gel led, respectively, to formation of composite fibers [99] and zeolite materials with a hierarchical micro/meso/macropore organization [100].

The template synthesis by the method of Mann *et al.* is limited to large objects because of the nanoparticle dimension. There is also an additional restriction on its widespread use caused by the nanoparticle being stable only in deionized water with pH between 10 and 11.

3.3
Biocompatible Approaches

The realization of the reasons for poor biocompatibility of general alkoxides with biopolymers led to the development of approaches to minimize or eliminate the problem of the detrimental effect of alcohols. This can be done in two ways: modification of the sol–gel processing or the silica precursor. This is considered in some detail below.

3.3.1
Modified Sol–Gel Processing

The alcohol evaporation under vacuum, the treatment of cells and microcapsules by silica precursor supplied by gas carrier and infiltration of silica nanoparticles, whose are discussed above, are the examples of such modified approaches. There are two original methods whose would be well to consider.

3.3.1.1 Method of Gill and Ballesteros

Gill and Ballesteros [101] suggested a chemical treatment of prehydrolyzed alkoxide. They performed first a partial hydrolysis of TEOS up to 50–75 %, preparing oligosilicates. According to Scheme 3.1, the authors stopped the process at the stage of sol formation. Then the oligomeric products were converted to polyol esters by transesterification by exchanging ethanol by various polyols. The chemically treated system was easily converted to the gel at near neutral pH even in the absence of a catalyst. The best results were obtained with glycerinated silicate, poly(glyceryl silicate). The authors entrapped various biopolymers successfully into a silica matrix, among them polysaccharides and proteins, including enzymes.

To my knowledge, this approach has not found further use. This could be due to the complicated procedure in which chemical treatment is applied to form pre-jelled silica. The solution is highly viscous presenting a severe experimental problem. It is also not easy to regulate the degree of polymerization of the silica.

However, this study is of great importance since Gill and Ballesteros demonstrated first by numerous examples [46,82,101] that the exchange of alcohol with polyols improves the compatibility of the sol–gel processing to biopolymers. This showed a method for modification of the silica precursors.

3.3.1.2 **Low-Molecular and Polymeric Organic Additives**

Wei *et al.* suggested a procedure in which the effect of alcohol on biopolymers was diminished by D-glucose, the sugar being introduced in a large amount, up to 50 wt. %. D-Glucose was initially applied as a dopant inducing optical rotation [102] and a porogenic agent for silica synthesized by the sol–gel technique with the help of TEOS [103,104]. It was found later that entrapped enzymes, acid and alkaline phosphatase, held their activity [105,106]. Moreover, it was 2–10 times higher than that in commonly synthesized nanocomposites (see Section 3). The observed increase in activity was accounted for by better accessibility of the entrapped enzyme in the substrate molecules because of the larger pores in the silica matrix. The stabilization of the protein structure, as followed from a systematic study of polyols, amino acids and methylamines by Eggers and Valentine, [107] is caused by their helix-enhancing effect that is observed in solution [108] and then retained in the silica matrix [107]. More recently, Brennan with collaborators [109] provided additional evidence, immobilizing a-chymotrypsin and ribonuclease T1 in the presence of sorbitol and *N*-methylglycine. They attributed the enzyme stabilization to the change in biomacromolecule hydration. It was also found that the removal of sugar and amino acid by washing resulted in a notable decrease in the enzyme activity. This is the decisive disadvantage of this approach because these water-soluble porogenic additives will leak inevitably into the surrounding aqueous medium in the course of the application of enzyme-loaded nanocomposites as biocatalysts.

It should be pointed out that the addition of substances, which could improve the biocompatibility of sol–gel processing and the functional characteristics of the silica matrix, is practiced rather widely. Poly(ethylene glycol) is one of such additives [110–113]. Enzyme stabilization was favored by formation of polyelectrolyte complexes with polymers. For example, an increase in the lactate oxidase and glycolate oxidase activity and lifetime took place when they were combined with poly(*N*-vinylimidazole) and poly(ethyleneimine), respectively, prior to their immobilization [87,114]. To improve the functional efficiency of entrapped horseradish peroxidase, a graft copolymer of polyvinylimidazole and polyvinylpyridine was added [115,116]. As shown in Refs. [117,118], the denaturation of calcium-binding proteins, cod III parvalbumin and oncomodulin, in the course of sol–gel processing could be decreased by complexation with calcium cations.

The addition of biopolymers, polysaccharides and proteins can also exert a good influence on the activity and lifetime of immobilized enzymes and cells. Alginate is applied rather widely (see, e.g., Refs. [70,96,119–123]) owing to its ability to fabricate microcapsules after a contact of microdrops containing polysaccharide with a solution containing a calcium salt. Enzymes and cells become entrapped in the hybrid polysaccharide-silica matrix formed in the inner volume. A sol–gel derived chitosan-silica hybrid film was used in Ref. [124] for successful immobilization of glucose oxidase. Gelatin addition brought about a great decrease in a lysis of *E. coli* bacteria [125].

These examples demonstrate that additives can have a beneficial effect on the entrapped biopolymers. Unfortunately, they are generally not universal. The additives need to be found for individual immobilized biopolymers and that is not so easy to do. For instance, lactate oxidase retained its activity in a silica matrix if the enzyme was taken as a complex with poly(*N*-vinylimidazole) prior to the immobilization, but the polymer did not stabilize glycolate oxidase [87,114]. Its stabilization was observed after an exchange of poly(*N*-vinylimidazole) for poly(ethyleneimine). This is a decisive disadvantage of the approaches because they do not offer a general solution that might be extended to any immobilized biopolymer.

3.3.2
Organically-modified Precursors

An alternative way of improving the biocompatibility of sol–gel processing is to use organically-modified silica precursors rather than organic additives. An organic group is linked directly to a silicon atom with no oxygen bridge (see Figure 3.2). Therefore, it is not susceptible to the hydrolysis (Equation (2)) which is why the organic groups are included in a silica network. The hybrid organic–inorganic materials so fabricated are called ***Or***ganically ***Mo***dified ***Si***licas ("ORMOSILs"). Some details of the organically-modified precursors, and the structure and properties of the hybrid nanocomposites synthesized by their use may be found in review articles [126,127].

A silica precursor with a covalently attached 3-aminopropyl group (Figure 3.2) was used in one of the first studies devoted to biopolymer entrapment [67]. The intent was to provide a better hold of biomacromolecules in the silica matrix through the electrostatic interactions of their negatively charged functional groups with introduced amino groups. This silane has gained wide acceptance for the immobilization of biomaterials by the sol–gel technique (see, e.g., Refs. [46,82,113,122,123,128,129]).

An example of the appropriate application of organically-modified silica precursors is alkoxides with an alkyl group. When methyltrimethoxy- or methyltriethoxysilane (Figure 3.2) was added in formulations to increase the hydrophobicity of ORMOSILs, it resulted in a better enzymatic activity of lipases immobilized in the alkyl-modified silica than in a hydrophilic matrix fabricated by means of TEOS alone [51,80,129–133]. Similarly, an increased stability of lipase from *Candida antarctica B* was observed after its immobilization in a silica matrix formed with the addition of octyltriethoxysilane [134]. Of great interest is that the enzyme from *Burkholderia cepacia* (previously called *Pseudomonas cepacia*) can catalyze the sol–gel process [135]. In its presence there was a substantial increase in the jellification rate. Furthermore, depending on the lipase concentration, the gel formation proceeded in neutral media with no introduced catalysts and the gel structure differed from that fabricated by the common procedure. More recently, it has been demonstrated that the lipase could catalyze gel formation even in such nonpolar media as isooctane [136]. When it was made in the presence of lauric acid and 1-octanol, there was a transesterification reaction that provided an addition of octyl radical to tetramethoxysilane, whereas in the presence of only lauric acid the lipase catalyzed complex reactions in which the alkoxide served as a substrate.

Organically-modified silica precursors have three alkoxy groups. As a result, alcohol is separated after the hydrolysis. Its appearance causes a detrimental effect on biopolymers typical to that of alkoxides. The next disadvantage of these precursors is the absence of universality. They are appropriate only in particular cases.

Progress has been made towards the solution of this problem when the silica precursors were modified by exchanging alkoxy groups for polyols. Shchipunov has used a silane in which molecule all the ethanol residues were exchanged for ethylene glycol (Figure 3.2) [52]. At the same time, Brennan with collaborators introduced an alkoxide with ethanol partially replaced by glycerol (Figure 3.2) [137,138]. There are also distinctions in their procedures for biopolymer entrapment. This results in notable differences in the structure and properties of the fabricated materials.

3.3.3
Biocompatible Precursors by Brennan *et al.*

It was suggested to use a silane containing glycerol residues. This made the chemical treatment of prehydrolyzed silica precursor in the procedure by Gill and Ballesteros (see Section 3.3.1.1) [46,82,101] unnecessary. Brennan *et al.* [113,138,139] replaced two ethanol groups in TEOS by glycerol, thus synthesizing a diglyceryl silane (Figure 3.2). The other two ethanol residues in the molecule remained unaltered. As a consequence of their presence, the hydrolysis of diglyceryl silane proceeds with separation of both glycerol and alcohol. This means that the use of this precursor does not eliminate completely the harmful effect of ethanol on biopolymers which are sensitive to its action.

It should be pointed out that the diglyceryl silane is a solid compound [113,138,139] thus sol–gel processing with its participation is performed through dissolution of the precursor in an aqueous solution. This leads, as shown experimentally in Ref. [140], to the hydrolysis and subsequent sol formation in accordance with Equations (2)–(4). The important consequence is that the diglyceryl silane can be involved only in the two-stage immobilization procedure shown schematically in Figure 3.3.

Of great importance is the volume shrinkage (syneresis) of the synthesized gel after its fabrication. The phenomenon is typical of the sol–gel derived materials fabricated through the two stages (Scheme 3.1, Figure 3.3). By using TEOS, one may find a 70 % reduction in the gel volume, with diglyceryl silane, the shrinkage is 53 % [138,141]. Such a decrease in volume results in deterioration of the functional characteristics of immobilized proteins. For example, the ligand-binding activity of human serum albumin was decreased by 45 % in comparison with the initial level after 45 days of aging, [141] whereas such enzymes as luciferase and tyrosine kinase were quite inactive [142,143]. The deterioration is attributable to the poor accessibility for substances owing to the diffusion restriction in small pores.

The syneresis of maturing sol-gel derived silica was decreased by adding an organically-modified precursor that contained a sugar group attached covalently to silicon. Brennan *et al.* [144] synthesized *N*-[(3-triethoxysilyl)propyl]maltonamide and *N*-[(3-triethoxysilyl)propyl]gluconamide, shown in Figure 3.2. The bulky sugar residue in their molecules brought about gel shrinkage at the level of 15 v/v%. [143] There were also larger pores in the silica matrix [141]. It is believed that these factors could favor the observed accessibility of entrapped enzymes to substrates. However, the sugar-containing precursors have usually been used as an additive to diglyceryl silane demonstrating better biocompatibility. Their mixing provided improvement in the functional characteristics of immobilized biopolymers [138,141–143].

The approach by Brennan with collaborators led to notable improvement in the biocompatibility of sol–gel processing. However, there are some disadvantages. Their approach does not exclude the hydrolytic separation of alcohol. Its presence is detrimental for sensitive biopolymers. Furthermore, the two-stage synthesis is accompanied by the significant shrinkage of sol-gel derived nanocomposites. This leads to a decrease in the pore size that sometimes can restrict the accessibility of enzymes to substrates.

3.4 One-Stage Approach Based on a Silica Precursor with Ethylene Glycol Residues

3.4.1 Precursor

The structure of the ethylene glycol-containing precursor, *tetrakis(2-hydroxyethyl) orthosilicate* (THEOS), can be seen in Figure 3.2. It was introduced by Hoffmann *et al.* [145]. They synthesized THEOS by substitution of ethanol groups with ethylene glycol as was suggested by Mehrotra and Narain in 1967 [146]. The Indian authors examined the product solubility in benzene, but its properties in aqueous solutions were not studied. This was done for the first time by Hoffmann and coworkers [145,147]. They noted that an advantage of THEOS over TEOS lies in its complete water solubility. The precursor can be mixed with water in any ratio and so there is no need to add an organic cosolvent. Furthermore, the THEOS itself and ethylene glycol produced in the course of hydrolysis, Equation (2) did not precipitate surfactants in the cases in which TEOS caused this effect and, which is also important, there was no restructuring of surfactant aggregates and their lyotropic mesophase [145]. This experimental evidence was indicative of the improved compatibility of the silica precursor with surfactants.

Ethylene glycol is a well-known biocompatible organic solvent which was taken into account when THEOS was first put together with polysaccharides to synthesize hybrid nanomaterials. The initial experiments were performed with iota-, kappa- and lambda-carrageenans [52]. TEOS could not be applied because the hydrolytically generated ethanol triggered precipitation of these polysaccharides from aqueous solution, [148,149] whereas carrageenans are soluble in ethylene glycol [150]. The admixing of THEOS with aqueous solutions of carrageenans proceeded in accordance with Scheme 3.2.

$$\text{THEOS} + H_2O + \text{Biopolymer} \xrightarrow[\text{Cross-linking}]{\text{Hydrolysis, Condensation}} \text{Gel}$$

Scheme 3.2

It did not give rise to phase separation or precipitation. Similar behavior was observed when other types of polysaccharides were examined [53,54]. By now all the commercially important polysaccharides have been applied to the fabrication of hybrid silica nanocomposites in accordance with Scheme 3.2. What is more, various proteins have been entrapped in silica by the same means. In all instances the THEOS demonstrated good biocompatibility with biopolymers, even though its amount in formulations was sometimes up to 60 wt.%. Biopolymer solutions after the precursor admixing remained homogeneous to the point of transition into a gel state.

3.4.2 Role of Biopolymers in Sol–Gel Processing

Most commercially important polysaccharides as well as proteins such as gelatin and albumin have been examined. They are listed in Table 3.1 together with structural formulas and some characteristics including those of the formed gels. The experiments were performed with anionic, uncharged and cationic polysaccharides. Most of them had a linear chain, but there were also samples with branched (starch, arabinogalactan) and cyclic (cyclodextrins) chains. Xanthan, locust bean gum, guar gum, hydroxyethylcellulose and its cationic derivatives are referred to as graft-polymers because of the short chains attached periodically to a linear backbone. Alginate is a block copolymer of L-guluronic and D-mannuronic acid residues [151,152]. For the most part the polysaccharides are applied as a thickener to regulate the viscosity of the solutions in view of the inability of carbohydrate chains to associate in solution [153–155]. The jellification of aqueous solutions occurs with xanthan [156–158], starch [159,160], and κ- and ι-carrageenans [148,149,161]. Alginates can form a gel, but only in the presence of cations of earth metals [151,152].

The examined proteins represent the main types of this class of biopolymers. One of them was albumin, another, gelatin. The former is a globular protein (MW 66.3 kDa) [162], whereas the latter is a fibrillar one, liable to form a helix from

Tab. 3.1 Polysaccharides and proteins involved in the synthesis of hybrid silica nanocomposite materials.

Biopolymer	**Structural Formula**	**Characteristic**	$C_{s\text{-}g.}$[a] (wt.%)	**Hydrogel**[b]
	Polysaccharides			
Alginate	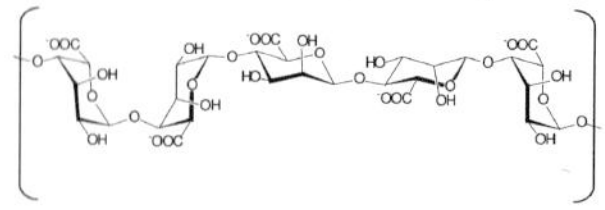	Gel-forming, linear, block copolymer,	~3	Turbid, monolith
Arabinogalactan	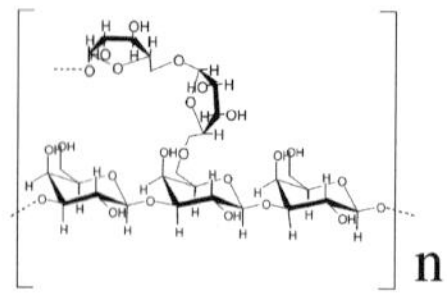	Non-jellified, branched, uncharged	~5	Turbid, syneresis
Carboxymethyl cellulose	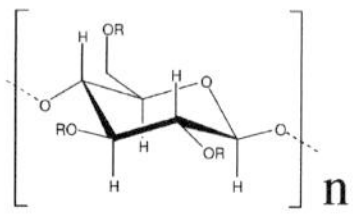 R=H or CH_2COO^-	Non-jellified, linear, anionic	nd[e]	Opalescent, monolith
Kappa-carrageenan	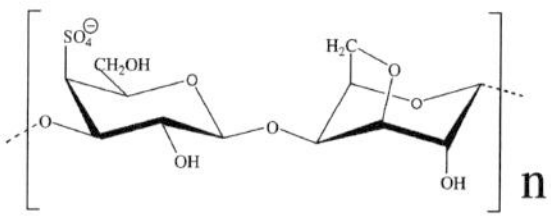	Gel-forming, linear, anionic	nd	Turbid, syneresis
Iota-carrageenan	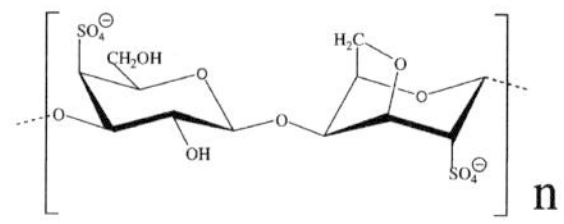			Turbid, monolith
Lambda-carrageenan	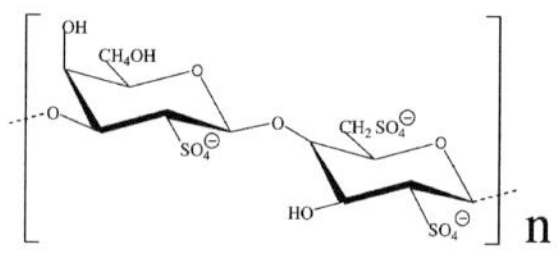	Non-jellified, linear, anionic		
Cationic derivative of hydroxyethylcellulose (cat-HEC)	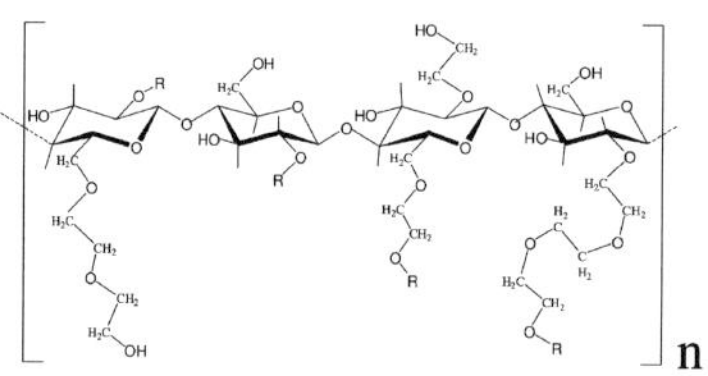 R=CH_2-CHOH-CH_2-$N^+(CH_3)_3$ Cl^-	Non-jellified, graft polymer, cationic	0.5	Transparent, monolith
Chitosan	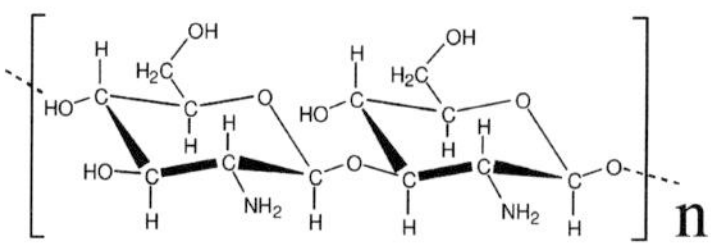	Non-jellified, linear, cationic	nd	Opalescent, monolith
Cyclodextrin	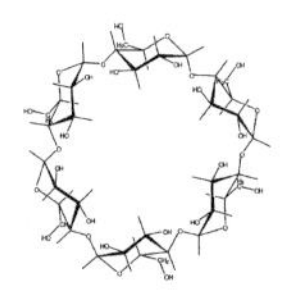	Non-jellified, cyclic, uncharged	~5	Opalescent, monolith

Tab. 3.1 (*Continued*)

Biopolymer	**Structural formula**	**Characteristic**	**$C_{s\text{-}g.}$[a] (wt.%)**	**Hydrogel[b]**
Fucoidan	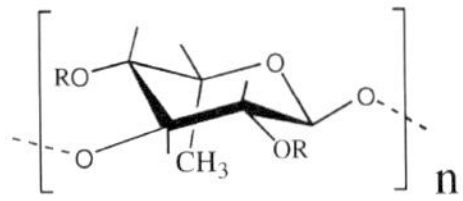 R=H or SO_3^-	Non-jellified, linear, anionic	nd	Turbid, monolith
Guar Gum	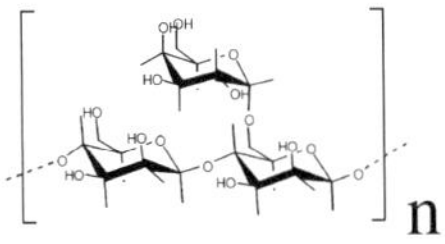	Non-jellified, graft polymer, uncharged	nd	Turbid, syneresis
Hyaluronic acid, sodium salt	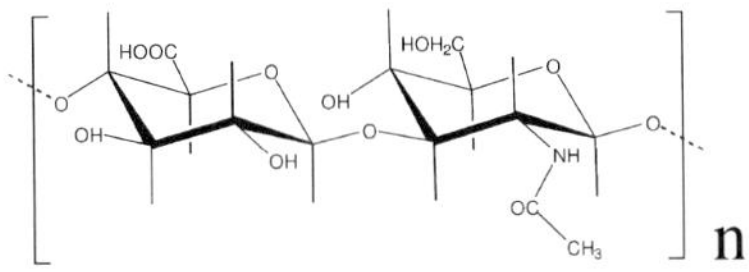	Non-jellified, linear, anionic	2	Opalescent, monolith
Hydroxyethyl-cellulose (HEC)	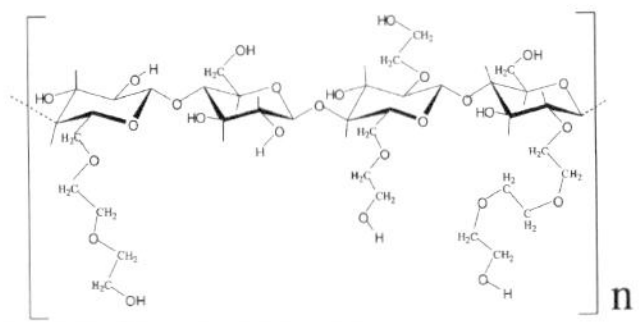	Non-jellified, graft polymer, uncharged	nd	Turbid, syneresis
Laminaran	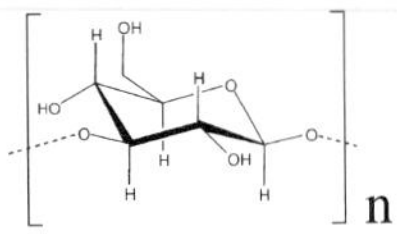	Non-jellified, linear, uncharged	nd	Turbid, syneresis
Locust bean gum	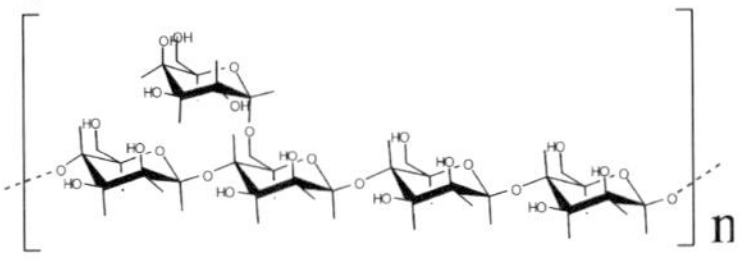	Non-jellified, graft polymer, uncharged	~5	Turbid, monolith
Starch	A mixture of amylase and amylopectin with ratio of about 1 : 4	Gel-forming, branched, uncharged	nd	Turbid, monolith
Xanthan	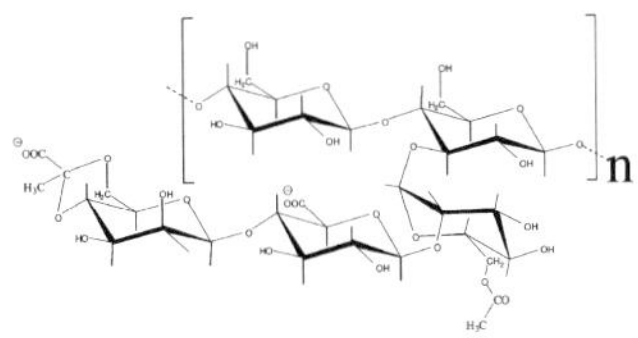	Gel-forming,graft polymer, anionic	~1	Turbid, monolith
	Proteins			
Albumin	—	Non-jellified, globular	~5	Transparent, monolith
Gelatin	—	Gel-forming, fibrillar	nd	Transparent, monolith

[a]Critical concentration of precursor at which gel formation is observed.
[b]Characteristics are given for gel fabricated by addition of 10 wt.% THEOS to a 1 wt.% solution of biopolymer.
[c]Not determined.

three macromolecules and experiencing a reversible phase transition from helix to coil at a certain temperature [163–165]. Gelatin solutions were in a gel state at ambient conditions when its concentration exceeded 1 wt.%. Albumin could jellify a solution after its denaturation at high temperature [162].

This short comparative characteristic gives some insight into the variety of the examined biopolymers, their structure and properties. However, they all had much in common in their influence on processes with THEOS when it was introduced in their aqueous solutions (Scheme 3.2). The first experiments were carried out with carrageenans. They revealed strong effects of polysaccharides on the jellification processes with THEOS [52]. It was suggested for the first time that biopolymers could participate in these processes. Attention was drawn to a drastic decrease in the time of gel setting and a change in the morphology of the nanocomposites. These effects were interpreted as catalysis of processes and mineralization of biomacromolecules serving as a template for silica [52–54].

One of the main reasons for the catalysis of reactions represented by Equations (2)–(4) is based on the fact that there was no jellification of the solutions with biopolymers in the neutral region after a month. When the processes could also proceed with the addition of an acid or alkali in the absence of biopolymers, one could observe a drastic acceleration. The gelation time could be decreased from hours to a few minutes once polysaccharides and proteins were added.

It is worth mentioning that a significant influence of the charged functional groups in biomacromolecules on the jellification kinetics was not mentioned. This distinguishes the fabrication of hybrids with the help of THEOS from the common sol–gel processing. By way of example let me cite a work by Coradin *et al.* [166] who studied the behavior of bovine serum albumin and lysozyme in sodium silicate solutions at various pH. It was demonstrated that a composite protein–silica gel was formed when inorganic sol nanoparticles and biomacromolecules bore charges opposite in sign. Therefore, they concluded that the mineralization mechanism is based on electrostatic interactions between protein and silica sol. Our unpublished results concerning the influence of albumin on the processes with THEOS demonstrated that nanocomposite materials were formed over a wide pH region from 2 to 10. In other words, the electrostatic interactions did not control the mineralization. A similar conclusion follows from experiments with polysaccharides, among which were samples with both oppositely charged and uncharged biomacromolecules [53,54].

The jellification at neutral pH and acceleration of processes in the acidic and alkaline regions points to catalysis of precursor hydrolysis as well as the condensation reactions represented by Equations (2)–(4), respectively. It was demonstrated that the effects of polysaccharides are associated with the presence of numerous hydroxy groups in their macromolecules [53]. Their participation can be attributed to the formation of hydrogen bonds with ethylene glycol residues in the precursor molecule and with hydrolytically generated sylanol groups. This is shown schematically in Figure 3.4. The hydrogen bonds formed could encourage the nucleophilic substitution of an ethylene glycol group for a hydroxyl group followed by condensation between neighboring sylanol groups in precursor molecules nucleating on biopolymer macromolecules. For some reasons it proceeds more easily than the processes in

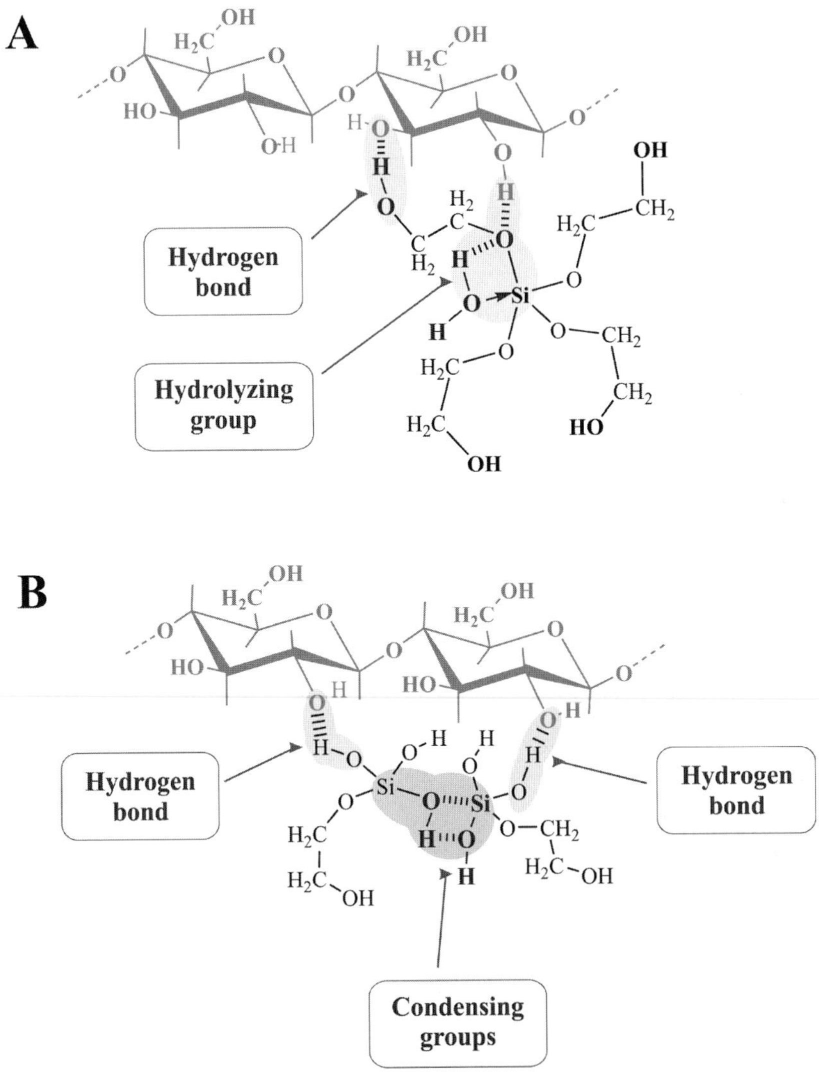

Fig. 3.4 Schematic drawings illustrating the nucleation of silica precursor and possible mechanisms of reactions leading to silica formation. (A) The precursor hydrolysis. THEOS nucleates on a polysaccharide macromolecule through hydrogen bonds that are formed with hydroxy groups in the biomacromolecule. Perhaps, this has a destabilizing effect, facilitating an exchange of the alkoxy group for the hydroxy group in accordance with Equation (2) (Section 3.2.1.1). (B) Condensation reaction. This can proceed between neighboring molecules containing sylanol groups. Their close location could favor the reaction.

the bulk solution. An actual catalytic mechanism is presently unknown, further study is needed.

The next important role of biopolymers in the processes after precursor nucleation is that their macromolecules can serve as a template for silica. This has been

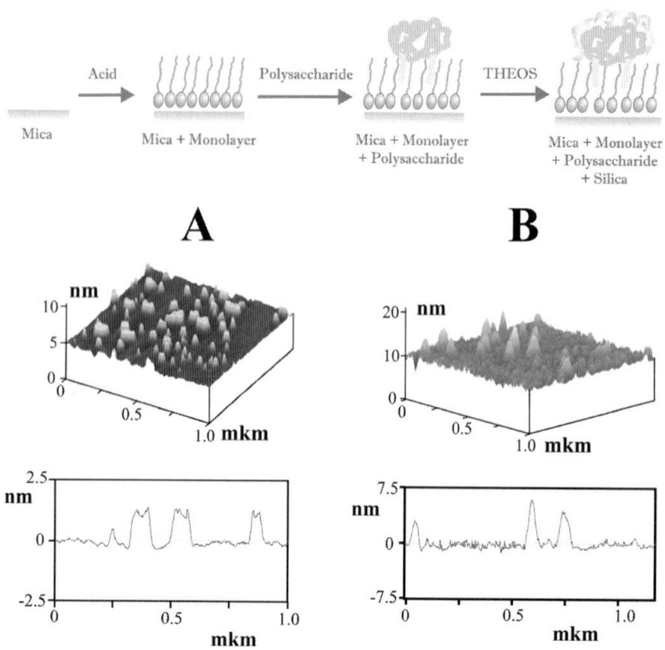

Fig. 3.5 Representation of a scheme of an experiment (upper set of drawings) and the obtained experimental results presented as AFM images (middle part) and cross-sectional profiles (bottom) that provides evidence of silica nucleation and shell formation on biopolymer macromolecules. *Scheme of experiment.* This includes the following main steps. 1. Protection of the mica surface against silica precipitation. It was covered with a fatty (arachidic) acid monolayer transferred from a water substrate with the Langmuir–Blodgett technique. This made the mica surface hydrophobic because of the orientation of the acid molecules with their hydrocarbon chains pointing outwards. 2. Adsorption of carbohydrate macromolecules. Hydrophobically modified cationic hydroxyethylcellulose was adsorbed from an aqueous solution. Hydrocarbon chains of polysaccharide served as anchors to fix the biomacromolecules firmly onto the acid monolayer. 3. Surface treatment by silica precursor. The mica covered with an acid monolayer with adsorbed polysaccharide was treated in THEOS solution for various times. If the silica nucleates on the biopolymers, one might find a silica precipitate on the biomacromolecules. *Experimental results.* (A) An AFM image of a mica surface covered by a monomolecular film of arachidic acid with adsorbed hydrophobically modified cationic hydroxyethylcellulose and the cross-sectional profile traced over it. Biomacromolecules are seen as separated "hills" on the surface. (B) An AFM image and cross-sectional profile traced over the mica surface after 5 s exposure in 10 wt.% aqueous solution of THEOS. It is obvious from comparison with the AFM image on the left side that even the short treatment with the silica precursor resulted in a notable increase in the height of the "hills" representing the adsorbed biomacromolecules. When the treatment was longer, the amount of precipitated silica increased and started to spread along the surface (see Ref. [167]). (Unpublished experimental results obtained with the participation of Prof. T. Imae and A. Kojima).

demonstrated with the help of two principal experiments. The scheme of one of them is presented in Figure 3.5. A sample for examination by *atomic force microscopy* (AFM) was prepared by a three-stage procedure [167]. First, a mica surface was covered with a monomolecular film of arachidic acid by means of a Langmuir–Blodgett technique to prevent silica precipitation on the surface. Thereupon hydrophobically modified cationic hydroxyethylcellulose was adsorbed in such a manner that there were both separated carbohydrate macromolecules and their aggregates. This is obvious from Figure 3.5A where an AFM image of a mica surface covered by an arachidic acid monolayer with hydrophobically modified polysaccharide and the cross-sectional profile traced over it are shown. When the sample was treated in a solution of THEOS, a notable increase in the height of adsorbed carbohydrate macromolecules is found (see Figure 3.5B). This is due to precipitation of silica. The height increase is observed in the initial stage after a short (5 s) treatment. Prolonged contact of the mica surface with the precursor solution led to the formation of a thick coating (shells) of biomacromolecules that caused spreading of the silica over the surface and merging of neighboring shells. As a result, a two-dimensional analog resembling a network structure was formed [167]. At the same time, the gel appearance was not noted in the bulk solution.

Additional evidence for silica nucleation on biopolymer macromolecules was furnished by experiments in which solutions of proteins were studied by dynamic light scattering. As an illustration, Figure 3.6 shows the relative intensity of light scattering versus the diameter of the scattering particles in solution with 1 wt.% of bovine serum albumin. Curve 1 presents the initial state where the protein was not yet treated with silica precursor. The measured

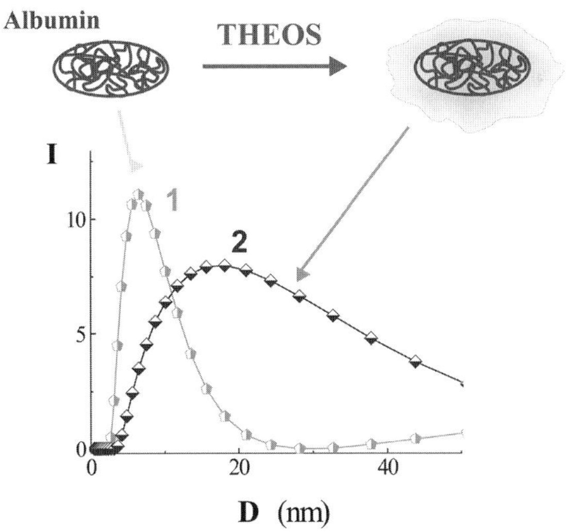

Fig. 3.6 The relative intensity of dynamic light scattering vs. the diameter of scattering particles for a solution with 1 wt.% bovine serum albumin (1) and the same solution after addition of 3 wt.% of THEOS (2). Before the measurements, the solutions were left at ambient temperature for a week. The drawings are a schematic representation of a protein macromolecule before and after the treatment by silica precursor.

diameter correlates well with the macromolecular dimension. Curve 2 demonstrates what happens to the biomacromolecules if 3 wt.% of THEOS is introduced into the solution. One may see the substantially increased dimensions of the scattering particles. This proceeded in a gradual manner, beginning from the state represented by curve 1 (unpublished results). The observed increase in particle diameter is indicative of encasement of albumin macromolecules in a silica shell. The following important features of the processes should be particularly emphasized.

1. The silica nucleated only on the biomacromolecules. By decreasing the THEOS amount in solution from 3 to 0.03 wt.%, we observed a curtailment of the particle diameter to the macromolecular dimension. This constitutes evidence that most of the silica generated *in situ* was used to form a shell.
2. The introduced THEOS did not bring about precipitation in protein solutions. This behavior differs from that observed with common silica precursors. For example, TEOS added in such small amounts caused precipitation. By using THEOS, we could prepare homogeneous mixtures. When its amount introduced into the albumin solution was less than 5 wt.%, there was no transition to a gel state (Table 3.1). A gradual increase in THEOS concentration resulted in a rise in the solution viscosity. The transition to a gel state took place as soon as a critical concentration was reached. Its value, as demonstrated in Ref. [168], is dependent on the biopolymer nature, varying to a rather large extent (see Table 3.1).

3.4.3 Advantages of One-Stage Processes

The ethylene glycol-containing silica precursor has been combined, as mentioned above, with most commercially important polysaccharides and two proteins listed in Table 3.1. In spite of the wide variety of their nature, structure and properties, the jellification processes on addition of THEOS to solutions of all of these biopolymers (Scheme 3.2) had a common feature, that is the formation of monolithic nanocomposite materials, proceeding without phase separation and precipitation. The syneresis mentioned in a number of cases in Table 3.1 was not more than 10 vol.%. It is worthwhile to compare it with common sol–gel processes. For example, the volume shrinkage of gels fabricated with the help of TEOS and diglyceryl silane was 70 and 53 %, respectively [138,141].

The striking difference in the syneresis of nanocomposite materials synthesized by various approaches is due, in my opinion, to the procedures of the synthesis. This is illustrated in the schematic drawings in Figure 3.7. The common sol–gel process for entrapping biomaterial consists of a two-stage synthesis (see Scheme 3.1 and Figure 3.3). The initial stage serves to prepare a sol. This is shown in Figure 3.7A.

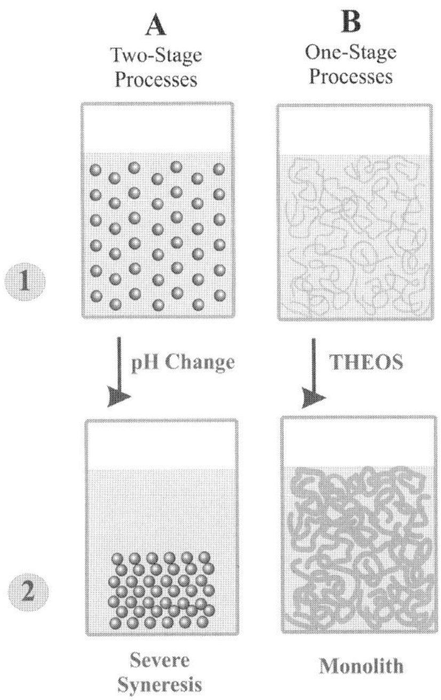

Fig. 3.7 Schematic drawings demonstrating the main features of two-stage (A) and one-stage (B) procedures leading to a difference in the morphology of the fabricated materials. (A) Sol nanoparticles initially prepared in the first stage (1, see also Figure 3.3) can self-assemble into a three-dimensional network when they are in direct contact with each other. For this reason, a gel formed after cross-linking (sol-gel transition) has a smaller volume (2). (B) The initial stage (1) is represented by a solution of entangled biopolymer macromolecules. The added silica precursor nucleates on the biomacromolecules (Figure 3.4). The following hydrolytic and condensation reactions lead to a shell formation from silica. The merging of the shells at their contact points results in the appearance of cross-links. A three-dimensional network thus generated represents the initial one consisting of entangled and now cross-linked biomacromolecules. The jellification in this case is not accompanied with gel shrinkage. There is a monolithic nanocomposite.

The next stage is a sol–gel transition that is accounted for by the formation of a three-dimensional network from cross-linked sol nanoparticles. It is apparent that it can happen only when the particles are in direct contact. The association of sol nanoparticles, as shown by Figure 3.7A, will inevitably result in shrinkage of the volume because of a decrease of the distance between them. The larger the initial distance and the denser the arrangement of particles in the gel, that is the larger the difference between the initial and final states in the system, the larger the syneresis.

The one-stage process with THEOS proceeds differently (Figure 3.7B). The difference is in the absence of sol nanoparticles in the initial solution. There are entangled macromolecular chains (stage 1, Figure 3.7B). The silica precursor is introduced in a biopolymer solution as a monomer. The experimental results available to date (see Section 3.4.2) demonstrate that instead of sol formation there is a nucleation of silane and a product of its hydrolysis on the biomacromolecules (Figure 3.4). The subsequent condensation reactions proceeding *in situ* result in encasement of the biopolymer macromolecules in silica shells. Biopolymers act as a template to build up the three-dimensional network. The cross-links are formed as a result of the merging of silica shells at the crossing points of entangled biomacromolecule chains. Figure 3.7B makes this mechanism clear. Since there is initial contact between entangled biomacromolecules, the necessity for condensing to provide the cross-linking is absent. Therefore, the transition to the gel state occurs without shrinkage. This is the principal difference between the approach based on using THEOS to entrap biopolymers and the common two-stage sol–gel process (Scheme 3.1, Figure 3.3).

3.4.4
Hybrid Biopolymer–Silica Nanocomposite Materials

From the above discussion one might expect that hybrid materials formed with the help of THEOS in accordance with Scheme 3.2 would differ in their characteristics from those synthesized by the common two-stage sol–gel process (Scheme 3.1, Figure 3.3). The main reason for that lies in the complete absence of, or insignificant, syneresis, while the common procedure leads to volume shrinkage up to 70 %. The important consequence of this difference is conspicuous in the structural organization of nanocomposites that is less dense for hybrids synthesized by using THEOS than for those fabricated by means of TEOS.

The peculiarities in the morphology of gels prepared by the one-stage procedure are obvious from Figure 3.8 representing pictures taken with a *scanning electron microscope* (SEM). One can see a cross-linked network from fibrils and spherical

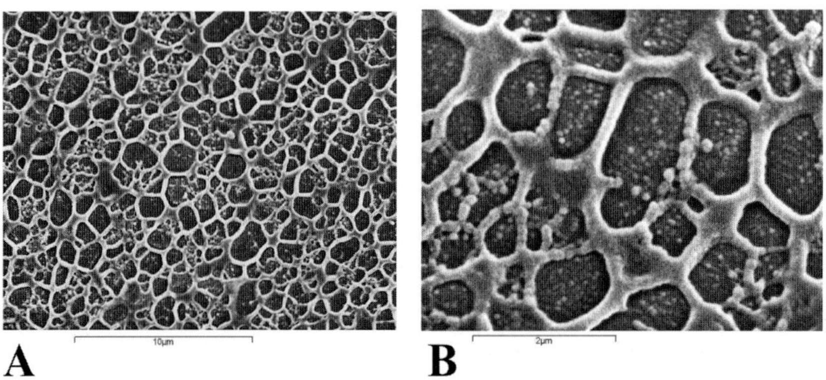

Fig. 3.8 SEM micrographs of a gel synthesized by adding 10 wt.% of THEOS to a 1.5 wt.%. aqueous solution of cationic hydroxyethylcellulose. The pictures were taken at various magnifications to demonstrate the main features of the material. (Unpublished results, Pictures taken by Dr. C. Abetz).

Fig. 3.9 SEM images of aerogels. They were prepared from initial gels synthesized by adding 10 wt.% THEOS to an aqueous solution containing 1 wt.% bovine serum albumin at the various pHs shown under the pictures. (Unpublished results obtained with the participation of N. Shipunova and D. Fomin).

particles filling the mesh space. The former forms a framework of nanocomposite material, providing mechanical strength, while the latter are not, in most cases, cross-linked. The gel morphology is dependent on the biopolymer nature, the concentration in the initial solution, the charge density on the macromolecules and the conditions of synthesis [52–54]. Their change opens up great opportunities for manipulation of the morphology, structure and porosity of hybrid biopolymer–silica nanocomposite materials. As an example, there is a set of SEM pictures of aerogels synthesized at various pH containing 1 wt.% of bovine serum albumin (see Figure 3.9). If the pH is shifted progressively from the isoelectric point at pH 4.8 in the acidic and alkaline regions, the protein macromolecule becomes more and more unfolded [169]. Its denaturation has been used to regulate the morphology of hybrid nanocomposites. It should be pointed out that this is possible because of the absence of the denaturation effect of THEOS. Otherwise, it would be impossible to realize this approach.

3.4.5 Enzyme Immobilization

The discussed advantages of the one-stage process and THEOS over commonly used silica precursors offer fresh possibilities for developing or improving functional materials. Here I would like to enlarge upon the considerable promise of enzyme immobilization. This is an area of great promise for developing biocatalysts for biotechnological and biosensor applications.

A list of entrapped enzymes is given in Table 3.2. There are two endo-1,3-β-D-glucanases from marine mollusks *Spisula sacchalinensis* and *Ch. albidus* and α-D-galactosidase from marine bacterium *Pseudoalteromonas* sp. MM 701. They were selected for immobilization for the following reasons:

1. The enzymes differ significantly in their optimal conditions for enzymatic activity. These include pH, ionic strength and temperature. The variety of conditions enabled us to verify if there could be limits to the approach.
2. Endo-1,3-β-D-glucanase L_o is a highly labile enzyme. It is especially sensitive to temperature, denaturating at 25 °C. The common approaches were unsuitable for its immobilization.
3. These enzymes exhibit specificity towards polysaccharides; that is, their substrates are high-molecular weight substances. For instance, laminarans, which were used as a substrate for endo-1,3-β-D-glucanases, have molecular weights ranging from 3 to 20 kDa. These values are of the order of the molecular weight of enzymes (Table 3.2). It was instructive to consider whether the endo-1,3-β-D-glucanases could retain their activity to such high-molecular weight substrates, being encapsulated in a silica matrix.

Tab. 3.2 Characteristics of studied enzymes (Adapted from Refs. [55,56]).

Enzyme	Type of hydrolyzing bond	Mw kDa	Optimum conditions for enzymatic activity		
			pH	T,°C	NaCl, M
endo-1,3-β-D-glucanase L_O	Glc. β-1 → 3	20–38	4.6	≤25	0.1–0.25
endo-1,3-β-D-glucanase L_{IV}	Glc. β-1 → 3	22–39	5.6	45	0.01–0.3
α-D-galactosidase	Gal, α-1 → 3	200	45	200	any

All the enzymes listed in Table 3.2 were successfully immobilized in THEOS-produced silica. Figure 3.10 presents data on the enzymatic activity of endo-1,3-β-D-glucanases. Further details on these enzymes and α-D-galactosidase can be found in Refs. [55,56]. As is obvious from Figure 3.10, the enzymes retained their activity. Sometimes it was even higher than that of the initial enzyme solutions. The functional properties depended strongly on the immobilization protocol, formulation, and composition of the final hybrid nanocomposites. What is important is that high-molecular wight laminarans could access the immobilized enzymes. The entrapment of endo-1,3-β-D-glucanases and α-D-galactosidase into a silica matrix favors long-term and thermal stability. The effect was very much more pronounced in the case of endo-1,3-β-D-glucanase L_o. Its lifetime in solution was a few hours prior to being immobilized by the one-stage procedure performed with THEOS.

The experiments with endo-1,3-β-D-glucanase L_o give evidence that the suggested approach is suitable for the entrapment of highly labile enzymes. This provides the basis on which novel types of biocatalysts could be developed. The approach has a few decisive advantages over the current immobilizing procedures performed in two stages (Figure 3.3).

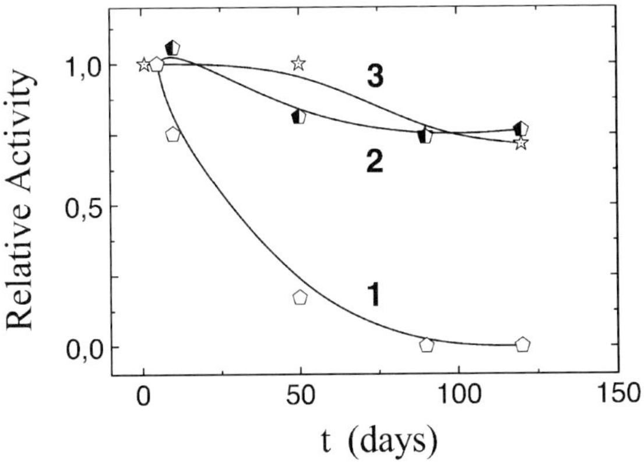

Fig. 3.10 Relative activities of endo-1,3-β-D-glucanase L_{IV} in aqueous solution (1) and in the immobilized state (2) as well as immobilized endo-1,3-β-D-glucanase L_O (3) vs. the time of testing. The enzyme entrapment was performed as described in Refs. [55,56]. The silica matrix was synthesized by using 10 wt.% THEOS and 1.5 wt.% locust bean gum (2) or xanthan (3). (Unpublished results obtained with the participation of T. Karpenko, Dr. Yu. Burtseva and Prof. T. Zvyagintseva).

1. The main advantage is that the entrapment conditions are dictated by the entrapped enzymes, but not the process. This includes such important denaturing factors as the solution pH, the temperature and the organic solvent released in the course of precursor hydrolysis. The immobilization by THEOS is performed at a pH and temperature that are optimal for encapsulated biomaterial [55,56]. The jellification processes are accomplished by the separation of ethylene glycol that possesses improved biocompatibility in comparison with alcohols.
2. The decreased denaturating action of the precursor and procedure enables one to immobilize reduced amounts of biomaterial. It was demonstrated in Ref. [55] that biocatalysts prepared by entrapping endo-1,3-β-D-glucanase and α-D-galactosidase in amounts comparable to that in living cells had a reasonable level of activity. When the TEOS is applied, the enzyme content in silica matrix can be up to 20–30 wt.% to counterbalance losses due to denaturation [50].
3. The porosity of hybrid nanocomposites provides access of the substrates to immobilized enzyme and their proper functioning. It is attributable to the absence of volume shrinkage of synthesized materials after their preparation. Although the compacting does not occur as in the common two-stage processes (Figure 3.7), enzyme macromolecules are held inside the silica matrix and not easily washed out of it.

3.5
Perspectives

The foregoing discussion and my own experience give valid reasons to believe that the developed one-stage approach to the entrapment of biomacromolecules in a silica matrix by using THEOS offers the greatest promise. It can be applied for simple regulation of rheological properties of solutions and the formation of hydrogels. Its advantage is that these materials are biocompatible. This allows their use in such areas as cosmetics, pharmacology, bioengineering and the food industry. If the hydrogels are dried to fabricate xerogels, aerogels or cryogels, these nanocomposites with regulated morphology and porosity can be applied as adsorbent, glasses, membranes, isolators, coatings. These materials of practical significance can also be prepared by using synthetic polymers or surfactants. In this case THEOS also has advantages over currently applied precursors [145,147,170–173]. However, the chief value of the one-stage approach and THEOS is in the entrapment of biopolymers such that they retain their functional properties as much as possible. It can be applied even for highly labile biomacromolecules which lose their activity when immobilized by the common sol–gel technique. The one-stage entrapment provides great scope for developing novel types of biocatalysts having considerable promise for biosensors and biotechnology. The approach can also be extended to the entrapment of microorganisms, cells and cell cultures. Our preliminary experiments have demonstrated that this is quite possible. Of great importance is the extension of the approach to the structure-driven synthesis of titania, zirconia, alumina, and so on by using biopolymers as a template. The challenge is to decrease the rates of hydrolysis and condensation reactions, bringing them under control. Our attempts have demonstrated that this is feasible by means of a radical change of procedure. Where the way to a nucleation of titania and other metal oxide on biomacromolecules is opened in lieu of the usual cluster formation, one may prepare nanocomposites like hybrid biopolymer–silica materials considered in this chapter. This gives promise that the developed approach will be further extended and find varied applications.

Acknowledgments

The work was supported by grants from the Russian Foundation for Basic Research (No. 06-03-96007-p_east_a) as well as from the Presidiums of the Russian Academy of Sciences (No. 06-I-OXHM-138) and the Far East Department of the Russian Academy of Sciences (No. 06-II-CO-04-018).

Notes

A set of articles devoted to hybrid protein–silica nanocomposites has been prepared for publication.

References

1 Antonietti, M. (2001) Surfactants for novel templating applications. *Current Opinion in Colloid and Interface Science*, **6**, 244–248.

2 Hentze, H.P. and Kaler, E.W. (2003) Polymerization of and within self-organized media. *Current Opinion in Colloid and Interface Science*, **8**, 164–178.

3 Patarin, J., Lebeau, B. and Zana, R. (2002) Recent advances in the formation mechanisms of organized mesoporous materials. *Current Opinion in Colloid and Interface Science*, **7**, 107–115.

4 Soler-Illia, G.J.A.A., Sanchez, C., Lebeau, B. and Patarin, J. (2002) Chemical strategies to design textured materials: from microporous and mesoporous cxides to nanonetworks and hierarchical structures. *Chemical Reviews*, **102**, 4093–4138.

5 Stein, A. (2003) Advances in microporous and mesoporous solids – Highlights of recent progress. *Advanced Materials*, **15**, 763–775.

6 Wight, A.P. and Davis, M.E. (2002) Design and preparation of organic-inorganic hybrid catalysts. *Chemical Reviews*, **102**, 3589–3614.

7 Kim, J.M., Sakamoto, Y., Hwang, Y.K., Kwon, Y.U., Terasaki, O., Park, S.E. and Stucky, G.D. (2002) Structural design of mesoporous silica by micelle-packing control using blends of amphiphilic block copolymers. *Journal of Physical Chemistry B*, **106**, 2552–2558.

8 Pierre, A.C. (1998) Introduction to Sol-Gel Processing, Kluwer, Boston.

9 Raman, N.K., Anderson, M.T. and Brinker, C.J. (1996) Template-based approaches to the preparation of amorphous, nanoporous silicas. *Chemistry of Materials*, **8**, 1682–1701.

10 Harada, M. and Adachi, M. (2000) Surfactant-mediated fabrication of silica nanotubes. *Advanced Materials*, **12**, 839–841.

11 Beck, J.S., Vartuli, J.C., Roth, W.J., Leonowicz, M.E., Kresge, C.T., Schmitt, K.D., Chu, C.T.W., Olson, D.H., Sheppard, E.W., McCullen, S.B., Higgins, J.B. and Schlenker, J.L. (1992) A new family of mesoporous molecular-sieves prepared with liquid-crystal templates. *Journal of the American Chemical Society*, **114**, 10834–10843.

12 Kresge, C.T., Leonowicz, M.E., Roth, W.J., Vartuli, J.C. and Beck, J.S. (1992) Ordered mesoporous molecular sieves synthesized by a liquid-crystal template mechanism. *Nature*, **359**, 710–712.

13 Simpson, T.L. and Volkani, B.E. (eds) (1981) Silicon and Siliceous Structures in Biological Systems, Springer-Verlag, New York.

14 Frausto da Silva, J.J.R. and Williams, R.J.P. (2001) The Biological Chemistry of the Elements. The Inorganic Chemistry of Life, Oxford University Press, Oxford.

15 Perry, C.C. and Keeling-Tucker, T. (2000) Biosilicification: The role of the organic matrix in structure control. *Journal of Biological Inorganic Chemistry*, **5**, 537–550.

16 Zaremba, C.M. and Stucky, G.D. (1996) Biosilicates and biomimetic silicate synthesis. *Current Opinion in Solid State and Materials Science*, **1**, 425–429.

17 Heuer, A.H., Fink, D.J., Laraia, V.J., Arias, J.L., Calvert, P.D., Kendall, K., Messing, G.L., Blackwell, J., Rieke, P. C., Thompson, D.H., Wheeler, A.P., Veis, A. and Caplan, A.I. (1992) Innovative materials processing strategies – A biomimetic approach. *Science*, **255**, 1098–1105.

18 Iler, R.K. (1979) The Chemistry of Silica: Solubility, Polymerization, Colloid and Surfaces Properties and Biochemistry, Wiley, New York.

19 Kroger, N., Lehmann, G., Rachel, R. and Sumper, M. (1997) Characterization of a 200-kDa diatom protein that is specifically associated with a silica-based substructure of the cell wall. *European Journal of Biochemistry*, **250**, 99–105.

20 Mayama, S. and Kuriyama, A. (2002) Diversity of mineral cell coverings and their formation processes: a review focused on the siliceous cell coverings. *Journal of Plant Research*, **115**, 289–295.

21 Morse, D.E. (1999) Silicon biotechnology: harnessing biological silica production to construct new materials. *Trends in Biotechnology*, **17**, 230–232.

22 Parkinson, J. and Gordon, R. (1999) Beyond micromachining: the potential of diatoms. *Trends in Biotechnology*, **17**, 190–196.

23 Shimizu, K., Cha, J., Stucky, G.D. and Morse, D.E. (1998) Silicatein alpha: Cathepsin L-like protein in sponge biosilica. *Proceedings of the National Academy of Sciences of the United States of America*, **95**, 6234–6238.

24 Sumper, M. and Brunner, E. (2006) Learning from diatoms: Nature's tools for the production of nanostructured silica. *Advanced Functional Materials*, **16**, 17–26.

25 Vrieling, E.G., Beelen, T.P.M., van Santen, R.A. and Gieskes, W.W.C. (1999) Diatom silicon biomineralization as an inspirational source of new approaches to silica production. *Journal of Biotechnology*, **70**, 39–51.

26 Cha, J.N., Chimizu, K., Zhou, Y., Christiansen, S.C., Chmelka, B.F., Stucky, G.D. and Morse, D.E. (1999) Silicatein filaments and subunits from a marine sponge direct the polymerization of silica and silicones in vitro. *Proceedings of the National Academy of Sciences of the United States of America*, **96**, 361–365.

27 Kroger, N., Deutzmann, R. and Sumper, M. (1999) Polycationic peptides from diatom biosilica that direct silica nanosphere formation. *Science*, **286**, 1129–1132.

28 Gordon, R. and Drum, R.W. (1994) The chemical basis of diatom morphogenesis. *International Review of Cytology-A Survey of Cell Biology*, **150**, 243–372.

29 Coffman, E.A., Melechko, A.V., Allison, D.P., Simpson, M.L. and Doktycz, M.J. (2004) Surface patterning of silica nanostructures using bio-inspired templates and directed synthesis. *Langmuir*, **20**, 8431–8436.

30 Coradin, T., Durupthy, O. and Livage, J. (2002) Interactions of amino-containing peptides with sodium silicate and colloidal silica: A biomimetic approach to silicification. *Langmuir*, **18**, 2331–2336.

31 Estroff, L.A. and Hamilton, A.D. (2001) At the interface of organic and inorganic chemistry: Bioinspired synthesis of composite materials. *Chemistry of Materials*, **13**, 3227–3235.

32 Jan, J.S., Lee, S.J., Carr, C.S. and Shantz, D.F. (2005) Biomimetic synthesis of inorganic nanospheres. *Chemistry of Materials*, **17**, 4310–4317.

33 Liu, Y.P., Shen, Z.R., Li, L.Y., Sun, P.C., Zhou, X.D., Li, B.H., Jin, Q.H., Ding, D.T. and Chen, T.H. (2006) Synthesis of hierarchically structured silica with polypeptide-based triblock copolymer as a template. *Microporous and Mesoporous Materials*, **92**, 189–194.

34 Naik, R.R., Whitlock, P.W., Rodriguez, F., Brott, L.L., Glawe, D.D., Clarson, S.J. and Stone, M.O. (2003) Controlled formation of biosilica structures in vitro. *Chemical Communications*, 238–239.

35 Naik, R.R., Brott, L.L., Clarson, S.J. and, Stone, M.O. (2002) Silica-precipitating peptides isolated from a combinatorial phage display peptide library. *Journal of Nanoscience and Nanotechnology*, **2**, 95–100.

36 Patwardhan, S.V., Clarson, S.J. and Perry, C.C. (2005) On the role(s) of additives in bioinspired silicification. *Chemical Communications*, 1113–1121.

37 Patwardhan, S.V., Mukherjee, N. and Clarson, S.J. (2001) The use of poly-L-lysine to form novel silica morphologies and the role of polypeptides in biosilicification. *Journal of Inorganic and Organometallic Polymers*, **11**, 193–198.

38 Sarikaya, M., Tamerler, C., Schwartz, D.T. and Baneyx, F.O. (2004) Materials assembly and formation using engineered polypeptides. *Annual Review of Materials Research*, **34**, 373–408.

39 Sumper, M., Lorenz, S. and Brunner, E. (2003) Biomimetic control of size in the polyamine-directed formation of silica nanospheres. *Angewandte Chemie-International Edition*, **42**, 5192–5195.

40 Wong, M.C., Cha, J.N., Choi, K.S., Deming, T.J. and Stucky, G.D. (2002) Assembly of nanoparticles into hollow spheres using block copolypeptides. *Nano Letters*, **2**, 583–587.

41 Avnir, D., Braun, S., Lev, O. and Ottolenghi, M. (1994) Enzymes and other proteins entrapped in sol-gel materials. *Chemistry of Materials*, **6**, 1605–1614.

42 Avnir, D., Coradin, T., Lev, O. and Livage, J. (2006) Recent bio-applications of sol-gel materials. *Journal of Materials Chemistry*, **16**, 1013–1030.

43 Coradin, T., Boissiere, M. and Livage, J. (2006) Sol-gel chemistry in medicinal science. *Current Medicinal Chemistry*, **13**, 99–108.

44 Dave, B.C., Dunn, B., Valentine, J.S. and Zink, J.I. (1994) Sol-gel encapsulation methods for biosensors. *Analytical Chemistry*, **66**, 1120A–1127A.

45 Dunn, B., Miller, J.M., Dave, B.C., Valentine, J.S. and Zink, J.I. (1998) Strategies for encapsulating biomolecules in a sol-gel matrix. *Acta Materialia*, **46**, 737–741.

46 Gill, I. and Ballesteros, A. (2000) Bioencapsulation within synthetic polymers (Part 1): Sol-gel encapsulated biologicals. *Trends in Biotechnology*, **18**, 282–296.

47 Kandimalla, V.B., Tripathi, V.S. and Ju, H.X. (2006) Immobilization of biomolecules in sol-gels: Biological and analytical applications. *Critical Reviews in Analytical Chemistry*, **36**, 73–106.

48 Kim, J., Grate, J.W. and Wang, P. (2006) Nanostructures for enzyme stabilization. *Chemical Engineering Science*, **61**, 1017–1026.

49 Livage, J., Coradin, T. and Roux, C. (2001) Encapsulation of biomolecules in silica gels. *Journal of Physics-Condensed Matter*, **13**, R673–R691.

50 Lu, Z.-L., Lindner, E. and Mayer, H.A. (2002) Applications of sol-gel-processed interphase catalysts. *Chemical Reviews*, **102**, 3543–3578.

51 Pierre, A.C. (2004) The sol-gel encapsulation of enzymes. *Biocatalysis and Biotransformation*, **22**, 145–170.

52 Shchipunov, Yu.A. (2003) Sol-gel derived biomaterials of silica and carrageenans. *Journal of Colloid and Interface Science*, **268**, 68–76.

53 Shchipunov, Yu.A. and Karpenko T.Yu. (2004) Hybrid polysaccharide-silica nanocomposites prepared by the sol-gel technique. *Langmuir*, **20**, 3882–3887.

54 Shchipunov, Yu.A., Karpenko, T.Yu. and Krekoten, A.V. (2005) Hybrid organic-inorganic nanocomposites fabricated with a novel biocompatible precursor using sol-gel processing. *Composite Interfaces*, **11**, 587–607.

55 Shchipunov, Yu.A., Karpenko, T.Yu., Bakunina, I.Yu., Burtseva, Yu.V. and Zvyagintseva, T.N. (2004) A new precursor for the immobilization of enzymes inside sol-gel derived hybrid silica nanocomposites containing polysaccharides. *Journal of Biochemical and Biophysical Methods*, **58**, 25–38.

56 Shchipunov, Yu.A., Burtseva, Yu.V., Karpenko, T.Yu., Shevchenko, N.M. and Zvyagintseva, T.N. (2006) Highly efficient immobilization of endo-1,3-β-D-glucanases (laminarinases) from marine mollusks in novel hybrid polysaccharide silica nanocomposites with regulated composition. *Journal of Molecular Catalysis B-Enzymatic*, **40**, 16–23.

57 Arkles, B. (1997) Kirk-Othmer Encyclopedia of Chemical Technology, 4th edn. John Wiley, New York, Vol. 22, 69.

58 Brinker, C.J. and Scherer, G.W. (1990) Sol-Gel Science. The Physics and Chemistry of Sol-Gel Processing, Academic Press, Boston.

59 Hench, L.L. (1998) Sol-Gel Silica. Properties, Processing and Technology Transfer, Noyes Publications, Westwood, New Jersey, USA.

60 Volkani, B.E. (1981) Silicon and Siliceous Structures in Biological Systems (eds T.L. Simpson and B.E. Volkani), Springer-Verlag, New York, Chapter 7.

61 Pierre, A.C. and Pajonk, G.M. (2002) Chemistry of aerogels and their applications. *Chemical Reviews*, **102**, 4243–4265.

62 Miller, J.M., Dunn, B., Valentine, J.S. and Zink, J.I. (1996) Synthesis conditions for encapsulating cytochrome c and catalase in SiO_2 sol-gel materials. *Journal of Non-Crystalline Solids*, **202**, 279–289.

63 Dave, B.C., Miller, J.M., Dunn, B., Valentine, J.S. and Zink, J.I. (1997) Encapsulation of proteins in bulk and thin film sol-gel matrices. *Journal of Sol-Gel Science and Technology*, **8**, 629–634.

64 Eggers, D.K. and Valentine, J.S. (2001) Molecular confinement influences protein structure and enhances thermal protein stability. *Protein Science*, **10**, 250–261.

65 Dickey, F.H. (1955) Specific adsorption. *Journal of Physical Chemistry*, **59**, 695–707.

66 Johnson, P. and Whateley, T.L. (1971) On the use of polymerizing silica gel systems for the immobilization of trypsin. *Journal of Colloid and Interface Science*, **37**, 557–563.

67 Venton, D.L., Cheesman, K.L., Chatterton, R.T. Jr., and Anderson, T.L. (1984) Entrapment of a highly specific antiprogesterone antiserum using polysiloxane copolymers. *Biochimica et Biophysica Acta*, **797**, 343–347.

68 Braun, S., Rappoport, S., Zusman, R., Avnir, D. and Ottolenghi, M. (1990) Biochemically active sol-gel glasses: the trapping of enzymes. *Materials Letters*, **10**, 1–5.

69 Braun, S., Shtelzer, S., Rappoport, S., Avnir, D. and Ottolenghi, M. (1992) Biocatalysis by sol-gel entrapped enzymes. *Journal of Non-Crystalline Solids*, **147**, 739–743.

70 Heichalsegal, O., Rappoport, S. and Braun, S. (1995) Immobilization in alginate-silicate sol-gel matrix protects beta-glucosidase against thermal and chemical denaturation. *Biotechnology*, **13**, 798–800.

71 Ellerby, L.M., Nishida, C.R., Nishida, F., Yamanaka, S.A., Dunn, B., Valentine, J.S. and Zink, J.I. (1992) Encapsulation of proteins in transparent porous silicate-glasses prepared by the sol-gel method. *Science*, **255**, 1113–1115.

72 Yamanaka, S.A., Nishida, F., Ellerby, L. M., Nishida, C.R., Dunn, B., Valentine,

J.S. and Zink, J.I. (1992) Enzymatic-activity of glucose-oxidase encapsulated in transparent glass by the sol-gel method. *Chemistry of Materials*, **4**, 495–497.

73 Wu, S.G., Ellerby, L.M., Cohan, J.S., Dunn, B., Elsayed, M.A., Valentine, J.S. and Zink, J.I. (1993) Bacteriorhodopsin encapsulated in transparent sol-gel glass – A new biomaterial. *Chemistry of Materials*, **5**, 115–120.

74 Zuhlke, J., Knopp, D. and Niessner, R. (1995) Sol-gel glass as a new support matrix in immunoaffinity chromatography. *Fresenius Journal of Analytical Chemistry*, **352**, 654–659.

75 Audebert, P. and Sanches, C. (1994) Modified electrodes from hydrophobic alkoxide silica gels – insertion of electroactive compounds as glucose oxidase. *Journal of Sol-Gel Science and Technology*, **2**, 809–812.

76 Collino, R., Therasse, J., Binder, P., Chaput, F. and Boilot, J.P. (1994) Thin films of functionalized amorphous silica for immunosensors application. *Journal of Sol-Gel Science and Technology*, **2**, 823–826.

77 Livage, J., Roux, C., Da Costa, J.M., Desportes, I. and Quinson, J.F. (1996) Immunoassays in sol-gel matrices. *Journal of Sol-Gel Science and Technology*, **7**, 45–51.

78 Pope, E.J.A. (1995) Gel encapsulated microorgansims: saccharomyces cerevisiae-silica gel biocomposites. *Journal of Sol-Gel Science and Technology*, **4**, 225–229.

79 Campostrini, R., Carturan, G., Caniato, R., Piovan, A., Filippin, R., Innocenti, G. and Cappelletti, E.M. (1996) Immobilization of plant cells in hybrid sol-gel materials. *Journal of Sol-Gel Science and Technology*, **7**, 87–97.

80 Reetz, M.T. (1997) Entrapment of biocatalysts in hydrophobic sol-gel materials for use in organic chemistry. *Advanced Materials*, **9**, 943–953.

81 Weetall, H.H. (1993) Preparation of immobilized proteins covalently coupled through silane coupling agents to inorganic supports. *Applied Biochemistry and Biotechnology*, **41**, 157–188.

82 Gill, I. (2001) Bio-doped nanocomposite polymers: Sol-gel bioencapsulates. *Chemistry of Materials*, **13**, 3404–3421.

83 Livage, J. (1997) Sol-gel processes. *Current Opinion in Solid State and Materials Science*, **2**, 132–138.

84 Golfen, H. and Mann, S. (2004) Higher-order organization by mesoscale self-assembly and transformation of hybrid nanostructures. *Angewandte Chemie-International Edition*, **42**, 2350–2365.

85 Mann, S. (1995) Biomineralization and biomimetic materials chemistry. *Journal of Materials Chemistry*, **5**, 935–946.

86 Das, T.K., Khan, I., Rousseau, D.L. and Friedman, J.M. (1998) Preservation of the native structure in myoglobin at low pH by sol-gel encapsulation. *Journal of the American Ceramic Society*, **120**, 10268–10269.

87 Chen, Q., Kenausis, G.L. and Heller, A. (1998) Stability of oxidases immobilized in silica gels. *Journal of the American Chemical Society*, **120**, 4582–4585.

88 Ferrer, M.L., Del Monte, F., Mateo, C. R., Gomez, J. and Levy, D. (2003) Denaturation and leaching study of horseradish peroxidase encapsulated in sol-gel matrices. *Journal of Sol-Gel Science and Technology*, **26**, 1169–1172.

89 Flora, K.K. and Brennan, J.D. (2001) Effect of matrix aging on the behavior of human serum albumin entrapped in a tetraethyl orthosilicate-derived glass. *Chemistry of Materials*, **13**, 4170–4179.

90 Ferrer, M.L., Del Monte, F. and Levy, D. (2002) A novel and simple alcohol-free sol-gel route for encapsulation of labile proteins. *Chemistry of Materials*, **14**, 3619–3621.

91 Ferrer, M.L., Yuste, L., Rojo, F. and Del Monte, F. (2003) Biocompatible sol-gel route for encapsulation of living bacteria in organically modified silica matrixes. *Chemistry of Materials* **15**, 3614–3618.

92 Ferrer, M.L., Garcia-Carvajal, Z.Y., Yuste, L., Rojo, F. and Del Monte, F. (2006) Bacteria viability in sol-gel materials revisited: Cryo-SEM as a suitable tool to study the structural integrity of encapsulated bacteria. *Chemistry of Materials* **18**, 1458–1463.

93 Cappelletti, E.M., Carturan, G. and Piovan, A. 7-12- (1999) US Patent 5, 998, 162.

94 Boninsegna, S., Dal Toso, R.D., Dal Monte, R. and Carturan, G. (2003) Alginate microspheres loaded with animal cells and coated by a siliceous layer. *Journal of Sol-Gel Science and Technology*, **26**, 1151–1157.

95 Boninsegna, S., Bosetti, P., Carturan, G., Dellagiacoma, G., Dal Monte, R. and Rossi, M. (2003) Encapsulation of individual pancreatic islets by sol-gel SiO_2: A novel procedure for perspective cellular grafts. *Journal of Biotechnology*, **100**, 277–286.

96 Carturan, G., Dal Toso, R., Boninsegna, S. and Dal Monte, R. (2004) Encapsulation of functional cells by sol-gel silica: actual progress and perspectives for cell therapy. *Journal of Materials Chemistry*, **14**, 2087–2098.

97 Muraca, M., Vilei, M.T., Zanusso, G.E., Ferraresso, C., Boninsegna, S., Dal Monte, R., Carraro, P. and Carturan, G. (2002) SiO_2 entrapment of animal cells: Liver-specific metabolic activities in silica-overlaid hepatocytes. *Artificial Organs*, **26**, 664–669.

98 Davis, S.A., Breulmann, M., Rhodes, K.H., Zhang, B. and Mann, S. (2001) Template-directed assembly using nanoparticle building blocks: A nanotectonic approach to organized materials. *Chemistry of Materials*, **13**, 3218–3226.

99 Zhang, B.J., Davis, S.A., Mendelson, N. H. and Mann, S. (2000) Bacterial templating of zeolite fibres with hierarchical structure. *Chemical Communications*, 781–782.

100 Zhang, B.J., Davis, S.A. and Mann, S. (2002) Starch gel templating of spongelike macroporous silicalite monoliths and mesoporous films. *Chemistry of Materials*, **14**, 1369–1375.

101 Gill, I. and Ballesteros, A. (1998) Encapsulation of biologicals within silicate, siloxane, and hybrid sol-gel polymers: An efficient and generic approach. *Journal of the American Chemical Society*, **120**, 8587–8598.

102 Wei, Y., Jin, D.L. and Ding, T.Z. (1997) Optical rotatory silica materials prepared via sol-gel processes. *Journal of Physical Chemistry B*, **101**, 3318–3323.

103 Wei, Y., Jin, D., Ding, T., Shih, W.H., Liu, X., Cheng, S.Z.D. and Fu, Q. (1998) A non-surfactant templating route to mesoporous silica materials. *Advanced Materials*, **10**, 313–316.

104 Wei, Y., Xu, J.G., Dong, H., Dong, J.H., Qiu, K.Y. and Jansen-Varnum, S.A. (1999) Preparation and physisorption characterization of D-glucose-templated mesoporous silica sol-gel materials. *Chemistry of Materials*, **11**, 2023–2029.

105 Wei, Y., Xu, J., Feng, Q., Lin, M., Dong, H., Zhang, W.-J. and Wang, C. (2001) A novel method for enzyme immobilization: direct encapsulation of acid phosphatase in nanoporous silica host materials. *Journal of Nanoscience and Nanotechnology*, **1**, 83–93.

106 Wei, Y., Xu, J., Feng, Q., Dong, H. and Lin, M. (2000) Encapsulation of enzymes in mesoporous host materials via the nonsurfactant-templated sol–gel process. *Materials Letters*, **44**, 6–11.

107 Eggers, D.K. and Valentine, J.S. (2001) Crowding and hydration effects on protein conformation: a study with sol-gel encapsulated proteins. *Journal of Molecular Biology*, **314**, 911–922.

108 Yansey, P.H., Clark, M.E., Hand, S.C., Bowlus, R.D. and Somero, G.N. (1982) Living with water stress: evolution of osmolyte systems. *Science*, **217**, 1214–1222.

109 Brennan, J.D., Benjamin, D., DiBattista, E. and Gulcev, M.D. (2003) Using sugar and amino acid additives to stabilize enzymes within sol-gel derived silica. *Chemistry of Materials*, **15**, 737–745.

110 Keeling-Tucker, T., Rakic, M., Spong, C. and Brennan, J.D. (2000) Controlling the material properties and biological activity of lipase within sol-gel derived bioglasses via organosilane and polymer doping. *Chemistry of Materials*, **12**, 3695–3704.

111 Altstein, M., Segev, G., Aharonson, N., Ben-Aziz, O., Turniansky, A. and Avnir, D. (1998) Sol-gel-entrapped cholinesterases: A microtiter plate method for monitoring anti-cholinesterase compounds. *Journal of Agricultural and Food Chemistry*, **46**, 3318–3324.

112 Besanger, T.R., Easwaramoorthy, B. and Brennan, J.D. (2004) Entrapment of highly active membrane-bound receptors in macroporous sol-gel derived silica. *Analytical Chemistry*, **76**, 6470–6475.

113 Hodgson, R.J., Chen, Y., Zhang, Z., Tleugabulova, D., Long, H., Zhao, X. M., Organ, M., Brook, M.A. and Brennan, J.D. (2004) Protein-doped monolithic silica columns for capillary liquid chromatography prepared by the sol-gel method: Applications to frontal affinity chromatography. *Analytical Chemistry*, **76**, 2780–2790.

114 Heller, J. and Heller, A. (1998) Loss of activity or gain in stability of oxidases upon their immobilization in hydrated silica: Significance of the electrostatic interactions of surface arginine residues at the entrances of the reaction channels. *Journal of the American Chemical Society*, **120**, 4586–4590.

115 Wang, B.Q. and Dong, S.J. (2000) Sol-gel-derived amperometric biosensor for hydrogen peroxide based on methylene green incorporated in Nafion film. *Talanta*, **51**, 565–572.

116 Zhang, J.Z., Li, B., Wang, Z.X., Cheng, G.J. and Dong, S.J. (1999) Functionalized inorganic-organic composite material derived by sol-gel for construction of a mediated amperometric hydrogen peroxide biosensor. *Analytica Chimica Acta*, **388**, 71–78.

117 Zheng, L.L., Flora, K. and Brennan, J.D. (1998) Improving the performance of a sol-gel-entrapped metal-binding protein by maximizing protein thermal stability before entrapment. *Chemistry of Materials*, **10**, 3974–3983.

118 Flora, K. and Brennan, J.D. (1998) Fluorometric detection of Ca^{2+} based on an induced change in the conformation of sol-gel entrapped parvalbumin. *Analytical Chemistry*, **70**, 4505–4513.

119 Chen, J.P. and Wang, H.Y. (1998) Improved properties of bilirubin oxidase by entrapment in an alginate-silicate sol-gel matrix. *Biotechnology Techniques*, **12**, 851–853.

120 Coradin, T., Mercey, E., Lisnard, L. and Livage, J. (2001) Design of silica-coated microcapsules for bioencapsulation. *Chemical Communications*, 2496–2497.

121 Kawakami, K. and Furukawa, S.Y. (1997) Alcohol-oxidation activity of whole cells of Pichia pastoris entrapped in hybrid gels composed of Ca-alginate and organic silicate. *Applied Microbiology and Biotechnology*, **67**, 23–31.

122 Sakai, S., Ono, T., Ijima, H. and Kawakami, K. (2003) Proliferation and

insulin secretion function of mouse insulinoma cells encapsulated in alginate/sol-gel synthesized aminopropyl-silicate/alginate microcapsule. *Journal of Sol-Gel Science and Technology*, **28**, 267–272.

123 Sakai, S., Ono, T., Ijima, H. and Kawakami, K. (2001) Synthesis and transport characterization of alginate/aminopropylsilicate/alginate microcapsule: application to bioartificial pancreas. *Biomaterials*, **22**, 2827–2834.

124 Tan, X.C., Tian, Y.X., Cai, P.X. and Zou, X.Y. (2005) Glucose biosensor based on glucose oxidase immobilized in sol–gel chitosan/silica hybrid composite film on Prussian blue modified glass carbon electrode. *Analytical and Bioanalytical Chemistry*, **381**, 500–507.

125 Nassif, N., Coiffier, A., Coradin, T., Roux, C., Livage, J. and Bouvet, O. (2003) Viability of bacteria in hybrid aqueous silica gels. *Journal of Sol-Gel Science and Technology*, **26**, 1141–1144.

126 Wen, J. and Wilkes, G.L. (1996) Organic/inorganic hybrid network materials by the sol-gel approach. *Chemistry of Materials*, **8**, 1667–1681.

127 Sanches, C., Ribot, F. and Lebeau, B. (1999) Molecular design of hybrid organic-inorganic nanocomposites synthesized *via* sol-gel chemistry. *Journal of Materials Chemistry*, **9**, 35–44.

128 Chong, A.S.M. and Zhao, X.S. (2004) Design of large-pore mesoporous materials for immobilization of penicillin G acylase biocatalyst. *Catalysis Today*, **93–95**, 293–299.

129 Kuncova, G., Czilva, J., Hetflejs, J. and Sabata, S. (2003) Catalysis in organic solvents with lipase immobilized by sol-gel technique. *Journal of Sol-Gel Science and Technology*, **26**, 1183–1187.

130 Kuncova, G., Guglielmi, M., Dubina, P. and Safar, B. (1995) Lipase immobilized by sol-gel technique in layers. *Collection of Czechoslovak Chemical Communications*, **60**, 1573–1577.

131 Reetz, M.T., Zonta, A., Simpelkamp, J. and Konen, W. (1996) In situ fixation of lipase-containing hydrophobic sol-gel materials on sintered glass – Highly efficient heterogeneous biocatalysts. *Chemical Communications*, 1397–1398.

132 Maury, S., Buisson, P., Perrard, A. and Pierre, A.C. (2004) Influence of the sol-gel chemistry on the activity of a lipase encapsulated in a silica aerogel. *Journal of Molecular Catalysis B-Enzymatic*, **29**, 133–148.

133 Reetz, M.T., Zonta, A. and Simpelkamp, J. (1995) Efficient heterogeneous biocatalysts by entrapment of lipases in hydrophobic sol-gel materials. *Angewandte Chemie-International Edition*, **34**, 301–303.

134 Blanco, R.M., Terreros, P., Fernandez-Perez, M., Otero, C. and az-Gonzalez, G. (2004) Functionalization of mesoporous silica for lipase immobilization: Characterization of the support and the catalysts. *Journal of Molecular Catalysis B-Enzymatic*, **30**, 83–93.

135 Buisson, P., El Rassy, H., Maury, S. and Pierre, A.C. (2003) Biocatalytic gelation of silica in the presence of a lipase. *Journal of Sol-Gel Science and Technology*, **27**, 373–379.

136 Pierre, A.C. and Buisson, P. (2006) Use of a lipase to synthesize silica gels in a hydrophobic organic solvent. *Journal of Sol-Gel Science and Technology*, **38**, 63–72.

137 Besanger, T.R., Chen, Y., Deisingh, A. K., Hodgson, R., Jin, W., Mayer, S., Brook, M.A. and Brennan, J.D. (2003) Screening of inhibitors using enzymes entrapped in sol-gel-derived materials. *Analytical Chemistry*, **75**, 2382–2391.

138 Brook, M.A., Chen, Y., Guo, K., Zhang, Z. and Brennan, J.D. (2004) Sugar-modified silanes: precursors for silica monoliths. *Journal of Materials Science*, **14**, 1469–1479.

139 Brook, M.A., Chen, Y., Guo, K., Zhang, Z., Jin, W., Deisingh, A., Cruz-Aguado, J. and Brennan, J.D. (2004) Proteins entrapped in silica monoliths prepared from glyceroxysilanes. *Journal of Sol-Gel Science and Technology*, **31**, 343–348.

140 Tleugabulova, D., Duft, A.M., Zhang, Z., Chen, Y., Brook, M.A. and Brennan, J.D. (2004) Evaluating formation and growth mechanisms of silica particles using fluorescence anisotropy decay analysis. *Langmuir*, **20**, 5924–5932.

141 Sui, X.H., Cruz-Aguado, J.A., Chen, Y., Zhang, Z., Brook, M.A. and Brennan, J.D. (2005) Properties of human serum albumin entrapped in sol-gel-derived silica bearing covalently tethered sugars. *Chemistry of Materials*, **17**, 1174–1182.

142 Cruz-Aguado, J.A., Chen, Y., Zhang, Z., Elowe, N.H., Brook, M.A. and Brennan, J.D. (2004) Ultrasensitive ATP detection using firefly luciferase entrapped in sugar-modified sol-gel-derived silica. *Journal of the American Chemical Society*, **126**, 6878–6879.

143 Cruz-Aguado, J.A., Chen, Y., Zhang, Z., Brook, M.A. and Brennan, J.D. (2004) Entrapment of Src protein tyrosine kinase in sugar-modified silica. *Analytical Chemistry*, **76**, 4182–4188.

144 Tleugabulova, D., Zhang, Z., Chen, Y., Brook, M.A. and Brennan, J.D. (2004) Fluorescence anisotropy in studies of solute interactions with covalently modified colloidal silica nanoparticles. *Langmuir*, **20**, 848–854.

145 Sattler, K., Gradzielski, M., Mortensen, K. and Hoffmann, H. (1998) Influence of surfactant on the gelation of novel ethylene glycol esters of silicic acid. *Berichte Der Bunsen-Gesellschaft-Physical Chemistry Chemical Physics*, **102**, 1544–1552.

146 Mehrotra, R.C. and Narain, R.P. (1967) Reactions of tetramethoxy- and triethoxysilanes with glycols. *Indian Journal of Chemistry*, **5**, 444–448.

147 Meyer, M., Fischer, A. and Hoffmann, H. (2002) Novel ringing silica gels that do not shrink. *Journal of Physical Chemistry B*, **106**, 1528–1533.

148 Guiseley, K.B., Stanley, N.F. and Whitehouse, P.A. (1980) Handbook of Water-Soluble Gums and Resins (ed R.L. Davidson), McGraw-Hill Book Company, New York.

149 Therkelsen, G.H. (1993) Industrial Gums: Polysaccharides and their Derivatives (eds R.L. Whistler and J.N. BeMiller), Academic Press, San Diego, Chapter 7.

150 Tye, R.J. (1998) Gums and Stabilisers for the Food Industry (eds G.O. Phillips, P.A. Williams, D.J. Wedlock), IRL Press, Oxford,**4**.

151 Clare, K. (1993) Industrial Gums: Polysaccharides and their Derivatives (eds R.L. Whistler and J.N. BeMiller), Academic Press, San Diego, Chapter 6.

152 Moe, S.T., Draget, K.I., Skjak-Braek, G. and Smidsrod, O. (1995) Food Polysaccharides and their Applications (ed A.M. Stephen), Marcel Dekker, New York, Chapter 9.

153 Gruber, J.V. (1999) Principles of Polymer Science and Technology in Cosmetics and Personal Care (eds E.D. Goddard and J.V. Gruber), Marcel Dekker, New York, Chapter 8.

154 Kaplan, D.L. (1998) Biopolymers from Renewable Resources, (ed D.L. Kaplan), Springer, Berlin, Chapter 1.

155 Sandford, P.A. and Baird, J. (1983) The Polysaccharides (ed G.O. Aspinall), Academic Press, New York, Chapter 7.

156 Garcia-Ochoa, F., Santosa, V.E., Casas, J.A. and Gomez, F. (2000) Xanthan gum: production, recovery, and properties. *Biotechnology Advances*, **18**, 549–579.

157 Cottrell, I.W., Kang, K.S. and Kovacs, P. (1980) Handbook of Water-Soluble Gums and Resins (eds R.L. Davidson), McGraw-Hill Book Company, New York.

158 Morris, V.J. (1990) Gums and Stabilizers for the Food Industry (eds G.O. Phillips, P.A. Williams, D.J. Wedlock), IRL Press, Oxford, **5**.

159 Zobell, H.F. and Stephen, A.M. (1995) Food Polysaccharides and their Applications (ed A.M. Stephen), Marcel Dekker, New York.

160 Shogren, R.L. (1998) Biopolymers from Renewable Resources (ed D.L. Kaplan), Springer, Berlin.

161 Piculell, L. (1995) Food Polysaccharides and their Applications (ed A.M. Stephen), Marcel Dekker, New York, Chapter 8.

162 Peters, T. Jr. (1985) Serum albumin. *Advances in Protein Chemistry*, **37**, 161–245.

163 Djabourov, M., Leblond, J. and Papon, P. (1988) Gelation of aqueous gelatin solutions. 1. Structural investigation. *Journal of Physics France*, **49**, 319–332.

164 Guenet, J.M. (1992) Thermoreversible Gelation of Polymers and Biopolymers, Academic Press, London.

165 Te Nijenhuis, K. (1997) Thermoreversible networks. Viscoelastic properties and structure of gels. *Advances in Polymer Science*, **130**, 1–267.

166 Coradin, T., Coupe, A. and Livage, J. (2003) Interactions of bovine serum albumin and lysozyme with sodium silicate solutions. *Colloids and Surfaces B-Biointerfaces*, **29**, 189–196.

167 Shchipunov, Yu.A., Kojima, A. and Imae, T. (2005) Polysaccharides as a template for silicate generated by sol–gel processes. *Journal of Colloid and Interface Science*, **285**, 374–380.

168 Shchipunov, Yu.A., Karpenko, T.Yu. Krekoten, A.V. and Postnova, I.V. (2005) Gelling of otherwise nongelable polysaccharides. *Journal of Colloid and Interface Science*, **287**, 373–378.

169 Clark, A.H. (1998) Functional Properties of Food Macromolecules (eds S.E. Hill, D.A. Ledward, J.R. Mitchell), Asper Publishers, Gaithersburg, Maryland, Chapter 3.

170 Brandhuber, D., Torma, V., Raab, C., Peterlik, H., Kulak, A. and Husing, N. (2005) Glycol-modified silanes in the synthesis of mesoscopically organized silica monoliths with hierarchical porosity. *Chemistry of Materials*, **17**, 4262–4271.

171 Brandhuber, D., Peterlik, H. and Husing, N. (2005) Simultaneous drying and chemical modification of hierarchically organized silica monoliths with organofunctional silanes. *Journal of Materials Chemistry*, **15**, 3896–3902.

172 Torma, V., Peterlik, H., Bauer, U., Rupp, W., Husing, N., Bernstorff, S., Steinhart, M., Goerigk, G. and Schubert, U. (2005) Mixed silica titania materials prepared from a single-source sol-gel precursor: A time-resolved SAXS study of the gelation, aging, supercritical drying, and calcination processes. *Chemistry of Materials*, **17**, 3146–3153.

173 Takahashi, S., Ikkai, Y., Rodriguez-Abreu, C., Aramaki, K., Ohsuna, T. and Sakamoto, K. (2007) Application of a water soluble alkoxysilane for the formation of mesoporous silica from nonionic surfactant micelles bearing cholesterol. *Chemistry Letters*, **36**, 182–183.

4
Immobilization of Biomolecules on Mesoporous Structured Materials

Ajayan Vinu, Narasimhan Gokulakrishnan, Toshiyuki Mori, Katsuhiko Ariga

4.1
Introduction

Various functional systems have been developed using artificially constructed structures and synthesized compounds, and the nanotechnology based on them has recently received much attention. However, current nanoscale technologies are still highly inferior to those seen in natural systems. The efficiency of the energy conversion that occurs in mitochondrial and photosynthetic systems far exceeds that obtained in artificial systems. A dog can smell and a bat can hear far more sensitively than most artificial sensors. The information processing exhibited by the brain and nervous systems is much more sophisticated than that exhibited by current computers. Nature developed superior nanotechnologies to our own several billion years ago. Therefore, the use of biomaterials for the development of various functional systems can be a very practical approach to reaching several goals in current nanotechnologies.

Unfortunately, biomolecules in most cases give their best performance under ambient conditions and are not mechanically and thermally strong enough to be used in harsh conditions. In order to overcome such difficulties, hybridization of biomolecules with supporting materials such as lipid assemblies, polymers and inorganic materials has been proposed to attain both biological functions and mechanical stability [1–30]. Biomolecules are frequently immobilized on silica supports through a sol–gel process. As advanced silica structures, mesoporous silica molecular sieves and related materials have received much attention as inorganic supports for biomolecules [31–40]. Mesoporous silica materials are chemically and mechanically stable and resistant to microbial attack. In addition, chemical modification to introduce organic functional groups to the inner silica pore wall is highly possible. Outstanding properties of the mesoporous structure also lie in their high specific surface area and huge specific pore volume, which are definitely advantageous for the efficient accommodation of biomolecules. In addition, the narrow pore size distribution could provide reliable immobilization of biomolecules. According to the

Bio-inorganic Hybrid Nanomaterials. Edited by E. Ruiz-Hitzky, K. Ariga, Y. M. Lvov

ISBN: 978-3-527-31718-9

IUPAC classification, mesoporous materials are defined as porous materials with diameters in the range 2–50 nm, which is rather close to the dimensions of functional biomolecules such as proteins. Therefore, unexplored phenomena and functions could be observed for biomolecules confined in mesopore channels due to their restricted motion and orientation. In this chapter, we briefly introduce recent developments on the immobilization of biomolecules in mesoporous media, where the use of mesoporous silica and mesoporous carbon are mainly discussed.

Prior to starting a discussion on biomolecule immobilization, mesoporous silica materials are briefly described below (see later for a description of mesoporous carbon). In 1990, Kuroda and coworkers first reported the preparation of mesoporous silica with uniform pore size distribution from the layered polysilicate kanemite (FSM-16, Folded Sheet Materials) [41,42] followed by a significant breakthrough in mesoporous materials research by Mobil scientists [43,44]. The latter researchers discovered ways to synthesize the M41S family of materials, which have large uniform pore structures, high specific surface areas and specific pore volumes, including hexagonal MCM-41 [43,44], cubic MCM-48 [45], and lamellar MCM-50 [46]. Mesoporous silica materials are basically prepared through silica formation around self-assembled micelles (template) followed by template removal by appropriate methods such as calcination (Figure 4.1). This concept has been generalized to produce various types of mesoporous silica. *Hexagonal mesoporous silica* (HMS) with

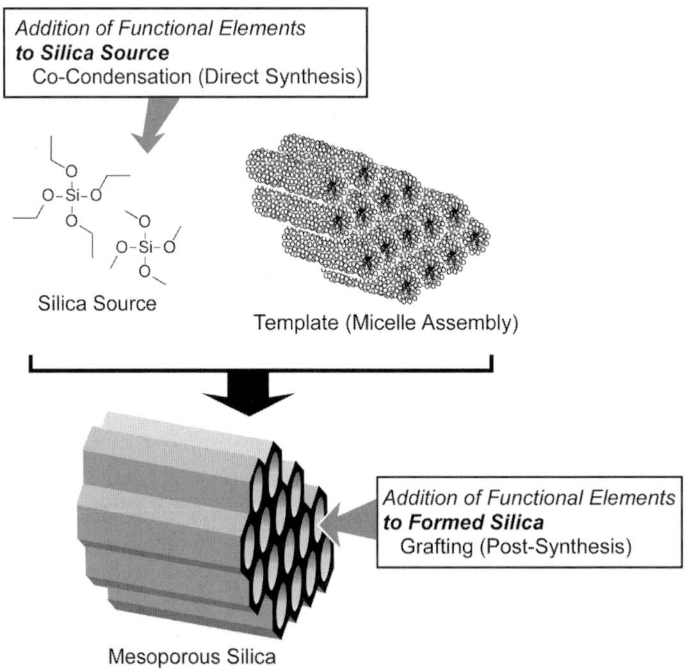

Fig. 4.1 Outline of mesoporous silica synthesis and modification.

slightly disordered hexagonal structure and thicker walls was synthesized using neutral amine as template and showed superior thermal stability upon calcination in air, and its smaller crystallite size affords complementary textural mesoporosity for improved access to the framework-confined mesopores [47]. The mesoporous silica called *Michigan State University-1* (MSU-1) prepared using *polyethylene oxide* (PEO) possesses large wall thickness and small particle size with considerable textural mesoporosity due to the pores being formed between relatively small particles [48]. The mesoporous silica called *Santa Barbara Amorphous-15* (SBA-15) with thicker pore walls and a highly ordered two-dimensional hexagonal structure, synthesized by using the amphiphilic triblock-copolymer of poly(ethylene oxide) and poly(propylene oxide) (Pluronic P123) in highly acidic media is now one of the most popular mesoporous silica materials [49,50]. It exhibits higher hydrothermal stability, than the other mesoporous silicas, and its pore diameters are well tunable in the range 5–30 nm. *Meso-cellular foam* (MCF) materials can be prepared by using triblock-copolymers stabilized in oil in water to result in an aerogel-like structure [51]. Not limited to these representative materials, novel mesoporous silica materials with unique structures and properties have been continuously proposed. In addition, functionalization of mesoporous silica through co-condensation (direct synthesis) and grafting (post-synthesis) has been widely researched (Figure 4.1) [52].

Various framework materials of mesoporous structures can also be designed. Incorporation of heteroatoms such as Cu, Zn, Al, B, Ga, Fe, Cr, Ti, V and Sn into the mesoporous silica framework has been widely investigated [53–71]. In addition, methodology to prepare mesoporous silica via the template synthesis has been extended to the synthesis of various mesoporous metal oxides [72–85] such as TiO_2, Ta_2O_5, Nb_2O_5, ZrO_2, Al_2O_3, and V_2O_5, as well as the synthesis of mesoporous aluminophosphate [86–89]. At the end of the 1990s, the synthesis of mesoporous carbon was proposed as a milestone discovery in mesoporous technology, the details are described later in this chapter. Another breakthrough finding was the synthesis of *periodic mesoporous organosilicates* (PMO), which was initiated independently by three groups (Inagaki group [90], Ozin group [91], and Stein group [92]) in 1999. PMO is prepared using organic molecules having multiple alkoxysilane groups such as bis(triethoxysilyl)ethene and bis(triethoxysilyl)benzene (Figure 4.2A). Chemical modification of the PMO pore wall was also demonstrated by several research groups [93–99]. One of the most fantastic findings on PMO materials is the formation of a crystalline pore wall (Figure 4.2B) [100]. Inagaki *et al.* synthesized PMO material using benzene-bridged organosilane, 1,4-bis(triethoxysilyl)benzene. The powder XRD patterns showed a set of peaks assignable to a two-dimensional hexagonal lattice. Four sharp peaks were detected in a small angle region, corresponding to a periodic structure with a spacing of 0.76 nm. Many lattice fringes, which stacked along the channel axes with a uniform spacing of 0.76 nm, were confirmed in the TEM image. These results are consistent with a crystal-like pore-wall structure, where hydrophilic silicate layers and hydrophobic benzene layers array alternately.

In this chapter, immobilization of proteins on these mesoporous silicas and PMO is first introduced, followed by a description of protein immobilization on mesoporous carbon materials. The adsorption behavior of other biopolymers such as

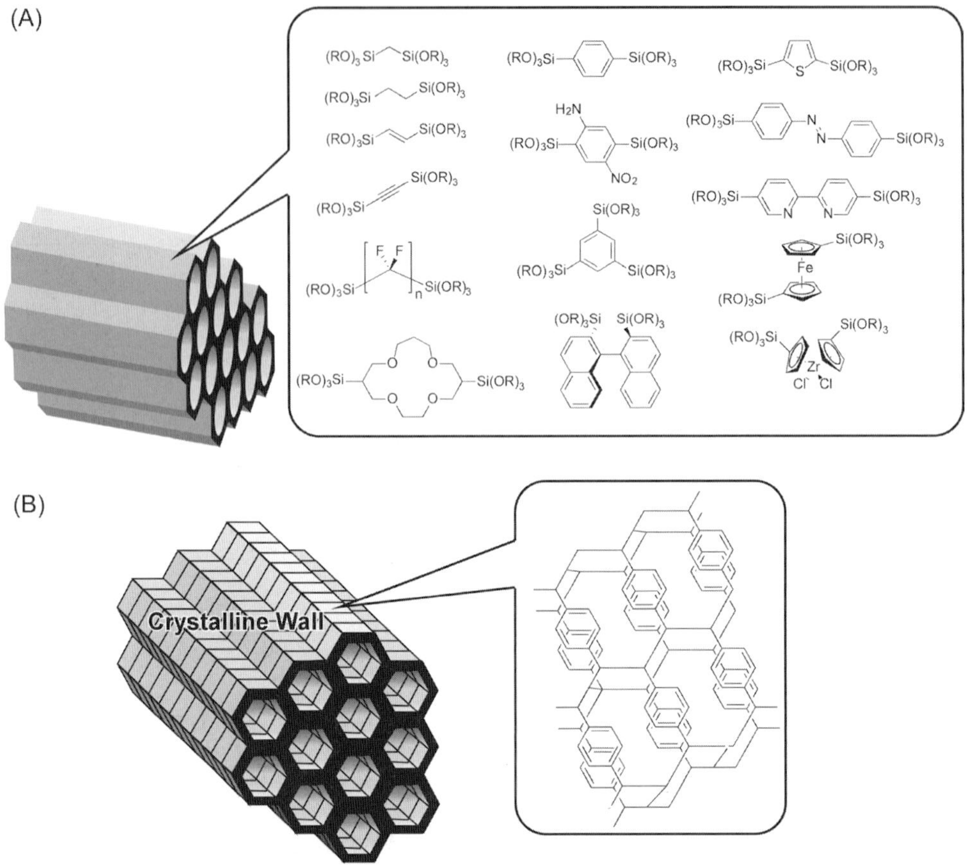

Fig. 4.2 *Periodic mesoporous organosilicates* (PMO): (A) various framework structures; (B) crystalline pore wall composed of phenyl rings and silica. Adapted from [52], A. Vinu *et al.*, *J. Nanosci. Nanotechnol.* **2005**, *5*, 347.

peptides, DNA and small biomolecules is also described. Finally, advanced functions related to nanohybrids of biomolecules in mesoporous structures are explained.

4.2 Immobilization of Protein on Mesoporous Silica

Since Diaz and Balkus first attempted to immobilize enzymes on mesoporous MCM-41 [101], several research groups have investigated the influence of various physical factors such as pore size, ambient pH, and ionic strength, on the adsorption efficiency of proteins [102–118]. This research revealed the general tendencies of protein adsorption behavior and outlines for successful immobilization of proteins onto mesoporous materials. As one of the representative examples, systematic

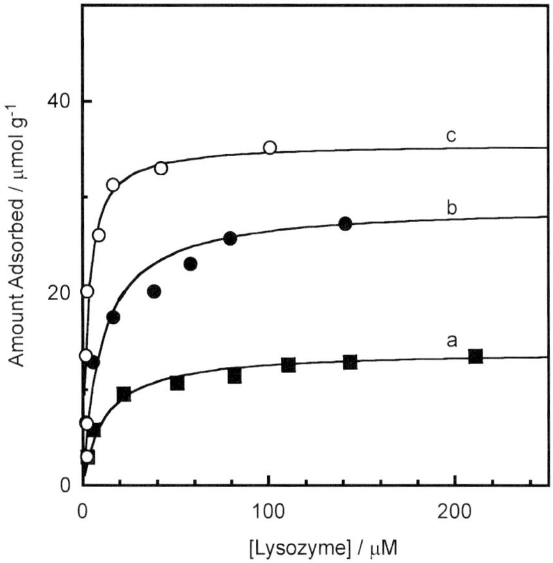

Fig. 4.3 Adsorption isotherms of lysozyme at pH 10.5 on mesoporous silica materials: (a) C_{12}-MCM-41; (b) C_{16}-MCM-41 (MCM-41); (c) SBA-15. Reprinted with permission from [119], A. Vinu *et al.*, *J. Phys. Chem. B* **2004**, *108*, 7323. 2004, American Chemical Society.

research on protein adsorption onto pore-engineered mesoporous silica, by Vinu and coworkers, is also introduced in this section.

Hen egg white lysozyme was used as a probe adsorbate in the systematic research. It is a small globular protein (molecular mass, 14 400 Da), and its size, in two characteristic cross-sections with dimensions of roughly $3.0 \times 4.5\,nm^2$ and an end of dimensions $3.0 \times 3.0\,nm^2$, is comparable to or a bit smaller than the pore sizes of common mesoporous silicates. This protein includes eighteen cationic amino acid residues (six lysyl groups including one at the *N*-terminal, eleven arginyl groups, and one histidyl group) and twelve anionic residues (two glutamates, nine aspartates, and one *C*-terminal carboxylate) and it has an isoelectric point at around 11. Figure 4.3 displays the adsorption isotherms of lysozyme at pH 10.5 [119] on three mesoporous silica materials, C_{12}-MCM-41, C_{16}-MCM-41 (conventional MCM-41), and SBA-15, of which C_{12}-MCM-41 and C_{16}-MCM-41 are MCM-41-type mesoporous silicates prepared using dodecyltrimethylammonium and hexadecytrimethylammonium surfactants as structure-directing reagents, respectively. The monolayer adsorption capacities derived from Langmuir-type adsorption analysis are 13.4, 28.1, and 35.3 $\mu mol\,g^{-1}$ for C_{12}-MCM-41, C_{16}-MCM-41, and SBA-15, respectively. These values increase with increasing pore volume (C_{12}-MCM-41, $0.70\,cm^3\,g^{-1}$; C_{16}-MCM-41, $0.86\,cm^3\,g^{-1}$; SBA-15, $1.25\,cm^3\,g^{-1}$) and pore diameter (C_{12}-MCM-41, 3.54 nm; C_{16}-MCM-41, 4.10 nm; SBA-15, 10.98 nm) of the mesoporous adsorbents. Comparison of the molecular volume of lysozyme ($17.8\,nm^3$) with the results of nitrogen adsorption/desorption measurement provides volume occupation by the adsorbed lysozyme as

20.1, 35.1, and 30.3 % for C_{12}-MCM-41, C_{16}-MCM-41, and SBA-15, respectively. The smaller value estimated for SBA-15 could be due to the presence of ultramicropores in SBA-15. The structural stability of the adsorbed proteins was confirmed using *diffuse reflectance Fourier transform infrared* (DRIFT) spectroscopy for the lysozyme molecules loaded onto mesoporous adsorbents. Representative peaks at amide I and amide II bands in the obtained spectra indicate that denaturation of lysozyme upon adsorption onto the mesoporous silicates is negligible. A similar observation was generalized to the other proteins such as cytochrome c (2.6 × 3.2 × 3.0 nm) [120] and myoglobin (3 × 4 × 5 nm) [121]. The adsorption capability of cytochrome c also increases in the order of C_{12}-MCM-41 C_{16}-MCM-41 SBA-15. The calculated volumes occupied by the cytochrome c molecules correspond to 13.7, 26.4, and 28.3 % of the total free volume of C_{12}-MCM-41, C_{16}-MCM-41, and SBA-15 adsorbents, respectively. The adsorption capacity of myoglobin on SBA-15 at pH 7.0 fell to 31.6 $\mu mol\,g^{-1}$, corresponding to 48 % of volume occupation of the pores.

These obtained data were compared with simplified pore-filling models, as illustrated in Figure 4.4 [121,122]. In these models, mesopore channels are represented as a cylinder of diameter D, while a sphere with diameter d is assigned for a protein molecule. The models include (a) separated single-molecular adsorption, (b) separated double-molecular adsorption, (c) separated triple-molecular adsorption, and (d) interdigitated triple-molecular adsorption where adjacent layers are interdigitated by 1/4 of the protein diameter through changing relative orientation. Simple mathematics assigned volume occupations as $2/3 \times (d/D)^2$, $4/3 \times (d/D)^2$, $2 \times (d/D)^2$, and $8/3 \times (d/D)^2$ for the models (a), (b), (c), and (d), respectively. When we assume 4 nm as a value of d and 10 nm as a value of D, which roughly corresponds to adsorption of the used proteins in the mesopores of SBA-15 materials, the pore occupation values by protein molecules are calculated as 11, 21, 32, and 43 % for the cases of (a), (b), (c), and (d), respectively. Comparison of the experimentally obtained values with these model estimations indicates that protein molecules adsorbed in mesopores are in a well packed state, as illustrated in models (c) and (d).

Physical conditions in solution are also a subject worthy of detailed investigation. In particular, the solution pH changes significantly the charged state of both proteins and silicates. Monolayer adsorption capacities of lysozyme on MCM-41 (C_{16}-MCM-41) under different pH conditions are plotted in Figure 4.5A as plot (a) [119]. The isoelectric point (pI) of lysozyme is around 11, while the isoelectric point of the silica surface of MCM-41 is around 3.6. One can expect that increasing the ambient pH would induce simple suppression of the adsorption efficiency due to reduced electrostatic attraction in the pH region between the two pI values. However, the obtained results (plot (a) in Figure 4.5A) did not obey this expectation. The adsorption capacity actually increases from pH 6.5 to 10.5 and decreases when the pH is increased to 12, resulting in the maximum adsorption capability of lysozyme on MCM-41 being near the isoelectric point of lysozyme. A similar tendency can be found in the adsorption behavior of lysozyme on SBA-15 adsorbent (plot (a) in Figure 4.5B). This tendency was experimentally generalized for adsorption of the other protein molecules on mesoporous silica adsorbents. For example, adsorption of cytochrome *c* on SBA-15 showed the maximum efficiency at around pH 9.6, which is

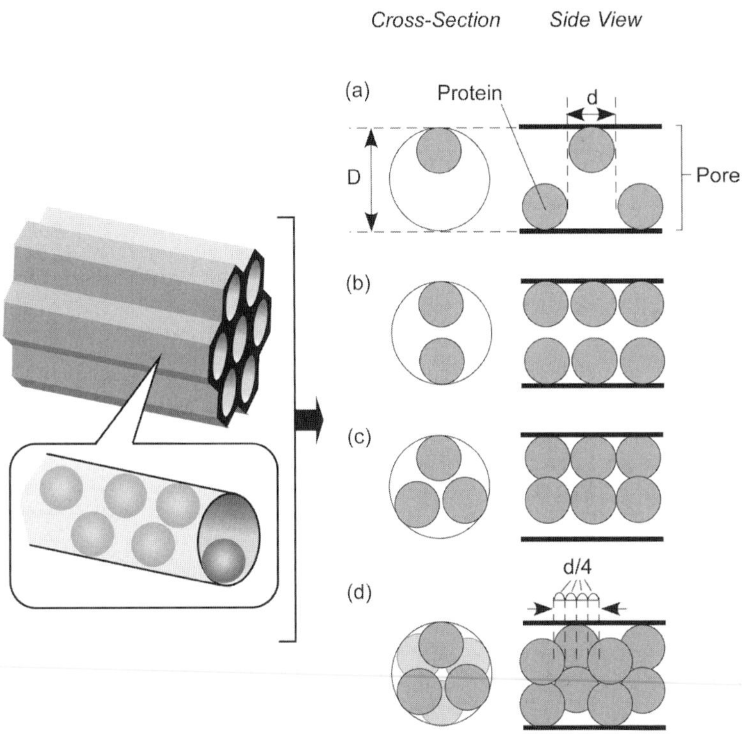

Fig. 4.4 Pore-filling models for protein adsorption in a mesopore channel: (a) separated single-molecular adsorption; (b) separated double-molecular adsorption; (c) separated triple-molecular adsorption; (d) interdigitated triple-molecular adsorption where adjacent layers are interdigitated by 1/4 of the protein diameter through changing the relative orientation. Adapted from [37], A. Vinu *et al.*, *J. Nanosci. Nanotechnol.* **2006**, *6*, 1510.

again close to the isoelectric point of cytochrome c. The maximum adsorption of myoglobin on SBA-15 was obtained at around pH 7, which is very close to the isoelectric point of myoglobin (about 7). Hence, these results suggest that the lateral repulsion between the protein molecules may play a more significant role in the adsorption process than the electrostatic interaction between the silica surface and the protein surface. The electric repulsion between the protein molecules is significantly suppressed, allowing dense packing of the proteins in confined spaces near the isoelectric point of the protein, where proteins are virtually neutral. Change in the protein size may also be another factor to explain the peculiar pH tendency of the protein adsorption. For example, the area per molecule of lysozyme in solution having a pH near the isoelectric point was reported to be similar to that in its crystallized state (13.5 nm^2), whereas it is 26.6 nm^2 at a solution pH of 4. It can be easily imagined that the smaller occupied area in the former case results in a larger monolayer adsorption capacity.

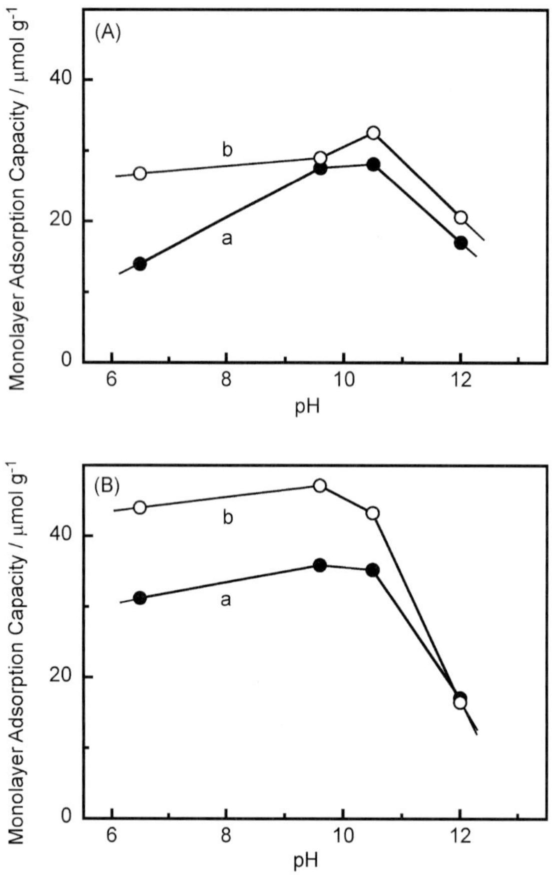

Fig. 4.5 Monolayer adsorption capacities of lysozyme onto (A) MCM-41-type mesoporous silicates ((a) MCM-41 and (b) AlMCM-41) and (B) SBA-15-type mesoporous silicates ((a) SBA-15 and (b) AlSBA-15). Reprinted with permission from [119], A. Vinu *et al.*, *J. Phys. Chem. B* **2004**, *108*, 7323. 2004, American Chemical Society.

Incorporation of heteroatoms into a mesoporous silica framework modifies the pH dependences of protein adsorption on mesoporous materials, as shown in Figure 4.5, for the adsorption of lysozyme on Al-substituted MCM-41 (AlMCM-41, see plots (b) in Figure 4.5A) and on Al-doped SBA-15 (SBA-15, see plot (b) in Figure 4.5B) [119]. The adsorption capacities of both the aluminum-substituted mesoporous materials are always higher than those of the corresponding non-substituted materials. In particular, the differences are more enhanced in the lower pH region, and the pH values for the maximum adsorption in the cases of AlMCM-41 and AlSBA-15 were a little shifted towards the lower pH region. Acidic sites at aluminum atoms would have strong electrostatic interaction with the protonated basic amino acid residues (lysine, arginine, histidine, *N*-terminal, etc.) of the protein (Figure 4.6). This interaction is

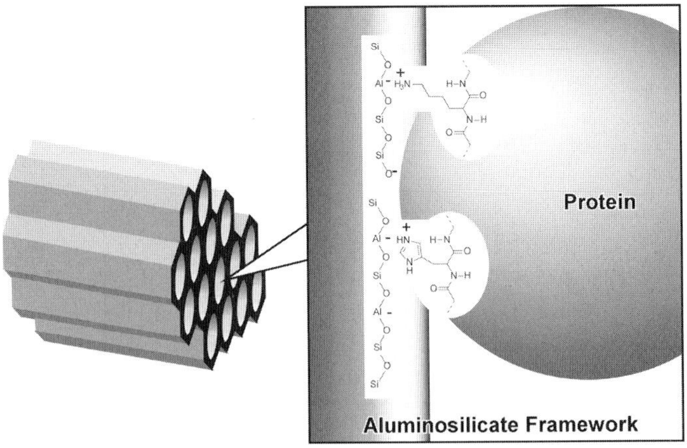

Fig. 4.6 Protein adsorption model on aluminum-substituted mesoporous silica. Adapted from [37], A. Vinu *et al.*, *J. Nanosci. Nanotechnol.* **2006**, *6*, 1510.

more apparent below the pK of the amino acid side chains, and, hence the protein adsorption is more promoted in lower pH regions.

Immobilization of proteins on mesoporous silica has also been investigated by other groups. Examples of recent research are briefly summarized below. Instead of adsorption of proteins on pre-synthesized mesoporous silica, direct incorporation of proteins during the mesoporous synthetic process was conducted, because proteins tend to denature under strongly acidic or basic conditions. So, synthesis of mesoporous silica under neutral conditions is required for this direct immobilization. Blin *et al.* demonstrated the synthesis of mesoporous silica under neutral pH conditions, using a nonionic fluorinated surfactant, $CF_3(CF_2)_7C_2H_4(OC_2H_4)_9OH$, in the presence of glucose oxidase [123]. Hexagonally arranged pore structures were successfully prepared by keeping the glucose oxidase concentration below 3.2 mg mL^{-1}. Galarneau and coworkers proposed the use of neutral lecithin and cationic dodecylamine as structure-directing reagents in the presence of lactose for direct protein immobilization [124]. Natural lecithin protects the enzyme and avoids direct interaction between the silica and the enzyme, and lactose also has the role of protecting the enzyme activity. Ying and coworkers reported a pressure-driven method for entrapping lipase onto siliceous mesocellular foam [125]. A high pressure (3000–5000 psi) was used to achieve high enzyme loading within the support (up to 275 mg per gram of silica). The adsorbed lipase demonstrated high activity with enhanced thermal stability, and enzyme leaching was not significant.

In addition to popular mesoporous silica materials, mesoporous silica supports with various morphologies have also been used for protein immobilization. Tang and coworkers synthesized lotus-leaf-like silica flakes with a three-dimensionally connected nanoporous structure and controllable thickness, which were used for immobilization of ribonuclease A [126]. The synthesized silica flakes have a thickness of 200 nm, and a diameter of 3 mm, showing a much higher initial adsorbing rate of ribonuclease A than their microspheric analogues with diameter of 1 mm. The adsorbed amount of ribonuclease A in the silica flakes reached 120 mg g^{-1} within 4 h, while it took the silica spheres more than 40 h to reach the same value, indicating that the short nanoporous channels and rich pore-openings of the silica flakes greatly facilitated the diffusion and adsorption of ribonuclease A molecules. Ma, Bao, and coworkers prepared nanoscale silica particles with ordered large mesopores (about 13 nm) by tuning the composition of the octane/water/P123 *triblock copolymer/ tetraethoxy orthosilicate* (TEOS) emulsion system in silica synthesis [127]. The mesoporous silica particles thus obtained have an unusual ultrafast enzyme adsorption speed, which has potential application in the fast separation of biomolecules. Ramón, Amorós, and coworkers demonstrated the capability of a hierarchical porous silica-based network with pore systems at two different length scales for protein immobilization [128]. Bimodal porous silica (UVM-7) was synthesized through the use of silicon atrane complexes as the hydrolytic inorganic precursor and surfactants as the porogen species. Small proteins such as lysozyme can be loaded in both pore systems, while large proteins such as α-L-arabinofuranosidase can be immobilized in the large interparticle voids.

Post-treatments of adsorbed proteins and/or enzyme-trapped mesoporous silica are often beneficial for stable immobilization of proteins. Kim, Hyeon, and coworkers used *hierarchically ordered mesocellular mesoporous silica materials* (HMMS) for immobilization of α-chymotrypsin [129]. In order to improve the retention of α-chymotrypsin in HMMS, the adsorbed α-chymotrypsin molecules were cross-linked using glutaraldehyde. The resulting cross-linked enzyme aggregates show an impressive stability with extremely high enzyme loadings. In contrast, adsorbed α-chymotrypsin without glutaraldehyde treatment resulted in a lower loading, which further decreased due to continuous leaching of adsorbed α-chymotrypsin. The activity of cross-linked α-chymotrypsin aggregates in HMMS was about 10 times higher than that of the simply adsorbed α-chymotrypsin. He *et al.* reported a method for stable entrapment of protein through modification of the mesoporous silica with polymers [130]. In their method, porcine pancreatic lipase was first immobilized in the mesoporous channels of the SBA-15 support through physical adsorption. Chemical modification was then performed to reduce the pore openings of SBA-15, in order to prevent the leaching of porcine pancreatic lipase through the surface grafting of the poly(methacrylate)-type polymer (Figure 4.7). The assay of enzyme activity showed that the reduction in pore size by chemical modification prevents the porcine pancreatic lipase leaching, as expected. Wang and Caruso proposed a very unique method to prepare a porous protein assembly using mesoporous silica as a hard template (Figure 4.8) [131]. In the first step of their method, proteins such as lysozyme, cytochrome c, and catalase were adsorbed in the channels of a mesoporous silica sphere. Next, an oppositely charged polyelectrolyte was infiltrated into the protein-loaded mesopores, to bridge the proteins. The silica template was finally removed by a solution of $HF–NH_4F$, resulting in the formation of free-standing nanoporous proteins. The nanoporous proteins can be prepared with protein contents as high as 83 wt%. Such materials are envisaged to find application in bio-sensing, catalysis, separations, and controlled drug delivery.

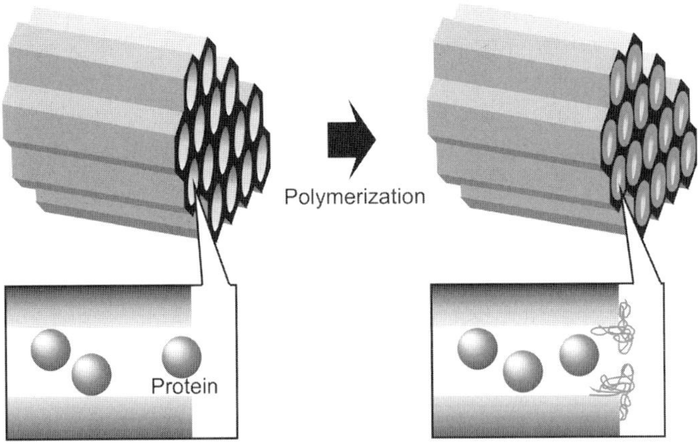

Fig. 4.7 Modification of the pore openings of SBA-15 through surface grafting of the poly(methacrylate)-type polymer to prevent the leaching of proteins.

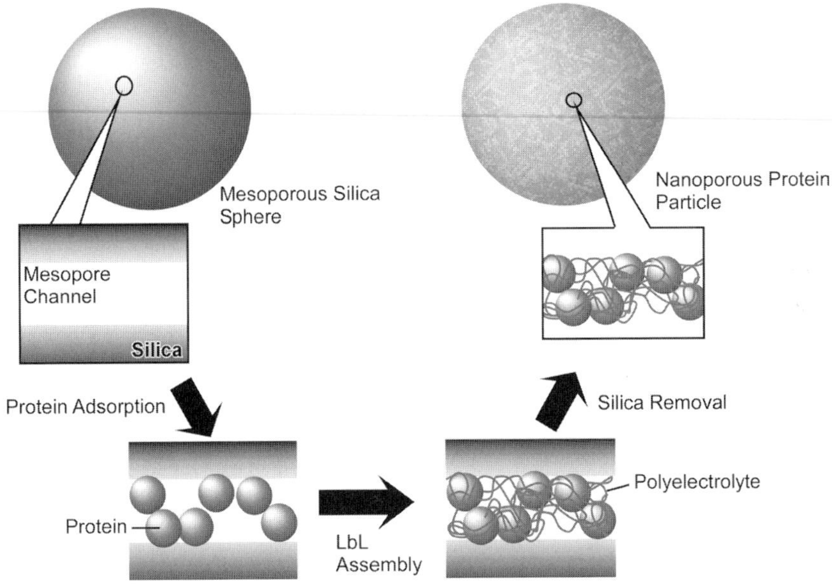

Fig. 4.8 Preparation of a porous protein assembly using mesoporous silica as a hard template.

Hudson, Hodanett, and coworkers discuss the driving forces for protein adsorption on mesoporous materials using two different materials, SBA-15 and MSE (PMO) [132]. Both materials were synthesized with the Pluronic surfactant, P123. While SBA-15 is a pure silica material, MSE consists of SiO_2 units with $–CH_2CH_2–$ groups interspersed periodically. Therefore, MSE and SBA-15 were found to have very similar physical properties but different chemical compositions. Cytochrome c and xylanase were adsorbed onto two mesoporous materials. Electrostatic attraction dominated the protein interactions with SBA-15, while weaker hydrophobic interactions are more prominent with MSE for both cytochrome c and xylanase. In their research cytochrome c immobilized on SBA-15 through electrostatic interaction showed resistance to leaching with an enhanced activity compared to free protein. Lu and coworkers similarly compared the protein adsorption capability of mesoporous silica and PMO, through an adsorption experiment with cytochrome c [133]. The adsorbed amount reached a maximum around the isoelectric point of cytochrome c, and the PMO materials did not have higher adsorption capacity than SBA-15 silica. The specific adsorption amounts of cytochrome c on PMO or pure silica decrease as the ionic strengths increase at all pH conditions. Their results suggest that the electrostatic interaction between cytochrome c and PMO/pure silica surface is more dominant than hydrophobic forces in the adsorption of cytochrome c. Such discussion seems still to be a source of confusion, hence, careful analyses are necessary in every case.

For application of protein-immobilized porous materials to sensor fields, use of an electroactive substance as the framework material is important. DeLouise and Miller demonstrated the immobilization of glutathione-*S*-transferase in electrochemically etched porous silicon films [134], which are attractive materials for the construction of biosensors and may also have utility for the production of immobilized enzyme bioreactors. Not limited to this case, practical applications of nanohybrids from biomolecules and mesoporous materials have been paid much attention. Examples of the application of such hybrids are summarized in a later section of this chapter.

4.3 Immobilization of Protein on Mesoporous Carbon and Related Materials

As seen in the comparison of mesoporous silica and PMO in protein adsorption behavior, the nature of the framework of mesoporous materials has unavoidable influence on the protein adsorption. Therefore, adsorption of protein on mesoporous structures composed of hydrophobic materials such as carbon is worthy of detailed investigation. In this section, systematic research on protein adsorption on mesoporous carbon materials by Vinu and coworkers is mainly introduced.

Before starting the discussion on protein adsorption on mesoporous carbon, an outline of mesoporous carbon materials is briefly given. From the late 1990s, regular carbon materials have been synthesized using regularly structured templates such as zeolite and mesoporous silica [135–139]. For example, Kyotani and coworkers successfully synthesized highly regular microporous carbon using the channels of Y zeolite as a template [140,141]. Mesoporous carbon material structures (CMK-x) were first reported by Ryoo using sucrose as the carbon source and mesoporous silicates such as MCM-48, SBA-1 and SBA-15 as templates [142–146]. Use of mesoporous silica MCM-48 in the cubic phase results in the representative mesoporous carbon, CMK-1 (Figure 4.9A). SBA-15 with a hexagonal array of cylindrical pores and the silica channels interconnected with micropores, leads to the formation of regular carbon structures (CMK-3, Figure 4.9B). Hyeon *et al.* reported, independently

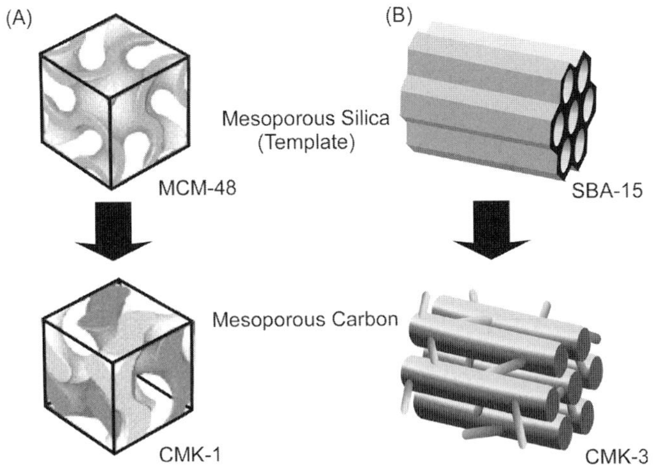

Fig. 4.9 Outline of mesoporous carbon syntheses: (A) CMK-1 synthesized using MCM-48; (B) CMK-3 synthesized using SBA-15.

and somewhat later, a similar approach for the well-ordered mesoporous carbon materials designated SNU-x [147–149].

Recently, Vinu *et al.* successfully prepared pore-tuned mesoporous carbon materials [150], where SBA-15 silicates synthesized at various temperatures [151] were used as hard templates. These SBA-15 silicates were named as SBA-15-x, where x represents the synthesis temperature, and have controlled pore sizes. The pore-engineered mesoporous silicates, SBA-15-100 (SBA-15), SBA-15-130, and SBA-15-150, provided CMK-3 type mesoporous carbon materials, CMK-3-100 (CMK-3), CMK-3-130, and CMK-3-150, respectively. XRD analyses revealed well-ordered hexagonal geometries for all the pore-tuned CMK-3 materials. Nitrogen adsorption and desorption isotherms of these materials are all categorized as type IV isotherms according to the IUPAC classification and exhibit an H1 hysteresis loop. Figure 4.10 displays their pore size distribution together with that of CMK-1, assigning averaged pore sizes of 2.3, 3.3, 4.3, and 5.4 nm for CMK-1, CMK-3, CMK-3-130, and CMK-3-150, respectively, based on *non-local density functional theory* (NLDFT) analysis. For CMK-1, CMK-3, CMK-3-130, and CMK-3-150, BET surface areas of 1675, 1260, 1250, and 1350 $m^2 g^{-1}$, and pore volumes of 1.05, 1.1, 1.3, and 1.6 $cm^3 g^{-1}$ were calculated, respectively. The relatively large surface area for CMK-1 as compared with its small pore diameter and pore volume may originate from an unavoidable contribution of micropores in the CMK-1 structure.

Adsorption of lysozyme on these mesoporous carbon materials was studied at a solution pH of 11 [152]. All the adsorption isotherms were of a Langmuir type (see Figure 4.11A), resulting in monolayer adsorption capacities of 3.8, 9.8, 15.9, and 22.9 $\mu mol\, g^{-1}$ for CMK-1, CMK-3, CMK-3-130, and CMK-3-150, respectively. These values are compared with the structural parameters of the CMK materials (Figure 4.11B). There is no clear relation between the adsorption capacity and the surface area (Figure 4.11B(a)). In sharp contrast, positive correlation between the adsorption

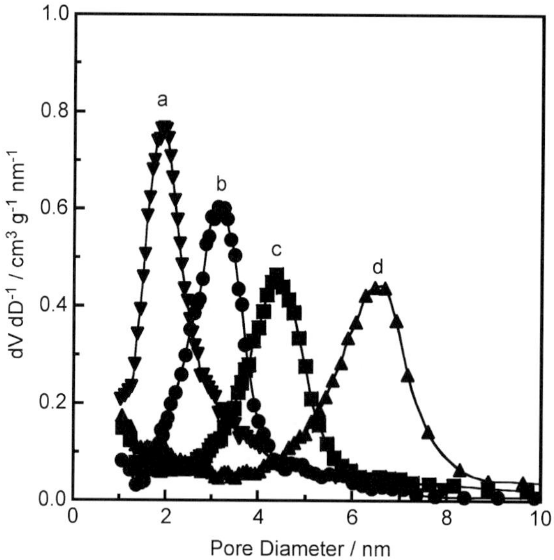

Fig. 4.10 Pore size distribution profiles: (a) CMK-1; (b) CMK-3; (c) CMK-3-130; (d) CMK-3-150. Reprinted with permission from [150], A. Vinu *et al.*, *J. Phys. Chem. B* **2003**, *107*, 8297. 2003, American Chemical Society.

capacity and the pore volume (Figure 4.11B(b)) and pore diameter (Figure 4.11B(c)) is clearly seen. This relation sounds reasonable from the viewpoint of pore filling by the lysozyme molecules, because adsorption of large substances such as proteins in size-limited mesopores would be regulated by the available pore volumes in the mesopores. The adsorbent was also characterized by XRD and nitrogen adsorption–desorption isotherms after the lysozyme adsorption in order to see whether the lysozyme molecule enters the mesopore of CMK-3 or not. The powder XRD patterns of CMK-3 both before and after the adsorption experiments consist of a strong (1 0 0) reflection at a low angle and two small peaks at a higher angle. Adsorption of lysozyme on CMK-3 did not cause shifts in the peak positions. These results indicate that the structural regularity of the host CMK-3 material was not altered upon lysozyme adsorption. In contrast, the intensity of the low angle (1 0 0) and the high angle peaks (110 and 200) of the lysozyme-loaded samples decreased as compared to the parent CMK-3 materials. This cannot be interpreted as a severe loss of structural order of CMK-3 itself, but it is likely that decreased contrast in density between the carbon walls and the lysozyme molecule inside the pores relative to that between the carbon walls and the open pores is responsible for the observed decrease in the XRD intensity. The presence of the lysozyme molecules inside the mesopores of CMK-3 was also supported by the results obtained by nitrogen adsorption–desorption profiles. Upon increasing the lysozyme loading (adsorbed at 280 μM), the specific surface area and pore volume of CMK-3 are decreased to 930 $m^2\,g^{-1}$ and 0.85 $cm^3\,g^{-1}$, respectively. The reduction in the specific mesopore volume after lysozyme adsorption clearly

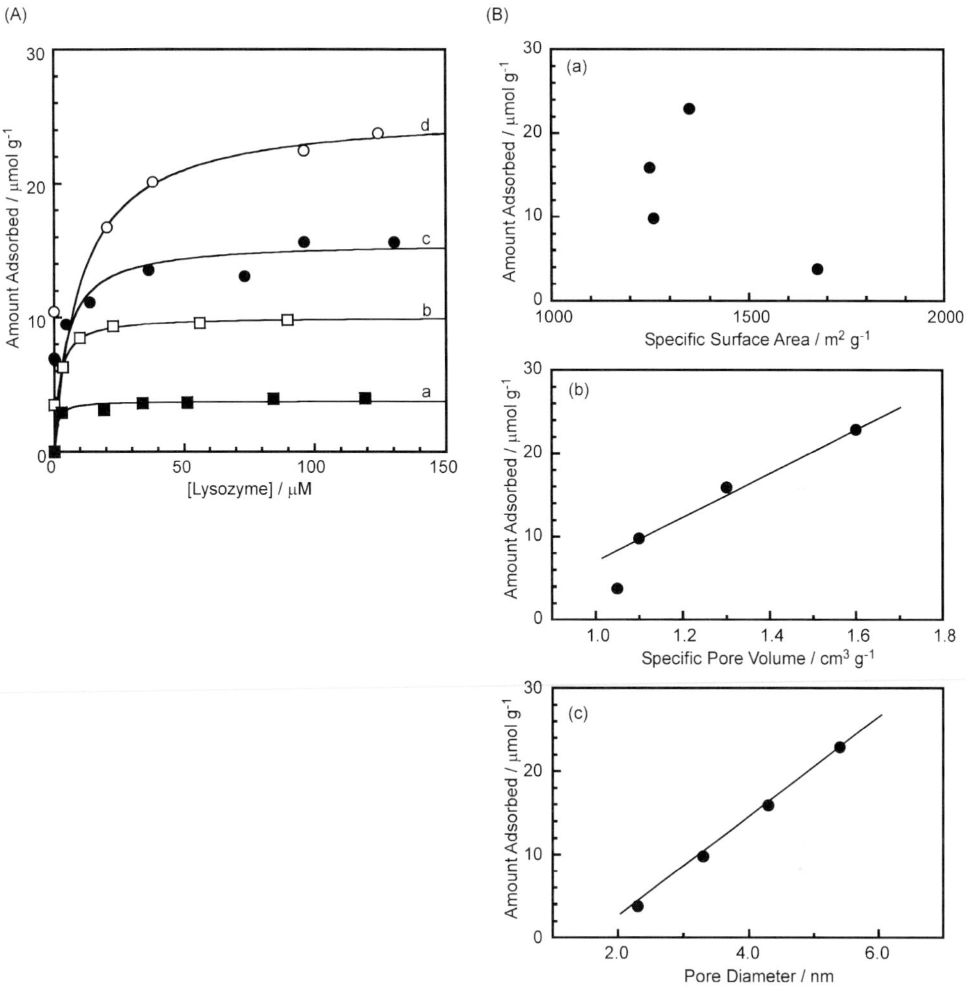

Fig. 4.11 (A) Adsorption isotherms of lysozyme on CMK-type mesoporous carbon materials at pH 11: (a) CMK-1; (b) CMK-3; (c) CMK-3-130; (d) CMK-3-150. (B) Dependence of the adsorption capacities of lysozyme on the structural parameters of the CMK materials: (a) specific surface area; (b) specific pore volume; (c) pore diameter. Reprinted with permission from [152], A. Vinu *et al.*, *J. Phys. Chem. B* **2005**, *109*, 6436. 2005, American Chemical Society.

indicates that the lysozyme molecules are adsorbed inside the mesopores of the CMK-3 adsorbent.

The effect of pH on the protein adsorption on CMK-3 was also investigated [152]. The monolayer adsorption capacities obtained under various pH conditions are plotted in Figure 4.12, where the maximum adsorption was observed in the pH region near the isoelectric point of lysozyme (pI of about 11). Near the isoelectric point, the net charges of the lysozyme molecule are minimized and would form the most compact assembly. A similar pH effect was also seen in the adsorption of cytochrome c on CMK-3. Although the nature of the surface of mesoporous silica and

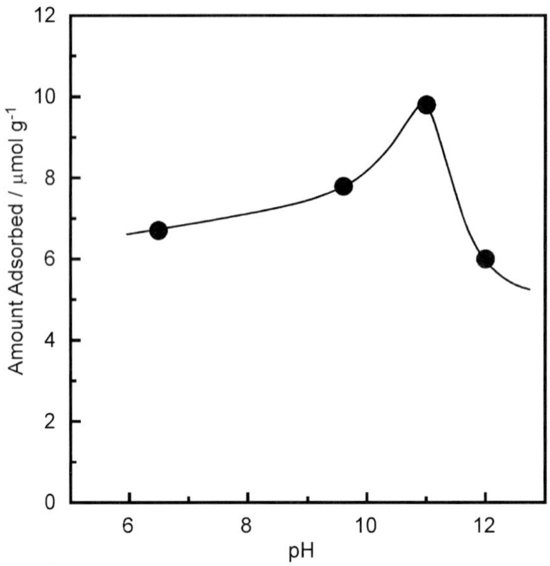

Fig. 4.12 Adsorption capacities of lysozyme on CMK-3 under various pH conditions. Reprinted with permission from [152], A. Vinu *et al.*, *J. Phys. Chem. B* **2005**, *109*, 6436. 2005, American Chemical Society.

mesoporous carbon is completely different, the protein adsorption behavior showed the same pH dependence. This suggests that lateral electrostatic repulsion between the charged protein molecules would be the most influential factor in protein adsorption on mesoporous materials, although several factors such as size selection and protein/surface interaction have also to be considered (Figure 4.13). FT-IR spectra were recorded for the lysozyme molecules loaded on the mesoporous adsorbent (CMK-3); these also confirmed the negligible denaturation, even upon adsorption onto highly hydrophobic mesoporous carbon materials.

If appropriate structures are selected as a hard template for mesoporous carbon syntheses, carbon materials with advanced geometries can be fabricated. Along with this concept, a novel nanocarbon material, a "carbon nanocage", has been invented by Vinu and coworkers [153–155]. The carbon nanocage can be prepared through replica synthesis using three-dimensional large cage-type face centered cubic mesoporous silica materials (KIT-5) as inorganic templates [156]. Schematic illustrations for the synthesis of the carbon nanocage are shown in Figure 4.14 that, however, show just a rough idea of the obtained materials and do not represent the correct geometry. Coincidentally, cage-shaped (hollow capsule) nanocarbons have the same name, carbon nanocage [157–162]. Therefore, our carbon nanocage should be assigned as a mesoporous-type carbon nanocage, which is, however, called "carbon nanocage" in the following just for convenience. Carbon nanocage materials with different pore diameter can be prepared by using different KIT-5 mesoporous silicas synthesized at different temperatures (from 100 to 150 °C) as templates and sucrose as the carbon source. Pore dimensions are also tunable through modification of the sucrose to

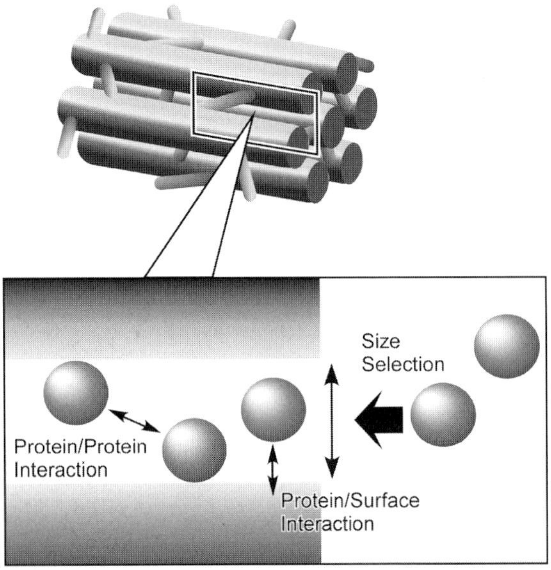

Fig. 4.13 Influence of factors on protein adsorption efficiency on mesoporous materials. Adapted from [37], A. Vinu *et al.*, *J. Nanosci. Nanotechnol.* **2006**, *6*, 1510.

silica weight ratios. *High-resolution transmission electron microscopy* (HRTEM) images of the selected carbon nanocage material are shown in Figure 4.15. The images were recorded along two different crystallographic directions ((A) longitudinal projection and (B) cross-sectional projection), and both confirm that the carbon nanocage possesses a highly ordered structure with uniform pore size distribution. A regular arrangement of bright spots suggests that the mesoporous material is of the three-dimensional cage type. The textural characteristics of the prepared carbon nanocage materials were further investigated by nitrogen adsorption–desorption measurements. The obtained structural parameters are summarized as a function of sucrose to silica ratio in the synthesis conditions in Figure 4.16. While the pore diameters (Figure 4.16a) change little with the sucrose to silica weight ratio, the specific surface area (Figure 4.16b) and specific pore volume (Figure 4.16c) can be controlled by tuning this ratio. Under selected conditions, the specific surface area and specific pore volume reached $1600\,m^2\,g^{-1}$ and $2.1\,cm^3\,g^{-1}$, respectively. These values are clearly larger than those reported for conventional mesoporous carbon, CMK-3 (surface area, $1260\,m^2\,g^{-1}$; pore volume, $1.1\,cm^3\,g^{-1}$). A further analysis with the method proposed by Ravikovitch *et al.* provided the cage diameter of 15 nm for the corresponding carbon nanocage, which has a pore diameter of 5.2 nm. Integrated structures with a large difference between pore size and cage size would result in huge values of the surface area and pore volume.

These excellent textural characteristics would be very useful for the adsorption of biomolecules. Figure 4.17 compares the adsorption behavior of lysozyme on CMK-3 (plot (a)), CMK-3-150 (plot (b)), and the carbon nanocage (plot (c)) at pH 11 [154].

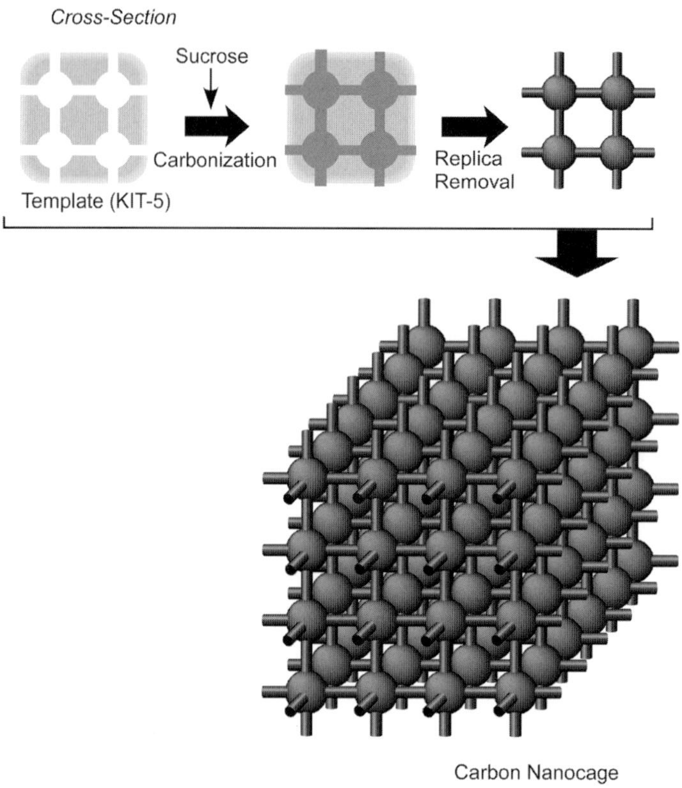

Fig. 4.14 Schematic illustration of the synthesis of the carbon nanocage.

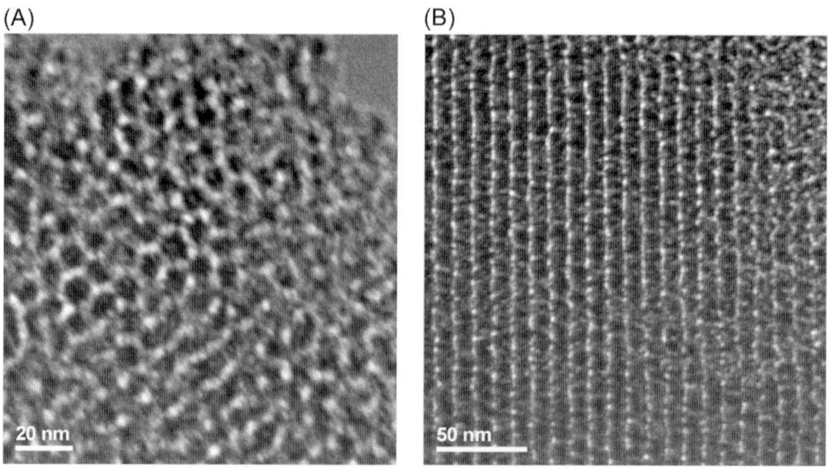

Fig. 4.15 HRTEM images of the selected carbon nanocage: (A) longitudinal projection; (B) cross-sectional projection. Reprinted with permission from [154], A. Vinu *et al.*, *J. Mater. Chem.* **2005**, *15*, 5122. 2005, Royal Society of Chemistry.

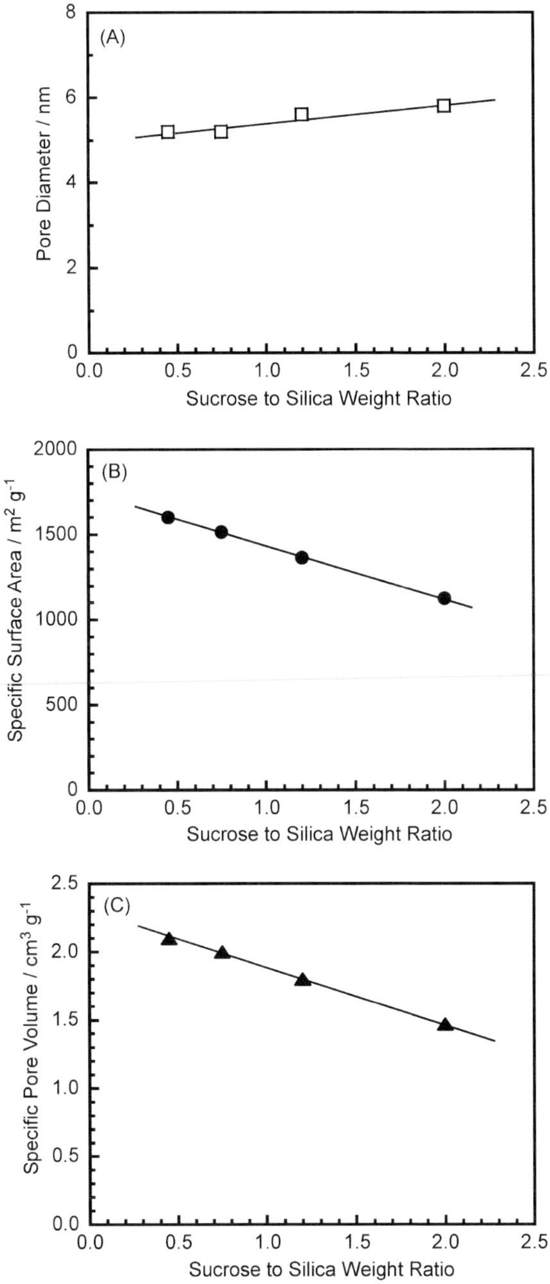

Fig. 4.16 Structural parameters of carbon nanocage materials as a function of the sucrose to silica ratio in their synthesis: (A) pore diameter; (B) specific surface area; (C) specific pore volume.

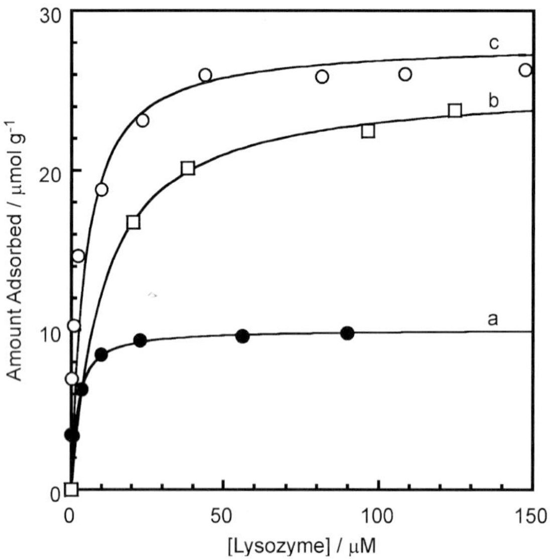

Fig. 4.17 Adsorption isotherms of lysozyme on various carbon materials: (a) CMK-3; (b) CMK-3-150; (c) carbon nanocage. Reprinted with permission from [154], A. Vinu *et al.*, *J. Mater. Chem.* **2005**, *15*, 5122. 2005, Royal Society of Chemistry.

Similarly to the CMK-3 materials, the adsorption isotherm of lysozyme showed Langmuir-type monolayer adsorption. However, the carbon nanocage showed apparently greater adsorption capacity (26.5 $\mu mol\,g^{-1}$) as compared with the former CMK-3 adsorbents, indicating that the superior structural features, such as high pore volume, high surface area, and large cage diameter, of the carbon nanocage are clearly advantageous for biomaterial adsorption. The IR spectrum of lysozyme adsorbed on the carbon nanocage is very similar to that of lysozyme before adsorption. In addition, the intensity ratio of the amide I and amide II band was not altered upon adsorption on the carbon nanocages. These IR characteristics confirmed that the adsorption on the carbon nanocage does not result in serious denaturation of lysozyme.

Recent reports describe the use of various porous carbon materials for protein adsorption. For example, Hyeon and coworkers summarized the recent development of porous carbon materials in their review [163], where the successful use of mesoporous carbons as adsorbents for bulky pollutants, as electrodes for supercapacitors and fuel cells, and as hosts for protein immobilization are described. Gogotsi and coworkers synthesized novel mesoporous carbon materials using ternary MAX-phase carbides that can be optimized for efficient adsorption of large inflammatory proteins [164]. The synthesized carbons possess tunable pore size with a large volume of slit-shaped mesopores. They demonstrated that not only micropores (0.4–2 nm) but also mesopores (2–50 nm) can be tuned in a controlled way by extraction of metals from carbides, providing a mechanism for the optimization of adsorption systems for selective adsorption of a large variety of biomolecules. Furthermore, Vinu and coworkers have successfully developed the synthesis of mesoporous materials composed of *carbon nitride* (CN) [165], *boron nitride* (BN) [166,167], and *boron carbon nitride* (BCN) [166,167]. These heteroatom-hybridized mesoporous carbon materials should have unique properties in protein adsorption and are now under further investigation.

4.4 Immobilization of Other Biopolymers on Mesoporous Materials

The superior structural characteristics of mesoporous materials, such as high surface area and large pore volume, are not limited to use with proteins but are also advantageous for hybridization of other biopolymers, as described below. The use of polypeptides as structure-directing reagents for nanostructured silica was pioneeringly demonstrated by Deming and coworkers [168]. They used synthetic cysteine-lysine block copolypeptides that self-assembled into structured aggregates and hydrolyzed TEOS. Oxidation of the cysteine sulfhydryl groups affected the assembly of the block copolypeptide, resulting in various silica morphologies such as silica hard spheres and well-defined columns of amorphous silica. Bellomo and Deming recently reported the formation of monoliths composed of silica-polypeptide hexagonal platelets using a nonionic, α-helical polypeptide in concentrated solutions [169]. The obtained platelets aligned normal to an applied magnetic field, yielding long-range order in the monoliths. Naik and coworkers found that poly-L-lysine promoted the precipitation of silica from a silicic acid solution within minutes and that the polypeptide secondary structure transition occurred during the silicification reaction [170]. The formation of the hexagonal silica platelets is attributed to the poly-L-lysine helical chains that are formed in the presence of monosilicic acid and phosphate (Figure 4.18). The backbone of the poly-L-lysine interacts with the silicate species in the solution via hydrogen bonding and electrostatic interactions. A condensation reaction between neighboring silicate precursors would in turn promote the formation of intramolecular hydrogen bonding on the polypeptide backbone, causing poly-L-lysine to zip into a helical conformation. The combination of the interactions results in the packing of the poly-L-lysine chains into a hexagonal lattice as the silicification reaction proceeds.

Ariga and coworkers demonstrated the use of polypeptide-functionalized surfactants for the synthesis of mesoporous silica hybrids with peptide segments confined in regular mesopores (Figure 4.19) [171,172]. This can be regarded as the preparation of a protein-like structure in an inorganic mesostructured framework, which was named proteosilica. Type I surfactants have a peptide residue with a hydrophobic chain at the *C*-terminal and a polar group at the *N*-terminal, and mesoporous silicates in both powder and film forms were successfully prepared with these templates. In particular, a spin-coating method effectively provided transparent mesoporous films, and the formation of regular structures depends significantly on hydrogen bonding between the peptide moieties. HRTEM images of the films indicate highly ordered regular pore arrays. Proteosilica provides an optically asymmetric environment for doped molecules. The photo-isomerization behavior of spiropyran doped in the

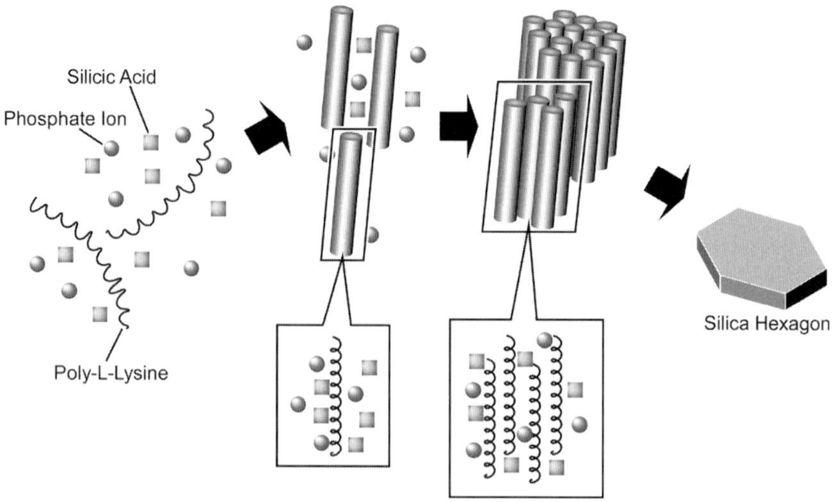

Fig. 4.18 Formation of hexagonal silica platelets with poly-L-lysine helical chains in the presence of monosilicic acid and phosphate. Reprinted with permission from [170], M.M. Tomczak *et al.*, *J. Am. Chem. Soc.* **2005**, *127*, 12577. 2005, American Chemical Society.

transparent proteosilica films was investigated spectroscopically, and control of circular dichroism signals upon photo-transformation was successfully demonstrated [173]. The presented materials are expected to be applied to memory devices with non-destructive read-out capability. The same research group have pioneered the preparation of mesoporous silica films filled with an amphiphilic polypeptide that was composed of a hydrophilic poly(ethylene glycol) chain (MW, about 3000) and a hydrophobic peptide segment with defined length (leucine 16mer) (type II surfactant in Figure 4.19) [174]. The structural regularity of the prepared transparent films was again verified by XRD and TEM measurements. Preservation of the α-helix structure of the poly(leucine) was confirmed by CD and FT-IR spectroscopies.

Nucleic acids, DNA and RNA, are attractive biopolymers that can be used for biomedical applications [175,176], nanostructure fabrication [177,178], computing [179,180], and materials for electron-conduction [181,182]. Immobilization of DNA and RNA in well-defined nanostructures would be one of the most unique subjects in current nanotechnology. Unfortunately, a silica surface cannot usually adsorb duplex DNA in aqueous solution due to the electrostatic repulsion between the silica surface and polyanionic DNA. However, Fujiwara *et al.* recently found that duplex DNA in protonated phosphoric acid form can adsorb on mesoporous silicates, even in low-salt aqueous solution [183]. The DNA adsorption behavior depended much on the pore size of the mesoporous silica. Plausible models of DNA accommodation in mesopore silica channels are depicted in Figure 4.20. Inclusion of duplex DNA in mesoporous silicates with larger pores, around 3.8 nm diameter, would be accompanied by the formation of four water monolayers on the silica surface of the mesoporous inner channel (Figure 4.20A), where sufficient quantities of Si–OH groups remained after solvent extraction of the template (not by calcination).

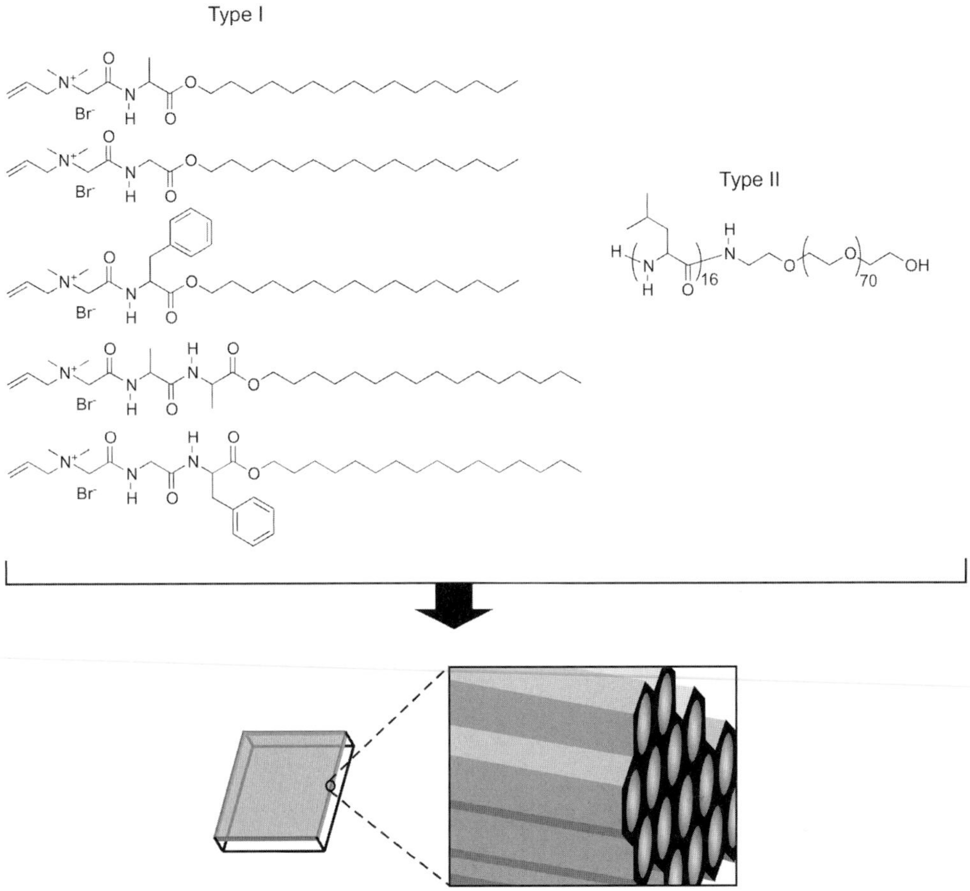

Fig. 4.19 Preparation of a protein-like structure in an inorganic mesostructured framework (Proteosilica) as a transparent film using polypeptide-functionalized surfactants.

Figure 4.20B illustrates the inclusion structure of duplex DNA in mesoporous silicates with a smaller pore size such as the sample. In this case, the pore diameter is slightly larger than duplex DNA that forms direct hydrogen bonds between the surface Si–OH groups and DNA. In this mode, adsorbed water layers are not important for the interaction between them. These modes suggest that both a suitable pore size and sufficient hydrogen bonds are required for effective incorporation of duplex DNA into the pores of mesoporous silicates. The effect of encompassing the phosphate groups of duplex DNA with the inner surface of mesoporous silica might be crucial to this DNA adsorption. Solberg and Landry also accomplished DNA adsorption onto acid-prepared mesoporous silica [184]. Mesoporous silica alone adsorbed a negligible amount of DNA, but exchanging divalent cations such as Mg^{2+} and Ca^{2+} into the pores of mesoporous silica prior to DNA uptake was found to cause a significant amount of DNA to be adsorbed. They also showed that postsynthetic

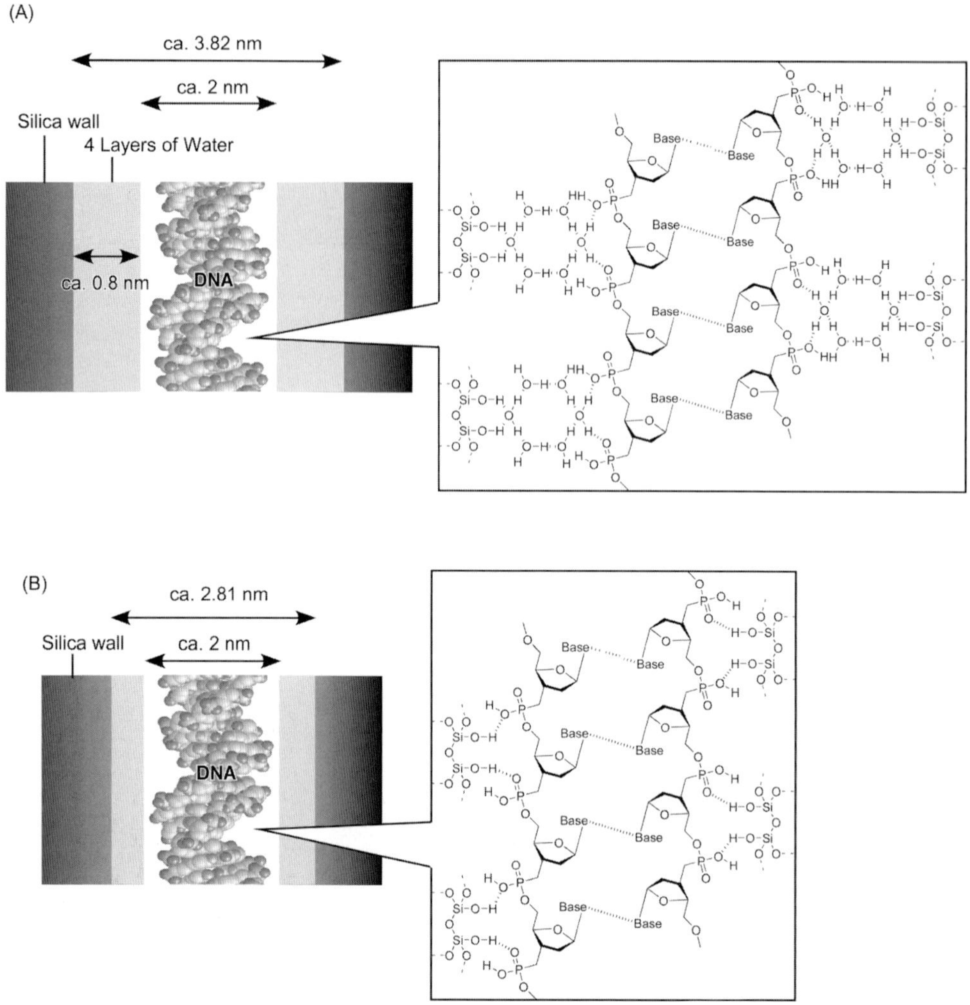

Fig. 4.20 Plausible models of DNA accommodation in mesopore silica channels: (A) within a larger pore; (B) within a smaller pore. Reprinted with permission from [183], M. Fujiwara *et al.*, *Anal. Chem.* **2005**, *77*, 8138. 2005, American Chemical Society.

modification of the silica surface with aminopropyl groups increased the DNA adsorption capacity to a maximum. The obtained value was 2–3 times larger than observed for the metal ion-exchanged mesoporous silica. The advantages of the modification of mesoporous silica with amino groups for DNA adsorption was also suggested in the report by Zhang *et al.* [185].

Mann and coworkers proposed a unique method to prepare nanohybrids of mesoporous silica and gold nanoparticles mediated by programmed DNA hybridization, as illustrated in Figure 4.21 [186]. MCM-41 nanoparticles were first functionalized with a 3′-amino-terminated 12-base single strand oligonucleotide

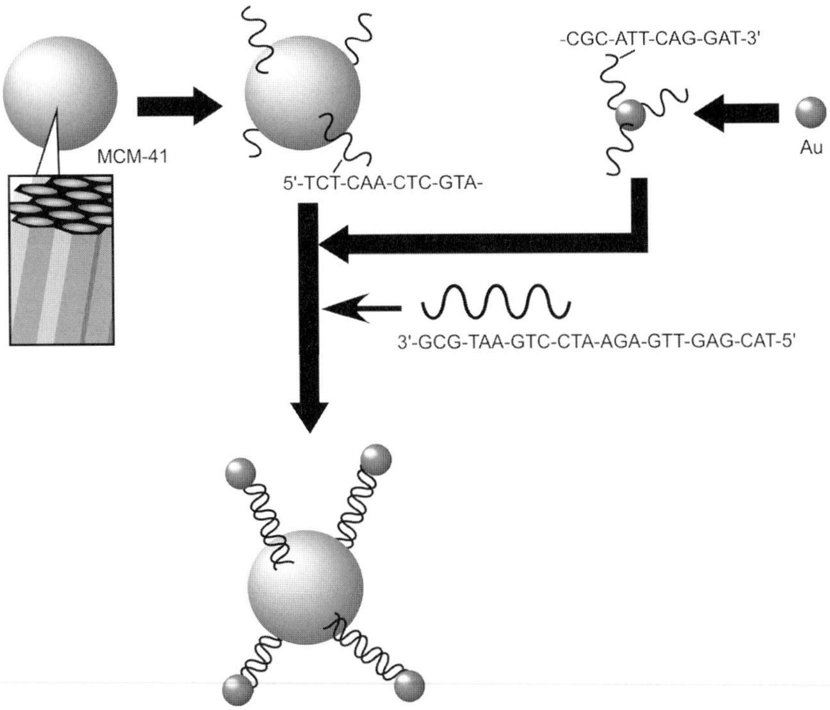

Fig. 4.21 Preparation of nanohybrids of mesoporous silica and gold nanoparticles mediated by programmed DNA hybridization. Adapted from [186], S. Sadasivan *et al.*, *Small* **2005**, *1*, 103.

(5′-TCT-CAA-CTC-GTACH(OH)-CH_2-CH(OH)-NH_2). Gold nanoparticles 11 nm in size were separately capped with a 5′-mercapto-terminated 12-base oligonucleotide (HS-propyl-CGC-ATT-CAG-GAT-3′). Self-assembly was induced by mixing DNA-capped MCM-41 silica and gold nanoparticles in the presence of a 24-base single strand target oligonucleotide (3′-GCG-TAA-GTCCTA-AGA-GTT-GAG-CAT-5′) that was simultaneously complementary to DNA fragments on MCM-41 and gold. This strategy should make it possible to extend this approach to the control of the assembly of a wide range of nanomaterials with increased levels of structural and compositional complexity, and tunable multi-functionality.

4.5 Immobilization of Small Biomolecules on Mesoporous Materials

Adsorption of small biomolecules such as amino acids, vitamins, and oligopeptides on mesoporous materials is also an important research target, because it certainly satisfies certain demands in material extraction, drug delivery, and pollutant removal. Vinu and coworkers recently reported systematic research on the adsorption of L-histidine on mesoporous materials [187,188]. Figure 4.22 compares the adsorption

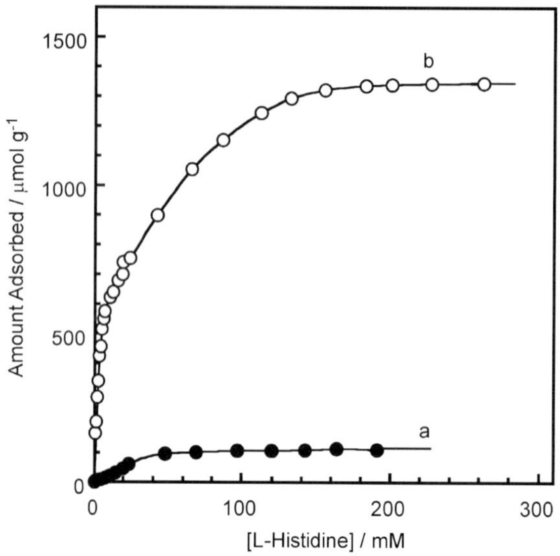

Fig. 4.22 Adsorption isotherms of L-histidine on mesoporous materials at pH 7.5: (a) SBA-15; (b) CMK-3. Adapted from [37], A. Vinu *et al.*, *J. Nanosci. Nanotechnol.* **2006**, *6*, 1510.

behavior of L-histidine on mesoporous silica SBA-15 (plot (a)) and mesoporous carbon CMK-3 (plot (b)) at pH 7.5. The adsorption capacities of these adsorbents are apparently different. CMK-3 registered an L-histidine adsorption capacity of about 1.35 mmol g^{-1}, which is more than 12 times higher than the total adsorption capacity of SBA-15 (0.11 mmol g^{-1}). This indicates that a hydrophobic surface is important for adsorption of the rather hydrophobic amino acid, histidine. Nitrogen adsorption isotherms of CMK-3 adsorbent before and after L-histidine adsorption were also investigated. The mesopore volume and the surface area of CMK-3 drastically decrease after L-histidine adsorption. After loading with 1.35 mmol g^{-1} of L-histidine, the specific surface area of CMK-3 is reduced from 1260 to 556 $m^2 g^{-1}$ which is a 55.9 % reduction of the total surface area, while the specific pore volume is reduced from 1.1 to 0.56 $cm^3 g^{-1}$ (49.1 %). This large reduction in the specific pore volume and the specific surface area could be attributed to tight packing of the L-histidine molecules in the mesopores of CMK-3.

Adsorption and hybridization of amino acids and peptides on mesoporous materials has also been investigated by other researchers. O'Connor *et al.* recently reported the adsorption of the basic amino acid, lysine, on MCM-41 molecular sieve which possesses acidic silanol groups [189]. Lysine was found to adsorb according to a Langmuir-type isotherm with a maximum capacity at pH 6 of 0.21 mmol g^{-1}. The extent of adsorption depends strongly on the pH and ionic strength of the adsorbate solution, due to a combination of ion exchange and electrostatic interactions governing the adsorption process. Wang and coworkers demonstrated chiral recognition using SBA-15 modified with chiral porphyrin that was introduced by the surface OH group on SBA-15 with an amino group of the porphyrin [190]. The investigation of chiral recognition of the porphyrin and the composite SBA-15 showed that the incorporated SBA-15 has a clear CD spectral signal of the chiral porphyrin and exhibits different chiral recognition abilities for small amino acids. The D-alanine is more tightly bound to the mesoporous composite SBA-15 than its L-isomer. Vega and coworkers used ^{2}H MAS NMR to determine the adsorption and desorption rates of a deuterium-labeled tetraalanine salt at the surface of the pores of MCM-41 [191]. The dynamic deuterium MAS NMR approach can also be applied to many other molecular adsorption–desorption systems, as long as the signal-to-noise ratio is sufficient. One of the most unique examples of amino acids in mesoporous media could be the chiral morphology control of mesoporous silica reported by Che *et al.* (Figure 4.23) [192–194]. By using synthesized surfactant, *N*-acetyl-type alanine-based surfactant as a structure-directing reagent, mesoporous silica materials with regularly twisted rod-like structures with diameter 130–180 nm and length 1–6 mm were synthesized. In the obtained structure, hexagonally aligned mesoscopic channels with diameter of 2.2 nm wound together in a particular direction.

Vinu and coworkers also reported systematic research on the adsorption of vitamin E (tocopherol) on activated carbon (plot (a)) and CMK-3 (plot (b)) in heptane solutions of vitamin E (Figure 4.24) [195,196]. Up to an initial solution concentration of vitamin

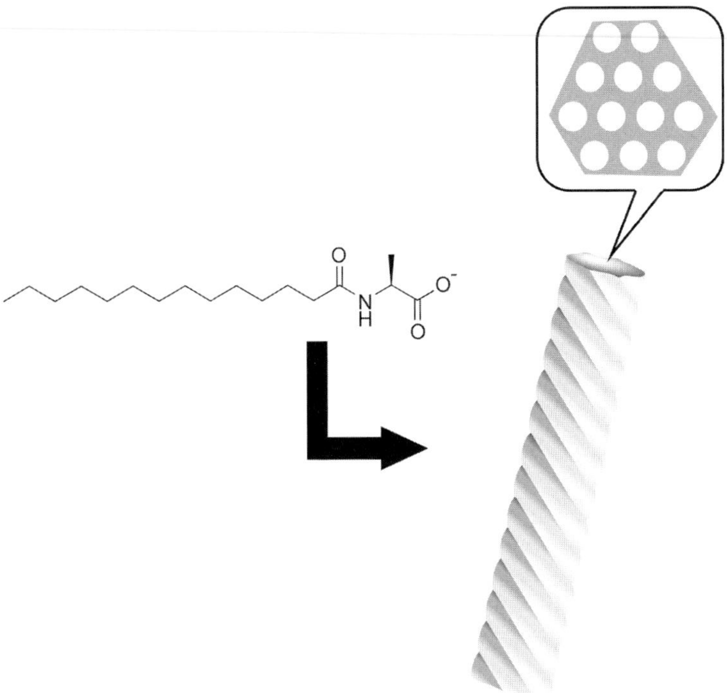

Fig. 4.23 Preparation of mesoporous silica materials with chirally twisted rod-like structures by using surfactant with a chiral amino acid moiety as a structure-directing reagent.

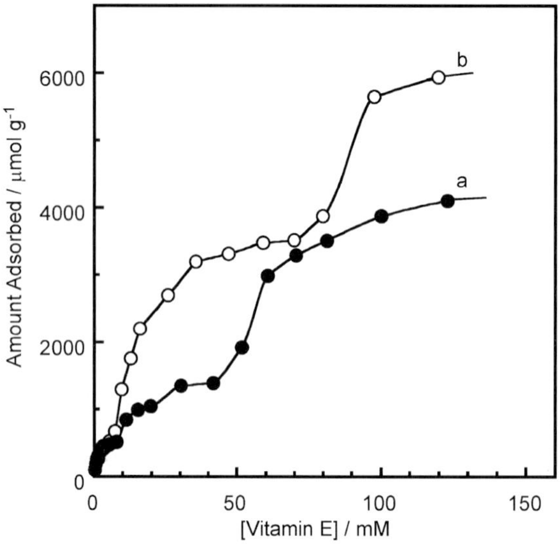

Fig. 4.24 Adsorption isotherms of vitamin E (tocopherol) on carbon adsorbents: (a) activated carbon; (b) CMK-3. Reprinted with permission from [195], M. Hartmann *et al.*, *Chem.Mater.* **2005**, *17*, 829. 2005, American Chemical Society.

E of about 2 g L^{-1}, the two adsorbents showed an almost similar amount of vitamin E adsorption, which could be attributed to a monolayer on the accessible surface of each adsorbent. Further increase in vitamin E concentration induced multi-layer adsorption of vitamin E. However the multi-layer adsorption apparently started at a lower concentration in the case of the CMK-3 adsorbent. This result indicates that pore filling by the vitamin E molecules would be more easily induced in mesoporous carbon materials CMK-3 than with activated carbon. The difference observed between mesoporous carbon and activated carbon may be attributed to the textural properties of the porous adsorbents other than surface area, such as pore volume and pore diameter. Vitamin immobilization on mesoporous structures was also investigated by others. For example, Qiu and coworkers researched the adsorption of vitamin B_{12} onto CMK-1, CMK-3 and their analogues modified with 10 wt% poly (methyl methacrylate) [197]. Compared with CMK-1, CMK-3 showed higher vitamin B_{12} adsorption due to a larger mesopore volume, a higher BET surface area and a larger pore diameter. After coating with poly(methyl methacrylate), both mesoporous carbons showed a higher adsorption capacity than the unmodified materials. These results indicate that the adsorption properties of vitamin B_{12} were influenced by the pore structure and surface properties of mesoporous carbons. Asaftei and Walder cross-linked the vitamin B_{12} derivatives equipped with multiple reaction sites in the TiO_2 mesopores [198]. The cross-linkers have multiple functions with one of them being to assist the electron transfer from TiO_2 to the Co centers through redox shuttling. The modified electrodes show high electrocatalytic reactivity toward organic halides as well as improved stability.

4.6
Advanced Functions of Nanohybrids of Biomolecules and Mesoporous Materials

Nanohybrids prepared from biomolecules and mesoporous materials can be used in various applications. In this section, examples of such applications are briefly introduced.

One of the most promising applications of enzyme-immobilized mesoporous materials is as microscopic reactors. Galarneau *et al.* investigated the effect of mesoporous silica structures and their surface natures on the activity of immobilized lipases [199]. Too hydrophilic (pure silica) or too hydrophobic (butyl-grafted silica) supports are not appropriate for the development of high activity for lipases. An adequate hydrophobic/hydrophilic balance of the support, such as a supported-micelle, provides the best route to enhance lipase activity. They also encapsulated the lipases in sponge mesoporous silicates, a new procedure based on the addition of a mixture of lecithin and amines to a sol–gel synthesis to provide pore-size control. This encapsulation procedure gives the highest activity for the lipases. The lecithin/amines mixture structuring the pore network leads to a suitable phospholipids bilayer-like environment, which avoids the necessity to create an interface by substrate assembly. Monduzzi and coworkers compared the activity of lipase that was immobilized on SBA-15 physically, or chemically with glutardialdehyde [200]. The prepared hybrids can be used both in aqueous or nonaqueous media. Chemical immobilization is preferred in aqueous solvents, while physical immobilization is sufficient in organic solvents. Hydrolytic activity measurements indicated that the chemically immobilized biocatalyst acquires size-selectivity, as deduced from a direct comparison between tributyrin and triolein assays. Lin, Mou, and coworkers investigated the spin state of cytochrome c immobilized in a silica mesopore channel [201]. Cytochrome c immobilized in mesoporous silica existed in both high and low spin states, while the native cytochrome c shows primarily the low spin state. The high spin state arises from the replacement of Met-80 ligands of heme Fe (III) by water or a silanol group on the silica surface. This replacement could open up the heme groove for easy access of oxidants and substrates to the iron center and facilitate the catalytic activity. Cytochrome c immobilized on mesoporous silica has high specific activity toward the oxidation of polycyclic aromatic hydrocarbons. Seelan, Kato and coworkers immobilized aldolase antibody 84G3 inside the mesoporous silica with 15–25 nm pore size, resulting in highly accelerated and stable aldol reactions compared to the free antibody [202]. The enhanced catalytic activities of the immobilized antibodies may result from a favored microenvironment that results in exceptionally high immobilization efficiency (because of size exclusion) with enhanced stability.

Addition of third components to nanohybrids of proteins and mesoporous materials sometimes brings advantages in their functions. Kim, Hyeon, and coworkers immobilized enzyme molecules together with magnetite (Fe_3O_4) nanoparticles in *hierarchically ordered, mesocellular, mesoporous silica* (HMMS) (Figure 4.25) [203]. In their method, after co-adsorption of enzymes and magnetite nanoparticles, enzyme molecules were cross-linked via glutaraldehyde treatment. The HMMS

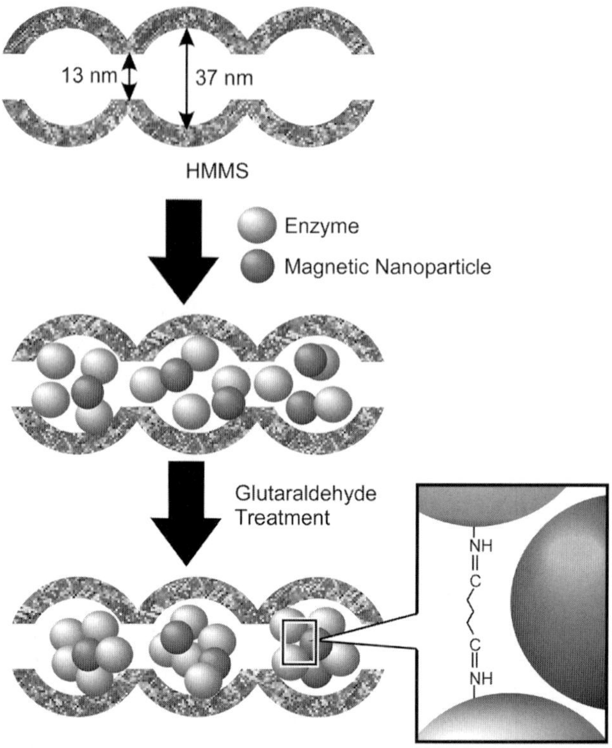

Fig. 4.25 Immobilization of enzyme molecules and magnetite nanoparticles in hierarchically ordered mesocellular mesoporous silica with the aid of cross-linking via glutaraldehyde treatment. Adapted from [203], J. Kim *et al.*, *Small* **2005**, *1*, 1203.

support consists of 37-nm cellular mesopores and 13-nm-sized mesoporous channels around the mesocellular pores. The composites of cross-linked enzyme aggregates and magnetite nanoparticles in the large mesocellular pores will not leach out from the HMMS through the narrow mesoporous channels. The synthesized three-component hybrids are magnetically separable, stable in enzyme activity, iteratively recyclable, and highly loaded with a large amount of enzymes. Caruso and coworkers applied *layer-by-layer* (LbL) techniques for an enzyme-loaded capsule by sequential processes of adsorption of enzymes in a mesoporous silica sphere, LbL coating of the sphere, and selective removal of the core sphere [204,205].

Instead of using real enzymes, well-organized assemblies of chemical functionalities sometimes lead to enzyme-like functions, which are called artificial enzymes [206–211]. A mesoporous nanospace can supply an environment where molecular motions and orientations are highly limited. This situation is similar to that in the confined spaces found in enzyme pockets, so that a mesoporous nanospace should be a good medium for the performance of biomimetic functions. Goettmann *et al.* immobilized a phosphanorbornadiene phosphonic acid derivative (1-phospha-4,5-dimethyl-3,6-diphenylnorbornadienyl phosphonic acid) grafted onto a zirconia-rich mesoporous material, which complexed to a rhodium(I) precursor, which showed high catalytic activity in the hydrogenation and hydroformylation of olefins [212]. The observed increase in regioselectivity, together with the increased activity, could be dependent on the complexation of the Lewis basic substrate onto the Lewis acidic zirconia wall. Pirngruber and coworkers immobilized the histidine and glutamic acid residues on the surface of mesoporous silica supports to construct an Fe(II) complex as a biomimetic catalyst [213]. The biomimetic iron complexes obtained were tested as catalysts for cyclohexane oxidation under mild reaction conditions with hydrogen peroxide as oxidant. The nature of the immobilized amino acid and its surface density has an influence on the catalytic activity. Fukuoka and coworkers modified the mesoporous silicates with sulfonic acid groups, either by a one-pot or a grafting method [214]. In the hydrolysis of sucrose and starch, the sulfonated mesoporous silicates work as water-tolerant recyclable solid acid catalysts, showing higher conversion and turnover frequency than conventional Amberlyst-15, Nafion-silica and HZSM-5 catalysts. Fukuoka and Dhepe also reported the preparation of two metal (platinum and ruthenium) catalysts on silica or alumina, to convert an aqueous mixture of cellulose and hydrogen gas into glucose at about 190 °C [215].

Selective binding and separation of biomolecules also play an important role in the functions of nanohybrids of biomolecules and mesoporous materials. Brennan and coworkers reported the application of protein-doped monolithic silica columns for immobilized enzyme reactor chromatography, which allowed the screening of enzyme inhibitors present in mixtures using mass spectrometry for detection [216]. In their method, the enzyme was immobilized by entrapment within a bimodal meso/macroporous silica material prepared by a biocompatible sol–gel processing route. Wall and coworkers investigated the adsorption of an anti-domoic acid *single-chain Fv* (scFv) antibody fragment on a range of mesoporous silicate supports [217]. The immobilized antibody was functional on all materials and bound its antigen, a naturally occurring neurotoxin produced by shellfish, in a rapidly saturating manner that suggested the antibody adsorbed in a multilayer on the mesoporous particles. While the studies were demonstrated with an antibody fragment specific for the neurotoxin domoic acid, the highly conserved domain architecture of antibody molecules and their derived fragments confer general relevance to scFv immobilization for sensor development. Mercuri, Jaroniec, and coworkers reported a study on the applicability of SBA-15 as an adjuvant where its efficiency to act as a vehicle for the bacterial 16.5 kDa recombinant protein and venom proteins of the snake were evaluated [218]. Katiyar and Pinto reported the first visualization of a protein–protein separation by a mesoporous molecular sieve, which was demonstrated on spherical SBA-15 particles suitable for chromatographic use [219].

The mesoporous materials possess highly reliable and controllable pore geometries that regulate the release of drugs from mesopores. Therefore, mesoporous materials have high potential in biomedical applications. Fujiwara and coworkers achieved photo-controlled regulation of drug storage and release using functionalized mesoporous silica (Figure 4.26) [220,221]. MCM-41 mesoporous silica was functionalized with photoactive coumarin that was grafted only at the pore outlet. Guest drugs such as cholestane were incorporated into the mesopores of template-free silica upon exposure

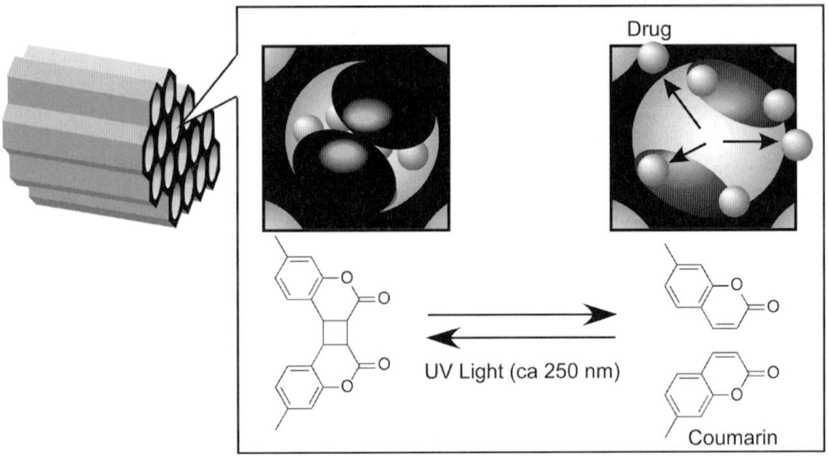

Fig. 4.26 Photo-controlled drug storage and release from mesoporous silica using photodimerization of coumarin grafted only at the pore outlet. Reprinted with permission from [221], M. Fujiwara *et al.*, *Chem. Mater.* **2003**, *15*, 3385. 2003, American Chemical Society.

to the guest solution. Irradiation by UV light (310 nm) causes dimerization of coumarin for trapping the guest cholestane in the stable form and blocks leaching of the guest from the mesopores. The dimerized coumarin was cleaved upon irradiation with a different wavelength of UV light (~250 nm) resulting in release of trapped cholestane. Lin and coworkers realized a mesoporous silica-based controlled-release delivery system using colloid capping (Figure 4.27) [222]. Their stimuli-responsive system consists of a 2-(propyldisulfanyl)ethylamine functionalized mesoporous silica nanosphere material with an average particle size of 200.0 nm and an average pore diameter of 2.3 nm. The mesopores were also used to store various drug molecules or neurotransmitters, such as vancomycin and *adenosine triphosphate* (ATP). The entrance of the mesopores of the drug/neurotransmitter-loaded mesoporous material was then capped *in situ* by allowing the pore surface-bound 2-(propyldisulfanyl)ethylamine functional groups to covalently capture the water-soluble mercaptoacetic acid-derivatized CdS nanocrystals. The resulting disulfide linkages could be chemically cleaved using various reducing agents, such as dithiothreitol or mercaptoethanol. Thus, stimuli-responsive release was realized while the system's *in vitro* biocompatibility with neuroglial cells (astrocytes) was also confirmed. Stoddart, Zink, and coworkers reported the tethering of pseudorotaxanes as gates at the entrances of the cylindrical pores in mesostructured silica, creating nanovalves capable of trapping luminescent molecules and releasing them on demand [223]. The same research group also realized a reversibly operating nanovalve that can be toggled on and off using redox chemistry in mesoporous silica functionalized with redox-activated bistable [2]rotaxane molecules [224].

Nanohybrids of biomolecules and mesoporous materials have also been used for sensor application. Lin and coworkers also demonstrated gate-controlled molecular

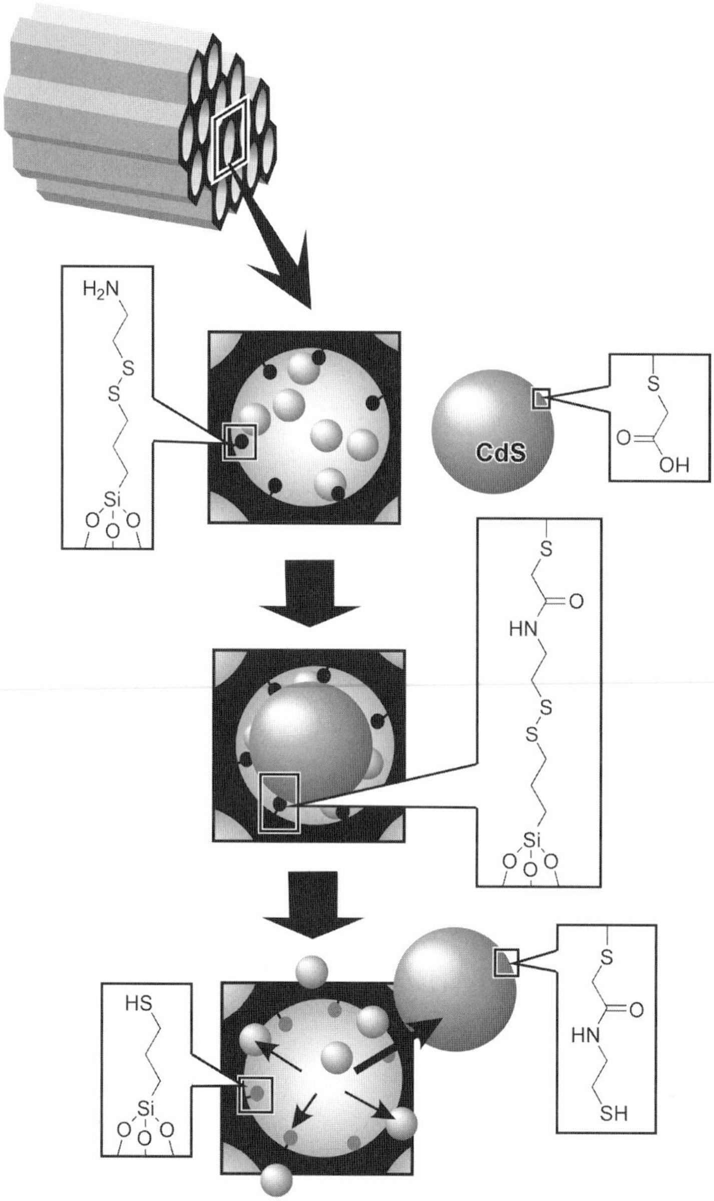

Fig. 4.27 Controlled drug release and delivery system using colloid capping of a mesoporous silica channel. Reprinted with permission from [222], C.-Y. Lai *et al.*, *J. Am. Chem. Soc.* **2003**, *125*, 4451. 2003, American Chemical Society.

recognition and fluorescence detection by selective functionalization of external and internal surfaces of mesoporous silicates (Figure 4.28) [225]. In the first step, thiol-modified MCM-41 type mesoporous silica – nanosphere silica was synthesized using a co-condensation route. Before removal of the surfactant template, epoxyhexyl groups were grafted selectively at the external surface of the mesoporous silica. The epoxy groups were then converted to dihydroxyhexyl groups and L-lactide was polymerized at the external surface followed by the immobilization of the *o*-phthalic hemithioacetal group at the internal thiol group. The *o*-phthalic hemithioacetal can operate as an optical sensor by fluorescing upon reaction with amines such as dopamine (a neurotransmitter), tyrosine, or glutaric acid. The polylactate layer at the exterior plays the role of gatekeeper, resulting in selectivity of detection for the neurotransmitters. Lee *et al.* fabricated a highly sensitive and fast glucose biosensor through simply immobilizing glucose oxidase in mesocellular carbon foam (MSU-F-C) [226]. A high enzyme loading without serious mass-transfer limitations owing to the unique structures of MSU-F-C (see Figure 4.29), led to a

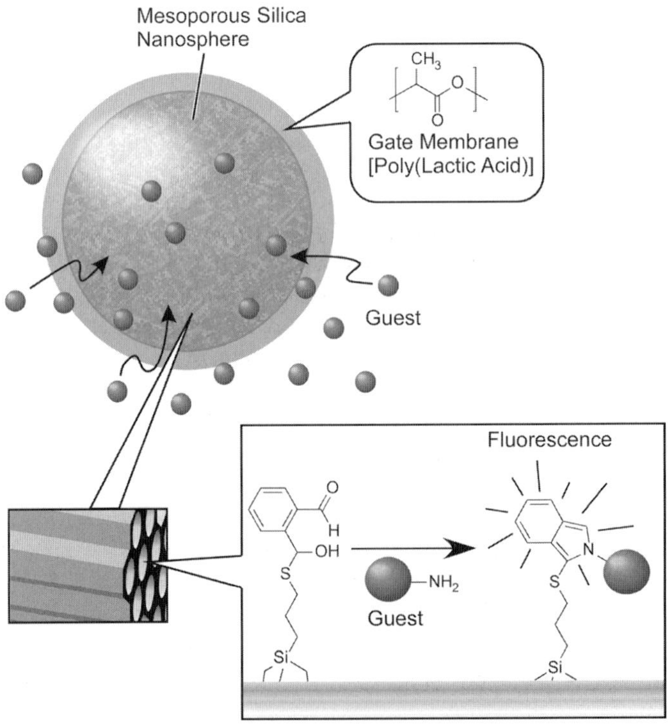

Fig. 4.28 Gate-controlled molecular recognition and fluorescence detection by selective functionalization of external and internal surfaces of mesoporous silicates. Reprinted with permission from [225], R. D. Radu *et al.*, *J. Am. Chem. Soc.* **2004**, *126*, 1640. 2004, American Chemical Society.

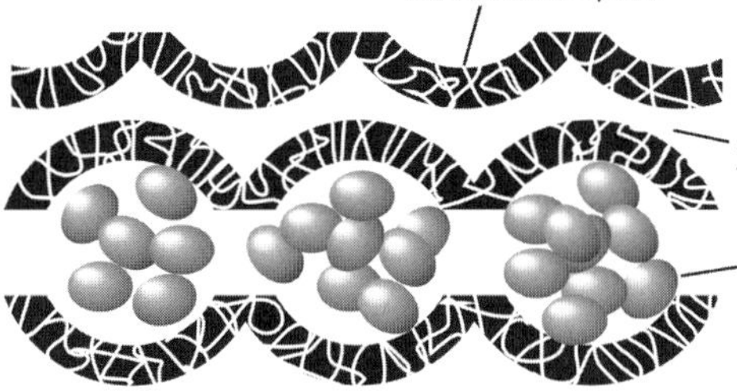

Fig. 4.29 Immobilization of glucose oxidase in mesocellular carbon foam for a sensitive and fast glucose biosensor. Adapted from [226], D. Lee *et al.*, *Adv. Mater.* **2005**, *17*, 2828.

high catalytic efficiency. Therefore, the glucose biosensor fabricated on a glassy carbon electrode showed high sensitivity and fast response.

Some kinds of biomolecules such as photosynthetic pigments have photoelectrical functions, which is useful for achieving photonic and electronic functions with nanohybrids of biomolecules in mesoporous structures. Itoh and coworkers prepared a chlorophyll-mesoporous silica (FSM) conjugate with high photo-stability [227,228]. The mesopore spaces of the FSM also allow nanoscale interactions between the absorbed chlorophyll molecules. The conjugate can evolve hydrogen gas efficiently upon visible light illumination in the presence of an electron donor, a carrier, and a catalyst. In another report, they investigated the arrangement of chlorophyll molecules within mesoporous nanospaces. The arrangement of chlorophylls and the distance between the porphyrin rings are both important in determining the molecular interaction that leads to the changes in absorption/fluorescence behavior, energy transfer efficiency, and charge separation probability. Their spectroscopic experiments suggest that the arrangement of chlorophyll molecules inside each nanoscale pore of FSM-22 (pore diameter, 4.0 nm) resembles that in light-harvesting or reaction-center complexes of living plants. That is, they exhibit red-shifted spectral forms and fast energy transfer. In contrast, FSM-10 with a smaller pore (pore diameter, 1.9 nm) is expected to have chlorophyll aggregates inside or outside the FSM, which seem to be too far apart to allow efficient energy transfer. These examples illustrate the importance of the selection of an appropriate mesoporous nanospace for the promoted functions. The interfacing of nanostructured semiconductor photo-electrodes with redox proteins is an innovative approach to the development of artificial photosynthetic systems. Kong and coworkers entrapped the bacterial photosynthetic reaction center from the *Rb. sphaeroides* strain RS601 on the tailored three-dimensional wormlike mesoporous WO_3–TiO_2 films [229,230]. Synthesis of the tailored three-dimensional wormlike mesoporous WO_3–TiO_2 films (pore size 7.1 nm) was carried out according to the acid–base pair concept (Figure 4.30A). The target film was achieved by spin-coating the mother solution on *indium tin oxide* (ITO) glass with thickness about 150 nm. Immobilization of the reaction center protein was achieved by immersing the freshly prepared WO_3–TiO_2

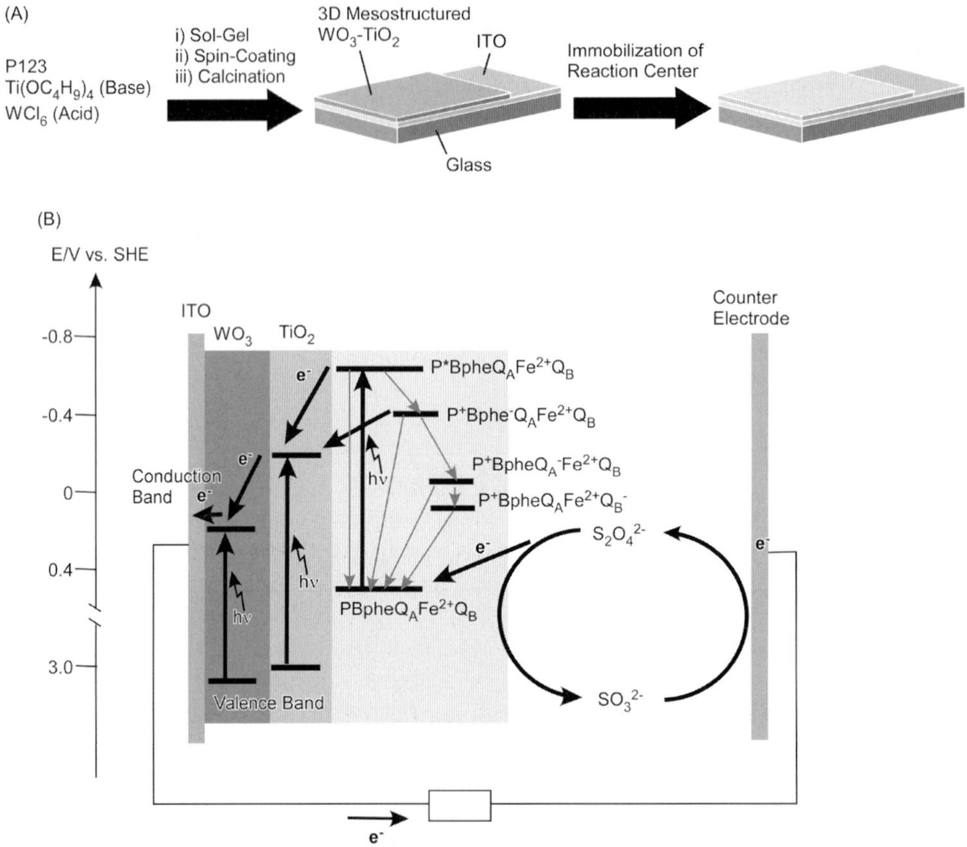

Fig. 4.30 Immobilization of the bacterial photosynthetic reaction center on tailored three-dimensional wormlike mesoporous WO_3–TiO_2 films for artificial photosynthetic systems: (A) procedure of film coating; (B) proposed scheme of photoelectric conversion. Reprinted with permission from [229], Y. Lu *et al.*, *Langmuir* **2005**, *21*, 4071. 2005, American Chemical Society.

films in the pH 8.0 Tris-HCl buffer solution of the reaction center for a few days. A reasonable mechanism that partly explains the enhancement of photoelectric conversion can be proposed, as illustrated in Figure 4.30B. The attractive features of mesoporous metal oxides modified with bacterial photosynthetic proteins may provide an alternative way to probe the photo-induced multiple pathway electron transfer of photosensitive chromophores and create a new perspective to develop versatile biophotoelectric devices. Durrant and coworkers reported the spectroelectrochemical investigation of proton-coupled electron transfer in flavodoxin D. vulgaris Hildenborough immobilized on nanocrystalline mesoporous SnO_2

electrodes with the aid of poly-L-lysine [231]. Such electrodes are suitable for addressing relatively slow redox events and, in this case, they applied the electrode to a study of the kinetics of deprotonation associated with the oxidation of flavodoxin semiquinone to its quinone state. The same research group investigated the photo-induced electron-transfer reactions of zinc-substituted cytochrome c, immobilized on mesoporous, nanocrystalline metal oxide electrodes [232]. Efficient electron injection from the triplet state of zinc-substituted cytochrome c into TiO_2 electrodes was observed, resulting in a long-lived charge-separated state (lifetime up to 0.4 s). The addition of ethylenediamine tetraacetic acid to the electrolyte as a sacrificial electron donor resulted in efficient photo-generation of molecular hydrogen, with a quantum yield per absorbed photon of 10 %.

4.7 Future Perspectives

In this chapter, recent examples of the adsorption, immobilization, and hybridization of various biomolecules on mesoporous structures are comprehensively introduced. These hybrid materials have high possibilities in a wide range of applications, as described in the previous section. Channels formed in mesoporous materials have structural dimensions comparable to biopolymers and their interactive partners. Therefore, activities of such substances in mesopores surely have limited freedom in their motion, conformation, and orientation. Such unnatural conditions would lead to the discovery of unexplored phenomena of entrapped biomolecules. Most researchers, who develop bio-functions in mesoporous media, rely on the advantages of the structural uniqueness of mesoporous materials such as high surface area, large pore volume, and controllable pore size. However, the use of pore arrangements and geometries has not been fully explored in functional developments of mesoporous hybrids. Mesoporous structures with different geometries such as hexagonal and cubic phases can be easily designed and synthesized, but the differences between these phases are not well reflected in bio-functions and applications. Mesoporous materials are not simple porous objects but surely have regularly ordered structures. Therefore, development of functions deeply based on structural regularity and geometry have to be seriously considered for meaningful use of mesoporous biohybrids in future.

Acknowledgment

The research described in this chapter was partially supported by the "Ground-based Research Program for Space Utilization" promoted by the Japan Space Forum and Grant-in Aid for Scientific Research on Priority Area (No. 18033059 "Chemistry of Coordination Space") from the Ministry of Education, Science, Sports, and Culture, Japan.

References

1 Okahata, Y., Tsuruta, T., Ijiro, K. and Ariga, K. (1988) *Langmuir*, **4**, 1373.

2 Okahata, Y., Tsuruta, T., Ijiro, K. and Ariga, K. (1989) *Thin Solid Films*, **180**, 65.

3 Bartlett, P.N. and Cooper J.M.(1993) *Journal of Electroanalytical Chemistry*, **362**, 1.

4 Lvov, Y., Ariga, K. and Kunitake, T. (1994) *Chemistry Letters*, 2323.

5 Lvov, Y., Ariga, K., Ichinose, I. and Kunitake, T. (1995) *Journal of the American Chemical Society*, **117**, 6117.

6 Lvov, Y., Ariga, K., Ichinose, I. and Kunitake, T. (1995) *Journal of the Chemical Society-Chemical Communications*, 2313.

7 Mrksich, M. and Whitesides G.M. (1996) *Annual Review of Biophysics and Biomolecular Structure*, **25**, 55.

8 Onda, M., Lvov, Y., Ariga, K. and Kunitake, T. (1996) *Biotechnology and Bioengineering*, **51**, 163.

9 Lvov, Y., Ariga, K., Ichinose, I. and Kunitake, T. (1996) *Thin Solid Films*, **285**, 797.

10 Onda, M., Lvov, Y., Ariga, K. and Kunitake, T. (1996) *Journal of Fermentation and Bioengineering*, **82**, 502.

11 Ariga, K., Onda, M., Lvov, Y. and Kunitake, T. (1997) *Chemistry Letters*, 25.

12 Onda, M., Lvov, Y., Ariga, K. and Kunitake, T. (1997) *Japanese Journal of Applied Physics*, **36**, L1608.

13 Lvov, Y., Onda, M., Ariga, K. and Kunitake, T. (1998) *Journal of Biomaterials Science-Polymer Edition*, **9**, 345.

14 Caruso, F., Furlong, D.N., Ariga, K., Ichinose, I. and Kunitake, T. (1998) *Langmuir*, **14**, 4559.

15 Walcarius, A.(1998) *Electroanalysis*, **10**, 1217.

16 Blawas, A.S. and Reichert W.M. (1998) *Biomaterials*, **19**, 595.

17 Cosnier, S.(1999) *Biosensors and Bioelectronics*, **14**, 443.

18 Kikuchi, J., Ariga, K., Miyazaki, T. and Ikeda, K. (1999) *Chemistry Letters*, 253.

19 Kikuchi, J., Ariga, K. and Ikeda, K. (1999) *Chemical Communications*, 547.

20 Onda, M., Ariga, K. and Kunitake, T. (1999) *Journal of Bioscience and Bioengineering*, **87**, 69.

21 Ariga, K., Tanaka, K., Katagiri, K., Kikuchi, J., Ohshima, E. and Hisaeda, Y. (2000) *Colloids and Surfaces A-Physicochemical and Engineering Aspects*, **169**, 47.

22 Ariga, K., Tanaka, K., Katagiri, K., Kikuchi, J., Shimakoshi, H., Ohshima, E. and Hisaeda, Y. (2001) *Physical Chemistry Chemical Physics*, **3**, 3442.

23 Gust, D., Moore, T.A. and Moore, A.L. (2001) *Accounts of Chemical Research*, **34**, 40.

24 Fukuda, K., Sasaki, Y., Ariga, K. and Kikuchi, J. (2001) *Journal of Molecular Catalysis B-Enzymatic*, **11**, 971.

25 Kikuchi, J., Ariga, K., Sasaki, Y. and Ikeda, K. (2001) *Journal of Molecular Catalysis B-Enzymatic*, **11**, 977.

26 Jin, W. and Brennan, J.D. (2002) *Analytica Chimica Acta*, **461**, 1.

27 Ginger, D.S., Zhang, H. and Mirkin, C. A. (2004) *Angewandte Chemie-International Edition*, **43**, 30.

28 Ariga, K., Sasaki, Y., Horiguchi, H., Horiuchi, N. and Kikuchi, J. (2001) *Defect and Diffusion Forum*, **191**, 35.

29 Katagiri, K., Hamasaki, R., Ariga, K. and Kikuchi, J. (2002) *Journal of the American Chemical Society*, **124**, 7892.

30 Katagiri, K., Hamasaki, R., Ariga, K. and Kikuchi, J. (2002) *Lamgmuir*, **18**, 6709.

31 Zhang, Q., Ariga, K., Okabe, A. and Aida, T. (2003) *Studies in Surface Science and Catalysis*, **146**, 465.

32 Zhang, Q., Ariga, K., Okabe, A. and Aida, T. (2004) *Journal of the American Chemical Society*, **126**, 988.

33 Yiu, H.H.P. and Wright P.A. (2005) *Journal of Materials Chemistry*, **15**, 3690.

34 Song, S.-W., Hidajat, K. and Kawi, S. (2005) *Langmuir*, **21**, 9568.

35 Zhao, X.S., Bao, X.Y., Guo, Y. and Lee, F.Y. (2006) *Materials Today*, **9**, 32.

36 Finnie, K.S., Jacques, D.A., McGann, M.J., Blackford, M.G. and Barbé, C.J. (2006) *Journal of Materials Chemistry*, **16**, 4494.

37 Vinu, A., Miyahara, M. and Ariga, K. (2006) *Journal of Nanoscience and Nanotechnology*, **6**, 1510.

38 Ariga, K., Vinu, A. and Miyahara, M. (2006) *Current Nanoscience*, **2**, 197.

39 Otani, W., Kinbara, K., Zhang, Q., Ariga, K. and Aida, T. (2007) *Chemistry – A European Journal*, **13**, 1731–1736.

40 Vinu, A., Mori, T. and Ariga, K. (2006) *Science and Technology of Advanced Materials*, **7**, 753.

41 Yanagisawa, T., Shimizu, T., Kuroda, K. and Kato, C. (1990) *Bulletin of the Chemical Society of Japan*, **63**, 988.

42 Inagaki, S., Fukushima, Y. and Kuroda, K. (1993) *Journal of the Chemical Society-Chemical Communications*, 680.

43 Kresge, C.T., Leonowicz, M.E., Roth, W.J., Vartuli, J.C. and Beck, J.S. (1992) *Nature*, **359**, 710.

44 Beck, J.S., Vartuli, J.C., Roth, W.J., Leonowicz, M.E., Kresge, C.T., Schmitt, K.D., Chu, C.T.W., Olson, D. H., Sheppard, E.W., McCullen, S.B., Higgins, J.B. and Schlenker, J.L. (1992) *Journal of the American Chemical Society*, **114**, 10834.

45 Vartuli, J.C., Schmitt, K.D., Kresge, C. T., Roth, W.J., Leonowicz, M.E., McCullen, S.B., Hellring, S.D., Beck, J. S., Schlenker, J.L., Olson, D.H. and Sheppard, E.W. (1994) *Chemistry of Materials*, **6**, 2317.

46 Dubois, M., Gulik-Krzywicki, T. and Cabane, B. (1993) *Langmuir*, **9**, 673.

47 Tanev, P.T. and Pinnavaia T.J. (1995) *Science*, **267**, 865.

48 Bagshaw, S.A., Prouset, E. and Pinnavaia, T.J. (1995) *Science*, **269**, 1242.

49 Zhao, D., Feng, J., Huo, Q., Melosh, N., Fredickson, G.H., Chmelka, B.F. and Stucky, G.D. (1998) *Science*, **279**, 548.

50 Zhao, D., Huo, Q., Feng, J., Chmelka, B.F. and Stucky, G.D. (1998) *Journal of the American Chemical Society*, **120**, 6024.

51 Schmidt-Winkel, P., Lukens, W.W., Zhao, D., Yang, P., Chmelka, B.F. and Stucky, G.D. (1999) *Journal of the American Chemical Society*, **121**, 254.

52 Vinu, A., Hossain, K.Z. and Ariga, K. (2005) *Journal of Nanoscience and Nanotechnology*, **5**, 347.

53 Morey, M.S., Davidson, A. and Stucky, G.D. (1998) *Journal of Porous Materials*, **5**, 195.

54 Khushalani, D., Hasenzahl, S. and Mann, S. (2001) *Journal of Nanoscience and Nanotechnology*, **1**, 129.

55 Hartmann, M., Vinu, A., Elangovan, S. P., Murugesan, V. and Böhlmann, W. (2002) *Chemical Communications*, 1238.

56 Leite, E.R., Carreno, N.L.V., Longo, E., Valentini, A. and Probst, L.F.D. (2002) *Journal of Nanoscience and Nanotechnology*, **2**, 89.

57 Vinu, A., Dědeček, J., Murugesan, V. and Hartmann, M. (2002) *Chemistry of Materials*, **14**, 2433.

58 Vinu, A. and Hartmann M. (2004) *Chemistry Letters*, **33**, 588.

59 Vinu, A., Ariga, K., Saravanamurugan, S., Hartmann, M. and Murugesan, V. (2004) *Microporous Mesoporous Materials*, **76**, 91.

60 Wu, S., Han, Y., Zou, Y.-C., Song, J.-W., Zhao, L., Di, Y., Liu, S.-Z. and Xiao, F.-S. (2004) *Chemistry of Materials*, **16**, 486.

61 Vinu, A., Krithiga, T., Murugesan, V. and Hartmann, M. (2004) *Advanced Materials*, **16**, 1817.

62 Vinu, A., Devassy, B.M., Halligudi, S. B., Böhlmann, W. and Hartmann, M. (2005) *Applied Catalysis A-General*, **281**, 207.

63 Vinu, A., Karthik, M., Miyahara, M., Murugesan, V. and Ariga, K. (2005) *Journal of Molecular Catalysis A-Chemical*, **230**, 151.

64 Vinu, A., Sawant, D.P., Ariga, K., Hartmann, M. and Halligudi, S.B. (2005) *Microporous and Mesoporous Materials*, **80**, 195.

65 Vinu, A., Kumar, G.S., Ariga, K. and Murugesan, V. (2005) *Journal of Molecular Catalysis A-Chemical*, **235**, 57.

66 Krithiga, T., Vinu, A., Ariga, K., Arabindoo, B., Palanichamy, M. and Murugesan, V. (2005) *Journal of Molecular Catalysis A-Chemical*, **237**, 238.

67 Vinu, A., Sawant, D.P., Ariga, K., Hossain, K.Z., Halligudi, S.B., Hartmann, M. and Nomura, M. (2005) *Chemistry of Materials*, **17**, 5339.

68 Sawant, D.P., Vinu, A., Jacob, N.E., Lefebvre, F. and Halligudi, S.B. (2005) *Journal of Catalysis*, **235**, 341.

69 Vinu, A., Krithiga, T., Balasubramanian, V.V., Asthana, A., Srinivasu, P., Mori, T., Ariga, K., Ramanath, G. and Ganesan, P.G. (2006) *Journal of Physical Chemistry B*, **110**, 11924.

70 Vinu, A., Srinivasu, P., Miyahara, M. and Ariga, K. (2006) *Journal of Physical Chemistry B*, **110**, 801.

71 Umamaheswari, V., Böhlmann, W., Poppl, A., Vinu, A. and Hartmann, M. (2006) *Microporous and Mesoporous Materials*, **89**, 47.

72 Antonelli, D.M. and Ying J.Y. (1995) *Angewandte Chemie-International Edition in English*, **34**, 2014.

73 Bagshaw, S.A. and Pinnavaia T.J. (1996) *Angewandte Chemie-International Edition in English*, **35**, 1102.

74 Antonelli, D.M. and Ying J.Y. (1996) *Angewandte Chemie-International Edition in English*, **35**, 426.

75 Antonelli, D.M., Nakahira, A. and Ying, J.Y. (1996) *Inorganic Chemistry*, **35**, 3126.

76 Antonelli, D.M. and Ying J.Y. (1996) *Chemistry of Materials*, **8**, 874.

77 Liu, P., Moudrakovski, I.L., Liu, J. and Sayari, A. (1997) *Chemistry of Materials*, **9**, 2513.

78 Bach, U., Lupo, D., Comte, P., Moser, J. E., Weissörtle, F., Salbeck, J., Spreitzer, H. and Gratzel, M. (1988) *Nature*, **395**, 583.

79 Wong, M.S. and Ying J.Y. (1998) *Chemistry of Materials*, **10**, 2067.

80 Kondo, J.N., Lu, L., Takahara, Y., Maruya, K., Domen, K., Igarashi, N. and Tatsumi, T. (2000) *Bulletin of the Chemical Society of Japan*, **73**, 1123.

81 Schüth, F.(2001) *Chemistry of Materials*, **13**, 3184.

82 Subramanian, V., Jiang, J.C., Smith, P. H. and Rambabu, B. (2004) *Journal of Nanoscience and Nanotechnology*, **4**, 125.

83 Kartini, I., Meredith, P., Zhao, X.S., Diniz da Costa, J.C. and Lu, G.Q. (2004) *Journal of Nanoscience and Nanotechnology*, **4**, 270.

84 Hayward, R.C., Chmelka, B.F. and Kramer, E.J. (2005) *Advanced Materials*, **17**, 2591.

85 Shibata, H., Ogura, T., Mukai, T., Ohkubo, T., Sakai, H. and Abe, M. (2005) *Journal of the American Chemical Society*, **127**, 16396.

86 Chakraborty, B., Pulikottil, A.C., Das, S. and Viswanathan, B. (1997) *Chemical Communications*, 911.

87 Zhao, D., Luan, Z. and Kevan, L. (1997) *Chemical Communications*, 1009.

88 Holland, B.T., Isbester, P.K., Blanford, C.F., Munson, E.J. and Stein, A. (1997) *Journal of the American Chemical Society*, **119**, 6796.

89 Kimura, T., Sugahara, Y. and Kuroda, K. (1999) *Chemistry of Materials*, **11**, 508.

90 Inagaki, S., Guan, S., Fukushima, Y., Ohsuna, T. and Terasaki, O. (1999)

Journal of the American Chemical Society, **121**, 9611.

91 Asefa, T., MacLachlan, M.J., Coombs, N. and Ozin, G.A. (1999) *Nature*, **402**, 867.

92 Melde, B.J., Holland, B.T., Blanford, C. F. and Stein, A. (1999) *Chemistry of Materials*, **11**, 3302.

93 Asefa, T., Kruk, M., MacLachlan, M.J., Coombs, N., Grondey, H., Jaroniec, M. and Ozin, A.G. (2001) *Journal of the American Chemical Society*, **123**, 8520.

94 Burleigh, M.C., Markowitz, M.A., Spector, M.S. and Gaber, B.P. (2001) *Chemistry of Materials*, **13**, 4760.

95 Burleigh, M.C., Markowitz, M.A., Spector, M.S. and Gaber, B.P. (2001) *Journal of Physical Chemistry B*, **105**, 9935.

96 Burleigh, M.C., Markowitz, M.A., Spector, M.S. and Gaber, B.P. (2001) *Langmuir*, **17**, 7923.

97 Yang, Q., Kapoor, M.P. and Inagaki, S. (2002) *Journal of the American Chemical Society*, **124**, 9694.

98 Wang, Z., Heising, J.M. and Clearfield, A. (2003) *Journal of the American Chemical Society*, **125**, 10375.

99 Kapoor, M.P., Yang, Q., Goto, Y. and Inagaki, S. (2003) *Chemistry Letters*, **32**, 914.

100 Inagaki, S., Guan, S., Ohsuna, T. and Terasaki, O. (2002) *Nature*, **416**, 304.

101 Diaz, J.F. and Balkus J.K. Jr. (1996) *Journal of Molecular Catalysis B-Enzymatic*, **2**, 115.

102 Washmon-Kriel, L., Jimenez, V.L. and Balkus, K.J. Jr. (2000) *Journal of Molecular Catalysis B-Enzymatic*, **10**, 453.

103 Takahashi, H., Li, B., Sasaki, T., Miyazaki, C., Kajino, T. and Inagaki, S. (2000) *Chemistry of Materials*, **12**, 3301.

104 Wei, Y., Xu, J., Feng, Q., Lin, M., Dong, H., Zhang, W.-J. and Wang, C. (2001) *Journal of Nanoscience and Nanotechnology*, **1**, 83.

105 Yiu, H.H.P., Botting, C.H., Botting, N.P. and Wright, P.A. (2001) *Physical Chemistry Chemical Physics*, **3**, 2983.

106 Yiu, H.H.P., Wright, P.A. and Botting, N.P. (2001) *Microporous and Mesoporous Materials*, **44–45**, 763.

107 Takahashi, H., Li, B., Sasaki, T., Miyazaki, C., Kajino, T. and Inagaki, S. (2001) *Microporous and Mesoporous Materials*, **44–45**, 755.

108 Sasaki, T., Kajino, T., Li, B., Sugiyama, H. and Takahashi, H. (2001) *Applied and Environmental Microbiology*, **67**, 2208.

109 Deere, J., Magner, E., Wall, J.G. and Hodnett, B.K. (2001) *Chemical Communications*, 465.

110 Deere, J., Magner, E., Wall, J.G. and Hodnett, B.K. (2002) *Journal of Physical Chemistry B*, **106**, 7340.

111 Han, Y.-J., Watson, J.T., Stucky, G.D. and Butler, A. (2002) *Journal of Molecular Catalysis B-Enzymatic*, **17**, 1.

112 Deere, J., Magner, E., Wall, J.G. and Hodnett, B.K. (2003) *Catalysis Letters*, **85**, 19.

113 Chong, A.S.M. and Zhao X.S. (2004) *Applied Surface Science*, **237**, 398.

114 Goradia, D., Cooney, J., Hodnett, B.K. and Magner, E. (2005) *Journal of Molecular Catalysis B-Enzymatic*, **32** (2005), 231.

115 Hartmann, M. (2005) *Chemistry of Materials*, **17**, 4577.

116 Moelans, D., Cool, P., Baeyens, J. and Vansant, E.F. (2005) *Catalysis Communications*, **6**, 307.

117 Di?az, J.F. and Balkus K.J. Jr. (2006) *Journal of Molecular Catalysis B-Enzymatic*, **2**, 115.

118 Han, Y., Lee, S.S. and Ying, J.Y. (2006) *Chemistry of Materials*, **18**, 643.

119 Vinu, A., Murugesan, V. and Hartmann, M. (2004) *Journal of Physical Chemistry B*, **108**, 7323.

120 Vinu, A., Murugesan, V., Tangermann, O. and Hartmann, M. (2004) *Chemistry of Materials*, **1**, 3056.

121 Miyahara, M., Vinu, A., Hossain, K.Z., Nakanishi, T. and Ariga, K. (2006) *Thin Solid Films*, **499**, 13.
122 Miyahara, M., Vinu, A. and Ariga, K. (2006) *Journal of Nanoscience and Nanotechnology*, **6**, 1765.
123 Blin, J.L., Gérardin, C., Carteret, C., Selve, R.C. and Stébé, M.J. (2005) *Chemistry of Materials*, **17**, 1479.
124 Mureseanu, M., Galarneau, A., Renard, G. and Fajula, F. (2005) *Langmuir*, **21**, 4648.
125 Han, Y., Lee, S.S. and Ying, J.Y. (2006) *Chemistry of Materials*, **18**, 643.
126 Shan, W., Wang, B., Zhang, Y. and Tang, Y. (2005) *Chemical Communications*, 1877.
127 Sun, J., Zhang, H., Tian, R., Ma, D., Bao, X., Su, D.S. and Zou, H. (2006) *Chemical Communications*, 1322.
128 Tortajada, M., Ramón, D., Beltra?n, D. and Amós, P. (2005) *Chemical Communications*, 3859.
129 Lee, J., Kim, J., Jia, H., Kim, M.I., Kwak, J.H., Jin, S., Dohnalkova, A., Park, H. G., Chang, H.N., Wang, P., Grate, J.W. and Hyeon, T. (2005) *Small*, **1**, 744.
130 He, J., Song, Z., Ma, H., Yang, L. and Guo, C. (2006) *Journal of Materials Chemistry*, **16**, 4307.
131 Wang, Y. and Caruso F. (2006) *Advanced Materials*, **18**, 795.
132 Hudson, S., Magner, E., Cooney, J. and Hodnett, B.K. (2005) *Journal of Physical Chemistry B*, **109**, 19496.
133 Qiao, S.Z., Yu, C.Z., Xing, W., Hu, Q.H., Djojoputro, H. and Lu, G.Q. (2005) *Chemistry of Materials*, **17**, 6172.
134 DeLouise, L.A. and Miller B.L.(2005) *Analytical Chemistry*, **77**, 1950.
135 Yang, N.Y.C., Jian, K., Külaots, I., Crawford, G.P. and Hurt, R.H. (2003) *Journal of Nanoscience and Nanotechnology*, **3**, 386.
136 Inagaki, M., Kaneko, K. and Nishizawa, T. (2004) *Carbon*, **42**, 1401.
137 Vinu, A. and Ariga K. (2005) *Chemistry Letters*, **34**, 674.
138 Yang, H. and Zhao D. (2005) *Journal of Materials Chemistry*, **15**, 1217.
139 Vinu, A. and Hartmann M. (2005) *Catalysis Today*, **102**, 189.
140 Kyotani, T., Nagai, T., Inoue, S. and Tomita, A. (1997) *Chemistry of Materials*, **9**, 609.
141 Kyotani, T. (2000) *Carbon*, **38**, 269.
142 Ryoo, R., Joo, S.H. and Jun, S. (1999) *Journal of Physical Chemistry B*, **103**, 7743.
143 Jun, S., Joo, S.H., Ryoo, R., Kruk, M., Jaroniec, M., Liu, Z., Ohsuna, T. and Terasaki, O. (2000) *Journal of the American Chemical Society*, **122**, 10712.
144 Ryoo, R., Joo, S.H., Kurk, M. and Jaroniec, M. (2001) *Advanced Materials*, **13**, 677.
145 Joo, S.H., Choi, S.J., Oh, I., Kwak, J., Liu, Z., Terasaki, O. and Ryoo, R. (2001) *Nature*, **412**, 169.
146 Solovyov, L.A., Zaikovskii, V.I., Shmakov, A.N., Belousov, O.V. and Ryoo, R. (2002) *Journal of Physical Chemistry B*, **106**, 12198.
147 Lee, J., Yoon, S., Hyeon, T., Oh, S.M. and Kim, K.B. (1999) *Chemical Communications*, 2177.
148 Lee, J., Yoon, S., Oh, S.M., Shin, C.-H. and Hyeon, T. (2000) *Advanced Materials*, **12**, 359.
149 Han, S., Kim, S., Lim, H., Choi, W., Park, H., Yoon, J. and Hyeon, T. (2003) *Microporous and Mesoporous Materials*, **58**, 131.
150 Vinu, A., Streb, C., Murugesan, V. and Hartmann, M. (2003) *Journal of Physical Chemistry B*, **107**, 8297.
151 Hartmann, M. and Vinu A. (2002) *Langmuir*, **18**, 8010.
152 Vinu, A., Miyahara, M. and Ariga, K. (2005) *Journal of Physical Chemistry B*, **109**, 6436.
153 Vinu, A., Miyahara, M. and Ariga, K. (2005) *Studies in Surface Science and Catalysis*, **158**, 971.
154 Vinu, A., Miyahara, M., Sivamurugan, V., Mori, T. and Ariga, K. (2005) *Journal of Materials Chemistry*, **15**, 5122.

155 Vinu, A., Miyahara, M., Mori, T. and Ariga, K. (2006) *Journal of Porous Materials*, **13**, 379.

156 Kleitz, F., Liu, D., Anilkumar, G.M., Park, I.-S., Solovyov, L.A., Shmakov, A. N. and Ryoo, R. (2003) *Journal of Physical Chemistry B*, **107**, 14296.

157 Oku, T., Hirano, T., Suganuma, K. and Nakajima, S. (1999) *Journal of Materials Research*, **14**, 4266.

158 Oku, T. and Suganuma K. (1999) *Chemical Communications*, 2355.

159 Oku, T., Hirata, T., Motegi, N., Hatakeyama, R., Sato, N., Mieno, T., Sato, N.Y., Mase, H., Niwano, M. and Miyamoto, N. (2000) *Journal of Materials Research*, **15**, 2182.

160 Erkoc, S. and Vural D.C. (2001) *International Journal of Modern Physics*, **12**, 685.

161 Kawai, T., Miyamoto, Y., Sugino, O. and Koga, Y. (2002) *Physical Review B-Condensed Matter*, **66**, 033404.

162 Wang, J.N., Zhang, L., Niu, J.J., Yu, F., Sheng, Z.M., Zhao, Y.Z., Chang, H. and Pak, C. (2007) *Chemistry of Materials*, **19**, 453.

163 Lee, J., Kim, J. and Hyeon, T. (2006) *Advanced Materials*, **18**, 2073.

164 Yushin, G., Hoffman, E.N., Barsoum, M.W., Gogotsi, Y., Howell, C.A., Sandeman, S.R., Phillips, G.J., Lloyd, A.W. and Mikhalovsky, S.V. (2006) *Biomaterials*, **27**, 5755.

165 Vinu, A., Ariga, K., Mori, T., Nakanishi, T., Hishita, S., Golberg, D. and Bando, Y. (2005) *Advanced Materials*, **17**, 1648.

166 Vinu, A., Terrones, M., Golberg, D., Hishita, S., Ariga, K. and Mori, T. (2005) *Chemistry of Materials*, **17**, 5887.

167 Murakami, M., Shimizu, T., Tansho, M., Vinu, A., Ariga, K. and Takegoshi, K. (2006) *Chemistry Letters*, **35**, 986.

168 Cha, J.N., Stucky, G.D., Morse, D.E. and Deming, T.J. (2000) *Nature*, **403**, 289.

169 Bellomo, E.G. and Deming T.J. (2006) *Journal of the American Chemical Society*, **128**, 2276.

170 Tomczak, M.M., Glawe, D.D., Drummy, L.F., Lawrence, C.G., Stone, M.O., Perry, C.C., Pochan, D.J., Deming, T.J. and Naik, R.R. (2005) *Journal of the American Chemical Society*, **127**, 12577.

171 Ariga, K., Zhang, Q., Niki, M., Okabe, A. and Aida, T. (2003) *Studies in Surface Science and Catalysis*, **146**, 427.

172 Ariga, K. (2004) *Chemical Record*, **3**, 297.

173 Ariga, K., Aimiya, T., Zhang, Q., Okabe, A., Niki, M. and Aida, T. (2002) *International Journal of Nanoscience*, **1**, 521.

174 Ariga, K., Zhang, Q., Okabe, A., Niki, M., Aida, T., Koga, T., Kojima, Y. and Kinoshita, T. (2004) *Transactions of Materials Research Society of Japan*, **29**, 3179.

175 Zakian, V.A. (1995) *Science*, **270**, 1601.

176 Gurunathan, S., Klinman, D.M. and Seder, R.A. (2000) *Annual Review of Immunology*, **18**, 927.

177 Winfree, E., Liu, F.R., Wenzler, L.A. and Seeman, N.C. (2003) *Nature*, **394**, 539.

178 Seeman, N.C. (2003) *Nature*, **421**, 427.

179 Adleman, L.M. (1994) *Science*, **266**, 1021.

180 Sakamoto, K., Gouzu, H., Komiya, K., Kiga, D., Yokoyama, S., Yokomori, T. and Hagiya, M. (2000) *Science*, **288**, 1223.

181 Fink, H.-W. and Schönenberger C. (1999) *Nature*, **398**, 407.

182 Kasumov, A.Y., Kociak, M., Guéron, S., Reulet, B., Volkov, V.T., Klinov, D.V. and Bouchiat, H. (2001) *Science*, **291**, 280.

183 Fujiwara, M., Yamamoto, F., Okamoto, K., Shiokawa, K. and Nomura, R. (2005) *Analytical Chemistry*, **77**, 8138.

184 Solberg, S.M. and Landry C.C. (2006) *Journal of Physical Chemistry B*, **110**, 15261.

185 Zhang, X., Guan, R.-F., Zhang, F. and Chan, K.-Y. (2006) *Scripta Materialia*, **54**, 1651.

186 Sadasivan, S., Dujardin, E., Li, M., Johnson, C.J. and Mann, S. (2005) *Small*, **1**, 103.

187 Vinu, A., Hossain, K.Z., Kumar, G.S. and Ariga, K. (2006) *Carbon*, **44**, 530.

188 Vinu, A., Hossain, K.Z., Kumar, G.S., Sivamurugan, V. and Ariga, K. (2005) *Studies in Surface Science and Catalysis*, **156**, 631.

189 OConnor, A.J., Hokura, A., Kisler, J.M., Shimazu, S., Stevens, G.W. and Komatsu, Y. (2006) *Separation and Purification Technology*, **48**, 197.

190 Fa, H.-B., Zhao, L., Wang, X.-Q., Yu, J.-H., Huang, Y.-B., Yang, M. and Wang, D.-J. (2006) *European Journal of Inorganic Chemistry*, 4355.

191 Pizzanelli, S., Kababya, S., Frydman, V., Landau, M. and Vega, S. (2005) *Journal of Physical Chemistry B*, **109**, 8029.

192 Che, S., Liu, Z., Ohsuna, T., Sakamoto, K., Terasaki, O. and Tatsumi, T. (2004) *Nature*, **429**, 281.

193 Jin, H., Liu, Z., Ohsuna, T., Terasaki, O., Inoue, Y., Sakamoto, K., Nakanishi, T., Ariga, K. and Che, S. (2006) *Advanced Materials*, **18**, 593.

194 Che, S.(2006) *Journal of Nanoscience and Nanotechnology*, **6**, 1557.

195 Hartmann, M., Vinu, A. and Chandrasekar, G. (2005) *Chemistry of Materials*, **17**, 829.

196 Chandrasekar, G., Vinu, A., Murugesan, V. and Hartmann, M. (2005) *Studies in Surface Science and Catalysis*, **158**, 1169.

197 Guo, Z., Zhu, G., Gao, B., Zhang, D., Tian, G., Chen, Y., Zhang, W. and Qiu, S. (2005) *Carbon*, **43**, 2344.

198 Asaftei, S. and Walder L. (2006) *Langmuir*, **22**, 5544.

199 Galarneau, A., Mureseanu, M., Atger, S., renard, G. and Fajula, F. (2006) *New Journal of Chemistry*, **30**, 562.

200 Salis, A., Meloni, D., Ligas, S., Casula, M.F., Monduzzi, M., Sonias, V. and Dumitriu, E. (2005) *Langmuir*, **21**, 5511.

201 Lee, C.-H., Lang, J., Yen, C.-W., Shih, P.-C., Lin, T.-S. and Mou, C.-Y. (2005) *Journal of Physical Chemistry B*, **109**, 12277.

202 Seelan, S., Sinha, A.K., Kato, K. and Yokogawa, Y. (2006) *Advanced Materials*, **18**, 3001.

203 Kim, J., Lee, J., Na, H.B., Kim, B.C., Youn, J.K., Kwak, J.H., Moon, K., Lee, E., Kim, J., Park, J., Dohnalkova, A., Park, H.G., Gu, M.B., Chang, H.N., Grate, J.W. and Hyeon, T. (2005) *Small*, **1**, 1203.

204 Yu, A., Wang, Y., Barlow, E. and Caruso, F. (2005) *Advanced Materials*, **17**, 1737.

205 Yu, A., Gentle, I., Lu, G. and Caruso, F. (2006) *Chemical Communications*, 2150.

206 Breslow, R.(1991) *Accounts of Chemical Research*, **24**, 159.

207 Ariga, K. and Anslyn E.V. (1992) *Journal of Organic Chemistry*, **57**, 417.

208 Smith, J., Ariga, K. and Anslyn, E.V. (1993) *Journal of the American Chemical Society*, **115**, 362.

209 Kneeland, D.M., Ariga, K., Lynch, V.M., Huang, C.Y. and Anslyn, E.V. (1993) *Journal of the American Chemical Society*, **115**, 10042.

210 Murakami, Y., Kikuchi, J., Hisaeda, Y. and Hayashida, O. (1996) *Chemical Reviews*, **96**, 721.

211 Motherwell, W.B., Bingham, M.J. and Six, Y. (2001) *Tetrahedron*, **57**, 4663.

212 Goettmann, F., Le Floch, P. and Sanchez, C. (2006) *Chemical Communications*, 2036.

213 Luechinger, M., Kienhöfer, A. and Pirgruber, G.D. (2006) *Chemistry of Materials*, **18**, 1330.

214 Dhepe, P.L., Ohashi, M., Inagaki, S., Ichikawa, M. and Fukuoka, A. (2005) *Catalysis Letters*, **102**, 163.

215 Fukuoka, A. and Dhepe P.L. (2006) *Angewandte Chemie-International Edition*, **45**, 5161.

216 Besanger, T.R., Hodgson, R.J., Green, J.R.A. and Brennan, J.D. (2006) *Analytica Chimica Acta*, **564**, 106.

217 Hu, X., Spada, S., Wjite, S., Hudson, S., Magner, E. and Wall, J.G. (2006) *Journal of Physical Chemistry B*, **110**, 18703.

218 Mercuri, L., Carvalho, L.V., Lima, F.A., Quayle, C., Fantini, M.C.A., Tanaka, G.S., Cabrera, W.H., Furtado, M.F.D., Tambourgi, D.V., Matos, J.do R., Jaroniec, M. and Sant'Anna, O.A. (2006) *Small*, **2**, 254.

219 Katiyar, A. and Pinto N.G. (2006) *Small*, **2**, 644.

220 Mal, N.K., Fujiwara, M. and Tanaka, Y. (2003) *Nature*, **421**, 350.

221 Mal, N.K., Fujiwara, M., Tanaka, Y., Taguchi, T. and Matsukata, M. (2003) *Chemistry of Materials*, **15**, 3385.

222 Lai, C.-Y., Trewyn, B.G., Jeftinija, D.M., Jeftinija, K., Xu, S., Jeftinija, S. and Lin, V.S.-Y. (2003) *Journal of the American Chemical Society*, **125**, 4451.

223 Hernandez, R., Tseng, H.-R., Wong, J. W., Stoddart, J.F. and Zink, J.I. (2004) *Journal of the American Chemical Society*, **126**, 3370.

224 Nguyen, T.D., Tseng, H.-R., Celestre, P.C., Flood, A.H., Liu, Y., Stoddart, J.F. and Zink, J.I. (2005) *Proceedings of the National Academy of Sciences, USA*, **102**, 10029.

225 Radu, D.R., Lai, C.-Y., Wiench, J.W., Pruski, M. and Lin, V.S.-Y. (2004) *Journal of the American Chemical Society*, **126**, 1640.

226 Lee, D., Lee, J., Kim, J., Kim, J., Na, H. B., Kim, B., Shin, C.-H., Kwak, J.H., Dohnalkova, A., Grate, J.W., Hyon, T. and Kim, H.-S. (2005) *Advanced Materials*, **17**, 2828.

227 Itoh, T., Yano, K., Inada, Y. and Fukushima, Y. (2002) *Journal of the American Chemical Society*, **124**, 13437.

228 Itoh, T., Yano, K., Kajino, T., Itoh, S., Shibata, Y., Mino, H., Miyamoto, R., Inada, Y., Iwai, S. and Fukushima, Y. (2004) *Journal of Physical Chemistry B*, **108**, 13683.

229 Lu, Y., Yuan, M., Liu, Y., Tu, B., Xu, C., Liu, B., Zhao, D. and Kong, J. (2005) *Langmuir*, **21**, 4071.

230 Lu, Y., Xu, J., Liu, Y., Liu, B., Xu, C., Zhao, D. and Kong, J. (2006) *Chemical Communications*, 785.

231 Astutli, Y., Topoglidis, E., Briscoe, P.B., Fantuzzi, A., Gilardi, G. and Durrant, J.R. (2004) *Journal of the American Chemical Society*, **126**, 8001.

232 Astuti, Y., Palomares, E., Haque, S.A. and Durrant, J.R. (2005) *Journal of the American Chemical Society*, **127**, 15120.

5
Bio-controlled Growth of Oxides and Metallic Nanoparticles

Thibaud Coradin, Roberta Brayner, Fernand Fiévet, Jacques Livage

5.1
Introduction

The remarkable magnetic, opto-electronic and catalytic properties of oxides and metallic nanoparticles (1–100 nm) are obtained by control of the nucleation and growth steps during the reaction. A variety of synthetic methodologies for nanoparticle synthesis with controlled shape and size and also a narrow size distribution are widely known [1–4]. In most cases, nanoparticles are synthesized by "wet chemistry" procedures, where reduction (to form metallic nanoparticles) or forced hydrolysis (to form oxides) proceeds in the presence of a germ (heterogeneous nucleation), a complexing agent (to control the growth step) or a surface capping agent such as carboxylic acids, thiols groups, phosphines and polymers (to avoid nanoparticle agglomeration).

Astonishing advances have been made in the biological sciences since the discovery of the double helix structure of DNA. To improve and to create new revolutionary materials, it seems crucial to combine biotechnology with materials science [5]. Fusing these important disciplines we can generate new advanced materials to solve biological problems. In nature, many situations can be found where living organisms and inorganic phases are in interaction, at scales ranging from nanometer (intracellular pool of metals) to micrometer (mineralized structures of marine organisms), up to human dimensions (bone) [6]. This ability of several living organisms to control the formation of inorganic phases and/or to survive in highly mineralized environments may be used to design new materials. One possibility lies in the understanding of biomineralization processes and its *in vitro* transcription to obtain biomimetic materials [7]. In parallel, it is possible to use the intrinsic self-organization properties and metal-binding capacities of cellular extracts, such as polysaccharides [8,9], and proteins/peptides [10,11], to synthesize non-biogenic (i.e., not formed naturally by living organisms) nanoparticles.

The scope of this chapter is to review such bio-controlled syntheses of metallic and oxide nanoparticles using biological molecular and macromolecular templates. The possible use of polymer templates to organize pre-formed nanoparticles has been

Bio-inorganic Hybrid Nanomaterials. Edited by E. Ruiz-Hitzky, K. Ariga, Y. M. Lvov

ISBN: 978-3-527-31718-9

recently reviewed [12] and will not be addressed here. We aim to demonstrate how it is possible to take advantage of the diversity of chemical functionality and structural features of biological systems to prepare a wide range of novel nanocomposites. In particular we wish to emphasize that successful approaches rely on one's ability to select the most suitable bio-template(s) to control the growth of a given inorganic phase and, conversely, to envision the metal/oxide materials that are more likely to benefit from the presence of a specific template. Such a "fit" is indeed dependent on the intrinsic properties of each "partner" and we have tried to provide some key information on biological template structure and properties as an introduction to each of the following sections. Moreover, the selection of the most appropriate systems should also take into account the targeted properties/applications of the resulting nanomaterials. Thus, we have tried to present several examples of bio-controlled functional nanocomposites, especially in the field of biotechnology and medical devices.

As a perspective to this review, we conclude with recent developments in the use of whole cells to control the growth of nanoparticles. Such approaches should allow one to take advantage of the whole "cellular machinery" and, combined with genetic engineering, appear very promising for the development of a "green" nanochemistry.

5.2
Biomimetic Approaches

As mentioned earlier, biological systems have developed optimized strategies to design materials with elaborate nanostructures [6]. A straightforward approach to obtaining nanoparticles with controlled size and organization should therefore rely on so-called biomimetic syntheses where one aims to reproduce *in vitro* the natural processes of biomineralization. In this context, a first possibility is to extract and analyze the biological (macro)-molecules that are involved in these processes and to use them as templates for the formation of the same materials. Such an approach has been widely developed for calcium carbonate biomimetic synthesis [13]. In the case of oxide nanomaterials, the most studied system so far is the silica shell formed by diatoms [14].

Diatoms are unicellular brown algae that build up a silica shell exhibiting a striking hierarchically-organized porosity [15] (Figure 5.1). The process of silica formation occurs within a specific vesicle where silica precursors and templating macromolecules interact in a still largely unknown manner, yielding synthetically-unchallenged silica morphologies. The first extractions of macromolecules from diatom cells were performed using mild chemical conditions that did not dissolve the silica material. Therefore, the recovered molecules were only weakly associated with the mineral network and should not be directly involved in the mineralization process [16]. This assumption was confirmed by *in vitro* experiments that showed that addition of these molecules to silica precursor solutions does not induce their condensation. In a further step, dissolution of the silica shell using hydrofluoric acid (HF) allowed the recovery of proteins called silaffins (for silica affinity),

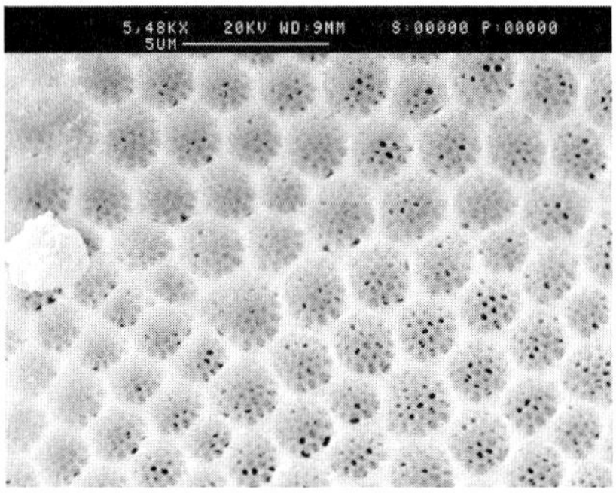

Fig. 5.1 SEM micrograph of the silica shell of a *Coscinodiscus* sp. diatom cell.

exhibiting a poly-cationic structure [17]. As these proteins are trapped in the mineral network, they are very likely to take part in the silicification process. In fact, it was shown that when added to aqueous solutions of pre-hydrolyzed tetraethoxysilane $Si(OEt)_4$ (TEOS), silaffins induced the formation of silica as aggregated nanoparticles. A variation in particle size was observed, depending whether purified proteins or protein mixtures were used. Further studies showed that the extraction protocol could be optimized to avoid the partial degradation of the proteins due to the hydrolytic activity of HF. This allowed the recovery of silaffins as phosphorylated and/or sulfated species [18]. In this case, the proteins are not poly-cations but exhibit a zwitterionic character, and present self-assembling properties in solution. The possibility for silaffins to be in fact glycoproteins (i.e., to bear some sugar moieties) was also mentioned. As a result of such post-transcriptional modifications, these macromolecules not only control the growth of silica nanoparticles but also have some influence on their organization into porous networks that bear some resemblance to diatom shells [19]. In addition to silaffins, poly-amines could also be extracted from many different diatom species [14]. The species-dependent variation in poly-amine chain length was recently shown to correlate nicely with the corresponding silica shell pore diameter [20]. In the presence of TEOS, these poly-amines have the ability to precipitate silica nanoparticles, whose size can be tailored by addition of phosphate ions. Finally, it was possible to identify a gene that is involved in the synthesis of silaffins and to obtain peptide fragments, termed pRn ($n = 1$–5), that can be considered as silaffin precursors [17]. The pR5 peptide was also shown to induce silica nanoparticle formation.

Overall, silaffins and extracted poly-amines are able to control the growth of silica nanoparticles of well-defined sizes and, to a lesser extent, to exert some morphological templating. However, these molecules are not easy to extract and can only be recovered in small amounts. Therefore, many studies have been devoted to the use of synthetic poly-amines or poly-cations to control the growth of silica nanoparticles [21,22]. A huge amount of literature can be found on the use of poly-lysine as a biomimetic template. The effect of the polymer chain length, the pH and stirring conditions and the nature of the silica precursors have been studied, leading to a wide

Fig. 5.2 SEM micrograph of a silica nanoparticles assembly resulting from the interaction between silicates and gelatine.

range of silica nanostructures [23]. Poly-arginine, poly-ethylene imine and biomimetically-designed poly-amines showed similar properties. These studies allowed the authors to propose a model of silica formation control via electrostatic interactions between cationic ammonium groups and anionic silicate species, favoring their condensation [24]. Amine-terminated dendrimers were found to exhibit a phosphate-dependent ability to control the size of silica nanoparticles, in good analogy with diatom poly-amines [25]. Poly-cationic proteins, such as lysozyme and gelatine, also allow the formation of colloidal silica [26,27] (Figure 5.2). In the latter case, it was shown that the size of these colloids can be varied from a few tens of nanometers to several microns by changing the gelatine concentration and/or the physical state (i.e., solution, thin films, bulk gel) [28]. This was attributed to the self-assembling properties of the protein that lead to a confinement of the silica growth process in the array of aggregated polymer chains.

In fact, such biomimetic molecules demonstrate the ability to tailor the growth of silica nanoparticles in a way that is very similar to diatom-extracted species. However, they demonstrate the same limitations in terms of morphological control of nanoparticle assembly. This is because the diatom shell architecture results not only from interactions of silica precursors with templating molecules but also benefits from a cell-driven molding of the vesicular compartment where silicification occurs [29]. Thus, it is very likely that diatom-like synthetic silica will only be achieved when such confinement/molding effects are taken into account in the design of biomimetic experiments [30].

Nevertheless, it is worth examining at this stage whether such bio-templated nanoparticles may have practical applications. In fact, silica nanoparticles are nowadays produced on a large scale using the well-known Stöber process [31]. This process is based on the hydrolysis/condensation of silicon alkoxides in ethanol/ammonia mixture and allows easy access to a wide range of colloid sizes. When compared to this method, the main advantage of biomimetic approaches would, in principle, lie in the use of conditions that follow the actual concerns for "green" processes (no organic solvents, bio-renewable reagents (especially if sodium silicate is used as the silica precursor instead of alkoxides), room-temperature conditions).

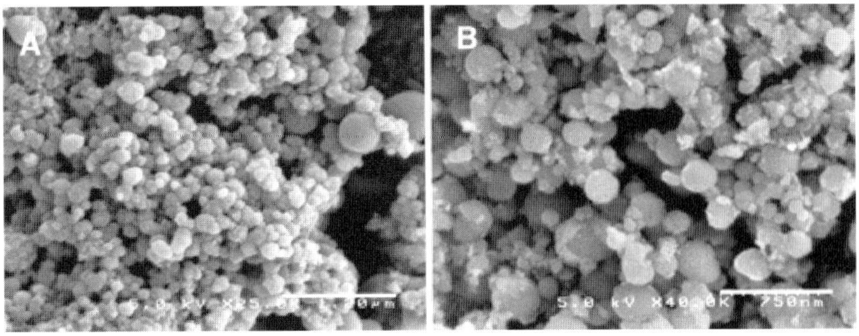

Fig. 5.3 SEM micrographs of TiO_2 nanoparticles obtained in the presence of pR5 (scale bar = 1.2 μm) (a) and poly-L-Lysine (scale bar = 750 nm). (b) (Reprinted with permission from [36]. Copyright (2006) American Chemical Society).

However, it has to be realized that biological templates remain inserted in the final nanoparticles and this is not acceptable for many applications. Nevertheless, some recent examples indicate that such biomimetic materials may be suitable for the design of biotechnological and medical devices [32]. For instance, it was shown that silica gels formed in the presence of p-R5 were excellent host matrices for enzyme encapsulation [33]. In parallel, biopolymer/silica hybrid macro-, micro- and nanocapsules were recently obtained via biomimetic routes and these exhibit promising properties for the design of drug delivery materials (see Section 3.1.1) [34,35].

In parallel, the possibility of using diatom-related macromolecules to elaborate inorganic materials differing from silica has been studied. In this context, the p-R5 peptide was shown to allow the elaboration of TiO_2 nanoparticles from a synthetic Ti (IV) complex [36] (Figure 5.3). Interestingly, it was found that poly-L-lysine has a similar ability, suggesting that the control of titanium oxide formation occurs in a way that is very similar to the proposed mechanism for silica nanoparticle formation. However, the success of this approach is largely due to the design of a suitable Ti precursor. In fact, due to their significant positive charge, silaffins and natural/synthetic poly-amines are expected to interact very efficiently with most metal alkoxides that are the traditional precursors of oxide phases in the sol–gel processes, presenting the possibility for control of their precipitation. For instance, germanium oxide is the only other example of a non-silica material reported so far whose formation from aqueous salts could be controlled by poly-lysine, probably due to the strong similarity of SiO_2 and GeO_2 solution chemistry [37].

In contrast, such approaches have been much more developed with proteins, termed silicateins, that have been extracted from some silicified sponges [38]. The success of these approaches probably originates from the fact that the reactivity of these proteins towards silica precursors differs significantly from the processes occurring in diatoms. Whereas silaffins and poly-amines activate silica formation via electrostatic interactions due to the presence of positively-charged ammonium

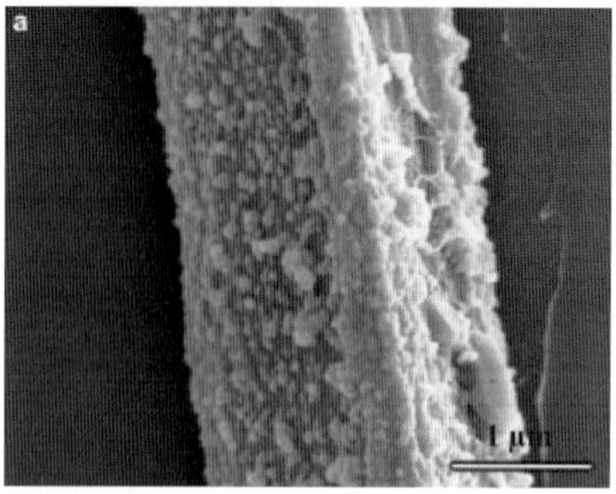

Fig. 5.4 SEM micrograph of Ga_2O_3-coated silicatein [43]. (Reproduced by permission of John Wiley & Sons).

groups, one of the extracted proteins, called silicatein α, exhibits an active-site-like amino acid triad containing only neutral groups (serine, asparagine, histidine). In this case, it was proposed that the protein influences silicon alkoxide hydrolysis, via nucleophilic activation of the histidine nitrogen, rather than condensation [39]. To support this assumption, it was shown that silicatein α does not activate silica formation from aqueous silicate solutions. Moreover, it was observed that silicatein addition to TEOS did not lead to the formation of well-defined silica nanoparticles, in contrast to diatoms or diatom-like macromolecules.

Small bifunctional molecules [40] and diblock copolypeptides [41] inspired by silicatein α were shown to activate silica formation and to control its morphology. Even more interesting results on these systems were obtained for other inorganic materials. Nanostructured TiO_2 and Ga_2O_3 could be obtained using silicatein α itself, using alkoxide-like molecular precursors [42,43] (Figure 5.4). This protein was also shown to catalyze the hydrolysis of $BaTiF_6$ to form a crystalline perovskite-like $BaTiOF_4$ phase, an intermediate in the formation of the piezo-electric $BaTiO_3$ structure, that could only previously be obtained in acidic or high pressure/temperature conditions [44]. To demonstrate further the great potential of this approach, a recombinant silicatein α was also fused with a peptide fragment, allowing it to anchor on the *Escherichia coli* bacteria cell surface [45]. In the presence of a Ti complex dissolved in the phosphate-containing cell culture medium, titanium phosphate nanocrystals, usually obtained at elevated temperatures, were formed on the bacteria surface.

In summary, it appears that the use of macromolecules extracted from biomineralizing organisms is a suitable approach to understanding how these organisms build up their associated inorganic phases. However, neither these molecules nor their biomimetic synthetic equivalents allow the replication of the natural materials, as molecular templating is only one of the multiple tools that living organisms use to control the mineral phase growth. It is nevertheless possible to profit from these two kinds of macromolecules to elaborate materials either constituted of the same inorganic phase or of some others, that may have no natural equivalent or are not accessible in mild reaction conditions. In the latter case, this is only possible if the mineral precursors have a similar reactivity or are designed to interact favorably with the templating molecules. In parallel, the recent developments in genetic engineering, together with actual exploration of the biomineralizing organism's genome, open the possibility of modifying the protein structure and functionality so that it may be adapted to the controlled growth of many different inorganic phases. This last approach would in fact represent a kind of "reverse" biomimetic approach as the protein modifications needed for a given material would probably partially replicate the actual strategies involved in the design of synthetic templates adapted to this specific phase.

5.3 *In vitro* Synthesis of Hybrid Nanomaterials

5.3.1 Polysaccharides

Natural polysaccharides such as alginates, carrageenans and chitosan (chitin derivative) have some important properties such as mucoadhesivity, biocompatibility and nontoxicity, which render them interesting biomaterials. From a physicochemical point of view, these polysaccharides have the special ability to gel upon contact with cations (alginate and carrageenan) and anions (chitosan) under very mild conditions and may therefore interact with a wide range of metal precursors, opening the route to the design of bio-nanocomposites.

5.3.1.1 Alginates

Alginates are a family of unbranched polysaccharides consisting of 1 → 4 linked, β-D-mannuronic acid (M) and α-L-guluronic acid (G) residues arranged in a block-wise, non-regular order along the chain (Figure 5.5A). They are produced by a number of microorganisms including *Pseudomonas aeruginosa*, *Azotobacter vinelandii* and *Azotobacter crococcum* and are the principal component of the cell walls of several brown macro-algae (Phaeophyceae) [46]. Depending upon their source, they can have different chemical (e.g., complexation) or physico-chemical (e.g., biopolymer rigidity) nature due to the relative proportions and sequences of the (M) and (G) residues [47,48]. The M/G ratio and their macromolecular conformation determine also the physical properties and the affinity of the alginate for divalent or trivalent cations. Poly-guluronic acid contains two diaxially linked α-L-guluronic acid residues in the chair form, producing a rod-like conformation, whereas poly-manuronic acid forms a flat ribbonlike chain and contains two diequatorially linked β-D-mannuronic acid residues in the chair form [46]. The difference in conformation between the two homopolymeric blocks may be responsible for their strong but variable affinity for divalent and trivalent cations. Another feature of these polymers lies in the very special process leading to gel formation. Alginates are able to adopt an ordered solution conformation through dimerization of the poly-G sequences in the

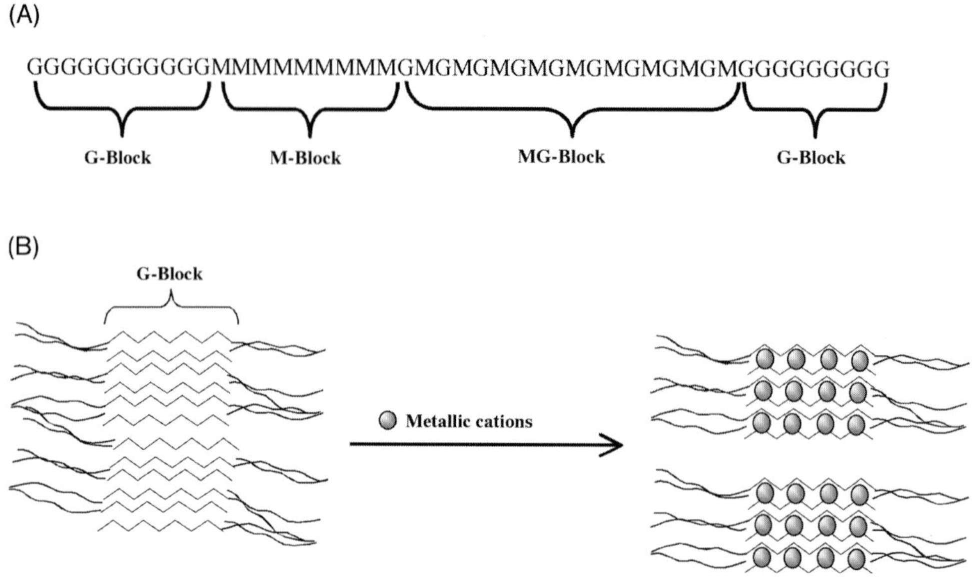

Fig. 5.5 (A) Alginate block copolymer structure with random sequences; (B) divalent cations induced gelation of alginate (formation of "egg-box" structure).

presence of Ca^{2+} or other divalent or trivalent cations, and this description is known as the "egg-box" model (Figure 5.5B) [47]. Dimerization regions are terminated by chain sequences of poly-M or mixed poly-MG and several different chains may be interconnected, promoting gel network formation. Thus, the gelation process involves the formation of ordered domains in which the polymer chain network defines a confined space for cation localization.

Such confinement is a key feature in numerous biomineralization processes and allows better control of particle growth, as demonstrated for Ni, Co and NiCo metallic nanoparticles synthesized inside different alginate matrixes [49,50]. The ability of three different sources of alginate (i.e., with different M/G ratio) to control the formation of Co, Ni and CoNi nanoparticles has also been studied. Variations in particle size, structure and magnetic properties were observed, depending on the alginate composition [50]. The more stable gels allowed the formation of Co_xNi_{100-x} nanoalloys over the whole solid solution domain (Figure 5.6). Moreover, these networks stabilize Ni nanoparticles in an unusual hexagonal compact phase (hcp). These results could be related to the strength of the metal-alginate gels and hence to the metal ion coordination within the polymer network. This suggests that the "egg-box" cavity intrinsic anisotropy may favor the growth of hcp platelets. In addition superconducting quantum interference device (SQUID) measurements indicated that hexagonal Ni metallic nanoparticles show superparamagnetic behavior

(A)

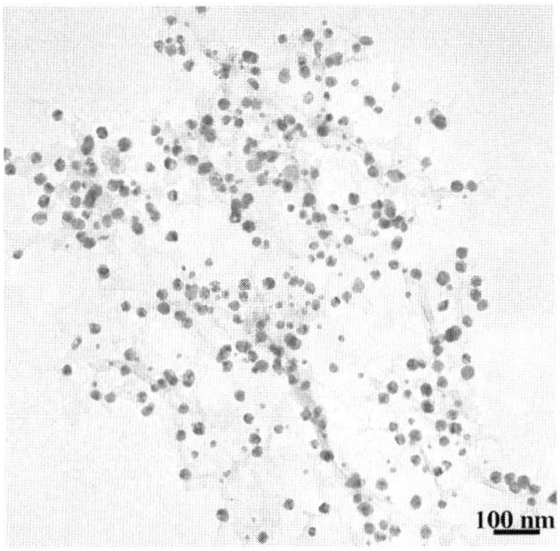

(B)

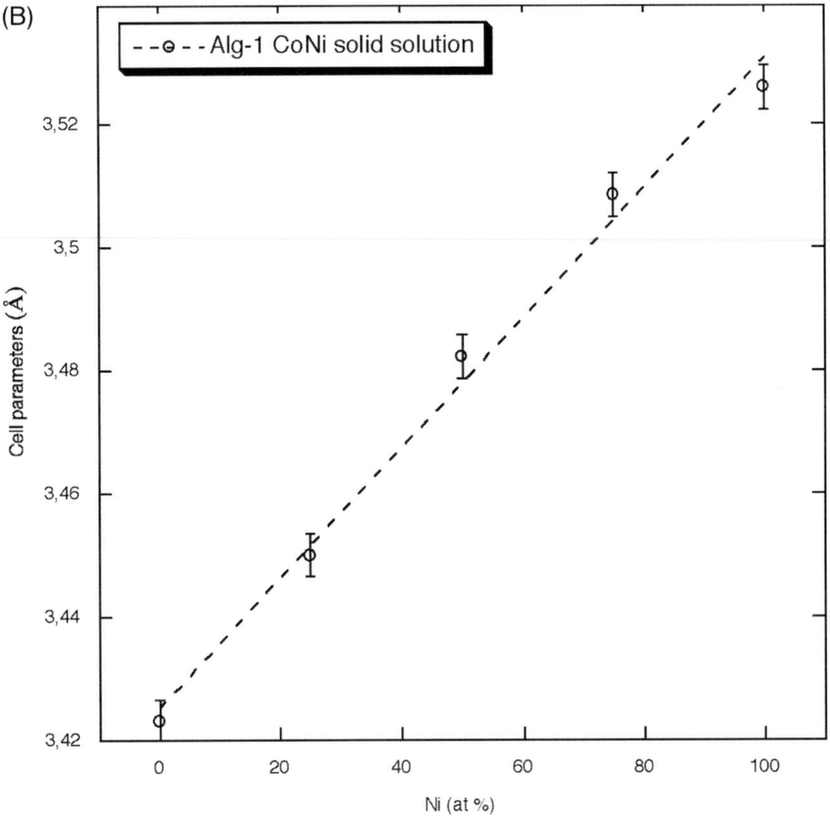

Fig. 5.6 (A) TEM micrograph of alginate-$Ni_{50}Co_{50}$ nanocomposites; (B) correlation between alginate-NiCo solid solution composition and face-centered cubic crystal parameters. (Adapted from [50]).

corresponding to the presence of a magneto-crystalline anisotropy [50]. The alginate polymeric matrix allows the control of nucleation and the growth steps during nanoparticle formation. In addition, the metallic nanoparticles were obtained after reduction under flowing H_2/N_2 at 350 °C and, in this case, alginate degradation was not observed. These new hybrid materials may find application, for example, in ultrahigh density magnetic media devices.

The application of small particles in *in vitro* diagnostics has been practised for many years due to a large number of beneficial factors including a large surface area/volume ratio, and the possibility of ubiquitous tissue accessibility. Recently, it has been possible to produce, characterize and specifically tailor the functional properties of magnetic nanoparticles for biomedical applications. Typical magnetic particles studied are composed of a magnetic core (usually magnetite Fe_3O_4, or maghemite γ-Fe_2O_3) with a biocompatible coating [51,52]. In this context, a new family of nanocomposites, named HYbrid MAgnetic Carriers (HYMAC), consisting of biopolymer/silica nanoparticles incorporating magnetic colloids, was recently proposed [53]. As a first step, alginate/silica hybrid nanomaterials could be easily obtained using a spray-drying technique [34]. These nanocomposites showed an enhanced thermal stability when compared to their biopolymer equivalents. They exhibit a homogeneous silica/alginate structure. As a consequence, the *in situ* reduction of Co^{2+} ions leads to the intra-capsular formation of magnetic cobalt silicate nanoparticles. When added to fibroblast cells, nanoparticles enter the intracellular space where they are degraded, suggesting that these new nanocomposites should be suitable for targeted drug delivery applications (Figures 5.7 and 5.8) [34].

First attempts to incorporate pre-formed magnetite colloids within alginate/silica nanocomposites via a spray-drying process have been described, but formation of lepidocrocite γ-FeOOH and fayalite Fe_2SiO_4 was observed, attributed to Fe^{2+} release during the aerosol thermal treatment [53].

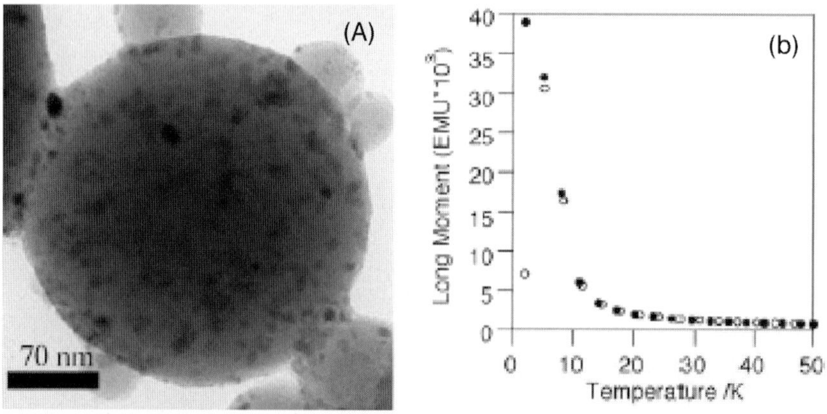

Fig. 5.7 (A) TEM micrograph of silica–alginate–Co hybrid nanocomposites and (B) corresponding Zero Field Cool-Field Cool (ZFC/FC) susceptibility curves ($H = 500$ Oe) indicating superparamagnetic behavior (Adapted from [34]).

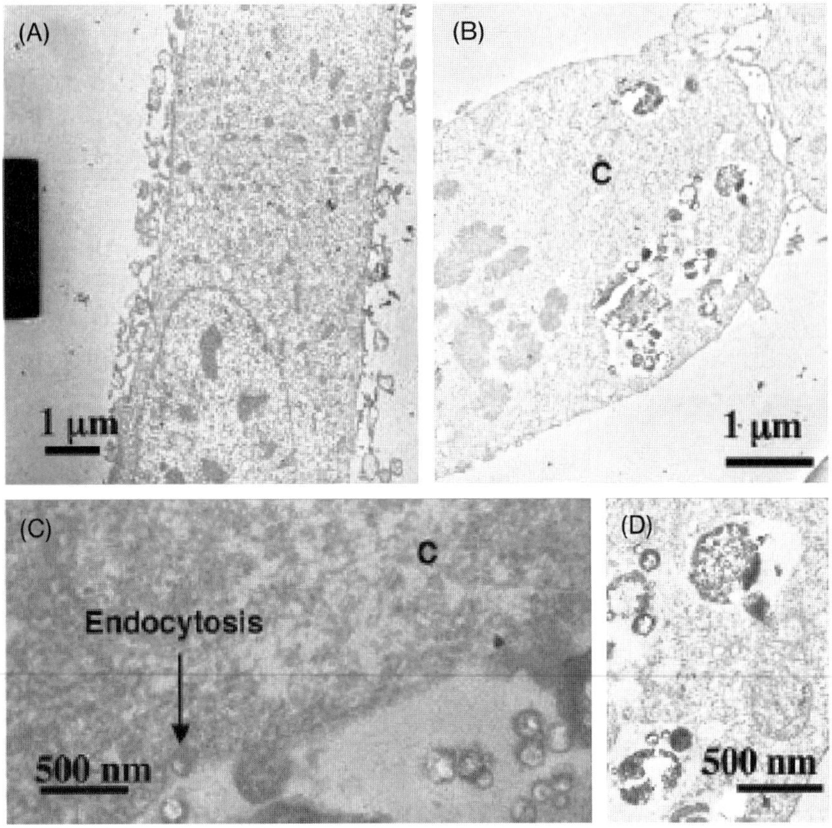

Fig. 5.8 TEM micrographs of fibroblast cells cultured in the absence (A) or presence (B–D) of silica/alginate nanocomposites (Adapted from [34]).

5.3.1.2 Carrageenans

Carrageenan is a collective term for polysaccharides prepared by alkaline extraction and modification from red macro-algae (Rhodophycae), mostly of genera *Chondrus, Eucheuma, Gigartina* and *Iridaea*. Different genera produce different carrageenans. They consist of alternating 3-linked-β-D-galactopyranose and 4-linked-α-D-galactopyranose units.

The ability of κ- and ι-carrageenan (Figure 5.9) to form higher order structures in solution, which sometimes occurs as part of a gelation process, has been investigated by several authors [54,55]. In contrast to alginates, the physical shape and strandedness associated with the ordered conformations have been, and still are, a matter of debate. Salt-induced conformational ordering of κ-carrageenan is usually accompanied by aggregation or gelation. However, by using iodide salts of non-gelling cations [e.g., Li^+, Na^+, $N(CH_3)_4{}^+$], an ordered conformation may be induced without further aggregation, due to hydrophobic interactions with κ-carrageenan double helices [56].

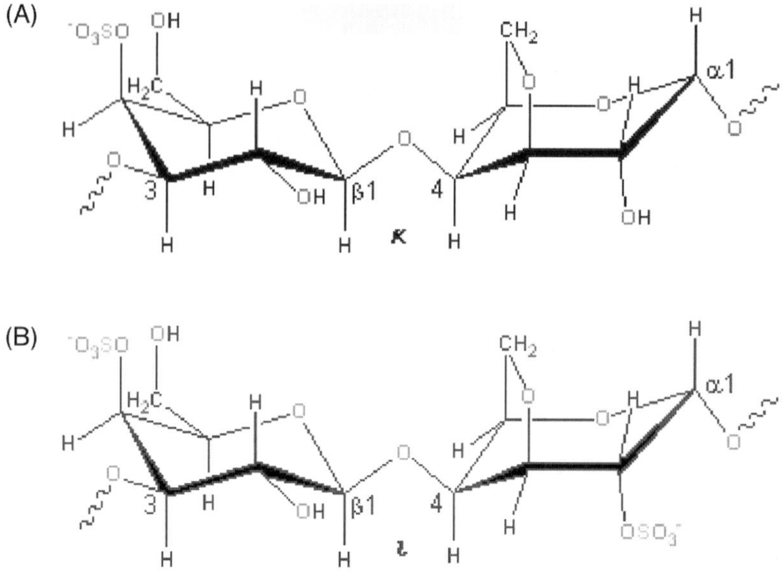

Fig. 5.9 Disaccharide repeating units in (A) κ-carrageenan and (B) ι-carrageenan.

Carrageenans and alginates present different conformations: "egg-box" structure (alginates) and double helices (carrageenan) but both natural biopolymers are able to form gels and consequently, to control nanoparticle growth.

Cölfen *et al.* showed that the functional polysaccharide κ-carrageenan acts as an efficient stabilizer during Fe, Ni and Co hydroxide nanoparticle formation up to very high pH values [57,58]. In this case, neutralization of iron salts in aqueous solutions of κ-carrageenan results in iron oxyhydroxide-polysaccharide hybrid nanocomposites with unusual pH stability up to pH 13 (Figure 5.10). It has been shown that κ-carrageenan forms a tight polymer layer (cross-linked by helical domains) surrounding the inorganic particles, forming a self-assembled nanoreactor that avoid particle agglomeration. The stabilized iron oxyhydroxide particles can undergo further chemical reactions to produce stabilized magnetite particles [57]. Repetition of the loading/neutralization steps in the reaction results in hybrids with iron contents much higher than the stoichiometric balance of iron and functional groups of the polymer (greater than 100 % Fe/SO_4^{2-}). This combination of high iron content with a natural polysaccharide stabilizer makes these colloids interesting for a number of applications, for example, for nutritional purposes or as contrasting agents for tomography.

It is also possible to synthesize metallic nanoparticles with κ- and ι-carrageenan [59]. Both Ag and Au nanoparticles have considerable potential for biochemical analysis [60]. The advantage of Ag nanoparticles is that the range of dyes which remain effective in biological media is much more extensive. On the other hand, in some biological systems, such as cell suspensions, Ag can react positively with the cell and it is well know as a bactericide. Au, Ag and Cu nanoparticles have the ability to

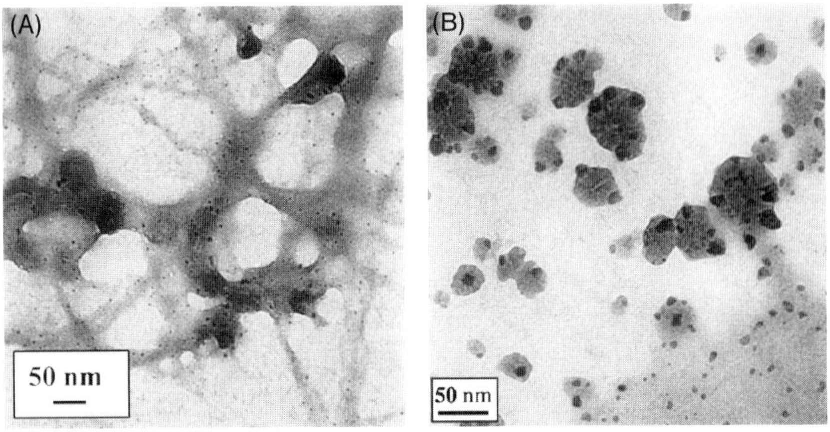

Fig. 5.10 TEM micrographs of gel and nanoparticle formation of iron-loaded κ-carrageenan at (A) pH 2 and (B) pH 13. In both cases the iron loading was 100%. ([57]. Copyright Springer-Verlag 2000. With kind permission of Springer Science and Business Media).

resonantly scatter visible and near-infrared light upon excitation of their surface plasmon oscillation. The scattering light intensity is extremely sensitive to the size and aggregation state of the nanoparticles [61]. Raman and SERRS techniques have the same sensitivities [62]. SERRS can be used, for example, for labeling both fluorophores and non-fluorophores. To test the sensitivity of the assay, the detection of 3,5-dimethoxy-4-(6′-azobenzotriazolyl)-phenylamine anchored on an Ag nanoparticle surface was studied, indicating a detection limit of about 10^{-10} M [63]. Carrageenan biopolymers have also been used as a template and as reduction agent for the synthesis of Ag metallic nanoparticles. In this case, homogeneous nanoparticles with a narrow size distribution were obtained inside κ- and ι-carrageenan biogels (Figure 5.11A) [59]. These biogels were then immersed in a dopamine neurotransmitter aqueous solution (used as a probe molecule) at a concentration of 10^{-5} M. An important SERRS effect for a dopamine molecule adsorbed on an Ag surface inside a carrageenan matrix was observed (Figure 5.11B). This polyfunctional hybrid material may be applied in bio-detection applications and also in drug delivery because carrageenans have a great protein loading capacity.

5.3.1.3 Chitosan

Chitin, poly-β-(1,4)-*N*-acetyl-D-glucosamine, is a cellulose-like biopolymer widely distributed in nature, especially in marine invertebrates, insects, fungi, and yeasts [64]. Chitin also has unique properties, including toughness, bioactivity, and biodegradability. Chitosan is a natural nontoxic biopolymer derived by the deacetylation of chitin. It is a copolymer of glucosamine and *N*-acetylglucosamine (Figure 5.12). Chitosan and its derivatives have attracted considerable interest due to their antimicrobial and antifungal activity [65].

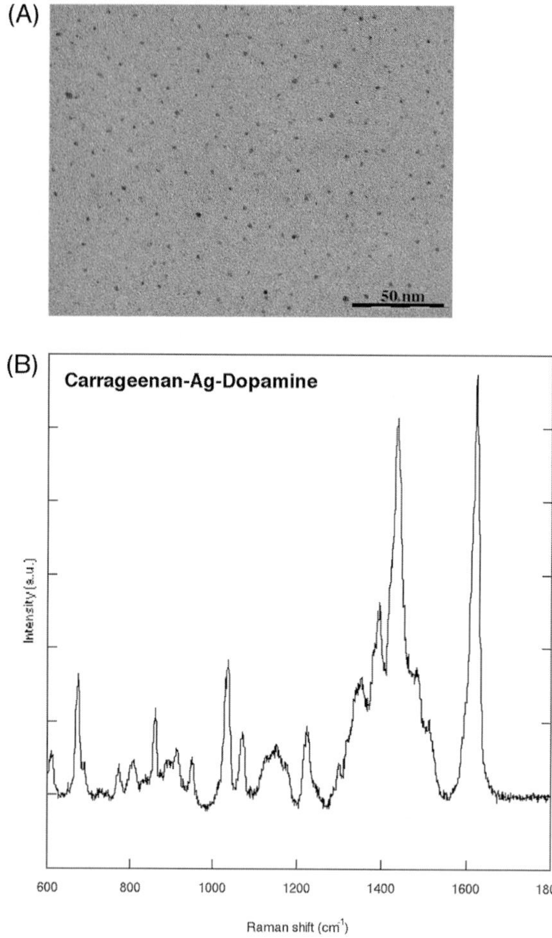

Fig. 5.11 (A) TEM micrograph of Ag-ι-carrageenan biogel; (B) Raman spectra of dopamine excited under Ag nanoparticles (SERRS effect).

Chitosan is insoluble at alkaline and neutral pH values but it can form salts with inorganic and organic acids such as hydrochloric acid, lactic acid, acetic acid and glutamic acid. In contrast to alginates and carrageenans in solution, the amino functions of chitosan are protonated and the resultant soluble polysaccharide is

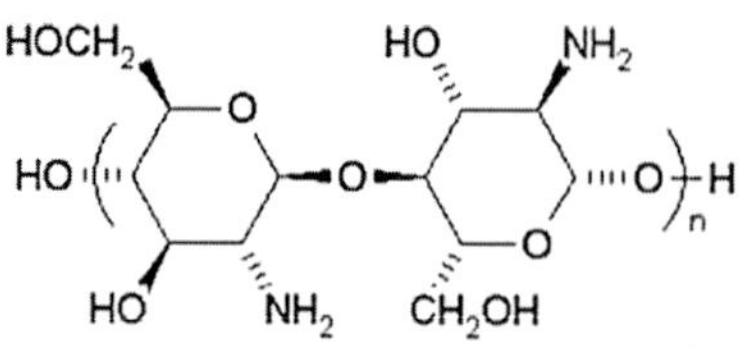

Fig. 5.12 Schematic representation of the chitosan molecule.

positively charged. Its cationic nature gives chitosan the ability to form complexes with various polyanions. For example, it interacts and easily forms complexes with negatively-charged DNA. Chitosan–DNA nanoparticles generally transfect cells more efficiently than naked DNA, but less than commercial liposome formulations [66]. It has been suggested that the strength of the interaction between chitosan and DNA results in highly stable particles, thereby preventing dissociation within the cell and ultimately precluding translation of the DNA, resulting in low transfection [67]. Another interest for chitosan is stimulated by possible applications in the field of self-organizing and growth step control of nanosized materials. Chitosan nanoparticles with magnetic properties can be potentially used as drug delivery carriers and separation materials. Zhi *et al.* showed the *in situ* preparation of magnetic chitosan/Fe_3O_4 nanocomposite materials with tiny water pools of water-in-oil microemulsion containing chitosan and ferrous salt as microreactors by adding NaOH to the microemulsion [68]. In this case, cubic-shaped magnetic Fe_3O_4 nanoparticles inside spherical chitosan capsules were obtained (Figure 5.13). The Fe_3O_4 particles vary in size from 10 to 80 nm, depending on the molecular weight of the chitosan. They present superparamagnetic behavior with a saturation magnetization higher than pure Fe_3O_4 nanoparticles. This hybrid material can be used in separation processes for nano-absorbent ions, such as Cu^{2+} and Co^{2+} [69].

Au–chitosan composites present antioxidant potential. All living organisms suffer from the damage caused by free-radical oxygen species which damage cells by attacking unsaturated fatty acids in the cell membrane. Fortunately, the superoxide dismutase enzyme can totally convert these free-radical species into two water molecules and oxygen. However, the enzyme activity decreases with organism aging, so Au–chitosan nanocomposites can act as a catalyst for the elimination of active radical species [70]. The particle size ratio and catalytic activity vary with chitosan concentration (Figure 5.14). This hybrid nanomaterial exhibits a catalytic activity 80

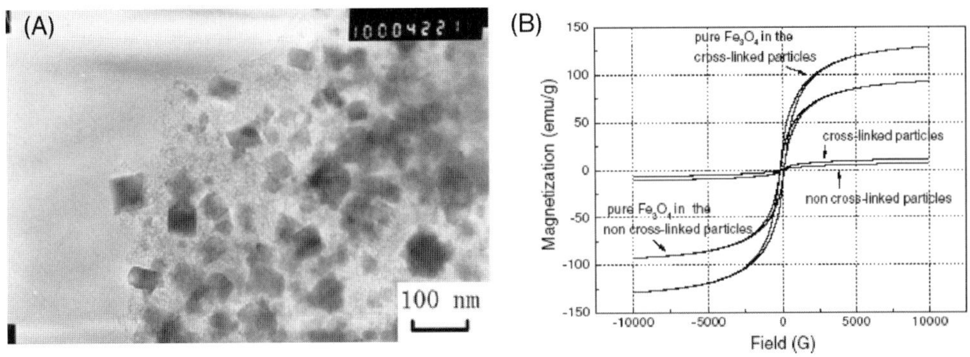

Fig. 5.13 Chitosan magnetic nanoparticles: (A) TEM micrograph, (B) magnetization hysteresis loops for chitosan and magnetite-loaded chitosan particles. (Reprinted from [68], Copyright (2006), with permission from Elsevier).

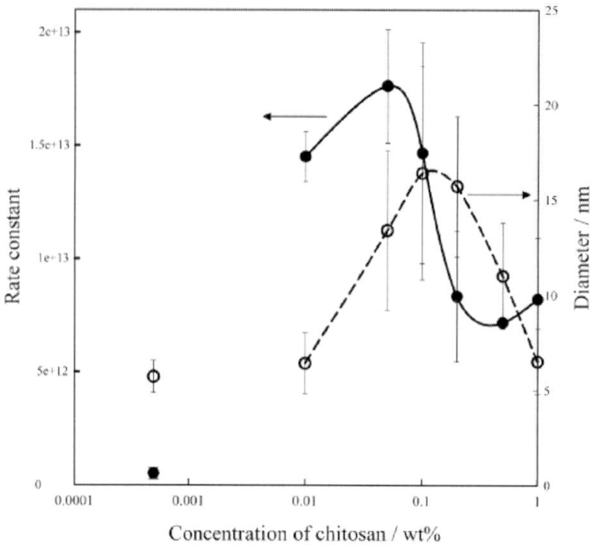

Fig. 5.14 Plots of rate constant of catalytic activity and diameter of Au nanoparticles versus the concentration of chitosan added. (Reprinted from [70], Copyright (2003), with permission from Elsevier).

times higher than that obtained with ascorbic acid, which is well-known as a powerful antioxidant.

5.3.2 Proteins

Traditional methods for fabricating nano-scaled arrays are usually based on lithographic techniques. Alternative new approaches rely on the use of self-organizing templates. Due to their intrinsic ability to adopt complex and flexible conformations, proteins have been used to control the size and shape, and also to form ordered two-dimensional arrays of nanoparticles. The following examples focus on the use of helical protein templates, such as gelatin and collagen, and protein cages such as ferritin-based molecules.

5.3.2.1 Gelatin

Gelatin aqueous solutions have long attracted chemists, physicists and food scientists, not only because of their wide range of applications in such as food, cosmetics, photography, pharmacology and glues, but also as a model system for the sol–gel transition [71,72]. Gelatin is a mixture of α (single helix), β (double helix) and γ (triple helix) obtained through collagen denaturation and contains a large number of glycine (almost one in three residues, arranged every third residue), proline and 4-hydroxyproline residues (Figure 5.15) [73,74].

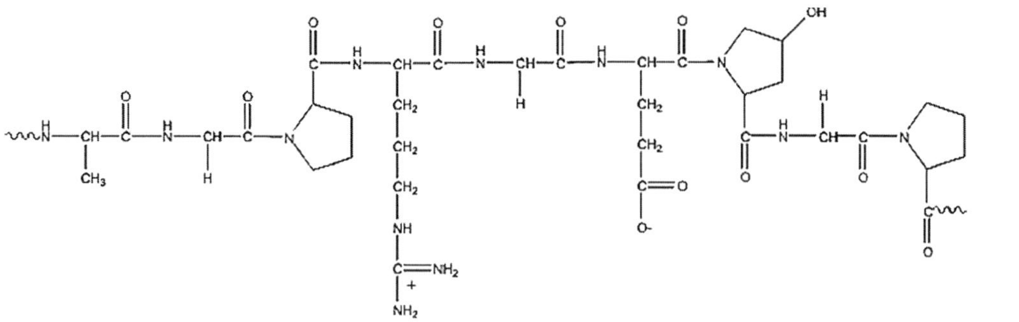

Fig. 5.15 Schematic structure of gelatine.

The interactions between dissolved ions and specific surfaces affect not only the particle size and habit of nucleating crystals but also the stability of intermediate phases by dropping the activation energy required for the formation of specific crystal faces [75]. The organic templates nucleate the mineral by controlling its crystallographic orientation and growth, by imitating the lattice of a two-dimensional (2D) face, or by the functional group stereochemistry at the interface [76]. Bauermann *et al.* showed that the presence of dissolved gelatin in the sol state during the crystallization of ZnO has an important influence on its morphology [77]. In this case, gelatin behaves as a matrix where the ZnO is heterogeneously nucleated. By the presence of dissolved gelatin (1.2 mg mL^{-1}), the morphology of ZnO is drastically changed to twinned hexagonal plates with (002) faces about 460 nm diameter and 40 nm thick (Figure 5.16).

Hybrid Au– and AuNi–gelatin nanocomposites have also been synthesized in mild conditions at 4 °C with 0.3 g L^{-1} of gelatin and initial pH 5 [78]. For the Au–gelatin sample, the gelatin 3D network was maintained, presenting Au close-packed linear assemblies along gelatin filaments forming a spider's-web-like network (Figure 5.17). In addition, AuNi–gelatin nanocomposites are ferromagnetic, presenting coercivity values (H_c) between 450 and 650 Oe.

5.3.2.2 Collagen

Type I collagen, more commonly designated as collagen, is the most common protein involved in the organic scaffold of mineralized connective tissues and therefore collagen-based composite materials may find application in biomedical devices [79,80]. The collagen triple helix is composed of three poly-(proline)-II strands that are held together by interchain hydrogen bonds [81]. This structure is similar to the double-helical structure of DNA. Both collagen and DNA exhibit melting transitions that reflect the stability and strength of the multiplex assemblies. Strand invasion in DNA by short DNA strands or peptide nucleic acids is well documented [82]. Although collagens are known to incorporate thermally unstable domains where small segments of the triple helix are thought to be partially unraveled [83], strand invasion by other collagen molecules or collagen analogues has not been reported to date. Thus, during the mineralization process of bone

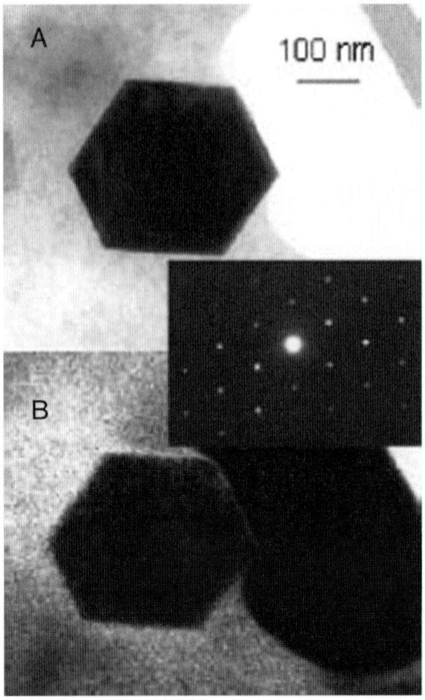

Fig. 5.16 (A) Bright-field TEM image and (B) element mapping carbon (brighter contrast corresponds to higher concentration of carbon) of ZnO synthesized in aqueous solution at 37 °C in pH 8 buffer for 4 h in the presence of 1.2 mg mL^{-1} of gelatin. The inset shows the electron diffraction pattern taken parallel to the platelet normal. (Reprinted with permission from [77]. Copyright (2006) American Chemical Society).

tissue, collagen is first synthesized, extruded from the cell, and then self-assembled in the extra-cellular space before hydroxyapatite precipitation occurs in the voids of the protein network [80].

Kim *et al.* showed for the first time the growth of hydroxyapatite on self-assembled Au–collagen nanoparticles by employing wet chemistry at ambient conditions [84]. In this case, collagen was immobilized on Au nanoparticles by an *in situ* chemical reduction process (Figure 5.18 A). Briefly, 2 mL of collagen solution was homogenized with 150 μL of 0.01 M aqueous $HAuCl_4$ solution for 1 h followed by addition of 60 μL of 1.0 M $NaBH_4$ in 0.3 M NaOH dropwise with vigorous stirring. The solution became fairly pink after 24 h stirring with the appearance of Au plasmon resonance (Figure 5.18B). This hybrid Au–collagen material was used for *in vitro* mineralization of hydroxyapatite in mild conditions. The possible mechanism of hydroxyapatite formation in an Au–collagen matrix is shown in Figure 5.18A and C.

The piezoelectric properties of collagen have been investigated in complex biological systems such as bone and tendon. The piezoelectric properties of bone have a great interest in view of their role in bone remodeling. The bone stress

A

B

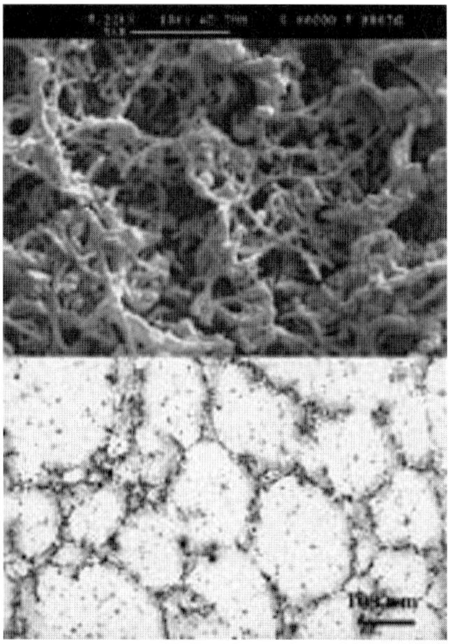

Fig. 5.17 Scanning (A) and transmission (B) electron micrographs of Au–gelatin nanocomposites (Adapted from [78]).

promotes osteogenesis, which is mediated by electrical current generated by piezoelectric materials [85]. Sombra *et al.* studied an interesting example of the electrical properties of hybrid collagen–nanoparticles materials [86]. They synthesized collagen membranes with iron as impurity to understand the role of iron in the collagen structure, aiming to increase the value of the piezoelectric strain tensor element d_{14}. The samples were prepared by two different routes: (i) by immersion of collagen in $FeCl_2$ solutions and (ii) by immersion of collagen directly in α-Fe_2O_3 colloid solution. In this study, the highest conductivity was obtained for collagen–Fe_2O_3 compared to collagen–$FeCl_2$ nanocomposite. At room temperature collagen–Fe_2O_3 conductivity was around ten times higher than that of pure collagen. However, collagen–Fe_2O_3 nanocomposite conductivity decreases by a factor of 10^{-4} when the Fe_2O_3 concentration decreases from 15 to 5 %.

5.3.2.3 Protein Cages and Viral Capsids

Ferritin is a class of iron storage proteins that possess a high structural similarity across a range of biological species [87]. The protein comprises 24 subunits of two types (H and L) that form a segmented hollow cage with an outer diameter of 12 nm (Figure 5.19). In the native ferritin, iron, in a ferric oxyhydroxide form, is stored within an 8 nm diameter cavity formed by the protein (Figure 5.19).

The mammalian protein has an enzymatic activity and catalyzes the oxidation of Fe^{2+} at "ferroxidase" sites present only in H subunits [88]. This reaction may proceed through a di-ferric-μ-peroxo species, which rapidly decomposes, eventually forming a ferric oxyhydroxide mineral core via an inorganic hydrolysis polymerization [89].

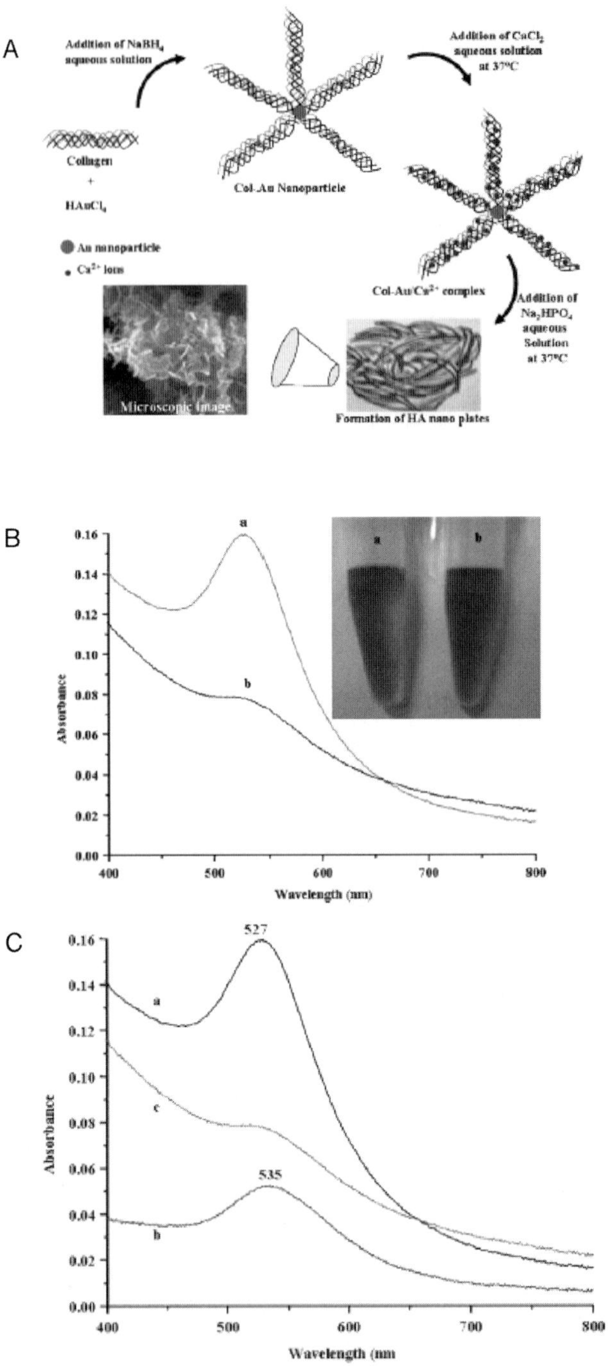

Fig. 5.18 (A) Schematic representation of the immobilization of collagen on Au nanoparticles and the formation of hydroxyapatite on the nanomatrix. (B) UV–Vis spectra of: (a) Au–collagen after 24 h; (b) immediately after addition of reducing agent. (C) Effect of addition of hydroxyapatite on the Au plasmon resonance: (a) Au–collagen solution; (b) after addition of $CaCl_2$; (c) after addition of Na_2HPO_4 [8]. (Reproduced by permission of The Royal Society of Chemistry).

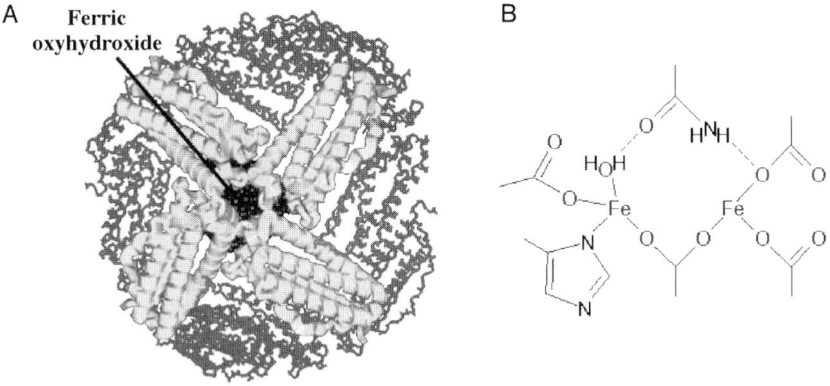

Fig. 5.19 (A) Ferritin 3D representation; (B) Human ferritin (H chain).

This reaction catalyzes the specific nucleation step on the interior surface of the protein cage [90]. The ferritin protein cage is very stable, able to withstand extremes of pH (2–10) and temperature (up to 70 °C). Previous work has shown that the cavity can be used as a confined reaction environment for the synthesis of protein-encapsulated nanoparticles of manganese oxide (MnOOH), magnetite (Fe_3O_4), iron sulfide, uranyl oxide and cadmium sulfide (CdS) [91–95]. In each case, the materials are synthesized with control of their size and shape, and encapsulated within the spherical protein cavity to give dispersed nanocomposites with potential biocompatible and bioactive properties. Kim *et al.* synthesized hollow cobalt oxide nanoparticles produced by a protein-regulated site-specific reconstitution process in aqueous solution [96]. In this work, ferritin was used as a biotemplate to produce hollow nanoparticles and also as a spacer between the hollow nanoparticles. The authors reported the mechanism of cobalt oxide growth in apoHoSF, which has a protein shell without a core, and the fabrication of the hollow nanoparticles by controlling the number of metal atoms in the ferritin. The reaction was carried out at pH 8.5 and the cobalt source used was $CoSO_4$. Cobalt salt was added very slowly to attain a certain loading of metal atoms equally distributed among the ferritins. In this case, reconstituted ferritins were prepared with 200 (Co_{200}) to 2000 Co (Co_{2000}) atoms per ferritin, in increments of 200 atoms. The cobalt oxide growth mechanism proposed by the authors is outlined in Figure 5.20A. Following the authors, the first step of metal core formation is the nucleation of small particles inside the ferritin protein shell. The metal ions form chemical bonds with the functional groups in the interior wall of the ferritin during the oxidation process. It seems that oxidized metal nanoparticles were associated with the carboxylate groups of glutamic acid residues on the interior of the protein wall. The cobalt mineral core is thus attached to the inside of the protein wall. If the interaction with the interior protein wall does not occur, a single nanocrystal would be formed in order to reduce surface tension. The nanoparticles continue to grow along the ferritin interior wall (step 2). In step 3, a hollow nanoparticle forms by coalescence of single nanocrystals. The hollow nanoparticle size is somewhat reduced due to the formation of metallic bonds. Finally, further Co addition results in hollow

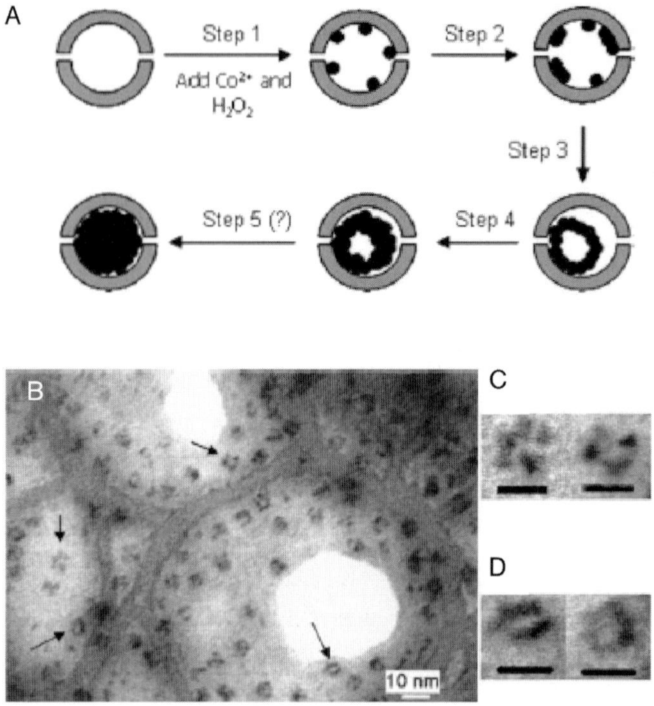

Fig. 5.20 (A) Scheme of cobalt oxide growth mechanism inside ferritin; (B) STEM image of Co_{1000}-cored ferritins; (C) and (D) STEM magnified images of Co_{1000} (arrowed particle (B)) metal oxide core in a ferritin. ([96], reproduced by permission of The Royal Society of Chemistry).

nanoparticles with thicker metal oxide walls (step 4). Figure 5.20B–D show STEM micrographs of Co_{1000}-cored ferritins and cobalt oxide hollow nanoparticles.

In contrast to ferritin, viral protein cages (or capsids) occur in a wide variety of sizes and shapes and have structural transitions that allow controlled access to the protein cage interior [97]. For example, the protein cage of the well characterized cowpea chlorotic mottle virus (CCMV), devoid of nucleic acid as a starting material for creating a new cage with specific chemical functionality, was used to mimic the iron storage in ferritin (L chain). Another example is the self-assembled virus-like particles having a brome mosaic virus protein coat and a functionalized gold core (Figure 5.21) [98].

5.3.3 Lipids

It is well known that liposomes are good candidates for controlled reactions. As lipid-based colloids, they have been used to mimic biomineralization processes [99,100], while as restricted volumes, they have been used as microreactors to synthesize

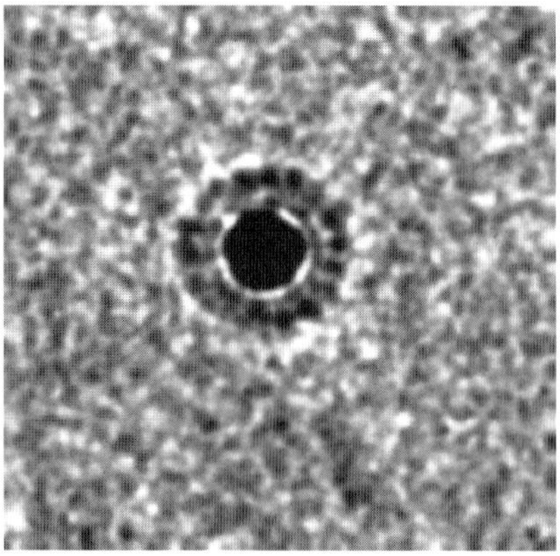

Fig. 5.21 Cryoelectron micrograph of a single virus-like particle showing the well-defined protein coating of the 12 nm diameter Au nanoparticle (black disk). (Reprinted with permission from [98]. Copyright (2006) American Chemical Society).

nanoparticles such as metal oxides [101], metal hydroxides [102], metal sulfides [103] or metallic nanoparticles [104]. For instance, Au nanoparticles can be generated into onion-type multilamellar vesicles (MLV) [105]. In this case, onion-type MLV were selected because they are suitable for drug delivery applications. Indeed, this multilamellar system has demonstrated some ability to supply oligonucleotides to cells and is able to encapsulate a large amount of enzymes [106,107]. The spontaneous *in situ* synthesis of Au nanoparticles within onion-type MLV was carried out in mild conditions [105]. Monoolein, one of the MLV components, was used as reducing agent. The Au growth mechanism within onion-type MLV is detailed in Figure 5.22. No external reducing agent was needed for Au metallic nanoparticle formation suggesting that the onions' surfactants themselves were responsible for the reduction of Au^{3+} ions.

5.3.4 **DNA Scaffolds**

Synthesis of DNA–Au hybrid nanocomposites holds promise for applications in nanotechnology. Following the pioneering work of Mirkin *et al.*, these modified Au nanoparticles can act as useful building blocks to form spatially well-defined superstructures, including nanocrystals [108], binary and multilayered nanoparticle assemblies [109,110] and also well-ordered 3D nanoclusters [111].

These nanocomposites open new horizons to develop highly sensitive biosensors such as selective colorimetric detection of DNA up to one single mismatch [112] and an ultrasensitive scanometric DNA array detection [113]. Another important result, reported later, was that aggregation of Au nanoparticles could be triggered by non-crosslinking-DNA hybridization also allowing single mismatch discrimination [114]. The wrapping of a negatively charged DNA around positively charged

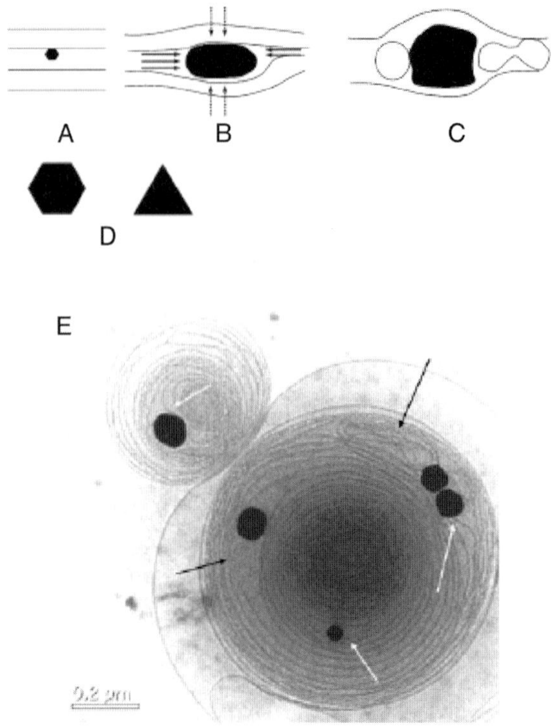

Fig. 5.22 (A–D) Scheme of Au nanoparticle growth between bilayers: (A) nucleation at bilayer interfaces and "bulk-like" growth when particle size is smaller than the lamellar *d* spacing; (B) once the "transversal" particle size is larger than the *d* spacing, growth is slowed by the bilayers, the transversal Au^{3+} flux being limited, leading to elongated particles shapes; (C) when the constraint exerted by the particles on the bilayers is too high, restructuring of lamellae in, for example, liposomes or oblate-shaped vesicles occurs; (D) classical multiply twinned particles essentially grown on the onion surface or within onion defect areas such as the onion core; (E) cryo-TEM of Au nanoparticles grown in onions dispersed in gold salt for 12 h. (Reprinted with permission from [105]. Copyright (2004) American Chemical Society).

ions/molecules primarily through electrostatic interaction is a recurrent fundamental biological process of DNA condensation [115]. Cationic Au and Ag nanoparticles become attractive materials for such studies. In addition, the application of Au and Ag nanoparticles as effective transfection agents adds further impetus to research activities in this direction. Here, we show for the first time the one-pot synthesis of Au–DNA nanowires and an Ag^+–DNA lamellar composite (Figure 5.23). In this case, Ag^+ and Au^{3+} salts were added to DNA solution (extracted from *E. coli* bacteria). The solutions were sonicated and a $NaBH_4$ solution added. The Au–DNA nanowires were formed by coalescence of Au nanoparticles around DNA helices during the reduction step. In contrast, Ag^+ ions react with DNA basic type units to form a stable complex [116].

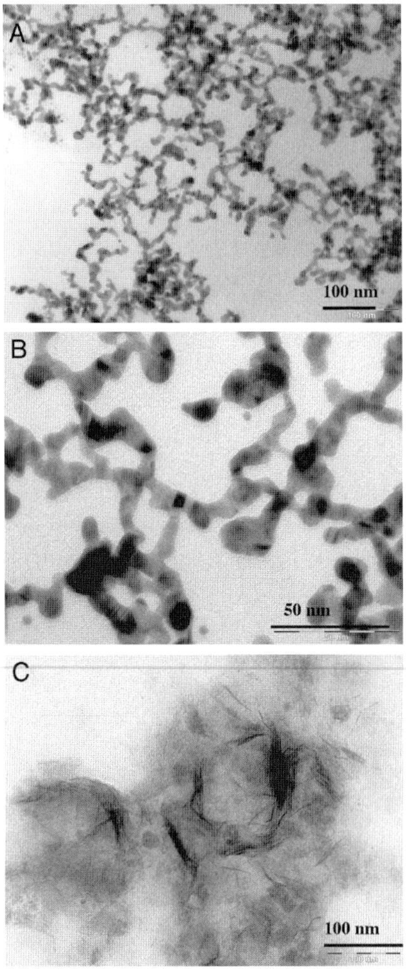

Fig. 5.23 TEM micrographs of (A), (B) Au–DNA nanowires and (C) Ag^+–DNA lamellar structure.

5.4
Perspectives: Towards a "Green Nanochemistry"

All the examples gathered here demonstrate the possibility to control the growth of metallic and oxide nanoparticles using biological templates. A wide variety of chemical composition, particle size and assemblage can be obtained via these approaches. Moreover, due to the biological nature of the template, applications in fields related to biotechnology and medicinal science can be envisioned.

However, when compared to the striking achievements of biomineralizing systems, *in vitro* macromolecular-based strategies still exhibit some limitations. First, these approaches imply the extraction of the biomolecules from the living organisms, sometimes including tedious separation/purification steps. Moreover, some variability may exist from one batch to another, especially for poly-saccharides, and this should influence the templating ability of the extracts [50]. In addition, complex natural morphologies, similar to diatoms shells or bone tissue, do not result from the influence of one macromolecule but involve the whole organism "machinery." To overcome these limitations, recent efforts have been devoted to the use of living cells to control the growth of inorganic nanoparticles [117].

The ability of organisms living in mineralized environments to incorporate metal salts and precipitate inorganic materials has been known for quite a long time [118]. However, the first *in vitro* approaches using whole cells to form nanoparticles took place only about ten years ago [119]. Since then, bacteria [120], fungi [121] and even human cells [122] have been evaluated not only for the growth of gold colloids, but also for silver [123], silica [124] and manganese oxide [125] nanoparticles. The particle growth process can occur extra-cellularly or within the cells and is supposed to involve specific reducing enzymes. Through these approaches, the size and the shape of the colloids can be varied and controlled. In a step forward, it was recently reported that several cyanobacteria cells were able to form Au, Ag, Pd and Pt nanoparticles intracellularly and to release these colloids in their culture medium [126]. Due to the presence of an external poly-saccharide coating, these nanoparticles do not aggregate and can be easily recovered. First steps in the setting-up of a bioreactor for the continuous production of metal nanoparticles have also been reported.

However, from our point of view, the most important result of this study is to demonstrate that the yield of the reduction reaction, and the size of the colloids, is highly species-dependent. In other words, it should be possible to select the most appropriate cell for a given material and, hence, for a given application. Moreover, it also suggests that further optimization of the process may be achieved by suitable genetic modification of the living organisms.

In addition, it must be underlined that the chemical routes to cell-based materials are intrinsically "green" reactions because they imply that the viability of the living organisms is maintained. They should therefore be performed in mild temperature conditions and avoid the presence of cytotoxic products (or at least at non-toxic dose). Moreover, they use living cells that can be considered from a chemical reactivity point of view, as bio-renewable reagents. Such approaches can therefore be considered as the first steps towards the development of a *"green" nanochemistry*.

References

1 Feldheim, D.L. and Foss, C.A., Jr. (eds) (2002) Metal Nanoparticles: Synthesis, Characterization and Application, Marcel Dekker,New York.

2 Schmid, G. (ed.) (2004) Nanoparticles: From Theory to Application,VCH, Weinheim.

3 Ozin, G. and Arsenault, A. (2005) Nanochemistry: A Chemical Approach to Nanomaterials, RSC Publishing, London.

4 Sugimoto, T. (ed.)(2000) Fine Particles: Synthesis, Characterization and Mechanisms of Growth, Surfactant

Science Series, Marcel Dekker, New York.

5 Wu, L.Q. and Payne, G.F. (2004) Biofabrication: The Use of Biological Materials and Biocatalysts to Construct Nanostructured Assemblies. *Trends in Biotechnology*, **22**, 593–599.

6 Lowenstam, H.A. and Weiner, S. (1989) On Biomineralization,Oxford University Press, New York.

7 Xu, A.-W., Ma, Y. and Cölfen, H. (2007) Biomimetic mineralization. *Journal of Materials Chemistry*, **17**, 415–449.

8 He, J., Kunitake, T. and Nakao, A. (2003) Facile *in situ* synthesis of noble metal nanoparticles in porous cellulose fibers. *Chemistry of Materials*, **15**, 4401–4406.

9 Sumar, T.K. and Chattopadhyay, A. (2004) Starch-mediated shape-selective synthesis of Au nanoparticles with tunable longitudinal plasmon resonance. *Langmuir*, **20**, 3520–3524.

10 Singh, A.V., Bandgar, B.M., Kasture, M., Prasad, B.L.V. and Sastry, M. (2005) Synthesis of gold, silver and their alloy nanoparticles using bovine serum albumin as foaming and stabilizing agent. *Journal of Materials Chemistry*, **15**, 5115–5121.

11 Yu, L., Banerjee, I.A. and Matsui, H. (2003) Direct growth of shape-controlled nanocrystals on nanotubes via biological recognition. *Journal of the American Chemical Society*, **125**, 14837–14840.

12 Shenhar, R., Norsten, T.B. and Rotello, V.M. (2005) Polymer-mediated nanoparticle assembly: structural control and applications. *Advanced Materials*, **17**, 657–669.

13 Mann, S., Archibald, D.D., Didymus, J. M., Douglas, T., Heywood, B.R., Meldrum, F.C. and Reeves, N.J. (1993) Crystallization at inorganic-organic interfaces-Biominerals and biomimetic synthesis. *Science*, **261**, 1286–1292.

14 Sumper, M. and Brunner, E. (2006) Learning from diatoms: Nature's tools for the production of nanostructured silica. *Advanced Functional Materials*, **16**, 17–26.

15 Round, F., Crawford, R. and Mann, D. (1990) The Diatoms: Biology and Morphology of the Genera,Cambridge University Press,Cambridge.

16 Swift, D.M. and Wheeler, A.P. (1992) Evidence of an organic matrix from diatom biosilica. *Journal of Phycology*, **28**, 202–209.

17 Kröger, N., Deutzmann, R. and Sumper, M. (1999) Polycationic peptides from diatom biosilica that direct silica nanosphere formation. *Science*, **286**, 1229–1132.

18 Kröger, N., Lorenz, S., Brunner, E. and Sumper, M. (2002) Self-assembly of highly phosporylated silaffins and their function in biosilica morphogenesis. *Science*, **298**, 584–586.

19 Poulsen, N., Sumper, M. and Kröger, N. (2003) Silica formation in diatoms: Characterization of native silaffin-2 and its role in silica morphogenesis. *Proceedings of the National Academy of Sciences of the United States of America*, **100**, 12075–12080.

20 Sumper, M. and Lehmann, G. (2006) Silica pattern formation in diatoms: Species-specific polyamine biosynthesis. *ChemBioChem*, **7**, 1419–1427.

21 Coradin, T. and Lopez, P.J. (2003) Biogenic silica patterning: simple chemistry or subtle biology? *ChemBioChem*, **4**, 251–255.

22 Patwardhan, S.V., Clarson, S.J. and Perry, C.C. (2005) On the role(s) of additives in bioinspired silicification. *Chemical Communications*, 1113–1121.

23 Lopez, P.J., Gautier, C., Livage, J. and Coradin, T. (2005) Mimicking biogenic silica nanostructures formation. *Current Nanoscience*, **1**, 73–83.

24 Coradin, T., Durupthy, O. and Livage, J. (2002) Interactions of amino-containing peptides with sodium silicate and colloidal silica: a biomimetic approach to silicification. *Langmuir*, **18**, 2331–2336.

25 Knecht, M.R. and Wright, D.W. (2004) Amine-terminated dendrimers as biomimetic templates for silica nanospheres formation. *Langmuir*, **20**, 4728–4732.

26 Coradin, T., Coupé, A. and Livage, J. (2003) Interactions of bovine serum albumin and lysozyme with sodium silicate solutions. *Colloids and Surfaces B-Biointerfaces*, **29**, 189–196.

27 Coradin, T., Bah, S. and Livage, J. (2004) Gelatine/silicate interactions: from nanoparticles to composite gels. *Colloids and Surfaces B-Biointerfaces*, **35**, 53–58.

28 Coradin, T., Marchal, A., Abdoul-Aribi, N. and Livage, J. (2005) Gelatine thin films as biomimetic surfaces for silica particles formation. *Colloids and Surfaces B-Biointerfaces*, **44**, 191–196.

29 Pickett-Heaps, J., Schmid, A.-M.M. and Edgar, L.A. (1990) *Progress in Phycology Research*, **7**, 1.

30 Gautier, C., Lopez, P.J., Hemadi, M., Livage and J., Coradin, T. (2006) Biomimetic growth of silica tubes in confined media. *Langmuir*, **22**, 9092–9095.

31 Stöber, W., Fink, A. and Bohn, E. (1968) Controlled growth of monodispersed silica spheres in the micron size range. *Journal of Colloid and Interface Science*, **26**, 62–69.

32 Coradin, T., Allouche, J., Boissière, M. and Livage, J. (2006) Sol-Gel biopolymer/silica nanocomposites in biotechnology. *Current Nanoscience*, **2**, 219–230.

33 Luckarift, H.R., Spain, J.C., Naik, R.R. and Stone, M.O. (2004) Enzyme immobilization in a biomimetic silica support. *Nature Biotechnology*, **22**, 211–213.

34 Boissière, M., Meadows, P.J., Brayner, R., Hélary, C., Livage, J. and Coradin, T. (2006) Turning biopolymer particles into hybrid capsules: the example of silica/alginate nanocomposites. *Journal of Materials Chemistry*, **16**, 1178–1182.

35 Allouche, J., Boissière, M., Hélary, C., Livage, J. and Coradin, T. (2006) Biomimetic core-shell gelatine/silica nanoparticles: a new example of biopolymer-based nanocomposites. *Journal of Materials Chemistry*, **16**, 3121–3131.

36 Sewell, S.S. and Wright, D.W. (2006) Biomimetic synthesis of titanium dioxide utilizing the R5 peptide derived from *Cylindrotheca fusiformis*. *Chemistry of Materials*, **18**, 3108–3113.

37 Patwardhan, S.V. and Clarson, S.J. (2005) Bioinspired mineralization: macromolecule mediated synthesis of amorphous germania structures. *Polymer*, **46**, 4474–4479.

38 Shimizu, K., Cha, J.N., Stucky, G.D. and Morse, D.E. (1998) Silicatein α: a cathepsin L-like protein in sponge biosilica. *Proceedings of the National Academy of Sciences of the United States of America*, **95**, 6234–6238.

39 Cha, J.N., Shimizu, K., Zhou, Y. and Christiansen, S.C., Chmelka, B.F., Stucky, G.D., Morse, D.E. (1999) Silicatein filaments and subunits from a marine sponge direct the polymerization of silica and silicone *in vitro*. *Proceedings of the National Academy of Sciences of the United States of America*, **96**, 361–365.

40 Roth, K.M., Zhou, Y., Yang, W. and Morse, D.E. (2005) Bifunctional small molecules are biomimetic catalysts for silica synthesis at neutral pH. *Journal of the American Chemical Society*, **127**, 325–330.

41 Cha, J.N., Stucky, G.D., Morse, D.E. and Deming, T.J. (2000) Biomimetic synthesis of ordered silica structures mediated by block copolypeptides. *Nature*, **403**, 289–292.

42 Summerel, J.L., Yang, W., Kisailus, D., Weaver, J.C., Choi, J.H. and Morse, D. E. (2003) Biocatalytically templated synthesis of titanium dioxide. *Chemistry of Materials*, **15**, 4804–4809.

43 Kisailus, D., Choi, J.H., Weaver, J.C., Yang, W. and Morse, D.E. (2005) Enzymatic synthesis and nanostructural control of gallium oxide at low temperature. *Advanced Materials*, **17**, 314–318.

44 Brutchey, R.L., Yoo, E.S. and Morse, D.E. (2006) Biocatalytic synthesis of a nanostructured and crystalline bimetallic perovskite-like barium oxofluorotitanate at low temperature. *Journal of the American Chemical Society*, **128**, 10288–10294.

45 Curnow, P., Bessette, P.H., Kisailus, D., Murr, M.M., Daugherty, P.S. and Morse, D.E. (2005) Enzymatic synthesis of layered titanium phosphates at low temperature and neutral pH by cell-surface display of silicatein α. *Journal of the American Chemical Society*, **127**, 15749–15755.

46 Moe, S.T., Draget, K.I., Skjaek-Braek, G. and Smidsrod, O. (1995) Alginates. *Food Science and Technology*, **67**, 245–286.

47 Davis, T.A., Llanes, F., Volesky, B. and Mucci, A. (2003) Metal selectivity of *Sargassum* spp. And their alginates in relation to their α-L-guluronic acid content and conformation. *Environmental Science & Technology*, **37**, 261–267.

48 Davis, T.A., Volesky, B. and Mucci, A. (2003) A review of the biochemistry of heavy metal biosorption by brown algae. *Water Research*, **37**, 4311–4330.

49 BraynerCoradin, T., Fiévet-Vincent, F., Livage, J. and Fiévet, F. (2005) Algal polysaccharide capsule-templated growth of magnetic nanoparticles. *New Journal of Chemistry*, **29**, 681–685.

50 Brayner, R., Vaulay, M.-J., Fiévet, F. and Coradin, T. (2007) Alginate-mediated growth of Co, Ni and CoNi nanoparticles: influence of the biopolymer structure.*Chemistry of Materials*, **19**, 1190–1198.

51 Arias, J.L., Gallardo, V., Gomez-Lopera, S.A., Plaza, R.C. and Delgado, A.V. (2001) Synthesis and characterization of poly(ethyl-2-cyanoacrylate) nanoparticles with a magnetic core. *Journal of Controlled Release*, **77**, 309–321.

52 Rudge, S.R., Kurtz, T.L., Vessely, C.R., Catterall, L.G. and Williamson, D.L. (2000) Preparation, characterization and performance of magnetic iron-carbon composite microparticles for chemotherapy. *Biomaterials*, **21**, 1411–1420.

53 Boissière, M., Allouche, J., Brayner, R., Chanéac, C., Livage, J. and Coradin, T. Design of iron oxide/silica/alginate HYbrid MAgnetic Carriers (HYMAC). *Journal of Nanoscience and Nanotechnology*, in press.

54 Morris, E.R., Rees, D.A. and Robinson, G. (1980) Cation-specific aggregation of carrageenan helices: Domain model of polymer gel structure. *Journal of Molecular Biology*, **138**, 349–362.

55 Smidsrod, O. and Grasdalen, H. (1984) Conformations of κ-carrageenan in solution. *Hydrobiologia*, **116/117**, 178–186.

56 Grasdalen, H. and Smidsrod, O. (1981) Iodide-specific formation of κ-carrageenan single helices- I-127 NMR spectroscopic evidence for selective site binding of iodide anions in the ordered conformations. *Macromolecules*, **14**, 1842–1845.

57 Jones, F., Cölfen, H. and Antonietti, M. (2000) Iron oxyhydroxide colloids stabilized with polysaccharides. *Colloid and Polymer Science*, **278**, 491–501.

58 Jones, F., Cölfen, H. and Antonietti, M. (2000) Interaction of κ-carrageenan with nickel, cobalt and iron hydroxides. *Biomacromolecules*, **1**, 556–563.

59 Brayner, R. *et al.* unpublished results.

60 Brayner, R. (2006) *Frontiers in Drug Design and Discovery*. **2**, 241.

61 Sokolov, K., Aaron, J., Hsu, B., Nida, D., Gillenwater, A., Follen, M., Macaulay, C., Adler-Storthz, K., Korgel, B., Discour, M., Pasqualini, R., Arap, W., Lam, W. and Richartz-Kortum, R. (2003) Optical systems for in vivo molecular imaging of cancer. *Technology in Cancer Research & Treatment*, **2**, 491–504.

62 Rodger, C., Smith, W.E., Dent, G. and Edmondson, M. (1996) Surface-enhanced resonance-Raman scattering: an informative probe of surfaces. *Journal of the Chemical Society Dalton Transactions*, 791–799.

63 Docherty, F.T., Clark, M., McNay, G., Graham, D. and Smith, W.E. (2004) Multiple labelled nanoparticles for biodetection. *Faraday Discussions*, **126**, 281–288.

64 Kumar, M.N.V.R., Muzzarelli, R.A.A., Muzzarelli, C., Sashiwa, H. and Domb, A.J. (2004) Chitosan chemistry and pharmaceutical perspectives. *Chemical Reviews*, **104**, 6017–6084.

65 Sudarshan, N.R., Hoover, D.G. and Knorr, D. (1992) Antibacterial action of chitosan. *Food Biotechnology*, **6**, 257–272.

66 Agnihori, S.A., Mallikarjuna, N.N. and Aminabhavi, T.M. (2004) Recent advances on chitosan-based micro and nanoparticles in drug delivery. *Journal of Controlled Release*, **100**, 5–28.

67 Maclaughlin, F.C., Mumper, R.J., Wang, J., Tagliaferri, J.M., Gill, I., Hinchcliffe, M. and Rolland, A.P. (1998) Chitosan and depolymerized chitosan oligomers as condensing carriers for in vivo plasmid delivery. *Journal of Controlled Release*, **56**, 259–272.

68 Zhi, J., Wang, Y., Lu, Y., Ma, J. and Luo, G. (2006) *In situ* preparation of magnetic chitosan/Fe_3O_4 composite nanoparticles in tiny pools of water-in-oil microemulsion. *Reactive and Functional Polymers*, **66**, 1552–1558.

69 Chang, Y.C., Chang, S.W. and Chen, D.H. (2006) Magnetic chitosan nanoparticles: studies on chitosan binding and adsorption of Co(II) ions. *Reactive and Functional Polymers*, **66**, 335–341.

70 Esumi, K., Takei, N. and Yoshimura, T. (2003) Antioxidant-potentiality of gold-chitosan nanocomposites. *Colloids and Surface Science B*, **32**, 117–123.

71 Texter, J. (2001) in Handbook of Applied Surface and Colloid Chemistry, (ed. Holmberg K.),Wiley, New York.

72 Shibayama, M. and Norisuye, T. (2002) Gel formation analyses by dynamic light scattering. *Bulletin of the Chemical Society of Japan*, **75**, 641–659.

73 Kenchington, A.W. and Ward, A.G. (1954) The titration curve of gelatin. *Biochemical Journal*, **58**, 202–207.

74 Bigi, A., Panzavolta, S. and Rubini, K. (2004) Relationship between triple-helix content and mechanical properties of gelatin films. *Biomaterials*, **25**, 5675–5680.

75 Cölfen, H. (2001) Double-hydrophilic block copolymers: synthesis and application as novel surfactants and crystal growth modifiers. *Macromolecular Rapid Communications*, **22**, 219–252.

76 Cölfen, H. and Mann, S. (2003) Higher-order organization by mesoscale self-assembly and transformation of hybrid nanostructures. *Angewandte*

Chemie-International Edition, **42**, 2350–2365.

77 Bauermann, L.P., DelCampo, A., Bill, J. and Aldinger, F. (2006) Heterogeneous nucleation of ZnO using gelatin as the organic matrix. *Chemistry of Materials*, **18**, 2016–2020.

78 Brayner, R., Coradin, T., Vaulay, M.-J., Mangeney, C., Livage, J. and Fiévet, F. (2005) Preparation and characterization of metal (Au)- and bimetallic alloys (AuNi)-gelatin nanocomposites. *Colloids and Surface Science A*, **256**, 191–197.

79 Lee, C.H., Singla, A. and Lee, Y. (2001) Biomedical applications of collagen. *International Journal of Pharmaceutics*, **221**, 1–22.

80 Giraud-Guille, M.M., Mosser, G., Hélary, C. and Eglin, D. (2005) Bone matrix-like assemblies of collagen: From liquid crystals to gels and biomimetic materials. *Micron*, **36**, 602–608.

81 Brodsky, B. and Ramshaw, J.A.M. (1997) The collagen triple-helix structure. *Matrix Biology*, **15**, 545–549.

82 Peffer, N.J., Hanvey, J.C., Bisi, J.E., Thomson, S.A., Hassman, C.F., Noble, S.A. and Babiss, L.E. (1993) Strand-invasion of duplex DNA by peptidic nucleic acid oligomers. *Proceedings of the National Academy of Sciences of the United States of America*, **90**, 10648–10652.

83 Miles, C.A. and Bailey, A.J. (2001) Thermally labile domains in the collagen molecules. *Micron*, **32**, 325–332.

84 Aryal, S., Bahadour, K.C.R., Bhattarai, J.R., Prabu, P. and Kim, H.Y. (2006) Immobilization of collagen on gold nanoparticles, preparation, characterization and hydroxyapatite growth. *Journal of Materials Chemistry*, **16**, 4642–4648.

85 Jianqing, F., Huipin, Y. and Xingdong, Z. (1997) Promotion of osteogenesis by a piezoelectronic biological ceramic. *Biomaterials*, **18**, 1531–1534.

86 Sousa Neto, V.O., Silva, C.C., Almeida, A.F.L., Figueiró, S.D., Góes, J.C., dePaiva, J.A.C., Magalhães, C.E.C. and Sombra, A.S.B. (2002) Study of the electrical conductivity and piezoelectricity in iron doped collagen films. *Solid State Sciences*, **4**, 43–51.

87 Gianazza, E. and Arosio, P. (1980) Electrophoretic analysis of horse tissue ferritins at different pH values. *Biochimica et Biophysica Acta*, **625**, 310–317.

88 Lawson, D.M., Treffry, A., Artymiuk, P. J., Harrison, P.M., Yewdall, S.J., Luzzago, A., Cesareni, G., Levi, S. and Arosio, P. (1989) Identification of the ferroxidase center in ferritin. *FEBS Letters*, **254**, 207–210.

89 Pereira, A.S., Small, W., Krebs, C., Tavares, P., Edmondson, D.E., Theil, E.C. and Huynh, B.H. (1998) Direct spectroscopic and kinetic evidence for the involvement of a peroxodiferric intermediate during the ferroxidase reaction in fast ferritin mineralization. *Biochemistry*, **37**, 9871–9876.

90 Lawson, D.M., Artymiuk, P.J., Yewdall, S.T., Smith, J.M.A., Livingstone, J.C., Treffry, A., Luzzago, A., Levi, S., Arosio, P., Cesareni, G., Thomas, C.D., Shaw, W.V. and Harrison, P.M. (1991) Solving the structure of human H-ferritin by genetically engineering intermolecular crystal contacts. *Nature*, **349**, 541–544.

91 Meldrum, F.C., Wade, V.J., Nimmo, D. L., Heywood, B.R. and Mann, S. (1991) Synthesis of inorganic nanophase materials in supramolecular protein cages. *Nature*, **349**, 684–687.

92 Meldrum, F.C., Heywood, B.R. and Mann, S. (1992) Magnetoferritin – *in vitro* synthesis of a novel magnetic protein. *Science*, **257**, 522–523.

93 Douglas, T., Dickson, D.P.E., Betteridge, S., Charnock, J., Garner, C.D. and Mann, S. (1995) Synthesis and structure of an iron(III) sulfide-ferritin bioinorganic nanocomposite. *Science*, **269**, 54–57.

94 Wong, K.K.W. and Mann, S. (1996) Biomimetic synthesis of cadmium sulfide-ferritin nanocomposites. *Advanced Materials*, **8**, 928–932.

95 Wong, K.K.W., Douglas, T., Gider, S., Awschalon, D.D. and Mann, S. (1998) Biomimetic synthesis and characterization of magnetic proteins (magnetoferritin). *Chemistry of Materials*, **10**, 279–285.

96 Kim, J.W., Choi, S.H., Lillehei, P.T., Chu, S.H., King, G.C. and Watt, G.D. (2005) Cobalt oxide hollow nanoparticles derived by bio-templating. *Chemical Communications*, 4101–4103.

97 Douglas, T. and Young, M. (1998) Host-guest encapsulation of materials by assembled virus protein cages. *Nature*, **393**, 152–155.

98 Chen, C., Daniel, M.C., Quinkert, Z.T., De, M., Stein, B., Bowman, V.D., Chipman, P.R., Rotello, V.M., Kao, C.C. and Dragnea, B. (2006) Nanoparticle-templated assembly of viral protein cages. *Nano Letters*, **6**, 611–615.

99 Mann, S., Hannington, J.P. and Williams, R.J.P. (1986) Phospholipid-vesicles as a model system for biomineralization. *Nature*, **324**, 565–567.

100 El-Rassy, H., Belamie, E., Livage, J. and Coradin, T. (2005) Onion phases as biomimetic confined media for silica nanoparticle growth. *Langmuir*, **21**, 8584–8587.

101 Gauffre, F. and Roux, D. (1999) Studying a new type of surfactant aggregate ("Spherulites") as chemical microreactors. A first example: Copper ion entrapping and particle synthesis. *Langmuir*, **15**, 3738–3747.

102 Bhandarkar, S. and Bose, A. (1990) Synthesis of nanocomposite particles by intravesicular coprecipitation. *Journal of Colloid and Interface Science*, **139**, 541–550.

103 Khramov, M.I. and Parmon, V.N. (1993) Synthesis of ultrafine particles of transition-metal sulfides in the cavities of lipid vesicles and the light-stimulated transmembrane electron-transfer catalyzed by these particles. *Journal of Photochemistry and Photobiology A-Chemistry*, **71**, 279–284.

104 Faure, C., Derré, A. and Neri, W. (2003) Spontaneous formation of silver nanoparticles in multilamellar vesicles. *Journal of Physical Chemistry B*, **107**, 4738–4746.

105 Regev, D., Backov, R. and Faure, C. (2004) Gold nanoparticles spontaneously generated in onion-type multilamellar vesicles bilayers. Particle coupling imaged by cryo-TEM. *Chemistry of Materials*, **16**, 5280–5285.

106 Miguet, N., Brun, A., Degert, C., Delord, B., Roux, D., Hélène, C., François, J.C. and Laversanne, L. (2000) The spherulites (TM): a promising carrier for oligonucleotide delivery. *Nucleic Acids Research*, **28**, 3134–3142.

107 Olea, D. and Faure, C. (2003) Quantitative study of the encapsulation of glucose oxidase into multilamellar vesicles and its effect on enzyme activity. *Journal of Chemical Physics*, **119**, 6111–6118.

108 Storhoff, J.J. and Mirkin, C.A. (1999) Programmed materials synthesis with DNA. *Chemical Reviews*, **99**, 1849–1862.

109 Mucic, R.C., Storhoff, J.J., Mirkin, C.A. and Letsinger, R.L. (1998) DNA-directed synthesis of binary nanoparticle network materials. *Journal of the American Chemical Society*, **120**, 12674–12675.

110 Taton, T.A., Mucic, R.C., Mirkin, C.A. and Letsinger, R.L. (2000) The DNA-mediated formation of supramolecular mono- and multilayered nanoparticle structure. *Journal of the American Chemical Society*, **122**, 6305–6306.

111 Mirkin, C.A., Letsinger, R.L., Mucic, R. C. and Storhoff, J.J. (1996) A DNA-based method for rationally assembling nanoparticles into macroscopic materials. *Nature*, **382**, 607–609.

112 Elghanian, R., Storhoff, J.J., Mucic, R. C., Letsinger, R.L. and Mirkin, C.A. (1997) Selective colorimetric detection of poly-nucleotides based on the distance-dependent optical properties of gold nanoparticles. *Science*, **277**, 1078–1081.

113 Taton, T.A., Mirkin, C.A. and Letsinger, R.L. (2000) Scanometric DNA array detection with nanoparticle probes. *Science*, **289**, 1757–1760.

114 Sato, K., Hosokawa, K. and Maeda, M. (2003) Rapid aggregation of gold nanoparticles induced by non-crosslinking DNA hybridization. *Journal of the American Chemical Society*, **125**, 8102–8103.

115 Alberts, B., Bray, D., Lewis, J., Raff, M., Roberts, K. and Watson, J.D. (1994) Molecular Biology of the Cell,Galland, New York.

116 Brayner R. *et al.* unpublished results.

117 Klaus-Joerger, T., Joerger, R., Olsson, E. and Granqvist, C. (2001) Bacteria as workers in the living factory: metal accumulating bacteria and their potential in material science. *Trends in Biotechnology*, **19**, 15–20.

118 Beveridge, T.J. and Doyle, R.J. (1989) Metal Ions and Bacteria, Wiley, New York.

119 Southam, G. and Beveridge, T.J. (1994) The in vitro formation of placer gold by bacteria. *Geochimica et Cosmochimica Acta*, **58**, 4527–4530.

120 Lengke, M. and Southam, G. (2006) Bioaccumulation of gold by sulfate-reducing bacteria cultured in the presence of gold(I)-thiosulfate complex. *Geochimica et Cosmochimica Acta*, **70**, 3646–3661.

121 Mukherjee, P., Ahmad, A., Mandal, D., Senapati, S., Sainkar, S.R., Khan, M.I., Ramani, R., Parischa, R., Ajayakumar, P.V., Alam, M., Sastry, M. and Kumar, R. (2001) Bioreduction of $AuCl_4^-$ ions by the fungus *Verticillium sp.* and surface trapping of the gold nanoparticles formed. *Angewandte Chemie-International Edition*, **40**, 3585–3588.

122 Venkataraman, A.J.S., Subramaniam, C., Kumar, R.R., Priya, S., Kumar, T.R. S., Omkumar, R.V., John, A. and Pradeep, T. (2005) Growth of gold nanoparticles in human cells. *Langmuir*, **21**, 11562–11567.

123 Bhainsa, K.C. and D'Souza, S.F. (2006) Extracellular biosynthesis of silver nanoparticles using the fungus *Aspergillus fumigatus*. *Colloids and Surfaces B*, **47**, 160–164.

124 Bansal, V., Ahmad, A., Sastry, M. (2006) Fungus-mediated biotransformation of amorphous silica in rice husk to nanocrystalline silica. *Journal of the American Chemical Society*, **128**, 14059–14066.

125 Kim, H.-S., Pasten, P.A., Gaillard, J.-F. and Slair, P.C. (2003) Nanocrystalline todorokite-like manganese oxide produced by bacterial catalysis. *Journal of the American Chemical Society*, **125**, 14284–14285.

126 Brayner, R., Barberousse, H., Hemadi, M., Djediat, S., Yéprémian, C., Coradin, T., Livage, J., Fiévet, F. and Couté, A. (2007) Cyanobacteria as bioreactors for the synthesis of Au, Ag, Pd and Pt nanoparticles via an enzyme-mediated route. *Journal of Nanoscience and Nanotechnology*, **7**, 2696–2708.

6
Biomineralization of Hydrogels Based on Bioinspired Assemblies for Injectable Biomaterials

Junji Watanabe, Mitsuru Akashi

6.1
Introduction

6.1.1
Biominerals as Nanomaterials

Biominerals have been present in Nature since ancient times. A volcano produced volcanic ash; subsequently, a river, a lake, and a sea were formed. In particular, the sea is well known as a source of living systems. The most primitive living thing is a magnetic bacterium, which contains nanoscale magnetic particles. It is really a bio-inorganic hybrid nanomaterial found in Nature. Furthermore, a diatom contains a silica compound, a coral reef is made from calcium carbonate, and a vertebrate contains bones as a frame, whose major component is hydroxyapatite (HAp). There are many kinds of biominerals, not only in Nature but also on the earth. In most cases, the biominerals in Nature are sophisticated and hierarchical microstructures, so the biominerals are very fine bio-inorganic hybrid nanomaterials indeed. In this chapter, we focus on HAp and calcium carbonate for their biofunctions as biomaterials. In most biominerals, HAp and calcium carbonate are significantly important inorganic compounds. Collagen and HAp form an excellent microstructure in bone tissue, and calcification of biomaterials is also a major problem which occurs at the implant site. The major component is calcium carbonate. Therefore, the bio-inorganic hybrid is closely connected with biomaterials, and will regulate biofunctions.

The stoichiometry of HAp is known to be $Ca_{10}(PO_4)_6(OH)_2$. However, slight imbalances in the stoichiometric ratio of calcium and phosphorus in HAp are generally observed (Figure 6.1). The HAp is a very adaptable crystalline structure, so alternative ions can be easily substituted for Ca^{2+}, PO_4^{3-}, and OH^-. Moreover, some of the ions are vacant without substitutions. The ideal HAp composition has a Ca/P molar ratio of roughly 1.67 (10 calcium atoms per 6 phosphorus atoms). However, in most cases, a higher Ca/P ratio (1.7–1.9) was observed due to the substitution of CO_3^{2-}, if the HAp was prepared in air. The only way we can obtain

Bio-inorganic Hybrid Nanomaterials. Edited by E. Ruiz-Hitzky, K. Ariga, Y. M. Lvov

ISBN: 978-3-527-31718-9

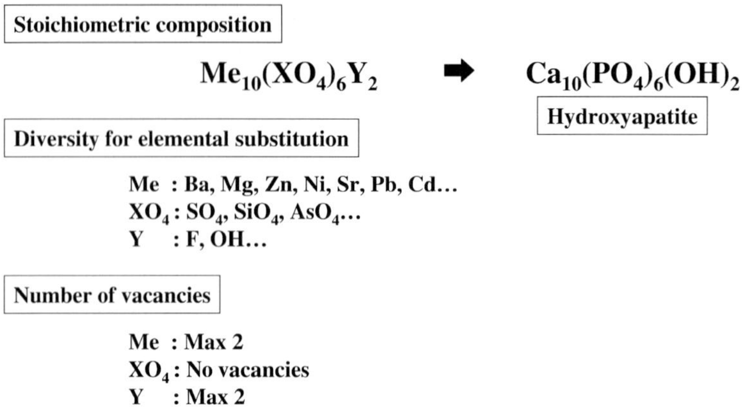

Fig. 6.1 Diversity of hydroxyapatite focusing on chemical composition. Hydroxyapatite is an adaptable crystalline structure, so substitution of other elements and vacancies are frequently observed.

the ideal HAp composition is by thermal processing, that is by a sintering process at over 900 °C using calcium and phosphate compounds. On the other hand, calcium carbonate shows unique crystalline morphologies. Calcite, aragonite, and vaterite are typical examples of its polymorphs, known as a limestone and a coral, respectively. The stability of the polymorphs is regulated by the thermodynamics of the system (Figure 6.2) [1]. Vaterite does not exist in nature due to its thermodynamic stability. However, we can control the polymorphs by addition of some divalent cations such as Mg^{2+}, Fe^{2+}, and Ba^{2+} [2]. Magnesium ions can be adsorbed onto a specific crystal face of aragonite, so that they enhance the growth of the aragonite crystal [3]. In the case of

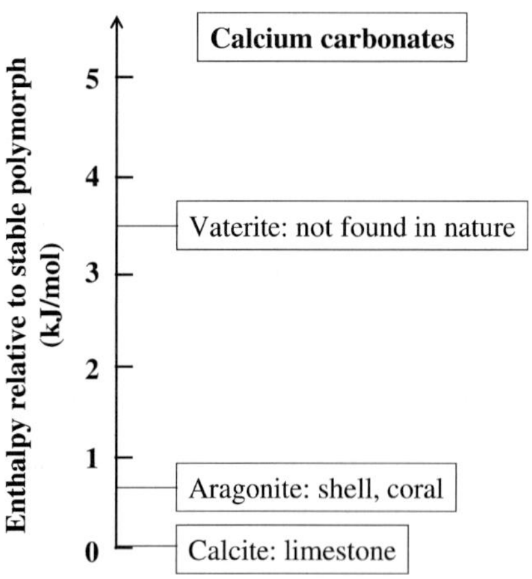

Fig. 6.2 Thermodynamic stability of calcium carbonates. Change in enthalpy is shown with polymorph.

barium ions, the ions suppress the transformation of calcite, so the formation of vaterite crystals can be induced. Thus, the resulting biominerals are easily varied by changes to the preparative conditions.

Hydroxyapatite powder itself has been widely prepared by, for example, precipitation from aqueous solutions and a mechanochemical-hydrothermal method [4,5]. LeGeros *et al.* reported tooth mineral formation, which contains HAp, and investigated the effect of fluoride in terms of stabilization against acid dissolution by caries and remineralization in the presence of fluoride solution [6]. With regard to biomineralization, Kokubo reported pioneering work in 1990; a desired thickness of HAp on a glass-ceramics substrate was prepared using a biomimetic process in simulated body fluid (SBF) [7]. A chemical bond between HAp crystals in the body milieu can give a fairly strong bond. The surface HAp layer provided an interface with a chemical reaction of calcium and silicate ions dissolved from the glass-ceramic. The SBF process is a very important methodology and is still considered to be a gold standard process. Therefore, much research on HAp formation is based on the SBF process. In 2006, Kokubo *et al.* summarized the history of SBF and the successful development of HAp formation in SBF as a leading opinion paper [8].

A template molecule assists biomineral formation and the crystalline structures of HAp and calcium carbonate have been regulated by a polymer template. Bertozi *et al.* reported a template-driven nucleation and biomineral growth for high integration of poly(2-hydroxyethyl methacrylate) and HAp [9,10]. The hydroxyethyl groups were converted into carboxyl groups by hydrolysis. The resulting carboxyl groups interact strongly with Ca^{2+}, and facilitate the nucleation two-dimensionally. Finally, the template surface was covered with HAp crystals. Mooney *et al.* reported a biodegradable polymer template that assisted the growth of carbonate apatite crystals [11]. Polyesters such as poly(lactic acid), poly(glycolic acid), and their copolymer could readily take part in biomineralization via carboxylic acid and hydroxy groups on the polymer surface. The morphologies of the minerals were regulated by changes in the functional groups present at the nucleation site, which were controlled by surface hydrolysis. Imai *et al.* studied calcium phosphate formation with a hierarchically laminated architecture in hydrogels [12]. The resulting calcium phosphate in hydrogels showed Liesegang periodic precipitation. This is a challenge for hierarchical architecture in our living system. Kato *et al.* reported the formation of an oriented calcium carbonate using polymer composites [13]. The oriented crystals of calcium carbonate were formed by using the specific combination of calcification-associated peptide and polysaccharides such as chitin as a matrix. The peptides were designed and synthesized to study their effect on mineral formation. These research reports are based on the template-assisted biomineralization process, which could regulate morphologies, crystallinity, and surface coverage. Yu *et al.* reported calcium carbonate formation at the air/water interface [14]. A block copolymer composed of poly(ethylene glycol) and poly(ethylene imine) played a key role in the mineral formation. The polymer formed blush architecture on the anionic substrate, and a polycation cavity was produced as a reservoir of carbonate and calcium ions. On the other hand, Aizenberg and coworkers focused on the use of functionalized self-assembled monolayers micropatterned by soft-lithography techniques [15]. The most favorable characteristics of the strategy provided a spatially constrained and chemically modified microenvironment for mineralization. The ordered crystalline minerals are critical inorganic solids, not a simple compound but a sophisticated material. The hierarchical mineral crystals are only produced by bottom-up fabrication. Therefore, the resulting materials are really "bio-inorganic hybrid nanomaterials".

6.1.2 Nanomaterials for Biofunctions

Biominerals such as HAp and calcium carbonate are common in Nature. However, their microstructure consists of excellently integrated organic and inorganic materials. A natural bone in a vertebrate is a typical example of a nanomaterial. The collagen fibers and HAp crystals are integrated at the nanoscale. Moreover, the crystalline structure of the bone can be spontaneously rearranged, the driving force being an adaptation in the living system. Nakano *et al.* investigated the alignment of the HAp in several kinds of bone using an X-ray diffractometer [16]. There are roughly 200 bones in our human body; typically, long bones, flat bones, and dentulous mandible. The alignment of the HAp crystals is different in each case to optimize the biofunction. In particular, the *c*-axis of the HAp crystal is sensitive to the moiety and shape in the calcified tissues. Perhaps, the diversity of the specific alignment depends on the history of bone formation and mechanical function to show biological function.

The mechanical strength of the bones is regulated by their alignment of HAp crystals. The alignment is continually remodeled by a coupling of osteoblast and osteoclast. The bone tissue is always maintained by a renewal process (Figure 6.3). An osteoblast contributes bone formation, for example the growth in the stature of

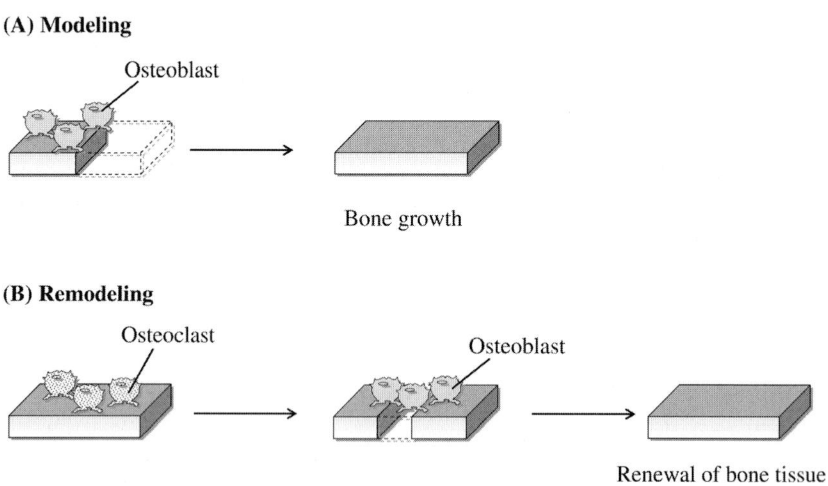

Fig. 6.3 Schematic illustration of bone modelling and remodelling. Bio-inorganic hybrid nanomaterials are kept fresh in bone tissue.

babies is based on bone growth. This process is called "Modeling," and the bone shape and alignment are changeable with the growth, however, it is limited until the growth generation (from baby to adult). On the other hand, an osteoclast could dissolve the bone tissue by reducing the surrounding pH (roughly pH 3). After dissolving the tissue, an osteoblast is induced and activated by a signal from cytokines. This coupling process is called "Remodeling," and the bone shape and alignment are maintained constantly. Moreover, the remodeling process continues forever, so it is really renewal. Thus, bio-inorganic hybrid nanomaterials such as bone tissue are a self-renewal system based on the osteoblast and osteoclast, and the nanomaterials are closely related to the biofunctions.

Considering calcium carbonate, a shellfish has a nice house made of shell for protection. Calcium carbonate is a typical inorganic material for shell and coral. However, through evolution of the living system we no longer have any calcium carbonate; we discarded the calcium carbonate, and took the HAp. Though these two compounds are fundamental inorganic materials, HAp is superior to calcium carbonate in terms of the theory of evolution. It is one of the mysteries, and the biominerals and their hybrid materials are closely related to the theory of evolution in our living system. Taking these sophisticated biofunctions into account, "bio-inorganic hybrid nanomaterials" are of great importance, not only to chemistry, science, and biology but also to the history of the living system.

6.2 Fundamental Concept of Bioinspired Approach

6.2.1 Bioinspired Approach to Materials

In this chapter, we will introduce a bioinspired approach to biomineralization. First, a fundamental concept of the bioinspired approach is compared with the biomimetic approach. Here is a question, "What is a bioinspired approach?" If we want to fly in the sky, when we look at flying birds what do we have to do? Two kinds of typical answer were considered, as shown in Figure 6.4. One is "Bird-Man," in the Edo period (AD 1600) in Japan, a man named Kokichi attempted to fly. He wore wide, large wings on both his arms, and fluttered his arms vigorously, but in vain. This is a typical *biomimetic* approach. He only mimicked the style of a flying bird, so his innovative idea ended in failure.

On the other hand, in 1903, the Wright brothers in the USA developed an "airplane" in order to fly in the sky. The airplane consisted of large wings and a propeller. The propeller-driven airplane could fly, and the fundamental mechanism was buoyancy by a propeller-driven driving force. This is an ideal *bioinspired* approach. The flying birds also utilize buoyancy produced by fluttering. The Wright brothers paid attention to the fundamental mechanism of a flying system; in particular, they clearly understood the principle of buoyancy generated by a driving force. Therefore, they picked up some important factors from the birds about how to

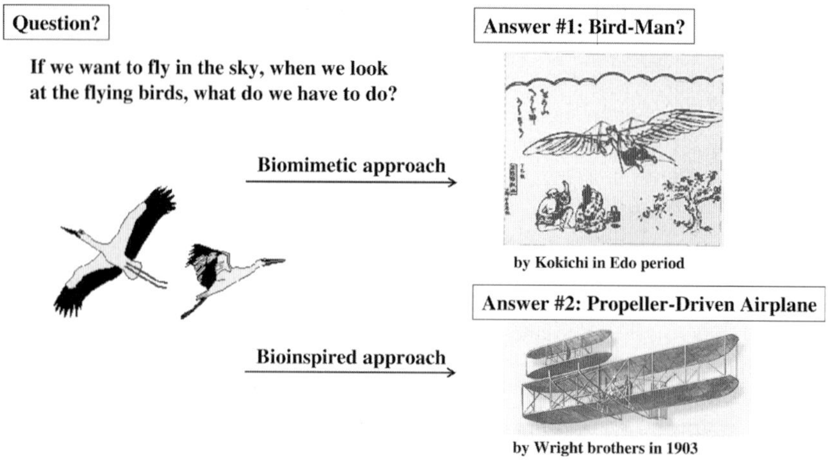

Fig. 6.4 Schematic illustration of a fundamental concept of a bioinspired approach relative to a biomimetic approach.

fly in the sky. The selected items were a wing and a propeller, and they assembled the items as a propeller-driven airplane. This is a fundamental strategy of the bioinspired approach. In the field of bio-inorganic hybrid nanomaterials, the bioinspired approach is of great importance in the development of innovative materials. This is the concept of this chapter, and we will introduce some concrete examples of the bioinspired approach.

6.2.2
Concrete Examples of the Bioinspired Approach

In a muscle tissue, the smooth and continuous sliding function is based on a pair of proteins, actin and myosin, and this is a device on the molecular scale. If we prepare a similar device using some molecules, what is the dominant molecular design? The key role of a muscle is a sliding function, therefore we have to design a sliding system. Yui *et al.* designed a sliding system using a polymer backbone with cyclic compounds, called a polyrotaxane structure [17]. The cyclic compounds such as β-cyclodextrins (β-CDs) were localized on the polymer backbone, which contains a triblock polymer segment. The CDs could be made movable on the polymer backbone by changing the temperature, showing the change in the molecular interaction between CDs and the polymer backbone. The molecular design of the polyrotaxane structure is diverse, and Yui *et al.* synthesized a pH-responsive movement system using cyclic cucurbit compounds and CDs on a diblock polymer backbone [18]. In this sliding function, the focus is on linear polymer and cyclic compounds. If the sliding function is synchronized, the resulting movement will be amplified enormously. This is a concrete example of a bioinspired approach.

Another example focused on cell membranes, which are sophisticated, nanostructured separators in living organisms. The membrane is mainly composed of phospholipid molecules, which play an important role in dividing intracellular cytoplasm from the outer environment, and glycoproteins as receptor and membrane penetrating proteins are also situated on the surface. If we prepare a similar interface artificially, what is the most important factor in the molecular design? The most obvious strategy is to use monolayer membranes, which are known as Langmuir–Blodgett (LB) membranes or self-assembled monolayers (SAMs), using appropriate chemical compounds with an alkyl chain on the air–water or air–solid interface. However, this strategy is a typical example of the biomimetic approach. If we design the interface using the bioinspired approach, further selection of important factors is necessary. In 1990, Ishihara *et al.* reported the design of a methacrylate monomer containing a phosphorylcholine group (2-methacryloyloxyethyl phosphorylcholine (MPC)) and its successful synthesis with excellent purity [19]. Originally, the monomer was prepared by Nakabayashi *et al.* [20]; however, the amount of the product was too small to examine for further application. The methacrylate monomer MPC could be synthesized with diverse methacrylate and vinyl monomers. Typically, a copolymer composed of MPC and *n*-butyl methacrylate was suitable for preparing a cell membrane-inspired interface [21]. The interface could provide excellent surface properties like the biomaterials, not only related to improvement of blood and cell compatibility but also to the fabrication of a micro-fluidic device [22,23]. The essential concept of the MPC is based on assemblies of the phosphorylcholine groups, thus the strategy is really a bioinspired approach.

6.3 Alternate Soaking Process for Biomineralization and their Bio-functions

6.3.1 Nanoassembly by Polyelectrolytes

In 1997, Decher reported a fundamental concept regarding the multilayer nanoassemblies composed of polycations and polyanions [24]. The polymer assemblies were prepared by an alternate adsorption process, as shown in Figure 6.5. If a substrate has an anionic charge on its surface then polycation could adsorb onto the surface as a first layer. Polyanion could subsequently adsorb onto the surface by immersing in a

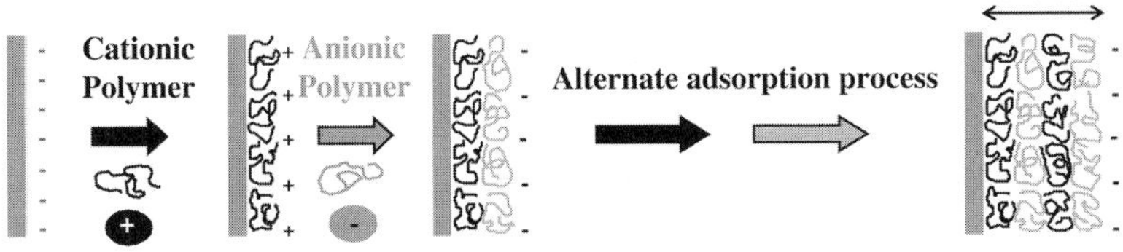

Fig. 6.5 Schematic illustration of the alternate adsorption process for nanoassemblies of polyelectrolytes onto a substrate.

polyanion aqueous solution. This process is called an alternate adsorption process. The process is capable of repetition and a multilayer of nanoassemblies would be formed. The thickness of the resulting ultra-thin membrane is variable in terms of ionic strength, which regulates the polymer conformation such as coil and globule. The driving force of the nanoassemblies is molecular interaction such as electrostatic interaction, hydrogen bonding, van der Waals forces, and charge-transfer complexation. This methodology itself is very simple, thus it can be adapted to further applications.

6.3.2
Alternate Soaking Process for Biomineralization

Biominerals such as HAp are a focus not only of materials science but also biomaterials science. The most favorable characteristic of HAp is its capability for use in bone regeneration. In the context of bone regeneration, stoichiometric HAp with a higher crystalline structure was utilized; however, this HAp could not dissolve in physiological conditions due to the high crystallinity. Though HAp is a good scaffold for bone regeneration, the HAp remains in the human body forever, therefore, bioabsorbable HAp is crucial when the aim is to produce implantable materials. A lower crystalline HAp is prepared by using calcium and phosphate ion solutions without any thermal processing. However, the resulting HAp does not have any mechanical strength, thus suitable carrier materials are necessary in order to utilize this material.

Akashi and coworkers investigated innovative bio-inorganic hybrid nanomaterials using agarose–HAp hybrid. The characteristics of agarose hydrogel are quite unique: (i) a physically cross-linked hydrogel, (ii) a bio-based material, (iii) non-immunogenicity, (iv) hydrolyzable under acidic conditions (similar pH condition relative to osteoclast behavior), and (v) non-toxicity. From these perspectives, agarose hydrogel is a good carrier material. In 1998, Taguchi and Akashi found an innovative biomineralization of agarose hydrogel using an alternate soaking process (Figure 6.6) [25]. Perhaps, they were inspired by Decher's alternate adsorption process as described above. The HAp is mainly composed of calcium and phosphate ions, therefore, the agarose hydrogel was soaked in calcium chloride and disodium hydrogen phosphate solutions repeatedly. In this process, each ion entered the agarose hydrogel by concentration gradient. The resulting materials contained HAp in the agarose hydrogel matrix. Moreover, the amount of HAp in the matrix was remarkably observed by changes in the turbidity of the hydrogel, as shown in Figure 6.7. After 12 repetitions, the agarose hydrogel became white, and bio-inorganic hybrid nanomaterials had then been prepared by the alternate soaking process. Figure 6.8 shows the total amount of HAp in the agarose hydrogel. The amount of HAp correlated well with the number of cycles. Therefore, the total content of HAp could be fully regulated by changing the preparative conditions, that is the number of cycles. Furthermore, the alternate soaking process is a very versatile methodology, thus the hydrogel and resulting biomineral can be varied. With respect to the hydrogel, poly(vinyl alcohol), poly(acrylamide), and collagen gels are adaptable

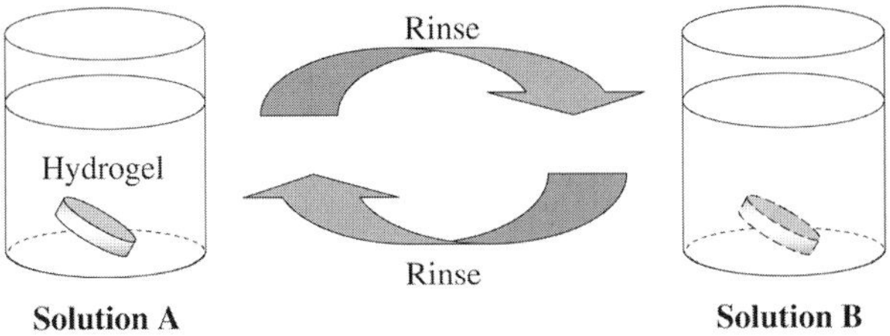

Diverse biomineralization of alternate soaking process

Combination of ionic solutions		Biomineral formation
Solution A	Solution B	
$CaCl_2$	Na_2HPO_4	$Ca_{10}(PO_4)_6(OH)_2$
$CaCl_2$	Na_2CO_3	$CaCO_3$

Fig. 6.6 Schematic illustration of the alternate soaking process for biomineralization of hydrogels and its diversity.

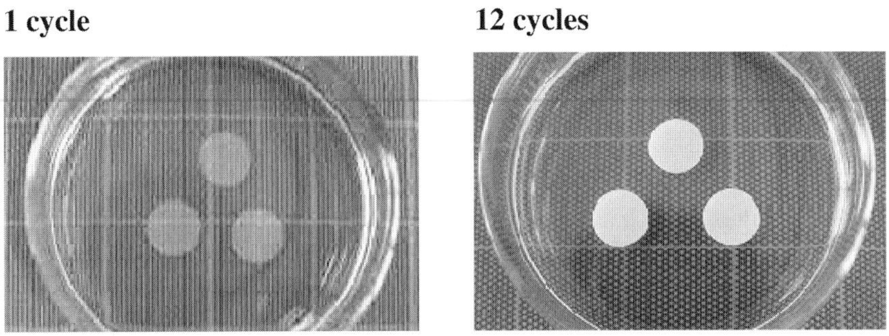

Fig. 6.7 Change in turbidity of hydrogels after 1 cycle and 12 cycles of the alternate soaking process. Round-shaped hydrogels were used (diameter: 10 mm, thickness 1 mm).

for biomineralization [26–28]. The combination of ionic solutions (calcium chloride and sodium carbonate to provide the calcium carbonate) may also be varied [29,30]. The process is quite simple and a bioinspired approach, so the methodology is capable of adaptation to another application.

6.3.3
Biomineralization of Hydrogels for Bio-functions

The resulting HAp in agarose hydrogel can be easily handled for injection, using a plastic syringe with a needle. Animal testing was carried out by a dental surgeon and/or an orthopedic surgeon. This kind of research has to involve collaboration with medical doctors, and all of the animal tests were approved by institution

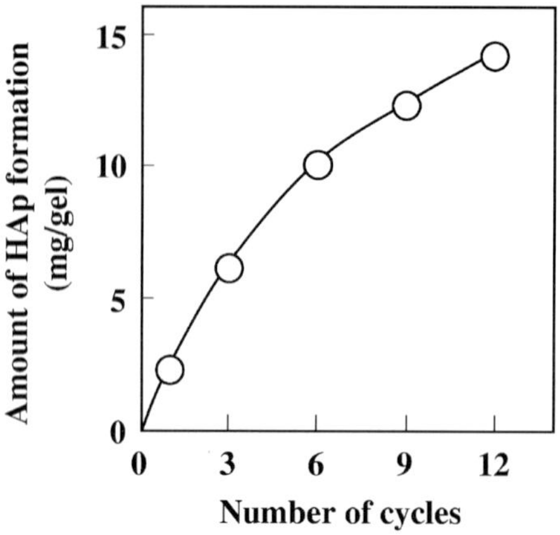

Fig. 6.8 The relationship between soaking cycles and amount of HAp formation. Round-shaped hydrogels were used (diameter: 10 mm, thickness 1 mm).

guidelines. Dental surgeons Tabata *et al.* carried out animal tests using HAp-agarose hybrid nanomaterials [31,32]. In this chapter, bio-functions are focused on using bio-inorganic hybrid nanomaterials. The biological functions involve many categories such as stimulation of cell differentiation, enhancement of cytokine production, and tissue regeneration. In particular, HAp is adapted for bone regeneration because bone tissue contains mainly HAp, roughly 70 %. Osteoconductivity is one of the biofunctions of the resulting HAp-agarose hybrid nanomaterials.

The biological behavior was investigated using monkeys (*Macasa fascicularis*). The defect model was prepared by extraction of the first premolar. The nanomaterials were injected at the defect site. The time of hemostasis was observed and it was finished within 1 min. The hemostasis property provided fine views at the surgical site in the operation, therefore it was quite an important property to use for biomedical materials. Moreover, the injectable nanomaterials still remained at the implant site, because the nanomaterials turned into a clot. Therefore, the nanomaterials provide a nice scaffold for new cells at the defect site. The bone defect was filled with regenerated bone tissue after 12 weeks implantation. The time course of the regeneration was faster than with conventional injectable materials composed of collagen and HAp particles. The implant site showed no inflammatory reaction and the implant nanomaterials disappeared, indicating bio-absorption. Moreover, six months postoperatively, neocementum was wider and cellular. These results indicated that the HAp–agarose hybrid nanomaterial is capable of conducting osteoblasts and osteoclasts. The excellent osteoconductivity is considered to be due to the bio-inorganic hybrid at the nano-scale.

6.4
Electrophoresis Process for Biomineralization

6.4.1
Innovative Methodology of Electrophoresis Process for Biomineralization

An alternate soaking process for biomineralization was introduced and the resulting bio-inorganic hybrid nanomaterials showed excellent bone regeneration, which was determined by collaboration with medical doctors. Though the alternate soaking process is a unique, simple, and adaptable methodology, the amount of HAp in the hydrogels was regulated by the diffusion of the calcium and phosphate ions subject to following equation [33]:

$$t = r^2/6D \tag{1}$$

where t is the time to reach complete diffusion, r is the diffusion distance, and D is the diffusion coefficient (in the case of immunoglobulin: 10^{-7} cm^2/s) (Figure 6.9). For example, a thick hydrogel (1 mm) needs roughly 70 min to diffuse in each solution. If the thickness is 1 cm of hydrogel, more than 115 h were needed to complete one cycle to prepare the HAp–hydrogel composite. Thus, it takes a long time to obtain homogeneous HAp formation in the hydrogel. In other words, the alternate soaking process is a time-consuming process, if the sample is a thick hydrogel.

Speeding up the ion diffusion is crucial for improvement of the process. Watanabe and Akashi discovered an innovative biomineralization using an electrophoresis process, as shown in Figure 6.10 [34]. The calcium ion and phosphate ion are positively and negatively charged, respectively. If the migration of ions into the hydrogels occurs by means of an electric field, the velocity of each ion diffusion would be accelerated. Therefore, an agarose electrophoresis apparatus was used for the biomineralization by

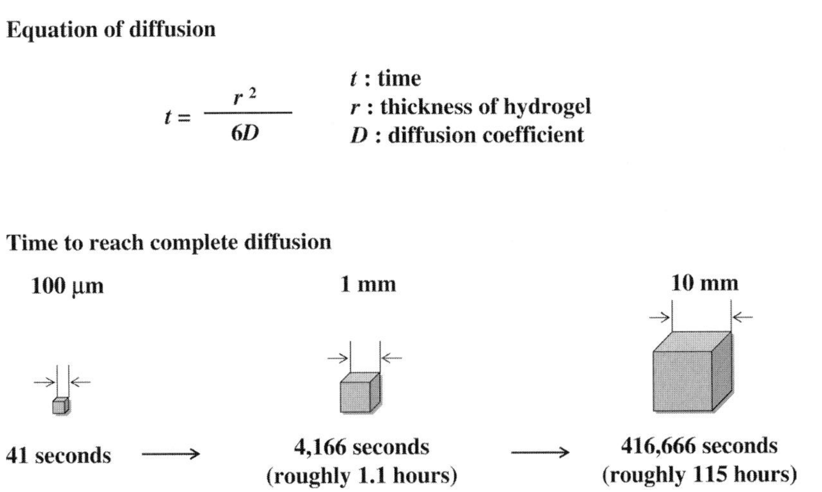

Fig. 6.9 Estimation of ion diffusion and a concrete example of time to reach complete diffusion.

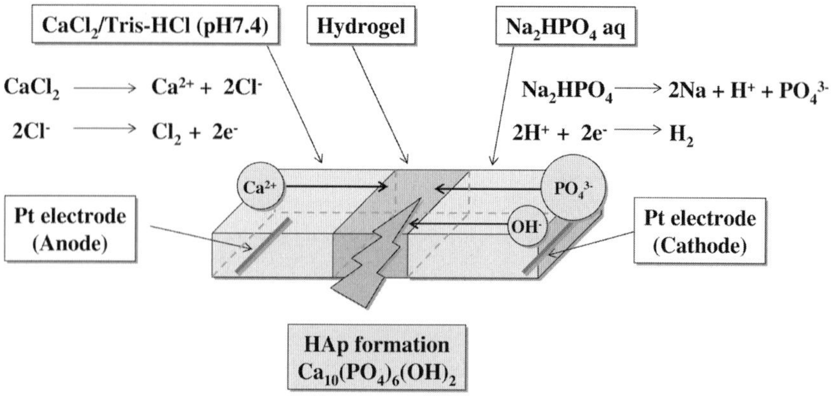

Fig. 6.10 Schematic illustration of innovative biomineralization using electrophoresis.

electrophoresis approach. The agarose hydrogel was set on the center part of the apparatus, and each cell was filled with calcium ion solution and phosphate ion solution. The electric field was applied at 100 V for 30 min, a turbid region was then observed at the center part of the agarose hydrogel. The resulting HAp–agarose hybrid contained a large homogeneous amount of HAp particles. Moreover, the particle diameter was 1 μm, with the particles being covered with needle-shaped crystals. The particle morphology was typical of HAp particles described in previous reports [35].

6.4.2 Application for Injectable Materials

The HAp–agarose hybrid nanomaterial prepared by the electrophoresis approach was also tested for orthopedic surgery implantation in rabbit bone. Figure 6.11 shows

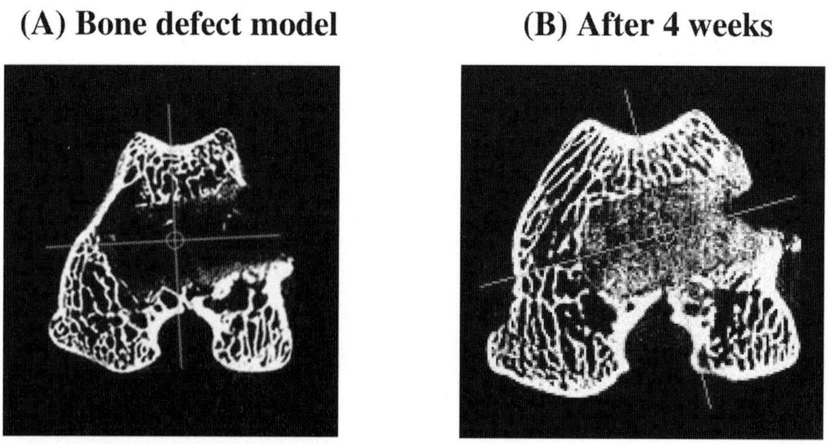

Fig. 6.11 μCT images of implantation site. Bone defect was filled with newly formed bone after 4 weeks implantation.

μ-CT images of the defect site and the same site 4 weeks after implantation. As a preliminary test, the nanomaterials could not show any high contrast images in μ-CT analysis due to the amorphous HAp and their smaller size in comparison with the μ-CT resolution (50 μm). On the other hand, some of the high contrast dots were observed in conventional high crystalline HAp containing hybrid materials, because the HAp shows high crystallinity and larger size (100 μm). From the μ-CT observation, the nanomaterial shows a stealth aspect (could not be detected by μ-CT) such as non-response to μ-CT analysis. Therefore, HAp–agarose hydrogel nanomaterials prepared by the electrophoresis approach are good candidates to research bone regeneration; the μ-CT images showed only bone regeneration without any problems attributed to HAp particles.

Figure 6.11 shows a bone defect model at the medial femoral condyle on rabbit. The drill hole did not penetrate to the end of the lateral side to suppress washing away of the composite. HAp–agarose hydrogel nanomaterials were injected into the drill hole, and μ-CT analysis was carried out. At the end of 4 weeks implantation, the interface between the HAp–agarose hydrogel nanomaterials and the edge of the drill hole was not clear, and the drill hole was filled with newly formed bone tissue. For a control sample, we injected a bare agarose hydrogel at the bone defect site. In this case, slight bone regeneration was observed by μ-CT analysis after 4 weeks implantation. However, the change in the contrast was significantly slower than with the HAp–agarose hydrogel nanomaterials. Therefore, the amorphous HAp particles may induce some osteoconduction leading to acceleration of the bone regeneration. The dominant factor for the acceleration is considered to be controlled release of the calcium ions from the amorphous HAp particles. Further results will be reported in a forthcoming paper.

The research regarding biominerals such as HAp is closely related to biomedical applications. In particular, the focus has been on bone regeneration in both dental and orthopedic surgery. There has been much basic material research on polymer materials, metal materials, and ceramic materials. The crucial strategy is to find some cross-links between fundamental research and medical therapeutics. However, clinical trials are severely regulated by the government in each country. There are a lot of stages necessary to progress the materials research. In this chapter, HAp–agarose hybrid nanomaterials for medical application were introduced as a typical example. The stages

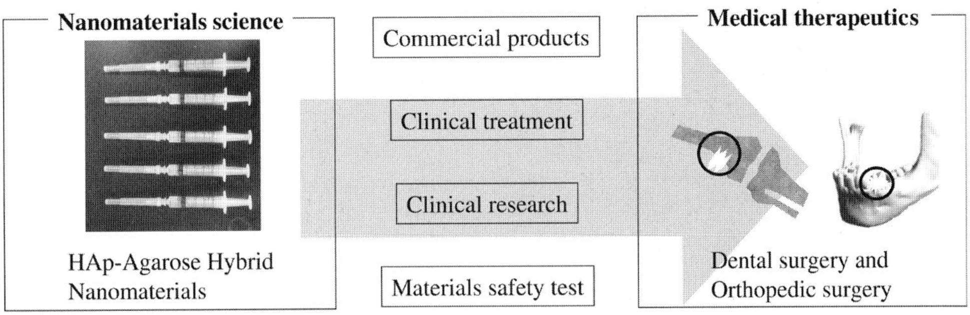

Fig. 6.12 Cross-links between nanomaterials science and medical therapeutics. There are several regulations to pass the cross-links.

to be cleared for this application were (i) materials safety data, (ii) clinical research, (iii) clinical treatment, and (iv) commercial products (Figure 6.12). Sometimes, the cross-links are called "Death valley;" however, researchers have to rise to the challenge to clear the several kinds of regulations. If not, the evolution of medical therapeutics using innovative materials will never be realized. Therefore, biomaterial researchers should collaborate well with each other through the "Bio-Inorganic Hybrid Nanomaterials".

6.5 Conclusions

In this chapter, biominerals and organic hydrogels were combined to create innovative hybrid nanomaterials; in particular, these nanomaterials target regenerative medicine for dental and orthopedic surgery. The strategy of the preparative process is mainly focused not on a biomimetic process but on a bioinspired one. If we are to discover innovative nanomaterials, we have to consider a bioinspired strategy. The bioinspired process is capable of preparing materials at the nano-scale, molecular level. Therefore, bio-inorganic hybrid nanomaterials with new bio-functions are produced.

Acknowledgements

A part of this study was financially supported by a Grant-in-Aid for the twenty first century COE Program "Center for Integrated Cell and Tissue Regulation" from The Ministry of Education, Culture, Sports, Science, and Technology, Japan. The authors would like to express their grateful acknowledgements to Dental Dr. Masashi Tabata, Miyakonojo National Hospital for animal testing using monkeys, and to Prof. Hideki Yoshikawa and Dr. Masafumi Kashii, Orthopedic Surgery, Graduate School of Medicine, Osaka University for animal testing using rabbits.

References

1 Navrotsky, A. (2004) Energetic clues to pathways to biomineralization: precursors, clusters, and nanoparticles. *Proceedings of the National Academy of Sciences of the USA*, **101**, 12096–12101.

2 Wada, N., Yamashita, K. and Umegaki, T. (1995) Effects of divalent cations upon nucleation, growth and transformation of calcium carbonate polymorphs under conditions of double diffusion. *Journal of Crystal Growth*, **148**, 297–304.

3 Wada, N., Yamashita, K. and Umegaki, T. (1999) Effects of carboxylic acids on calcite formation in the presence of Mg^{2+} ions. *Journal of Colloid and Interface Science*, **212**, 357–364.

4 Puajindanetra, S., Best, S.M. and Bonfield, W. (1994) Clinical applications of hydroxyapatite reinforced materials. *British Ceramic Transactions*, **93**, 104–107.

5 Chen, C.-W., Oakes, C.S., Byrappa, K., Riman, R.E., Brown, K., TenHuisen, K. S. and Janas, V.F. (2004) Synthesis, characterization, and dispersion properties of hydroxyapatite prepared by mechanochemical–hydrothermal

methods. *Journal of Materials Chemistry*, **14**, 2425–2432.

6 LeGeros, R.Z., Silverstone, L.M., Daculsi, G. and Kerebel, L.M. (1983) *In vitro* caries-like lesion formation in F-containing tooth enamel. *Journal of Dental Research*, **62**, 138–144.

7 Kokubo, T. (1990) Surface chemistry of bioactive glass-ceramics. *Journal of Non-Crystalline Solids*, **120**, 138–151.

8 Kokubo, T. and Takadama, H. (2006) How useful is SBF in predicting *in vivo* bone bioactivity? *Biomaterials*, **27**, 2907–2915.

9 Song, J., Saiz, E. and Bertozzi, C.R. (2003) A new approach to mineralization of biocompatible hydrogel scaffolds: an efficient process toward 3-dimensional bonelike composites. *Journal of the American Chemical Society*, **125**, 1236–1243.

10 Song, J., Malathong, V. and Bertozzi, C.R. (2005) Mineralization of synthetic polymer scaffolds: a bottom-up approach for the development of artificial bone. *Journal of the American Chemical Society*, **127**, 3366–3372.

11 Murphy, W.L. and Mooney, D.J. (2002) Bioinspired growth of crystalline carbonate apatite on biodegradable polymer substrata. *Journal of the American Chemical Society*, **124**, 1910–1917.

12 Imai, H., Tatara, S., Furuichi, K. and Oaki, Y. (2003) Formation of calcium phosphate having a hierarchically laminated architecture through periodic precipitation in organic gel. *Chemical Communications*, 1952–1953.

13 Sugawara, A., Nishimura, T., Yamamoto, Y., Inoue, H., Nagasawa, H. and Kato, T. (2006) Self-organization of oriented calcium carbonate/polymer composites: effects of a matrix peptide isolated from the exoskeleton of a crayfish. *Angewandte Chemie International Edition*, **45**, 2876–2879.

14 Gao, Y.-X., Yu, S.-H. and Guo, X.-H. (2006) Double hydrophilic block copolymer controlled growth and self-assembly of $CaCO_3$ multilayered structures at the air/water interface. *Langmuir*, **22**, 6125–6129.

15 Aizenberg, J. (2004) Crystallization in patterns: a bio-inspired approach. *Advanced Materials*, **16**, 1295–1302.

16 Nakano, T., Kaibara, K., Tabata, Y., Nagata, N., Enomoto, S., Marukawa, E. and Umakoshi, Y. (2002) Unique alignment and texture of biological apatite crystallites in typical calcified tissues analyzed by microbeam X-ray diffractometer system. *Bone*, **31**, 479–487.

17 Fujita, H., Ooya, T. and Yui, N. (1999) Thermally induced localization of cyclodextrins in a polyrotaxane consisting of β-cyclodextrins and poly (ethylene glycol)-poly(propylene glycol) triblock copolymer. *Macromolecules*, **32**, 2534–2541.

18 Ooya, T., Inoue, D., Choi, H.S., Kobayashi, Y., Loethen, S., Thompson, D.H., Ko, Y.H., Kim, K. and Yui, N. (2006) pH-responsive movement of cucurbit[7]uril in a diblock polypseudorotaxane containing dimethyl β-cyclodextrin and cucurbit[7]uril. *Organic Letters*, **8**, 3159–3162.

19 Ishihara, K., Ueda, T. and Nakabayashi, N. (1990) Preparation of phospholipid polymers and their properties as polymer hydrogel membranes. *Polymer Journal*, **22**, 355–360.

20 Kadoma, Y., Nakabayashi, N., Masuhara, E. and Yamauchi, J. (1978) Synthesis and hemolysis test of the polymer containing phosphorylcholine groups. *Kobunshi Ronbunshu*, **35**, 423–427.

21 Ishihara, K., Oshida, H., Endo, Y., Ueda, T., Watanabe, A. and Nakabayashi, N. (1992) Hemocompatibility of human whole blood on polymers with a phospholipid polar group and its mechanism. *Journal of Biomedical Materials Research*, **26**, 1543–1552.

22 Ishihara, K. (2000) Bioinspired phospholipid polymer biomaterials for making high performance artificial organs. *Science and Technology of Advanced Materials*, **1**, 131–138.

23 Sakai-Kato, K., Kato, M., Ishihara, K. and Toyo-oka, T. (2004) An enzyme immobilization method for integration of biofunctions on a microchip using a water-soluble amphiphilic phospholipid polymer having a reacting group. *Lab on a Chip*, **4**, 4–6.

24 Decher, G. (1999) Fuzzy nano-assemblies: toward layered polymeric multicomposites. *Science*, **277**, 1232–1237.

25 Taguchi, T., Kishida, A. and Akashi, M. (1998) Hydroxyapatite formation on/in poly(vinyl alcohol) hydrogel matrices using an alternate soaking process. *Chemistry Letters*, **27**, 711–712.

26 Taguchi, T., Shiraogawa, M., Kishida, A. and Akashi, M. (1999) A study on hydroxyapatite formation on/in the hydroxyl groups-bearing nonionic hydrogels. *Journal of Biomaterials Science–Polymer Edition*, **10**, 19–32.

27 Taguchi, T., Kishida, A. and Akashi, M. (1999) Apatite formation on/in hydrogel matrixes using an alternate soaking process: II. Effect of swelling ratios of poly(vinyl alcohol) hydrogel matrixes on apatite formation. *Journal of Biomaterials Science–Polymer Edition*, **10**, 331–339.

28 Taguchi, T., Kishida, A. and Akashi, M. (1999) Apatite formation on/in hydrogel matrices using an alternate soaking process (III): effect of physico-chemical factors on apatite formation on/in poly (vinyl alcohol) hydrogel matrices. *Journal of Biomaterials Science–Polymer Edition*, **10**, 795–804.

29 Ogomi, D., Serizawa, T. and Akashi, M. (2003) Bioinspired organic-inorganic composite materials prepared by an alternate soaking process as a tissue reconstitution matrix. *Journal of Biomedical Materials Research*, **67A**, 1360–1366.

30 Ogomi, D., Serizawa, T. and Akashi, M. (2005) Controlled release based on the dissolution of a calcium carbonate layer deposited on hydrogels. *Journal of Controlled Release*, **103**, 315–323.

31 Tabata, M., Shimoda, T., Sugihara, K., Ogomi, D., Serizawa, T. and Akashi, M. (2003) Osteoconductive and hemostatic properties of apatite formed on/in agarose gel as a bone-grafting material. *Journal of Biomedical Materials Research Part B: Applied Biomaterials*, **67B**, 680–688.

32 Tabata, M., Shimoda, T., Sugihara, K., Ogomi, D., Ohgushi, H. and Akashi, M. (2005) Apatite formed on/in agarose gel as a bone-grafting material in the treatment of periodontal infrabony defect. *Journal of Biomedical Materials Research Part B: Applied Biomaterials*, **75B**, 378–386.

33 Berg, H.C. (1993) Random Walks in Biology,Princeton University Press,NJ, USA, p. 5.

34 Watanabe, J. and Akashi, M. (2006) Novel biomineralization for hydrogels: electrophoresis approach accelerates hydroxyapatite formation in hydrogels. *Biomacromolecules*, **7**, 3008–3011.

35 Zhang, R. and Ma, P.X. (1999) Porous poly(L-lactic acid)/apatite composites created by biomimetic process. *Journal of Biomedical Materials Research*, **45**, 285–293.

7
Bioinspired Porous Hybrid Materials via Layer-by-Layer Assembly

Yajun Wang, Frank Caruso

7.1
Introduction

Hybrid porous materials are attracting considerable interest for applications such as biocatalysis, biosensing, and bioseparation, because they combine benefits originating from their unique pore structure and intrinsic composite properties. These materials are generally composed of a porous inorganic framework, which acts as a robust support for functionalization with organic species or biological molecules. The unique pore channels make it possible to use the functional sites situated in the inorganic framework. This chapter discusses recent trends in using porous inorganic materials with pore sizes from several nanometers to micrometers as substrates for biologically oriented applications, emphasizing the "layer-by-layer" (LbL) method. The flexibility of the LbL strategy will be demonstrated by a number of examples, where various species are deposited in porous substrates with different morphologies and pore structures. We will focus on porous materials in which new biological properties have been engineered, and highlight our recent investigations into the bioapplication of mesoporous silica particles. Additionally, we will briefly discuss the potential applications of the novel bio-hybrids with different pore structures.

7.2
Porous Materials

Materials with uniform pore structures offer a wide range of applications, including catalysis, adsorption, and separation. These materials have the benefit of both specific pore systems and intrinsic chemical properties [1–3]. The pores in the materials are able to host guest species and provide a pathway for molecule transportation. The skeletal pore walls provide an "active" and/or "affinity" surface to associate with guest molecules. According to the International Union of Pure and Applied Chemistry (IUPAC), porous materials can be classified into three main categories based on the diameters of their pores, that is, microporous, mesoporous, and macroporous

Bio-inorganic Hybrid Nanomaterials. Edited by E. Ruiz-Hitzky, K. Ariga, Y. M. Lvov

ISBN: 978-3-527-31718-9

materials with pore diameters of <2, 2–50, and >50 nm, respectively [4]. In this section, we will introduce microporous materials, particularly zeolites, and discuss recent developments in mesoporous silica materials. Finally, several representative materials with macroporosity will be presented.

7.2.1 Microporous Materials

The preparation of ordered porous materials began with the preparation of the first synthetic zeolites in the 1950s [5]. Zeolites are crystalline materials with wide commercial applications as absorbents, molecular sieves, and shape-selective catalysts because of their extremely well-defined structures, high specific surface areas, good thermal/hydrothermal stability, and uniform microporous channels. Due to the relatively small pores in zeolites (<2 nm), conventional applications for them are where small molecules (5–12 Å) are present. Recently, they have been used in the purification of nucleic acids and proteins, based on electrostatic and hydrophobic interactions between the biomacromolecules and the zeolite substrates [6,7]. Nygaard *et al.* reported that certain proteins were able to distinguish between the crystallographic planes of aluminum silicate zeolites, binding to their cognate zeolite surfaces with high affinity and allowing enzyme immobilization without loss of specific activity [8]. Hayakawa *et al.* used surfactant-modified P- and X-type zeolites as drug carriers for the release of chloroquin [9]. Enhanced properties have been reported for zeolite materials with nanosized crystals in enzyme immobilization [10] and protein separation [11,12], due to the dramatically increased external surface area of zeolite nanocrystals. However, in all of these studies, the biological molecules only associate with the external surface of the zeolite crystals, not the abundant internal surface of the micropores, which constitute most of the total surface area of the zeolite materials. For applications involving larger molecules, such as biological compounds or polymers, materials with larger pores are often more useful. In the following sections, we will focus on materials with pore sizes in the mesoporous to macroporous range and their use in biological applications.

7.2.2 Mesoporous Material

The ability to produce materials with larger pores has led to new opportunities for the application of porous materials. This began with the discovery of mesoporous silicas (MSs) by Kresge *et al.* from the Mobil company in 1992 [13]. In the past 15 years, MS materials have attracted increasing research interest, and have opened up new possibilities in many areas of chemistry and materials science. Nowadays, the popular term "nanoporous" silica is often used, which can be defined as materials having pores in the nanometer range. The MS materials are typically prepared using self-assembled molecular arrays (e.g., surfactant micelles) as templates (Figure 7.1), in which the silica precursors are solidified to form a silica–surfactant composite [14].

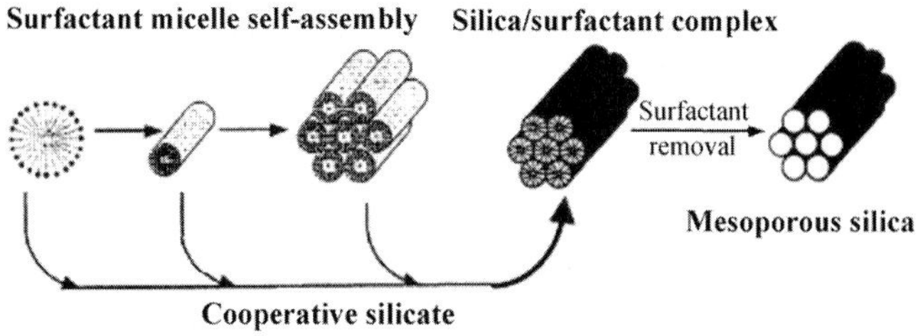

Fig. 7.1 Mechanistic pathway for the formation of mesoporous silica. (Adapted from [14] with permission of the American Chemical Society, Copyright 1997 American Chemical Society).

After removal of the surfactant template, a regular arrangement of mesopores with a narrow pore size distribution is obtained in the replicated silicas.

Outstanding features of MS materials include their high specific surface areas (up to about 1500 m^2/g), high specific pore volumes (up to about 2 cm^3/g), and well-ordered pore structures. This renders MS materials ideal candidates as porous substrates for loading with large quantities of drug and biological molecules. MS materials also fulfill some of the additional requirements for application as substrates for biological molecules, such as adjustable porosity and particle morphology, good chemical, thermal and mechanical stability, and low-toxicity (for cells). Details on the application of MS materials as supports for protein and drug loading can be found in recent reviews by Hartmann [15] and Yiu and Wright [16].

Recent research shows that it is critical to decrease the MS particle size to the nanometer range to deliver drug-loaded MS particles through cell membranes to various intracellular targets, because most cell uptake occurs in this size range [17]. For example, MS with an average particle size of 250 nm was reported as a gene transfection system. Lin *et al.* applied nanosized MSs loaded with fluorescent dyes and gadolinium as markers in cell tracking and novel contrast agents for magnetic resonance imaging (MRI), respectively [18,19]. The biocompatibility of these novel particles to cells has also been demonstrated. Lin and coworkers demonstrated that the uptake of MS nanoparticles by cells can be regulated by surface functionalization of the nanoparticles [20]. In that work, the efficiency of endocytosis of MS nanoparticles with different charge profiles on human cervical cancer cells was investigated. In addition to the surface charge, folate groups on the nanoparticles could also play an active role in facilitating receptor-mediated endocytosis, since the membranes of human cancer cells are abundant with folate receptors.

7.2.3 Macroporous Materials

Compared with microporous and mesoporous materials, the larger, interconnected voids in macroporous materials potentially provide easier molecule transportation through the materials. This is of particular interest for the transport of large biomolecules (e.g., proteins and cells). The pore sizes in macroporous materials are usually from tens to hundreds of nanometers, and the pores are typically distributed and interconnected. Macroporous materials can be prepared with various morphologies including particles, fibers, membranes, and three-dimensional ordered macroporous (3DOM) materials. Here, we briefly introduce some macroporous materials that are employed as substrates for LbL assembly.

Anodized alumina oxide (AAO) membranes are materials with macropores commonly used in templating synthesis. They are formed by anodizing aluminum in different electrolytes, by applying a direct current with potentials from tens to hundreds of volts. The pores in the membrane are packed hexagonally, with amorphous boehmite forming the pore wall [21]. Pore densities in the membrane can be as high as 10^{11} pores cm^{-2}, while the bulk porosity remains the same ($\sim$30 %) because the small pores are packed more densely than the larger pores [22]. The pore diameter can be controlled by the voltage of the anodic potential, with the pore diameter increasing by 1.5 nm for every volt applied [23]. The pore length, typically up to $\sim$100 μm, is proportional to the time of anodic oxidation. Increasing this time leads to an increase in the oxide layer thickness until equilibrium is reached between oxide formation and dissolution of the porous layer in the electrolyte. AAO membranes with submicrometer-sized pores are often used as templates to prepare nanowires and nanotubes of various compositions and morphologies for biological applications [24–26].

Three-dimensionally ordered macroporous (3DOM) materials are prepared by utilizing colloidal crystals as templates [27]. These are similar to an opal in structure and consist of ordered arrays of monodisperse colloidal spheres, such as emulsion droplets, polymer or silica microspheres with diameters in the submicrometer to micrometer range. The interstices of the opal template are filled with suitable precursors such as monomers, preformed polymers, and inorganic precursors, followed by polymerization of the precursors. 3DOM materials are obtained after removal of the opal template from the composite by chemical etching or calcination. The products, often called "inverse opals," are a close-packed array of macropores with a regular, three-dimensional spacing and long-range order. The size of the macropores can be tuned by the particle size in the opal template. The macropores are commonly connected with "windows" induced by the former contact points between neighboring spheres. To date, 3DOM materials of diverse composition, including polymers [28–31], metals [32], and inorganic oxides [33–36] have been prepared. A review paper by Schroden and Stein has summarized the synthesis, characterization, and application of 3DOM materials [37].

Many natural materials are intrinsically porous. For example, diatomite, also known as diatomaceous earth or kieselguhr, possesses unique macroporous channel structures with pore sizes ranging from submicrometers to tens of micrometers [38]. Diatomite is a sedimentary rock primarily composed of the fossilized remains of unicellular fresh algae known as diatoms [38]. Diatomite is an inert, lightweight, highly porous, super absorbent material, and has wide industrial applications, including as a sound and heat insulator, absorbent, filter, filler, and as a lightweight building material. Owing to its inherent macroporous structure, low cost and availability, diatomite has been recently used as a template to fabricate macroporous carbon [39], zirconia [40], magnesium oxide [41], and zeolite [38,42,43] materials with intricate three-dimensional structures.

7.3 LbL Assembly

A convenient method to prepare hybrid materials is to exploit self-assembly, by which different species ranging from inorganic to organic materials, can be assembled together at the molecular level. Due to the combination of benefits from different materials, hybrid assemblies often have enhanced properties compared with the single components. Electrostatic attraction, hydrophilicity–hydrophobicity, and H-bonding are common driving forces in self-assembly processes. Langmuir-Blodgett deposition, self-assembly of monolayers (SAMs), and LbL assembly have been the dominant techniques developed in this area.

The origins of LbL assembly can be traced back to the work of Iler in the 1960s on sequentially-deposited charged particles [45]. However, for more than two decades, there was limited progress made in the area of LbL assembly. In the early 1990s, Decher and Hong demonstrated that positively and negatively-charged polyelectrolytes (PEs) could be LbL-assembled on substrates for PE multilayer construction [46]. Since then, the LbL technique has rapidly expanded to become a popular method to prepare layered thin films with tailored properties on the nanometer scale. The versatility of the technique has allowed a broad range of charged materials (e.g., polymers, enzymes, proteins, nanoparticles and dye molecules) to be assembled into nanostructured materials. Further details on the progress of the LbL technique can be found in a number of review papers [47–52].

The benefit of the LbL technique is that the properties of the assemblies, such as thickness, composition, and function, can be tuned by varying the layer number, the species deposited, and the assembly conditions. Further, this technique can be readily transferred from planar substrates (e.g., silicon and quartz slides) [53,54] to three-dimensional substrates with various morphologies and structures, such as colloids [55] and biological cells [56]. Application of the LbL technique to colloids provides a simple and effective method to prepare core–shell particles, and hollow capsules, after removal of the sacrificial core template particles. The properties of the capsules prepared by the LbL procedure, such as diameter, shell thickness and permeability, can be readily adjusted through selection of the core size, the layer number, and the nature of the species deposited [57]. Such capsules are ideal candidates for applications in the areas of drug delivery, sensing, and catalysis [48–51,57].

Recently, the LbL technique has been extended from conventional nonporous substrates to macroporous substrates, such as 3DOM materials [58,59], macroporous membranes [60–63], and porous calcium carbonate microparticles [64,65], to prepare porous PE-based materials. LbL-assembly of polyelectrolytes can also be performed on the surface of MS particles preloaded with enzymes [66,67] or small molecule drugs [68], and, under appropriate solution conditions, within the pores of MS particles to generate polymer-based *nanoporous* spheres following removal of the silica template [69].

In the next section, we will highlight recent developments in the engineering of mesoporous and macroporous substrates via the LbL procedure to produce porous, hybrid materials for various bioapplications.

7.4
LbL Assembly on MS Substrates

7.4.1
Encapsulation of Biomolecules in MS Particles

The development of versatile and efficient encapsulation technologies is of interest in medicine, pharmaceutics, and cosmetics [70,71]. Liposomes, microemulsion droplets, and hollow capsules are currently used to encapsulate materials, including drugs, pesticides, fragrances, and biomolecules [72]. Recent progress has demonstrated that the LbL technique can be used to prepare hollow capsules with various dimensions, compositions, and properties [73–78]. There are two dominant encapsulation approaches based on the LbL technique. The first involves the direct coating of a crystal core of the material to be encapsulated (e.g., low molecular weight compounds, enzymes) with PE multilayers [73–76]. A limitation of this method is the limited number of materials that can be crystallized for encapsulation. The second approach is to load preformed LbL-assembled polymer capsules by altering the shell permeability via changing the environmental solution conditions (e.g., pH) [77,78]. However, low degrees of loading (particularly for larger biomacromolecules) and, often poorly reproducible data, limit the application of this capsule loading method [56].

Recently, we proposed an alternative process for encapsulating biomacromolecules within PE microcapsules. This approach involves using nanoporous particles as sacrificial templates for both enzyme immobilization and PE multilayer capsule formation (Figure 7.2, route (I)) [66,67]. Unlike previous LbL encapsulation strategies, this approach is not limited to species that undergo crystallization, and is not dependent upon adjustments in electrostatic interactions within PE microcapsules to alter shell permeability characteristics. The salient feature of this method is that it is applicable to a wide range of materials for encapsulation.

MS materials are ideal porous substrates for such an encapsulation strategy for several reasons. First, the MS materials have a very large surface area (and pore volume), which enables a high degree of loading. Second, the silica materials are robust in biological environments, which is important for applications in enzyme immobilization and protein separations. Third, MS materials exhibit low toxicity to biological systems. Finally, MS materials can be prepared with tunable pore sizes in the nanometer range, and with controllable morphologies such as spheres, fibers, rods, and membranes, to satisfy specific applications. An example of MS materials with spherical morphology is shown in Figure 7.3. To achieve high protein loadings, bimodal mesoporous silica (BMS) spheres with sufficiently large pore sizes and volumes ($\sim$1.7 mL g^{-1}) have been used [80]. BMS spheres have a relatively wide particle size distribution of 2–4 μm (Figure 7.3A), and their relatively homogeneous

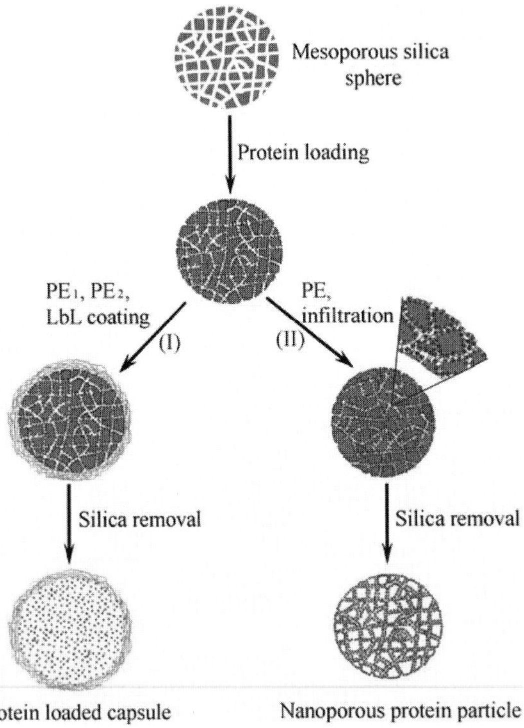

Fig. 7.2 Schematic representation of the procedure for the encapsulation of enzyme in PE microcapsules (I) and preparing nanoporous protein particles (II) using MS spheres as templates.

inner structure can be seen from the ultramicrotomed samples (~90 nm thick slices) in Figure 7.3B. At higher magnification, the disordered pore structure with pore sizes of 10–40 nm becomes apparent (Figure 7.3C).

A range of enzymes with different molecular sizes and isoelectric points (pI) (e.g., catalase, peroxidase, cytochrome C, lysozyme, transferrin, urease and bovine serum albumin (BSA)) have been immobilized in BMS spheres [67,87–89]. High enzyme loadings (20–40 wt%, 200–400 $mg\,g^{-1}$) and rapid uptake (several minutes) were observed in BMS spheres for enzymes of size $\leq$3 nm and pI $\geq$10 (e.g., cytochrome C, lysozyme and protease) [67]. Catalase has a relatively large molecular size (about 10 nm) and a net negative charge at pH 7.0 (pI 5.4). Approximately 75 $mg\,g^{-1}$ was immobilized within BMS particles, which was about 10 times more than that immobilized in MS spheres (denoted as HMS spheres) with only small mesopores (~2–3 nm) (Figure 7.3D) [81]. The significantly different loading ability of the MS particles is further evidenced by confocal laser scanning microscopy (CLSM) images of the particles after dye-labeled protein loading [67]. The bright spheres (Figure 7.4A) are due to the homogeneous distribution of fluorescein isothiocyanate-labeled peroxidase (FITC-POD) in the BMS particles with high enzyme loadings. For the HMS spheres (with only small mesopores), distinct fluorescent rings with

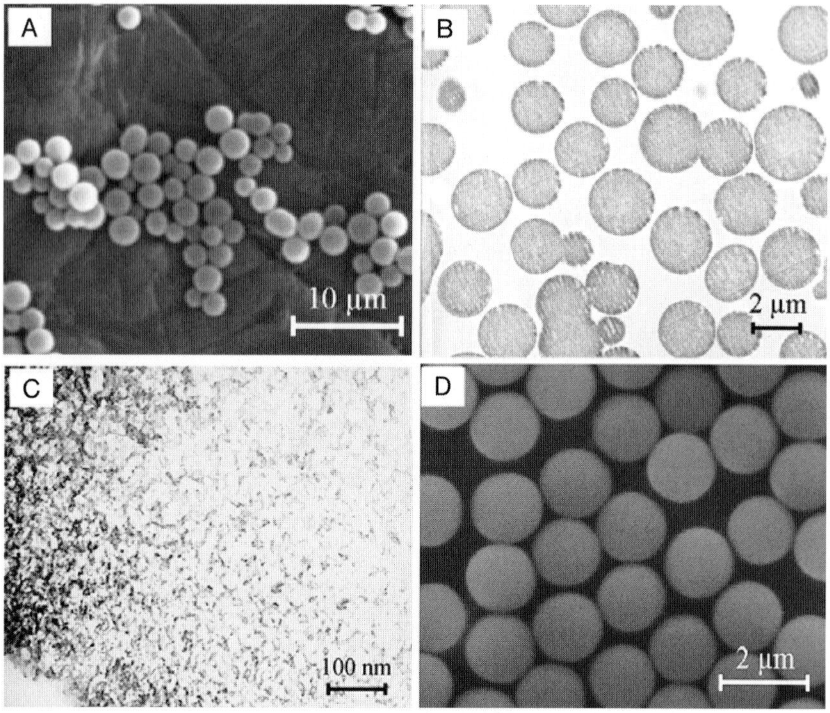

Fig. 7.3 SEM images of BMS spheres (A) and TEM images of ultramicrotomed BMS spheres (B, C), and SEM image of HMS spheres (D). (Adapted from [67] with permission of the American Chemical Society, Copyright 2005 American Chemical Society).

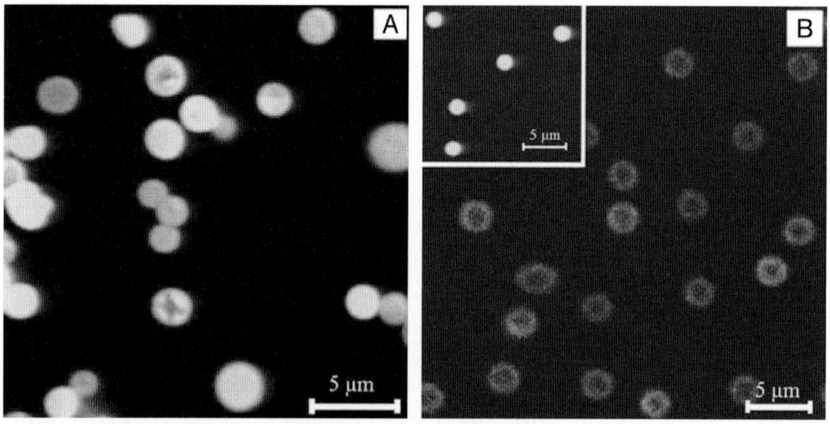

Fig. 7.4 CLSM images of BMS spheres (A) and HMS spheres (B) loaded with FITC-labeled POD. The inset in (B) corresponds to Rhodamine 6G-loaded HMS spheres. (Reproduced from [67] with permission of the American Chemical Society, Copyright 2005 American Chemical Society).

significantly less fluorescence intensity are observed, indicating localization of most of the POD on the sphere surface (Figure 7.4B). The CLSM image of bright spheres after dye (Rhodamine 6G, M_w 479) loading (Figure 7.4B, inset) suggests interconnectivity of the mesopores in the HMS spheres.

After enzyme loading, a multilayer shell is assembled on the MS sphere surface to encapsulate the protein in the nanoporous particles via the LbL procedure. Shells with various compositions and different properties have been deposited [66–68,84,87–89]. To date, the reported layer pairs assembled on MS particles include (but are not limited to): poly(diallyldimethylammonium chloride) (PDDA)/poly(sodium 4-styrenesulfonate (PSS); poly(allylamine hydrochloride) (PAH)/PSS; poly(L-glutamic acid) (PGA)/poly(L-lysine) hydrobromide (PLL); chitosan/dextran sulfate; poly(vinylpyrrolidone) (PVON)/poly(methacrylic acid) (PMA); and PDDA/silica nanoparticles. During the assembly of the multilayer shells (especially the first polyelectrolyte layer), the loaded protein is partially released from the MS particle because of competitive adsorption between the polyelectrolyte and protein at the silica surface. The conditions (e.g., pH, salt concentration) of the PE assembly solution also influence protein desorption. A small amount (<5 %) of catalase is lost from catalase-loaded BMS particles when a PLL/PGA shell is assembled at pH 5.5 in a 0.05 M 2-(N-morpholino)ethanesulfonic acid (MES) buffer. In contrast, enzyme loss of around 40 % is observed when PLL/PGA coatings are deposited at pH 7 [82]. The leakage of protein dramatically decreases during subsequent deposition steps, and is negligible after three bilayers of PDDA/PSS are deposited [67]. This indicates that a PE multilayer coating of six layers is sufficient to encapsulate the loaded protein in the MS particles.

There is a direct relationship between enzyme activity and the number of layers deposited on the MS particles, due to an increase in shell thickness and a decrease in shell permeability. When encapsulating the protein catalase in BMS particles, its activity decreased with an increase in layer number [67]. The type and structure of the coatings also influence the activity of encapsulated enzymes [67]. For example, with the same layer number, catalase has less activity when encapsulated in PDDA/PSS-coated BMS particles compared with PDDA/silica nanoparticle-coated particles. This is attributed to the PDDA/PSS coatings being less permeable. Enzyme stability is enhanced with an increase in the number of layers [67]. For example, catalase-loaded BMS spheres coated with two, four, and six bilayers of PDDA/PSS yielded an activity of 15, 29, and 47 %, respectively, after 120 min. Another advantage of enzymes encapsulated in BMS spheres is their reusability with substantial retention of enzyme activity. The activity of catalase encapsulated in the BMS particles with four PDDA/silica nanoparticle bilayer coatings decreased by only about 4 % after five successive batch reactions. Significantly, more activity (about 30 %) was lost if the enzyme was only loaded in the BMS particles without a coating.

The multilayer shells can also provide a protective barrier for the loaded enzyme in environments where enzyme-degrading substrates such as proteases may be present [67]. Dissolved catalase was inactivated immediately by protease, losing its entire activity within 60 min in solution. For catalase loaded in BMS spheres, inactivation is slower, with an activity loss of about 20 % in 60 min. Notably, a negligible decrease in activity is found for catalase encapsulated in BMS with a four bilayer PDDA/PSS shell. In this system, the PE multilayers form a dense shell with micropores, through which only small molecules (molecular weight below ~4000) can penetrate [83]. The permeability of a PE shell deposited on the surface of BMS particles can be probed via a molecular beacon (MB) approach [84]. This technique involves immobilizing MBs [85] within BMS particles (BMS_{MB}) and encapsulating the BMS_{MB} particles within the PE multilayer shell to be probed (e.g., PAH/PSS). The encapsulated BMS_{MB} particles are then incubated with DNA target sequences of different lengths. Permeation of the DNA targets through the shell into the BMS_{MB} particles causes the immobilized MBs to open, due to hybridization of the DNA targets with the complementary loop region of the MBs, resulting in an increase in the MB fluorescence.

Encapsulation of small drugs in MS has also been demonstrated by a similar strategy [68]. In that work, the MS spheres were loaded with ibuprofen and then encapsulated within eight layers of PAH and PSS on the particle surface to cap the mesopore openings [68]. The encapsulated drug molecules were subsequently released from the MS particles under the influence of solution pH and salt concentration.

7.4.2
MS Spheres as Templates for the Preparation of Hollow Capsules

Sacrificial solid cores (e.g., silica and polymer microspheres) are usually used as template particles for hollow capsule preparation via the LbL method [55,56]. In this process, the cores are first coated with multiple layers of polyelectrolytes, preformed inorganic nanoparticles, or inorganic molecular precursors, utilizing electrostatic attraction for construction of the layers on the particles. Hollow capsules are obtained after removal of the core template, and loading of the capsules is possible with some species. As we have shown, a broad range of materials can be encapsulated at high concentrations in MS spheres by a shell that is LbL-assembled on the surface of the MS spheres. Accordingly, after the MS substrate is dissolved, polymeric capsules with pre-loaded proteins can be prepared. This yields capsules with high enzyme loadings, retained enzyme activity, and, notably for release studies, enzymes in a "free" (non-immobilized) state [82]. The porous silica template cores can be removed by exposure to a diluted HF solution. Silica particles readily decompose in HF to form $[SiF_6]^{2-}$, which can easily diffuse through PE multilayers [79]. To retain protein functionality, the pH of the HF solution can be adjusted by adding ammonium fluoride (NH_4F) to form a HF/NH_4F buffer [86]. The effective removal of silica from the samples is shown by various techniques, including energy-dispersive X-ray analysis [82], Fourier transform infrared spectroscopy [69], and inductively coupled plasma analysis [88].

The morphology of the capsules obtained shows some deformation/collapse upon drying (Figure 7.5A) [82], similar to that of polymeric capsules prepared using nonporous particles as templates. CLSM examination of FITC-labeled catalase encapsulated in $(PLL/PGA)_3$ microcapsules showed that the catalase is uniformly distributed throughout the capsules (Figure 7.5B). The homogeneous fluorescence suggests that most of the catalase is in its free state inside the capsule [82]. Further

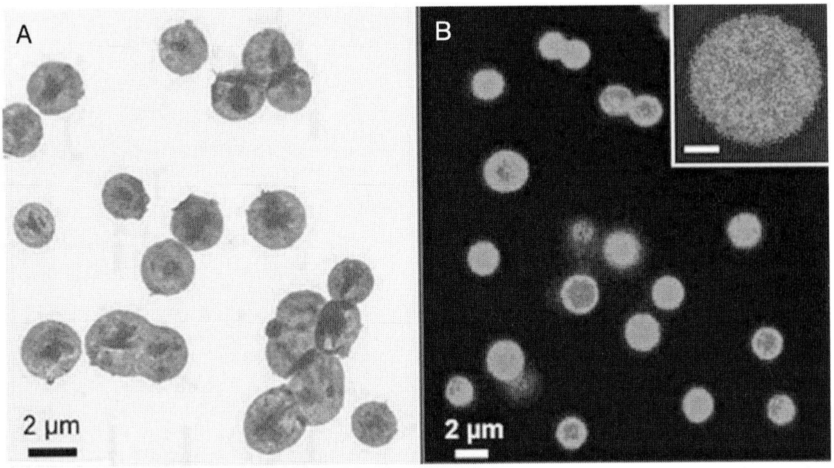

Fig. 7.5 TEM image of microcapsules prepared by LbL assembly of three bilayers of a PLL/PGA shell on catalase-loaded BMS spheres, following removal of the BMS particle template (A). CLSM images of $(PLL/PGA)_3$ microcapsules loaded with FITC-labeled catalase (B). The scale bar in the inset corresponds to 800 nm. PLL/PGA layers were assembled from a 0.05 M MES, pH 5.5 buffer. The MS spheres were dissolved using HF/NH_4F at pH ~5. (Adapted from [82] with permission of Wiley-VCH).

evidence for this is provided by CLSM images of catalase-loaded $(PLL/PGA)_3$ microcapsules prepared using FITC-labeled PLL, which show that the fluorescence intensity from PLL only arises from the capsule shell and not the interior.

Release of encapsulated drugs and proteins can be achieved by changing the permeability of the capsule shells through various methods, such as environmental solution changes (e.g., pH, and salt) [82], chemical triggers [87], and enzymatic degradation [88] of the capsule shells. For instance, weak PEs (e.g., PLL and PGA) are sensitive to changes in solution pH and salt concentration. Proteins encapsulated in the $(PLL/PGA)_3$ capsules can be released through changing the physicochemical properties of the constituent polymers forming the membrane shell, which undergo conformational changes upon changing the conditions of the medium [82]. For example, "burst" release (about 80 %) of enzyme from PLL/PGA capsules is observed when the solution pH is changed from 5.5 to 2. Capsules composed of hydrogen-bonded multilayer shells of $PVPON/PMA_{SH}$, with additional stabilization of the shell via thiol-disulfide chemistry, have also been studied as a model for the chemically triggered release of encapsulated contents [87]. Transferrin-loaded capsules are prepared by LbL assembly of $PVPON/PMA_{SH}$ multilayers on BMS particles pre-loaded with transferrin, followed by removal of the porous silica. The loaded capsules are stable at physiological pH, however, with addition of the triggering agent (dithiothreitol, DTT, a thiol-disulfide exchange reagent), the capsule shells are disassembled. The encapsulated protein is quickly released into the bulk solution due to the deconstruction of the $PVPON/PMA_{SH}$ multilayers after cleavage of the S–S bonds by DTT. This offers a promising way to design polymeric capsules for delivery applications, as protein-assisted thiol-disulfide exchange leading to deconstruction of the capsules and triggered release of the cargo drugs may occur in an intracellular environment. Sustained release of the encapsulated cargo is achieved by enzymatic degradation of the hollow capsule membranes, where the permeability of the polymeric shell is gradually increased. As an example of this, Itoh *et al.* prepared biodegradable hollow capsules via the LbL assembly of chitosan and dextran sulfate on protein-loaded MS particles, followed by the removal of the silica particle templates [88]. To release the encapsulated proteins, an enzyme (chitosanase) is added to the capsule solution. The chitosan component on the capsule shell is degraded by chitosanase, and the capsules deform and are finally destroyed. The release profile of the encapsulated protein due to enzymatic degradation of the hollow capsules can be controlled over a period of several days.

Capsules with high enzyme loading and activity prepared by templating BMS spheres can also be used as biomimetic reactors [89]. For example, PGA/PLL capsules with pre-loaded urease (prepared via a BMS sacrificial template) are capable of catalyzing the hydrolysis of urea and have been shown to induce the exclusive formation of $CaCO_3$ particles inside the capsules [89].

In summary, the encapsulation technique of templating MS spheres is significant for a number of reasons: (i) a range of enzymes can be encapsulated in microcapsules comprised of a diverse range of PEs; (ii) the encapsulated enzyme can be readily released by altering the permeability of the polymeric shells; and (iii) this technique is readily transferable to encapsulate various species, ranging from (bio)macromolecules, drugs, fragrances and nanoparticles, due to the immobilization capacity of MS for various materials and the versatility of the LbL method.

7.4.3
Preparation of Protein Particles via MS Sphere Templating

Biomaterials designed with different morphologies are of interest for applications in biosensing, catalysis, separations, and drug delivery. For example, peptides and proteins in the form of nanofibrils [90,91], nanotubes [92–94], and spherical assemblies [95–98], have been prepared by various methods, including self-assembly, acid–base chemistry, AAO membrane templating, ultrasonic emulsification, and precipitation in organic solvents. Suslick and coworkers prepared micron-sized core–shell protein microspheres, with an outer shell made from disulfide cross-linked protein roughly 50 nm thick and a core containing air or a nonaqueous liquid [97]. These microspheres can be easily prepared during ultrasonic emulsification, particularly from serum albumin. The prepared protein particles have been used in biomedical imaging and as vesicles to encapsulate hydrophobic material [97]. Another promising application of the protein particles is in targeted delivery, following LbL modification of the protein microspheres with integrin receptors that are overexpressed in several tumor types [97].

Recently, a general method to prepare protein-based particles with tailored properties through MS sphere templating was reported [98]. In this approach, polyelectrolytes are infiltrated into the mesopores of MS spheres pre-loaded with proteins to "cross-link" the proteins. Following removal of the MS template, self-standing nanoporous protein-based particles (NPPs) are obtained (Figure 7.2, route II). Unlike the encapsulation technique using MS particles, where positively-charged PEs are electrostatically adsorbed onto the negatively-charged silica surface as the first layer, the PEs used to form the NPPs are negatively-charged. The negatively-charged PEs (e.g., PAA, PGA, PSS) complex with the oppositely-charged proteins pre-loaded in the MS spheres, rather than the silica surface which has the same charge, thereby forming a protein/polyelectrolyte complex. The stability of the protein/PE complex can be further enhanced through the formation of amide bonds between the $-COOH$ groups of the PE (PAA or PGA) and the $-NH_2$ moieties of the pre-loaded protein, using 1-ethyl-3-(3-dimethylaminopropyl) carbodiimide hydrochloride (EDC) as a catalyst [99]. Negligible enzyme desorption ($<0.1\,\%$) is observed from cross-linked lysozyme-loaded MS spheres after exposing the samples to an aqueous solution for 48 h, while about 25 % of the immobilized lysozyme is desorbed under the same conditions when the lysozyme is not cross-linked.

After removal of the MS template by exposing the samples to a buffer solution of HF/NH_4F, free-standing NPPs are obtained (Figure 7.6). The protein loadings in the MS spheres play a key role in the morphology of the NPPs. For relatively small proteins (e.g., lysozyme, cytochrome C), the NPPs retain the original spherical shape of the MS templates (Figure 7.6), and do not show signs of collapse, as is typically observed for PE capsules. The prepared lysozyme-NPPs have a protein content of about 80 wt% and diameters ranging from 1.6–2.4 μm [98], about 20 % smaller than the MS template particles (Figure 7.6A). Disordered nanopores (pore sizes of ~10–50 nm) are clearly seen by TEM of ultramicrotomed spheres (~90 nm thick slices) (Figure 7.6B). For the relatively large catalase, non-spherical ("collapsed") particles are also observed.

The influence of complexation and cross-linking with the PE on enzyme activity was assessed using the protein catalase [98]. The activity of catalase loaded in the BMS particles was normalized at 100 %. The activity decreased slightly to ~90 % after PAA infiltration. After EDC-induced cross-linking, the MS sphere-immobilized and cross-linked protein retained 75 % of its activity, which is significantly higher than that for catalase cross-linked by glutaraldehyde (GA) in chitosan beads (~1 % activity) [100]. EDC-induced cross-linking, therefore, provides a simple and effective pathway to immobilize and stabilize enzymes/proteins in porous matrices through the formation of protein networks that are bridged by polymers. The catalase activity increased from 75 to 86 % after removal of the MS template, suggesting increased substrate accessibility [98].

7.4.4
Template Synthesis of Nanoporous Polymeric Spheres

Porous polymer materials, especially in particulate form, are of interest in a diverse range of applications, including controlled drug delivery, enzyme immobilization, molecular separation technology, and as hosts for chemical synthesis [101–104]. MS materials have been used as hosts for the template synthesis of nanoporous polymer replicas through *in situ* polymerization of monomers in the mesopores [105–108].

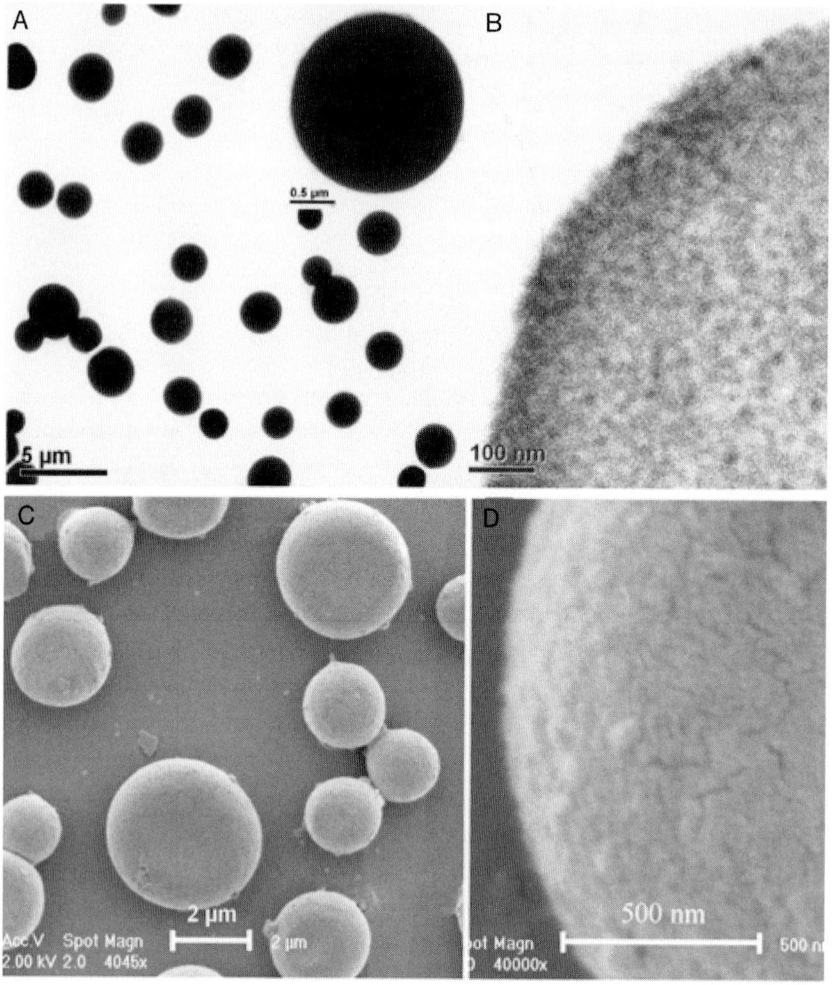

Fig. 7.6 TEM (A, B) and SEM (C, D) images of NPP-lysozyme prepared using 8000 Da PAA as the bridging molecule. (Image B is from an ultramicrotomed sample.) The lysozyme/PAA was cross-linked using EDC, after which the MS template was decomposed using HF/NH_4F at pH ~5. (Adapted from [98] with permission of Wiley-VCH).

Recently, a new procedure was reported for the preparation of nanoporous polymeric spheres (NPSs) with well-controlled structure via the LbL infiltration and coating of MS spheres with preformed polymers (Figure 7.7) [69]. A main advantage of this approach is that it offers a general and versatile route to the preparation of NPSs of different and tailored compositions, as it is based on electrostatic assembly [69,109].

A schematic representation of the formation of PAA/PAH NPSs is shown in Figure 7.7. The BMS particles are modified with a layer of 3-aminopropyltriethoxysilane (APTS) on the BMS skeleton. This process introduces amine ($-NH_2$) surface

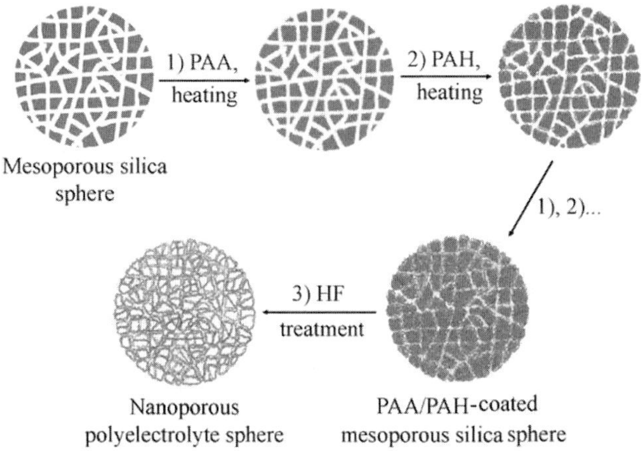

Fig. 7.7 Schematic illustration showing the preparation of NPS. APTS-modified BMS spheres are LbL coated with PEs of opposite charge (PAA and PAH) (steps 1 and 2). The samples were heated (160 °C for 2 h) after deposition of each PE to partially cross-link the layers. The BMS template is then dissolved by exposure to HF (step 3), yielding replicated NPS. (Reprinted from [69] with permission of Wiley-VCH).

functionality to the BMS spheres [110], which promotes specific adsorption of the subsequently deposited PAA through interaction of the $-NH_2$ and $-COOH$ groups. Chemical bonds between the MS templates and the PE building blocks can be selectively formed through cross-linking of amine groups (in the template) and the carboxylic groups (in the PE deposited), by chemical or thermal means. Under this treatment, amide bonds are formed by the $-COOH$ groups (in PAA) and the $-NH_2$ moieties (in PAH or on the APTS-BMS template), thus enhancing the structural stability of the layers. Self-supported NPSs are obtained for samples comprised of two or more PAA/PAH layers. The diameters of the resulting NPSs were found to depend on the number of PE layers deposited on the template APTS-BMS particles [69]. For example, the NPSs prepared with two and seven layers of PE had diameters ranging from 0.8–1.3 and 1.4–2.1 μm, respectively, indicating that less shrinkage occurred with NPSs containing more PE layers. SEM imaging revealed the NPSs to be individual particles with no obvious aggregation (Figure 7.8A). At higher magnification, the roughness and porosity of the spheres is apparent (Figure 7.8B), with pores in the range of ~10–50 nm (Figure 7.8C). Homogeneous nanoporosity with a pore size distribution of ~10–50 nm is clearly seen in the ultramicrotomed spheres (Figure 7.8E and F). The accessibility of the nanopores to proteins is evidenced by the homogeneous and bright spheres seen in the CLSM images of the NPSs after loading with FITC-labeled lysozyme through adsorption from solution.

Given that the primary step in the formation of NPSs is the infiltration of macromolecules into the porous silica matrix, the polymer size is an important parameter. Low molecular weight PEs are preferred for infiltration into the pores of the APTS–BMS spheres, and this process can be promoted by tuning the adsorption

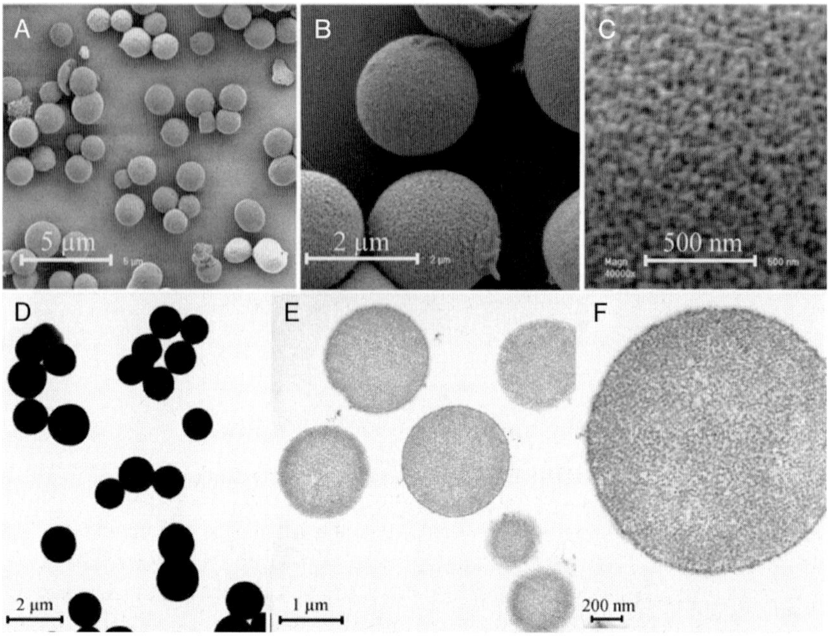

Fig. 7.8 SEM (A–C) and TEM (D–F) images of $(PAA/PAH)_2/PAA$ NPS at different magnification. PAA (2000 g mol^{-1}) and PAH (15 000 g mol^{-1}) were deposited from 5 mg mL^{-1} solutions containing 0.7 M NaCl at pH 2.9 and 4.3, respectively. Cross-linking was applied after the deposition of each layer by heating the samples at 160 °C for 2 h. Images (C) and (F) show the porosity of the NPS. The difference in the diameters seen in image (E) is a result of preparation of the samples for the ultramicrotoming process. (Adapted from [69] with permission of Wiley-VCH).

solution conditions. Since PAA and PAH are weak PEs, their ionization depends on the pH [111,112]. PEs with a reduced charge density adopt a more coiled conformation (due to the weaker intramolecular electrostatic repulsions) and are adsorbed in greater amounts. PEs with a more extended conformation are more likely to be adsorbed outside the nanopores, as they tend to be spatially restricted by the mesopores. NPSs can be prepared at pH $2 \sim 6$ for PAA ($pK_a \sim 4.3$) [111], with the pH of PAH fixed at 4.5. If the PAA deposition solution has a pH <1 or >7, only capsule-like products are obtained after silica removal [109]. The pH influence of PAH ($pK_a \sim 8$–10) has also been investigated by fixing the pH of the PAA solution at 4.5 [113]. Intact PAA/PAH NPSs could be prepared provided the PAH solution had a pH between ~4 and ~8.5 [109]. In contrast, capsule-like products are obtained at a PAH solution pH of 2.5. Like solution pH, the ionic strength of the deposition solutions influences the conformation and charge density of the weak PEs. The salt facilitates PE coiling [114] and, hence, penetration into the mesopores of the BMS sphere templates. Intact NPSs can be prepared from PAA (M_w 30 000 g mol^{-1}) and PAH (M_w 70 000 g mol^{-1}) deposition solutions that contain 0.5 M NaCl [109]. The density of the NPSs is gradually decreased if PE solutions of lower ionic strength are used. Also, only capsule-like products are obtained when no salt is added to the PAH deposition solution, due largely to the adsorption of the PEs onto the particle surfaces [109].

The pore size and distribution in the porous particles play essential roles in NPS synthesis. For example, only hollow capsules are obtained when MS spheres with only small mesopores (<3 nm) are used as the templates [69]. This suggests that the PE has difficulty infiltrating mesopores in this size range, and is primarily restricted to the surface of the spheres. The density and homogeneity of the pores in the sacrificial particles is also important to prepare intact NPSs. In a separate study, employing $CaCO_3$ microparticles with radial channel-like pore structures (surface area $\sim 8.8\,m^2\,g^{-1}$) as sacrificial templates resulted in PE microcapsules that collapse when dried, which is in stark contrast to the free-standing NPSs described above [64].

NPSs show an excellent capacity to sequester proteins because of their nanoporosity and large number of functional groups. They exhibit a high capacity for enzyme loading (about $470\,mg\,mL^{-1}$ lysozyme for $(PAA/PAH)_2/PAA$ NPS), with stimuli-responsive loading and release of the protein triggered by changes in solution pH. Reversible loading and release of the NPSs is achieved simply by varying the pH to mediate the electrostatic association between the proteins and the NPS host [109]. For example, at pH 7, lysozyme is positively-charged and is expected to electrostatically associate with the negatively-charged NPS (pI $\sim$5.3). When the particles are dispersed in a pH 2.0 solution, almost 90 % of the lysozyme is released from the NPSs within 60 min due to the diminished electrostatic interactions, as both the protein and NPSs are positively-charged at this pH. The protein loading ability can be fully recovered after washing the NPSs with pH 7.0 buffer [109].

The LbL-template synthesis strategy can be generalized to prepare a broad range of NPS of diverse compositions and morphologies (e.g., spheres, fibers, and membranes) with various properties, including pH-reversible loading and release characteristics for drugs and proteins [109]. Various other species such as copolymers, peptides, proteins, and small molecules can also be sequentially assembled in MS particles for the preparation of NPSs with different composition and function. For example, the ability to form NPSs from copolymers represents a significant step, as monomer units with different properties and diverse functional groups can be readily incorporated into the chain, and the proportion of each unit varied by changing the copolymer composition [109]. Biocompatible peptide particles (e.g., PGA/PLL NPS) are appealing supports for drug loading and delivery [109]. NPSs composed of polymer species can also be prepared through the alternate deposition of carboxyl group-containing polymers (e.g., PAA or PGA) and small molecules with diamine groups (e.g., ethylenediamine) via covalent bonds [109]. This provides another level of control over the composition and function of the NPS materials.

7.5
LbL Assembly on Macroporous Substrates

Macroporous materials with various shapes such as particles, tubes, rods, fibers, membranes, and 3DOM have been designed to cater for different applications. Three examples of macroporous material templating via LbL assembly will be detailed, to demonstrate their potential use in bioapplications. These include the formation of nanotubes, functionalization of macroporous membrane materials, and modification of naturally occurring porous materials (e.g., diatomite).

7.5.1
LbL Assembly on Tubular Substrates

Nanotubes are alternatives to spherical particles that might offer some advantages for biotechnological applications. For example, nanotubes have open ends, which make it possible to incorporate species inside them. They also have much larger inner surface areas compared with hollow spheres of the same diameter, rendering them superior in terms of loading capacity for guest molecules [115,116]. Several approaches have been developed to prepare nanotubes of various compositions and morphologies using nanorods [117], nanofibers [118], and porous membranes as templates [60–63]. The LbL assembly technique has proven to be an efficient method to control the thickness, composition, and functionalization of nanotubes. Mayya *et al.* have applied the LbL method to modify the surface of nickel nanorods with eight layers of PDDA/PSS [117]. Nanotubes of PDDA/PSS were obtained after dissolving the nickel core particles using diluted hydrochloric acid. Similarly, electrospun polystyrene (PS) fibers with high aspect ratios ($>100\,000:1$) have been functionalized with composite layers of PEs, deoxyribonucleic acid (DNA) oligonucleotides, and PEs/gold nanoparticles [118]. After dissolving the inner PS fibers with tetrahydrofuran, hollow PE fibers are obtained. An alternative method to produce tubes is by using porous membranes as templates. This involves the LbL assembly of polyelectrolytes and nanoparticles in an organic filter membrane such as cellulose acetate and polycarbonate. Zeolitic tubes [63] and hybrid nanotubes of metal [60], semiconductor nanoparticles [60], and polymer [60] are prepared by alternately depositing PEs and/or nanoparticles in the filter membrane, then removing the organic membrane with dichloromethane or by calcination.

Recently, inorganic AAO membranes with submicrometer pores have also been widely used in nanotube preparation. The outside diameter of these tubes is determined by the pore diameter in the template membrane used, and the length of the tube is determined by the thickness of the template membrane. The wall thickness and components can be adjusted by the number of layers and the species deposited, respectively. Ai *et al.* introduced a pressure-filter-template technique during the preparation of nanotubes to overcome the problem of pore blockage during LbL assembly [61]. In this technique, the assembly solutions are squeezed through a porous membrane by external pressure. Pressurized flow of solutions of polymers with alternating charges (e.g., PAH and PSS) results in the LbL deposition of a PE multilayer on the inner walls of the membrane. Much thicker multilayer coatings are formed in the porous walls when the PEs are LbL-assembled by the pressure-filter-template method, compared to when the multilayers are LbL-assembled on flat surfaces. The multilayer thickness increase is likely caused by more PE being deposited under pressure in the macropores of AAO membranes, since regular thickness growth of multilayers was found when no pressure is applied to the polymer deposition solutions [60]. Finally, the PE nanotubes are liberated into solution after removal of the AAO membrane by NaOH solution (1 M) or by H_3PO_4 solution (1.5–5 wt%).

Protein nanotubes composed of human serum albumin (HSA) can be prepared via the alternate adsorption of HSA of opposite charge (by varying the pH) onto the inner surfaces of AAO templates [93]. The positively-charged HSA (pI ~4.8) solution at pH 3.8 is injected into the filter through a syringe and then passed through the porous AAO template (negatively-charged) under pressure. Afterwards, pure water is used to wash the pores in the template. A negatively-charged HSA solution at pH 7.0 is then used to form a second layer. Alternating adsorption for five cycles leads to the formation of a multilayer protein coating in the AAO membrane. The membrane is subsequently removed, resulting in free-standing protein nanotubes. The prepared HSA nanotubes have smooth surfaces with a wall thickness of around 30 nm (Figure 7.9A). The aligned HSA nanotubes exhibit good stability (Figure 7.9B), and their length is about 60 μm, analogous to the thickness of the AAO template (Figure 7.9C). The HSA nanotubes are capable of bending and are highly flexible (Figure 7.9D). Composite nanotubes of protein/lipid can also be prepared via sequential adsorption of HSA and L-R-dimyristoylphosphatidic acid in aqueous or

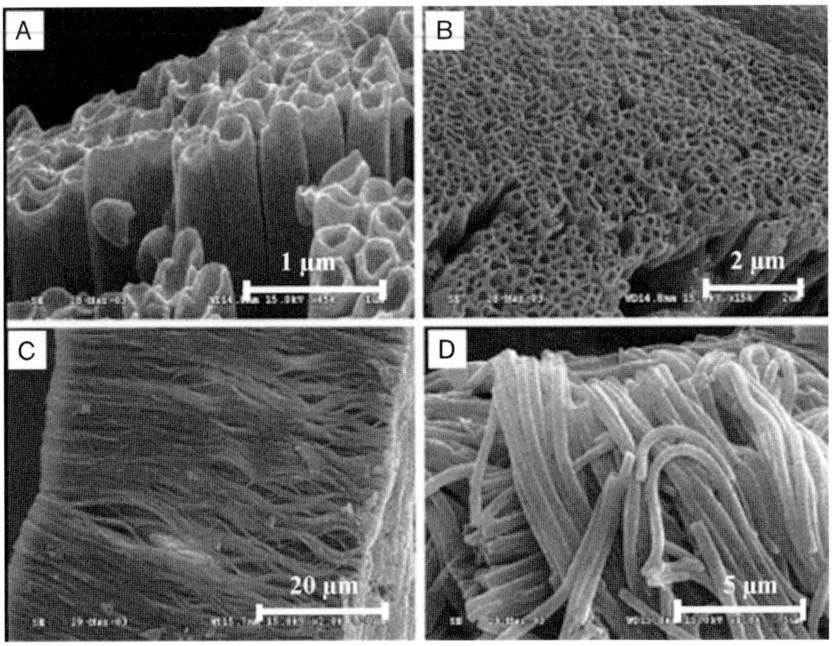

Fig. 7.9 SEM images of HSA nanotubes prepared by self-assembly in AAO membranes. (A) HSA nanotube arrays after the removal of the AAO template. (B) Highly ordered HSA nanotubes with a wall thickness of around 30 nm. (C) The length of the resulting nanotubes is about 60 μm. (D) Highly flexible HSA nanotubes. (Reproduced from [93] with permission of the American Chemical Society, Copyright 2005 American Chemical Society).

chloroform solutions [93]. The incorporation of lipophilic components such as channels or receptors could lead to biogenic nanotubes that may serve as bioreactors, biosensors, or drug carriers in a living system. Cross-linked protein nanotubes are prepared by alternately exposing the AAO membrane to a solution of the desired proteins and then to a solution of glutaraldehyde (GA), which acts as a cross-linking agent to hold the protein layers together [94,119]. The AAO template is immersed first into a solution of 3-aminopropylphosphonic acid (APA), resulting in attachment of an APA monolayer to the pore walls via the phosphonate group [94]. The amino groups reacted with an excess quantity of GA, leaving unreacted aldehyde groups on the pore walls. The membrane is subsequently exposed to a solution of the desired protein, which reacts via free amino sites with the aldehyde groups on the pore wall. The membrane is then alternately exposed to GA and then the protein solution until the desired number of layers is reached. The prepared glucose oxidase nanotubes retain their heme electroactivity, with an increase in enzymatic activity observed as the nanotube wall thickness increases. Tian *et al.* reported the preparation of protein nanotubes based on the LbL electrostatic assembly of cytochrome C and PSS through electrostatic adsorption on AAO templates [119]. The assembled cytochrome C tubes retained their protein biochemical activity and electrical activity, as shown by cyclic voltammograms.

Fluorescent nanotubes of polyethyleneimine (PEI) and 3,4,9,10-perylenetetra-carboxylicdianhydride (PTCDA) have been prepared through the alternating deposition of polymers and small functional molecules that form covalent bonds (Figure 7.10) [120]. The nanotube synthesis starts with the deposition of PEI in the pores of an AAO membrane as the first layer. The PTCDA solution is then used to bind to the PEI via covalent bonding (Figure 7.10). The electro-optical properties of the small molecule (PTCDA) are retained in the multilayer films of PEI/PTCDA. The prepared nanotubes retain their fluorescent properties for up to 10 months without

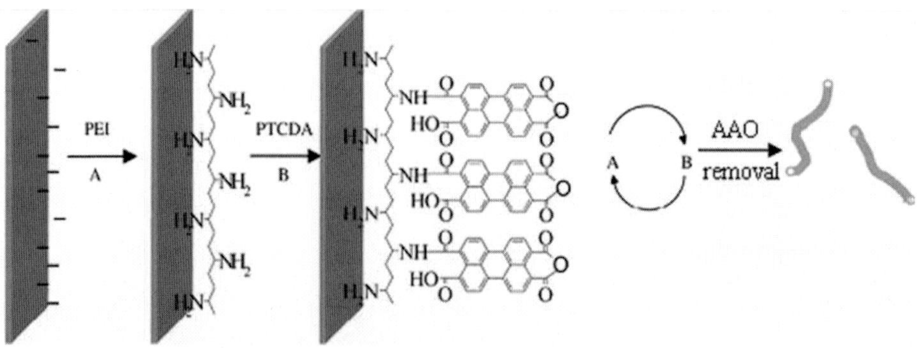

Fig. 7.10 Schematic representation of fluorescent nanotubes prepared through the alternate deposition of PEI and PTCDA in AAO membrane templates. (Reproduced from [119] with permission of the American Chemical Society, Copyright 2006 American Chemical Society).

morphological variations. Given their optoelectronic response, such nanotubes are expected to be applied as biological labels.

It should be noted that LbL assembly also occurs on the surface of AAO membranes when the multilayers are deposited in the pore walls [61]. The film deposited on the top and bottom surfaces of the membrane can be wiped off from the surface with carborundum sand paper after the multilayers have formed [61]. Sputter coating of both faces of the AAO membrane with ultrathin (~5 nm) films of Au has been used to prevent the adsorption of multilayers on the surface in the preparation of α, ω-diorganophosphonate/Zr(IV) nanotubes [121]. These Au films are too thin to block the pores at the membrane surfaces, but they prevent the adsorption of the first layer of the α, ω-diorganophosphonate to the faces of the membrane. As a result, layered α, ω-diorganophosphonate/Zr(IV) films are deposited only along the pore walls and not on the faces of the membrane. Without these Au films, the faces become preferentially coated, and nanotubes are not obtained. When adsorption on the outer membrane surfaces was prevented, the wall thickness of the α, ω-diorganophosphonate/Zr(IV) nanotubes increased by 1.7 nm per layer, which is similar to the thickness of films deposited on planar surfaces [121]. Composite DNA nanotubes have also been prepared by LbL assembly of DNA segments in AAO membranes pre-coated with a layer of α, ω-diorganophosphonate/Zr(IV) [122]. In this case, hybridization between DNA segments is the driving force to assemble the multilayers [122].

Recent progress in AAO templating synthesis includes the preparation of "nano-test-tubes" [123–125]. Unlike the nanotubes that are open at both ends (resembling pipes), silica nano-test-tubes are prepared through a simple modification of the template-synthesis method [123]. The nano-test-tubes are able to be corked with polystyrene latex nanoparticles [124] or gold nanoparticles [125] through chemical self-assembly. A potential application for these novel nanotubes is as vehicles for the delivery of drugs, DNA, proteins or other biomolecules. This test-tube geometry would allow for convenient filling of the nanotube with the biomolecule of interest and, by applying a cap to the open end, the biomolecule could be kept "bottled-up" inside until it is ready to be delivered [124,125].

7.5.2
LbL Assembly on 3DOM Materials

Due to their high surface area, ordered and well-connected macropores (through connecting windows), 3DOM materials are also attracting increasing attention for potential use in biological applications. A 3DOM material-based potentiometric biosensor has been reported for the detection of analytes [30], where an enzyme, creatinine deiminase, is doped within ordered arrays of hollow poly(pyrrole) 3DOMs. Macroporous thin film transducers for optical affinity biosensing that are capable of biospecific recognition through simple monitoring of the bandgap peak position, have also been prepared [31]. A macroporous bioactive glass of CaO/SiO_2 composite has been reported as a bone graft substitute, since it can bond to bone through the growth of biocompatible hydroxycarbonate apatite when placed in simulated body fluid [126]. Also, a macroporous material prepared by templated precipitation of a precursor solution of calcium nitrate and phosphoric acid was shown to be capable of adsorbing antibiotics and then releasing them in simulated body fluid [127].

The application of LbL assembly on 3DOM substrates was first reported for the preparation of heterogeneous macroporous materials through the sequential electrostatic deposition of PAH and cadmium telluride (CdTe) nanocrystals in macroporous titania walls [58]. Biomacromolecule-functionalized macroporous zeolitic membrane bioreactors have also been prepared via the LbL electrostatic assembly of PEs and an enzyme (catalase) on silicalite-1 3DOM [59]. The 3DOM substrate is prepared by LbL electrostatically seeding mesoporous silica sphere self-assembled films with silicalite-1 nanoparticles, followed by hydrothermal treatment. Figure 7.11A shows a cross-section of the free-standing membrane with a thickness of about 140 μm. The periodic structure extends across areas up to 100 μm^2, as shown in Figure 7.11B. From the higher magnification images (Figure 7.11C and D), it is evident that the adjacent macropores are interconnected through "windows," which have a size of 100–140 nm. The enzyme-functionalized membranes with interconnected macroporous structures display enzyme loading amounts and activities that are one order of magnitude higher than corresponding zeolite films with closed macropores, and three orders of magnitude higher than non-porous planar films assembled on glass slides [59]. The enzyme loading amounts and activities were found to be approximately linearly dependent on the thickness of the membranes,

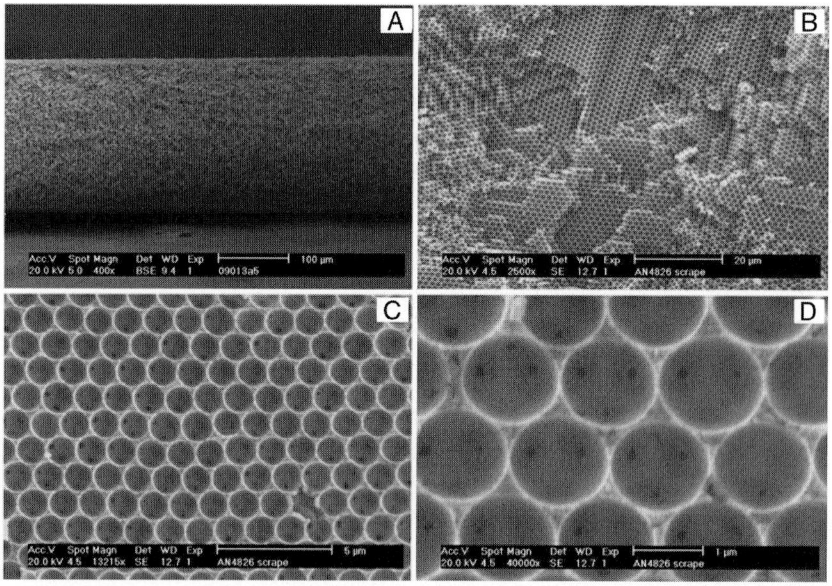

Fig. 7.11 SEM images of the 3D macroporous zeolitic membrane used as a support for enzyme immobilization by the LbL procedure. Images (A–D) are cross-sections of the membrane at different magnifications. A biomacromolecule-functionalized macroporous zeolitic membrane bioreactor was prepared via the LbL electrostatic assembly of PEs and enzyme (catalase) on the 3D macroporous membrane. (Reprinted from [59] with permission of Wiley-VCH).

suggesting the effective utilization of the whole membrane due to the interconnected macropores. The immobilized enzymes also exhibit enhanced reaction stability in comparison with enzymes in bulk solution.

3DOM materials also provide a unique cellular scaffold system for investigation of cell–cell interactions due to their highly ordered 3D geometry. Lee *et al.* modified the surface of macroporous hydrogel scaffolds (pore size ~25–30 μm) through LbL assembly of a clay/PDDA multilayer [128]. The clay particles are biocompatible and their coating created nanoscale roughness, increased charging on the surface, and led to much stiffer films than the unmodified poly(acrylamide) hydrogel, which rarely supports cell adhesion. Ten layers of clay/PDDA altered the surface property from cell repulsive to cell adhesive and, as a result, thymic epithelial cells could attach to the hydrogel scaffold.

7.5.3 LbL Assembly on Naturally Occurring Porous Substrates

Many of the materials currently under development draw their inspiration from structures found in nature. That is, by mimicking the supramolecular architecture of natural materials, one can prepare complex materials capable of highly sophisticated functions. An important aspect of this work involves the selection of microorganism templates (e.g., diatomite) based on specific porous structures that may benefit targeted applications.

A novel zeolite material possessing an inherent hierarchical structure with good mechanical and chemical strength has been prepared by the LbL assembly of zeolite nanocrystals and PDDA on the diatomite substrates [129]. The diatomite used has a disk-like morphology (Figure 7.12A) and exhibits abundant and uniform macropores (about 300–500 nm) in the diatomite plates (Figure 7.12B). The zeolite–diatomite (ZD)

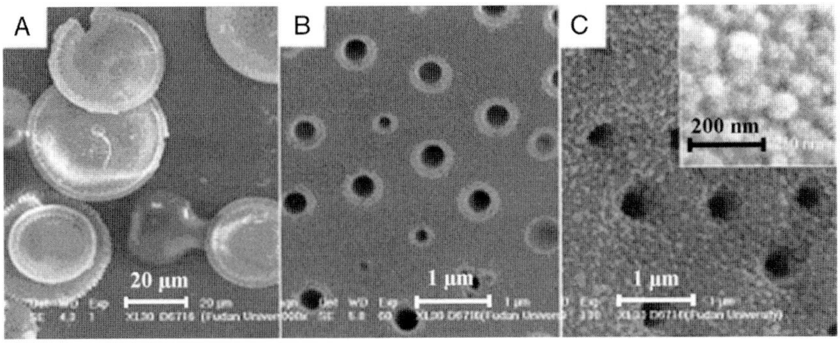

Fig. 7.12 SEM images of the diatomite substrate at low (A) and high (B) magnification. The diatomite has a disk-like shape about 1.2 μm in thickness and 20–40 μm in diameter with a nearly regular array of submicrometer-sized pores (about 300–500 nm). SEM images of diatomite coated with three layers of zeolite β nanocrystals through sequential assembly of positively-charged PDDA and negatively-charged zeolite nanocrystals (C). The inset in (C) shows an enlarged view of the nanoparticle-modified surface. (Reprinted from [129] with permission of Wiley-VCH.)

composite possesses a high outer surface area due to the deposition of zeolite nanocrystals [130], and the inherent macroporous structure of the diatomite substrate is well-preserved after the assembly of three bilayers of PDDA/zeolite nanocrystals (Figure 7.12C). When applied to separation processes, this combination provides less flow resistance to the effluent, and allows fast diffusion. The abundant cationic exchange sites in the zeolitic framework endow the material with good cation (e.g., Co^{2+}) immobilization and regeneration ability. The Co^{2+}-ion-immobilized nanozeolite β-diatomite composite is capable of separating histidine-containing polypeptides by immobilized metal-ion affinity chromatography (IMAC) [131]. When this material is used to separate proteins in a living system, Se-containing protein with two histidine-enriched domains [132] coordinates with the Co-ZD composite and is thus separated by IMAC from the other two, structurally different, Se-containing proteins, albumin and extracellular glutathione peroxidase, in mouse plasma [133].

Rosi *et al.* [134] covalently functionalized diatom cell walls with DNA and then used the DNA modified diatomite as templates to build hierarchical structure and complexity into the templated materials. Three-dimensional microscale metallic materials with patterned nanometer-scale features have been prepared through the LbL assembly of DNA-functionalized nanoparticles. Multilayers of gold nanoparticles were deposited onto the diatom surfaces by the alternating assembly of oligonucleotide (3′-HS-C3H6-Cy3-ATCCTTATCAATATT-5′13)-functionalized Au nanoparticles (13 nm) and the complementary oligonucleotide (3′-HS-C3H6-AATATTGATAAG-GAT-5′)-functionalized Au nanoparticles. Here, the build up of multilayers of nanoparticles was programmed by DNA, through hybridization-directed LbL assembly. The nanostructured surfaces of diatom-templated metallic materials also render them useful candidates as surface enhanced Raman scattering substrates [135].

7.6 Summary and Outlook

This chapter has summarized recent advances in the synthesis of bioinspired porous hybrid materials prepared by LbL assembly. The LbL encapsulation of MS materials has proven to be a general approach for the loading and encapsulation of biomacromolecules. The high protein loadings possible using designed structures are of particular interest in protein drug delivery, especially for improving drug efficacy and decreasing side effects. It is of widespread interest to develop methods to controllably load and release small molecules and drugs; however, the LbL encapsulation of such materials is still under development. Further work in this area is likely to focus on the design of intelligent particles with tailored optical, magnetic, catalytic, and biological properties for functional applications.

As part of the next generation of complex colloidal containers, NPSs show promise for the facile uptake and controlled delivery of high quantities of biomolecules. The broad range of MS materials available with tunable size, morphology and porosity will enable the preparation of tailored NPS materials. NPSs are promising for the uptake of various species, including small drugs, and conjugated drugs. The results from clinical trials with polymer-anticancer-drug conjugates [136] bode well for the future design and development of ever more sophisticated NPSs that are capable of targeted delivery. Controlled drug delivery systems may be improved further by functionalizing NPSs with superparamagnetic nanoparticles [137], integrin receptors [97], and antibodies for receptor-mediated endocytosis [138].

Macroporous substrates with interconnected voids can be used as platforms for biomacromolecule separation and enzyme immobilization. These assemblies are likely to find application in biocatalysis and bioassays. The inorganic framework can provide a robust substrate, while their large and abundant pores allow the transportation of biomolecules. The availability of various morphologies for macroporous materials provides another level of control over the function of the hybrids.

Given the significant recent developments in the burgeoning areas of porous materials and self-assembly, we anticipate that, with further development, bioinspired porous hybrid materials will find their way into a range of therapeutic and diagnostic applications.

Acknowledgements

Funding from the Australian Research Council (Federation Fellowship and Discovery Project schemes) and the Victorian State Government (STI Initiative) is gratefully acknowledged. Robert Johnson and Alexandra Angelatos are thanked for critical reading of the manuscript.

References

1 Soler-illia, G.J.D., Sanchez, C., Lebeau, B. and Patarin, J. (2002) *Chemical Reviews*, **102**, 4093–4138.

2 Davis, M.E. (2002) *Nature*, **417**, 813–821.

3 Stein, A. (2003) *Advanced Materials*, **15**, 763–775.

4 Sing, K.S.W., Everett, D.H., Haul, R.A.W., Moscou, L., Pierotti, J., Rouquerol, J. and Siemieniewska, T. (1985) *Pure and Applied Chemistry*, **57**, 603–619.

5 Breck, D.W., Eversole, W.G., Milton, R. M., Reed, T.B. and Thomas, T.L. (1956) *Journal of the American Chemical Society*, **78**, 5963–5971.

6 Matsui, M., Kiyozumi, Y., Yamamoto, T., Mizuchina,Y., Mizukami, F. and Sakaguchi, K. (2001) *Chemistry-A European Journal*, **7**, 1555–1560.

7 Munsch, S., Hartmann, M. and Ernst, S. (2001) *Chemical Communications*, 1978–1979.

8 Nygaard, S., Wendelbo, R. and Brown, S. (2002) *Advanced Materials*, **14**, 1853–1856.

9 Hayakawa, K., Mouri,Y., Maeda, T., Satake, I. and Sato, M. (2000) *Colloid and Polymer Science*, **278**, 553–558.

10 Shan, W., Yu, T., Wang, B., Hu, J.K., Zhang, Y.H., Wang, X.Y. and Tang, Y. (2006) *Chemistry of Materials*, **18**, 3169–3172.

11 Zhang, Y.H., Wang, X.Y., Shan, W., Wu, B.Y., Fan, H.Z., Yu, X.J., Tang, Y. and Yang, P.Y. (2005) *Angewandte Chemie International Edition*, **44**, 615–617.

12 Huang, Y., Shan, W., Liu, B.H., Liu, Y., Zhang, Y.H., Zhao, Y., Lu, H.J., Tang, Y.

and Yang, P.Y. (2006) *Lab on a Chip*, **6**, 534–539.

13 Kresge, C.T., Leonowicz, M.E., Roth, W.J., Vartuli, J.C. and Beck, J.S. (1992) *Nature*, **359**, 710–712.

14 Corma, A. (1997) *Chemical Reviews*, **97**, 2373–2419.

15 Hartmann, M. (2005) *Chemistry of Materials*, **17**, 4577–4593.

16 Yiu, H.H.P. and Wright, P.A. (2005) *Journal of Materials Chemistry*, **15**, 3690–3700.

17 Radu, D.R., Lai, C.Y., Jeftinija, K., Rowe, E.W., Jeftinija, S. and Lin, V.S.Y. (2004) *Journal of the American Chemical Society*, **126**, 13216–13217.

18 Lin, Y.S., Tsai, C.P., Hung, H.Y., Kuo, C.T., Hung, Y., Huang, D.M., Chen, Y.C. and Mou, C.Y. (2005) *Chemistry of Materials*, **17**, 4570–4572.

19 Lin, Y.S., Hung, Y., Su, J.K., Lee, R., Chang, C., Lin, M.L. and Mou, C.Y. (2004) *Journal of Physical Chemistry B*, **108**, 15608–15611.

20 Slowing, I., Trewyn, B.G. and Lin, V.S.Y. (2006) *Journal of the American Chemical Society*, **128**, 14792–14793.

21 O'Sullivan, J.P. and Wood, G.C. (1970) *Proceedings of the Royal Society of London*, **317** (1531), 511.

22 Wefers, K. and Misra C. (1987) Oxides and Hydroxides of Aluminium, Alcoa Laboratories.

23 Lohrengel, M.M. (1993) *Materials Science and Engineering*, **11**, 243–291.

24 Martin, C.R. (1994) *Science*, **266**, 1961–1966.

25 Mitchell, D.T., Lee, S.B., Trofin, L., Li, N., Nevanen, T.K., Söderlund, H. and Martin, C.R. (2002) *Journal of the American Chemical Society*, **124**, 11864–11865.

26 He, B., Son, S.J. and Lee, S.B. (2006) *Langmuir*, **22**, 8263–8265.

27 Xia, Y., Gates, B., Yin, Y. and Lu, Y. (2000) *Advanced Materials*, **12**, 693–713.

28 Míguez, H., Meseguer, F., López, C., López-Tejeira, F. and Sánchez-Dehesa, J. (2001) *Advanced Materials*, **13**, 393–397.

29 Edrington, A.C., Urbas, A.M., DeRege, P., Chen, C.X., Swager, T.M., Hadjichristidis, N., Xenidou, M., Fetters, L.J., Joannopoulos, J.D., Fink, Y. and Thomas, E.L. (2001) *Advanced Materials*, **13**, 421–425.

30 Cassagneau, T. and Caruso, F. (2002) *Advanced Materials*, **14**, 1837–1841.

31 Cassagneau, T. and Caruso, F. (2002) *Advanced Materials*, **14**, 1629–1633.

32 Velev, O.D., Tessier, P.M., Lenhoff, A. M. and Kaler, E.W. (1999) *Nature*, **401**, 548–548.

33 Velev, O.D., Jede, T.A., Lobo, R.F. and Lenhoff, A.M. (1997) *Nature*, **389**, 447–448.

34 Holland, B.T., Blanford, C.F. and Stein, A. (1998) *Science*, **281**, 538–540.

35 Wijnhoven, J.E.G.J. and Vos, W.L. (1998) *Science*, **281**, 802–804.

36 Dong, A.G., Wang, Y.J., Tang, Y., Zhang, Y.H., Ren, N. and Gao, Z. (2002) *Advanced Materials*, **14**, 1506–1509.

37 Schroden, R. and Stein, A. (2004) in Colloids and Colloid Assemblies (ed.F. Caruso),Wiley-VCH,Weinheim, Chapter 15.

38 Anderson, M.W., Holmes, S.M., Hanif, N. and Cundy, C.S. (2000) *Angewandte Chemie International Edition*, **39**, 2707–2710.

39 Holmes, S.M., Graniel-Garcia, B.E., Foran, P., Hill, P., Roberts, E.P.L., Sakakini, B.H. and Newton, J.M. (2006) *Chemical Communications*, 2662–2663.

40 Zhao, J.P., Gaddis, C.S., Cai, Y. and Sandhage, K.H. (2005) *Journal of Materials Research*, **20**, 282–287.

41 Sandhage, K.H., Dickerson, M.B., Huseman, P.M., Caranna, M.A., Clifton, J.D., Bull, T.A., Heibel, T.J., Overton, W.R. and Schoenwaelder, M.E.A. (2002) *Advanced Materials*, **14**, 429–433.

42 Anderson, M.W., Holmes, S.M., Mann, R., Foran, P. and Cundy, C.S. (2005) *Journal of Nanoscience and Nanotechnology*, **5**, 92–95.

43 Wang, Y., Tang, Y., Dong, A., Wang, X., Ren, N., Shan, W. and Gao, Z. (2002) *Journal of Materials Chemistry*, **12**, 1812–1819.

44 Tang, Z., Wang, Y., Podsiadlo, P. and Kotov, N.A. (2006) *Advanced Materials*, **18**, 3203–3224.

45 Iler, R.K. (1966) *Journal of Colloid and Interface Science*, **21**, 569–594.

46 Decher, G. and Hong, J.D. (1991) *Berichte Der Bunsen-Gesellschaft-Physical Chemistry*, **95**, 1430–1434.

47 Decher, G. (1997) *Science*, **277**, 1232–1237.

48 Peyratout, C.S. and Dähne, L. (2004) *Angewandte Chemie International Edition*, **43**, 3762–3783.

49 Caruso, F. (2001) *Advanced Materials*, **13**, 11–22.

50 Angelatos, A.S., Katagiri, K. and Caruso, F. (2006) *Soft Matter*, **2**, 18–23.

51 Johnston, A.P.R., Cortez, C., Angelatos, A.S. and Caruso, F. (2006) *Current Opinion in Colloid & Interface Science*, **11**, 203–209.

52 Quinn, J.F., Johnston, A.P.R., Such, G.K., Zelikin, A.N. and Caruso, F. (2007) Chemical Society Reviews, **36**, 707.

53 Tedeschi, C., Caruso, F., Möhwald, H. and Kirstein, S. (2000) *Journal of the American Chemical Society*, **122**, 5841–5848.

54 Mamedov, A.A., Belov, A., Giersig, M., Mamedova, N.N. and Kotov, N.A. (2001) *Journal of the American Chemical Society*, **123**, 7738–7739.

55 Caruso, F., Caruso, R.A. and Möhwald, H. (1998) *Science*, **282**, 1111–1114.

56 Ai, H., Fang, M., Jones, S.A. and Lvov, Y.M. (2002) *Biomacromolecules*, **3**, 560–564.

57 Antipov, A.A. and Sukhorukov, G.B. (2004) *Advances in Colloid and Interface Science*, **111**, 49–61.

58 Wang, D. and Caruso, F. (2001) *Chemical Communications*, 489–490.

59 Wang, Y. and Caruso, F. (2004) *Advanced Functional Materials*, **14**, 1012–1018.

60 Liang, Z., Susha, A.S., Yu, A. and Caruso, F. (2003) *Advanced Materials*, **15**, 1849–1854.

61 Ai, S., Lu, G., He, Q. and Li, J. (2003) *Journal of the American Chemical Society*, **125**, 11140–11141.

62 Hou, S., Harrell, C., Trofin, L., Kohli, P. and Martin, C.R. (2004) *Journal of the American Chemical Society*, **126**, 674–675.

63 Wang, Y., Tang, Y., Dong, A., Wang, X., Ren, N., Shan, W. and Gao, Z. (2002) *Advanced Materials*, **14**, 994–998.

64 Volodkin, D.V., Petrov, A.I., Prevot, M. and Sukhorukov, G.B. (2004) *Langmuir*, **20**, 3398–3406.

65 Sukhorukov, G.B., Volodkin, D.V., Günther, A.M., Petrov, A.I., Shenoy, D.B. and Möhwald, H. (2004) *Journal of Materials Chemistry*, **14**, 2073–2081.

66 Wang, Y. and Caruso, F. (2004) *Chemical Communications*, 1528–1529.

67 Wang, Y. and Caruso, F. (2005) *Chemistry of Materials*, **17**, 953–960.

68 Zhu, Y.F., Shi, J.L., Shen, W.H., Dong, X.P., Feng, J.W., Ruan, M.L. and Li, Y.S. (2005) *Angewandte Chemie International Edition*, **44**, 5083–5087.

69 Wang, Y., Yu, A. and Caruso, F. (2005) *Angewandte Chemie International Edition*, **44**, 2888–2894.

70 Langer, R. (1998) *Nature*, **392**, 5–10.

71 Park, K. (1997) Controlled Drug Delivery: Challenges and Strategies, American Chemical Society, Washington DC.
72 Kreuter, J. (1994) Colloidal Drug Delivery Systems, Marcel Dekker, New York.
73 Caruso, F., Trau, D., Möhwald, H. and Renneberg, R. (2000) *Langmuir*, **16**, 1485–1492.
74 Jin, W., Shi, X. and Caruso, F. (2001) *Journal of the American Chemical Society*, **123**, 8121–8122.
75 Shi, X. and Caruso F. (2001) *Langmuir*, **17**, 2036–2043.
76 Qiu, X., Leporatti, S., Donath, E. and Möhwald, H. (2001) *Langmuir*, **17**, 5375–5382.
77 Sukhorukov, G.B., Antipov, A.A., Voigt, A., Donath, E. and Möhwald, H. (2001) *Macromolecular Rapid Communications*, **22**, 44–50.
78 Lvov, Y., Antipov, A.A., Mamedov, A., Möhwald, H. and Sukhorukov, G.B. (2001) *Nano Letters*, **1**, 125–129.
79 Dähne, L., Baude, B. and Voigt, A. (2003) Patent WO 2004/014540 A1.
80 Schulz-Ekloff, G., Rathouský, J. and Zukal, A. (1999) *International Journal of Inorganic Materials*, **1**, 97–102.
81 Grun, M., Buchel, C., Kumar, D., Schumacher, K., Bidingmaier, B. and Unger, K.K. (2000) *Studies in Surface Science and Catalysis*, **128**, 155–162.
82 Yu, A., Wang, Y., Barlow, B. and Caruso, F. (2005) *Advanced Materials*, **17**, 1737–1741.
83 Klitzing, R. and Möhwald H. (1996) *Macromolecules*, **29**, 6901–6907.
84 Angelatos, A.S., Johnston, A.P.R., Wang, Y. and Caruso, F. (2007) *Langmuir*, **23**, 4554.
85 Tyagi, S. and Kramer F.R. (1996) *Nature Biotechnology*, **14**, 303–308.
86 Shimizu, K., Cha, J., Stucky, G.D. and Morse, D.E. (1998) *Proceedings of the National Academy of Sciences*, **95**, 6234–6238.
87 Zelikin, A.N., Quinn, J.F. and Caruso, F. (2006) *Biomacromolecules*, **7**, 27–30.
88 Itoh, Y., Matsusaki, M., Kida, T. and Akashi, M. (2006) *Biomacromolecules*, **7**, 2715–2718.
89 Yu, A., Gentle, I., Lu, G.Q. and Caruso, F. (2006) *Chemical Communications*, 2150–2152.
90 Mesquida, P., Ammann, D.L., MacPhee, C.E. and McKendry, R.A. (2005) *Advanced Materials*, **17**, 893–897.
91 Kaneko, T., Higashi, M., Matsusaki, M., Akagi, T. and Akashi, M. (2005) *Chemistry of Materials*, **17**, 2484–2490.
92 Couet, J., Samuel, J.D.J.S., Kopyshev, A., Santer, S. and Biesalski, M. (2005) *Angewandte Chemie International Edition*, **44**, 2–6.
93 Lu, G., Ai, S. and Li, J. (2005) *Langmuir*, **21**, 1679–1682.
94 Hou, S., Wang, J. and Martin, C.R. (2005) *Nano Letters*, **5**, 231–234.
95 Cha, J.N., Birkedal, H., Bartl, M.H., Deming, T.J. and Stucky, G.D. (2003) *Journal of the American Chemical Society*, **125**, 8285–8286.
96 McKenna, B.J., Birkedal, H., Bartl, M. H., Deming, T.J. and Stucky, G.D. (2004) *Angewandte Chemie International Edition*, **43**, 5652–5656.
97 Toublan, F.J.J., Boppart, S. and Suslick, K.S. (2006) *Journal of the American Chemical Society*, **128**, 3472–3473.
98 Wang, Y. and Caruso, F. (2006) *Advanced Materials*, **18**, 795–800.
99 Hermanson, G.T., Mallia, A.K. and Smith, P.K. (1992) Immobilized Affinity Ligand Techniques, Academic Press, London.
100 Cetinus, S.A. and Öztop, H.N. (2003) *Enzy. Microbial Technology*, **32**, 889–900.
101 Meyer, U., Larsson, A., Hentze, H.P. and Caruso, R.A. (2002) *Advanced Materials*, **14**, 1768–1772.
102 Shchukin, D.G. and Caruso, R.A. (2003) *Chemical Communications*, 1478–1479.

103 Norrlöw, O., Glad, M. and Mosbach, K. (1984) *Journal of Chromatography*, **299**, 29–41.

104 Mayes, A.G. and Mosbach, K. (1996) *Analytical Chemistry*, **68**, 3769–3774.

105 Goltner, C.G., Henke, S., Weissenberger, M.C. and Antonietti, M. (1998) *Angewandte Chemie International Edition*, **37**, 613–616.

106 Kageyama, K., Tamazawa, J.-I. and Aida, A. (1999) *Science*, **285**, 2113.

107 Kim, J.Y., Yoon, S.B., Kooli, F. and Yu, J. (2001) *Journal of Materials Chemistry*, **11**, 2912–2914.

108 Yilmaz, E., Ramsrtöm, O., Möller, P., Sanchez, D. and Mosbach, K. (2002) *Journal of Materials Chemistry*, **12**, 1577.

109 Wang, Y. and Caruso, F. (2006) *Chemistry of Materials*, **18**, 4089–4100.

110 Zhao, X.S. and Lu, G.Q. (1998) *Journal of Physical Chemistry B*, **102**, 1556–1561.

111 Kato, N., Schuetz, P., Fery, A. and Caruso, F. (2002) *Macromolecules*, **35**, 9780–9787.

112 Burke, S.E. and Barrett, C.J. (2003) *Langmuir*, **19**, 3297–3303.

113 Harris, J.J., DeRose, P.M. and Bruening, M.L. (1999) *Journal of the American Chemical Society*, **121**, 1978–1979.

114 Shiratori, S.S. and Rubner, M.F. (2000) *Macromolecules*, **33**, 4213–4219.

115 Martin, C.R. (2003) *Nature Reviews Drug Discovery*, **2**, 29–37.

116 Kohli, P. and Martin, C.R. (2005) *Journal of Drug Delivery Science and Technology*, **15**, 49–57.

117 Mayya, S., Gittins, D.I. and Caruso, F. (2001) *Nano Letters*, **1**, 727–730.

118 Muller, K., Quinn, J.F., Johnston, A.P. R., Becker, M., Greiner, A. and Caruso, F. (2006) *Chemistry of Materials*, **18**, 2397–2403.

119 Tian, Y., He, Q., Tao, C., Cui, Y. and Li, J. (2006) *Biomacromolecules*, **7**, 2539–2542.

120 Tian, Y., He, Q., Tao, C. and Li, J. (2006) *Langmuir*, **22**, 360–362.

121 Hou, S., Harrell, C.C., Trofin, L., Kohli, P. and Martin, C.R. (2004) *Journal of the American Chemical Society*, **126**, 5674–5675.

122 Hou, S., Wang, J. and Martin, C.R. (2005) *Journal of the American Chemical Society*, **127**, 8586–8587.

123 Gasparac, R., Kohli, P., Mota, M.O., Trofin, L. and Martin, C.R. (2004) *Nano Letters*, **4**, 513–516.

124 Hillebrenner, H., Buyukserin, F., Kang, M., Mota, M.O., Stewart, J.D. and Martin, C.R. (2006) *Journal of the American Chemical Society*, **128**, 4236–4237.

125 Lee, S.B., Willis, A.C. and Webster, R.D. (2006) *Journal of the American Chemical Society*, **128**, 9332–9333.

126 Yan, H., Zhang, K., Blanford, C.F., Francis, L.F. and Stein, A. (2001) *Chemistry of Materials*, **13**, 1374–1382.

127 Melde, B.J. and Stein, A. (2002) *Chemistry of Materials*, **14**, 3326–3331.

128 Lee, S., Shanbhag, N.A. and Kotov, N. A. (2006) *Journal of Materials Chemistry*, **16**, 3558–3564.

129 Xu, F., Wang, Y., Wang, X., Zhang, Y., Tang, Y. and Yang, P. (2003) *Advanced Materials*, **15**, 1751–1755.

130 Ravishankar, R., Kirscnhock, C., Schoeman, B.J., Vanoppen, P., Grobet, P.J., Storck, S., Maier, W.F., Martens, J. A., De Schryver, F.C. and Jacobs, P.A. (1998) *Journal of Physical Chemistry B*, **102**, 2633–2639.

131 Porath, J., Carlsson, J., Olsson, I. and Belfrage, G. (1975) *Nature*, **258**, 598–599.

132 Hill, K.E., Lloyd, R.S. and Burk, R.F. (1993) *Proceedings of the National Academy of Sciences USA*, **90**, 537–541.

133 Deagen, J.T., Butler, J.A., Zachara, B.A. and Whanger, P.D. (1993) *Analytical Biochemistry*, **208**, 176–181.

134 Rosi, N.L., Thaxton, S. and Mirkin, C. A. (2004) *Angewandte Chemie International Edition*, **43**, 5500–5503.

135 Payne, E.K., Rosi, N.L. and Xue, C. (2005) *Angewandte Chemie International Edition*, **44**, 5064–5067.

136 Duncan, R. (2003) *Nature Reviews Drug Discovery*, **2**, 347–360.

137 Ding, Y., Hu, Y., Zhang, L., Chen, Y. and Jiang, X. (2006) *Biomacromolecules*, **7**, 1766–1772.

138 Cortez, C., Tomaskovic-Crook, E., Johnston, A.P.R., Radt, B., Cody, S.H., Scott, A.M., Nice, E.C., Heath, J.K. and Caruso, F. (2006) *Advanced Materials*, **18**, 1998–2003.

8
Bio-inorganic Nanohybrids Based on Organoclay Self-assembly

Avinash J. Patil, Stephen Mann

8.1
Introduction

A core focus of the burgeoning field of nanomaterials science is to develop innovative synthesis protocols that enable controlled and integrated organization of functional organic, inorganic and biological building blocks [1,2]. In this context, research on the synthesis of clay-based organic–inorganic hybrid materials has received considerable attention because these lamellar materials not only possess ordered and constrained structures in 2D but also furnish interesting chemical intercalation and surface properties. As a result, entrapment and orientation of various functional guest molecules into layered inorganic solids has been well exploited [3–6].

Considering the technological importance of layered materials, there is a great deal of interest in the rational design of hybrid bio-inorganic composites on the nanometer to micrometer length scales. These hybrids encompass highly selective recognition properties associated with biological species, combined with catalytic, optical and electronic properties of the inorganic lamellar framework. Such biomaterials have numerous biotechnological applications ([5,6] and references therein). However, as naturally occurring clay minerals suffer from heterogeneity and poor swelling ability in polar/nonpolar solvents, both of which hamper their application, current research has placed increasing emphasis on the development of low temperature protocols for synthetic clay minerals [7]. Recently, a new class of synthetic organic-inorganic layered materials have emerged that are derivatives of 2 : 1 trioctahedral phyllosilicates with covalently bonded organic moieties occupying the interlayer regions. Significantly, the structural and chemical properties of these organically modified phyllosilicates can be utilized for the integration of functional guest biomolecules to produce bio-inorganic hybrids with potential applications in biosensing, biocatalysis and pharmaceutical/biomedical devices.

This chapter reviews recent work on the fabrication and characterization of bio-inorganic nanomaterials based on organically functionalized magnesium phyllosilicate materials. We begin with the general procedures used to synthesize and characterize these organoclays (Section 8.2), and then describe how higher-order

Bio-inorganic Hybrid Nanomaterials. Edited by E. Ruiz-Hitzky, K. Ariga, Y. M. Lvov

ISBN: 978-3-527-31718-9

arhitectures of these hybrid materials can be prepared using self-assembled organic/biomolecular templates, such as emulsion droplets, helical lipid tubules and colloidal crystals (Section 8.3). The intercalation of proteins/enzymes, DNA, or drug molecules within the interlayers of aminopropyl-functionalized magnesium phyllosilicate by re-assembly of exfoliated organoclay aqueous dispersions is discussed in Section 8.4. This approach is contrasted with a novel strategy for wrapping single protein/enzyme or DNA molecules in an ultrathin inorganic shell by using low molecular weight aminopropyl-functionalized organoclay oligomers (Section 8.5). Finally, in Section 8.6, we highlight how nanothin sheets of organoclay or organosilica can be intercalated within the interlayer spaces of stacked lipid bilayers containing the membrane-bound protein bacteriorhodopsin to produce nanocomposites with exceptional photo-active properties. A short summary of the chapter is presented in Section 8.7.

8.2 Synthesis and Characterization of Organically Functionalized 2:1 Magnesium Phyllosilicates

To synthesize hybrid materials with exceptional properties, various inter disciplinary approaches such as organic synthesis, sol–gel chemistry and supramolecular assembly must be integrated and intelligently employed to design and construct multicomponent materials for specific tasks [8,9]. Among various soft chemical routes, sol–gel chemistry provides excellent control over reaction kinetics such that multiple components can be incorporated and co-assembled in single materials by tuning the reaction conditions [10]. Fukushima and coworkers first demonstrated a one step sol–gel synthetic route in which the reaction of 3-[(methacryloxy)propyl] methoxysilane with magnesium chloride in methanol under mild conditions produced a lamellar magnesium organosilicate material in which the organic groups were covalently attached to the inorganic framework through Si–C linkages [11,12]. This report illustrated the potential of preparing a wide range of synthetic lamellar inorganic-organic hybrid materials, and as a consequence a wide range of metal salts (aluminum, copper, zinc and calcium) and organotrialkoxysilane precursors have been used to synthesize numerous organophyllosilicates (Table 8.1) [13–33]. The resulting materials have a 2 : 1 trioctahedral phyllosilicate smectite-like structure (talc, $Si_8Mg_6O_{20}(OH)_4$) with approximate compositions of $Si_8R_8Mg_6O_{16}(OH)_4$ (Figure 8.1). Typically, powder X-ray diffraction (PXRD) patterns showed broader reflections compared to those observed for the parent talc structure, suggesting significant intralayer disorder due to the organic moieties (Figure 8.2). However, the intralayer reflection at $2\theta = 59°$ ($d_{060,330} = 0.156$ nm), which is a characteristic reflection of the 2 : 1 trioctahedral phyllosilicate structure, remained unchanged in the synthetic derivatives. Significantly, the interlayer d_{001} spacings for the organically modified phyllosilicates were larger than the basal spacing of naturally occurring talc (~0.96 nm), and increased with the steric size of the organosiloxane side chains

Tab. 8.1 Examples of organically modified 2 : 1 trioctahedral phyllosilicates.

Metal cations	Organic moieties covalently attached to phyllosilicate framework	References
Mg	(3-methacryloxy)propyl, methyl, phenyl, 3-aminopropyl, 3-mercaptopropyl, [n-(2-aminoethyl)-3-aminopropyl], [(10-amino)-4,7-diazanonyl], hexadecyl, n-propyl-ethylenediamine, n-propyltriethylenediamine, 3-chloropropyl, 5-amino-1,3,4-thiadiazol-2-thiol, $(CH_3CH_2O)Si(CH_2)_3NHC(O)NH$-R (R = 3-propyl, 3-pentyl, 3-heptyl)	[11–13,15,16,18,19,21–23,26,33]
Al–Mg	n-dodecyl, n-octyl, n-pentyl, (3-methacryloxy)propyl, isobutyl, phenyl, 3-mercaptopropyl	[14,24,25,28,31,32]
Ca	methyl, ethyl, n-butyl, n-hexyl, n-octyl, n-dodecyl, phenyl, n-octadecyl	[29]
Ni	n-propyltriethylenediamine, n-propyldiethylenetriamine, 3-aminopropyl	[11,12,20,27]
Cu–Mg	3-mercaptopropyl	[17]
Zn	3-aminopropyl, *N*-propylethylenediamine	[30]

covalently linked to the phyllosilicate framework, which confirmed that the organic groups were directed into the interlayer spaces of the inorganic–organic composite material. In the case of magnesium organophyllosilicates produced using long chain (C_{12}) alkyl- or amine-functionalized trialkoxysilanes, XRD studies revealed basal spacings that were consistent with an interlayer structure comprising an interdigitated array of the organic moieties.

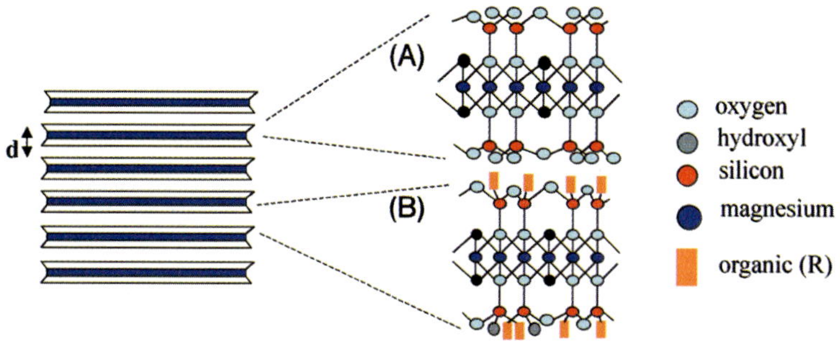

Fig. 8.1 Schematic representation of the layered structures of (A) 2 : 1 trioctahedral phyllosilicate (talc; $[Si_8Mg_6O_{20}(OH)_4]$), and (B) organically functionalized magnesium phyllosilicate ($[Si_8R_8Mg_6O_{16-x/2}(OH)_{4+x}]$).

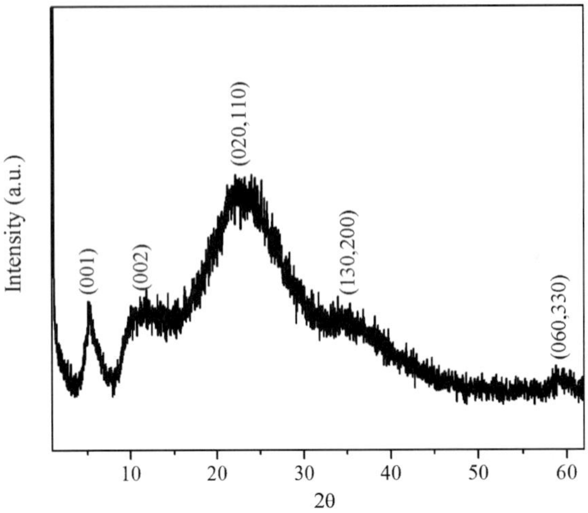

Fig. 8.2 PXRD pattern of ethlyenediamine-functionalized magnesium phyllosilicate showing reflections indexed according to the 2 : 1 trioctahedral phyllosilicate structure of talc.

FTIR spectra showed all the characteristic features associated with the 2:1 phyllosilicate structure, and indicated that the Si–C bond remained intact in the synthesized organoclays. Peak assignments in solid state ^{13}C CP MAS NMR spectra also confirmed the presence of intact and unreacted organic functionalities [13]. Corresponding ^{29}Si NMR spectra showed fully condensed T^3 organosilicate centres, as well as hydroxylated T^1 and T^2 species, indicating domains of incomplete condensation that were presumably associated with geometric constraints imposed on the organosilanol linkages, as well as slow condensation of the organosilsequioxanes. As a result, both inter- and intra-layer regions of the resulting materials exhibited significant structural disorder. It seems likely that accommodation of the organic functionalities within the gallery spaces is governed in part by packing constraints, which in turn increase the epitaxial mismatch at the interface between the brucite and organosilicate layers. In addition, structural disorder arises directly from replacing the four-oxygen bridged Si centers in the parent talc structure with T^n ($n = 1$–3) sites. This reduces the connectivity of the brucite sheet, which in turn leads to distortion and puckering of the tetrahedral sheets [13–15].

The processes of formation of synthetic magnesium organophyllosilicates have been described in terms of several mechanisms. Formation of silicate clays in nature is dependent on silica concentration, pH and the type of metal ions that constitute the octahedral sheets. Similarly, synthesis studies have shown that low-temperature co-precipitation of silica and selective metal hydroxides results in the crystallization of a variety of minerals. In general, increasing the pH causes gelation due to metal hydroxide formation and, as a result, brucite sheets formed by 2-D condensation act as incipient templates for subsequent condensation of the silicate layers [34]. In this regard, magnesium organophyllosilicate synthesis is usually carried out under basic reaction conditions (pH ~ 12), in which the hydrolysis of magnesium cations is facile, but the rate of organotrialkoxysilane hydrolysis is slow relative to condensation [14]. Moreover, the presence of bulky organic substituents can sterically hinder the rate of hydrolysis, which under basic conditions proceeds either by a conventional S_N2-type mechanism, or via a five-coordinated negatively charged intermediate. Once formed, O/OH sites on the brucite layers are available for rapid condensation with silanol derivatives and subsequent nucleation and growth of the tetrahedral sheets takes place on the either side of the central octahedral layer [15]. In addition, hydrophobic interactions between adjacent organic moieties and associated microphase separation could also facilitate formation of the lamellar structure. Thus, the octahedral coordination of magnesium is considered to be the primary structure determinant in the formation of 2 : 1 trioctahedral phyllosilicates at high pH. This is analogous to other self-assembled layered materials such as zirconium organophosphates and organophosphonates, where structure formation is driven principally by the coordination chemistry of the inorganic components [13].

Other studies, in which aluminum/magnesium organophyllosilicates were prepared using long chain trialkoxysilanes under acidic pH conditions, suggested an alternative mechanism of formation [14]. Under acidic reaction conditions, the rate of organosilane hydrolysis was significantly faster than condensation, with the consequence that lamellar intermediates consisting of self-assembled amphiphilic organosilanol molecules were produced. As the partially condensed silanol groups were acidic, the lamellar sheets were considered to be suitable negatively charged templates for Al(III) binding and subsequent deposition of the inorganic framework. This mechanism is therefore the converse of that proposed for the synthesis of magnesium phyllo(organo)silicates under basic conditions. In contrast, ethylenediamine- or aminopropyl-functionalized magnesium phyllosilicates can be prepared at neutral pH [15]. Under these conditions, the rates of organosilane and Mg^{2+} hydrolysis are negligible, whereas intermolecular condensation is favorable due to the presence of significant concentrations of both protonated and deprotonated organosilanol groups. However, the basic character of the pendent amine groups shifts the local pH towards higher values, thereby catalyzing the hydrolysis reactions of alkoxysilane and magnesium ions simultaneously. Complex interactions appear prior to the coupling of the organosilanols on to the inorganic framework such that co-operative assembly of the organic and inorganic components leads to formation of the 2 : 1 organophyllosilicate structure.

8.3 Magnesium Organophyllosilicates with Higher-order Organization

The ability to manipulate the morphogenesis of inorganic materials through controlled chemical synthesis for the design of complex architectures on various length scales remains challenging [8,9,35]. Among various strategies, inorganic replication of organized reaction fields such as bicontinuous microemulsions [36,37] and emulsion foams or droplets [38,39], has proved to be an interesting approach to

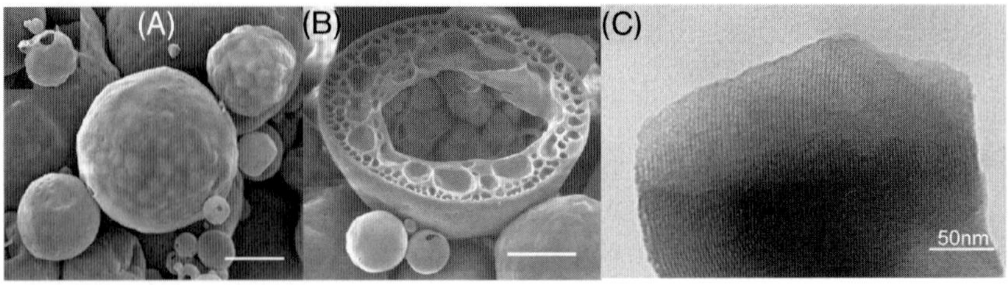

Fig. 8.3 SEM images of hexadecyl-functionalized magnesium phyllosilicate showing (A) intact spheroids (scale bar = 20 μm) and (B) fractured spheroid with foam like interior (scale bar = 20 μm). (C) TEM image of a wall fragment showing lattice fringes corresponding to a periodic lamellar structure (scale bar = 50 nm).

the synthesis of inorganic materials with intricate morphologies [37–42]. Except for a handful of studies, however, the higher-order organization of layered materials into novel and hierarchical forms has not been investigated in detail. Notably, anion exchange resins [43], cationic miniemulsions and latex spheres [44], and layer-by-layer co-assembly [45], have been used to produce hollow clay spheres. In each case, preformed plate-like clay particles were used to produce the higher-ordered microstructures. Mann and coworkers demonstrated a bottom-up approach for the direct synthesis of organically functionalized magnesium phyllosilicate microspheres with sponge-like interiors [46]. In this emulsion-based synthesis, addition of an alkaline solution of hydrolyzed hexadecyltrimethoxysilane precursor to sodium dodecylsulfate stabilized Mg^{2+}-containing water droplets in oil resulted in the formation of 5–25 μm-diameter spherical particles (Figure 8.3A). The microspheres consisted of a sponge-like interior (Figure 8.3B), and transmission electron microscopy (TEM) of fragments revealed that the interior walls were composed of highly-ordered organoclay lamellae with an interlayer spacing ~3.5 nm (Figure 8.3C), which was in good agreement with PXRD data. The complex morphology was produced by the diffusion-controlled reaction of aqueous OH^- with the prehydrolysed long chain organotrialkoxysilane precursor specifically at the oil/water droplet interface to produce primary hollow microspheres that subsequently aggregated into larger spheroids with foam-like interiors. This method clearly demonstrated the flexibility of using sol–gel protocols in emulsion media to generate hierarchically organized organoclay-based materials with potential applications in the controlled release of molecules such as drugs, pesticides, dyes or inks, entrapment and display of functional biomolecules such as proteins, enzymes or DNA, and sequestration and extraction of heavy metal ions from contaminated fluids.

Organoclay materials with higher-order organization can also be prepared by template-directed methods involving self-assembled supramolecular structures. In this approach, preformed organic architectures in the form of tubes, fibers, hollow shells, gyroids, helicoids, and so on are transferred into hybrid materials exhibiting structural hierarchy, complex form and ordered mesoporosity [47–55]. For example, synthetic lipids such as 1,2-bis- (10,12-tricosadiynoyl)-*sn*-glycerol-3-phosphatidylcholine ($DC_{89}PC$) form tubular/helical microstructures with diameters of approximately 0.5 μm and lengths ranging from 50 to 200 μm [55], and these architectures can be used as templates for the controlled deposition and organization of a variety of inorganic materials [56–63], including the fabrication of organoclay pipes [64]. In the latter case, reaction of magnesium chloride and 3-(2-aminoethyl)-3-aminopropyl trimethoxysilane was undertaken in the presence of preformed lipid tubules. TEM revealed that the product consisted of intact helical or tubular microstructures coated with a continuous and smooth organoclay overlayer, 25–50 nm in thickness. PXRD patterns of the as-synthesized lipid–organoclay materials showed reflections corresponding to a well ordered lamellar lipid mesostructure, along with characteristic peaks for a disordered layered magnesium organophyllosilicate with basal spacing of 1.8 nm. Addition of excess ethanol dissolved the lipid template to produce well ordered microscopic tubular replicas consisting of a smooth organoclay surface with helical decoration and external diameter of 0.85 μm (Figure 8.4). Furthermore, by extending the immersion time in ethanol, organoclay pipes with thicker walls, up to 60 nm in width, could be produced by prolonged deposition on the tubule surface.

The results indicated that nanosized organoclay clusters bind specifically to the zwitterionic phosphocholine headgroups present along the inner and outer walls of the lipid tubules. At neutral pH, this is facilitated by electrostatic interactions between the protonated ethylenediaminopropyl groups ($pK_a \sim 10.5$) and negatively charged phosphate headgroups, which together provide an interfacial layer that facilitates

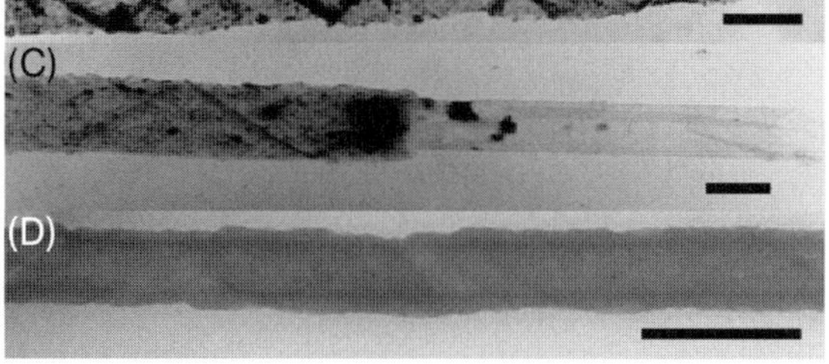

Fig. 8.4 TEM images showing ethylenediaminopropyl-functionalized magnesium phyllosilicate microtubules produced by lipid helicoid templating. In each case the template was removed by washing with ethanol. (A) intact open-ended tubule, (B) tubule with helical-ribbon morphology, (C) tubule-within-tubule morphology, and (D) micropipe with increased wall thickness produced after 10 days ethanol treatment (scale bar = 500 nm).

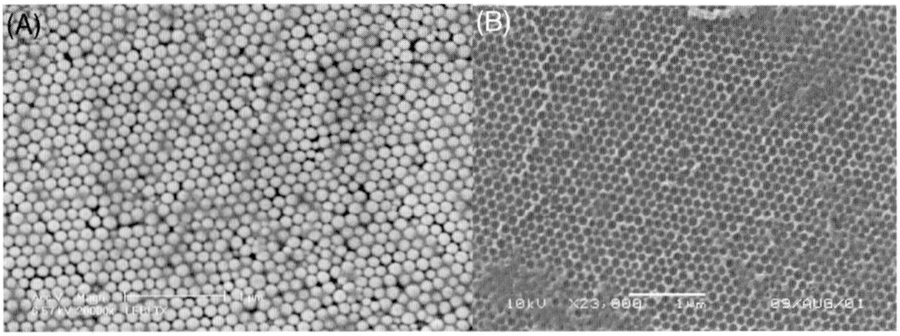

Fig. 8.5 SEM images of (A) close packed array of latex beads (scale bar = 1 μm) and (B) macroporous aminopropyl-functionalized magnesium phyllosilicate monolith obtained after infiltration and extraction of colloidal template (scale bar = 1 μm).

nucleation specifically on the tubule surface by partial hydrolysis and co-condensation in the presence of magnesium ions to produce coherent overlayers that remain intact after dissolution of the lipid template. Other studies highlighted the potential utility of the ethylenediamine-functionalized organoclay pipes as metal-ion sorbents or supports for inorganic nanoparticles. For example, when hollow organoclay micropipes were immersed in a sol of mercaptoacetic acid-stabilized CdS nanoparticles the entire surface of the hybrid microstructure was coated in quantum dots [64].

The above investigations illustrate how discrete tubular architectures of layered magnesium organophyllosilicates can be fabricated using appropriate self-assembled microstructures. Similar approaches can be developed to produce extended 3-D organized organoclay architectures, provided that suitable templates are available. In this regard, Lebeau and coworkers demonstrated a simple strategy to produce highly ordered macroporous networks of organically modified 2 : 1 trioctahedral phyllosilicates using latex colloidal crystal templates (Figure 8.5A) [65]. In this method, an exfoliated dispersion of aminopropyl-functionalized magnesium phyllosilicate clay (see Section 8.4.1) was infiltrated into the void spaces of the preformed colloidal crystal. Extensive penetration of the exfoliated organoclay into the array was facilitated by the presence of negatively charged sulfonate groups on the polystyrene spheres. Subsequently, dissolution of the latex spheres in toluene produced an intact, highly ordered macroporous microstructure consisting of walls of layered organo-functionalized magnesium phyllosilicate (Figure 8.5B).

8.4
Intercalation of Biomolecules within Organically Modified Magnesium Phyllosilicates

A key feature of the chemistry of layered materials is their ability to locate and orient guest molecules of interest within constrained interlayer regions. Once intercalated, the guest molecules can demonstrate a range of interesting properties [3,66]. In general, insertion of guest molecules into the layered hosts can be achieved via two types of interaction processes. First, ions and molecules can be intercalated into the gallery spaces of as-synthesized layered materials by exchange of the native guest species. This process is usually slow, and often limited to intercalands with low molecular mass and size. In many cases, exchange proceeds with retention of the host framework, even though the interlayer spacing may undergo substantial expansion due to attractive interactions between the intercalated guest and flanking sheets of the host matrix [67]. Secondly, the as-synthesized layered material can be exfoliated by increasing the interlayer repulsion forces such that the individual sheets are physically separated in the form of colloidal dispersions of discrete lamellae. The lamellae can be subsequently restacked in the presence of appropriate guest molecules to produce the intercalated nanocomposites [68]. Although a wide range of layered host–guest structures can be prepared by this approach, periodic re-stacking of the sheets can be notoriously difficult, with the consequence that the nanocomposites often exhibit increased levels of disorder compared with the as-synthesized materials.

In the next section, we demonstrate how exfoliation and ordered restacking of aminopropyl-derivatized magnesium phyllosilicates in the presence of proteins, enzymes or DNA can be used to prepare new types of bio-inorganic layered nanocomposites.

8.4.1 Protein–Organoclay Lamellar Nanocomposites

Recent studies have shown that organically modified magnesium phyllosilicates can be used as platforms to prepare functionalized nanocomposites by utilizing the chemical reactivity and binding properties of their pendent moieties. With regard to as-synthesized materials, *in situ* crosslinking of epoxide-functionalities with *m*-phenylenediamine [13], or binding of $[AuCl_4]^-$ to pendent thiol groups [15], have been used to produce polymer–organoclay or gold nanoparticle-intercalated layered materials, respectively. Similarly, thiol- or ethylenediamine-functionalized magnesium organophyllosilicates have been exploited for the sequestration of heavy metal ions [21–23,31,64]. Alternatively, as described below, exfoliated magnesium organophyllosilicates can be used to prepare a range of new bio-inorganic layered nanocomposites that have significant bionanotechnological potential.

In general, biomolecules such as proteins and enzymes display sophisticated recognition abilities but their commercial viability is often hampered by their fragile structure and lack of long term stability under processing conditions [69]. These problems can be partially overcome by immobilization of the biomolecules on various supports, which provide enhanced stability, repetitive and continuous use, potential modulation of catalytic properties, and prevention of microbial contaminations. Sol–gel and synthetic polymer-based routes for biomolecule encapsulation have been studied extensively and are now well established [70–72]. Current research is also concerned with improving the stability of the immobilized biomolecules, notably enzymes, to increase the scope for exploitation in various biotechnological applications such as affinity separations, biosensing, bioreactors and in the fabrication of biofuel cells [73]. To date several inorganic supports such as metal oxide thin films [74,75], molecular sieves [76–78], colloidal nanoparticles [79,80], microcrystals [81], and magnetic γ-Fe_2O_3 nanoparticles [82] have been investigated in bio-immobilization studies.

Layered materials are of special interest for bio-immobilization due to the accessibility of large internal and external surface areas, potential to confine biomolecules within regularly organized interlayer spaces, and processing of colloidal dispersions for the fabrication of protein–clay films for electrochemical catalysis [83–90]. These studies indicate that layered materials can serve as efficient support matrices to maintain the native structure and function of the immobilized biomolecules. Current trends in the synthesis of functional biopolymer nano composites based on layered materials (specifically layered double hydroxides) have been discussed in excellent reviews by Ruiz-Hitzky [5] and Duan [6]; herein we focus specifically on the fabrication of bio-inorganic lamellar nanocomposites based on the exfoliation and ordered restacking of aminopropyl-functionalized magnesium phyllosilicate (AMP) in the presence of various biomolecules [91].

An AMP clay with an approximate unit cell composition of $[H_2N(CH_2)_3]_8$ $[Si_8Mg_{\sim 6}O_{16}(OH)_4]$ was synthesized in ethanol and then exfoliated by water-induced protonation of the aminopropyl side chains. TEM investigations indicated that the exfoliated particles were in the form of ultrathin sheets, 30 to 150 nm in size (Figure 8.6A). Exfoliated sheets viewed edge-on indicated that the particles were only ~2 nm in thickness (Figure 8.6B), corresponding to a single layer of the organoclay framework. Subsequent additions of negatively charged proteins, such as myoglobin (Mb), hemoglobin (Hb) and glucose oxidase (GOx), to a suspension of the exfoliated AMP clay dispersion resulted in the immediate formation of reddish-brown (Mb-AMP, Hb-AMP) or yellow (GOx-AMP) bulk precipitates.

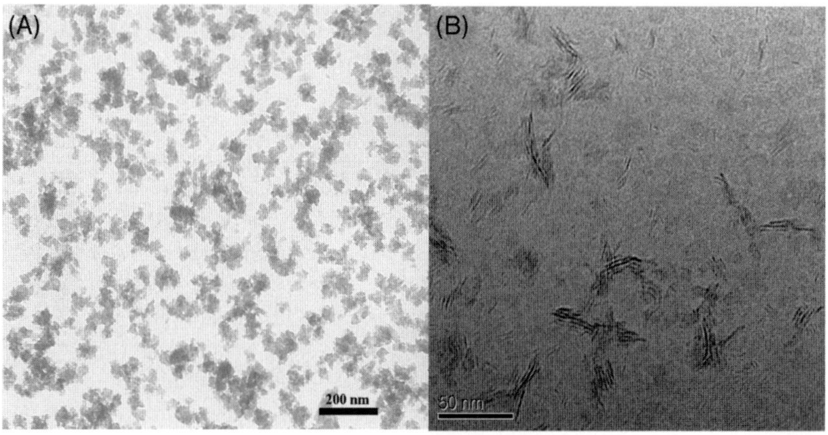

Fig. 8.6 TEM images of (A) exfoliated sheets of AMP in water (scale bar = 200 nm), and (B) cryo-TEM image of exfoliated AMP organoclay sheets viewed edge-on (scale bar = 50 nm).

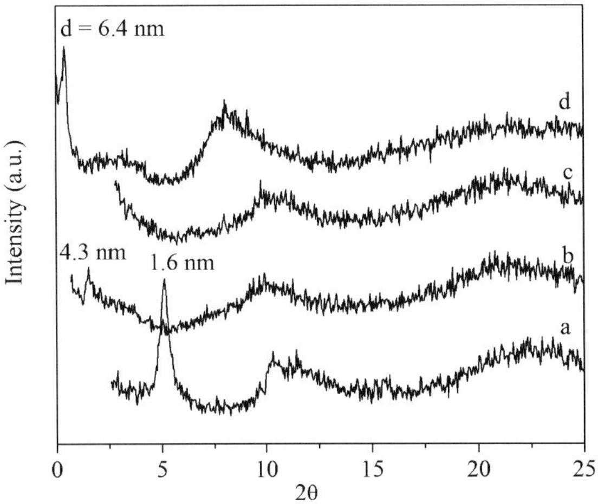

Fig. 8.7 PXRD profiles for (a) as-synthesized AMP, and (b–d) intercalated bio-nanocomposites; (b) myoglobin–AMP, (c) haemoglobin–AMP and (c) glucose oxidase–AMP.

Powder XRD patterns of Mb-AMP and GOx-AMP samples showed that the d_{001} reflection of the parent AMP clay at 1.6 nm was completely eliminated and replaced by a d_{001} peak at 4.3 or 6.4 nm, respectively (Figure 8.7). The associated expansions of 2.7 and 4.8 nm in the interlayer spacing were consistent with the molecular size of Mb ($2.5 \times 3.5 \times 4.5$ nm) and GOx ($3.7 \times 5.2 \times 6.0$ nm), confirming immobilization of the proteins in the gallery regions during re-assembly of the exfoliated AMP sheets, and formation of well defined protein–organoclay mesolamellar nanocomposites. Considering the molecular dimensions of Mb, the PXRD data suggested that the intercalated protein molecules were preferentially aligned within the interlayer spaces. In contrast, XRD patterns of precipitates prepared in the presence of Hb, which is significantly larger than Mb (Hb molecular size; $6.5 \times 5.4 \times 5.3$ nm), did not show reflections for the parent organoclay basal spacing or an expanded gallery spacing corresponding to a putative Hb–AMP lamellar nanocomposite.

Under optimized experimental conditions, the above negatively charged biomolecules and exfoliated organoclay sheets spontaneously co-assemble due to electrostatic interactions between the countercharged host and guest species. This was consistent with other observations that indicated that hybrid nanocomposites were not produced in the presence of the 3.5 nm-sized positively charged protein, cytochrome *c*. However, the absence of regular stacking in the Hb-AMP precipitate suggested that the size and local surface charge density of the guest molecules also influenced the intercalation process. Indeed, other studies have indicated that different proteins may interact diversely with the same substrate [90], and that exfoliated nanocomposites comprising randomly arranged dispersed inorganic sheets and immobilized guest molecules are often produced in polymer–clay materials [92].

Structural characterization of the biomolecule–AMP lamellar nanocomposites using FTIR and circular dichroism (CD) spectroscopies showed no change in the

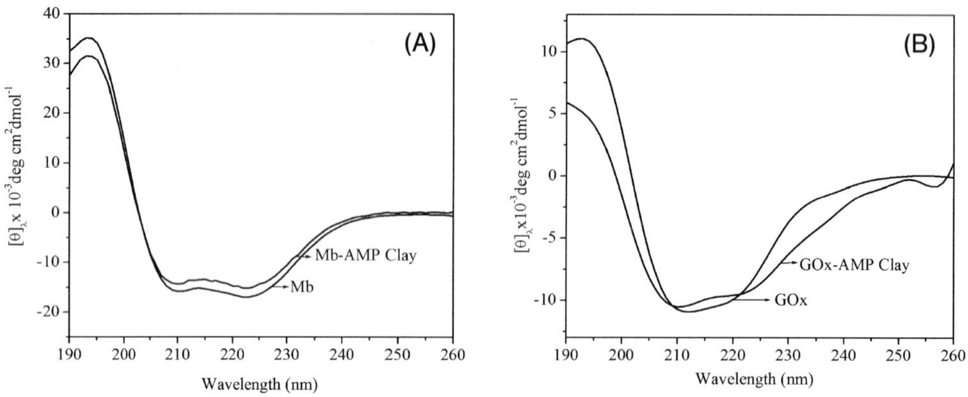

Fig. 8.8 CD spectra of native or intercalated biomolecules in solution or within re-assembled AMP, respectively. (A) Native myoglobin and myoglobin–AMP, (B) native glucose oxidase and glucose oxidase–AMP.

polypeptide amide I (C=O str, 1653 cm^{-1}) and amide II (N−H def; C−N str, 1542 cm^{-1}) bands, indicating that Mb and GOx were intercalated, and Hb immobilized, within the AMP clay without loss of secondary structure. The results were consistent with the CD spectra of Mb–AMP and Hb–AMP, which showed typical α-helical profiles with characteristic bands for π-π^* and n-π^* amide transitions at 190, 209 and 220 nm, respectively. Similar measurements with GOx–AMP gave slightly broadened CD bands compared with the native enzyme (Figure 8.8), suggesting that the protein secondary structure was influenced to a small extent by intercalation.

The biochemical activity and accessibility of biomolecule-intercalated AMP clays to small molecules was retained in the hybrid nanocomposites. For example, the absorption spectrum of the intercalated Mb–AMP nanocomposite showed a characteristic soret band at 408 nm associated with the intact prosthetic heme group of the oxidised protein (Fe(III), met-myoglobin) (Figure 8.9). Treatment of Mb with sodium dithionite solution resulted in a red shift of the soret band from 408 to 427 nm, consistent with the formation of intercalated deoxy-Mb. Reversible binding of CO under argon to the deoxy-Mb–AMP lamellar nanocomposite was demonstrated by a shift in the soret band from 427 to 422 nm. Subsequent dissociation of CO from the heme centre due to competitive O_2 binding shifted the soret band to 416 nm on formation of intercalated oxy-Mb.

The enzymatic activities of intercalated GOx–AMP layered nanocomposites at various pH values and temperatures were compared with the native enzyme in aqueous solution. In both cases, characteristic linear plots consistent with Michalis–Menton kinetics were obtained. The Lineweaver–Burk plots indicated that the reaction rates (V_{max}) for free and intercalated GOx (3.3 and 4.0 μM min^{-1} respectively), were comparable, suggesting that the turnover rate at substrate saturation was only marginally influenced by entrapment between the re-assembled organoclay sheets. However, the dissociation constant (K_m) associated with the activity of the enzyme was higher for intercalated GOx (6.63 mM) compared to native GOx (2.94 mM), suggesting

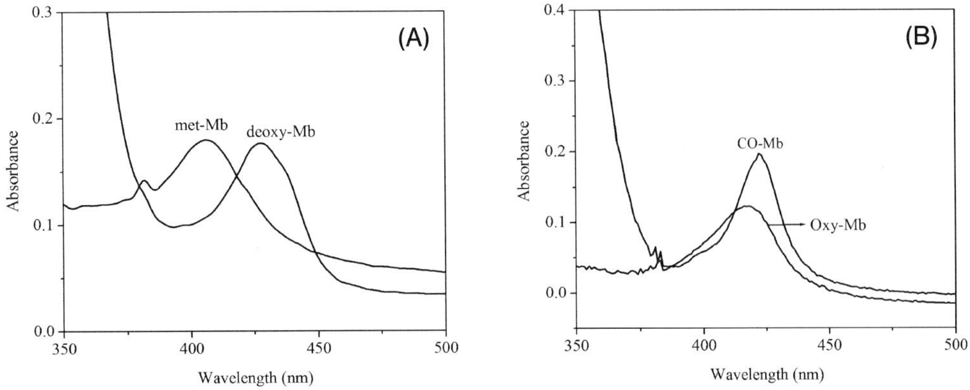

Fig. 8.9 UV–Vis spectra of intercalated biomolecules in re-assembled AMP. Soret band absorptions for (A) oxidized myoglobin (met-Mb) and after dithionite reduction (deoxy-Mb), and (B) after CO (CO-Mb) and O_2 binding to intercalated deoxy-Mb.

that glucose binding at intermediate substrate concentrations was partially inhibited, perhaps due to restricted diffusion of the substrate through the gallery spaces of the lamellar nanocomposite. Enzyme kinetic studies over a pH range of 4–9 showed that the relative activities of the intercalated GOx molecules were maintained at a level commensurate with that determined for the native enzyme at an optimum pH of 6 (Figure 8.10). As enzymes are very pH sensitive and usually active only within narrow pH ranges [93], the results were of significant interest. Moreover, a similar stabilization of the intercalated GOx molecules was observed when relative activities were determined over a temperature range of 25–85 °C (Figure 8.10B).

The above studies indicated that intercalation of GOx during re-assembly of exfoliated AMP sheets produced lamellar nanocomposites that retain their

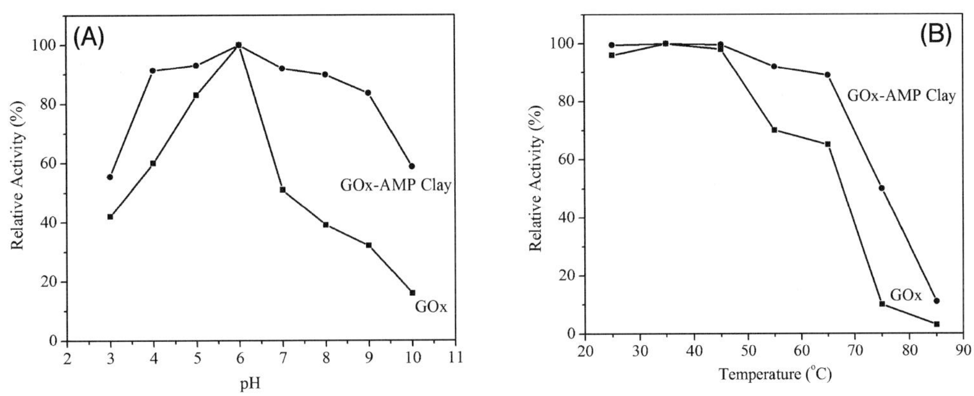

Fig. 8.10 Plots of relative activity (%) of GOx in solution or after intercalation within AMP clay) against (A) pH and (B) temperature.

bio-activity under conditions that conventionally result in unfolding and deactivation of the protein molecules. The improved thermal stability could arise from multipoint electrostatic interactions between the intercalated enzyme molecules and cationic aminopropyl side chains of the clay lamellae, as suggested previously for other systems [94,95]. In addition, preservation of enzyme activity over an extended pH range could be due to interlayer buffering of H^+ and OH^- ions, and/or the effect of confinement on local ionic strength that facilitates additional interactions such as salt-bridge formation with the host matrix. Alternatively, molecular confinement in the layered galleries may act as a significant diffusion barrier, such that considerable changes in the bulk solution pH cause only small pH shifts in the environment associated with the intercalated enzyme molecules [94]. Such interactions could be exploited for the generation of microscopic organoclay-based biosensors operating efficiently under detrimental environmental conditions.

8.4.2
DNA–Organoclay Lamellar Nanostructures

Recent advances in biomedical applications have demonstrated the target-specific efficiency of viral vectors for the introduction of desirable genetic sequences or drugs for the treatment of genetic disorders and promotion of tissue engineering [96]. Because of concerns over the safety and carrier capacity of this approach, non-viral vectors based on polycationic lipids, polypeptides and biopolymers have been developed, although significant physical, chemical, and toxicity barriers remain to be overcome [2]. Recent studies involving functional inorganic/metal nanoparticles [97–101], or clays such as layered double hydroxides [102–105], indicate that the use of synthetic inorganic materials as gene/drug delivery vehicles should be seriously considered. In recent studies we demonstrated a simple method by which intercalative DNA–organoclay nanocomposites could be prepared by addition of stoichiometric amounts of a DNA (700–1000 base-pairs) solution to an exfoliated dispersion of an aminopropyl-functionalized magnesium phyllosilicate clay [106]. Confinement of the DNA within the interlayer regions during co-assembly produced a layered composite with a basal spacing of 3.6 nm that was 2 nm larger than the parent organoclay. This was in good agreement with the diameter of the double helical strand of DNA, suggesting that a monolayer of the biomacromolecules was entrapped within the gallery spaces. Significantly, structural integrity and helicity of the intercalated DNA were retained within the layered nanohybrid. Melting temperature curves for the intercalated DNA showed a characteristic sigmoidal melting curve (Figure 8.11), suggesting cooperative denaturing of the double helix. The melting temperature was 83 °C, compared to a value of 63 °C for native DNA in solution, indicating a substantial increase in the thermal stability of the intercalated DNA, possibly due to electrostatic interactions between the aminopropyl moieties of the organoclay side chains and negatively charged phosphate groups of the biomacromolecule. These preliminary investigations suggest that DNA–organoclay layered nanocomposites could be useful vectors for the protection, storage and transport of gene-based materials.

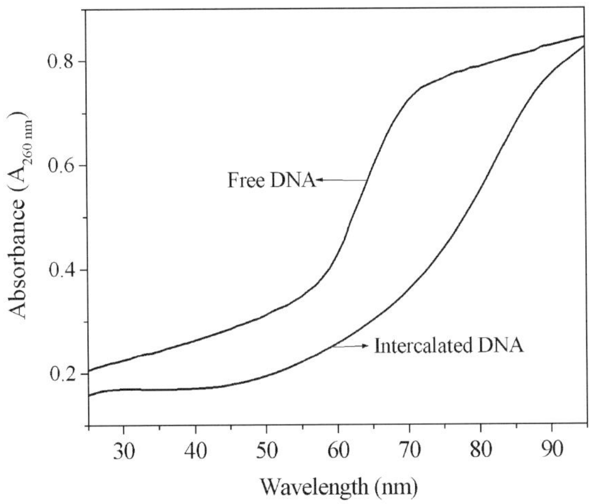

Fig. 8.11 Melting curves obtained for native DNA and DNA intercalated within AMP.

8.4.3 Drug–Organoclay Layered Nanocomposites

Along with the rational design of effective drug molecules, the constant search for novel formulation techniques for targeted and more efficient delivery of drugs is a major goal of pharmaceutical research [107]. Recent advances in nanotechnology have shown that ceramic based nano- and micro-particulate drugs can be used as therapeutic carriers [108,109]. Such materials can be used in a range of delivery methods (oral, nasal, parenteral, etc.) and enable drug loading and targeted delivery to be controlled for extended therapeutic effects. As a result, a range of nanostructural and biocompatible inorganic composites doped with pharmaceutically active compounds have been investigated for effective drug administration [5,6,110–116]. As part of an ongoing study, we used an AMP clay as a storage matrix for the anti-inflammatory drug ibuprofen, and the antioxidant molecule epigallocatechin (EGCG) [117]. EGCG is a polyphenol from green tea and has been shown to inhibit the growth of cancer cells [118]. Ibuprofen–AMP and EGCG–AMP organoclays were fabricated from exfoliated clay sheets by re-assembly in the presence of the drug molecules. Increases in the basal spacing of the precipitated materials indicated that the drug molecules were intercalated within the interlamellar spaces, and suggested that the ibuprofen or EGCG molecules were anchored to the cationic aminopropyl side chains of the organoclay sheets through carboxylate and phenolate groups, respectively. The increase in basal spacing from 1.48 nm in the parent organoclay to 2.75 nm in the ibuprofen–AMP nanocomposite suggested intercalation of a monolayer of ibuprofen molecules oriented parallel to the stacking direction. In contrast, the interlayer spacing increased by only 0.47 nm in the EGCG–AMP material, suggesting that the guest molecules were aligned parallel to the organoclay sheets. Significantly, auxiliary polymers, such as polystyrene sulfonate, polyacrylic acid and polymethyl-acrylamido propane sulfonic

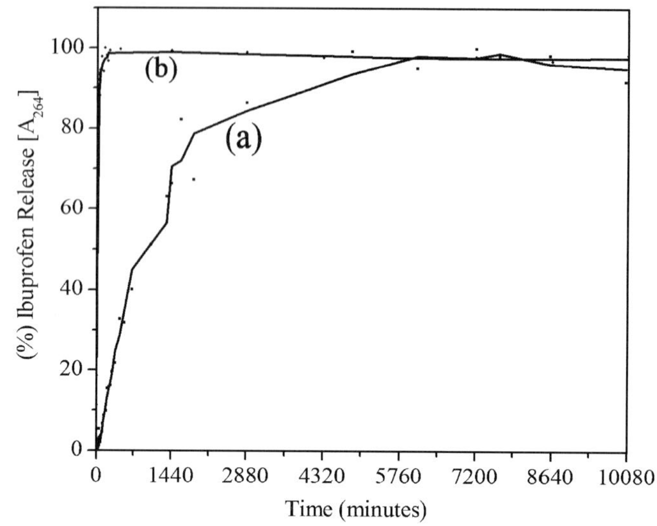

Fig. 8.12 Comparison of drug-release profiles recorded in simulated gastric medium (pH = 2) for (a) ibuprofen–AMP nanocomposite and (b) control sample (ibuprofen–talc suspension).

acid, could be co-intercalated with the drug molecules within the AMP clay gallery regions. In each case, the basal d_{001} spacing was increased compared with the polymer-intercalated composites alone, indicating that more than one type of guest molecule could be entrapped by the exfoliation/re-assembly procedure. Other studies showed that the release of ibuprofen from the organoclay nanocomposites suspended in water or simulated gastric medium (pH 2) was delayed compared with control samples (Figure 8.12). This was attributed to electrostatic interactions between the protonated aminopropyl groups and carboxylate anions of the ibuprofen molecules. PXRD of the composite after drug release showed a basal spacing commensurate with the parent organoclay, confirming complete release of ibuprofen molecules from the gallery regions. These results suggested that modification of the organic functionalities attached to the phyllosilicate framework could be used to control the release of pharmaceutically active molecules from the interlayer spaces.

8.5 Hybrid Nanostructures Based on Organoclay Wrapping of Single Biomolecules

8.5.1 Organoclay-wrapped Proteins and Enzymes

Recent advances in nanotechnology have shown that self-assembled cage structures of nanometer dimensions can be used as constrained environments for the encapsulation of guest molecules with potential applications in drug delivery, catalysis and sensory devices [119–121]. Kunitake *et al.* succeeded in encapsulating the fluorescent dye fluorescein within an ultrathin layer of TiO_2 gel with retention of monochromic properties [122], and reported the electrostatically induced wrapping of single molecules of a positively charged linear polymer with oligomeric silicate anions [122–125]. Similarly, single enzyme molecules were encapsulated within a nanometer-thick organic–inorganic porous composite network using selective modification of the enzyme surface followed by controlled silicate polymerization [126,127]. As a result, unlike the free enzymes, the encapsulated proteins exhibited significantly more stable catalytic activity. Recently, we demonstrated a simple method for the wrapping of single molecules of proteins such as Mb, Hb and GOx with an ultrathin sheath of aminopropyl-functionalized magnesium phyllosilicate oligomers of molecular mass 400–700 [128]. For this, exfoliated and partially disintegrated aminopropyl-functionalized clay layers were fractionated by gel chromatography to produce stable sols of organoclay oligomers that were subsequently added to the biomolecule solutions. TEM images of these samples revealed discrete electron dense nanoparticles that were randomly arranged across the support film (Figure 8.13). Typically, the nanoparticles were spheroidal and monodisperse in size with mean dimensions of 4.0 nm ($\sigma = 0.6$ nm), 7.8 nm ($\sigma = 0.8$ nm) and 6.4 nm ($\sigma = 0.9$ nm) for samples prepared in the presence of Mb, Hb and GOx, respectively. The variation in nanoparticle size showed a direct correlation with the respective molecular dimensions of the different proteins, suggesting that each nanoparticle comprised a single biomolecule enveloped by a continuous condensed sheet of organoclay oligomers, one unit cell layer in thickness. On wrapping, a significant decrease in the sedimentation coefficient and surface charge were observed, presumably due to tessellation of the biomolecules by the organoclay sheets, which was estimated to involve approximately thirty organoclay unit cells per biomolecule. Condensation of the cationic aminoproyl-functionalized magnesium phyllosilicate oligomers was driven by electrostatic interactions with negatively charged amino acids clustered on the protein surface. Thus, no organoclay-wrapped nanoparticles were produced in the presence of the positively charged redox protein, cytochrome *c*.

Higher-order organization of the organoclay-wrapped protein nanoparticles was achieved by tailoring the organic functionalities attached to the organoclay oligomers. For example, replacing about 50% of the aminopropyl groups in the clay with covalently attached hexadecyl groups significantly increased the hydrophobicity of the oligomeric species produced by gel chromatography and, as a consequence, the resulting organoclay-wrapped myoglobin molecules readily self-assembled into closed packed layers of discrete 4.5 nm-sized nanoparticles that were separated by an interparticle spacing of 4–5 nm (Figure 8.14). As shown by XRD studies of the corresponding bulk synthetic clay, this spacing was commensurate with an interparticle bilayer of interdigitated hexadecyl/aminopropyl chains.

The structural and functional integrity of organoclay-wrapped Mb and Hb molecules was demonstrated by retention of the secondary protein structure as well as distinctive shifts in the absorption spectra associated with oxygen or carbon monoxide binding to the heme metallocenter. The latter indicated that the wrapped

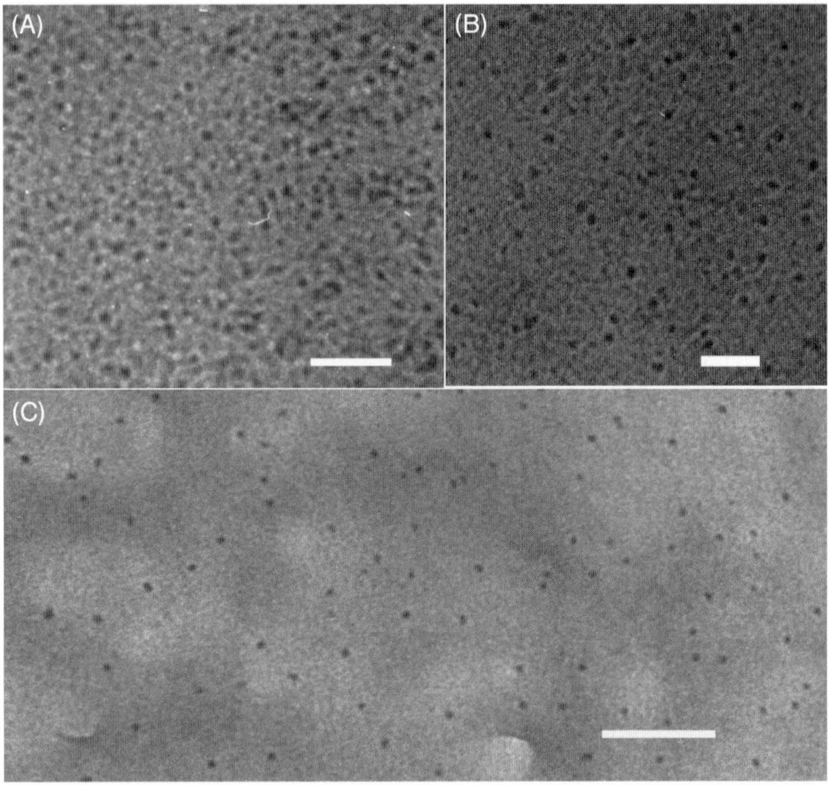

Fig. 8.13 TEM images showing single protein molecules wrapped with an ultrathin shell of AMP. (A) myoglobin, (B) hemoglobin and (C) glucose oxidase, Scale bars; 50, 100 and 100 nm, respectively.

protein molecules remained accessible to small molecules, presumably by diffusion through defects in the organoclay shell. Enzyme activity studies showed that at substrate concentrations between 3.5 and 50 μM the turnover rates for free GOx and GOx–organoclay-wrapped nanoconjugates (2.7 and 2.9 μmol min^{-1} respectively) were not significantly affected by entrapment within the organoclay shell. However, the dissociation constant associated with the activity of GOx–organoclay nanoparticles was increased, suggesting that glucose binding was partially inhibited at intermediate substrate concentrations, possibly by restricted diffusion through the organoclay shell of the hybrid conjugate. Similar experiments on relative enzyme activities at different pH values showed a nonlinear dependence with a small enhancement under acid or alkaline conditions compared with the free enzyme in solution (Figure 8.15). Significantly, the organoclay-conjugated proteins/enzymes showed improved thermal stability compared with native proteins in free solution.

In general, single molecule encapsulation techniques could provide integrated biogenic materials for the isolation and stabilization of functional bio-nanostructures.

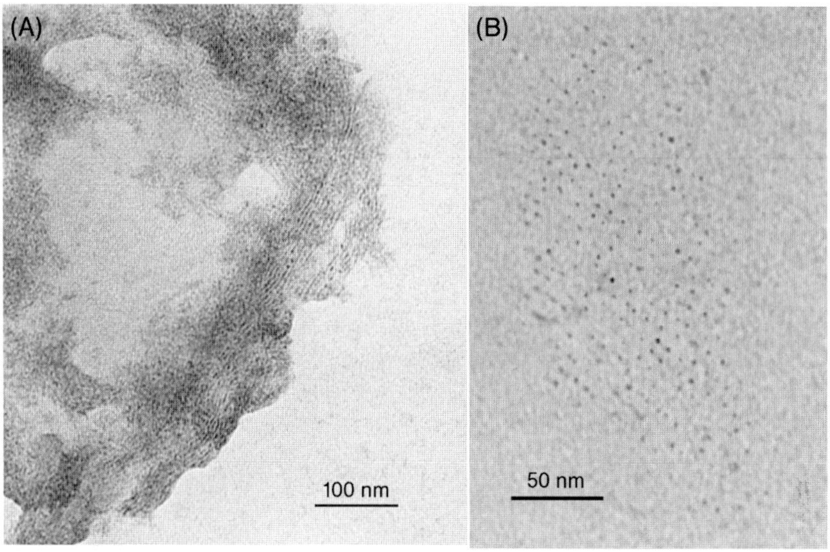

Fig. 8.14 TEM images showing (A) self-assembled layers of individual myoglobin molecules wrapped by condensed oligomers of AMP; (B) top view of single sheet showing ordered superstructure of the hydrophobic organoclay wrapped myoglobin molecules.

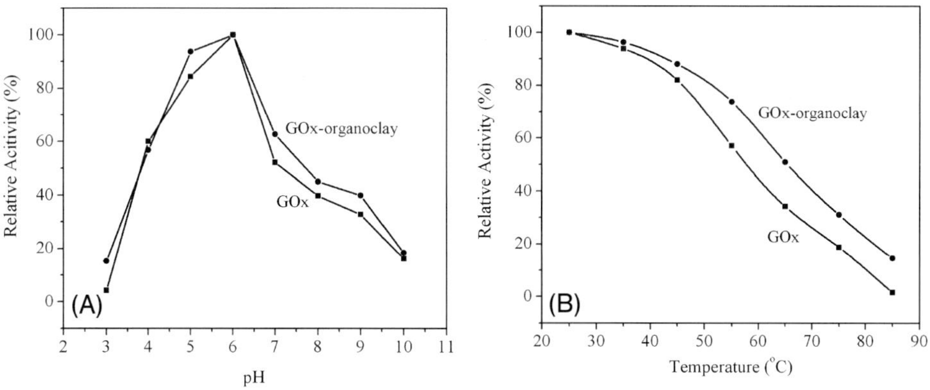

Fig. 8.15 Plots of relative activity (%) of GOx in solution or after wrapping of single molecules within an ultrathin shell of AMP against (A) pH and (B) temperature.

Thus, the organoclay armoured-proteins should have diverse applications in areas such as biosensing, biocatalysis and bioelectronics. Moreover, the experimental procedure is extremely facile and could, in principle, be applied to many types of biomolecules, including bio-macromolecules such as DNA.

8.5.2
Organoclay-wrapped DNA

Among a variety of biomolecules, DNA has proved to be a most attractive and promising programmable "assembler" for nanostructure scaffolds [129,130]. The unique molecular recognition capabilities, physico-chemical stabilities and mechanical rigidity of DNA templates have been exploited, for example, in the growth of a wide range of inorganic and polymer nanowires [131–135]. Recently, we have shown the unprecedented encasing of single λ-DNA molecules of 2 nm thickness and approximately 16 µm in length with cationic organoclay oligomers [106]. The entire surface of single DNA molecules was coated in an ultrathin layer of condensed organoclay oligomers to produce 15 nm-thick DNA–organoclay nanowires (Figure 8.16). Similarly, single strands of plasmid DNA were also successfully wrapped with organoclay oligomers to produce 12–14 nm-thick supercoiled nanowires. Molecular encasing of the DNA was confirmed by agarose-gel electrophoresis, which showed reversal in the DNA surface charge due to coating of the double helix with the cationic organoclay oligomers.

Wrapping of the individual DNA molecules was driven by electrostatic interactions between the anionic phosphodiester groups and cationic organoclay oligomers. This occurred with retention of the structural integrity of the organoclay wrapped λ-DNA/plasmid molecules. The effect of organoclay wrapping on DNA denaturation was examined by recording the thermal melting profiles. As shown in Figure 8.17, λ-DNA in solution showed a sharp rise in the abosorption intensity at 260 nm at around 46 °C, corresponding to the dissociation temperature associated with unwinding of the double helix. In contrast, organoclay-wrapped DNA nanowires exhibited a melting temperature of 61 °C, indicating that association of the organoclay oligomers with the DNA backbone inhibited dehybridization of the double helix.

In conclusion, we have highlighted in this and the preceding section two versatile synthetic strategies to bio-inorganic layered nanocomposites based on the self-assembly of organically functionalized magnesium phyllosilicates (Figure 8.18).

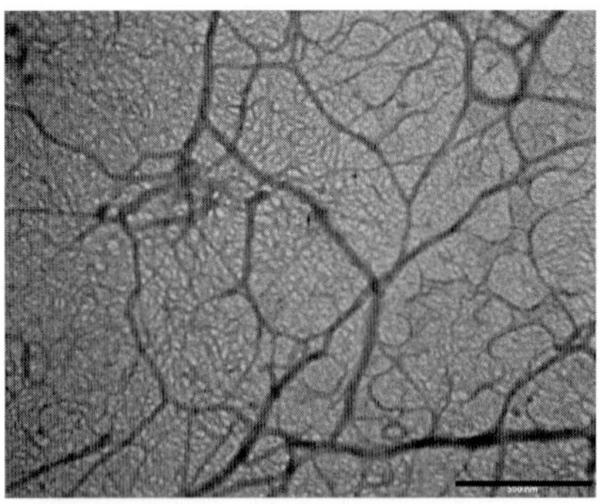

Fig. 8.16 TEM image showing nanowires formed by wrapping λ-DNA with an outer nanoshell of AMP (scale bar = 500 nm).

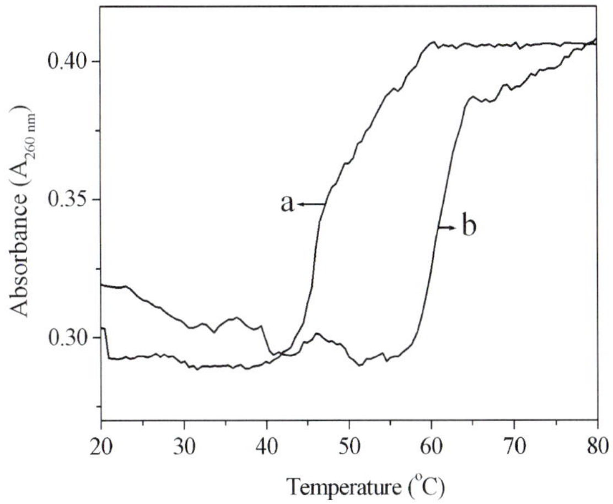

Fig. 8.17 Melting temperature curves recorded for (A) native λ-DNA and (B) λ-DNA molecules wrapped in AMP.

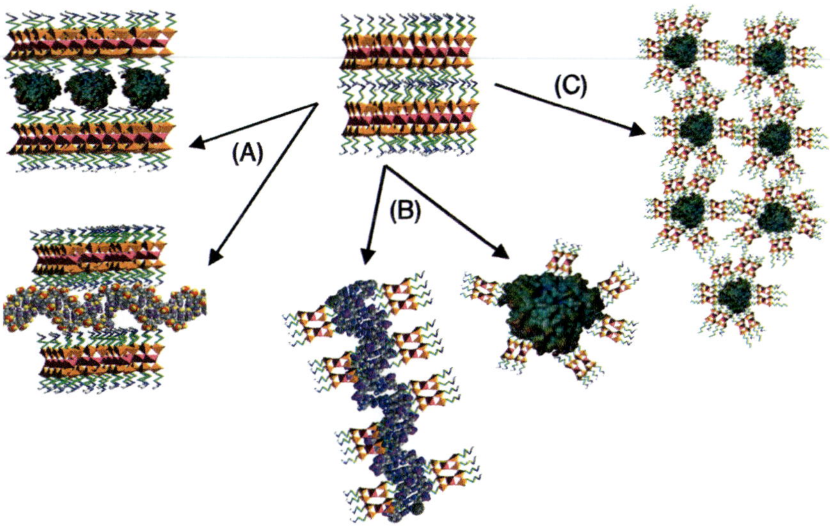

Fig. 8.18 Schematic diagram showing the potential scope of organically functionalized magnesium phyllosilicate (shown in top centre of figure) for the preparation of functional bio-inorganic nanomaterials. (A) biomolecule-induced co-assembly of exfoliated aminopropyl-functionalized organoclay sheets to produce layered nanocomposites containing functional protein molecules (top left) or DNA (bottom left). (B) molecular wrapping approach in which single molecules of negatively charged proteins (bottom centre right) or DNA (bottom centre left) are encased in cationic organoclay oligomers to produce core–shell nanoparticles or nanowires, respectively. (C) formation of partially ordered superstructures of organoclay-wrapped biomolecules by hydrophobic modification of covalently linked organic groups attached to the phyllosilicate framework.

Nanohybrids can be prepared in the form of intercalated layered nanocomposites produced by co-assembly of guest biomolecules in the presence of exfoliated organoclay sheets (Section 8.4), or by wrapping single biomolecules in ultrathin layers of condensed organoclay oligomers (Section 8.5). Such approaches should provide new general routes towards the development of functional biomaterials with numerous applications.

8.6
Functional Mesolamellar Bio-inorganic Nanocomposite Films

In the above sections we have shown that organically functionalized magnesium phyllosilicates can serve as host materials for a range of important biological molecules. Recently, we demonstrated that organoclay or organosilica nanosheets can be intercalated within the interlayer spaces of stacked arrays of naturally occurring bacterial cell membranes to produce self-assembled mesolamellar bio-nanocomposites with exceptional properties [136]. For this, we used micron-sized fragments of a purple membrane (PM) extracted from halophilic bacteria. The PM fragments consist of a 2-D crystalline array of unidirectionally oriented bacteriorhodopsin (BR) molecules embedded in a 5 nm thick lipid bilayer [137,138] BR is a photoactive protein-retinal complex that acts as a light activated proton pump and, as a result, the extracted PM flakes exhibit photoelectric, photochromic and photoinduced proton transport properties that have been exploited in holography, non-linear optics, and optical information processing [139–141].

Typically, we prepared the hybrid bio-inorganic films by swelling electrophoretically deposited PM films with aqueous droplets of prehydrolysed aminopropyltriethoxysilane or aminopropyl-functionalized magnesium phyllosilicate oligomers followed by drying the intercalated films in air. Alternatively, aqueous polymers such as chitosan, polyvinyl alcohol or polylysine could be used as the swelling solutions. Unlike the native PM films, which were extremely brittle, the resulting biocomposite films were mechanically more stable. SEM images showed smooth and continuous films with thicknesses between ~5 and 15 μm depending on the conditions of deposition (Figure 8.19). PXRD studies gave broad low angle reflections corresponding to a mesolamellar structure with interlayer spacings of 7.6, 5.9–6.3, or 6.9 nm for PM–organosilica, PM–organoclay, and PM–polylysine composites, respectively. In contrast, mesolamellar films prepared from stacked arrays of non-intercalated native PM fragments had an interlayer spacing of 5 nm, which was commensurate with the known PM layer thickness. Thus the results were consistent with the formation of uniformly oriented PM lipid bilayers that were interlaminated with 2–2.5 nm-thick sheets of amorphous organosilica, organoclay or polymer macromolecules.

Significantly, the bio-inorganic and polymer-containing PM nanocomposites showed no significant shift in the protein amide I and II vibration bands, or in the characteristic 567 nm optical absorption band of the retinal chromophore of BR, indicating that the structural and dynamical properties of the membrane-bound

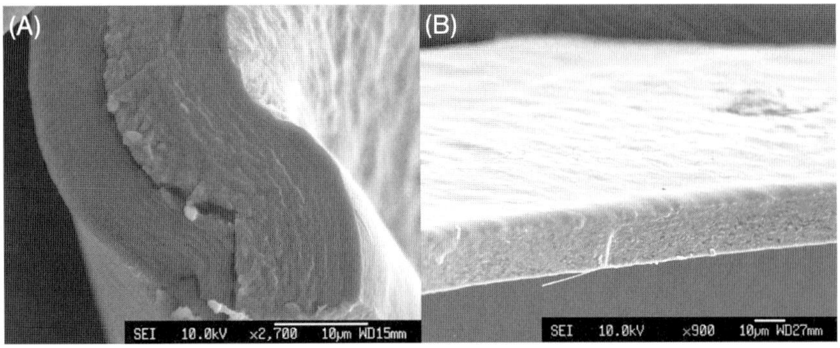

Fig. 8.19 SEM images of mesolamellar thin films produced by intercalation of nanosheets of (A) aminopropyl-functionalized silica or (B) AMP between stacked purple membrane fragments containing bacteriorhodopsin (scale bars = 10 μm).

protein were retained within the mesolamellar hybrid films. This was further supported by preliminary studies of the photochromic properties of the hybrid films prepared under alkaline conditions, which displayed reversible bleaching of the purple ground state to produce a yellow coloration associated with the relatively long-lived M state intermediate. Moreover, when compared with native PM films, the PM–organoclay and PM–organosilica films exhibited prolonged stabilization of the M state, possibly due to the high pK_a of the covalently linked amino moieties which facilitate the interlayer buffering of H^+ and OH^- ions.

Photovoltaic characterization of the nanocomposite PM films showed typical light-on and light-off photoelectric signals similar to those recorded for native PM films, indicating that the BR molecules remained functionally active within the mesolamellar structure. Interestingly, when the relative humidity was 25 %, the magnitude of the photovoltaic signals for PM–silica and PM–organoclay nanocomposite films was twice that recorded for the native PM film. Moreover, at 50 % relative humidity the light-on signal for the native PM films was significantly reduced, whereas negligible changes were observed for the nanocomposite films.

The enhanced photo-induced properties observed for the layered bio-nanocomposite films probably originate from the presence of hydrogen bonded water molecules associated with silanol groups of the entrapped organosilica or organoclay layers, which act as a proton reservoir [142]. As a result, in the presence of an electric field, proton hopping between trapped water molecules and SiOH groups could facilitate proton transfer across the layered nanocomposite. In addition, the ion back-current resulting from water molecules located at defect sites in the stacked PM array could be structurally hindered by the presence of the intercalated nanometer-thick inorganic silica layers. The above results clearly demonstrate the viability of this new approach, which offers significant promise for the development of robust BR-based mesolamellar nanocomposites for integration into a wide range of novel electronic devices and optical storage systems.

8.7 Summary

In this chapter we have introduced an emerging class of synthetic lamellar nanomaterials, based on organic modification of synthetic magnesium phyllosilicates, that offers several advantages over naturally occurring clay minerals. The broad range and flexibility in covalent tethering of organic moieties to the phyllosilicate framework furnishes opportunities for the development of novel bottom-up approaches to higher-order hybrid bio-inorganic materials with a variety of morphologies and functionalities for use in biosensing, biocatalysis and bioelectronic/optical applications. Along with the use of conventional intercalation procedures involving exfoliated layered organoclays to produce new biomolecule-entrapped layered nanocomposites, we have described a novel approach for the synthesis of ultra-small hybrid core–shell nanoparticles formed by the wrapping/encapsulation of single biomolecules with retained functionality. We also have highlighted a novel approach to the fabrication of mesolamellar bio-inorganic nanocomposite films comprising alternating stacks of organoclay/organosilica nanosheets interspaced with photoactive BR/lipid purple membrane bilayers. Finally, we envisage that the design, construction and higher-order assembly of organoclay-based bio-inorganic layered nanocomposites will generate new insights and opportunities for the wide ranging development of novel functional nanohybrid materials with diverse properties and applications.

Acknowledgements

The authors would like to thank EPSRC for financial support, and Dr E. Muthusamy, Dr A.M. Seddon, Dr M. Li, K.M. Bromley and S.C. Holmstrom for their valuable contributions to the studies cited in this chapter.

References

1 Mirkin, C. and Niemeyer C.M. (2004) *Nanobiotechnology: principles, concepts and applications*,Wiley-VCH, Weinheim.
2 Kumar, C. (2006) *Biological and Pharmaceutical Nanomaterials*,Wiley-VCH,Weinheim.
3 Ogawa, M. and Kuroda K. (1995) Photofunctions of intercalation compounds. *Chemical Reviews*, **75**, 399–438.
4 Schollhorn, R. (1996) Intercalation systems as nanostructural functional materials. *Chemistry of Materials*, **8**, 1747–1757.
5 Ruiz-Hitzky, E., Dardar, M. and Aranda, P. (2005) Functional biopolymer nanocomposites based on layered solids. *Journal of Materials Chemistry*, **15**, 3650–3662.
6 Evans, D.G. and Duan X. (2006) Preparation of layered double hydroxides and their application as additives, polymers, as precursors to magnetic materials and in biology and

medicine. *Chemical Communications*, 485–496.

7 Carrado, K.A. (2000) Synthetic organo- and polymer-clays: preparation, characterization and materials applications. *Applied Clay Science*, **17**, 1–23.

8 Mann, S. (2001) *Biomineralization: principles and concepts in bioinorganic materials chemistry*,Oxford University Press,Oxford.

9 Jones, W. and Rao, C.N.R.(eds)(2002) *Supramolecular organization and materials design*,Cambridge University Press,Cambridge.

10 Soler-Ilia, G.J.A.A., Sanchez, C., Lebeau, B. and Patarin, J. (2002) Chemical strategies to design textured materials: from microporous and mesoporous oxides to nanonetworks and hierarchical structures. *Chemical Reviews*, **102**, 4093–4138.

11 Fukushima, Y. and Tani M. (1995) An organic/inorganic hybrid layered polymer:methacrylate-magnesium (nickel) phyllosilicate. *Chemical Communications*, 241–242.

12 Fukushima, Y. and Tani M. (1996) Synthesis of 2 : 1 type 3-(methacryloxy) propyl magnesium (nickel) phyllosilicate. *Bulletin of the Chemical Society of Japan*, **69**, 3667–3671.

13 Burkett, S.L., Press, A. and Mann, S. (1997) Synthesis, characterization and reactivity of layered inorganic-organic nanocomposites based on 2 : 1 trioctahedral phyllosilicates. *Chemistry of Materials*, **9**, 1071–1073.

14 Ukrainczyk, L., Bellman, R.A. and Anderson, A.B. (1997) Template synthesis and characterization of layered Al- and Mg-silsesquioxanes. *Journal of Physical Chemistry B*, **101**, 531–539.

15 Whilton, N.T., Burkett, S.L. and Mann, S. (1998) Hybrid lamellar nanocomposite based on organically functionalized magnesium phyllosilicate clays with interlayer reactivity. *Journal of Materials Chemistry*, **8**, 1927–1932.

16 da Fonseca, M.G. and Airoldi C. (1999) Phyllosilicate-like structure anchored silylating agents: calorimetric data on divalent cation-aminated centre interactions in the lamellar cavity. *Journal of the Chemical Society, Dalton Transactions*, 3687–3692.

17 da Fonseca, M.G. and Airoldi C. (2000) New layered inorganic-organic nanocomposites containing n-propylmercapto copper phyllosilicates. *Journal of Materials Chemistry*, **10**, 1457–1463.

18 Carrado, K.A., Xu, L., Csencsits, R. and Muntean, J.V. (2001) Use of organo- and alkoxysilanes in the sysnthesis of grafted and pristine clays. *Chemistry of Materials*, **13**, 3766–3773.

19 da Fonseca, M.G., Silva, C.R. and Airoldi, C. (1999) Aminated phyllosilicates synthesized via a sol-gel process. *Langmuir*, **15**, 5048–5055.

20 da Fonseca, M.G., Silva, C.R., Barone, J.S. and Airoldi, C. (2000) Layered hybrid nickel phyllosilicates and reactivity of the gallery space. *Journal of Materials Chemistry*, **10**, 789–795.

21 Fonseca, M.G. and Airoldi C. (2000) Mercaptopropyl magnesium phyllosilicate-thermodynamic data on the interaction with divalent cations in aqueous solution. *Thermochimica Acta*, **359**, 1–9.

22 da Fonseca, M.G. and Airoldi C. (2001) New amino-inorganic hybrid from talc silylation and copper adsorption properties. *Materials Research Bulletin*, **36**, 277–287.

23 da Fonseca, M.G., Barone, J.S. and Airoldi, C. (2000) Self-organized inorganic-organic hybrid induced by silylating agents with phyllosilicate-like structure and the influence of the adsorption of cations. *Clays and Clay Minerals*, **48**, 638–647.

24 Jaber, M., M-Brendle, J., Roux, M., Dentzer, J., Le Dred, R. and Louis Guth, J. (2002) A new Al, Mg-organoclay. *New Journal of Chemistry*, **26**, 1597–1600.

25 Jaber, M., M-Brendle, J. and Le Dred, R. (2002) Mercaptopropyl Al-Mg phyllosilicate:synthesis and characterization by XRD, IR, and NMR. *Chemistry Letters*, 954–955.

26 Silva, C.R., Fonseca, M.G., Barone, J. S. and Airoldi, C. (2002) Layered inorganic-organic talc-like nanocomposites. *Chemistry of Materials*, **14**, 175–179.

27 Plouet, M.R., Vilminot, S., Guillot, M. and Kurmoo, M. (2002) Canted antiferromagnetism in an organo-modified layered nickel phyllosilicate. *Chemistry of Materials*, **14**, 3829–3836.

28 Jaber, M., M-Brendle, J., Delmotte, L. and Le Dred, R. (2003) New range of Al-Mg organoclays with tailored hydrophobicity: incorporation of fluoride as a local probe to study the octahedral character. *Microporous and Mesoporous Materials*, **65**, 155–163.

29 Minet, J., Abramson, S., Bresson, B., Sanchez, C., Montouillout, V. and Lequeux, N. (2004) New layered calcium organosilicate hybrids with covalently linked organic functionalities. *Chemistry of Materials*, 3955–3962.

30 da Fonseca, M.G., Silva Filho, C.R., Machado, R.S.A., Jr., Arakaki, L.N.H., Espinola, J.G.P. and Airoldi, C. (2004) Zinc phyllosilicates containing amino pendent groups. *Journal of Solid State Chemistry*, 2316–2322.

31 Jaber, M., M-Brendle, J., Michelin, L. and Delmotte, L. (2005) Heavy metal retention by organoclays: synthesis, applications, and retention mechanism. *Chemistry of Materials*, **17**, 5275–5281.

32 Jaber, M., M-Brendle, J., Delmotte, L. and Le Dred, R. (2005) Formation of organoclays by a one step synthesis. *Solid State Sciences*, **7**, 610–615.

33 Sales, J.A.A., Petrucelli, G.C., Oliveira, F.J.V.E. and Airoldi, C.A. (2006) Some features associated with organosilane groups grafted by the sol-gel process onto synthetic talc-like phyllosilicate. *Journal of Colloid and Interface Science*, **297**, 95–103.

34 Mizutani, T., Fukushima, Y. and Kamigaito, O. (1990) Mechanism of the copolymerization of silicic acid and metal ions in aqueous media. *Bulletin of the Chemical Society of Japan*, **63**, 618–619.

35 Mann, S. (2000) The chemistry of form. *Angewandte Chemie-International Edition*, **39**, 3392–3406.

36 Walsh, D., Hopwood, J.D. and Mann, S. (1994) Crystal tectonics: construction of reticulated calcium-phosphate frameworks in bicontinuous reverse microemulsions. *Science*, **264**, 1576–1578.

37 Sims, S.D., Walsh, D. and Mann, S. (1998) Morphosynthesis of macroporous silica frameworks in bicontinuous microemulsions. *Advanced Materials*, **10**, 151–154.

38 Walsh, D. and Mann S. (1995) Fabrication of hollow porous shells of calcium carbonate from self-organizing media. *Nature*, **377**, 320–323.

39 Imhof, A. and Pine D.J. (1997) Ordered macroporous materials by emulsion templating. *Nature*, **389**, 948–951.

40 Walsh, D., Lebeau, B. and Mann, S. (1999) Morphosynthesis of calcium carbonate (vaterite) microsponges. *Advanced Materials*, **11**, 324–328.

41 Fowler, C.E., Khushalani, D. and Mann, S. (2001) Facile synthesis of hollow silica microspheres. *Journal of Materials Chemistry*, **11**, 1968–1971.

42 Schacht, S., Huo, Q., Voight-Martin, I. G., Stucky, G.D. and Schuth, F. (1996) Oil-water interface templating of mesoporous macroscale structures. *Science*, **273**, 768–771.

43 Bourlinos, A.B., Karakassides, M.A. and Petridis, D. (2001) Synthesis and characterization of hollow microspheres through a resin template approach. *Chemical Communications*, 1518–1519.

44 zu Pulitz, B., Landfester, K., Fischer, H. and Antonietti, M. (2001) The generation of "Armoured Latexes" and hollow inorganic shells made of clay sheets by templating cationic miniemulsions and latexes. *Advanced Materials*, **13**, 500–503.

45 Caruso, R.A., Susha, A. and Caruso, F. (2001) Multilayered titania, silica, and laponite nanoparticles coating on polystyrene colloidal templates and resulting inorganic hollow spheres. *Chemistry of Materials*, **13**, 400–409.

46 Muthusamy, E., Walsh, D. and Mann, S. (2002) Morphosynthesis of organoclay microspheres with sponge-like or hollow interiors. *Advanced Materials*, **14**, 969–972.

47 Mann, S. (1993) Molecular tectonics in biomineralization and biomimetic materials chemistry. *Nature*, **365**, 499–505.

48 Archibald, D.D. and Mann S. (1993) Template mineralization of self-assembled anisotropic lipid microstructures. *Nature*, **364**, 430–433.

49 Kim, S.S., Zhang, W. and Pinnavaia, T. J. (1998) Ultrastable mesostructured silica. *Science*, **282**, 1302–1305.

50 Kleitz, F., Marlow, F., Stucky, G.D. and Schuth, F. (2001) Mesoporous silica fibres: synthesis, internal structure and growth kinetics. *Chemistry of Materials*, **13**, 3587–3595.

51 Fowler, C.E., Khushlani, D. and Mann, S. (2001) Interfacial synthesis of hollow microspheres of mesostructured silica. *Chemical Communications*, 2028–2029.

52 van Bommel, K.J.C., Friggeri, A. and Shinkai, S. (2003) Organic templates for generation of inorganic materials. *Angewandte Chemie-International Edition*, **42**, 981–999.

53 Davis, S.A., Dujardin, E. and Mann, S. (2003) Biomolecular inorganic materials chemistry. *Current Opinion in Solid State and Materials Science*, **7**, 273–281.

54 Shimizu, T., Masuda, M. and Minamikawa, H. (2005) Supramolecular nanotube architectures based on amphiphilic molecules. *Chemical Reviews*, **105**, 1401–1443.

55 Schnur, J.M. (1993) Lipid tubules: A paradigm for molecularly engineered structures. *Science*, **262**, 1669–1676.

56 Behroozi, F., Orman, M., Reese, R., Stockton, W., Calvert, J., Rachford, F. and Schoen, P. (1990) Interaction of metallized tubules with electromagnetic-radiation. *Journal of Applied Physics*, **68**, 3688–3693.

57 Krebs, J.J., Rubinstein, M. Lubitz, P., Harford, M.Z., Baral, S., Shashidhar, R., Ho, Y.S., Chow, M. and Qadri, F S. (1991) Magnetic properties permalloy-coated organic tubules. *Journal of Applied Physics*, **70**, 6404–6406.

58 Baral, S. and Schoen P. (1993) Silica-deposited phospholipids tubules as a precursor to hollow submicron-diameter silica cylinders. *Chemistry of Materials*, **5**, 145–147.

59 Burkett, S.L. and Mann S. (1996) Spatial organization and patterning of gold nanoparticles on self-assembled biolipid tubular templates. *Chemical Communications*, 321–322.

60 Goren, M., Qi, Z. and Lennox, R.B. (2000) Selective templated growth of polypyrrole strands on lipid tubule edges. *Chemistry of Materials*, **12**, 1222–1228.

61 Lvov, Y.M., Price, R.R., Selinger, J.V., Singh, A., Spector, M.S. and Schnur, J. M. (2000) Imaging nanoscale patterns on biologically derived microstructures. *Langmuir*, **16**, 5932–5935.

62 Seddon, A.M., Patel, H.M., Burkett, S. L. and Mann, S. (2002) Chiral templating of silica-lipid lamellar mesophase with helical tubular architecture. *Angewandte Chemie-International Edition*, **41**, 2988–2991.

63 Price, R.R., Dressick, W.J. and Singh, A. (2003) Fabrication of nanoscale metallic spirals using phospholipid microtubule organizational templates. *Journal of the American Chemical Society*, **125**, 11259–11263.

64 Patil, A.J., Muthusamy, E., Seddon, A. M. and Mann, S. (2003) Higher-order synthesis of organoclay pipes using self-assembled lipid templates. *Advanced Materials*, **15**, 1816–1819.

65 Lebeau, B., Brendle, J., Marichal, C., Patil, A.J., Muthusamy, E. and Mann, S. (2006) One-step synthesis of hybrid organic-inorganic phyllosilicate-like materials. *Journal of Nanoscience and Nanotechnology*, **6**, 352–359.

66 Bruce, D.W. and and O'Hare D. (1996) *Inorganic materials*, 2nd edn, John Wiley & Sons, England.

67 Kaschak, D.M., Jhonson, S.A., Hooks, D.E., Kim, H.-N., Ward, M.D. and Mallouk, T.E. (1998) Chemistry on the edge: A microscopic analysis of the intercalation exfoliation, edge functionalization and monolayer surface tiling reactions of alpha-zirconium phosphate. *Journal of the American Chemical Society*, **120**, 10887–10894.

68 Krishnamoorti, R., Vaia, R.A. and Giannelis, E. (1996) Structure and dynamics of polymer-layered silicate nanocomposites. *Chemistry of Materials*, **8**, 1728–1734.

69 Bornscheuer, U.T. (2003) Immobilizing enzymes: how to create more suitable biocatalysts. *Angewandte Chemie-International Edition*, **42**, 3336–3337.

70 Ellerby, L.M., Nishida, C.R., Nishida, F., Yamanaka, S.A., Dunn, B., Valentine, J.S. and Zink, J.I. (1992) Encapsulation of proteins in transparent porous silicate-glasses prepared by the sol-gel method. *Science*, **255**, 1113–1115.

71 Gill, I. and Ballesteros A. (2000) Bioencapsulation within synthetic polymers (Part 1): sol-gel encapsulated biologicals. *Trends in Biotechnology*, **18**, 282–296.

72 Gill, I. and Ballesteros A. (2000) Bioencapsulation within synthetic polymers (Part 2): non sol-gel protein-polymer biocomposites. *Trends in Biotechnology*, **18**, 469–479.

73 Katz, E. and Willner I. (2004) Integrated nanoparticle-biomolecule hybrid system: synthesis, properties and applications. *Angewandte Chemie-International Edition*, **43**, 6042–6108.

74 Topogildis, E., Cass, C.J., Cass, A.E.G., Gilardi, G., Sadeghi, S., Beaumont, N. and Durrant, J.R. (1998) Protein adsorption on nanocrystalline TiO_2 films: An immobilization strategy for bioanalytical devices. *Analytical Chemistry*, **70**, 5111–5113.

75 Topogildis, E., Cass, C.J., Cass, A.E.G. and Durrant, J.R. (2001) Factors that affect protein adsorption on nanostructured titania films: A novel spectrochemical application to sensing. *Langmuir*, **17**, 7899–7906.

76 Nygaard, S., Wendelbo, R. and Brown, S. (2002) Surface-specific zeolite-binding proteins. *Advanced Materials*, **14**, 1853–1856.

77 Fan, J., Lei, J., Wang, L., Yu, C., Tu, B. and Zhao, D. (2003) Rapid and high-capacity immobilization of enzymes based on mesoporous silicas with controlled morphologies. *Chemical Communications*, 2140–2141.

78 Yiu, H.H.P. and Wright P.A. (2005) Enzymes supported on ordered mesoporous solids: a special case of an inorganic-organic hybrid. *Journal of Materials Chemistry*, **15**, 3690–3700.

79 Caruso, F. and Schuler C. (2000) Enzyme multilayers on colloid particles: Assembly, stability and enzyme activity. *Langmuir*, **16**, 9595–9603.

80 Lvov, Y. and Caruso F. (2001) Biocolloids with ordered urease multilayer shells as enzymatic reactors. *Analytical Chemistry*, **73**, 4212–4217.

81 Kreiner, M., Moor, B.D. and Parker, M. C. (2001) Enzyme coated micro-crystals: a 1-step method for high activity biocatalyst preparation. *Chemical Communications*, 1096–1097.

82 Dyal, A., Loos, K., Noto, M., Chang, S., Spagnoli, C., Shafi, K.V.P.M., Ulman, A., Cowman, M. and Gross, R.A. (2003) Activity of Candida rugosa lipase immobilized on γ-Fe_2O_3 magnetic nanoparticles. *Journal of the American Chemical Society*, **125**, 1684–1685.

83 Kumar, C.V. and Choudhari A. (2000) Proteins immobilized at the galleries of layered alpha-zirconium phosphate: structure and activity studies. *Journal of the American Chemical Society*, **122**, 830–837.

84 Kelleher, B.P., Oppenheimer, S.F., Han, F.X., Willeford, K., Simpson, M. J., Simpson, A.J. and Kingery, W.L. (2003) Dynamical systems and phase plane analysis of protease-clay interactions. *Langmuir*, **19**, 9411–9417.

85 Kumar, C.V. and Chaudhari A. (2001) Efficient renaturation of immobilized met-hemoglobin at the galleries of alpha-zirconium phosphonate. *Chemistry of Materials*, **13**, 238–240.

86 Kumar, C.V. and Chaudhari A. (2003) Unusual thermal stabilities of some proteins and enzymes bound in the galleries of layered alpha-Zr(IV) phosphate/phosphonates. *Microporous and Mesoporous Materials*, **57**, 181–190.

87 Kumar, C.V. and Chaudhari A. (2002) High temperature peroxidase activities of HRP and hemoglobin in the galleries of layered Zr(IV) phosphate. *Chemical Communications*, 2382–2383.

88 Whilton, N.T., Vickers, P.J. and Mann, S. (1997) Bioinorganic clays: synthesis and characterization of amino- and polyaminacid intercalated double hydroxides. *Journal of Materials Chemistry*, **7**, 1623–1629.

89 Coradin, T., Coupe, A. and Livage, J. (2003) Intercalation of biomolecules in the $MnPS_3$ layered phase. *Journal of Materials Chemistry*, **13**, 705–707.

90 Zhou, Y., Hu, N., Zeng, Y. and Rusling, J.F. (2002) Heme protein-clay films: Direct electrochemistry and electrochemical catalysis. *Langmuir*, **18**, 211–219.

91 Patil, A.J., Muthusamy, E. and Mann, S. (2005) Fabrication of functional protein-organoclay lamellar nanocomposites by biomolecule-induced assembly of exfoliated aminopropyl-functionalized magnesium phyllosilicates. *Journal of Materials Chemistry*, **15**, 3838–3843.

92 Giannelis, E.P., Krishnamoorthy, R. and Manias, E. (1999) Polymer-silicate nanocomposites: Model systems for confined polymers and polymer brushes. *Advances in Polymer Science*, **138**, 107–147.

93 Copeland, R.A. (1996) *Enzymes*,VCH Publishers,New York.

94 Klibanov, A.M. (1983) Immobilized enzymes and cells as practical catalysts. *Science*, **219**, 722–727.

95 Onda, M., Ariga, K. and Kunitake, T. (1999) Activity and stability of glucose oxidase in molecular films assembled alternatively with polyions. *Journal of Bioscience and Bioengineering*, **87**, 69–75.

96 Verma, I.M. and Somia N. (1997) Gene therapy-promises, problems and prospects. *Nature*, **389**, 239–242.

97 Shen, H., Tan, J. and Saltzman, W.M. (2004) Surface-mediated gene transfer from nanocomposites of controlled texture. *Nature Materials*, **3**, 569–574.

98 Kneur, C., Sameti, M., Haltner, E.G., Schiestel, T., Schirra, H. and Schmidt, H.K. (2000) Silica nanoparticles modified with aminosilanes as carriers for plasmid DNA. *International Journal of Pharmaceutics*, **196**, 257–261.

99 Roy, I., Ohulchanskyy, T.Y., Bharali, D.J., Pudavar, H.E., Mistretta, R.A., Kaur, N. and Prasad, P.N. (2005) Organically modified silica nanoparticles: A non-viral vector for in-vivo gene delivery and expression. *Proceedings of the National Academy of Sciences of the United States of America*, **102**, 11539–11544.

100 Salem, A.K., Searson, P.C. and Leong, K.W. (2003) Multifunctional nanorods for gene delivery. *Nature Materials*, **2**, 668–671.

101 Xu, Z.P., Zeng, Q.H., Lu, G.Q. and Yu, A.B. (2006) Inorganic nanoparticles as carriers for efficient cellular delivery. *Chemical Engineering Science*, **61**, 1027–1040.

102 Choy, J.H., Kwak, S.Y., Jeong, Y.J. and Park, J.S. (2000) Inorganic layered double hydroxides as nonviral vectors. *Angewandte Chemie-International Edition*, **39**, 4042–4045.

103 Kwak, S.Y., Jeong, Y.J., Park, J.S. and Choy, J.H. (2002) Bio-LDH nanohybrid for gene therapy. *Sold State Ionics*, **151**, 229–234.

104 Tyner, K.M., Roberson, M.S., Berghorn, K.A., Li, L., Gilmour, R.F., Jr., Batt, C.A. and Giannelis, E.P. (2004) Intercalation, delivery and expression of the gene encoding green fluorescence protein utilizing nanobiohybrids. *Journal of Controlled Release*, **100**, 399–409.

105 Desigaux, L., Belkacem, M.B., Richard, P., Cellier, J., Leone, P., Cario, L., Leroux, F., Taviot-Gueho, C. and Pitard, B. (2006) Self-assembly and characterization of layered double hydroxide/DNA hybrids. *Nano Letters*, **6**, 199–204.

106 Patil, A.J., Li, M., Dujardin, E. and Mann, S. (2007) Novel bioinorganic nanostructures based on mesolamellar intercalation or single-molecule wrapping of DNA using organoclay building blocks. *Nano Letters* (in Press).

107 Sahoo, S.K. and Labhasetwar V. (2003) Nanotech approach to delivery and imaging drug. *Drug Dicovery Today*, **24**, 1112–1120.

108 Edwards, D.A., Hanes, J., Caponetti, G., Hrkach, J., Ben-Jebria, A., Eskew, M., Mintzes, J., Deaver, D., Lotan, N. and Langer, R. (1997) Large porous particle for pulmonary drug delivery. *Science*, **276**, 1868–1871.

109 Tsapis, N., Benent, D., Jackson, B., Weitz, D.A. and Edward, D.A. (2002) Trojan particles: large porous carriers of nanoparticles for drug delivery. *Proceedings of the National Academy of Sciences of the United States of America*, **99**, 12001–12005.

110 Hall, S.R., Bolger, H. and Mann, S. (2003) Morphosynthesis of complex inorganic forms using pollen grain templates. *Chemical Communications*, 2784–2785.

111 Kulak, A., Hall, S.R. and Mann, S. (2004) Single-step fabrication of drug-encapsulated inorganic microspheres with complex form by sonication-induced nanoparticle assembly. *Chemical Communications*, 576–577.

112 Roy, I., Ohulchanskyy, T.Y., Pudavar, H. E., Bergey, E.J., Oseroff, A.R., Morgan, J., Dougherty, T.J. and Prasad, P.N. (2003) Ceramic-based nanoparticles entrapping water-insoluble photosensitizing anticancer drugs: A novel drug-carrier system for photodynamic therapy. *Journal of the American Chemical Society*, **125**, 7860–7865.

113 Barbe, C., Barlett, J., Kong, L., Finnie, K., Lin, H.Q., Larkin, M., Calleja, S., Bush, A. and Calleja, G. (2004) Silica particles: A novel drug-delivery system. *Advanced Materials*, **16**, 1959–1966.

114 Vallet-Regi, M. (2006) Revisiting ceramics for medical applications. *Dalton Transactions*, 5211–5220.

115 Horcajada, P., Serre, C., Vallet-Regi, M., Sebban, M., Tauelle, F. and Ferry, G. (2006) Metal-organic frameworks as efficient materials for drug delivery. *Angewandte Chemie-International Edition*, **56**, 5974–5978.

116 Perkin, K.K., Turner, J.L., Wooley, K.L. and Mann, S. (2005) Fabrication of hybrid nanocapsules by calcium phosphate mineralization of shell cross-linked polymer micelles and nanocages. *Nano Letters*, **5**, 1457–1461.

117 Holmström, S.C., Patil, A.J., Butler, M. and Mann, S. (2007) Influence of polymer co-intercalation on guest release from aminopropyl-functionalized magnesium phyllosilicate mesolamellar nanocomposites. *Journal of Materials Chemistry*, **17**, 3894–3900.

118 Hou, Z., Sang, S., You, H., Lee, M.J., Hing, J., V-Chin, K. and Yang, C.S. (2005) Mechanism of action of (-)-epigallocatechin-3-gallate: Auto-oxidation-dependent inactivation of epidermal growth factor receptor and direct effects on growth inhibition in human oesophageal cancer KYSE 150 cells. *Cancer Research*, **65**, 8049–8056.

119 Grunes, J., Zhu, J. and Somorjai, G. (2003) Catalysis and nanoscience. *Chemical Communications*, 2257–2260.

120 Douglas, T. and Young M. (1998) Host-guest encapsulation of materials by assembled virus protein cages. *Nature*, **393**, 152–155.

121 Muller, A., Das, S.K., Kogerler, P., Bogge, H., Schunemann, V., Krickmeyer, E. and Preetz, W. (2000) A new type of supramolecular compound: molybdenum-oxide-based composites consisting of nanocapsules with encapsulated keggin-ion electron reservoirs cross-linked to a two-dimensional network. *Angewandte Chemie-International Edition*, **39**, 3413–3417.

122 Ichinose, I. and Kunitake T. (2001) Molecular wrapping of a fluorescent dye with TiO_2-gel and capping reagents. *Chemistry Letters*, **7**, 626–627.

123 Ichinose, I. and Kunitake T. (2002) Wrapping individual chains of a viologen polymer with an ultrathin silicate sheath. *Advanced Materials*, **14**, 344–346.

124 Ichinose, I. and Kunitake T. (2002) Wrapping and inclusion of organic molecules with ultrathin, amorphous metal oxide films. *Chemical Record*, **2**, 339–351.

125 Ichinose, I., Hashimoto, Y. and Kunitake, T. (2004) Wrapping of bio-macromolecules (dextran, amylopectin and horse heart cytochrome *c*) with ultrathin silicate layer. *Chemistry Letters*, **33**, 656–657.

126 Kim, J. and Grate J.W. (2003) Single-enzyme nanoparticle armoured by a nanometre-scale organic/inorganic network. *Nano Letters*, **3**, 1219–1222.

127 Kim, J., Grate, J.W. and Wang, P. (2006) Nanostructures for enzyme stabilization. *Chemical Engineering Science*, **61**, 1017–1026.

128 Patil, A.J., Muthusamy, E. and Mann, S. (2004) Synthesis and self-assembly of organoclay-wrapped biomolecules. *Angewandte Chemie-International Edition*, **43**, 4928–4933.

129 Mirkin, C.A., Letsinger, R.L., Mucic, R. C. and Storhoff, J.J. (1996) A DNA-based method for rationally assembling nanoparticles into macroscopic materials. *Nature*, **382**, 607–609.

130 Alivisatos, A.P., Johson, K.P., Oeng, X., Wilson, T.E., Loweth, C.J., Bruchez, M. P. and Schulthz, P.G. (1996) Organization of nanocrystal molecules using DNA. *Nature*, **382**, 609–611.

131 Braun, E., Eichen, Y., Sivan, U. and Ben-Yoseph, G. (1998) DNA-templated assembly and electrode attachment of conducting silver wire. *Nature*, **391**, 775–778.

132 Patolsky, F., Weizmann, Y., Lioubashevski, O. and Willner, I.

(2002) Au-nanoparticle nanowires based on DNA and polylysine templates. *Angewandte Chemie-International Edition*, **41**, 2323–2327.

133 Richter, J., Seidel, R., Kirsch, R., Mertig, M., Pompe, W., Plaschke, J. and Schackert, K. (2000) Nanoscale palladium metallization of DNA. *Advanced Materials*, **12**, 507–510.

134 Monson, C.F. and Woolley A.T. (2003) DNA templated construction of copper nanowires. *Nano Letters*, **3**, 359–363.

135 Ma, Y., Zhang, J. and He, H. 2004 Polyaniline nanowires on Si surfaces fabricated with DNA templates. *Journal of the American Chemical Society*, **126**, 7097–7101.

136 Bromley, K.M., Patil, A.J., Seddon, A. M., Booth, P. and Mann, S. (2007) Bio-functional mesolamellar nanocomposite based on inorganic/ polymer intercalation in purple membrane (Bacteriorhodopsin) films. *Advanced Materials*, **19**, 2433–2438.

137 Hampp, N. (2000) Bacteriorhodopsin as photochromic retinal protein for optical memories. *Chemical Reviews*, **100**, 1755–1776.

138 Brauchle, C., Hampp, N. and Oesterhelt, D. (1991) Optical applications of bacteriorhodopsin and its mutated variants. *Advanced Materials*, **3**, 420–428.

139 He, J.A., Samuelson, L., Li, L., Kumar, J. and Tripathy, S.K. (1999) Bacteriorhodopsin thin film assemblies-immobilization, properties and applications. *Advanced Materials*, **11**, 435–446.

140 Wu, S.G., Ellerby, L.M., Cohan, J.S., Dunn, B., El-Sayed, M.A., Valentine, J. S. and Zink, J.I. (1993) Bacteriorhodopsin encapsulated in transparent sol-gel glass-A new biomaterial. *Chemistry of Materials*, **5**, 115–120.

141 Luo, T.J.M., Soong, R., Lane, E., Dunn, B. and Montemagno, C. (2005) Photo-induced proton gradients and ATP biosynthesis produced by vesicles encapsulated in a silica matrix. *Nature Materials*, **4**, 220–224.

142 Nogami, M., Nagao, R. and Wong, C.J. (1998) Proton conduction in porous silica glasses with high water content. *Journal of Physical Chemistry B*, **102**, 5772–5775.

9
Biodegradable Polymer-based Nanocomposites: Nanostructure Control and Nanocomposite Foaming with the Aim of Producing Nano-cellular Plastics

Masami Okamoto

9.1
Introduction

The increased importance of renewable resources for raw materials and recyclability or biodegradability of the material at the end of its useful life is demanding a shift from petroleum-based synthetics to agro-based materials in industrial applications. Increased social awareness of environmental problems posed by the non-degradable, non-recyclable content of their products is forcing manufacturers to enhance the biodegradable content, which in turn favors a switch to biomaterials [1].

Hence, with this background, there is an urgent need to develop renewable source-based environmentally benign polymeric materials (biopolymers [2]). Such materials would not involve the use of toxic or noxious components in their manufacture, and could be naturally degraded by composting.

In today's commercial environment, biopolymers have proven to be relatively expensive and available only in small quantities. This has led to limitations on their application to date. However, there are signs that this is changing, with increasing environmental awareness and more stringent legislation regarding recyclability and restrictions on waste disposal. Cargill Dow has a polylactide (PLA) in production (with the Trade Name Natureworks). Metabolix has been working on polyhydroxyalkanoate (PHA) (Trade Name Biopol) [2].

Thus, the increasing application of the various intrinsic properties of biopolymers, coupled with the knowledge of how such properties can be improved to achieve compatibility with thermoplastics processing, manufacturing, and end-use requirements, and has fueled technological and commercial interest in biopolymers.

Of particular interest are the recently developed nanocomposites consisting of a polymer and layered silicate because they often exhibit remarkably improved mechanical and other properties [3] when compared with pure polymer or conventional composites (both micro- and macro-composites). A primary progress in

Bio-inorganic Hybrid Nanomaterials. Edited by E. Ruiz-Hitzky, K. Ariga, Y. M. Lvov

ISBN: 978-3-527-31718-9

polymer/layered filler nanocomposites (PLFNCs), a Nylon 6/layered silicate hybrid [4] reported by Toyota Central Research & Development Co. Inc. (TCRD), was successfully prepared by *in situ* polymerization of ε-caprolactam in a dispersion of montomorrillonite (MMT). The silicate can be dispersed in liquid monomer or a solution of monomer. It has also been possible to melt-mix polymers with layered silicates, thus avoiding the use of organic solvents. The latter method permits the use of conventional processing techniques such as injection molding and extrusion. The extensive literature on nanocomposite research is covered in recent reviews [5–7]. The study of nanocomposites has gained greater momentum. This new class of materials is now being introduced in structural applications, such as a gas barrier film, a flame retardant product, and other load-bearing applications [8].

This chapter intends to highlight the biodegradable polymer-based nanocomposites with the primary focus on nanostructure control and the foam processing operation with the aim of producing nano-cellular plastics. Development of nanocomposite foams is one of the latest evolutionary technologies of polymeric foam through a pioneering effort by Okamoto and his colleagues [9,10].

9.2 Nano-structure Development

9.2.1 Melt Intercalation

Since the possibility of direct melt intercalation was first demonstrated [11], melt intercalation has become a method of preparation of the intercalated polymer/layered silicate nanocomposites (PLSNCs). This process involves annealing, statically or under shear, a mixture of the polymer and organically modified layered fillers (OMLFs) above the softening point of the polymer. During annealing, the polymer chains diffused from the bulk polymer melt into the nano-galleries between the layered fillers.

In order to understand the thermodynamic issues associated with the nanocomposite formation, Vaia *et al.* have applied the mean-field statistical lattice model and found that conclusions based on the mean field theory agreed nicely with the experimental results [12,13]. The entropy loss associated with confinement of a polymer melt is not prohibited to nanocomposite formation because an entropy gain associated with the layer separation balances the entropy loss of polymer intercalation, resulting in a net entropy change near to zero. Thus, from the theoretical model, the outcome of nanocomposite formation via polymer melt intercalation depends on energetic factors, which may be determined from the surface energies of the polymer and OMLF.

Nevertheless, we have often faced the problem that the nanocomposite shows fine and homogeneous distribution of the nanoparticles in the polymer matrix (e.g., poly(L-lactide)) without a clear peak shift of the mean interlayer spacing of the (001) plane, as revealed by wide-angle X-ray diffraction (WAXD) analysis [14]. Furthermore we sometimes encounter a decrease in the interlayer spacing compared with that of pristine OMLF, despite very fine dispersion of the silicate particles. For this reason, information on the structure of the surfactant (intercalant)–polymer interface is necessary to understand the intercalation kinetics that can predict the final nanocomposite morphology and overall materials properties.

9.2.2
Interlayer Structure of OMLFs and Intercalation

9.2.2.1 Nano-fillers

In characterizing layered silicate, including layered titanate (HTO), the surface charge density is particularly important because it determines the interlayer structure of the intercalants as well as the cation exchange capacity (CEC). Lagaly proposed a method of calculation consisting of total elemental analysis and the dimensions of the unit cell [15]:

$$\text{Surface charge}: e^-/nm^2 = \xi/ab \qquad (1)$$

where ξ is the layer charge (1.07 for HTO, 0.66 for synthetic fluorine hectrite (*syn*-FH) and 0.33 for MMT. *a* and *b* are cell parameters of HTO ($a = 3.782$ Å, $b = 2.978$ Å [16]), *syn*-FH ($a = 5.24$ Å, $b = 9.08$ Å [17]), and MMT ($a = 5.18$ Å, $b = 9.00$ Å [18]). For *syn*-FH, however, about 30 % of the interlayer Na^+ ions are not replaced quantitatively by intercalants since they do not take part in ion-exchange reactions [17]. For HTO, only 27 % of the interlayer H^+ (H_3O^+) is active for ion-exchange reactions [14]. Thus the incomplete replacement of the interlayer ions is ascribed to the intrinsic chemical reactivity. The characteristic parameters of two nano-fillers are also summarized in Table 9.1 [18]. HTO has a high surface charge density of 1.26 e^-/nm^2 compared with those of *syn*-FH (0.971 e^-/nm^2) and MMT (0.780 e^-/nm^2). From these results, we can estimate the average distance between exchange sites as 0.888 nm for HTO, 1.014 nm for *syn*-FH and 1.188 nm for MMT. This estimation assumes that the

Tab. 9.1 Characteristic parameters of nano-fillers.

Parameters	HTO	syn-FH	MMT
chemical formula	$H_{1.07}Ti_{1.73}O_{3.95}\cdot 0.5H_2O$	$Na_{0.66}Mg_{2.6}Si_4O_{10}(F)_2$	$Na_{0.33}(Al_{1.67}Mg_{0.33})Si_4O_{10}(OH)_2$
particle size/nm	~100–200	~100–200	~100–200
BET area/m^2/g	~2400	~800	~700
CEC/meq/100 g	~200 (660)	~120 (170)	~90 (90)
e^-/charge/nm^2	1.26	0.971	0.708
density/g/cm^3	2.40	2.50	2.50
refractive index (n_D^{20})	2.3	1.55	1.55
pH	4–6	9–11	7.5–10

[a] Methylene blue adsorption method. The values in parentheses are calculated from the chemical formulae of the nano-fillers.
Source: Reprinted from [18].

cations are evenly distributed in a cubic array over the silicate surface and that half of the cations are located on the one side of the platelet and the other half reside on the other side.

9.2.2.2 **Molecular Dimensions and Interlayer Structure**

The calculated models of the intercalant structures are presented in Figure 9.1. For octadecylammonium ($C_{18}H_3N^+$), the molecular length, thickness and width are 2.466, 0.301 and 0.301 nm, respectively. Since the length of all alkyl units is more than 2 nm, these spacings (distance between exchange sites) of 0.888–1.188 nm do not allow parallel layer arrangement like flat-lying chains [15] in each gallery space of the nano-fillers. All the intercalants are oriented with some inclination to the host layer in the interlayer space to form an interdigitated layer. This is suggested by the paraffin-type layer structure proposed by Lagaly, especially in the case of clay minerals with high surface charge [15].

WAXD patterns for three OMLF powders are presented in Figure 9.2. The mean interlayer spacing of the (001) plane ($d_{(001)}$) for the HTO intercalated with $qC_{14}(OH)$ (HTO-$qC_{14}(OH)$) obtained by WAXD measurements is 2.264 nm (diffraction angle, $2\Theta \cong 3.90°$). The appearance of small peaks observed at $2\Theta \cong 7.78$, 11.78 and 15.74° confirmed that these reflections were due to the (002) to (004) planes of HTO-$qC_{14}(OH)$. HTO-$qC_{14}(OH)$ has a surprisingly well-ordered suprastructure, as proved by WAXD, with diffraction maxima up to the fourth order due to the high surface charge density of the HTO layers. On the other hand, *syn*-FH and MMT, which have low surface charge density compared with that of HTO, show a less-ordered interlayer structure, that is, the coherent order of the silicate layers is much lower for *syn*-FH and MMT intercalated with surfactants.

From the WAXD results the interlayer opening is estimated after subtraction of the layer thickness value of 0.374 nm for HTO [16], 0.98 nm for *syn*-FH [17] and 0.96 nm for MMT [15]. This is an important point for the following discussion on the interlayer structure. The illustration of a model of the interlayer structure of the $qC_{14}(OH)$ in the gallery space of the HTO is shown in Figure 9.3. For nano-fillers with a high surface charge density, the intercalants can adopt a configuration with an orientation where the alkyl chains are tilted under the effect of the van der Waals forces, which decreases the chain–chain distance. For this reason, the angle α should be directly related to the packing density of the alkyl chains. The value of α decreases until close contact between the chains is attained, giving an increase in the degree of crystallinity of the intercalants in the nano-galleries. To estimate the tilt angle α, we combined the molecular dimensions, interlayer spacing and loading amount of the intercalant in the layers, which was calculated from thermogravimetry analysis (TGA). The characteristic parameters are summarized in Tables 9.2 and 9.3. Note that HTO exhibits a large value of layer opening accompanied by large values of α and endothermic heat flow ΔH, due to the melting of the intercalants in the galleries, when compared with those of *syn*-FH and MMT. This indicates that HTO leads to a highly interdigitated layer structure and the interlayer opening becomes more uniform compared with MMT and *syn*-FH (possessing lower surface charge density).

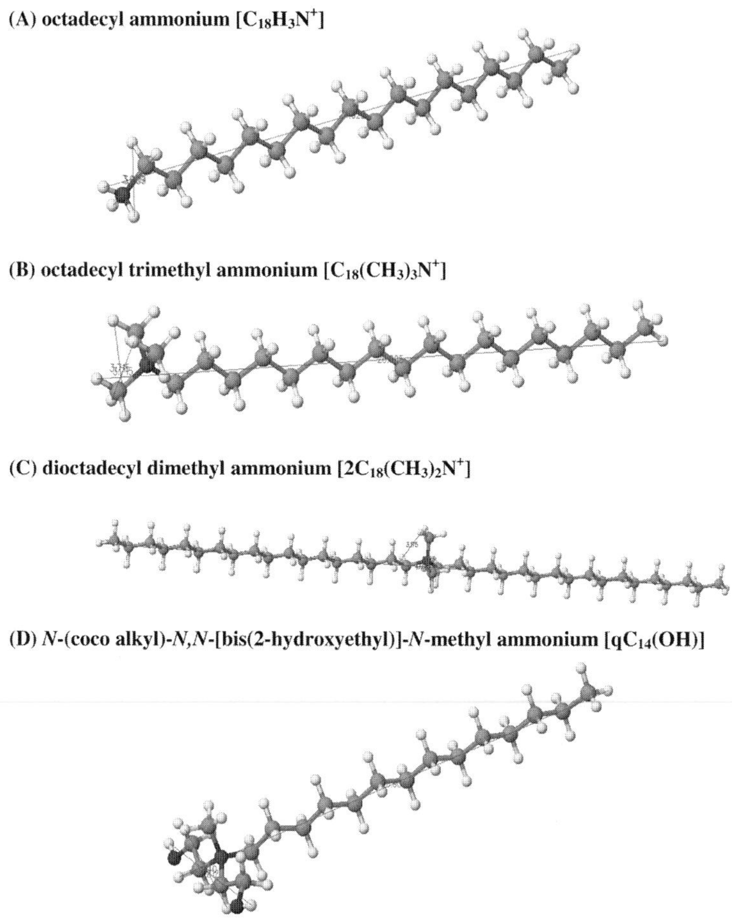

Intercalant	$C_{18}H_3N^+$	$C_{18}(CH_3)_3N^+$	$2C_{18}(CH_3)_2N^+$	$qC_{14}(OH)$
Length /nm	2.466	2.601	4.766	2.090
Thickness /nm	0.301	0.372	0.434	0.374
Width /nm	0.301	0.372	0.318	0.881

Fig. 9.1 Molecular dimensions of intercalants: (A) octadecylammonium ($C_{18}H_3N^+$), (B) octadecyltrimethylammonium ($C_{18}(CH_3)_3N^+$), (C) dioctadecyldimethylammonium ($2C_{18}(CH_3)_2N^+$), and (D) *N*-(cocoalkyl)-*N,N*-[bis(2-hydroxyethyl)]-*N*-methyl ammonium cations ($qC_{14}(OH)$). Reprinted from [18], © 2006, Wiley-VCH.

This explains why the well-defined diffraction peaks up to the (004) plane are observed (see Figure 9.2). The entropic contribution of the intercalants, which leads to the entropy gain associated with the layer expansion after intercalation of the polymer chains, may not be significant because of the interdigitated layer structure.

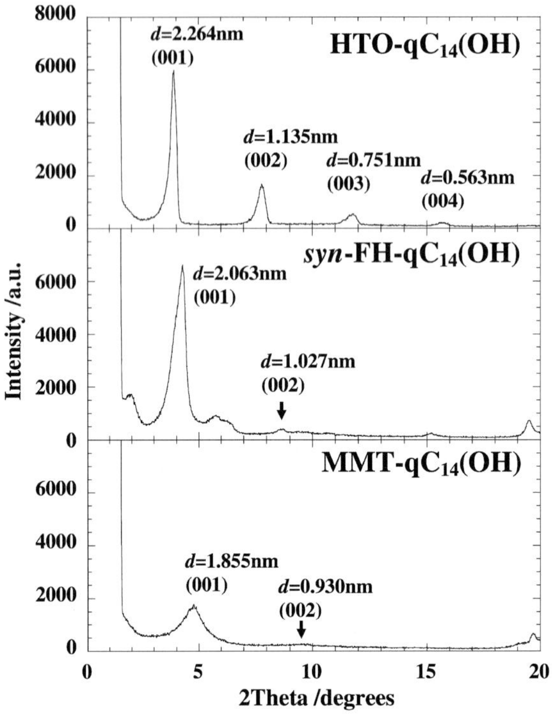

Fig. 9.2 WAXD patterns of HTO, *syn*-FH and MMT intercalated with $qC_{14}(OH)^+$. Reprinted from [18], © 2006, Wiley-VCH.

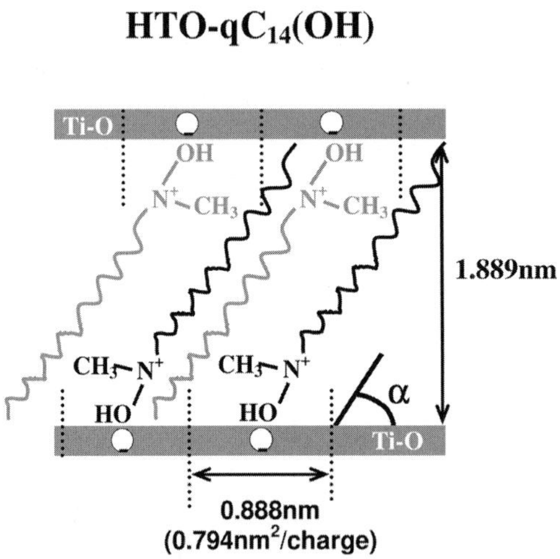

Fig. 9.3 Illustration of a model of interlayer structure of intercalant *N*-(cocoalkyl)-*N*,*N*-[bis (2-hydroxyethyl)]-*N*-methyl ammonium cation ($qC_{14}(OH)$) in the gallery space of layered titanate (HTO). The average distance between exchange sites is 0.888 nm, calculated from the surface charge density of $1.26\,e^-/nm^2$. For $qC_{14}(OH)$, the obtained molecular length, thickness and width are 2.09, 0.881 and 0.374 nm, respectively. The tilt angle α of the intercalants can be estimated by combination of the interlayer spacing, molecular dimensions and loading amount of intercalants when the alkyl chains adopted an all-trans conformation. Reprinted from [18], © 2006, Wiley-VCH.

Tab. 9.2 Comparison of characteristic parameters of HTO, *syn*-FH and MMT prepared with $qC_{14}(OH)$.

Parameter	HTO-$qC_{14}(OH)$	*syn*-FH-$qC_{14}(OH)$	MMT-$qC_{14}(OH)$
layer opening/nm	1.889	1.083	0.895
tilt angle α/°	64.4	31.1	25.3
organic content/wt%	39.6	30.4	32.5
T_m/°C	108.3	111.3	97.7
ΔH/J/g	214.5	141.2	138.6

[a] The melting and heat flow of $qC_{14}(OH)^+Cl^-$ are 35.8 °C and 69.8 J/g, respectively.
Source: Reprinted with permission from [18] © 2006, Wiley-VCH.

9.2.2.3 **Correlation of Intercalant Structure and Interlayer Opening**

For the interdigitated layer structure in MMT, the alkyl chain length (i.e. $C_{18}H_{37}$, CH_3 and $(CH_2)_2OH$ in the amine structure) changes the interlayer opening. That is, when we compare different intercalants having the same long alkyl chain (i.e., $C_{18}H_3N^+$ and $C_{18}(CH_3)_3N^+$), three methyl (CH_3) substituents instead of hydrogen (H) disturb the contact with the silicate surfaces. The value of α decreases until close contact between the ammonium cations and silicate surfaces is attained, giving a decreasing in the interlayer opening ($d_{(001)}$) (see Table 9.3 and Figure 9.2).

In the case of an intercalant having two long alkyl chains (i.e., $2C_{18}(CH_3)_2N^+$), the packing density of the alkyl chains is reduced and sterically limited to the nano-galleries. Consequently, MMT-$2C_{18}(CH_3)_2N^+$ exhibits a large interlayer opening accompanied by low crystallinity of the intercalant (ΔH ~130 J/g) compared with MMT-$C_{18}H_3N^+$ and MMT-$C_{18}(CH_3)_3N^+$. Accordingly, the disordered diffraction peak of the (001) plane of MMT-$2C_{18}(CH_3)_2N^+$ is observed in the WAXD analysis (see Figure 9.1 in Ref. [19]). We have to pay attention to the molecular size of the substituents instead of H attached to the nitrogen to get a better understanding of the interdigitated layer structure and direct polymer melt intercalation. This feature has

Tab. 9.3 Comparison of characteristic parameters of MMT-based OMLF prepared with $C_{18}H_3N^+$, $C_{18}(CH_3)_3N^+$ and $2C_{18}(CH_3)_2N^+$.

Parameter	$C_{18}H_3N^+$	$C_{18}(CH_3)_3N^+$	$2C_{18}(CH_3)_2N^+$
layer opening/nm	1.350	1.011	1.540
tilt angle α/°	33.2	22.9	40.1
organic content/wt%	35.5	29.5	39.8
T_m/°C	69.9	69.5	44.0
ΔH/J/g	177.7	189.6	129.7

[a] The melting and heat flow of $C_{18}H_3N^+$, $C_{18}(CH_3)_3N^+$ and $2C_{18}(CH_3)_2N^+$ are 83.8 °C and 95.6 J/g; 103.5 °C and 161.2 J/g; and 37.0 °C and 54.6 J/g, respectively.
Source: Reprinted with permission from [18] © 2006, Wiley-VCH.

been observed in the results of OMLFs intercalated with various intercalants (such as octadecyl di-methylbenzylammonium, n-hexadecyl tri-n-butyl phosphonium, n-hexadecyl tri-phenyl phosphonium cations) [20].

9.2.2.4 **Nanocomposite Structure**

Figure 9.4 shows the results of transmission electron microscopy (TEM) bright field images of PLA-based nanocomposites, in which the dark entities are the cross section of the intercalated MMT layers. The organically modified MMT content in all nanocomposites was 4 wt%. From the TEM images, it becomes clear that there are some intercalated-and-stacked silicate layers in the nanocomposites. Yoshida *et al.* estimate the form factors obtained from the TEM images, that is the average value of the particle length (L) of the dispersed particles and the correlation length (ξ) between them [21]. From the WAXD patterns, the crystallite size (D) of the intercalated stacked silicate layers of each nanocomposite is calculated by using the Scherrer equation. The calculated value of D ($\cong$ thickness of the dispersed particles) and other parameters for each nanocomposite are presented in Table 9.4.

For PLA/MMT-$C_{18}(CH_3)_3N^+$, L and D are in the range (200 ± 25) nm and 10.7 nm. On the other hand, PLA/MMT-$C_{18}H_3N^+$ exhibits a large value of L (450 ± 200 nm) with a large level of stacking of the silicate layers ($D{\sim}21$ nm). The ξ value of the PLA/MMT-$C_{18}(CH_3)_3N^+$ (80 ± 20 nm) is lower than that of PLA/MMT-$C_{18}H_3N^+$ (260 ± 140 nm), suggesting that the intercalated layers are more homogeneously and finely dispersed in the case of PLA/MMT-$C_{18}(CH_3)_3N^+$. The number of stacked individual silicate layers ($\equiv D/d_{(001)} + 1$) is 5 for PLA/MMT-$C_{18}(CH_3)_3N^+$ and the ξ value for this nanocomposite is one order of magnitude lower than those of PLA/MMT-$C_{18}H_3N^+$ and PLA/MMT-$2C_{18}(CH_3)_2N^+$, suggesting that the intercalated silicate layers are more homogeneously and finely dispersed.

Although the (initial) interlayer opening of MMT-$C_{18}(CH_3)_3N^+$ at 1.011 nm is smaller than that of MMT-$C_{18}H_3N^+$ at 1.350 nm and that of MMT-$2C_{18}(CH_3)_2N^+$ at 1.540 nm, the intercalation of the PLA in these different OMLFs gives almost the same basal spacing after preparation of the nanocomposites. Note that the existence of a sharp Bragg peak in the PLA-based nanocomposites after melt extrusion clearly indicates that the dispersed silicate layers still retain an ordered structure after melt extrusion.

Table 9.4 summarizes the layer expansions after preparation ($=\Delta$ opening) of three nanocomposites, or after subtraction of the initial layer opening. For the same MMT with different intercalants (e.g., comparison between MMT-$C_{18}(CH_3)_3N^+$ and MMT-$2C_{18}(CH_3)_2N^+$), the layer expansion of the former (0.879 nm) exhibits a large value compared with that of the latter (0.45 nm) in PLA-based nanocomposites. In other words, the smaller interlayer opening caused by the configuration with small tilt angle ($\alpha = 22.9°$ for $C_{18}(CH_3)_3N^+$) promotes a large amount of intercalation of the polymer chains. Accordingly, PLA/MMT-$C_{18}(CH_3)_3N^+$ exhibits finer dispersion of the nano-fillers than do PLA/MMT-$2C_{18}(CH_3)_2N^+$ and PLA/MMT-$C_{18}H_3N^+$, as discussed before (see Figure 9.4).

One more interesting feature is the absolute value of Δ opening. According to the molecular modeling, the width and thickness of the PLA are 0.76 and 0.58 nm (see Figure 9.5). This may suggest that the polymer chains could not penetrate into the

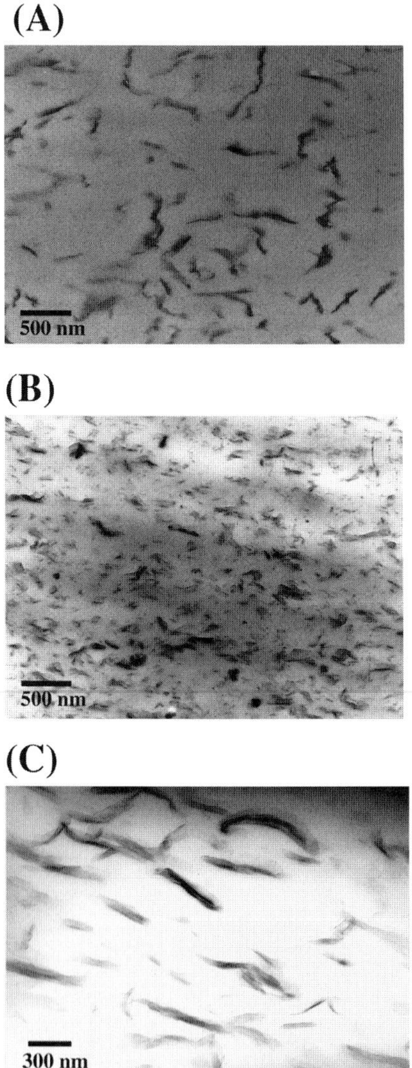

Fig. 9.4 Bright field TEM images of PLA-based nanocomposites prepared with (A) MMT-$C_{18}H_3N^+$, (B) MMT-$C_{18}(CH_3)_3N^+$ and (C) MMT-$2C_{18}(CH_3)_2N^+$. The dark entities are the cross section and/or face of the intercalated-and-stacked silicate layers and the bright areas are the matrix. Reprinted from [18], © 2006, Wiley-VCH.

galleries in the case of MMT-$2C_{18}(CH_3)_2N^+$ when we compare the apparent interlayer expansion (=Δ opening).

Now it is necessary to understand the meaning of the interlayer expansion in the intercalated nanocomposites. As discussed before, we have to take the interdigitated layer structure into consideration. This structure may suggest that a different

Tab. 9.4 Form factors of three nanocomposites obtained from WAXD and TEM observations.

Form factor	PLA/MMT-$C_{18}H_3N^+$	PLA/MMT-$C_{18}(CH_3)_3N^+$	PLA/MMT-$2C_{18}(CH_3)_2N^+$
d_{001}/nm	3.03	2.85	2.95
Δ opening/nm	0.72	0.879	0.45
final layer opening/nm	2.07	1.89	1.99
D/nm	20.9	10.73	14.71
$(D/d_{001})+1$	7.9	4.8	6.0
L/nm	450 ± 200	200 ± 25	655 ± 121
ξ/nm	260 ± 140	80 ± 20	300 ± 52

Source: Reprinted with permission from [18] © 2006, Wiley-VCH.

orientation angle could be adopted when the polymer chains penetrate into the galleries, giving a decrease in the basal spacing after intercalation. At the same time, this structure apparently provides a balance between polymer penetration and the different orientation angles of the intercalants. That is, we have to pay attention to the polymer chains intercalation into the galleries from the change in the basal spacing, as revealed by WAXD. Presumably the penetration of the polymer chain is prevented or reduced by the steric-limitation of the configuration with a large value of α (e.g., $\alpha = 40.1°$ for MMT-$2C_{18}(CH_3)_2N^+$. Accordingly, we sometimes observe small inter-layer expansion and encounter a decrease in the interlayer spacing after melt intercalation. As seen in Table 9.4, the initial interlayer opening depends on the interlayer expansion (=Δ opening) after melt intercalation. The smaller initial opening leads to the larger interlayer expansion, and gives almost the same final

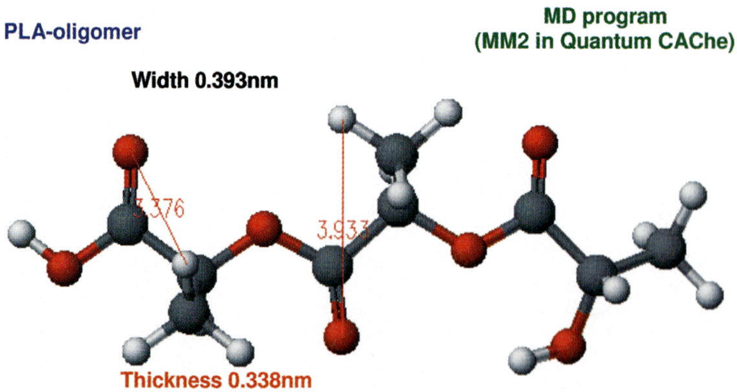

Fig. 9.5 Molecular dimensions of the PLA-backbone using the molecular dynamics program (MM2 in Quantum CAChe) with van der Waals radii taken into consideration. Optimization of structure is based on minimization of the total energy of the molecular system.

interlayer opening. This feature has been observed in the results of other nanocomposites prepared from different OMLFs intercalated with different surfactants [22]. From this result, the entropic contribution of the intercalants, which leads to the entropy gain associated with the layer expansion after intercalation of the small molecules and/or polymer chains, may not be significant due to the interdigitated layer structure. Presumably the penetration takes place by pressure drop within the nano-galleries, nano-capillary action, generated by the two platelets.

As reported in the literature [20], the pressure drop (Δp) into the nano-galleries, which makes the polymer penetration more difficult, should be discussed. The estimated pressure difference ($\sim$24 MPa) is much larger than the shear stress ($\sim$0.1 MPa) during melt compounding [20]. This suggests that the shear stress has little effect on the delamination (exfoliation) of the layer. This reasoning is consistent with the intercalated structure reported by so many nanocomposite researchers, who can prepare only intercalated (not exfoliated) nanocomposites via a simple melt extrusion technique [5]. A novel compounding process is currently in progress. Solid-state shear processing may be an innovative technique to delaminate the layered fillers [23].

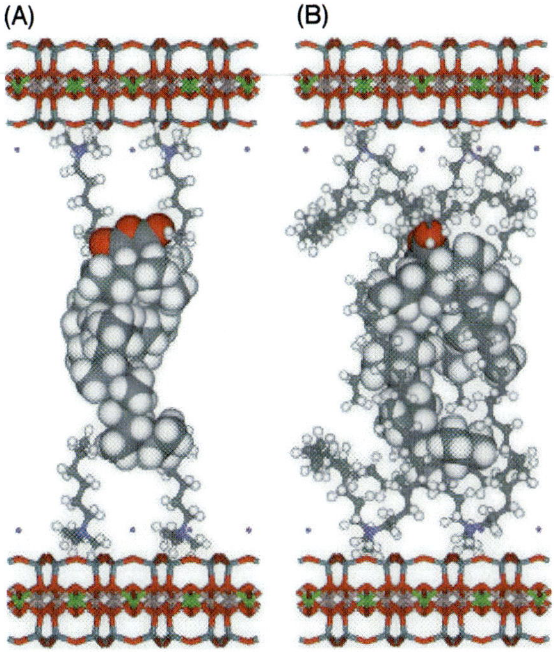

Fig. 9.6 Three-component model used for basal spacing simulations, consisting of two layers of MMT with K^+ cations (stick model), four molecules of trimethylammonium cation (A) or dimethylstearylammonium cation (B) (stick and ball model), and one molecule of maleated PP (PP-MA) (ball model). Reprinted from [24], © 2006, Elsevier Science.

Compared to OMLFs, nanocomposite structure is difficult to model using atomic scale molecular dynamics (MD) because the intercalated polymer chain conformation is complex and can hardly be the equilibrium state. However, Pricl *et al.* [24] explored and characterized the atomic scale structure to predict binding energies and basal spacing of PLFNCs based on polypropylene (PP) and maleated (MA) PP (PP-MA), MMT, and different alkylammonium ions as intercalants (see Figure 9.6). From a global interpretation of all these MD simulation results, they concluded that intercalants with smaller volume are more effective for clay modification as they improve the thermodynamics of the system by increasing the binding energy. On the other hand, intercalants with longer tails are more effective for intercalation and exfoliation processes, as they lead to higher basal spacing. Additional information is necessary to predict a more reasonable nanostructure of PLFNCs; some studies using coarse-grained MD simulation of the confined polymer chains within the silicate galleries have been reported [25–28].

9.3 Control of Nanostructure Properties

9.3.1 Flocculation Control and Modulus Enhancement

Most nanocomposite researchers obdurately believe that the preparation of a completely exfoliated structure is the ultimate target for better overall properties. However, these significant improvements are not observed in every nanocomposite system, including systems where the silicate layers are near to exfoliated [29]. While, from the barrier property standpoint, the development of exfoliated nanocomposites is always preferred, Nylon 6-based nanocomposite systems are completely different from other nanocomposite systems, as discussed [3,8].

In Figure 9.7, Okamoto summarized the clay content dependence of the dynamic storage modulus (G') of various types of nanocomposites obtained at well below the T_g of the matrices. The Einstein coefficient (k_E), derived by using Halpin and Tai's theoretical expression modified by Nielsen, is shown in the figure, and represents the aspect ratio (L_{filler}/d_{filler}) of dispersed MMT particles without intercalation. From Figure 9.7, it is clearly observed that poly(butylene succinate)(PBS)-based nanocomposites (PBSNCs) show a very high increment in G' compared to other nanocomposites having the same content of clay in the matrix. N6NCs are well-established exfoliated Nylon 6-based nanocomposites.

PLANCs are becoming established intercalated-and-flocculated PLA-based nanocomposites, while PBSNCs are intercalated-and-extended flocculated nanocomposite systems [30,31]. Due to the strong interaction between hydroxylated edge–edge groups, the MMT particles are sometimes flocculated in the polymer matrix. As a result of this flocculation the length of the MMT particles increases enormously and hence the increase in the overall aspect ratio. For the preparation of high molecular weight PBS, di-isocyanate [OCN-(C_6H_{12})-NCO] type end-groups are generally used as

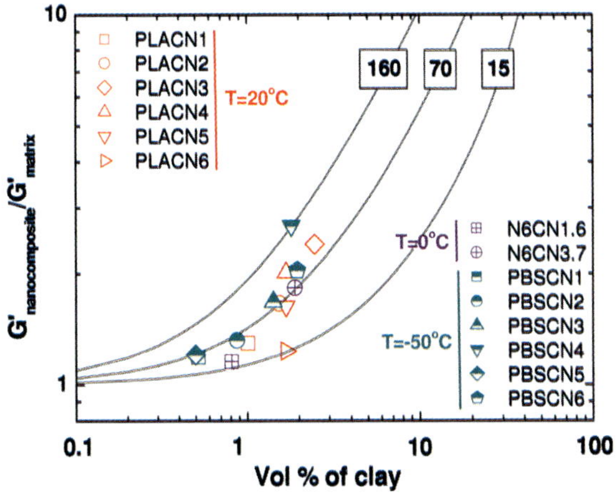

Fig. 9.7 Plots of $G'_{\text{nanocomposite}}/G'_{\text{matrix}}$ vs. vol.% of MMT for various nanocomposites. The Einstein coefficient k_E is shown by the number in the box. The lines show the calculated results from Halpin and Tai's theory with various k_E.

a chain extender. These isocyanate end-group chain extenders form urethane bonds with hydroxy-terminated low molecular weight PBS, and each high molecular weight PBS chain contains two such bonds. These urethane-type bonds lead to strong interaction with the silicate surface by forming hydrogen bonds and hence to strong flocculation (see Figure 9.8). For this reason, the aspect ratio of dispersed clay particles is much higher in the case of PBSNCs than for all nanocomposites, and hence the high enhancement of the modulus.

This behavior can be explained with the help of the classical rheological theory of suspension of conventional filler reinforced systems. According to this theory [32], rotation of the filler is possible when the volume fraction of clay $\phi_{\text{filler}} < \phi_{\text{critical}} \cong (\text{aspect ratio})^{-1}$. All PBSNCs studied here follow this relation except PBSNC4 (MMT = 3.6 wt%), in which $\phi_{\text{filler}} \gg (\text{aspect ratio})^{-1}$. For this reason, in

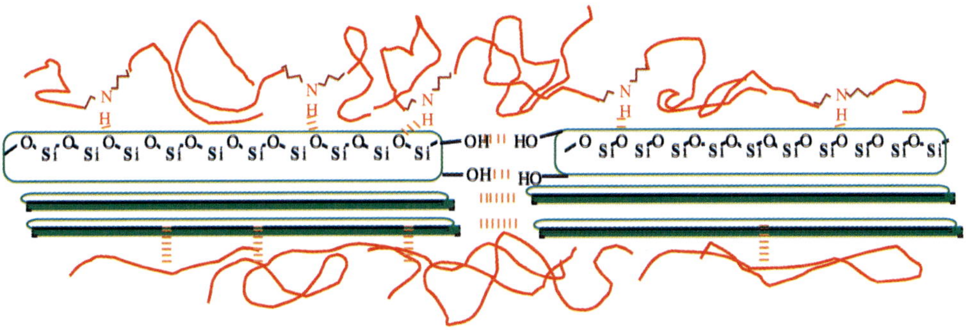

Fig. 9.8 Formation of hydrogen bonds between PBS and MMT, which leads to the flocculation of the dispersed silicate layers. Reprinted from [30], © 2003 American Chemical Society.

PBSNC4, rotation of dispersed intercalated with flocculated stacked silicate layers is completely hindered and only translational motion is available. Hence PBSNC4 exhibits a very high modulus. This behavior is clearly observed in dynamic storage modulus measurements in the molten state [30]. In the case of N6NC3.7 (MMT = 3.7 wt%) we can see the same high increment in G' as for the PBSNCs. The development of the flocculated structure occurs even though N6NCs are well-established exfoliated nanocomposite systems.

9.3.2 Linear Viscoelastic Properties

The measurement of rheological properties of the PLFNCs in the molten state is crucial in order to gain a fundamental understanding of the nature of the processability and the structure–property relationships for these materials.

Dynamic oscillatory shear measurements of polymeric materials are generally performed by applying a time dependent strain of $\gamma(t) = \gamma_o \sin(\omega t)$ and the resultant shear stress is $\sigma(t) = \gamma_o[G' \sin(\omega t) + G'' \cos(\omega t)]$, with G' and G'' being the storage and loss modulus, respectively.

Generally, the rheology of polymer melts depends strongly on the temperature at which the measurement is carried out. It is well known that for thermorheological simplicity, isotherms of storage modulus ($G'(\omega)$), loss modulus ($G''(\omega)$) and complex viscosity ($\eta^*(\omega)$) can be superimposed by horizontal shifts along the frequency axis:

$$b_T G'(a_T\omega, T_{ref}) = b_T G'(\omega, T);$$

$$b_T G''(a_T\omega, T_{ref}) = b_T G''(\omega, T)$$

$$|\eta^*|(a_T\omega, T_{ref}) = |\eta^*|(\omega, T)$$

where a_T and b_T are the frequency and vertical shift factors, and T_{ref} is the reference temperature. All isotherms measured for pure polymer and for various nanocomposites can be superimposed along the frequency axis.

In the case of polymer samples, it is expected that, at the temperatures and frequencies at which the rheological measurements were carried out, the polymer chains should be fully relaxed and exhibit characteristic homo-polymer-like terminal flow behavior (i.e., the curves can be expressed by a power-law of $G' \propto \omega^2$ and $G'' \propto \omega$).

The rheological properties of *in-situ* polymerized nanocomposites with end-tethered polymer chains were first described by Krisnamoorti and Giannelis [33]. The flow behavior of PCL- and Nylon 6-based nanocomposites differed extremely from that of the corresponding neat matrices, whereas the thermorheological properties of the nanocomposites were entirely determined by the behavior of the matrices [33]. The slope of $G'(\omega)$ and $G''(\omega)$ versus $a_T\omega$ is much smaller than 2 and 1, respectively. Values of 2 and 1 are expected for linear mono-dispersed polymer melts, and the large deviation, especially in the presence of a very small amount of layered silicate loading, may be due to the formation of a network structure in the molten state. However, such nanocomposites based on the *in-situ* polymerization technique exhibit fairly broad molar mass distribution of the polymer matrix, which hides the structure-relevant information and impedes the interpretation of the results.

To date, the melt state linear dynamic oscillatory shear properties of various kinds of nanocomposites have been examined for a wide range of polymer matrices including Nylon 6 with various matrix molecular weights [34], polystyrene (PS) [35], PS-polyisoprene (PI) block copolymers [36,37], poly(ε-caprolactone) (PCL) [38], PLA [39,40], PBS [30,41], and so on [42].

The linear dynamic viscoelastic master curves for the neat PLA and various PLA-based nanocomposites (PLANCs) are shown in Figure 9.9 [40]. The linear dynamic viscoelastic master curves were generated by applying the time–temperature superposition principle and shifted to a common temperature T_{ref} using both the frequency shift factor a_T and the modulus shift factor b_T. The moduli of the nanocomposites increase with increasing clay loading at all frequencies ω. At high ωs, the qualitative behavior of $G'(\omega)$ and $G''(\omega)$ is essentially the same and unaffected by frequency. However, at low frequencies $G'(\omega)$ and $G''(\omega)$ increase monotonically with increasing clay content. In the low frequency region, the curves can be expressed by the power-law of $G'(\omega) \propto \omega^2$ and $G''(\omega) \propto \omega$ for neat PLA, suggesting that this is similar to those of the narrow M_w distribution homopolymer melts. On the other hand, for $a_T < 5\ \text{rad s}^{-1}$, the viscoelastic response [particularly $G'(\omega)$] for all the nanocomposites displays significantly diminished frequency dependence as compared to the matrices. In fact, for all PLANCs, $G'(\omega)$ becomes nearly independent at low $a_T\omega$ and exceeds $G''(\omega)$, characteristic of materials exhibiting a pseudo-solid-like behavior [33]. The values of the terminal zone slopes of both neat PLA and PLANCs are estimated in the lower $a_T\omega$ region ($<10\ \text{rad s}^{-1}$), and are presented in Table 9.5. The lower slope values and the higher absolute values of the dynamic moduli indicate the formation of a "spatially-linked" structure in the PLANCs in the molten state [36]. Because of this structure or high geometric constraints, the individual stacked silicate layers are incapable of rotating freely and hence, by imposing small $a_T\omega$, the relaxations of the structure are prevented almost completely. This type of prevented relaxation due to the high geometric constraints of the stacked and intercalated silicate layers leads to the presence of the pseudo-solid-like behavior, as observed in PLANCs. This behavior probably corresponds to the shear-thinning tendency, which appears strongly in the viscosity curves ($a_T\omega < 5\ \text{rad s}^{-1}$) ($\eta^*$ vs $a_T\omega$). Such a feature depends strongly on the shear rate in the dynamic measurement because of the formation of shear-induced alignment of the dispersed MMT particles [43].

The temperature dependence frequency shift factor (a_T, Williams–Landel–Ferry type [44]) used to generate the master curves shown in Figure 9.9 is shown in Figure 9.10. The dependence of the frequency shift factors on the silicate loading suggests that the temperature-dependent relaxation process observed in the viscoelastic measurements is somehow affected by the presence of the silicate layers [33]. In the case of the Nylon 6-based nanocomposite, where there is hydrogen bonding to the silicate surface, the system exhibits a high flow activation energy nearly one order of magnitude greater than that of neat Nylon 6 [45].

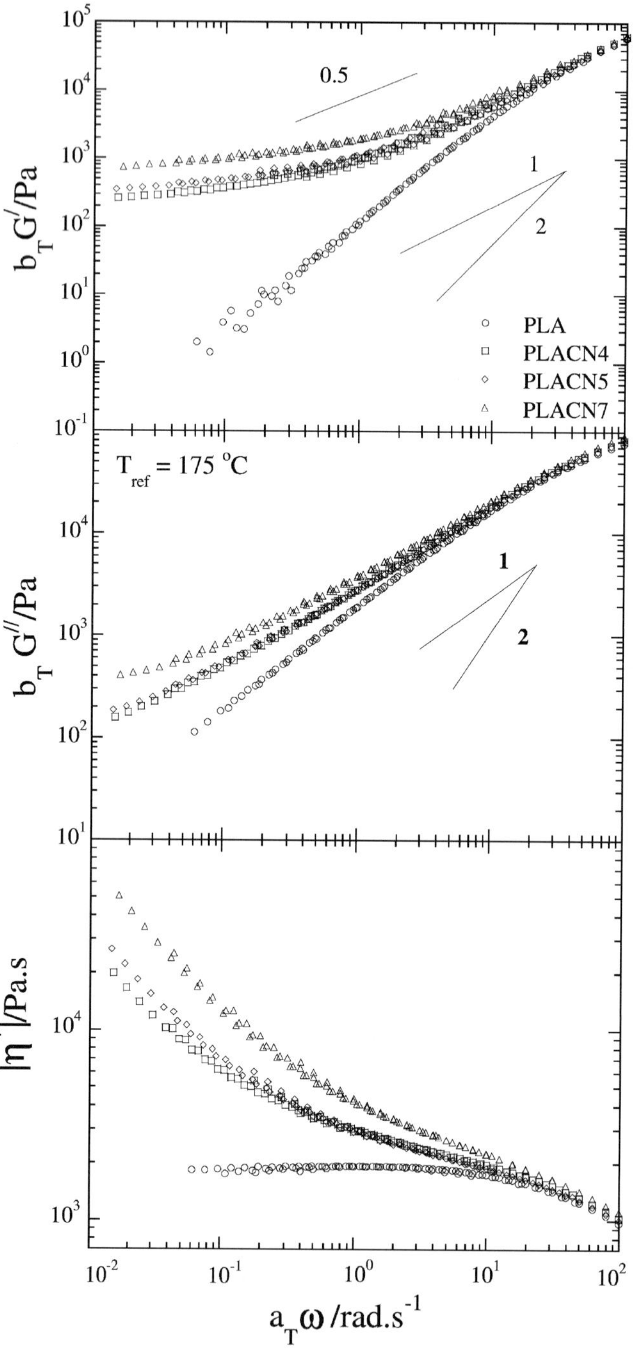

Fig. 9.9 Reduced frequency dependence of storage modulus, loss modulus and complex viscosity of neat PLA and various nanocomposites (PLANCs). Reprinted from [40], © 2003, Elsevier Science.

Tab. 9.5 Terminal slopes of G′ and G″ vs. $a_T\omega$ for PLA and various PLACNs.

System	G′	G″
PLA	1.3	0.9
PLACN4	0.2	0.5
PLACN5	0.18	0.4
PLACN7	0.17	0.32

Source: Reprinted from [40], © 2003, Elsevier Science.

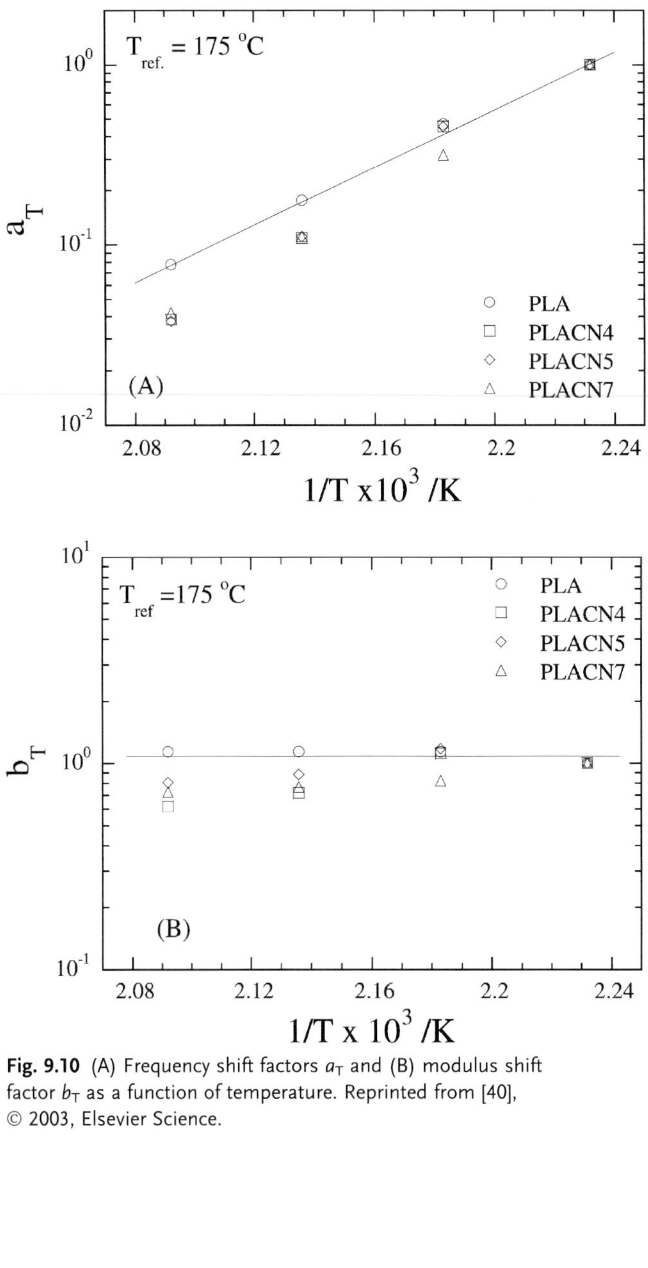

Fig. 9.10 (A) Frequency shift factors a_T and (B) modulus shift factor b_T as a function of temperature. Reprinted from [40], © 2003, Elsevier Science.

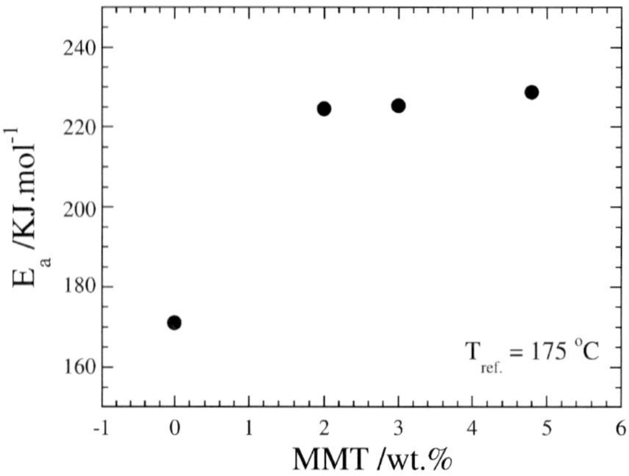

Fig. 9.11 Flow activation energy of pure PLA and various PLA-based nanocomposites as a function of MMT content. Reprinted from [47], © 2003 Wiley-VCH Verlag GmbH & Co.

The shift factor b_T shows a large deviation from a simple density effect, it would be expected that the values would not vary far from unity [44]. One possible explanation is an internal structure development occurring in PLANCs during measurement (shear process). The alignment of the silicate layers probably enables the PLFNC melts to withstand the shear force, thus leading to the increase in the absolute values of $G'(\omega)$ and $G''(\omega)$.

Galgali and his colleagues [46] have also shown that the typical rheological response in nanocomposites arises from frictional interactions between the silicate layers and not from the immobilization of confined polymer chains between the silicate layers. They have also shown a dramatic decrease in the creep compliance for the PP-based nanocomposite with 9 wt% MMT. They showed a dramatic three orders of magnitude drop in the zero shear viscosity beyond the apparent yield stress, suggesting that the solid-like behavior in the quiescent state is a result of the percolated structure of the layered silicate.

Figure 9.11 represents the MMT content-dependent (wt%) flow activation energy (E_a) of pure PLA and various PLANCs obtained from an Arrhenius fit of the master curves [47]. It is clearly observed that the E_a value increases significantly for the nanocomposite containing 3 wt% MMT and then is almost unchanged with increasing MMT content. This result indicates that, in the presence of MMT, it is very difficult for the materials to flow. This behavior is also ascribed to the formation of a spatially linked structure in the nanocomposite in the molten state.

9.3.3
Elongational Flow and Strain-induced Hardening

Okamoto *et al.* [48] first conducted an elongation test on PP-based nanocomposites in the molten state at constant Hencky strain rate, $\dot{\varepsilon}_0$ using elongation flow optorheometry [49]. They also attempted to control the alignment of the dispersed silicate layers with nanometer dimensions of an intercalated PP-based nanocomposite under uniaxial elongational flow.

Figure 9.12 shows double-logarithmic plots of transient elongational viscosity (η_E) against time (t) observed for the PLA-based nanocomposite (PLANC) (MMT = 4 wt%) at 170 °C with different Hencky strain rates ($\dot{\varepsilon}_0$) ranging from 0.01 to 1 s^{-1} [47]. A very strong tendency for *strain-induced hardening* in the nanocomposite melt was observed. In the early stage, η_E gradually increases with t but is almost independent of $\dot{\varepsilon}_0$; this is called the *linear region* of the viscosity curve. After a certain time, t_{η_E} which is called the *up-rising* time (marked with the upward arrows in Figure 9.12(A)), strongly dependent on $\dot{\varepsilon}_0$, rapid upward deviation of η_E from the curves of the linear region is seen. The very low shear viscosity of pure PLA was given as the main reason for this,

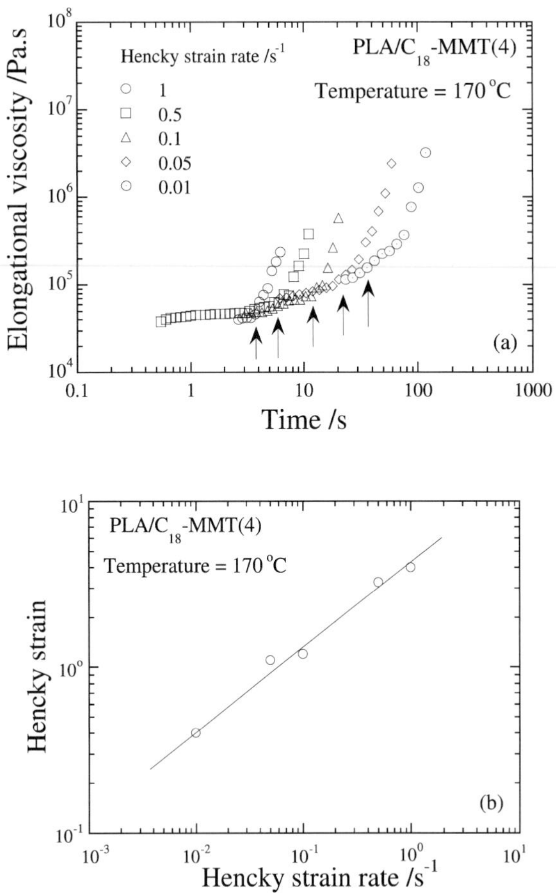

Fig. 9.12 (A) Time variation of elongational viscosity for PLA-based nanocomposite (MMT = 4 wt%) melt at 170 °C; (B) Strain rate dependence of up-rising Hencky strain. Reprinted from [47], © 2003 Wiley-VCH Verlag GmbH & Co.

because the minimum viscosity range of their instrument was greater than 10^4 Pa s. However, they confirmed that neither strain-induced hardening in elongation nor rheopexy in shear flow took place with pure PLA having the same molecular weight and polydispersity as the nanocomposite.

As in PP-based nanocomposite systems, the extended Trouton rule, $3\eta_0(\dot{\gamma};t) \cong \eta_E(\dot{\varepsilon}_0;t)$, also does not hold for PLANC melts, in contrast to the melt of pure polymers. These results indicate that in the case of PLANC, the flow induced internal structural changes also occur in elongation flow [48], but the changes are quite different in shear flow. The strong rheopexy observed in the shear measurements for the PLA-based nanocomposite at very slow shear rate reflects the fact that the shear-induced structural change involved a process with an extremely long relaxation time.

Regarding elongation-induced structure development, Figure 9.12(B) shows the Hencky strain rate dependence of the up-rising Hencky strain $(\varepsilon_{\eta_E}) = \dot{\varepsilon}_0 \times t_{\eta_E}$ recorded for PLANC at 170 °C. The ε_{η_E} increases systematically with the $\dot{\varepsilon}_0$. The lower the value of $\dot{\varepsilon}_0$, the smaller the value of ε_{η_E}. This tendency probably corresponds to the rheopexy of PLANC under slow shear flow.

9.4
Physicochemical Phenomena

9.4.1
Biodegradability

Another most interesting and exciting aspect of nanocomposite technology is the significant improvements in biodegradability of biodegradable polymers after nanocomposite preparation with OMLS. Aliphatic polyesters are among the most promising materials for the production of environmentally friendly biodegradable plastics. Biodegradation of aliphatic polyester is well known, in that some bacteria degrade it by producing enzymes, which attack the polymer. Tetto and his colleagues [50] first reported some results on the biodegradability of nanocomposites based on PCL, where they found that the PCL-based nanocomposites showed improved biodegradability compared to pure PCL. They suggest that the improved biodegradability of PCL after nanocomposite formation may be due to the catalytic role of the OMLS in the biodegradation mechanism. But it is still unclear how the clay increases the biodegradation rate of PCL.

In 2002, Lee *et al.* [51] reported the biodegradation of aliphatic polyester-based nanocomposites under compost. Figure 9.13(A, B) represent the clay content dependence of biodegradation of APES-based nanocomposites prepared with two different types of MMT clays. They assumed that the retardation of biodegradation was due to the improvement of the barrier properties of the aliphatic APSE after nanocomposite preparation with clay. However, there are no data about permeability.

Recently, Yamada and Okamoto *et al.* [52–54] first reported the biodegradability of neat PLA and PLA-based nanocomposites prepared with trimethyl octadecylammonium-modified MMT (MMT-$C_{18}(CH_3)_3N^+$) with a detailed mechanism. The compost

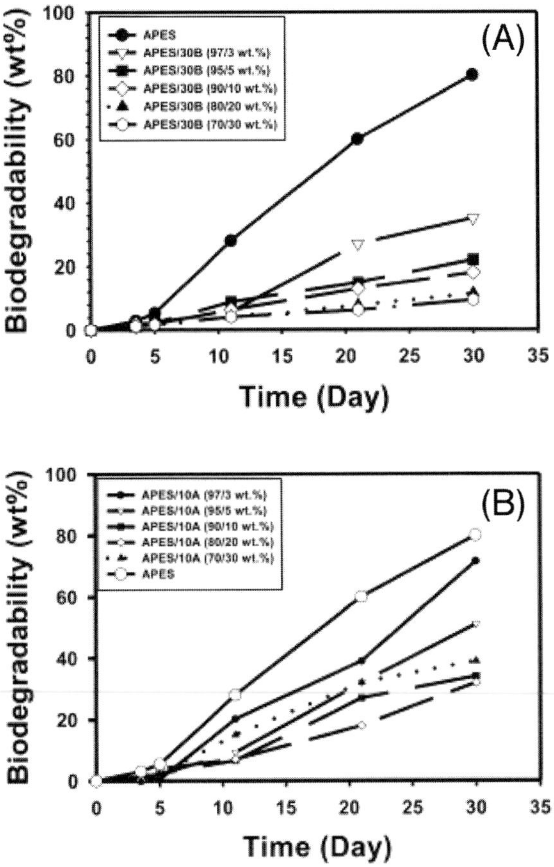

Fig. 9.13 Biodegradability of APES nanocomposites with: (A) Closite 30B and (B) Closite 10A. Reprinted from [51], © 2002, Elsevier Science.

used was prepared from food waste and tests were carried out at temperature of $58 \pm 2\,^{\circ}C$. Figure 9.14(A) shows photographs of the samples of neat PLA and nanocomposite (MMT = 4 wt%) recovered from the compost at various times. The decreased molecular weight M_w and residual weight percentage R_w of the initial test samples with time is also reported in Figure 9.14(B). The biodegradability of neat PLA is significantly enhanced after nanocomposite preparation. Within one month, both the M_w and the weight loss are almost the same for both PLA and the nanocomposite. However, after one month, a sharp change occurs in the weight loss of nanocomposite, and within two months it is completely degraded into compost. The degradation of PLA in compost is a complex process involving four main phenomena, namely: water absorption, ester cleavage and formation of oligomer fragments, solubilization of oligomer fragments, and finally diffusion of soluble oligomers by bacteria [55]. Therefore, the factor which increases the hydrolysis tendency of PLA ultimately

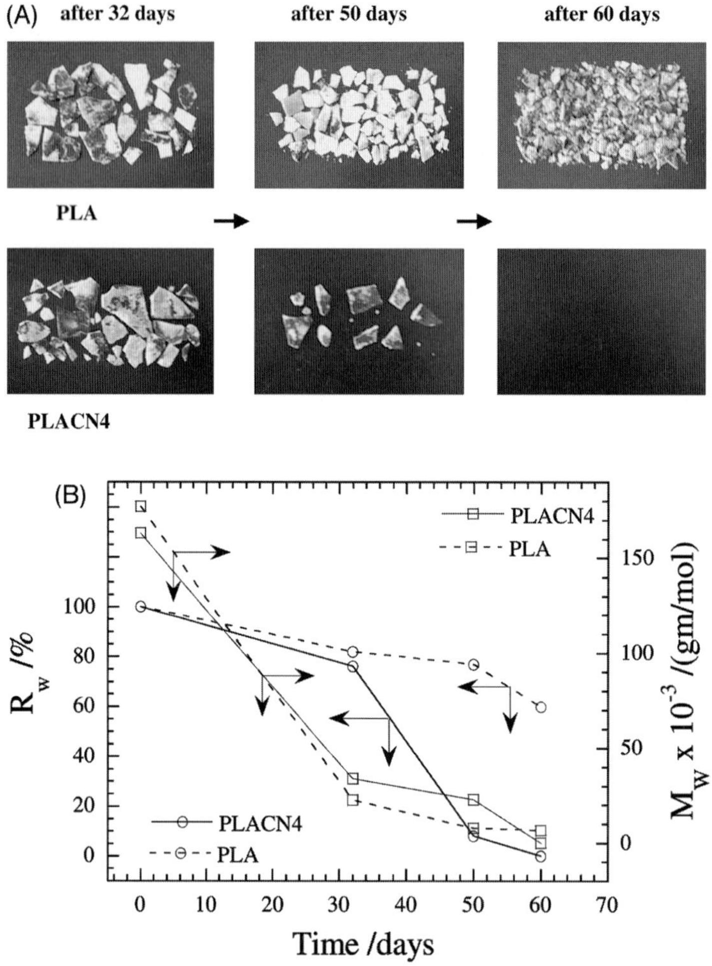

Fig. 9.14 (A) Photographs of biodegradability of neat PLA and PLA-based nanocomposite recovered from compost with time. Initial size of the crystallized samples was $3 \times 10 \times 0.1\,\text{cm}^3$. (B) Time dependence of residual weight, R_w and of matrix, M_w of PLA and PLANC4 under compost at $58 \pm 2\,^\circ$C. Reprinted from [52], © 2002, American Chemical Society.

controls the degradation of PLA. The authors expect that the presence of terminal hydroxylated edge groups on the silicate layers may be one of the factors responsible for this behavior. In the case of PLA-based nanocomposite, the stacked (~4 layers) and intercalated silicate layers are homogeneously dispersed in the PLA matrix (as seen from the TEM image [54]) and these hydroxy groups start heterogeneous hydrolysis of the PLA matrix after absorbing water from the compost. This process takes some time to start which explains why the weight loss and degree of hydrolysis of PLA and nanocomposite is almost the same in the first month (see Figure 9.14(B)). However, after one month there is a sharp weight loss with the nanocomposite compared to that with PLA. This means that one month is the critical time after which heterogeneous hydrolysis starts and the matrix breaks into very small fragments and disappears as compost. This assumption was confirmed by conducting same type of experiment with PLANC prepared by using $2C_{18}(CH_3)_2N^+$ modified *syn*-FH which has no terminal hydroxylated edge group, the degradation pattern was almost the same as with neat PLA [53].

They also conducted a respirometric test to study the degradation of the PLA matrix in a compost environment at $58 \pm 2\,°C$. For this test the compost used was made from bean-curd refuse, food waste, and cattle feces. Unlike weight loss, which reflects the structural changes in the test sample, CO_2 evolution provides an indicator of the ultimate biodegradability of PLA in a PLA-based nanocomposite (prepared with $qC_{14}(OH)$ modified *syn*-FH,), that is mineralization of the samples. Figure 9.15 shows the time dependence of the degree of biodegradation of neat PLA and nanocomposite, indicating that the biodegradability of PLA in the nanocomposite is enhanced significantly. The presence of OMLF may thus cause a different mode of attack on the PLA component, which might be due to the presence of hydroxy groups. Details of the mechanism of biodegradability are presented in the relevant literature [53,54].

K. Okamoto and M. Okamoto also investigated the biodegradability of neat PBS before and after nanocomposite preparation with three different types of OMLF. They used alkylammonium or alkylphosphonium salts for the modification of pristine layered silicates, and these surfactants are toxic for microorganisms [56].

Figure 9.16(A) shows photographs of samples of neat PBS and various nanocomposites recovered from the compost after 35 days. It can clearly be seen that many

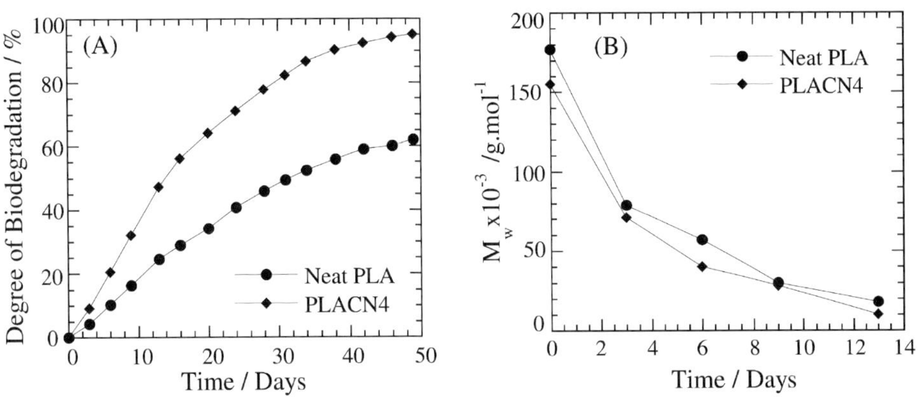

Fig. 9.15 (A) Degree of biodegradation (i.e., CO_2 evolution), and (B) time-dependent change of matrix M_w of neat PLA and PLA-based nanocomposite (*syn*-FH = 4 wt%) under compost at $58 \pm 2\,°C$. Reprinted from [53], © 2004 WILEY-VCH.

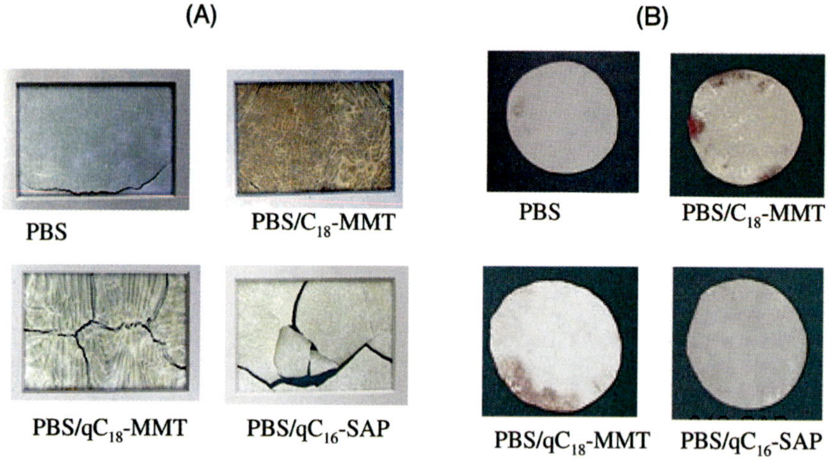

Fig. 9.16 Biodegradability of neat PBS and various nanocomposite sheets (A) under compost, and (B) under soil field. Reprinted from [56], © 2003 John Wiley & Sons, Inc.

cracks have appeared in the nanocomposite samples compared to that of neat PBS. This indicates the improved degradability of nanocomposites in compost. This kind of fracture has an advantage for biodegradation because it is easy to mix with the compost and create much more surface area for further attack by microorganisms. It should be noted that the extent of fragmentation is directly related to the nature of OMLF used in the nanocomposite preparation. Gel permeation chromatography (GPC) was performed on the samples recovered from the compost, and it was found that the extent of molecular weight loss was almost the same for all samples. This indicates that the extent of hydrolysis of PBS in the pure state or in OMLF filled systems is the same in compost.

Except for the PBS/SAP-qC_{16} (n-hexadecyl tri-n-butyl phosphonium cation modified saponite) system, the degree of degradation is the same for other samples. This indicates that MMT or alkylammonium cations, and at the same time other properties, have no effect on the biodegradability of PBS. The accelerated degradation of PBS matrix in the presence of SAP-qC_{16} may be due to the presence of alkylphosphonium surfactant. This kind of behavior is also observed in the case of PLA/MMT-based nanocomposite systems.

The nature of the degradation of PBS and various nanocomposites under a soil field was also observed. This experiment was conducted for one, two, and six months. After one and two months, there was no change in the nature of the surface of the samples, but after six months black or red spots appeared on the surface of the nanocomposite samples. Figure 9.16(B) represents the results of degradation of neat PBS and various nanocomposite sheets recovered from a soil field after six months. The spots on the sample surface were reported as being due to fungus attack because when they were put into a slurry clear growth of fungus was observed. These results also indicate that nanocomposites exhibit the same or a higher level of biodegradability compared with the PBS matrix.

9.4.2 Photodegradation

Hiroi and Okamoto *et al.* [57] first reported the photodegradability of neat PLA and the corresponding PLA-based nanocomposite prepared by using organically modified layered titanate (HTO) as a new nano-filler. One of the features of this material is its photocatalytic reactivity, like that of titania (TiO_2). The photocatalytic reactions of anatase-TiO_2, such as the evolution of hydrogen gas from water or the oxidative degradation of organic compounds, have attracted intense research interest because of their possible application to the conversion of solar energy into chemical energy [58].

Figure 9.17 shows the UV/VIS transmission spectra of pure PLA and nanocomposite (PLANC1.7: HTO = 1.7 wt%). The spectra show change in the VIS region (>~400 nm) with increasing absorption in the presence of titanate layers compared with neat PLA. For UV wavelengths, there is strong absorption up to 320 nm, resulting in 0 % transmittance. This significant change in the spectra may indicate photodegradation of the PLA matrix. To confirm this, some preliminary experiments on the photodegradation of PLA-based nanocomposites under a sunshine weathermeter at 60 °C were carried out. After 300 h, there was no change in the nature of the sample surfaces of neat PLA, however, the surface color of the nanocomposite samples became yellow and/or light brown. Table 9.6 shows the GPC measurement of samples recovered from the test. The drop in M_w accompanied by broadening of M_w/M_n indicates that enhancement of degradation of PLA in the titanate-filled system has occurred.

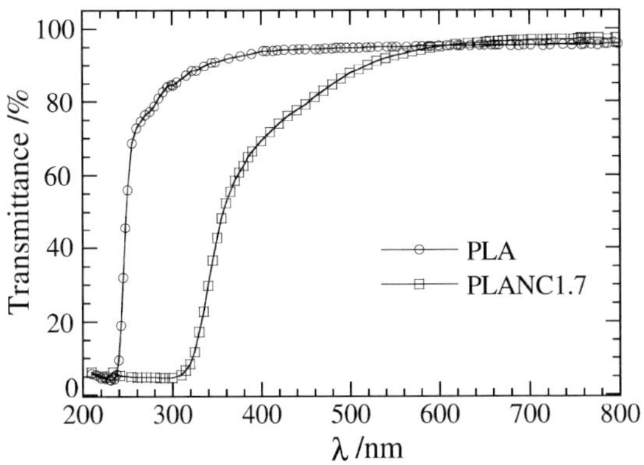

Fig. 9.17 UV/VIS transmission spectra of neat PLA and nanocomposite (PLANC1.7). Reprinted from [57], © 2004 WILEY-VCH.

Tab. 9.6 GPC results of sample recovered from weathermeter after 300 h.

Samples	$M_w \times 10^{-3}$ (g/mol)	M_w/M_n	M_w/M_w^0 [a]
PLA	198	1.53	0.94
PLANC1.7	93.7	1.89	0.68
PLANC3.9	86.3	1.86	0.76

[a] M_w^0 is molecular weight before test.
Source: Reprinted from [57], © 2004 WILEY-VCH.

9.5
Foam Processing using Supercritical CO_2

9.5.1
PLA-based Nanocomposite

PLA-based nanocomposites have already been shown to exhibit a tendency toward strong strain-induced hardening. This strain-induced hardening behavior is an indispensable characteristic for foam processing due to its capacity to withstand the stretching force experienced during the latter stages of bubble growth. On the basis of this result, the first successful nanocomposite foam, processed by using supercritical (sc)-CO_2 as a physical foaming agent, appeared through a pioneering effort by Okamoto *et al.* [9,10]. A small amount of nano-fillers in the polymer matrix serve as nucleation sites to facilitate bubble nucleation during foaming. Novel nanocomposite foams based on the combination of new nano-fillers and sc-CO_2 have led to a new class of materials. The process consists of four stages: (i) saturation of CO_2 in the sample at the desired temperature, (ii) cell nucleation when the release of CO_2 pressure starts (supersaturated CO_2), (iii) cell growth to an equilibrium size during the release of CO_2, and (iv) stabilization of the cell via cooling of the foamed sample. The set-up of the autoclave used in the study is shown in Figure 9.18.

Figure 9.19 shows typical scanning electron microscope (SEM) images of the fracture surfaces of the PLA-based nanocomposite (PLA/MMT-ODA) and neat PLA without MMT foamed in the temperature range 100–140 °C under different isobaric saturation conditions (14, 21 and 28 MPa) [59]. All foams exhibit nicely the closed-cell structure. We note here that homogeneous cells were formed in the case of nanocomposite foams, while neat PLA foams show a rather non-uniform cell structure having large cell size. The nanocomposite foams show smaller cell size (*d*) and larger cell density (N_c) compared to the neat PLA foam, suggesting that the dispersed silicate particles act as nucleating sites for cell formation [9].

For both foam systems, the calculated distribution functions of the cell size from the SEM images are presented in Figure 9.20. The nanocomposite foams nicely obeyed a Gaussian distribution. In the case of PLA/ODA foamed at 150 °C under a high pressure of 24 MPa, we can see that the width of the distribution peaks, which indicates the dispersity for cell size, became narrow, accompanied by a finer dispersion of silicate particles.

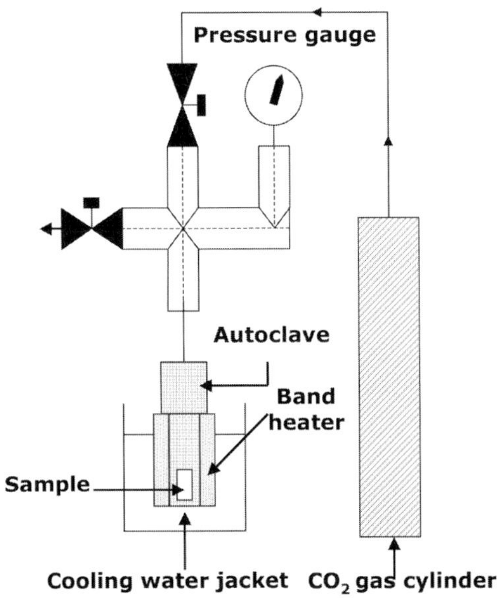

Fig. 9.18 Schematic representation of autoclave set-up. Reprinted with permission from [10], © 2002, Society of Plastic Engineers.

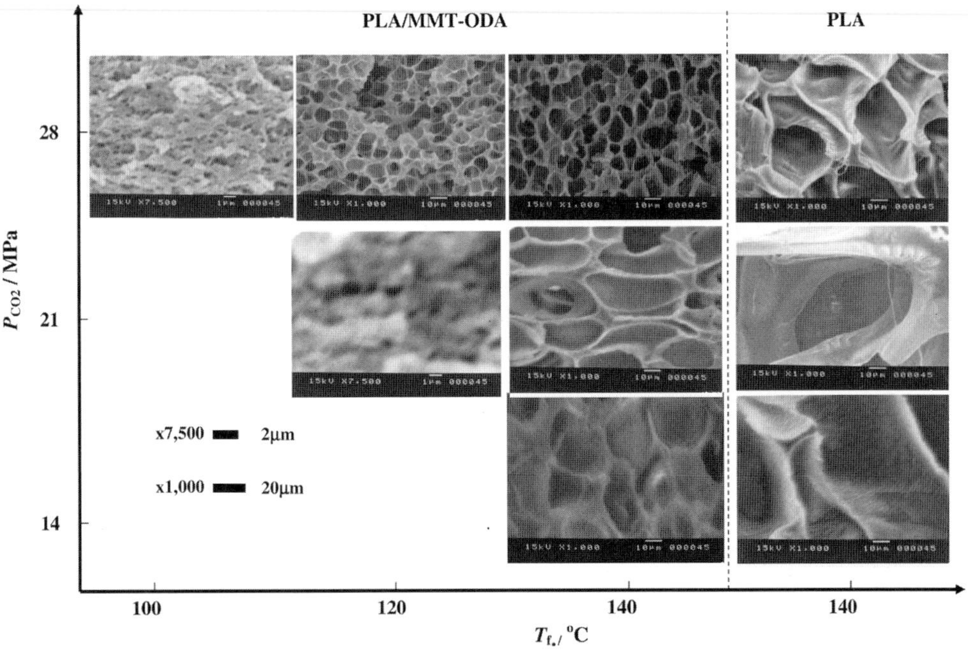

Fig. 9.19 Typical results of SEM images of the fracture surfaces of PLA/MMT-ODA and neat PLA foamed in the temperature range 100–140 °C under different isobaric saturation conditions (14, 21 and 28 MPa). Reprinted from [59], © 2006, Elsevier Science.

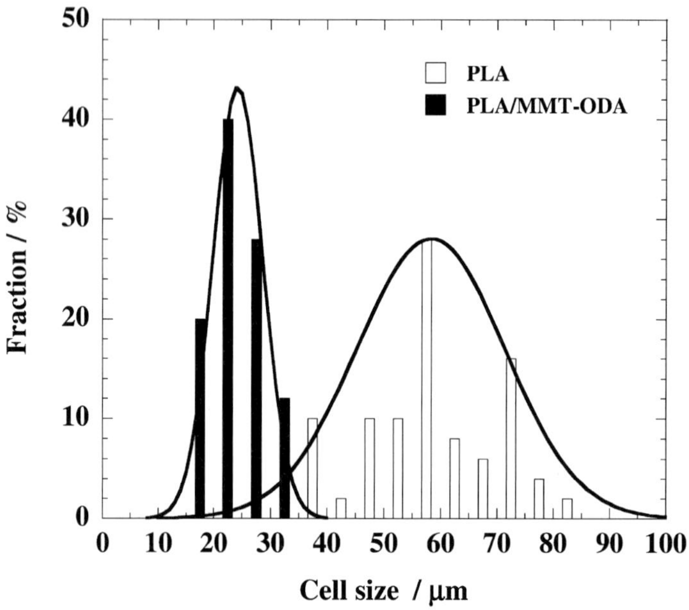

Fig. 9.20 Typical example of cell size distribution of foamed PLA/MMT-ODA and neat PLA in experiment at 150 °C under 24 MPa. Average values of *d* in μm and variances σ_d^2 in μm^2 in the Gaussian fit through the data are 24.2 and 19.1 for PLA/MMT-ODA foam and 58.3 and 171.0 for PLA foam. Reprinted from [59], © 2006, Elsevier Science.

Obviously, with decreasing saturation pressure (~140 °C and 14 MPa), both foams exhibit large cell size due to the low supply of CO_2 molecules, which can subsequently form a small population of cell nuclei upon depressurization. The incorporation of MMT induces heterogeneous nucleation because of a lower activation energy barrier compared with homogeneous nucleation [60]. However, the competition between homogeneous and heterogeneous nucleation is no longer discernible.

9.5.2 Temperature Dependence of Cellular Structure

The dependence of the foam density (ρ_f) at the T_f under different CO_2 pressures is shown in Figure 9.21. Throughout the whole CO_2 pressure range, the mass density of PLA/MMT-ODA foams remains constant at low foaming temperature (T_f) range, abruptly decreases beyond a certain T_f, and then attains a minimum constant value up to 150 °C. It can be said that such behavior of mass density is due to the competition between cell nucleation and cell growth. In the low T_f range (~110 °C), where a large supply of CO_2 molecules is provided, cell nucleation is dominant, while at high T_f, (~140 °C), the cell growth and coalescence of the cell predominate due to the low viscosity of the systems compared with the low T_f range (~110 °C). This

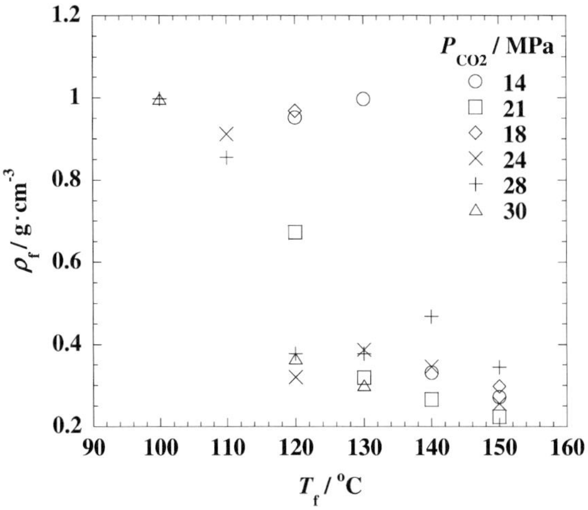

Fig. 9.21 Foaming temperature dependence of mass density for PLA/MMT-ODA foamed under different CO_2 pressures. Reprinted from [59], © 2006, Elsevier Science.

behavior is clearly seen in the plots of the cell size (≅2*d*), the cell density (N_c), and the mean cell wall thickness (δ) versus T_f under various pressure conditions. As seen in Figure 9.22, with increasing T_f all nanocomposite foams show an increasing tendency for 2*d* and/or δ to attain a maximum. On the other hand, the temperature dependence of N_c shows the opposite behavior compared with the tendency of 2*d* due to cell growth and coalescence. Both 2*d* and N_c affect the mass density of the foams.

Using the glass transition temperature (T_g) depressions (corresponding to ΔT_g), plots of ρ_f versus $T_f + \Delta T_g$ were constructed from the data of Figure 9.21. The results are shown in Figure 9.23. All the data, including those for neat PLA and different types of PLA-based nanocomposites (PLA/MMT-SBE), conform well to a reduced curve with $\rho_f \sim 1.0 \pm 0.1\,g/cm^3$ at $T_f + \Delta T_g < 140 \pm 4\,°C$ (nano-cellular region), whereas ρ_f values approach $0.3 \pm 0.15\,g/cm^3$ as the reduced temperature ($T_f + \Delta T_g$) is increased well above 150 °C (microcellular region). The critical temperature is thus 140 ± 4 °C, above which cell growth prevails. Below the critical temperature, cell nucleation dominates and cell growth is suppressed due to the high modulus and viscosity, as revealed by the temperature dependence of the storage, $G'(\omega)$ and loss, $G''(\omega)$ moduli ($G' = 162$ MPa and viscosity component; $G''/\omega \cong 2$ MPa s at 140 °C).

Figure 9.24 shows temperature-reduced plots of 2*d*, N_c and δ versus $T_f + \Delta T_g$. All the data conform well to a reduced curve like Figure 9.23. Interestingly, when the authors used both T_g and the melting temperature (T_m) [61] depressions to conduct superposition, they recognized that the reduced curve is nicely constructed but there is no significant difference compare with the case of $T_f + \Delta T_g$. This indicates that T_g depression is important in optimizing foam processing conditions but T_m depression is not a significant factor for processing because the T_f range is still below T_m after CO_2 saturation.

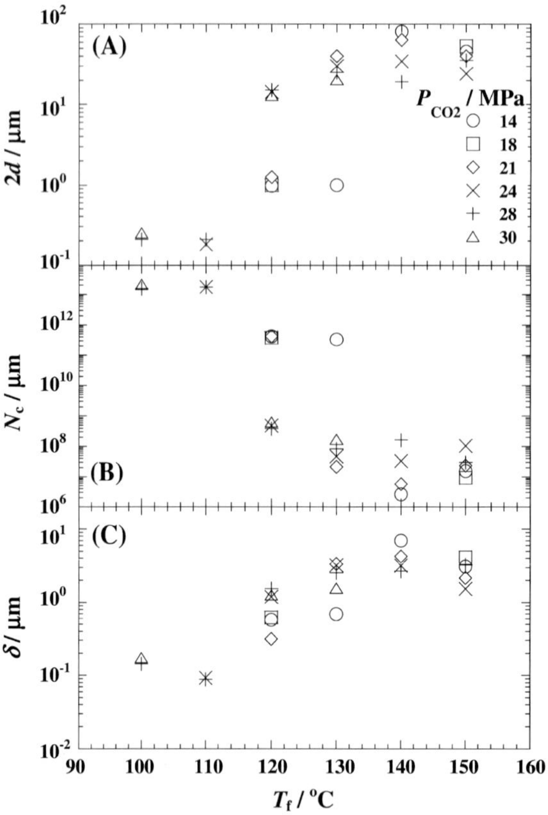

Fig. 9.22 Foaming temperature dependence of (A) cell size, (B) cell density and (C) mean cell wall thickness under different CO_2 pressures. Reprinted from [59], © 2006, Elsevier Science.

In Figure 9.25, the relations between $2d$ and N_c, and δ and $2d$ are shown. The relation obeys well Equations (2) and (3).

$$N_c = 10^4 \frac{3[1-(\rho_f/\rho_p)]}{4\pi d^3} \tag{2}$$

$$\delta = d(1/\sqrt{1-(\rho_f/\rho_p)}-1) \tag{3}$$

where ρ_p and ρ_f are the mass density of pre-foamed and post-foamed samples, respectively.

But the deviation occurs beyond the value of $N_c{\sim}10^{12}$ cell/cm^{-3} for panel (A) and below $2d \sim 1$ μm for panel (B). The downward and upward deviations indicate that the heterogeneous cell distribution mechanism due to the rigid crystalline phases in the PLA matrix is caused by the high degree of crystallinity (~49 wt%) in the low foaming temperature range (~100 °C). As seen in Figure 9.26, the PLANC foams exhibit

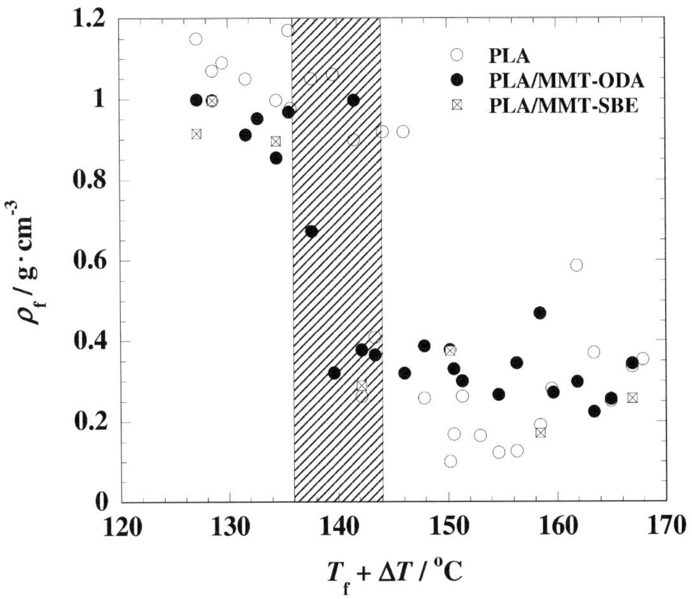

Fig. 9.23 Plot of mass density for PLA/MMT-ODA, PLA/MMT-SBE and neat PLA versus reduced foaming temperature ($T_f+\Delta T_g$). The critical temperature (140 ± 4 °C) is shaded. Reprinted from [59], © 2006, Elsevier Science.

heterogeneous cell distribution. The PLA foam reduces the value of N_c accompanied by a large value of δ compared with that of PLANC foams. In the case of PLANC foams, the controlled structure is from microcellular ($2d \cong 30\,\mu m$ and N_c 3.0×10^7 cells/cm^3) to nano-cellular ($2d \cong 200$ nm and $N_c \cong 2.0 \times 10^{13}$ cells/cm^3).

9.5.3
CO_2 Pressure Dependence

At high pressure, both homogeneous and heterogeneous nucleation mechanisms may be of comparable significance. All systems demonstrate that N_c increases systematically with increasing CO_2 pressure in the low T_f region (~100–120 °C). For PLA/MMT-ODA foams, it appears that heterogeneous nucleation is favored at high pressure. Cell nucleation in a heterogeneous nucleation system such as PLA/MMT-ODA foams takes place in the boundary between the matrix and the dispersed MMT (nano-filler) particles. Accordingly, the cell size decreases without individual cell coalescence for PLA/MMT-ODA and neat PLA systems, as seen in Figure 9.26. To investigate whether the addition of internal surfaces of the dispersed nano-clay may hinder CO_2 diffusion by creating a more tortuous diffusive pathway [21], characterization of the interfacial tension between bubble and matrix was carried out using the modified classical nucleation theory [60].

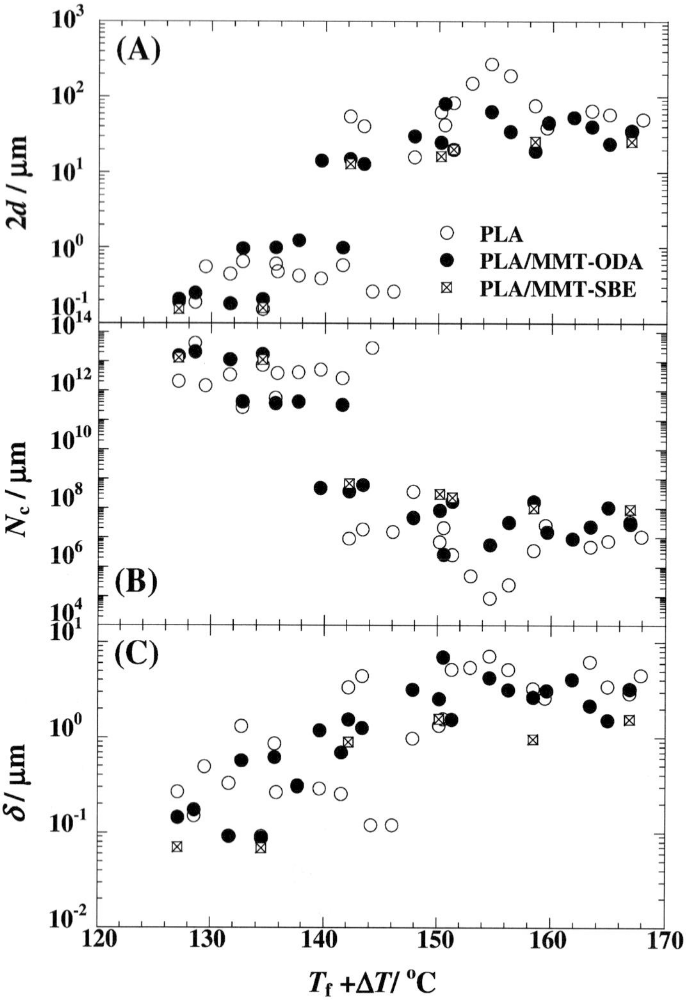

Fig. 9.24 Temperature-reduced plots of (A) $2d$, (B) N_c and (C) δ versus $T_f + \Delta T_g$ for PLA/MMT-ODA, PLA/MMT-SBE and neat PLA. Reprinted from [59], © 2006, Elsevier Science.

According to the theory proposed by Suh and Colton, the rate of nucleation of cells per unit volume ($\dot{N}$) can be written as

$$\dot{N} \sim Cf \exp\left[\frac{-16\pi\gamma^3 S(\theta)}{3(\Delta P_{CO2})^2 k_B T}\right] \tag{4}$$

where C is the concentration of CO_2 and/or the concentration of heterogeneous nucleation sites, f is the collision frequency of CO_2, γ is the interfacial tension between bubble and matrix, $S(\theta)$ is the energy reduction factor for the heterogeneous

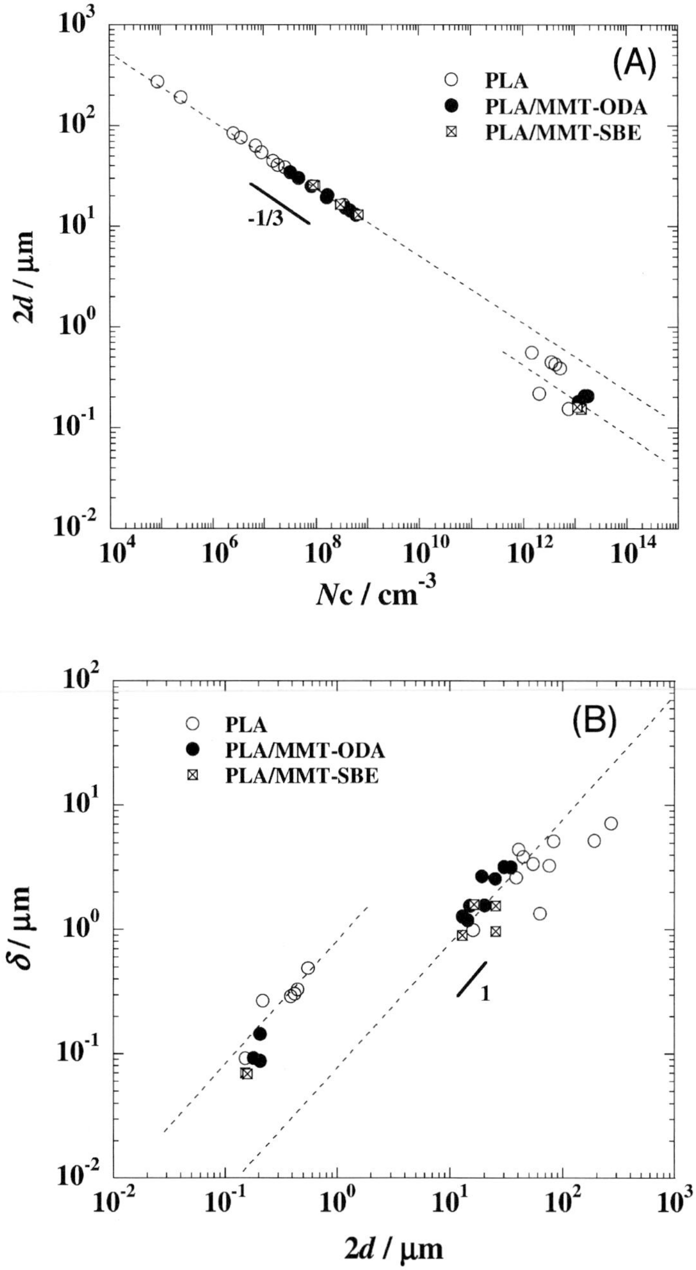

Fig. 9.25 Relation between (A) Cell size and cell density and (B) cell wall thickness and cell size for all foams. Reprinted from [59], © 2006, Elsevier Science.

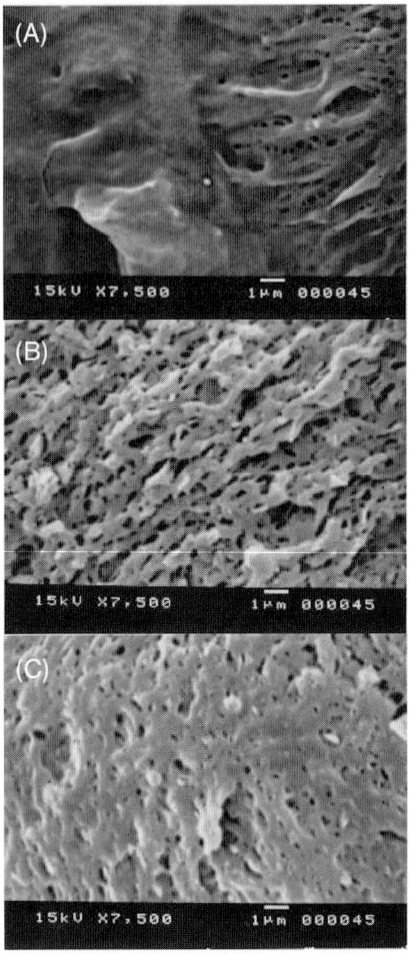

Fig. 9.26 SEM images of the fracture surfaces of (A) neat PLA, (B) PLA/MMT-ODA and (C) PLA/MMT-SBE foamed at 100 °C under 28 MPa. Reprinted from [59], © 2006, Elsevier Science.

nucleation (i.e., PLA/MMT-ODA), ΔP_{CO_2} is the magnitude of the pressure quench during depressurization, k_B is the Boltzmann constant, and T is the absolute temperature.

The theoretical cell density is given by

$$N_{\text{theor}} = \int_0^t \dot{N} \mathrm{d}t \tag{5}$$

where t is the foaming time, approximately 3 s. Assuming no effect of the coalescence of the cell on the value of N_c, we estimate the interfacial tension of the systems calculated using Equations (9.4) and (9.5), that is, the slope of the plots (N_c, versus

Tab. 9.7 Characteristic interfacial parameters of two systems.

	T_f/°C	$\gamma S(\theta)^{1/3}$/mJ/m^2	$S(\theta)$	θ/°
PLA/CO_2	110	7.43		
PLA/MMT-ODA/CO_2		6.65	0.717	107.3
PLA/CO_2	120	7.08		
PLA/MMT-ODA/CO_2		5.38	0.439	85.3

Source: Reprinted from [59], © 2006, Elsevier Science.

$1/\Delta P_{CO_2}$). The characteristic parameters of two systems are shown in Table 9.7. The interfacial tension of PLA/MMT-ODA and neat PLA are 6.65 and 7.43 mJ/m^2 at 110 °C, respectively. These estimated γ values are in good agreement with that of the poly(methyl methacrylate) (PMMA)-CO_2 system (~10 mJ/m^2) [62]. The PLA/MMT-ODA system has a low value compared to that of neat PLA. This trend reflects the relative importance of heterogeneous nucleation, which dominates over homogeneous nucleation when the amount of CO_2 available for bubble nucleation is limited because of a lower activation energy barrier, as mentioned before. That is, in heterogeneous nucleation (PLA/MMT-ODA), we have to take the reduction of the critical energy into consideration because of the inclusion of nucleants, which is a function of the PLA–gas–nano-clay contact angle (θ) and the relative curvature (W) of the nucleant surface to the critical radius of the nucleated phase [63]. When $W \geq 10$, the energy reduction factor $S(\theta)$ can be express by

$$S(\theta) = (1/4)(2 + \cos\theta)(1 - \cos\theta)^2 \qquad (6)$$

In the case of homogeneous nucleation $S(\theta)$ is unity ($\theta = 180°$). The obtained values of the contact angle are 107.3° at 110 °C and 85.3° at 120 °C. The estimated reduction factor ($S(\theta) = 0.4$–0.7) was not so small when compared with the other nano-filler (e.g., carbon nanofingers, $S(\theta) = 0.006$) [64]. However, experimentally, nano-clay particles lead to an increase in N_c.

For PLA/MMT-SBE foams prepared under the condition with low T_f (~100–110 °C) and high pressure (~28 MPa), the nanocomposite foams exhibit no significant difference in N_c compared with PLA/MMT-ODA foams. This reasoning is consistent with the large value of W in both systems.

9.5.4
TEM Observation

To confirm the heterogeneous nucleation and the nanocellular feature in the foam processing, TEM observation of the cell wall in the PLA/MMT-ODA foam was conducted.

Figure 9.27 shows the TEM micrograph of the cell wall foamed at 100 °C under 28 MPa. Interestingly, the grown cells with a diameter of ~200 nm are localized along the dispersed nano-clay particles in the cell wall. In other words, the dispersed nano-filler particles act as nucleating sites for cell formation and the cell growth occurs on the surfaces of the MMTs, that is, the cellular structure has an oval-faced morphology rather than a spherical cellular structure for the high T_f condition (~140 °C). In Figure 9.27, in addition to the nano-cellular structure formation, a lammellar pattern beside the nano-clay (MMT) particles is observed. This behavior appears to arise from the formation of the α-phase of the PLA crystal in the presence of nano-clay particles [65]. This is a unique observation of the epitaxial crystallization of PLA grown from MMT (clay) surfaces due to the nucleation effect of the dispersed nano-clays.

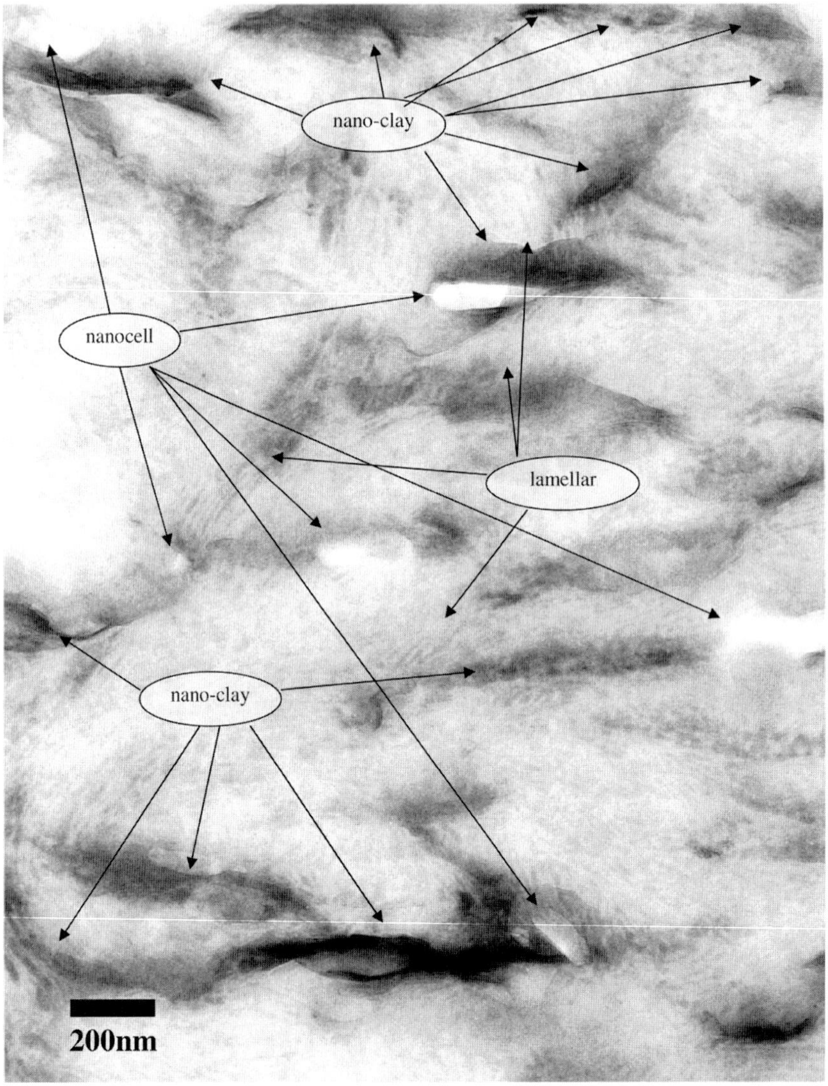

Fig. 9.27 TEM micrograph for the structure of PLA/MMT-ODA cell wall foamed at 100 °C under 28 MPa. Reprinted from [59], © 2006, Elsevier Science.

9.5.5
Mechanical Properties of Nanocomposite Foams

Figure 9.28 shows the relation of the relative modulus (K_f/K_p) to the relative density (ρ_f/ρ_p) of neat PLA and PLA-based nanocomposite foams, taken in the directions parallel (A) and perpendicular (B) to the flow.

To clarify whether the modulus enhancement of the nanocomposite foams was reasonable, Equation (9.7) proposed earlier by Kumar [66] was applied to estimate the relative moduli with various foam densities:

$$\frac{K_f}{K_p} = \left(\frac{\rho_f}{\rho_p}\right)^4 - \left(\frac{\rho_f}{\rho_p}\right)^2 + \left(\frac{\rho_f}{\rho_p}\right) \qquad (7)$$

where K_p and K_f are the modulus of pre-foamed and post-foamed samples, respectively. The solid line in the figure represents the fit to Equation (9.7). The neat PLA foams do not show any difference between the two moduli. On the other hand, for nanocomposite foams, the relative moduli exhibit a large value compared with the theoretical one. The dispersed MMT particles in the cell wall align along the thickness direction of the sample. In other words, the MMT particles arrange due to the biaxial flow of material during foaming. The MMT particles seem to act as a secondary cloth layer to protect the cells from being destroyed by external forces. In the direction perpendicular to the flow, the relative moduli of PLA/MMT-ODA and PLA/MMT-SBE foams appear higher than the predicted values even though at the same relative mass density, in the range 0.7–0.85 (see Figure 9.28(a)). This upward deviation suggests that the small cell size with large cell density enhances the material property, as predicted by Weaire [67]. This may create an improvement in the mechanical properties of polymeric foams through polymeric nanocomposites.

9.6
Porous Ceramic Materials via Nanocomposites

A new route for the preparation of porous ceramic material from thermosetting epoxy/MMT nanocomposite was first demonstrated by Brown *et al.* [68]. This route offers attractive potential for diversification and application of the PLFNCs. Okamoto and coworkers have reported the results on a novel porous ceramic material formed via burning of the PLA-based nanocomposite [69]. The nanocomposite containing 3.0 wt% inorganic MMT. The SEM image of the fracture surface of the porous ceramic material prepared from simple burning of the PLANC in a furnace at up to 950 °C is shown in Figure 9.29. After complete burning the nanocomposite becomes a white mass with a porous structure. The bright lines in the SEM image correspond to the edge of the stacked silicate layers. In the porous ceramic material, the silicate layers form a 'house of cards' structure, which consists of large plates of length ~1000 nm and thickness ~30–60 nm. This implies that the further stacked platelet structure is formed during burning. The material exhibits an open-cell type structure having a 100–1000 nm diameter void, a BET surface area of 31 $m^2 g^{-1}$ and a low

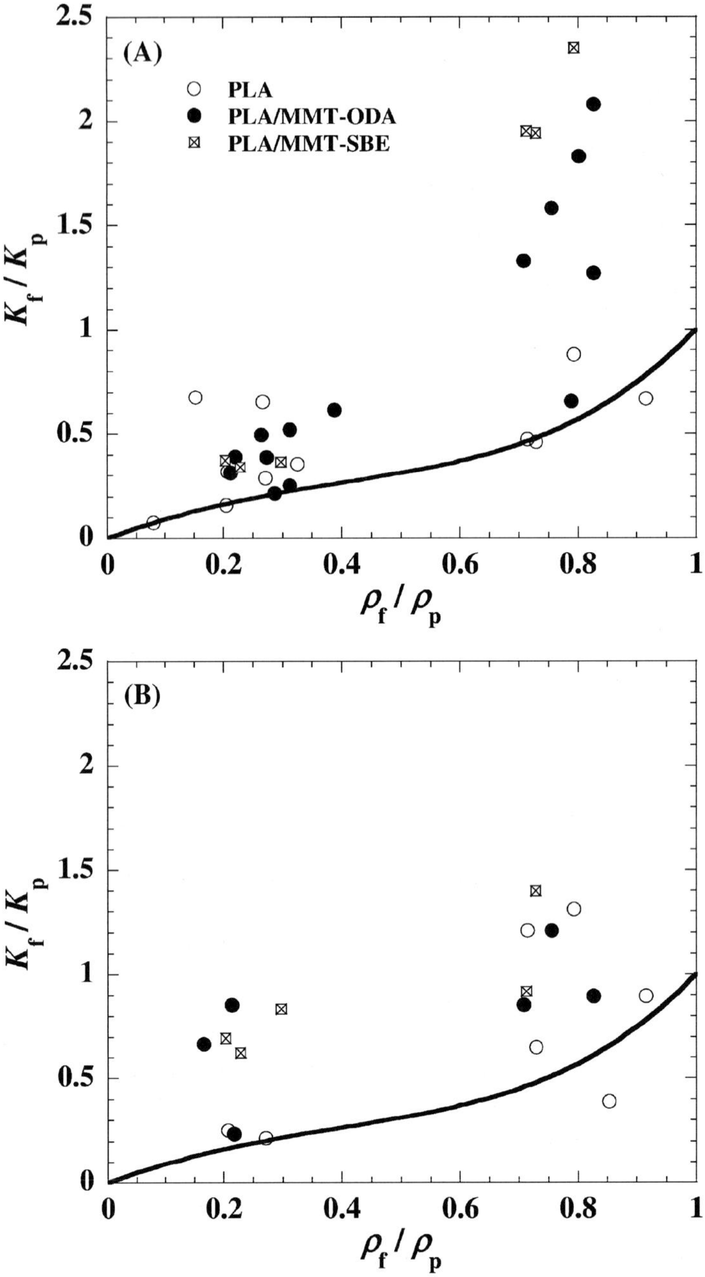

Fig. 9.28 The relation of the relative modulus (K_f/K_p) and the relative density (ρ_f/ρ_p) of neat PLA and PLA-based nanocomposite foams, taken in the directions parallel (A) and perpendicular (B) to the flow.

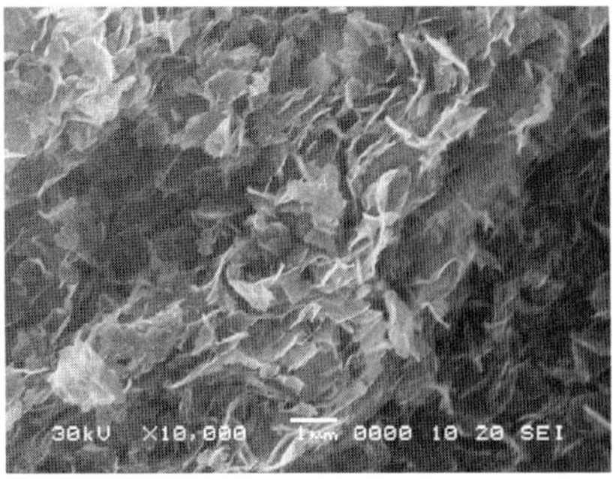

Fig. 9.29 SEM image of porous ceramic material after coating with a platinum layer (~10 nm thickness). Reprinted from [69], © 2002, American Chemical Society.

density of porous material of 0.187 g ml^{-1}, estimated by the buoyancy method. The BET surface area value of MMT is 780 m^2/g and that of the porous ceramic material is 31 m^2/g, suggesting that there are about 25 MMT plates stacked together. When MMT is heated above 700 °C (but below 960 °C) first all the OH groups are eliminated from the structure and thus MMT is decomposed into a non-hydrated aluminosilicate. This transformation radically disturbs the crystalline network of the MMT, and the resulting diffraction pattern is indeed often typical of an amorphous (or non-crystalline) phase. The estimated value of the compression modulus (K) is of the order of ~1.2 MPa, which is five orders of magnitude lower than the bulk modulus of MMT (~10^2 GPa) [3]. In the stress–strain curve, the linear deformation behavior is nicely described in the early stage of the deformation, that is, the deformation of the material closely resembles that of ordinary polymeric foams [70]. This open-cell type porous ceramic material consisting of the 'house of cards' structure is expected to provide strain recovery and an excellent energy dissipation mechanism after unloading in the elastic region up to 8 % strain, probably each plate bends like a leaf spring. This porous ceramic material is a new material possessing elasticity and is very lightweight. This new route for the preparation of porous ceramic material via burning of nanocomposites can be expected to pave the way for a much broader range of applications of the PLSNCs. The porous ceramic material closely relates to an excellent insulator property for flame reterdant of PLFNCs [3]. The flame behavior must derive from the morphological control of the shielding properties of the graphitic clay created during polymer ablation.

9.7 Future Prospects

Development of the PLFNCs is one of the latest evolutionary steps in polymer technology. The PLFNCs offer attractive potential for diversification and application of conventional polymeric materials. Some of the PLFNCs are already commercially available and applied in industrial products.

Biodegradable polymers based on nanocomposites and foams have great future promise for potential applications as high-performance biodegradable materials. They are entirely new types of materials based on materials from plants and nature. When disposed of in compost, they are safely decomposed into CO_2, water, and humus through the activity of microorganisms. The CO_2 and water will become corn or sugarcane again through plant photosynthesis. Undoubtedly, their unique properties originating from their controlled nanostructure pave the way to a much broader range of applications (some already commercially available through Unitika Ltd., Japan), and open a new dimension for plastics and composites. The major impact is at least a decade away.

References

1 Smith, R. (2005) *Biodegradable Polymer for Industrial Applications*, CRC Press, New York.

2 Johnson, R.M., Mwaikambo, L.Y. and Tucker, N. (2003) "Biopolymer", Rapra Review Report No 159, 148 pp Rapra Technology Ltd., London.

3 Okamoto, M. (2003) "Polymer/Layered Silicate Nanocomposites", Rapra Review Report No 163, 166 pp Rapra Technology Ltd., London.

4 Usuki, A., Kojima, Y., Okada, A., Fukushima, Y., Kurauchi, T. and Kamigaito, O. (1993) *Journal of Materials Research*, **8**, 1174.

5 Sinha Ray, S. and Okamoto, M. (2003) *Progress in Polymer Science*, **28**, 1539.

6 Gao, F. (2004) *Materials Today*, **7**, 50.

7 Okada, A. and Usuki, A. (2006) *Macromolecular Materials and Engineering*, **291**, 1449.

8 Okamoto, M. (2006) Recent advances in polymer/layered silicate nanocomposites: an overview from science to technology. *Materials Science and Technology*, **22** (7), 756–779.

9 Okamoto, M., Nam, P.H., Maiti, M., Kotaka, T., Nakayama, T., Takada, M., Ohshima, M., Usuki, A., Hasegawa, N. and Okamoto, H. (2001) *Nano Letters*, **1**, 503.

10 Nam, P.H., Okamoto, M., Maiti, P., Kotaka, T., Nakayama, T., Takada, M., Ohshima, M., Hasegawa, N. and Usuki, A. (2002) *Polymer Engineering and Science*, **42** (9), 1907.

11 Vaia, R.A., Ishii, H. and Giannelis, E.P. (1993) *Chemistry of Materials*, **5**, 1694.

12 Vaia, R.A. and Giannelis, E.P. (1997) *Macromolecules*, **30**, 7990.

13 Vaia, R.A. and Giannelis, E.P. (1997) *Macromolecules*, **30**, 8000.

14 Hiroi, R., Sinha Ray, S., Okamoto, M. and Shiroi, T. (2004) *Macromolecular Rapid Communications*, **25**, 1359.

15 Lagaly, G. (1970) *Clay Minerals*, **16**, 1.

16 Nakano, S., Sasaki, T., Takemura, K. and Watanabe, M. (1998) *Chemistry of Materials*, **10**, 2044.

17 Teyama, H., Nishimura, S., Tsunematsu, K., Jinnai, K., Adachi, Y. and Kimura, M. (1992) *Clays and Clay Minerals*, **40**, 180.

18 Yoshida, O. and Okamoto, M. (2006) *Macromolecular Rapid Communications*, **27**, 751.

19 Sinha Ray, S., Yamada, K., Okamoto, M. and Ueda, K. (2003) *Journal of Nanoscience and Nanotechnology*, **3**, 503.

20 Yoshida, O. and Okamoto, M. (2006) *Journal of Polymer Engineering*, **26**, 919.

21 Sinha Ray, S., Yamada, K., Okamoto, M., Ogami, A. and Ueda, K. (2003) *Chemistry of Materials*, **15**, 1456.

22 Saito, T., Okamoto, M., Hiroi, R., Yamamoto, M. and Shiroi, T. (2006) *Macromolecular Materials and Engineering*, **291**, 1367.

23 Saito, T., Okamoto, M., Hiroi, R., Yamamoto, M. and Shiroi, T. (2006) *Macromolecular Rapid Communications*, **27**, 1472.

24 Totha, R., Coslanicha, A., Ferronea, M., Fermeglia, M., Pricl, S., Miertus, S. and Chiellini, E. (2004) *Polymer*, **45**, 8075.

25 Sinsawat, A., Anderson, K.L., Vaia, R. A. and Farmer, B.L. (2003) *Journal of Polymer Science Part B: Polymer Physics*, **41**, 3272.

26 Kuppa, V., Menakanit, S., Krishnamoorti, R. and Manias, E. (2003) *Journal of Polymer Science Part B: Polymer Physics*, **41**, 3285.

27 Zeng, Q.H., Yu, A.B., Lu, G.Q. and Standish, R.K. (2003) *Chemistry of Materials*, **15**, 4732.

28 Sheng, N., Boyce, M.C., Parks, D.M., Rutledge, G.C., Abes, J.I., and Cohen, R.E. (2004) *Polymer*, **45**, 487.

29 Chen, J.S., Poliks, M.D., Ober, C.K., Zhang, Y., Wiesner, U. and Giannelis, E.P. (2002) *Polymer*, **43**, 4895.

30 Sinha Ray, S., Okamoto, K. and Okamoto, M. (2003) *Macromolecules*, **36**, 2355.

31 Okamoto, K., Sinha Ray, S. and Okamoto, M. (2003) *Journal of Polymer Science Part B: Polymer Physics*, **41**, 3160.

32 Utracki, L.A. (1990) *Polymer Alloys and Blends: Thermodynamics and Rheology*, Hasser Publishers, New York.

33 Krishnamoorti, R. and Giannelis, E.P. (1997) *Macromolecules*, **30**, 4097.

34 Fornes, T.D., Yoon, P.J., Keskkula, H. and Paul, D.R. (2001) *Polymer*, **42**, 9929.

35 Hoffman, B., Dietrich, C., Thomann, R., Friedrich, C. and Mulhaupt, R. (2000) *Macromolecular Rapid Communications*, **21**, 57.

36 Ren, J., Silva, A.S. and Krishnamoorti, R. (2000) *Macromolecules*, **33**, 3739.

37 Mitchell, C.A. and Krishnamoorti, R. (2002) *Journal of Polymer Science Part B: Polymer Physics*, **40**, 1434.

38 Lepoittevin, B., Devalckenaere, M., Pantoustier, N., Alexandre, M., Kubies, D., Calberg, C., Jerome, R. and Dubois, P. (2002) *Polymer*, **43**, 1111.

39 Sinha Ray, S., Maiti, P., Okamoto, M., Yamada, K. and Ueda, K. (2002) *Macromolecules*, **35**, 3104.

40 Sinha Ray, S., Yamada, K., Okamoto, M. and Ueda, K. (2003) *Polymer*, **44**, 857.

41 Okamoto, K., Sinha Ray, S. and Okamoto, M. (2003) *Journal of Polymer Science Part B: Polymer Physics*, **41B**, 3160.

42 Krishnamoorti, R. and Yurekli, K. (2001) *Current Opinion in Colloid and Interface Science*, **6**, 464.

43 Okamoto, M., Taguchi, H., Sato, H., Kotaka, T. and Tatayama, H. (2000) *Langmuir*, **16**, 4055.

44 Williams, M.L., Landel, R.F. and Ferry, J.D. (1955) *Journal of the American Chemical Society*, **77**, 3701.

45 Nam, P.H. (2001) Master Thesis, Toyota Technological Institute.

46 Galgali, G., Ramesh, C. and Lele, A. (2001) *Macromolecules*, **34**, 852.

47 Sinha Ray, S. and Okamoto, M. (2003) *Macromolecular Rapid Communications*, **24**, 815.

48 Okamoto, M., Nam, P.H., Maiti, P., Kotaka, T., Hasegawa, N. and Usuki, A. (2001) *Nano Letters*, **1**, 295.

49 Kotaka, T., Kojima, A. and Okamoto, M. (1997) *Rheologica Acta*, **36**, 646.

50 Tetto, J.A., Steeves, D.M., Welsh, E.A. and Powell, B.E. (1999) ANTEC'99, 1628.

51 Lee, S.R., Park, H.M., Lim, H.L., Kang, T., Li, X., Cho, W.J. and Ha, C.S. (2002) *Polymer*, **43**, 2495.

52 Sinha Ray, S., Okamoto, M., Yamada, K. and Ueda, K. (2002) *Nano Letters*, **2**, 1093.

53 Sinha Ray, S., Yamda, K., Ogami, A., Okamoto, M. and Ueda, K. (2002)

Macromolecular Rapid Communications, **23**, 943.

54 Sinha Ray, S., Yamada, K., Okamoto, M. and Ueda, K. (2003) *Macromolecular Materials and Engineering*, **288**, 936.

55 Liu, J.W., Zhao, Q. and Wan, C.X. (2001) *Space Medicine & Medical Engineering*, **14**, 308.

56 Okamoto, K., Sinha Ray, S. and Okamoto, M. (2003) *Journal of Polymer Science Part B: Polymer Physics*, **41**, 3160.

57 Hiroi, R., Sinha Ray, S., Okamoto, M. and Shiroi, T. (2004) *Macromolecular Rapid Communications*, **25**, 1359.

58 Fujishima, A. and Honda, K. (1972) *Nature*, **37**, 238.

59 Ema, Y., Ikeya, M. and Okamoto, M. (2006) *Polymer*, **47**, 5350.

60 Colton, J.S. and Suh, N.P. (1987) *Polymer Engineering and Science*, **27**, 485.

61 Takada, M. (2004) PhD Thesis, Crystallization control and foam processing of semi-crystalline polymers via supercritical CO_2, Kyoto University.

62 Goel, S.K. and Beckman, E.J. (1994) *Polymer Engineering and Science*, **34**, 1137.

63 Fletcher, N.H. (1958) *Journal of Chemical Physics*, **29**, 572.

64 Shen, J., Zeng, C. and Lee, L.J. (2005) *Polymer*, **46**, 5218.

65 Nam, J.Y., Sinha, S.R. and Okamoto, M. (2003) *Macromolecules*, **36**, 7126.

66 Kumar, V. and Weller, J.E. (1991) ANTEC,1401.

67 Weaire, D. and Fu, T.L. (1988) *Journal of Rheology*, **32**, 271.

68 Brown, J.M., Curliss, D.B. and Vaia, R. A. (2000)Proceedings of PMSE Spring Meeting, San Francisco, California, 278.

69 Sinha Ray, S., Okamoto, K., Yamada, K. and Okamoto, M. (2002) *Nano Letters*, **2**, 423.

70 Gibson L.J. and Ashby M.F. (eds) (1988) Cellular Solids, Pergamon Press, New York, p. 8.

10
Biomimetic and Bioinspired Hybrid Membrane Nanomaterials

Mihail Barboiu

10.1
Introduction

Many fundamental biological processes appear to depend on unique properties of molecular recognition or self-assembled domains of the biomolecules. Such behavior is illustrated by the functional complexity of self-organized membrane proteins, which may assist in proton and ion translocation through membranes [1,2]. Gramicidin A [3], KCsA K^+ channel [4], aquaporin water channels [5] are well known non-exclusive examples of functional systems in which protons, ions and water molecules are envisioned to diffuse along a directional pathway according to different relaying and migration mechanisms [1–7]. The 3D structure of such transporting proteins is determined by a delicate balance of weak interactions (hydrogen bonds, salt bridges, hydrophobic and cation–π interactions), directing selective molecular recognition processes. The final functional framework is acting as a *dynamic, adaptative, evolutive* sum of *collective* and *cooperative* sub-devices allowing the *directional translocation* of millions of ions per second with a remarkable ability to distinguish between different ions [1].

Multiple molecular recognition controls or initiates specific biophysical functions and is the essence of biological functions. Examples include: antibody–antigen interactions, biochemical catalytic reactions, the DNA double helix, protein assembly, and metabolic molecular transport of sugars and amino acids [4,8]. An elucidation of the rules and restrictions which govern these intermolecular interactions is important for understanding and manipulating these processes [2]. Molecular recognition and self-assembly processes could constitute the basis for selective membrane transport and separation [9–11].

Hybrid organic–inorganic materials produced by sol–gel processes are the subject of various investigations, offering the opportunity to achieve nanostructured materials, first from robust organogel systems or secondly from self-organised supramolecular silsesquioxane systems [12–19]. These compounds reveal great potentialities both at the level of their chemical composition or organization and for their applications. [12] Shinkai and coworkers made an important advance and provided

Bio-inorganic Hybrid Nanomaterials. Edited by E. Ruiz-Hitzky, K. Ariga, Y. M. Lvov

ISBN: 978-3-527-31718-9

useful insights in this field. Their studies are confined to the use of organogels resulting from multiple self-assembly processes and acting as robust templates during the siloxane polycondensation process on their supramolecular surface [13]. Many groups, including our own, have found new methods for the elaboration of hybrid anisotropic self-organized materials based on silsesquioxanes in which the functional organic and siloxane inorganic networks are covalently connected. Rigid aromatic molecules [14] and urea H-bonding ribbons [15–19], are used to transcribe an oriented supramolecular self-organization in a siloxane matrix by a sol–gel process.

Of special interest is the structure-directed function of biomimetic and bioinspired hybrid materials and the control of their build-up from suitable units by self-organisation. The main interest focuses on functional biomimetic membranes in which the recognition-driven transport properties could be ensured by a well-defined incorporation of receptors of specific *molecular recognition and self-organization* functions, incorporated in a hybrid solid of dense or mesoporous materials. A renewed interest in new membrane systems was generated by intriguing proposals using molecular recognition and self-assembly for the construction of new functional nanomaterials [20–34].

We therefore propose to review in this chapter the membrane transport properties of such supramolecular organic–inorganic hybrid membrane materials. The first part is a survey of different methods and processes which can be used for the generation of molecular recognition-based hybrid materials. Then the basic working principles of self-organized membranes are provided in order to better understand the requirements in material design for the generation of functional nanostructured materials. Finally, actual and potential applications of such self-organized systems presenting the combined features of structural adaptation in a specific hybrid nanospace will be presented.

10.2
Molecular Recognition-based Hybrid Membranes

10.2.1
Multiple Molecular Recognition Principles

The multiple molecular recognition and self-organization observed in liquid media or in bilayer membranes could be transferred to solid dense hybrid membrane materials. One way towards this objective is to develop functional hybrid membrane materials that form selective molecular recognition patterns so as to enable efficient translocation events. As a part of the program concerned with the description of membrane systems effecting the selective transport of biologically relevant species, our group has developed during the last decade new hybrid *fixed-site complexant membrane systems* [20–34]. In this way, grafted polytopic fixed-site complexants could improve the selectivity and the solute flux across the hybrid siloxane matrix due to a more specific complexation, by combining different non-covalent interactions.

Molecular recognition of neutral or charged organic molecules has only recently received attention, in sharp contrast to the far more advanced development of the corresponding coordination chemistry of cations [20–22]. These substrates are bulkier with miscellaneous geometries, and are thus more polarizable and more strongly solvated than cations [23–25]. For these reasons, *artificial receptors should have a specific geometry with a correct geometrical disposition of binding sites, in order to allow a strong entrapment of the organic guest* [25,26]. Much interest in the field of molecular recognition of organic compounds has been focused on the amino acids. In this case, it is generally easier to complex either the hydrophobic protonated ammonium ($H_3N^+{-}R{-}COOH$) or the deprotonated carboxylate ($H_2N{-}R{-}COO^-$) forms, rather than the zwitterionic natural form of the amino acids ($H_3N^+{-}R{-}COO^-$). The selective heterocomplexation of the zwitterionic form via three hydrogen bonds of protonated ammonium moiety or the charge interactions of the carboxylate form (as counter-ion of the supramolecular species) can be completed by lateral specific interactions, allowing an enantiomeric molecular recognition of the amino acid [23]. However, examples of this specific molecular recognition and membrane transport system remain rare.

Polytopic macrocyclic receptors **1**, **2** (Figure 10.1) are able to complex the zwitterionic form of the amino acids by a double non-covalent charge interaction [28,29]. The unsymmetrical benzocrown sulfonamide derivative, **2** which contains benzo-18-crown-6 and benzo-15-crown-5 moieties was used as a ditopic receptor for multiple molecular recognition of the amino acids, by combining two non-covalent interactions: ammonium-crown hydrogen bonding and carboxylate- complexed Na^+-benzo-15-crown-5 charge interactions [28,33].

Multiple molecular recognition of amino acids can be transferred into dense solid materials by anchoring the macrocycle in the hybrid organic–inorganic siloxane matrix [27,29]. By combining the different non-covalent forces (hydrogen bonding, charge and hydrophobic interactions) the fixed macrocyclic receptors may recognize the organic solutes in a solid dense polymeric matrix. Here, the macrocyclic complexant does not play the role of carrier, but selectively improves the partition of the solute in the membrane, yielding a dual mode of transport. This results in a *new*

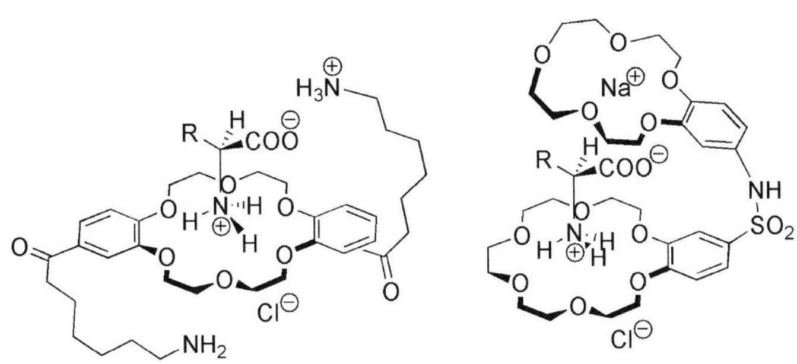

1 **2**

Fig. 10.1 Multiple molecular recognition of the zwitterionic amino acids by macrocyclic polytopic receptors **1**, **2** [28,33].

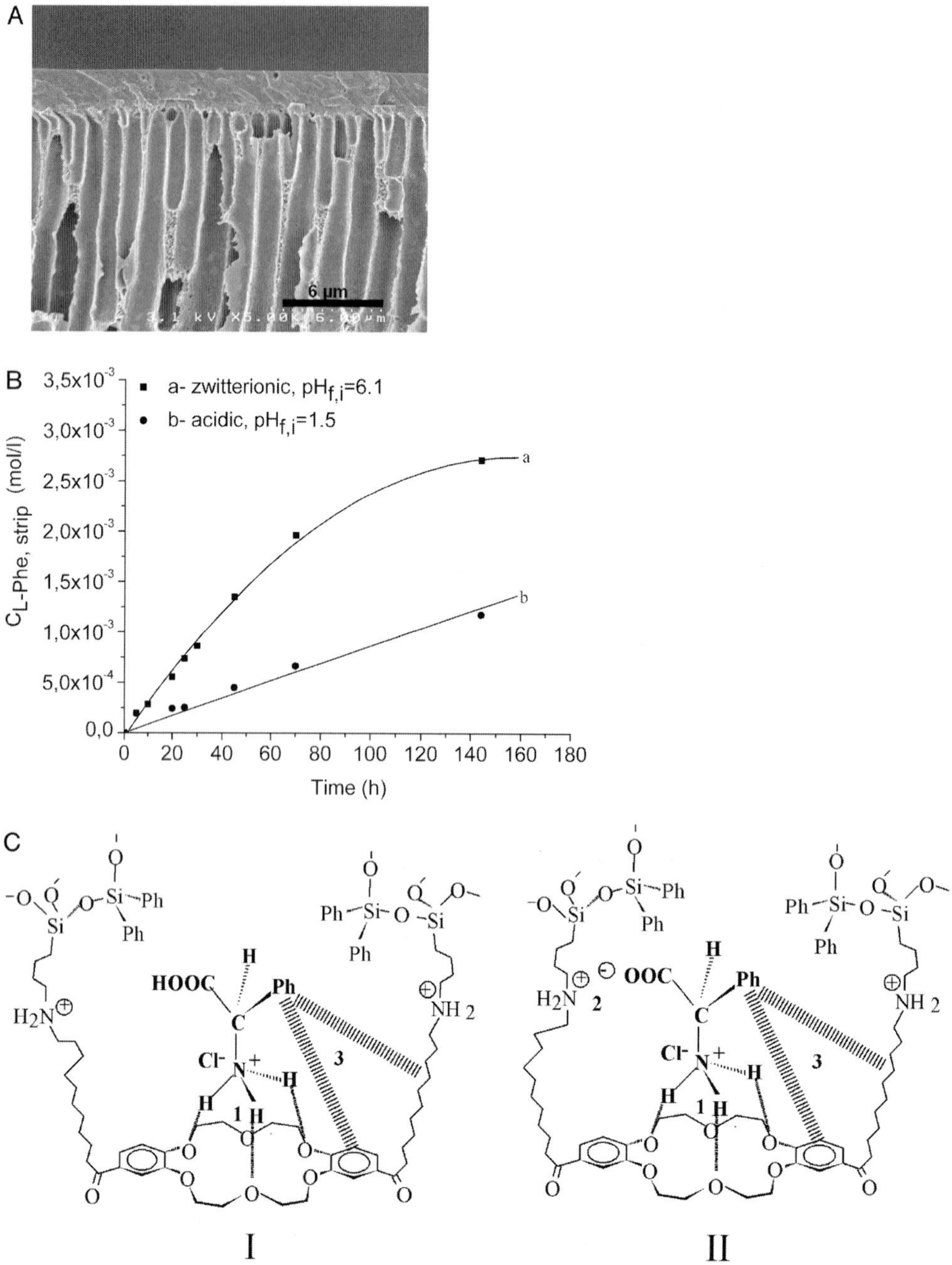

Fig. 10.2 (A) Cross-section SEM micrograph of the hybrid membrane containing the receptor **1**, (B) membrane transport concentration profiles and (C) molecular recognition principles of acidic **I** and zwitterionic **II** L-phenylalanine in the heteropolysiloxane material membrane (1-hydrogen bonding, 2-charge interaction, 3-Van der Waals forces) [29].

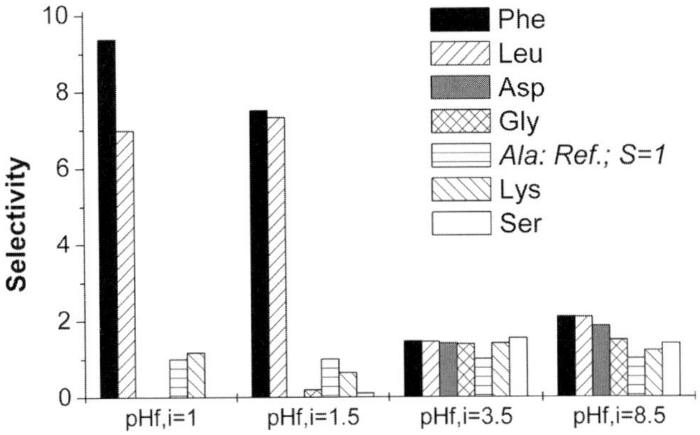

Fig. 10.3 Selectivity in facilitated transport of amino acids as a function of pH value [29,32].

multiple supramolecular recognition principle of the zwitterionic form of amino acids in solid dense material (Figure 10.2) [24–28]. Complexation–decomplexation reactions from macrocyclic sites seem to be kinetically rapid but thermodynamically slow. In fact, the transport of the amino acid across the membrane is due to a kinetic effect, but when the macrocyclic sites are blocked by the amino acid, because the resulting supramolecular complex is too stable, the transport is slowed due to the thermodynamics of the supramolecular complex formation.

Further, facilitated membrane separations by fixed-site heteropolysiloxane membranes for a mixture of seven amino acids provided a good selectivity ($S = 9$–10, calculated as flux ratio) for lypophilic amino acids (L-phenylalanine and L-leucine) (Figure 10.3) [30,31].

A good selectivity was also obtained in the facilitated transport of different organic acids of very similar molecular structure (Figure 10.4) by these solid membranes, due

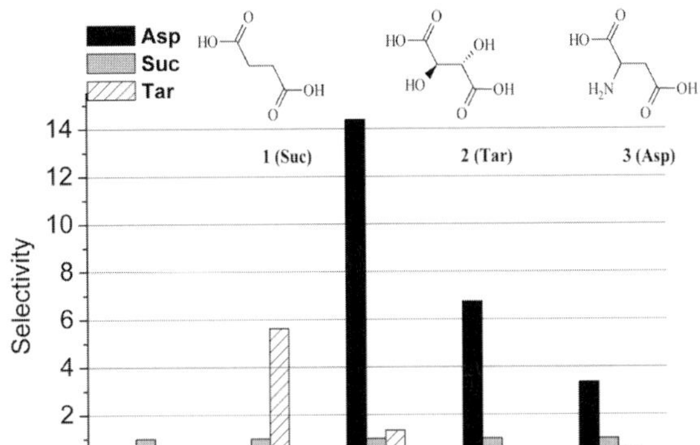

Fig. 10.4 Selectivity in facilitated transport of amino acids as a function of pH value [17].

to the combination of non-covalent interactions (van der Waals and ionic interactions) between the solute and the solid dense functionalized material [31].

10.3
Self-organized Hybrid Membrane Materials

10.3.1
Ionic-conduction Pathways in Hybrid Membrane Materials

Numerous artificial transport systems utilizing carriers, channel-forming or self-organized polymeric superstructures able to orient, to select and to pump the ionic transport across membranes have been developed in the last decades. Crown ethers [11], cyclic peptides [35], oligophenyl barrel-stave structures [36] and oligoester bola-amphiphiles [37] have all been used in this context, mostly in lipid bilayer membranes. Crystalline organic [38] and inorganic materials [39] progressed the field of bio-inspired supramolecular frameworks, generating synthetic water- and ion-channels or reservoirs on the nanometric scale. Convergent multidimensional self-assembly strategies have been used for the synthesis of non-covalent self-organized polymeric [40] or dendrimeric devices [41], designed to mimic natural ion channel proteins on the micrometric scale.

Based on this strategy three heteroditopic classes of receptors have been reported by our group: crown-ether ureido-silsesquioxanes, **3** [16–19] amino acid conjugates **4** [44,45] and nucleobase ureido-silsesquioxanes [46] **A**, **U**, **G** derivatives (Figure 10.5). They generate self-organized continual superstructures in solution and in the solid state based on three encoded features: (i) the molecular recognition, (ii) the supramolecular H-bond directing interactions and (iii) the covalently bonded triethoxysilyl groups. The inorganic precursor moiety allows us, by sol–gel processes, to transcribe the solution self-organized dynamic superstructures in the solid heteropolysiloxane materials. This represents an intermediate approach between

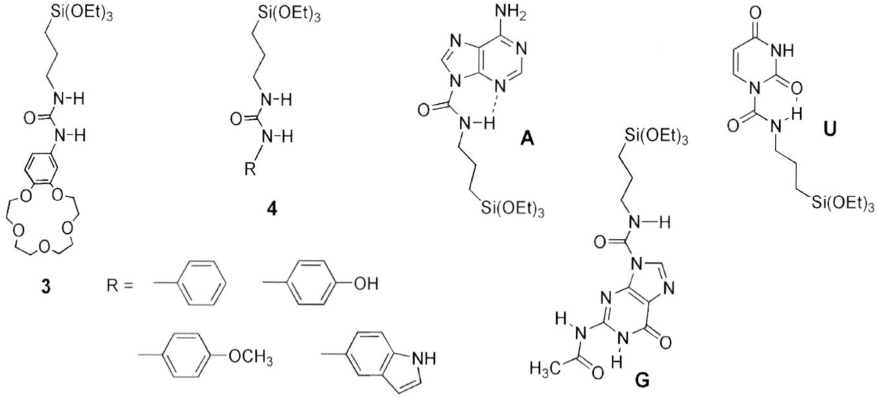

Fig. 10.5 Molecular structures of the molecular receptors **3**, **4**, **A**, **U** and **G**.

the previously reported methods to form self-organized hybrid materials using a TEOS organogel template and appropriate silylated organic molecules.

10.3.1.1 **Ionic-conduction Pathways in Macrocyclic Hybrid Materials**

Of particular interest is the potential ability of hybrid thin-layer membrane films to present polyfunctional properties such as solute molecular recognition and the generation of directional diffusion pathways by self-assembling at the supramolecular level. This would be a rational approach for building molecular channels in hybrid organic–inorganic materials via the inorganic (sol–gel) transcription of dynamic self-assembled superstructures. The basic and specific molecular information encoded in the molecular precursors (crown ether, amino acid and guanine ureido-silsesquioxanes) results in the generation of tubular and continual superstructures in solution and in the solid state which *can be "frozen" in a polymeric hybrid matrix by the sol–gel process*. These systems have been successfully employed to design solid dense membranes, functioning as ion-channels, and illustrate how a self-organized hybrid material performs interesting and potentially useful transporting functions. The urea-based head-to-tail motifs were used for the self-assembly of macrocyclic receptors of type **3**, resulting in the formation of tubular ion-channel-type superstructures (Figure 10.6a). We therefore studied the membrane transport properties in solid materials of such supramolecular and organic–inorganic hybrid membrane materials resulting from the dynamic self-assembly of the hydrogen-bonded urea-crown ethers or organic functional molecules. Our published results [16,17] showed that the self-organization properties in the membrane phase may provide the first evidence for the possible hybrid transport carrier vs. channel mechanisms, in correlation with self-assembly properties of the heteroditopic receptors. *These dynamic self-organized systems can be transferred into a polymeric hybrid matrix by a sol–gel process, allowing us to design a novel class of hybrid nanomembranes* [23]. The hybrid membranes successfully formed transport patterns so as to enable efficient translocation events. Moreover, this system was the first example of a hybrid nanomaterial where the concept of self-organization and a specific function (generation of specific translocation ionic pathways in a hybrid solid) might, in principle, be associated. As an example, a schematic representation of the dynamic self-assembling of crown ethers in solution and the sol–gel transcription in the hybrid membrane is illustrated in Figure 10.6b.

10.3.1.2 **Ionic-conduction Pathways in Peptido-mimetic Hybrid Materials**

Convergent multidimensional self-assembly strategies have been used for the synthesis of non-covalent self-organized devices, designed to mimic natural ion channel proteins. Intermolecular interactions involving aromatic rings are key processes in both chemical and biological recognition. Among these processes, cation–π interactions between positively charged species (alkali, ammonium, etc.) and aromatic systems with delocalized π-electrons are now recognized as important non-covalent binding forces of increasing relevance. The importance of interactions between alkali cations and the aromatic side-chains of aromatic amino acids of

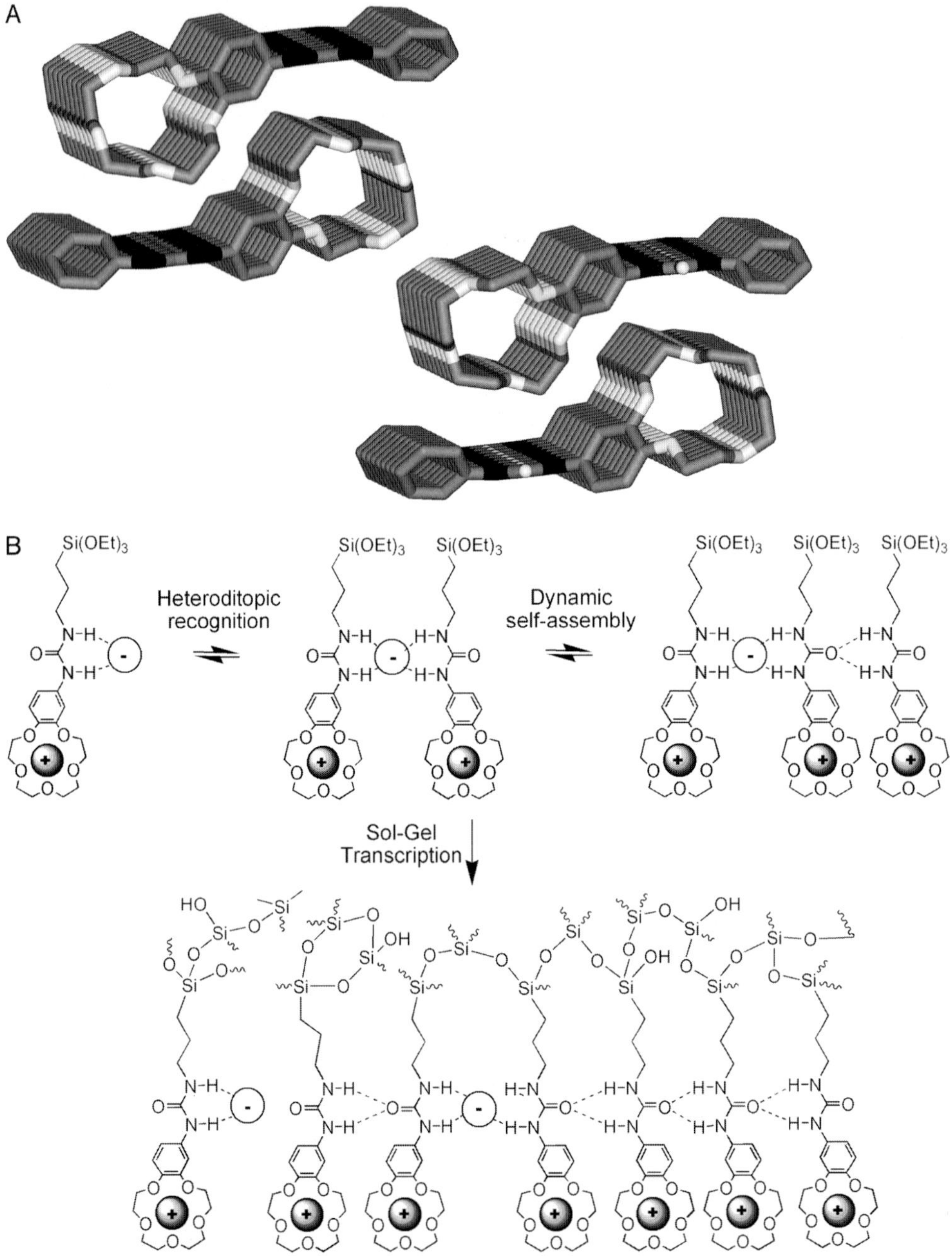

Fig. 10.6 (A) Stick representation of the packing of 3 in the crystal, showing the formation of directional tubular conduction pathways; (B) schematic representation of the hierarchically organized system **3**: (top) self-organization in solution and (bottom) sol–gel transcription of encoded molecular features into a hybrid heteropolysiloxane matrix [18].

proteins has been known for many years and they are of particular biological significance [1,2].

New heterocomplex structures emphasizing particular K^+–π contacts with phenyl, phenol and indole rings have been recently reported by many groups including our own [4,11,45]. These results have now led us to focus on designing functional amino acid ureido-silsesquioxane precursors to design suitable molecular pathways in hybrid membranes. Several novel findings related to this field are related to new hybrid organic–inorganic dense materials in which protons and ions are envisioned to diffuse along the *directional hydrophilic pathways*. For this reason the heteroditopic 3-(ureiodoarene)triethoxypropylsilane ($ArNHCONH(CH_2)_3Si(EtO)_3$) compounds of type **4** were reported. We restricted such specific studies to benzene, phenol and indole, the arene (Ar) termini of phenylalanine, tyrosine and tryptophan, three aromatic essential amino acids. They generate self-organized superstructures in solution, sols, gels and in the solid state, based on three encoded features: (i) they contain a functional aromatic part; the importance of interactions between alkali cations and side chains of aromatic aminoacids has been known for many years and they are of particular biological significance; (ii) the supramolecular guiding interaction is the urea head-to-tail H-bond association; reminiscent of the H-bonding of the amide moiety in proteins: (iii) these *ureido aromatic amino acids* are covalently bonded to triethoxysilyl groups allowing, by means of a sol–gel process, the transcription of their solution self-organized superstructures in a solid heteropolysiloxane material (Figure 10.7).

Successive H-bond urea self-assembly of **4** and sol–gel transcription steps yield preferential conduction pathways within the hybrid membrane materials. Crystallographic, microscopic and transport data confirm the formation of self-organized molecular channels transcribed in solid dense thin-layer membranes. The ionic transport across the organized domains illustrates the power of the supramolecular approach for the design of continual hydrophilic transport devices in hybrid membrane materials by self-organization (Figure 10.8) [42–44].

The hydrophilic pathways in hybrid membranes are generally defined by the presence of the stable −Si−OH groups in the hydrophobic condensed siloxane −Si−O−Si− network. Hydrophilic organic moieties (phenol, methoxyphenyl or indole) of **4** presenting a tight H-bond contact with an inorganic matrix enhance the hydrophobicity of the resulting hybrid material. Taking the hydrophobic phenyl-derived material as reference, where the transport of ions occurs mainly via hydrophilic silica pathways, the transport of protons and the hydrated alkali cations $Na(H_2O)_n^+$ and $K(H_2O)_n^+$ is amplified when phenolic groups are introduced into the membrane. Moreover, in the two cases, single or competitive transport, the smallest $Na(H_2O)_n^+$ ions are selectively transported and the transport rate is related to the diffusion coefficients of the hydrated ions along the hybrid matrix. Amazingly, the transport rates with indole-derived membranes decrease, even compared with the reference membrane and, moreover, the membranes selectively transport K^+ ions. Certainly, a substantial contribution to that selectivity arises from features of exchange between the hydrated ions and the indole moiety: the complexation of ions is thermodynamically more favorable for K^+ ions than Na^+. The NH indole

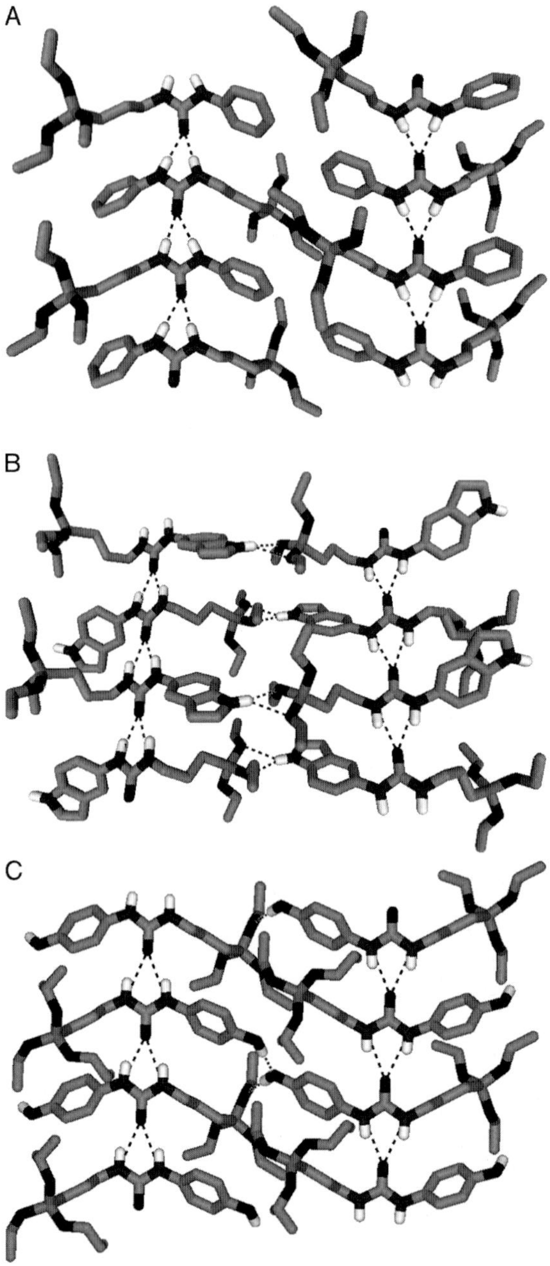

Fig. 10.7 Crystal structure of compounds **4**: stick representation of H-bond (dotted lines) superstructures of (A) phenyl, (B) indole, (C) phenol ureido-silsesquioxanes; H atoms were omitted for clarity.

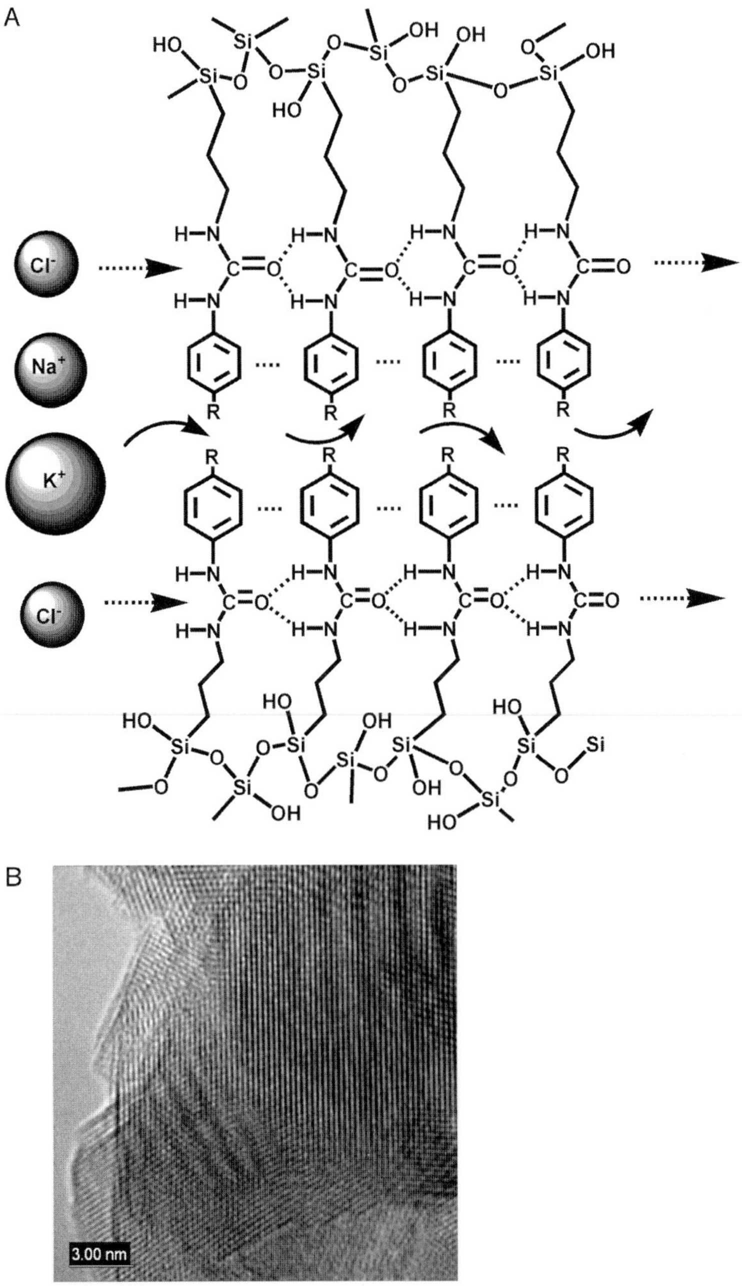

Fig. 10.8 (A) Sol–gel transcription of the self-organized molecular channels in solid dense hybrid membranes; (B) crystalline self-organized domains at the surface of the hybrid membrane containing compounds **4** (TEM image).

moiety is H-bonded with the oxygen of a vicinal siloxane network, thus the indole pathway is hydrophobic. This explains the reduced transport rates and suggests possible cation–π interactions to confer K^+ selectivity. It has been found the pyrolo subunit of the indole (i.e., $C_2{=}C_3$ bond) is a strong π-donor group for K^+ cation [11,44,45]. This suggests that voluminous hydrated $Na(H_2O)_n{}^+$ may be slowed by the "indole pathway" while dehydrated K^+ ions are transported along this pathway more easily.

10.3.2
Self-organization in Hybrid Supramolecular Polymers

Nanosized supramolecular materials have received increasing attention during the last two decades. Their properties have been surveyed for their ability to form aggregates in the solution phase, which form channel-like arrays in the solid state, and which ultimately form single channels in planar bilayer membranes. These systems therefore illustrate, in general, the convergence of supramolecular self-organization and supramolecular function.

Supramolecular synthesis provides a powerful tool for the non-covalent generation of such functional supramolecular architectures [8]. The supramolecular polymers offer solutions for material molding at the macroscopic level but their manipulation at the molecular (supramolecular) and nanoscopic levels is still difficult [47]. They may be divided into two partially overlapping classes: (i) *supramolecular* polymers formed by spontaneous polyassociation of a large number of components, via non-covalent bonding (H-bonding, van der Waals, metal ion coordination, etc.) or reversible covalent bonding, into the large polymolecular architectures; (ii) *supramacromolecular* polymers resulting from specific intermolecular self-organization of molecular components covalently connected to a polymeric network or to polymerizable groups as precursors for polymeric materials (Figure 10.9A) [48].

Considerable challenges still lie ahead with the second class of materials synthesised by classical polymerization reactions or by chemical grafting to a polymeric backbone. One of the more significant challenges is the "*dynamic marriage*" between the non-covalent supramolecular self-assembly and covalent polymerization processes which kinetically and sterochemically might communicate in order to converge to the right supramolecular self-organization. Moreover, during the generation of such superstructures, multiple outputs resulting from the processing of different interaction algorithms might coexist in equilibrium, leading, after polymerization, to combinatorial polymorphic or amorphous solids on the nanometric scale (Figure 10.9B).

The weak supramolecular interactions (H-bonds, coordination or van der Waals interactions, etc.) positioning the molecular components to give the supramolecular architectures are typically several orders of magnitude less robust than the cross-linked covalent bonds formed in a specific polymerization process. Accordingly, the sole solution to overcome these difficulties is to improve the binding (association) efficiency of the molecular components generating supramolecular assemblies. At least in theory, an increased number of interaction moieties and the selection of the

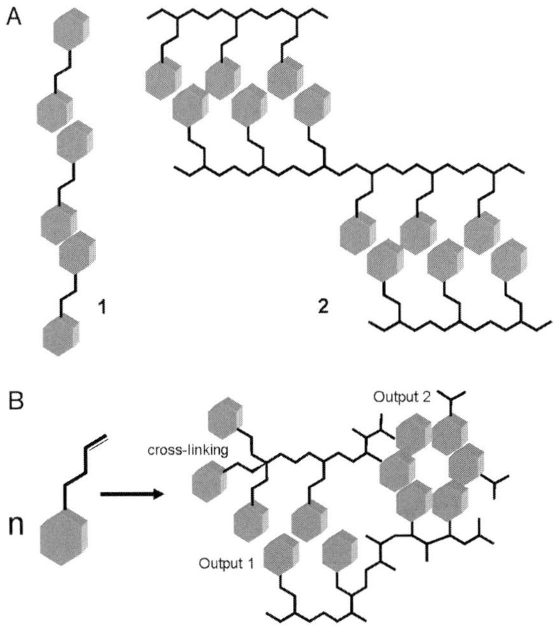

Fig. 10.9 (A) *Supramolecular* **1** and *supramacromolecular* **2** polymers resulted from specific intermolecular and intermacromolecular self-organization; (B) cross-linking and multiple output generation resulting after polymerization during the generation of the *supramacromolecular* polymers.

right solvent might improve the stability of the templating supramolecular architecture, communicating with the inorganic siloxane network. Among these systems the nucleobases [49] and nucleosides [9] as well as DNA [50] are well known fascinating compounds having a high ability to form directionally controlled multiple intermolecular H-bonding of a complementary nature, C–H···O, hydrophobic and stacking interactions.

10.3.2.1 **Self-organization by Base Pairing in Hybrid Supramolecular Polymers**

The remarkable single self-association properties of nucleobases and nucleosides via Watson–Crick and Hoogsteen pairing play a critical role in the stabilization of higher-order RNA hairpin loops, double or triple helix DNA [48] and G-quartets or G-quadruplexes [9]. Therefore during the last decades several studies have reported the preparation of discrete supramolecular assemblies, synthetic polymers and hybrid materials possesing nucleobases or nucleic acids as side groups or chain-ends. Even though the Watson–Crick mode of bonding is prevalent in natural systems other hydrogen-bonding motifs are present in natural and artificial systems: reverse Watson–Crick, Hoogsteen, reverse Hoogsteen, Wobble or reverse Wobble [48].

Amazingly, a very diverse set of interconverting supramolecular and macromolecular entities may be generated by using only two nucleobases. The adenine–uracil A=U interaction, which involves two hydrogen bonds ($K_a \cong 10^2\,M^{-1}/CDCl_3$) is weak and nonspecific compared with the guanine–cytosine G≡C interaction involving three hydrogen bonds ($K_a \cong 10^3$–$10^5\,M^{-1}/CDCl_3$) [49]. Homo- and heteropairing of adenine and uracil compounds appear inadequate for them to function in any predefined recognition scheme; formation of interconverting dimers, trimers and oligomeric assemblies is possible via the combination of Watson–Crick (WC) interactions reverse Watson–Crick (rWC), Hoogsteen (H) and reverse Hoogsteen (rH) pairing (Figure 10.10).

Oligomerization of nucleobases can be advantageous to reinforce the H-bonding supramolecular motifs when *supramacromolecular* polymers are desired. Moreover the different interconverting outputs that may form by oligomerization define a dynamic polyfunctional diversity which may be "extracted selectively" under the intrinsic stability of the system or by interaction with external factors by polymerization in the solid state.

For all these reasons, in this study, the alkoxysilane derivatives of the nucleobases are used as molecular precursors to conceive self-organized hybrid materials on the nanometric scale. Due to the diversity of the libraries of oligomers formed by self-assembly, we have limited our efforts to the self-assembly of the adenine, **A** and uracil **U** alkoxysilane precursors (Figure 10.5). The **A** and **U** molecules were designed as rigid H-bonding modules. For instance, by introducing bulky blocking alkoxysilanepropylcarboxamide groups in the N9 (**A**) and N1 (**U**) positions we can limit the interactions to only the Watson–Crick and the Hoogsteen interactions as preferential H-bonding motifs. Moreover, the electronic energies show the intermolecular H-bonded structures **A** and **U** to be favored over the non-H-bonded ones by 1.1 and 1.4 kcal/mol. By this strategy the alkoxy silanepropyl moiety will be directionally fixed, avoiding specific disorders in the structure of the hybrid material.

The precursor **A** recrystallized in ethanol, presents two well-resolved Bragg diffraction peaks corresponding to two crystallographically distinct phases: $\mathbf{(A)}_{nWC\text{-}H}$ and $\mathbf{(A)}_{nH}$ oligomers. The crystallized precursor **U** presents one diffraction peak corresponding to the Watson–Crick $\mathbf{(U)}_{2WC}$ dimer. The small-angle XRPD pattern of the hybrid material $\mathbf{M}_{A\text{-}U}$ presents a unique Bragg diffraction peak corresponding to a characteristic Watson–Crick $\mathbf{A}_{2WC}\mathbf{U}_{2H}$ oligomer (Figure 10.11). Another possible structure, the oligomer $\mathbf{A}_{2H}\mathbf{U}_{2WC}$, does not correlate with the experimentally determined distances. Amazingly, the structures correlating with the experimental results are consistent with the formation of the Hoogsteen motif between U and A and the Watson–Crick motif between two adenine molecules. Early contributions by Etter [51], Diederich [52] and others, have been recently confirmed by Zimmermann's [53] calculations giving nearly exclusive preference for Hoogsteen binding within 1 : 1 basepairing complexes between alkyladenine and alkylthimine derivatives. Although many have recognized the preference for Hoogsteen binding in the solid state for 1 : 1 complexes with adenine these results have confirmed that the sol–gel technique can be considered as a powerful technique for reading biomolecular information.

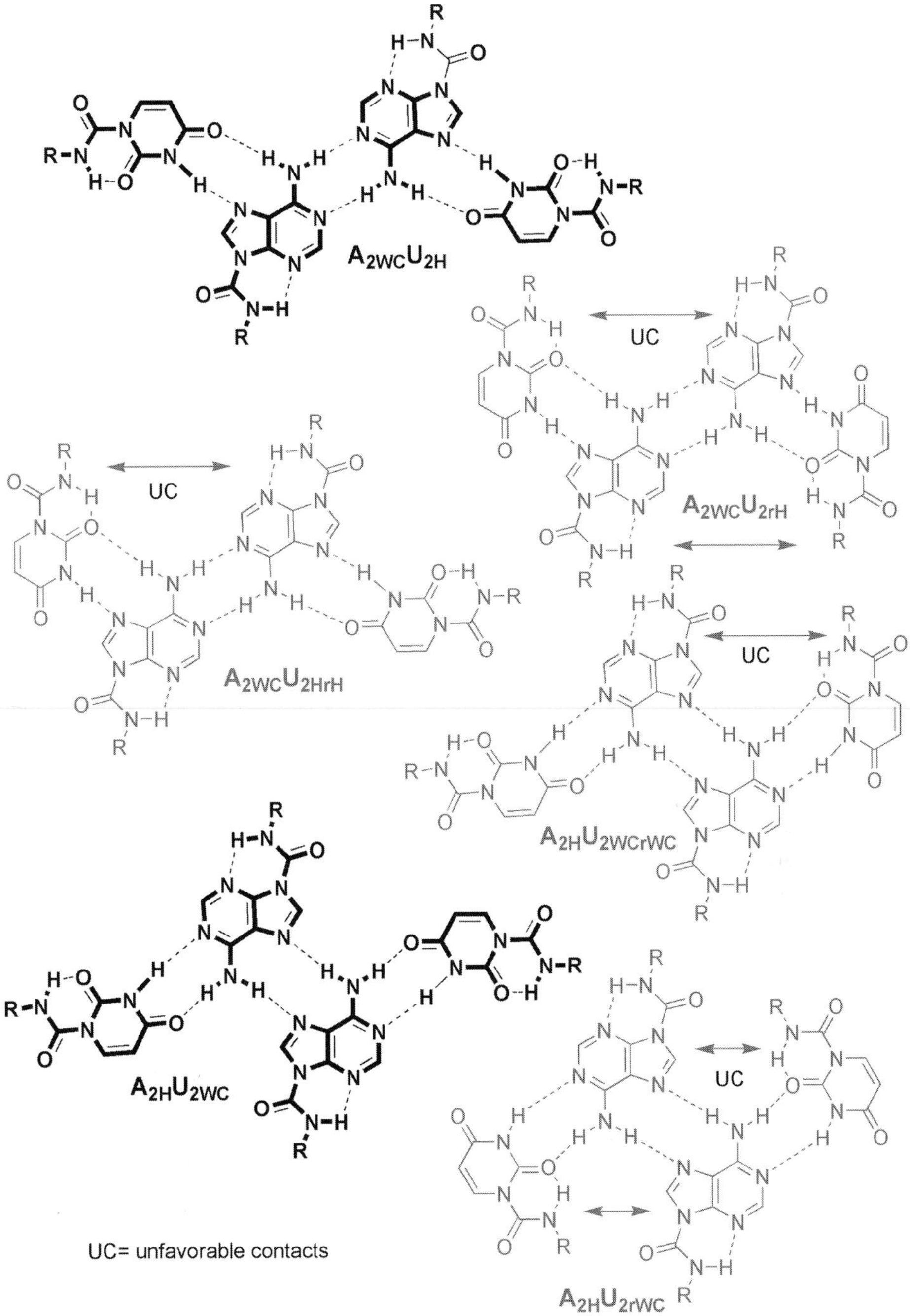

Fig. 10.10 Diverse sets of interconverting supramolecular and macromolecular polymers which may be generated by using only two adenine and uracil nucleobases (the gray formulas are stereochemically unstable).

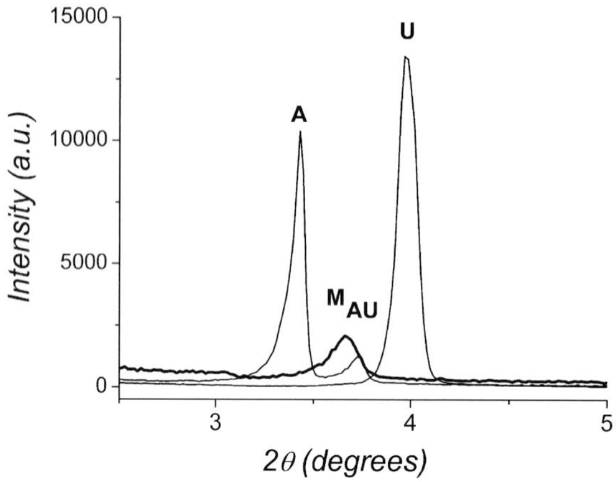

Fig. 10.11 The small-angle XRPD patterns of the precursors **A**, **U** and their hybrid material $\mathbf{M}_{A\text{-}U}$.

10.3.2.2 Self-Organization of the Guanine Quadruplex in Hybrid Supramolecular Polymers

The G-quartet, the hydrogen-bonded macrocycle formed by cation-templated assembly of guanosine, and the G-quadruplex, the columnar device formed by the vertical stacking of four G-quartets, represent examples of ion-channel supramolecular systems of functional complexity. The G-quadruplex represents an example of a chiral supramolecular architecture that is structurally unstable and dynamic in solution when guanine and guanosine molecules are used. It plays a very important role in biology, in particular in nucleic acid telomers of potential interest for cancer therapy [9]. The G-quadruplex represents an example of a chiral twisted supramolecular architecture that is structurally unstable when guanine and guanosine molecules are used. Moreover, the G-quadruplex represents a nice example of a very dynamic supramolecular system of multicomponent architectures in dynamic equilibria in solution or in lipid bilayer membranes. Only very recently, a new strategy based on dynamic constitutional chemistry was successfully used by Davis *et al.* [54] to generate a rich array of interconverting ion-channel conductance states and to make the functional dynamic G-quadruplex along a 40 Å thickness of the lipid bilayers.

The guaninesiloxane monomer **G** self-assembles in a G-quartet and then G-quadruplex supramolecular hybrid architectures, which can be transcribed into an inorganic silica material (Figure 10.12). The main strategy consists in generating (amplifying) dynamic G-quartets and G-quadruplexes from a pool of supramolecular dynamic architectures (Figure 10.12a) and subsequently fixing these by the sol–gel process, leading to soft chiral materials (Figure 10.12b). We reported a long-range amplification of the G-quadruplex supramolecular chirality into hybrid organic–inorganic twisted nanorods followed by transcription into inorganic silica microsprings [47]. Amazingly, these materials are, on the nanometric or micrometric scale,

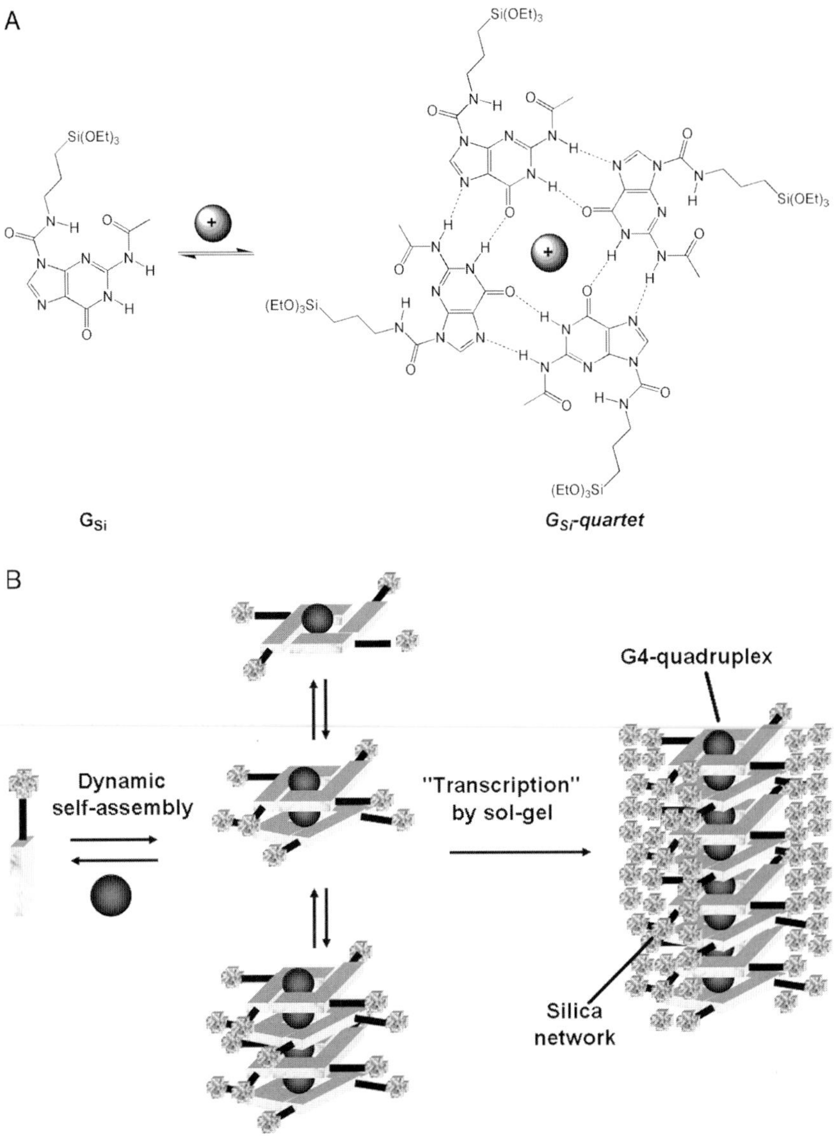

Fig. 10.12 (A) The cation-templated hierarchic self-assembly of guanine alkoxysilane gives the G-quartet and (B) the G-quadruplex transcribed in solid hybrid materials by the sol–gel process.

topologically analogous to their G-quadruplex supramolecular counterpart. The assembly behavior of G-quadruplex and the transcription of structural information into polymeric, hybrid and inorganic materials have been amply described before. However, the multicomponent chiral self-assembly of the G-quadruplex is difficult to preserve along long distances, when guanine and guanosine molecules are used. These findings show a new way to transcribe the supramolecular chirality of a

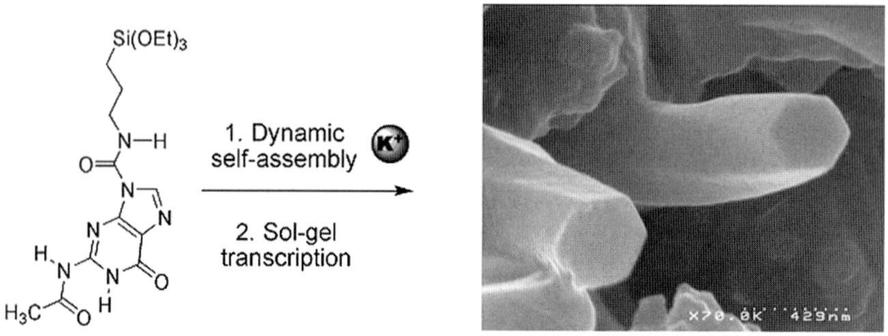

Fig. 10.13 The guanine-silylated conjugate **5** self-associates in the presence of alkali-metal cations to form hexagonal nanorod hybrids, presumably formed by G-quartet-based interactions.

dynamic supramolecular architecture; the transfer of the supramolecular chirality of the G-quadruplex on the nanometric and micrometric scale is reported, thereby creating nanosized hybrid or microsized inorganic superstructures, respectively. Moreover, we obtain *chiral materials* by using a starting *achiral* guaninesiloxane **G** as precursor of the *achiral* G-quartet and of the *chiral supramolecular* G-quadruplex.

Figure 10.13 shows the first picture of the *dynamic* G-quadruplex transcribed at the nanometric level; it unlocks the door to a new materials world paralleling that of biology.

10.4
Dynamic Site Complexant Membranes

Our interest has focused also on hybrid solid membranes in which the molecular recognition-driven transport function could be ensured by a dynamic incorporation of specific organic receptors, non-covalently linked in a hydrophobic dense siloxane inorganic matrix. Of particular interest is the potential ability of such solid membranes to combine functional properties such as solute molecular recognition and generation by self-assembly of the directional conduction pathways at the supramolecular level. New dynamic self-organized hybrid membranes have been prepared (Figure 10.14) by embedding self-organized lipophilic ureidocrown-ethers 15-crown-5, **5** or 18-crown-6, **6** [19] into silica mesoporous hybrid materials, regularly oriented along the pores of the alumina Anodisc 47 (0.02 μm) membranes (AAMs) as support. In a first step, the selective recognition functions of alkali metal ions and self-organization inside regular nanochannels of about 40 Å (Figure 10.14) have been revealed by using NMR, FTIR and X-ray diffraction techniques. The MCM41-type mesostructured powders and membranes were used as hydrophobic host matrices for physically or chemically entrapped 15-crown-5 and 18-crown-6 self-organized receptors (Figure 10.15).

In this way, based on hydrophobic and specific hydrogen bonds such as urea–urea or urea–anion interactions, molecular carriers can be non-covalently trapped in an

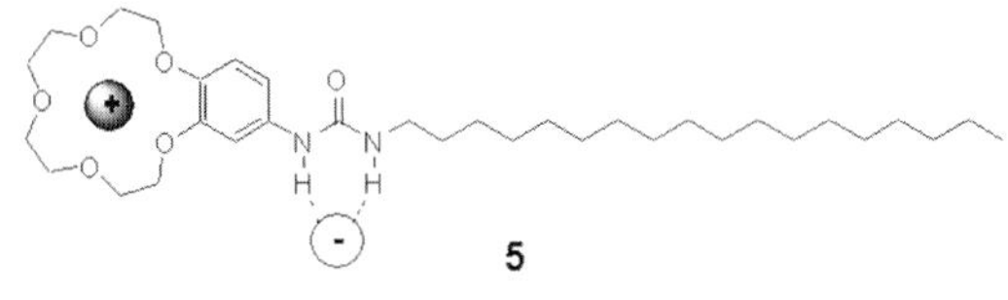

5

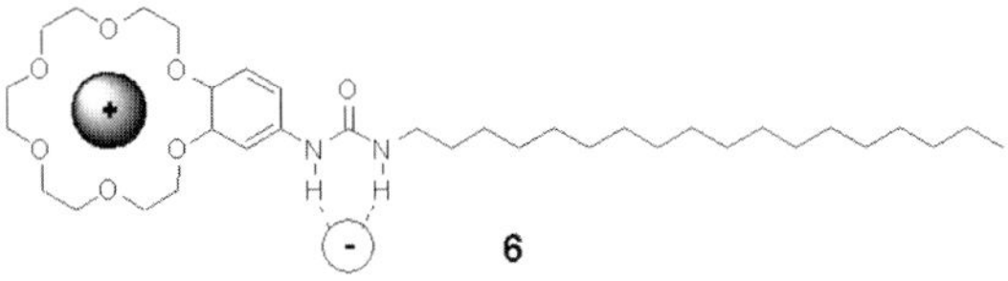

6

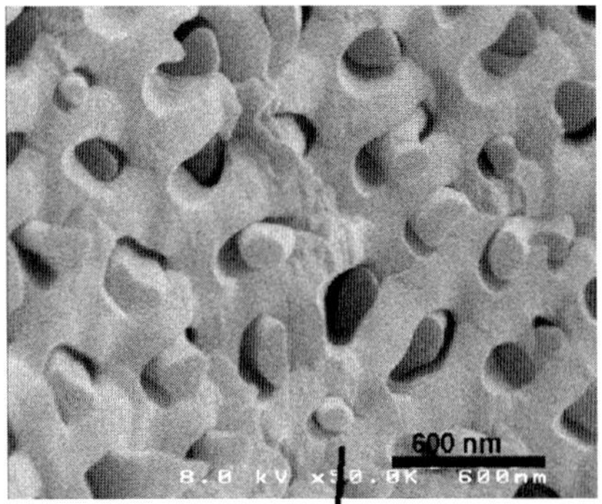

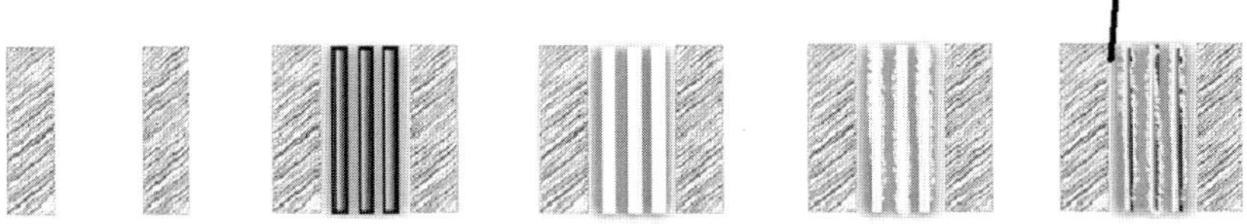

A B C D E

Fig. 10.14 Schematic representation of the synthetic route to dynamic functionalized mesostructured silica membranes in the AAMs: (A) anodic alumina membrane (pore diameter ~200 nm, thickness = ~60 μm, diameter of membrane = 47 mm), mesostructured silica–surfactant before (B) and after (C) calcination, ODS-hydrophobized silica before (D) and after (E) inclusion of the hydrophobic carriers **5** or **6**. SEM micrograph of alumina membrane.

inorganic matrix, which allows us to prepare very promising dynamic-site complexant molecular channel membranes [19,42,43]. Subsequently, hybrid organic–inorganic membranes have been prepared by filling a porous alumina membrane coupled with the sol–gel process. The MCM41-type functionalized materials were successfully oriented along the alumina membrane pores and characterized by SEM microscopy. These membranes have been tested in selective Na^+/K^+ transport. Periodic mesoporous materials have attracted considerable attention during the last decade because of their promising applications as membranes or as hosts for functional membrane materials [55,56]. Many of these applications benefit from arrangements of preferentially aligned, ordered arrays of certain mesostructures. The evaporation-induced self-assembly method has been established as an efficient process for the preparation of thin films with mono-oriented materials. However, the most frequently obtained films display hexagonally ordered channels that are aligned in a nonfavorable orientation, parallel to the surface of the substrate. Recently, the synthesis of mesoporous materials within the regular 200 nm channels of Anodisc alumina membranes (AAMs) has been explored, with the aim of attaining greater control over the morphology (orientation) of the mesoporous system [57]. It was then demonstrated that porous anodic alumina can serve as a support material to form silica–surfactant nanocomposites with a desirable orientation of nanochannels, perpendicular to the surface of the support and, consequently, parallel to (along) the alumina pores.

We also applied this method for the preparation of our membranes, in order to allow preferential transport nano-paths for molecules. In the first step, the AAMs

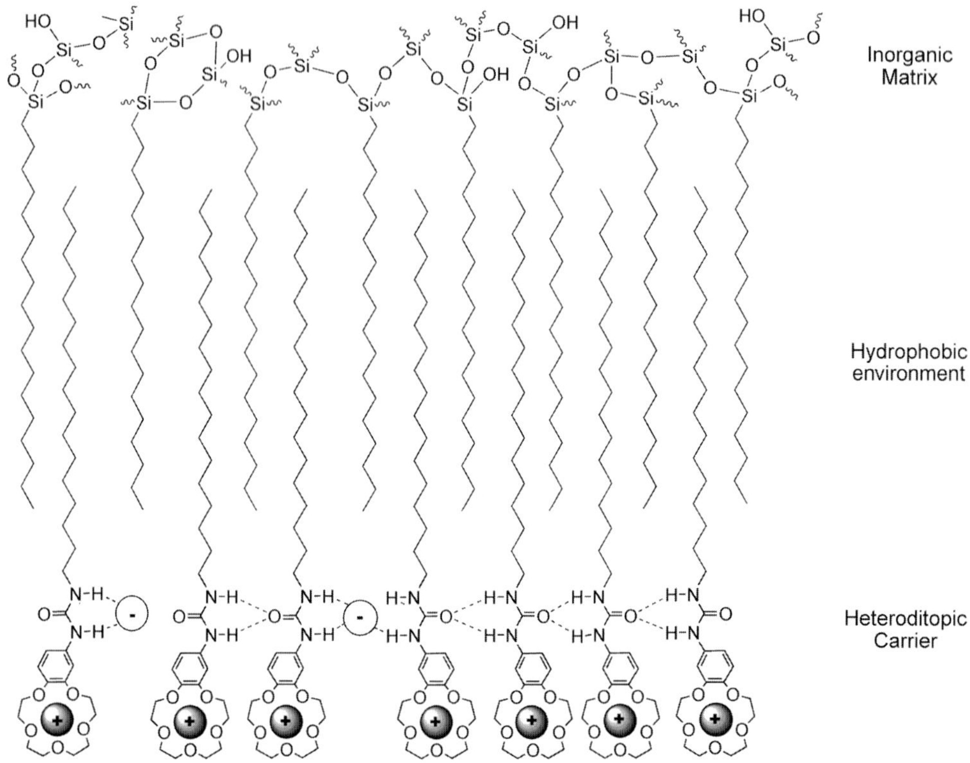

Fig. 10.15 Schematic representation of the dynamic transcription of encoded molecular features of the hierarchical organized system **5** into a hydrophobic heteropolysiloxane matrix.

were filled with surfactant (CTAB)-template silica and then calcinated to remove CTAB. The silica was then reacted with ODS followed by the incorporation of long chain hydrophobic carriers [25,26]. In the absence of the silica–surfactant–receptor nanocomposite in the alumina membrane, Na^+ and K^+ cations are transported through the membrane in a similar proportion. In contrast, the hybrid crown-alumina membranes, including the silica–surfactant composite, show a selective transport of ions depending on the receptor selectivity, **5** or **6**, respectively. In every case, we can distinguish two stages for the transport mechanism: (i) a simple and (ii) a facilitated diffusion. In the first one, the membranes are functioning like a "*sponge*", and the simple rapid diffusion through the membrane is accompanied by the selective complexation of the most suitable cation (Na^+ for **5** and K^+ for **6**, respectively); this is the so-called "membrane self-preparing step".The selective transport of the specific cation occurs at a much faster rate in the second stage. Thus, one can conclude that the membrane with the **15C5** receptor facilitates in the second step the transport of Na^+, whereas the **18C6** receptor facilitates the transport of K^+. These experimental results suggest that the self-assembly of receptors inside the surfactant–templated silica nanochannels of the columnar alumina pores can reorganize during the molecular transport, the mechanism being characterized by an initial self-preparing step.

10.5 Conclusions

Controlled formation of three-dimensional functional devices in silica makes the hybrid membrane materials presented here of interest for the development of a new supramolecular approach to nanoscience and nanotechnology through self-organization, towards systems of increasing behavioral and functional addressabilities (catalysis, optical and electronic applications, etc.).

The above results describe a simple synthetic hybrid material which successfully formed molecular recognition devices with transport patterns that can enable efficient translocation events. To our knowledge, the present systems are the first example of a hybrid nanomaterial where the concept of self-organization and a specific function (generation of specific translocation pathways in a hybrid solid) might, in principle, be associated. Finally, these findings show that the sol–gel process may be used to read biomolecular information. This is a new way of embedding basepairing or transcribing the supramolecular chirality in materials, of interest for developing a new sol–gel approach for nanoscience and nanotechnology.

The combined features of structural adaptation in a specific hybrid nanospace and of a dynamic supramolecular selection process make the dynamic-site membranes, presented in the third part, of general interest for the development of a specific approach toward nanomembranes of increasing structural selectivity. From the conceptual point of view these membranes express a synergistic adaptative behavior: the addition of the most suitable alkali ion drives a constitutional evolution of the membrane toward the selection and amplification of a specific transport crown-ether superstructure in the presence of the solute that promoted its generation in the first place. It embodies a constitutional self-reorganization (self-adaptation) of the membrane configuration producing an adaptative response in the presence of its solute. This is the first example of dynamic "smart" membranes where a solute induces the preparation of its own selective membrane.

Acknowledgement

This work, conducted as part of the award "Dynamic Adaptative Materials for Separation and Sensing Microsystems" made under the European Heads of Research Councils and European Science Foundation EURYI (European Young Investigator) Awards scheme in 2004, was supported by funds from the Participating Organisations of EURYI and the EC Sixth Framework Programme. See www.esf.org/euryi.

I thank the past and present members of my laboratory and the many individuals with whom we have collaborated in these studies: Adinela Cazacu, Mathieu Michau,

Remi caraballo, Carole Arnal-Herault, Andreea Pasc-Banu, Andre AYRAL, Arie van der Lee, Eddy Petit, Anca Meffre and Yves Marie Legrand.

References

1 Hucho, F. and Weise, C. (2001) Ligand-gated ion channels. *Angewandte Chemie-International Edition*, **40**, 3100–3116.

2 Vaitheeeswaran, S., Yin, H., Raisaiah, J.C. and Hummer, G. (2004) Water clusters in nonpolar cavities. *Proceedings of the National Academy of Sciences of the United States of America*, **101**, 17002–17005.

3 Cukierman, S. (2000) Proton mobilities in water and in different stereoisomers of covalently linked gramicidin A channels. *Biophysical Journal*, **78**, 1825–1834.

4 MacKinnon, R. (2004) Potassium channels and the atomic basis of selective ion conduction (Nobel Lecture). *Angewandte Chemie-International Edition*, **43**, 4265–4289.

5 Agre, P. (2004) Aquaporin water channels. *Angewandte Chemie-International Edition*, **43**, 4278–4290.

6 Dougherty, D.A. (1996) Cation-π interactions in chemistry and biology: A new view of benzene, Phe, Tyr, and Trp. *Science*, **271**, 163–168.

7 Gallivan, J.P. and Dougherty, D.A. (1999) Cation-π interactions in structural biology. *Proceedings of the National Academy of Sciences of the United States of America*, **96**, 9459–9464.

8 Lehn, J.-M. (1995) Supramolecular Chemistry-Concepts and Perspectives, VCH,Weinheim.

9 Davis, J.F. (2004) G-quartets 40 years later, from 5′GMP to molecular biology and supramolecular chemistry. *Angewandte Chemie-International Edition*, **43**, 668–698.

10 Zhang, X.X., Bradshaw, J.S. and Izatt, R.M. (1997) Enantiomeric recognition of amine compounds by chiral macrocyclic receptors. *Chemical Reviews*, **97**, 3313–3361.

11 Gokel, G.W. and Abel, E. (1996) in Comprehensive Supramolecular Chemistry, (eds Atwood, J.L. Davies, J.E.D. MacNicol, D.D. Vögtle, F. Suslick K.S.), Vol **1**, Pergamon, Oxford, 511–534.

12 Sanchez, C., Julian, B., Belleville, P. and Popall, M. (2005) Applications of hybrid organic-inorganic nanocomposites. *Journal of Materials Chemistry*, **15**, 3559–3592.

13 van Bommel, K.J.C., Frigerri, A. and Shinkai, S. (2003) Organic templates for the generation of inorganic materials. *Angewandte Chemie-International Edition*, **42**, 980–999.

14 Corriu, R.J.P. (2001) The control of nanostructured solids: A challenge for molecular chemistry. *European Journal of Inorganic Chemistry*, **5**, 1109–1121.

15 Moreau, J.J.E., Pichon, B.P., Pritzkow, H., Wong Chi Man, M., Bied, C., Bantignies, J.L., Dieudonné, P. and Sauvajol, J.L. (2004) A better understanding of the self-structuration of bridged silsesquioxanes. *Angewandte Chemie-International Edition*, **43**, 203–206.

16 Barboiu, M., Vaughan, G. and van der Lee, A. (2003) Self-organized heteroditopic macrocyclic superstructures. *Organic Letters*, **5**, 3073–3076.

17 Barboiu, M. (2004) Supramolecular polymeric macrocyclic receptors-hybrid carrier vs. channel transporters

in bulk liquid membranes. *Journal of Inclusion Phenomena and Molecular Recognition*, **49**, 133–137.

18 Barboiu, M., Cerneaux, S., Vaughan, G. and van der Lee, A. (2004) Ion-driven ATP-pump by self-organized hybrid membrane materials. *Journal of the American Chemical Society*, **126**, 3545–3550.

19 Cazacu, A., Tong, C., van der Lee, A., Fyles, T.M. and Barboiu, M. (2006) Columnar self-assembled ureidocrown-ethers – an example of ion-channel organization in lipid bilayers. *Journal of the American Chemical Society*, **128** (29), 9541–9548.

20 Barboiu, M., Luca, C., Guizard, C., Hovnanaian, N., Cot, L. and Popescu, G. (1997) Hybrid organic-inorganic fixed site dibenzo-18-crown complexant membranes. *Journal of Membrane Science*, **129**, 197–207.

21 Villamo, O., Barboiu, C., Barboiu, M., Yau-Chun-Wan, W. and Hovnanian, N. (2002) Hybrid organic-inorganic membranes containing a fixed thioether complexing agent for the facilitated transport of silver ions. *Journal of Membrane Science*, **204**, 97–110.

22 Lachowicz, E., Rózànska, B., Teixidor, F., Meliani, H., Barboiu, M. and Hovnanian, N. (2002) Comparison of sulphur and sulphur–oxygen ligands as ionophores for liquid–liquid extraction and facilitated transport. *Journal of Membrane Science*, **210**, 279–290.

23 Barboiu, M., Hovnanian, N., Luca, C., Popescu, G. and Cot, L. (1998) Functionalized derivatives of benzocrown-ethers, III, New macrocyclic derivatives containing chiral and linear lateral amino-acid moieties. *European Journal of Organic Chemistry*, 1705–1708.

24 Barboiu, M., Supuran, C.T., Scozzafava, A., Briganti, F., Luca, C., Popescu, G., Cot, L. and Hovnanian, N. (1997) Supramolecular complexes of L-amino acids as efficient activators of zinc enzyme carbonic anhydrase. *Liebigs Annalen der Chemie*, 1853–1859.

25 Barboiu, M., Supuran, C.T., Luca, C., Popescu, G. and Barboiu, C. (1996) Functionalized derivatives of benzocrown-ethers. Part I. Ionophores containing amino and pyridinium groups derived from benzo-15-crown-5, benzo-18-crown-6 and dibenzo-18-crown-6. *Liebigs Annalen der Chemie*, 959–963.

26 Barboiu, M., Luca, C., Popescu, G. and Cot, L. (1996) Facilitated transport of L-amino-acids I. New supramolecular complexes of L-amino-acids. *Roumaine Biotechnology Letters*, **1** (2), 77–86.

27 Barboiu, M., Luca, C., Popescu, G., Cot, L., Guizard, C., Hovnanian, N. and Barboiu, C. (1996) Facilitated transport of L-amino-acids II. Transport of L-phenylalanine by macrocyclic fixed-site carrier membranes. Preliminary report. *Roumaine Biotechnology Letters*, **1** (2), 87–97.

28 Barboiu, M., Hovnanian, N., Luca, C. and Cot, L. (1999) Functionalized derivatives of benzo-crown-ethers, V, Multiple molecular recognition of zwitterionic phenylalanine. *Tetrahedron*, **55**, 9221–9232.

29 Barboiu, M., Guizard, C., Luca, C., Albu, B., Hovnanian, N. and Palmeri, J. (1999) A new alternative to amino acid transport: Facilitated transport of L-phenylalanine by hybrid siloxane membrane containing a fixed site macrocyclic complexant. *Journal of Membrane Science*, **161**, 193–206.

30 Guizard, C., Bac, A., Barboiu, M. and Hovnanaian, N. (2000) Organic-inorganic hybrid materials with specific solute and gas transport properties for membrane and sensors applications. *Molecular Crystals and Liquid Crystals*, **354**, 91–106.

31 Barboiu, M., Guizard, C., Luca, C., Hovnanian, N., Palmeri, J. and Cot, L.

(2000) Facilitated transport of organics of biological interest II. Selective transport of organic acids by macrocyclic fixed site complexant membranes. *Journal of Membrane Science*, **174**, 277–286.

32 Barboiu, M., Guizard, C., Hovnanian, N., Palmeri, J., Reibel, C., Luca, C. and Cot, L. (2000) Facilitated transport of organics of biological interest I. A new alternative for the amino acids separations by fixed-site crown-ether polysiloxane membranes. *Journal of Membrane Science*, **172**, 91–103.

33 Barboiu, M., Guizard, C., Hovnanian, N. and Cot, L. (2001) New molecular receptors for organics of biological interest for the facilitated transport in liquid and solid membranes. *Separation and Purification Technology*, **25**, 211–218.

34 Guizard, C., Bac, A., Barboiu, M. and Hovnanian, N. (2001) Hybrid organic-inorganic membranes with specific transport properties. Applications in separation and sensors technologies. *Separation and Purification Technology*, **25**, 167–180.

35 Bong, D.T., Clark, T.D., Granja, J.R. and Ghadiri, M.R. (2001) Self-assembling organic nanotubes. *Angewandte Chemie-International Edition*, **40**, 988–1011.

36 Matile, S. (2001) En route to supramolecular functional plasticity: synthetic β-barrels, the barrel-stave motif, and related approaches. *Chemical Society Reviews*, **30**, 158–167.

37 Eggers, P.K., Fyles, T.M., Mitchell, K.D. D. and Sutherland, T. (2003) Ion channels from linear and branched bolaamphiphiles. *Journal of Organic Chemistry*, **68**, 1050–1058.

38 Cheruzel, L.E., Pometum, M.S., Cecil, M.R., Mashuta, M.S., Wittebort, R.J. and Buchanan, R.M. (2003) Structures and solid-state dynamics of one-dimensional water chains stabilized by imidazole channels. *Angewandte Chemie-International Edition*, **42**, 5452–5455.

39 Cronin, L. (2006) Inorganic molecular capsules: from structure to function. *Angewandte Chemie-International Edition*, **45**, 3576–3578.

40 Chemical Reviews (2005), 105, Special Issue on Functional Nanostructures.

41 Percec, V., Dulcey, A.E., Peterca, M., Ilies, M., Nurnmelin, S., Sienkowska, M.J. and Heiney, P.A. (2006) Principles of self-assembly of helical pores from dendritic dipeptides. *Proceedings of the National Academy of Sciences of the United States of America*, **103**, 2518–2523.

42 Cazacu, A., Pasc-Banu, A. and Barboiu, M. (2006) Molecular and supramolecular dynamics – a versatile tool for self-organization of polymeric membranes systems. *Macromolecular Symposia*, **245–246**, 435–438.

43 Cazacu, A., Michau, M., Arnal-Herault, C., Pasc-Banu, A., Meffre, A., Caraballo, R., Pasc, A. and Barboiu, M. (2006) Hybrid supramolecular dynamic membranes as selective information transfer devices. *Desalination*, **199**, 521–522.

44 Michau, M., Arnal-Herault, C., Caraballo, R., van der Lee, A. and Barboiu, M. (2007) Ionic pathways in dense and functionalized hybrid membranes, manuscript in preparation.

45 Arnal-Herault, C., Barboiu, M., Petit, E., Michau, M. and van der Lee, A. (2005) Cation-π interaction: A case for macrocycle-cation π-Interaction by its ureidoarene counteranion. *New Journal of Chemistry*, **29**, 1535–1539.

46 Arnal-Herault, C., Banu, A., Barboiu, M., Michau, M. and Van der Lee, A. (2007) Amplification and transcription of the dynamic supramolecular chirality of the guanine quadruplex,

Angewandle Chemie International Edition, **46**, 4268–4272.

47 Ciferi A. (ed.)(2005) Supramolecular Polymers Chemistry-Scope and Perspectives, 2nd edn. CRC Taylor and Francis,Boca Raton.

48 Sivakova, S. and Rowan, S.T. (2005) Nucleobases as supramolecular motifs. *Chemical Society Reviews*, **34**, 9–21.

49 Sessler, J.L. and Jayawickramarajah, J. (2005) Functionalized base-pairs: Versatile scaffolds for self-assembly. *Chemical Communications*, 1939–1949.

50 Numata, M., Sugiyasu, K., Hasegawa, T. and Shinkai, S. (2004) Sol-gel reaction using DNA as a template: An attempt toward transcription of DNA into inorganic materials, S. *Angewandte Chemie-International Edition*, **43**, 3279–3283.

51 Etter, M.C., Reutzel, S.M. and Choo, C. G. (1993) Self-organization of adenine and thymine in the solid state. *Journal of the American Chemical Society*, **115**, 4411–4412.

52 Castellano, R.K., Gramlich, V. and Diederich, F. (2002) Rebek imides and their adenine complexes: Preferences for Hoogsteen binding in the solid state and in solution. *Chemistry – A European Journal*, **8**, 118–129 and references therein.

53 Quinn, J.R., Ziemmerman, S.C., Del Bene, J.E. and Shavitt, I. (2007) Does the A·T or G·C base-pair possess enhanced stability? Quantifying the effects of CH·O interactions and secondary interactions on base-pair stability using a phenomenological analysis and ab initio calculations. *Journal of the American Chemical Society*, **129**, 934–941.

54 Kaucher, M.S., Harrell, W.A. and Davis, J.T. (2006) A unimolecular G-quadruplex that functions as a synthetic transmembrane Na^+ transporter. *Journal of the American Chemical Society*, **128**, 38–39.

55 Bosc, F., Ayral, A., Albouy, P.A., Datas, L. and Guizard, C. (2004) Mesostructure of anatase thin films prepared by mesophase templating. *Chemistry of Materials*, **16** (11), 2208–2214.

56 Klotz, M., Ayral, A., Guizard, C. and Cot, L. (2000) Synthesis conditions for hexagonal mesoporous silica layers. *Journal of Materials Chemistry*, **10** (3), 663–669.

57 Yamaguchi, A., Uejo, F., Yoda, T., Uchida, T., Tanamura, Y., Yamashita, T. and Teramae, N. (2004) Self-assembly of a silica-surfactant nanocomposite in a porous alumina membrane. *Nature Materials*, **3**, 337–341.

11
Design of Bioactive Nano-hybrids for Bone Tissue Regeneration

Masanobu Kamitakahara, Toshiki Miyazaki, Chikara Ohtsuki

11.1
Introduction

Techniques that repair damaged bones are important for patients to keep their quality of life (QOL), because bones support our body and enable us to perform various motions. When the damaged area is relatively small, the tissue and its function can be spontaneously repaired by the self-reconstructing ability of the tissues. However, when the damaged bone area is too large for self-reconstruction, the damaged bones must be repaired by using alternative materials. As such a material, autograft, which is transferred from other healthy parts of the bones of the same patient, is usually used. Although autograft shows high performance, it has the problems of only a limited amount of tissue being available and damage to the body because the bone tissue is extracted from the patient. Another candidate material is allograft, which is transferred from other people. Allograft has problems related not only to the limited amount available but also to overcoming reaction to the foreign body and the risk of infection.

Therefore, artificial materials that are safe and free from the amount limitation are needed to repair bone defects. Now, artificial bones made of ceramics or metals are used for the reconstruction of bone. These biomaterials have played important roles in the reconstruction of bones, joints and teeth. Among these materials, ceramics have been used because some of them show bone-bonding ability, that is bioactivity. In general, artificial materials implanted in the bony defects are encapsulated by a fibrous tissue to be isolated from the surrounding bone. Hench *et al.* [1,2] first developed glasses which show bioactivity in the Na_2O–CaO–SiO_2–P_2O_5 system and named these glasses Bioglass®. Since the discovery of Bioglass®, many researchers have developed bioactive ceramics, such as sintered hydroxyapatite ($Ca_{10}(PO_4)_6(OH)_2$) [3,4] and the glass-ceramic A-W [5,6]. These bioactive ceramics are used to repair bone defects because they can bond directly to living bone [7]. In spite of the specific biological activity of bioactive ceramics with bone tissue, these bioactive ceramics do not satisfy every clinical application because the mechanical properties of the ceramic materials are different from those of natural bone. The brittleness and higher elastic modulus than natural bone may cause

Bio-inorganic Hybrid Nanomaterials. Edited by E. Ruiz-Hitzky, K. Ariga, Y. M. Lvov

ISBN: 978-3-527-31718-9

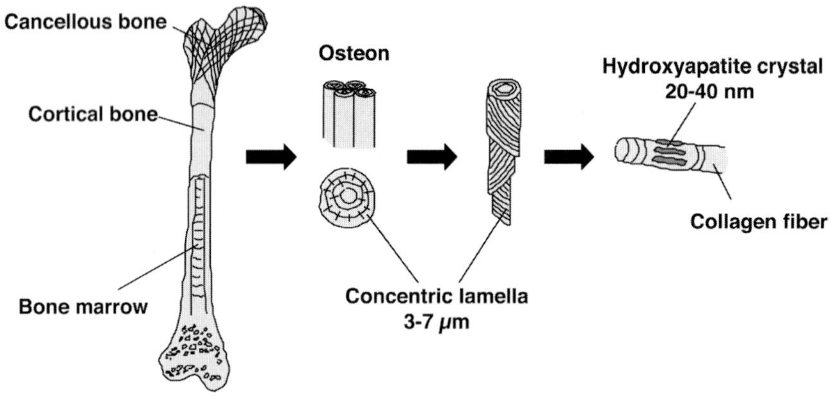

Fig. 11.1 Structure of bone [8].

so-called stress shielding, as well as fracture of the materials. Therefore, autografts and allografts are still widely used. There is a need for the development of bioactive materials which show not only bioactivity but also mechanical properties similar to living bone.

Living bone is a composite of 70 mass % of the inorganic component, hydroxyapatite, and 30 mass % of the organic component, collagen (Figure 11.1) [8]. Because of its unique composition and structure, natural bone can demonstrate not only high strength and high fracture toughness but also deformability and low elastic modulus. Therefore, the combination of an organic substance with bioactive inorganic components would produce a novel bone-repairing material showing not only bioactivity but also mechanical properties similar to living bone. This chapter presents an overview of studies of novel nano-hybrids for bone tissue regeneration.

11.2 Composite of Bioactive Ceramic Particles and Polymers

Bonfield *et al.* [9] developed a composite of hydroxyapatite particles with high-density polyethylene in the 1980s, in order to obtain an artificial bone that shows not only bioactivity but also mechanical compatibility. The model structure is shown in Figure 11.2. Hydroxyapatite particles, which show bioactivity, are dispersed in the polyethylene matrix, which shows high ductility, high toughness and high chemical durability. Although the polyethylene itself implanted in the body is encapsulated by fibrous tissue, the resultant composite begins to show bioactivity when the hydroxyapatite content is greater than 20 vol %. The exposure of the hydroxyapatite particles on the surface of the composite is important. As hydroxyapatite shows a higher Young's modulus than the polyethylene, the Young's modulus of the composite increases with increasing volume fraction of hydroxyapatite and approaches the value of cortical bone [10]. These composites show ductility when the hydroxyapatite content is less than 40 vol %, although they lose ductility and become brittle when the hydroxyapatite content exceeds 45 vol %. The ductility is advantageous not only

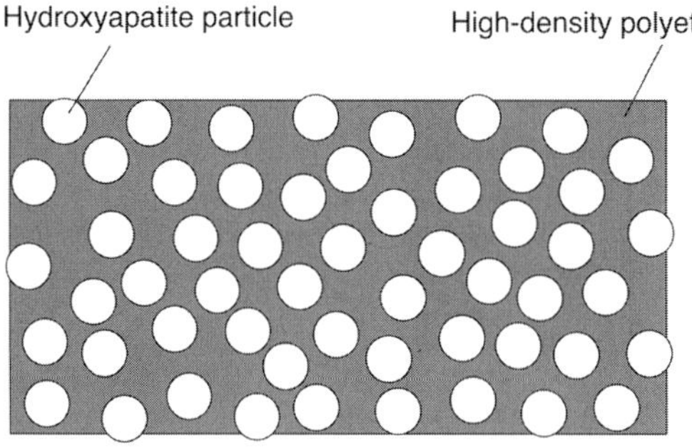

Fig. 11.2 Structural model of the composite of hydroxyapatite particles and high density polyethylene.

for the mechanical properties in use in the body but also for the ease of handling during an operation when the composite can be shaped by the clinicians.

The hydroxyapatite–polyethylene composite with hydroxyapatite content between 20 and 40 vol% shows bioactivity, ductility and comparable Young's modulus to bone. The optimum composition is 40 vol% of hydroxyapatite because the fracture toughness of the composite is still significantly higher than cortical bone and shows high bioactivity. The design of the combination of hydroxyapatite and polyethylene gives a novel bone-repairing material matching the mechanical properties of bone. This composite material is used as a trimmable orbital implant for either orbital floor fractures or volume augmentation, and as a trimmable shaft in a middle ear implant.

A composite of polyethylene and Bioglass® has also been developed [11]. Bioglass® is known to be a highly bioactive glass and can bond not only to hard tissue but also to soft tissue, such as soft connective tissue and cartilage. Therefore, a suitable composite is expected to show the ability to bond to both hard and soft tissue. The composite of polyethylene and glass-ceramic A-W [5,6] which shows high mechanical strength and high bioactivity, has also been reported [12].

The combination of bioactive ceramic particles and a polymer matrix gives bioactive materials which show mechanical properties analogous to those of human cortical bone. However, the bioactivity is not so high because the filler content is limited due to the brittleness, and the weak bonding between the filler and matrix may induce problems.

11.3 Bone-bonding Mechanism of Bioactive Materials

11.3.1 Interface between Bone and Bioactive Material

It has been shown that the bioactive materials bond to living bone through a hydroxyapatite layer which is formed on their surfaces in the body [13]. Therefore, it is believed that the essential requirement for an artificial material to bond to living bone is the formation of a hydroxyapatite layer on their surfaces in the living body [14]. This hydroxyapatite layer consists of calcium-deficient and carbonate-containing nanosized hydroxyapatite with low crystallinity and defect structure [15]. This hydroxyapatite is similar to the bone mineral in composition and structure, and hence the bone-producing cell, called an osteoblast, can proliferate preferentially on this surface apatite layer, and differentiate to produce bone matrix composed of hydroxyapatite and collagen. As a result, the surrounding bone can come into direct contact with the surface hydroxyapatite layer. When this occurs, a tight chemical integration is formed between the bone and hydroxyapatite.

11.3.2 Simulated Body Fluid

Kokubo *et al.* [16,17] showed that the hydroxyapatite formation on the surfaces of bioactive materials in the living body can be reproduced even in an acellular protein-free simulated body fluid (SBF) with ion concentrations nearly equal to those of human blood plasma. This indicates that the hydroxyapatite layer is formed through chemical reaction of the bioactive glass with the surrounding body fluids. The formed layer consists of carbonated hydroxyapatite with small crystallites and low crystallinity, which is similar to bone hydroxyapatite. Hence the bioactivity of a material can be evaluated even *in vitro* by examining the hydroxyapatite formation on its surface in SBF.

SBF is a solution that has inorganic ion concentrations similar to those of human blood plasma but does not contain any cells or protein. A brief summary of SBF, introduced by Cho *et al.* [17], follows. The ion concentrations of SBF are given in Table 11.1 [17]. The pH of SBF is typically adjusted to 7.25 or 7.40 at 36.5 °C. This fluid is a metastable solution containing calcium and phosphate ions supersaturated with respect to hydroxyapatite. SBF is prepared by successively dissolving the reagent-grade chemicals in ultra-pure water in the order given in Table 11.2 [17]. Each new chemical is added after the previous one has completely dissolved. The temperature

Tab. 11.1 Ion concentrations of human blood plasma and SBF [17].

	Concentration/mM ($mol \cdot m^{-3}$)	
Ion	**Blood plasma**	**SBF**
Na^+	142.0	142.0
K^+	5.0	5.0
Mg^{2+}	1.5	1.5
Ca^{2+}	2.5	2.5
Cl^-	103.0	147.8
HCO_3^-	27.0	4.2
HPO_4^{2-}	1.0	1.0
SO_4^{2-}	0.5	0.5

Tab. 11.2 Reagents for preparing one liter of SBF [17].

Order	Reagent	Amount (SBF)
1	NaCl	7.996 g
2	$NaHCO_3$	0.350 g
3	KCl	0.224 g
4	$K_2HPO_4 \cdot 3H_2O$	0.228 g
5	$MgCl_2 \cdot 6H_2O$	0.305 g
6	$1\,kmol \cdot m^{-3}$ HCl	$40\,cm^3$
7	$CaCl_2$	0.278 g
8	Na_2SO_4	0.071 g
9	$NH_2C(CH_2OH)_3$	6.057 g

of the solution in the bottle is adjusted to 36.5 °C in a water bath, and the pH of the solution is adjusted to 7.25 or 7.40 by titrating with $1\,kmol \cdot m^{-3}$ HCl solution. After the pH has been adjusted, the solution is transferred to a glass volumetric flask, and then ultra-pure water is added to give a total solution volume of $1\,dm^3$. Recently, SBF was subjected to round robin testing [18] and the recipe for preparing SBF was refined by Kokubo *et al.* [19].

11.3.3
Hydroxyapatite Formation on Bioactive Materials

Human body fluid and SBF contain calcium and phosphate ions that are already supersaturated with respect to hydroxyapatite [20]. However, these fluids do not spontaneously deposit hydroxyapatite under normal conditions. This is because the activation energy barrier for hydroxyapatite nucleation is very high. Therefore, the ability of substrates to induce heterogeneous nucleation of hydroxyapatite and the degree of supersaturation of SBF with respect to hydroxyapatite are important factors for hydroxyapatite formation on materials in the body fluid and SBF.

In order to design novel bioactive materials, we must know what compositions can act as the basic components of bioactive ceramics. Ohtsuki *et al.* [21] examined the compositional dependence of hydroxyapatite formation on the glasses in the system $CaO–SiO_2–P_2O_5$ in SBF. The results are summarized in Figure 11.3. It is noted that surface hydroxyapatite formation is restricted to the compositional regions based on CaO and SiO_2 and that hydroxyapatite formation cannot be observed on glasses in the compositional regions based on CaO and P_2O_5. This occurred although both the calcium ions released from CaO and SiO_2-based glasses and the phosphate ions released from CaO and P_2O_5-based glasses increased almost equally the degree of supersaturation of the surrounding fluid with respect to the hydroxyapatite [22]. The CaO,SiO_2-based glasses formed a surface hydroxyapatite layer but the CaO,P_2O_5-based glasses did not. This indicates that the surfaces of the CaO,SiO_2-based glasses exclusively provide favorable sites for hydroxyapatite nucleation.

Ohtsuki *et al.* [22] assumed the hydroxyapatite formation mechanism to be as follows. The formation of the hydroxyapatite layer is triggered by release of Ca^{2+} ions

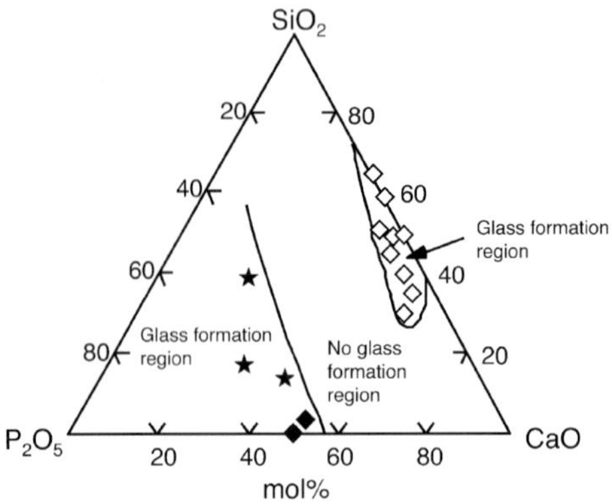

Fig. 11.3 Compositional dependence of hydroxyapatite formation on the glasses in the system $CaO–SiO_2–P_2O_5$. Soaking period: 30 days. ◇ hydroxyapatite formation, ◆ no hydroxyapatite formation, ★ severe dissolution.

from the glass and subsequent formation of Si–OH groups on the surface. Release of Ca^{2+} increases the degree of supersaturation with respect to hydroxyapatite, and Si–OH groups induce heterogeneous nucleation of hydroxyapatite on the surface [23]. Once hydroxyapatite nuclei form, crystals of hydroxyapatite grow spontaneously by consuming calcium and phosphate ions from SBF, because SBF is already supersaturated with respect to hydroxyapatite, to form a layer of hydroxyapatite. The schematic representation of the mechanism of hydroxyapatite formation is shown in Figure 11.4. Consequently, the hydroxyapatite layer fully covers the surface. From the findings described above, it can be concluded that hydroxyapatite deposition from SBF onto materials is governed by (i) the ability to selectively induce crystal

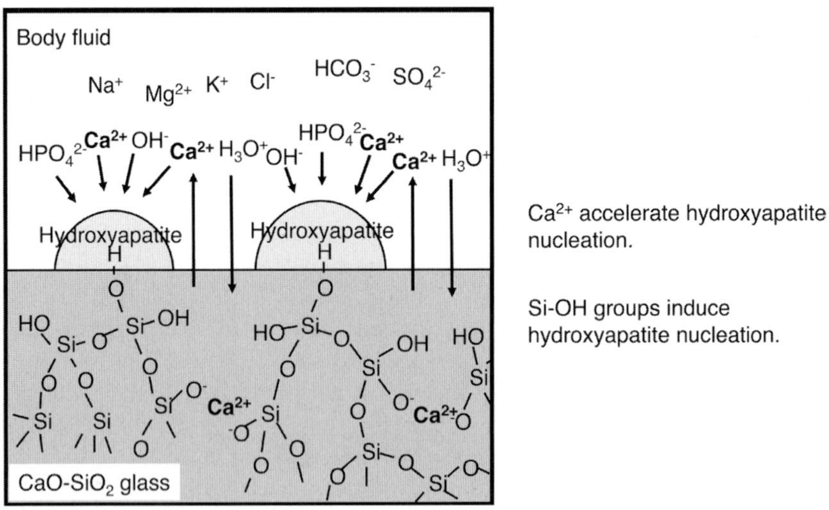

Fig. 11.4 Mechanism of hydroxyapatite formation on $CaO–SiO_2$ glass in the body environment.

nucleation (heterogeneous nucleation) of hydroxyapatite on the surface of the materials, and (ii) the increase in the degree of supersaturation of SBF with respect to hydroxyapatite. If we can design substrates that fulfil the above conditions, we can obtain bioactive materials.

11.4
Sol–Gel-derived Bioactive Nano-hybrids

11.4.1
Silicate-based Nano-hybrids

The sol–gel process is a popular process for the preparation of nano-hybrids of inorganic and organic components because the process can be conducted at low temperature. It has already been reported that organically modified silicates can be synthesized through hydrolysis and polycondensation of tetraethoxysilane (TEOS) and polydimethylsiloxane (PDMS) [24]. The synthesized organic–inorganic nano-hybrid has the potential to show the properties of both organic and inorganic components. Based on the findings that a $CaO–SiO_2$ glass is effective for bioactivity [21], Tsuru *et al.* [25] synthesized PDMS–$CaO–SiO_2$ nano-hybrids by a sol–gel method from TEOS, PDMS and calcium nitrate ($Ca(NO_3)_2$), and revealed that the nano-hybrid obtained shows hydroxyapatite-forming ability in SBF. In this nano-hybrid, the inorganic component which shows bioactivity is homogeneously distributed at the molecular level, and chemically bonds to the organic component (Figure 11.5).

Kokubo *et al.* [26–28] then reported the synthesis of bioactive and flexible organic–inorganic nano-hybrids with mechanical properties analogous to those of human cancellous bone through sol–gel processing in the PDMS–$CaO–SiO_2$ and PDMS–$CaO–SiO_2–TiO_2$ systems by controlling the synthesis condition. Scanning electron microscope (SEM) photographs of the surfaces of the PDMS–$CaO–SiO_2$ nano-hybrids before and after soaking in SBF for 7 days are shown in Figure 11.6. The amount of the precipitate increases with decreasing PDMS content when the calcium content is constant. The X-ray diffraction (XRD) analysis indicates that these

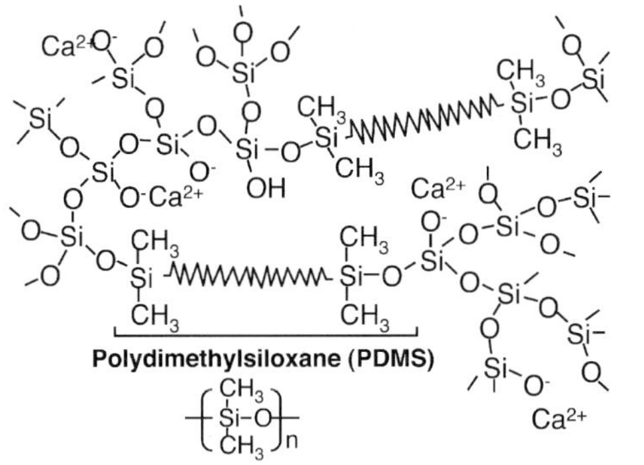

Fig. 11.5 Structural model of PDMS–$CaO–SiO_2$ nano-hybrid.

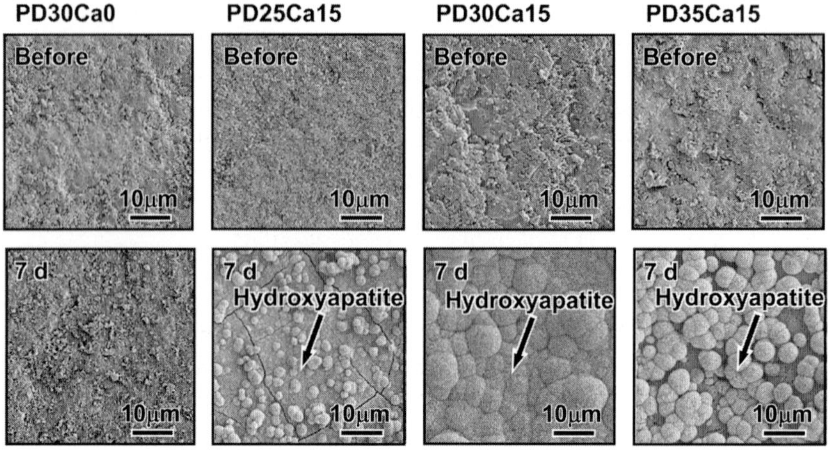

Fig. 11.6 SEM photographs of surfaces of PDMS–CaO–SiO_2 nano-hybrids before and after soaking in SBF for 7 days. PD*x*Ca*y*: $x = 100 \times$ PDMS/(PDMS + TEOS), $y = 100 \times Ca(NO_3)_2$/TEOS in starting compositions.

precipitates can be assigned to hydroxyapatite. The morphology of the deposited particles on the surface is quite similar to that of the hydroxyapatite layer formed on bioactive glasses and glass-ceramics. That is, the hydroxyapatite-forming ability of the PDMS–CaO–SiO_2 nano-hybrids increases with decreasing PDMS content. This should be attributed to the decrease in the amount of silanol groups which leads to a decrease in hydroxyapatite nucleation. Once the hydroxyapatite nuclei are formed, they can grow spontaneously by consuming calcium and phosphate ions from the surrounding SBF. On the other hand, the CaO-free nano-hybrid (PD30CaO) did not form apatite on its surface. The calcium-containing nano-hybrids release calcium ions and the released calcium ions accelerate the nucleation of hydroxyapatite.

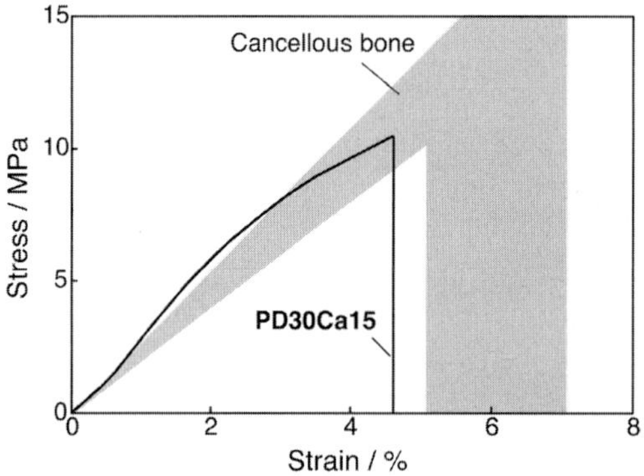

Fig. 11.7 Representative stress–strain curve of nano-hybrid PD30Ca15, in comparison with that reported for human cancellous bone.

A representative stress–strain curve of one of the PDMS–CaO–SiO_2 nano-hybrids is shown in Figure 11.7, in comparison with that reported for human cancellous bone [29]. Unlike the usual brittle ceramics, the nano-hybrid was deformable and showed mechanical properties analogous to those of human cancellous bone.

The design of biodegradable nano-hybrids has also been reported by Rhee *et al.* by the combination of a CaO–SiO_2 component with the biodegradable polymer, poly (epsilon-caprolactone) [30]. These nano-hybrids show hydroxyapatite-forming ability and are expected to show bioactivity and biodegradability.

11.4.2
Nano-hybrids Starting from Methacryloxy Compounds

Modification of an organic polymer by incorporation of methacryloxypropyl triethoxysilane (MPS) and calcium ions [31,32] has been reported. Ohtsuki *et al.* attempted to synthesize organic–inorganic nano-hybrids starting from MPS and 2-hydroxyethylmethacrylate (HEMA) with addition of a calcium salt. MPS has alkoxysilane groups that provide silanol groups after hydrolysis, whereas HEMA provides a hydrophilic polymer matrix in the nano-hybrid. Mixtures of MPS and HEMA with various molar ratios were dissolved in 100 cm^3 of ethanol at a total concentration of 1 kmol· m^{-3}. The mixtures were polymerized by heat treatment at 75 °C for 3 hours with benzoylperoxide (BPO) as initiator. Then 20 cm^3 of ethanol solution containing 0.01 mol calcium chloride ($CaCl_2$) was added to the obtained polymer solution. The obtained solution was cast in a polypropylene container and dried at room temperature to obtain a bulk gel. A homogeneous and transparent gel was obtained at the composition of MPS: HEMA = 0.1 : 0.9. Higher concentrations of MPS led to the formation of many cracks in the specimens during the drying process, crack-free specimens were not obtained when the molar ratio of MPS was more than 0.1. Figure 11.8 shows the appearance of the MPS–HEMA nano-hybrid with MPS: HEMA = 0.1 : 0.9. Figure 11.9 shows SEM photographs of the surface of the MPS–HEMA nano-hybrid synthesized at the composition MPS: HEMA = 0.1 : 0.9 before and after soaking in SBF for 7 days. A sample containing no MPS, remarkably, dissolved in SBF, the specimen was not able

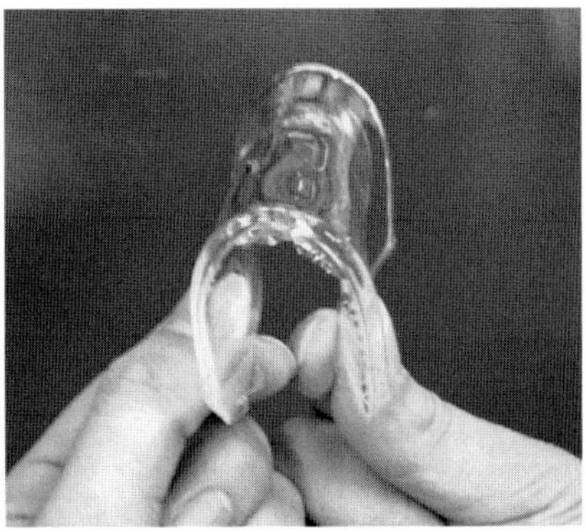

Fig. 11.8 Flexibility of the MPS–HEMA nano-hybrid (MPS: HEMA = 0.1 : 0.9).

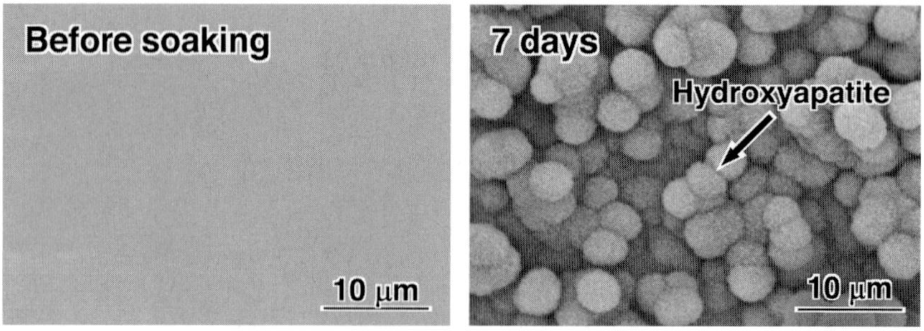

Fig. 11.9 SEM images of the surface of the MPS–HEMA nano-hybrid synthesized at the composition MPS: HEMA = 0.1 : 0.9 before and after soaking in SBF for 7 days.

to be analysed. Fine particles, which were assigned to hydroxyapatite by XRD, completely covered the surface of the sample soaked for 7 days.

Ohtsuki *et al.* [33–35] also applied organic modification of Si–OH and Ca^{2+} to a poly(methyl methacrylate) (PMMA) bone cement consisting of PMMA powder and methyl methacrylate (MMA) liquid in order to induce bioactivity in the cement. One of the significant problems of the PMMA bone cement is the loosening at the interface between the bone and the cement due to the lack of bioactivity. Although several attempts were made to obtain bioactive PMMA bone cement by mixing bioactive ceramic powders [36], large amounts of ceramic powder had to be incorporated into the cement in order to show any bioactivity. Moreover, the existence of an interface between the bioactive ceramic powders and the PMMA matrix might be a problem because the bond between the bioactive ceramic powders and the PMMA matrix is not so strong. Therefore, Ohtsuki *et al.* attempted the preparation of bioactive PMMA bone cements by incorporation of MPS and various calcium salts. In this system, MPS and MMA can be copolymerized and Si–OH groups are chemically bonded to the PMMA matrix. They used PMMA powder with a molecular weight of about 100 000 and an average particle size of about 14 μm. The PMMA powder was mixed with a calcium salt, such as $CaCl_2$, calcium acetate ($Ca(CH_3COO)_2$), calcium hydroxide ($Ca(OH)_2$), calcium lactate ($Ca(CH_3CHOHCOO)_2$), calcium benzoate ($Ca(C_6H_5COO)_2$) or calcium methacrylate ($Ca(CH_2{=}C(CH_3)COO)_2$) at 20 mass % of the powder. BPO was then added to the powder as a polymerization initiator. MMA liquid was mixed with MPS at 20 mass % of the liquid. *N*-dimethyl-p-toluidine (NDT) was then added to the liquid as a polymerization accelerator. The powder was mixed with the liquid at a powder to liquid ratio of 1g/0.5 g at $23 \pm 2\,^{\circ}C$. At half the setting time of the specimens, they were soaked in SBF to examine their hydroxyapatite-forming ability. Figure 11.10 shows SEM photographs of the surfaces of the cements modified with MPS and various kinds of calcium salts after soaking in SBF for 7 days. "Reference" is the sample unmodified by MPS or calcium salt. Hydroxyapatite fine particles were observed on the cements modified with $CaCl_2$, $Ca(CH_3COO)_2$, $Ca(OH)_2$ and $Ca(CH_2{=}C(CH_3)COO)_2$ after the soaking. Modification of PMMA cement by incorporation of MPS and appropriate calcium salts makes the cement form hydroxyapatite in the body environment. The solubility of the calcium salt in water decreases in the order:

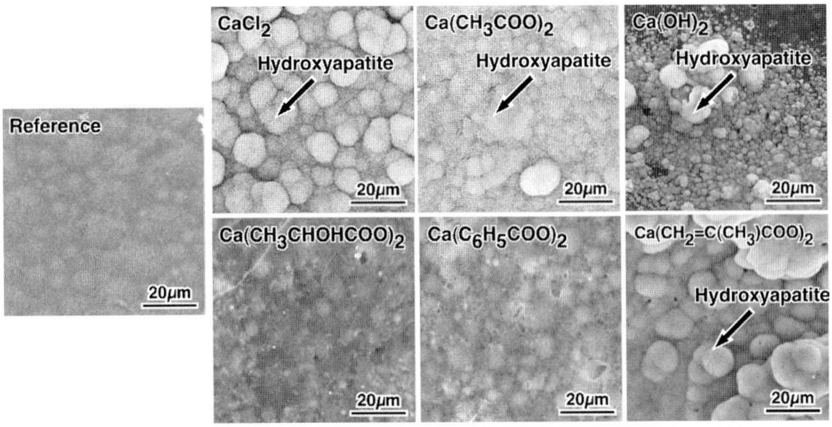

Fig. 11.10 SEM photographs of the surfaces of the cements modified with MPS and various kinds of calcium salts after soaking in SBF for 7 days.

$CaCl_2 > Ca(CH_3COO)_2 > Ca(CH_3CHOHCOO)_2 > Ca(C_6H_5COO)_2 > Ca(OH)_2$. The cements modified with calcium salts that are highly soluble in water have the greatest tendency to form the hydroxyapatite in SBF. However, it is noted that the cement modified with $Ca(OH)_2$ formed the hydroxyapatite in SBF even though the solubility of $Ca(OH)_2$ is the lowest among the calcium salts used in this study. The pH of the surrounding solution remarkably increased after soaking the cement modified with $Ca(OH)_2$ in SBF. The increase in pH would accelerate the hydroxyapatite nucleation, since OH^- is a component of the apatite.

The compressive strength of the modified cements decreased after exposure to SBF, except for the cement modified with $Ca(OH)_2$. This is attributed to release of Ca^{2+} ions from the cement into the solution. Among the cements examined in this study, those modified with $Ca(CH_3COO)_2$, $Ca(OH)_2$ or $Ca(CH_3CHOHCOO)_2$ showed compressive strength near the lower limit (~70 MPa) required by ISO5833. When Ca^{2+} ions are rapidly released, pores are formed inside the cement, leading to a decrease in compressive strength. An appropriate rate of release of Ca^{2+} is necessary to maintain the high mechanical strength of the cement.

11.4.3
Nano-hybrids Based on Other than Silicate

Miyazaki *et al.* [37] synthesized organic–inorganic hybrids from tetraisopropyl titanate, hydroxyethylmethacrylate and $CaCl_2$, and revealed that the hybrids form hydroxyapatite in SBF. Kamitakahara *et al.* [38] found that the CaO-containing nano-hybrids showed a decrease in mechanical strength in SBF in the PDMS–CaO–SiO_2–TiO_2 system. On the other hand, CaO-free PDMS–SiO_2–TiO_2 nano-hybrids do not form a hydroxyapatite layer on their surfaces in SBF. Uchida *et al.* [39] reported that anatase, one of the crystalline phases of TiO_2, is highly effective for hydroxyapatite formation in the body environment. Nano-hybrids are expected to show high hydroxyapatite-forming ability, even if they contain no CaO, when TiO_2 adopts an anatase structure in them. Kamitakahara *et al.* [40] attempted the preparation of bioactive CaO-free PDMS- modified nano-hybrids that can form hydroxyapatite on their surfaces. A PDMS–TiO_2 nano-hybrid was synthesized and its structure was controlled by hot-water treatment. A PDMS–TiO_2 nano-hybrid was prepared by a sol–gel method from PDMS ($M = 550$) and tetraethylorthotitanate with PDMS/tetraethylorthotitanate molar ratios at 1.35. The as-prepared nano-hybrid took an amorphous structure and did not form hydroxyapatite on its surface after soaking in SBF for 7 days. The PDMS–TiO_2 nano-hybrid was then soaked in hot water at 80 °C for various periods. The structural change was examined by thin film X-ray diffraction (TF-XRD) and transmission electron microscopy (TEM) (Figure 11.11). Anatase peaks were observed in the TF-XRD patterns for the PDMS–TiO_2 nano-hybrid treated with hot water at 80 °C within 1 day and the peak intensities of anatase increased with increasing length of the hot-water treatment. TEM observation showed that the nano-hybrid adopted a homogeneous amorphous structure before the hot-water treatment, and precipitated anatase particles, 10–20 nm in size, after the hot-water treatment at 80 °C for 7 days. The hot-water-treated nano-hybrids were soaked in SBF. Hydroxyapatite was formed on the nano-hybrid treated with hot water at 80 °C for periods longer than 1 day, and the amount of hydroxyapatite formed increased with increasing time of hot-water treatment (Figure 11.12). This indicates that the anatase precipitated in the nano-hybrid induces hydroxyapatite formation on the surface of the nano-hybrid. The representative stress–strain curves of the PDMS–TiO_2 nano-hybrid before and after the hot-water treatment, and after both hot-water treatment and SBF treatment are shown in Figure 11.13. Both the tensile strength and the Young's modulus of the nano-hybrid decreased, and its strain to failure increased after the hot-water treatment. It is considered that the matrix of the nano-hybrid was

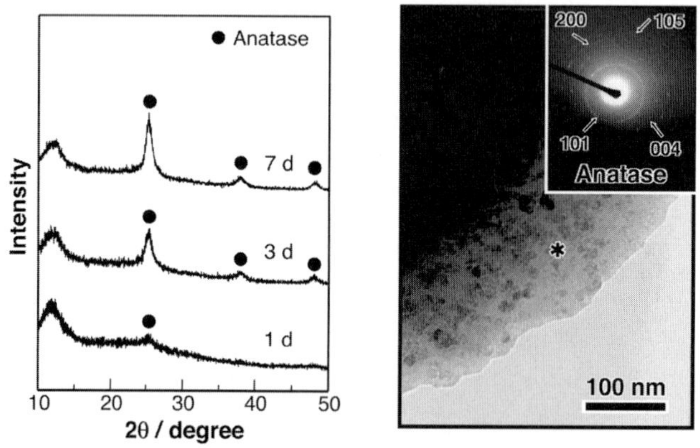

Fig. 11.11 TF-XRD patterns of the surfaces of the PDMS–TiO_2 nano-hybrid after hot-water treatment at 80 °C for various periods (left), and TEM photographs of the PDMS–TiO_2 nano-hybrid after hot-water treatment at 80 °C for 7 days. (*: center of electron diffraction) (right).

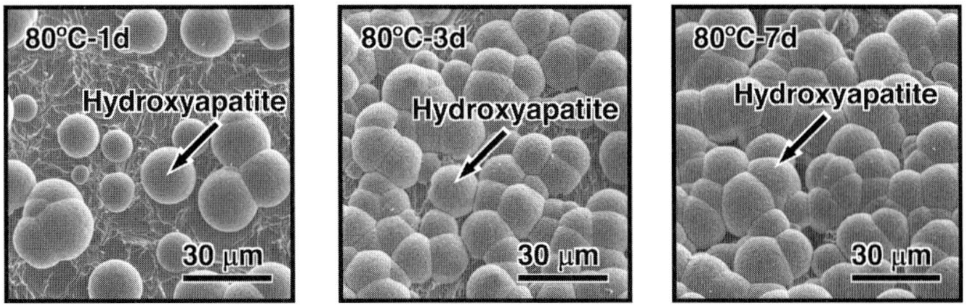

Fig. 11.12 SEM photographs of surfaces of the PDMS–TiO_2 nano-hybrid treated with hot water at 80 °C for various periods and soaked in SBF for 7 days.

enriched with PDMS due to the segregation of TiO_2 by the hot-water treatment and hence the mechanical properties of the nano-hybrid subjected to the hot-water treatment became close to the matrix rich in PDMS, since the continuous phase predominantly governs the overall mechanical behavior of composite materials. The observation that the strain energies stored until failure were not significantly changed by the hot-water treatment might indicate that rearrangement of the nano-hybrid by the hot-water treatment does not cause deterioration of the nano-hybrid. The obtained material showed rubber elasticity even after the hot-water treatment. After soaking in SBF, its strain to failure decreased, but its tensile strength did not change and its Young's modulus increased slightly. A highly deformable PDMS–TiO_2 nano-hybrid with hydroxyapatite-forming ability and durability in SBF was obtained by this method.

In order to obtain a nano-hybrid with higher mechanical properties, PDMS was replaced with poly(tetramethylene oxide) (PTMO) [41]. PTMO–TiO_2 nano-hybrids were prepared by a sol–gel method from triethoxysilane functionalized PTMO (Si-PTMO) and tetraisopropyltitanate with weight ratios of 30/70, 40/60 and

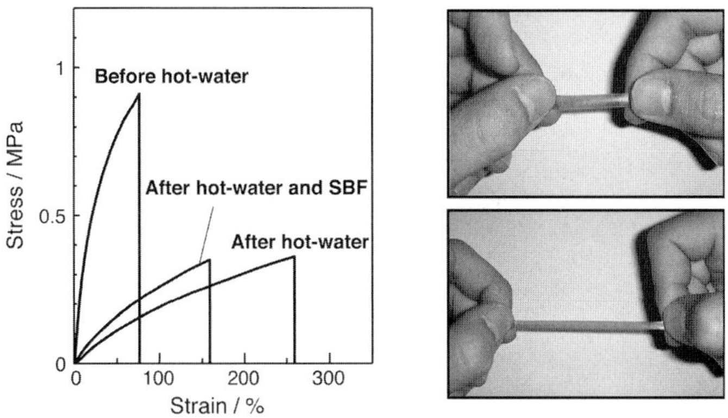

Fig. 11.13 Representative stress–strain curves of the PDMS–TiO_2 nano-hybrid before and after hot-water treatment at 80 °C for 7 days, and that after both hot-water treatment and SBF treatment for 28 days (left), and appearance of PDMS–TiO_2 nano-hybrid after the hot-water treatment (right).

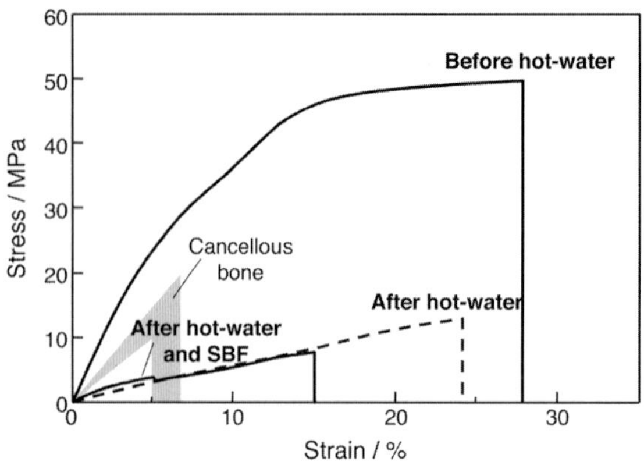

Fig. 11.14 Representative stress–strain curves of the PTMO–TiO_2 nano-hybrid before and after hot-water treatment, and after both hot-water treatment and SBF treatment for 28 days, in comparison with that reported for human cancellous bone.

50/50 (nano-hybrids TiPT30, TiPT40 and TiPT50, respectively), and subsequently subjected to a hot-water treatment at 95 °C for 2 days [42]. All of the obtained nano-hybrids were amorphous before the hot-water treatment, and precipitated nanosized anatase after the hot-water treatment. The amount of precipitated anatase increased with decreasing PTMO content. Hydroxyapatite was not formed on the surfaces of the nano-hybrids in SBF before the hot-water treatment, but was formed after treatment, and its amount increased with decreasing PTMO content. This indicates that hydroxyapatite formation was induced by the anatase precipitated by hot-water treatment, as in the case of the PDMS–TiO_2 nano-hybrid. Figure 11.14 shows representative stress–strain curves of nano-hybrid TiPT40 before and after the hot-water treatment, and after both the hot-water treatment and SBF treatment for 28 days, in comparison with that reported for human cancellous bone. Nano-hybrid TiPT40 showed strength and Young's modulus analogous to those of human cancellous bones, and high ductility after the hot-water treatment. The mechanical strength and Young's modulus of the nano-hybrid were improved by changing the organic component from PDMS to PTMO. After soaking in SBF, the Young's modulus increased, and the bending strength and strain to failure decreased. The decreases in bending strength and strain to failure after soaking in SBF might be due to the degradation of the nano-hybrid. These decreases were, however, much lower than those of a CaO-containing PTMO–CaO–SiO_2–TiO_2 nano-hybrid [38]. As the environment in hot water is more severe than in the body environment, the unreacted reagents or solvents remaining in the nano-hybrid might be released into hot water, and therefore the resultant nano-hybrid is expected to be non-toxic and stable in the living body at around 36.5 °C.

Miyazaki *et al.* [43] reported that a sol–gel-derived Ta_2O_5 gel shows hydroxyapatite-forming ability in SBF even in the amorphous state. Therefore, PTMO–Ta_2O_5 nano-hybrids prepared by a sol–gel method can be expected to form hydroxyapatite in SBF without any treatment. PTMO–CaO–Ta_2O_5 nano-hybrids were prepared by

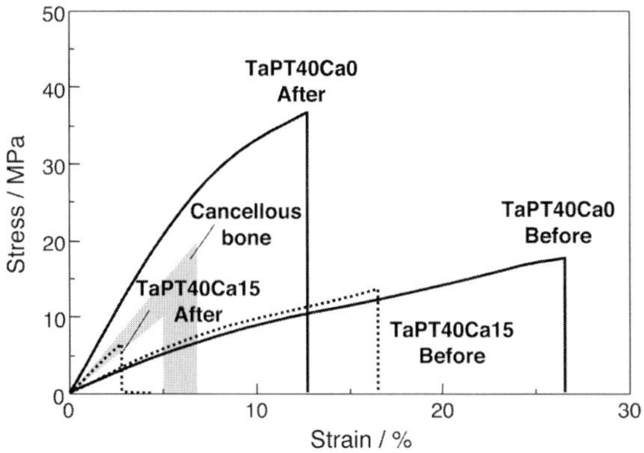

Fig. 11.15 Stress–strain curves of nano-hybrids TaPT40Ca0 and TaPT40Ca15 before and after soaking in SBF for 28 days, in comparison with that reported for human cancellous bone.

hydrolysis and polycondensation of Si-PTMO, tantalum ethoxide ($Ta(OEt)_5$) and $CaCl_2$ [44]. A CaO-free PTMO–Ta_2O_5 nano-hybrid from Si-PTMO/$Ta(OEt)_5$ weight ratio of 40/60 (TaPT40Ca0), and a PTMO-CaO-Ta_2O_5 from Si-PTMO/$Ta(OEt)_5$ weight ratio of 40/60 and $CaCl_2$/$Ta(OEt)_5$ mole ratio of 0.15 (TaPT40Ca15) were prepared. Even CaO-free nano-hybrids PT40Ca0 formed hydroxyapatite on their surfaces in SBF within 14 days. Nano-hybrid TaPT40Ca0 showed higher mechanical strength, which was increased by soaking in SBF, and larger strain to failure than human cancellous bone (Figure 11.15). The CaO-containing nano-hybrid (TaPT40Ca15) showed higher hydroxyapatite-forming ability than the CaO-free hybrids, but its mechanical strength was lower than PT40Ca0 and was decreased by soaking in SBF. Thus the flexible bioactive CaO-free PTMO–Ta_2O_5 hybrids are expected to be useful as bone substitutes.

11.4.4 Nano-hybrids Combined with Calcium Phosphates

In order to obtain bioactive nano-hybrids with high mechanical strength, Uchino *et al.* [45] focused on calcium phosphates with adequate solubility as a source of calcium ions, because they could also provide reinforcement to the nano-hybrid. The MPS–HEMA nano-hybrid was combined with α-tricalcium phosphate porous body or calcium phosphates powder (CPP) consisting of anhydrous dicalcium phosphate and tetracalcium phosphate as the Ca^{2+} sources, to improve the mechanical strength of the nano-hybrid. The composites were named TCP-MPS-HEMA and CPP-MPS-HEMA, respectively. These composites formed hydroxyapatite on their surfaces in SBF. When the calcium phosphates were not added, these materials did not form hydroxyapatite in SBF. The calcium ions might have been released from calcium phosphates and accelerated hydroxyapatite deposition by increasing the ionic activity with respect to hydroxyapatite. The mechanical strength of the nano-hybrid was improved by the reinforcement with calcium phosphates (Figure 11.16). When CPP

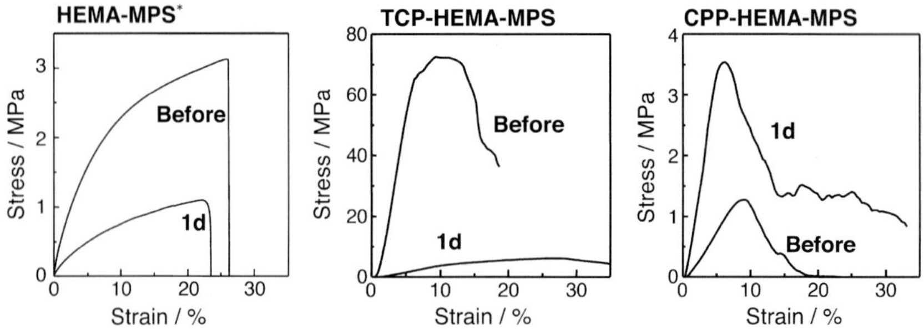

Fig. 11.16 Representative stress–strain curves of HEMA–MPS nano-hybrid, TCP-HEMA-MPS and CPP–HEMA–MPS before and after soaking in SBF for 1 day. *Tensile test, the others: compressive test.

was used, the compressive strength of the composite increased after soaking in SBF for 1 day. This is because the self-setting reaction of CPP occurred in SBF and increased the mechanical strength of the composite. The combination of the nano-hybrid and calcium phosphates offers a novel design for bioactive materials.

These sol–gel-derived nano-hybrids have a potential to show flexibility and bioactivity. As the biological and mechanical properties of nano-hybrids can be tailored by selecting the combination of the organic and inorganic components, these nano-hybrids may be a useful design to obtain flexible and bioactive bone-repairing materials.

11.5 Nano-hybrid Consisting of Bone-like Hydroxyapatite and Polymer

11.5.1 Biomimetic Process

The essential requirement for artificial materials to show bioactivity is to form bone-like hydroxyapatite on their surfaces in the body. Therefore, a material coated with bone-like hydroxyapatite is expected to bond to living bone. The structure of bone-like hydroxyapatite is different from the sintered stoichiometric hydroxyapatite. The XRD patterns of the bone hydroxyapatite and the sintered stoichiometric hydroxyapatite are shown in Figure 11.17. The bone hydroxyapatite gives broader peaks than the sintered stoichiometric hydroxyapatite. This is because the bone hydroxyapatite shows low crystallinity due to substitution with other ions, such as Mg^{2+}, HPO_4^{2-}, CO_3^{2-} and Cl^-, and consists of nano-sized crystallites. Living bone is a composite of nano-sized hydroxyapatite and collagen fibers. Therefore, if nano-hybrid materials, organic polymers coated with nano-sized bone-like hydroxyapatite, can be synthesized using a process mimicking biomineralization to produce bone as in living tissues, it would be possible to fabricate novel nano-hybrid materials which show high bioactivity and mechanical properties similar to those of human bone.

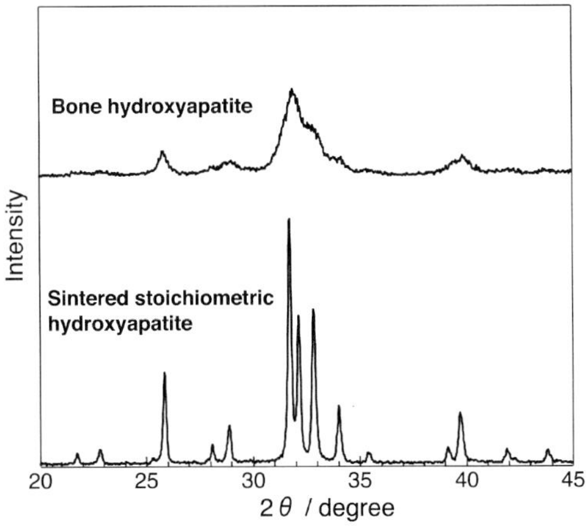

Fig. 11.17 XRD patterns of bone hydroxyapatite and sintered stoichiometric hydroxyapatite. (Bone hydroxyapatite was obtained by heating bone of pig at 600 °C).

Kokubo *et al.* [16] revealed that the hydroxyapatite formed on the bioactive ceramics in SBF is similar to bone hydroxyapatite. This indicates that a bone-like hydroxyapatite coating can be achieved by using SBF. From the mechanism of the formation of bone-like hydroxyapatite on the surface of bioactive glass, it was revealed that silanol (Si–OH) groups are effective for inducing heterogeneous nucleation of hydroxyapatite on the surfaces of the substrates. Abe *et al.* [46] reported that a bone-like hydroxyapatite layer can be formed on the surfaces of a substrate using a CaO,SiO_2-based glass as the source of the hydroxyapatite nucleating agent. First, a substrate is apposed to the CaO,SiO_2-based glass in SBF. The silicate ions released from the glass attach to the surface of the substrate and induce hydroxyapatite nucleation [47]. The substrates are then soaked in another SBF or 1.5SBF to induce crystal growth of hydroxyapatite.

It has been reported that functional groups other than Si–OH also induce heterogeneous nucleation of hydroxyapatite in SBF. Ti–OH groups on titania (TiO_2) gels also form hydroxyapatite in SBF [48], and the Ti–OH groups on TiO_2 with anatase structure are more effective for hydroxyapatite nucleation than those with rutile or amorphous structures [39]. Ta–OH [43], Zr–OH [49] and Nb–OH [50] groups have also been reported to induce hydroxyapatite nucleation. To find functional groups effective for inducing the formation of hydroxyapatite, Tanahashi *et al.* [51] prepared self-assembled layers containing various types of functional groups on the surface of gold to investigate their hydroxyapatite-forming ability. In this report, carboxyl (–COOH) groups acted as a functional group that induces heterogeneous nucleation of hydroxyapatite. Therefore, a nano-hybrid composed of nano-sized hydroxyapatite and polymer would be obtained when an organic polymer containing carboxyl groups on the surface as well as releasing Ca^{2+} ions into the fluid is soaked in SBF. In order to accelerate the hydroxyapatite formation, a solution with ion concentrations 1.5-fold greater than those of SBF (1.5SBF) can also be used.

11.5.2
Hydroxyapatite Deposition on Polymers Modified with Silanol Groups

Attempts to introduce silanol groups covalently to organic polymers in order to induce hydroxyapatite nucleation have also been reported. Kim *et al.* [52] applied a vapor-phase photografting of vinyltrimethoxysilane and subsequent hydrolysis in a hydrochloric acid solution to produce silanol groups on a polyethylene substrate. The polyethylene modified with silanol groups formed a dense and homogeneous bone-like hydroxyapatite layer in 1.5SBF. In order to obtain the polymer with high hydroxyapatite-forming ability, Oyane *et al.* [53] treated an ethylene–vinyl alcohol copolymer (EVOH) with a silane coupling agent and then calcium silicate solutions. A uniform bone-like hydroxyapatite layer was formed within 2 days in SBF on EVOH treated by the above method. They also reported the modification with Ti–OH groups by a silane coupling treatment followed by treatment with a titania solution [54]. After the arrangement of the Ti–OH groups formed on EVOH was changed from amorphous into an anatase structure by a subsequent HCl solution treatment, bone-like hydroxyapatite was formed on the modified EVOH in SBF. If these surface modifications can be applied to a three-dimensional fabric with a structure similar to the collagen fibers of natural bone, the resultant composite is expected to exhibit bioactivity and mechanical properties analogous to those of natural bone.

Nano-hybrids consisting of bone-like hydroxyapatite and biodegradable polymers are attractive materials for bone repair since they are expected to show bioactivity and be resorbed with bone repair. Hosoya *et al.* [55] have reported a hydroxyapatite–alginate nano-hybrid using SBF. It is known that alginate can act as a scaffold material for the repair of skin and nerves. Alginate was reacted with 3-aminopropyltriethoxysilane (APES), which gives silanol groups after hydrolysis by dehydration condensation using water-soluble carbodiimide to form gels. Modification of alginate with APES gave a gel, because the alginate could be cross-linked by dehydration of silanol groups derived from APES. The gels obtained were soaked in a 1 kmol $\cdot m^{-3}$ $CaCl_2$ solution and subsequently soaked in SBF. Hydroxyapatite was formed on and inside the alginate gels modified with APES (Figure 11.18). On the other hand, alginate gel

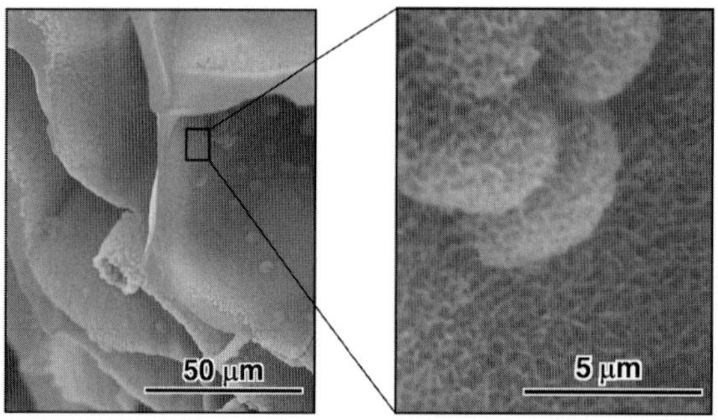

Fig. 11.18 SEM photographs of the alginate gel modified with silanol groups and calcium ions after soaking in SBF for 7 days.

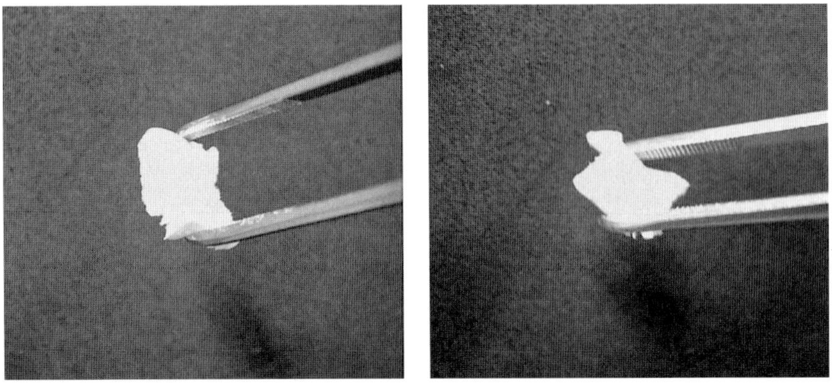

Fig. 11.19 Flexibility of the hydroxyapatite–polypeptide nano-hybrid.

cross-linked with ethylenediamine did not form hydroxyapatite, even via the same procedure. Modification of alginate with silanol groups induced not only gel formation but also the hydroxyapatite-forming ability on and inside the alginate gel in SBF. A hydroxyapatite–alginate nano-hybrid can be produced by modification of alginate with silanol groups, and subsequent soaking in $CaCl_2$ solution and SBF. Kamitakahara *et al.* [56] reported the nano-hybrid consisting of collagen-like polypeptide and hydroxyapatite. The freeze-dried collagen-like polypeptide was treated with isocyanatopropyltriethoxysilane (IPTS) and a $CaCl_2$ solution, and subsequently soaked in 1.5SBF. Hydroxyapatite formation was observed on its surface after soaking in 1.5SBF and a hydroxyapatite–polypeptide nano-hybrid was obtained. The appearance of the hydroxyapatite–polypeptide nano-hybrid is shown in Figure 11.19. The nano-hybrid shows flexibility and can be cut with scissors or a knife. Such a material is expected to be useful for bone repair since it is expected to show bioactivity, biodegradability, flexibility and workability.

11.5.3
Hydroxyapatite Deposition on Natural Polymers

There are several reports on the coating of bone-like hydroxyapatite onto natural polymer substrates. Kawashita *et al.* [57] reported that carboxymethylated chitin and gellan gum gels, which have carboxyl groups, can form hydroxyapatite on their surfaces in SBF if they are treated with a saturated $Ca(OH)_2$ solution in advance, while curdlan gel, which has no carboxyl group, does not form hydroxyapatite in SBF, even if it is treated with $Ca(OH)_2$ solution. These results support the hypothesis that carboxyl groups induce hydroxyapatite nucleation. Kokubo *et al.* [58,59] reported that non-woven fabrics of carboxymethylated chitin and alginate fibers also form hydroxyapatite on their surfaces in SBF if they are treated with $Ca(OH)_2$ solution.

Takeuchi *et al.* [60] reported that cloth made of raw silk (R-silk) fiber forms hydroxyapatite in 1.5SBF but cloth made of normal silk (N-silk) fiber does not. The surfaces of raw silk and normal silk fibers consist of sericin and fibroin, respectively.

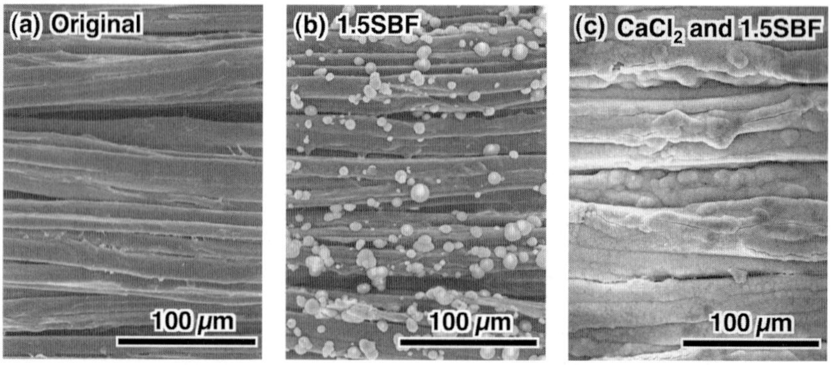

Fig. 11.20 SEM photographs of surfaces of R-silk fibers. (a) Original fibers before soaking in 1.5SBF. (b) Fibers soaked in 1.5SBF. (c) Fibers soaked in 1 kmol · m^{-3} $CaCl_2$ solution and subsequently in 1.5SBF.

The N-silk is prepared by removing the sericin from the R-silk surface. Sericin has approximately 20 mol % of acidic amino acids (aspartic and glutamic acids), while fibroin contains only 3 mol % of acidic amino acids [61,62]. This indicates that the surface of the raw silk fiber is richer in carboxyl groups than that of normal silk, and the high content of carboxyl groups in sericin is effective for heterogeneous hydroxyapatite nucleation. The hydroxyapatite-forming ability of the sericin fiber can be enhanced by soaking in $CaCl_2$ solutions before soaking in 1.5SBF (Figure 11.20). They then studied the effect of the structure of sericin on hydroxyapatite formation [63]. Hydroxyapatite formation on sericin films made by different processes was examined. The sericin films were prepared from the solution extracted from raw silk fibers under various conditions. Amongst these films only the sericin film with the highest β sheet content showed hydroxyapatite formation in 1.5SBF. This suggests that hydroxyapatite nucleation on sericin depends not only on the content of carboxyl groups but also on the content of β sheet. This indicates that not only the number of functional groups but also their arrangement are important factors governing hydroxyapatite-forming ability.

11.5.4
Hydroxyapatite Deposition on Synthetic Polymers

Synthetic polymer substrates are also useful for the preparation of hydroxyapatite-polymer nano-hybrids. Miyazaki *et al.* [64] reported that aromatic polyamides containing carboxyl groups (polyamide-COOH) [65] show hydroxyapatite-forming ability in 1.5SBF when they contain $CaCl_2$ (Figure 11.21). The hydroxyapatite layer was tightly attached to the polymer and was not peeled off by the tape test. The amount of hydroxyapatite formed increased with increasing amounts of carboxyl groups and $CaCl_2$. Kawai *et al.* [66] found that aromatic polyamides containing sulfonic ($-SO_3H$) groups (polyamide-SO_3H) also have the ability to induce hydroxyapatite formation after soaking in 1.5SBF when they are incorporated with $CaCl_2$. The amount of hydroxyapatite formed and the rate of formation increased

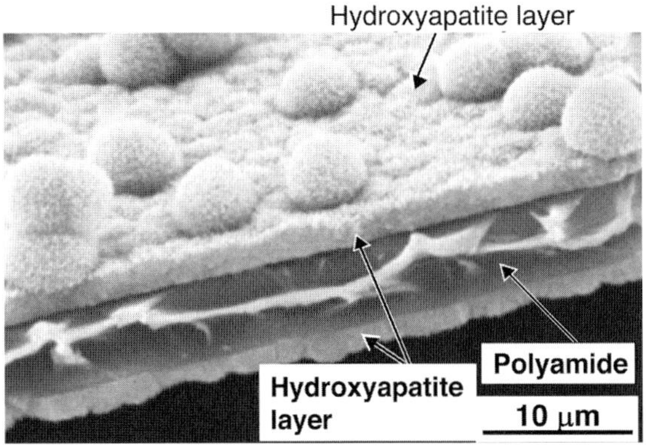

Fig. 11.21 SEM photograph of the cross-section of a polyamide-COOH film containing 40 mass % of $CaCl_2$ after soaking in 1.5SBF.

with increasing amounts of sulfonic groups in the polyamides. These results indicate that sulfonic groups can act as an effective functional group for hydroxyapatite nucleation.

Kawai *et al.* [67] then compared the induction period for the nucleation of hydroxyapatite, the rate of crystal growth, and the adhesive strength of hydroxyapatite formed on polyamide-COOH or polyamide-SO_3H incorporated with calcium chloride by soaking these materials in 1.5SBF. The induction period for hydroxyapatite nucleation on polyamide-SO_3H was shorter than for polyamide-COOH. The initial stage of the nucleation induced by silanol groups was described by Takadama *et al.* [68]. They examined the process of hydroxyapatite nucleation induced by silanol groups in SBF and reported that silanol groups do not induce hydroxyapatite nucleation directly but form amorphous calcium silicate compounds such as $\equiv SiOCa^+$ and $(\equiv SiO)_2Ca$ by incorporation of calcium ions. The amorphous calcium silicate compounds then incorporate phosphate ions to form hydroxyapatite nuclei. Considering that functional groups effective for hydroxyapatite nucleation are negatively charged in physiological solutions, the calcium incorporation at the initial stage is very important for hydroxyapatite nucleation. Therefore, easier association of sulfonic groups with calcium ions may lead to a shorter induction period for hydroxyapatite nuclei than that for carboxyl groups. On the other hand, the rate of crystal growth did not depend on the kind of functional groups, but depended on the degree of supersaturation of the surrounding solution. The adhesive strength of the hydroxyapatite layer with polyamide-SO_3H was significantly lower than that with polyamide-COOH, depending not only on the chemical interactions but also on the mechanical properties of the polyamide film.

11.5.5 Control of the Structure of Hydroxyapatite

The composition and nature of the hydroxyapatite can also be changed by modification of the solutions. Kim *et al.* [69] reported that both the Ca/P atomic ratio and the lattice constant c of the formed hydroxyapatite decrease with increasing ion concentrations from 0.75 to 2.00 times those of SBF. When the hydroxyapatite nuclei were grown by soaking the substrate in SBF under an atmosphere with a CO_2 partial pressure of 14.8 kPa to increase the content of carbonate ion to a level nearly equal to that of human blood plasma, the hydroxyapatite formed contained a higher carbonate content than the hydroxyapatite formed under air [70]. The composition and structure of the hydroxyapatite formed can be arranged to be more similar to those of bone hydroxyapatite. These results mean that we may be able to control the biological properties of hydroxyapatite, since the microstructure of hydroxyapatite can be controlled. The rate of hydroxyapatite crystal growth can also be increased by increasing solution temperature and the degree of supersaturation [71]. This indicates that we can also control the thickness of the formed hydroxyapatite layer.

11.6
Nano-hybrid Consisting of Hydroxyapatite and Protein

There have also been attempts to prepare hydroxyapatite layers functionalized by incorporation of proteins. Uchida *et al.* [72,73] reported a hydroxyapatite–laminine nano-hybrid coating on substrates to give cell-adhesive properties. A substrate which can induce hydroxyapatite nucleation is soaked in a laminine-containing metastable calcium phosphate solution which is supersaturated with hydroxyapatite (Figure 11.22). The coexistence of laminine in the formation of the hydroxyapatite coating provides efficient incorporation of laminine into the hydroxyapatite layer and the resultant coating is favorable to cell adhesion. Leonor *et al.* [74] also reported that the enzyme is able to be incorporated in the hydroxyapatite layer without losing the activity of the enzyme when hydroxyapatite was formed in SBF with the enzyme added. Although it is known that proteins are easily denatured, the biomimetic process using a solution mimicking body fluid enables us to prepare hydroxyapatite layers containing proteins without losing their activities. This is because the biomimetic process is conducted under mild synthetic conditions, for example,

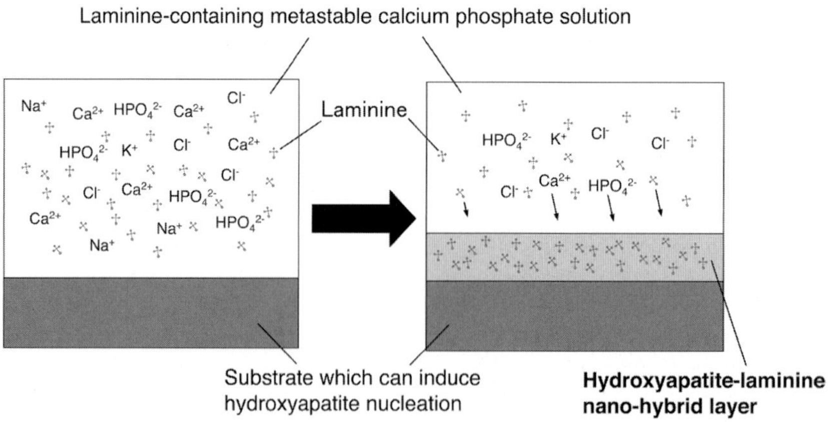

Fig. 11.22 Preparation of hydroxyapatite–laminine nano-hybrid layer.

neutral conditions and low temperature. This technique promises the preparation of hydroxyapatite with various functionalities.

11.7 Conclusion

Bioactive ceramics have already played an important role in bone repair in the clinical fields because of their ability to bond to living bone. However, the use of these bioactive ceramics is limited because of their brittleness and higher Young's modulus than bone. The combination of ceramic and organic components provides us with a design for novel tissue-repairing materials. The establishment of these techniques promises a novel bone repairing material.

References

1 Hench, L.L., Splinter, R.J., Allen, W.C. and Greenlee, T.K. Jr. (1971) Bonding mechanism at interface of ceramic prosthetic materials. *Journal of Biomedical Materials Research Symposium*, **2**, 117–141.

2 Hench, L.L. (1991) Bioceramics: from concept to clinic. *Journal of the American Ceramic Society*, **74**, 1487–1510.

3 Jarcho, M., Kay, J.L., Gumaer, R.H. and Drobeck, H.P. (1977) Tissue, cellular, and subcellular events at a bone-ceramic apatite interface. *Journal of Bioengineering*, **1**, 79–92.

4 LeGeros, R.Z. and LeGeros, J.P. (1993) An Introduction to Bioceramics (eds L. L. Hench and J. Wilson),World Scientific, Singapore, pp. 139–180.

5 Kokubo, T., Shigematsu, M., Nagashima, Y., Tashiro, M., Nakamura, T., Yamamuro, T. and Higashi, S. (1982) Apatite- and wollastonite-containing glass-ceramics for prosthetic application. *Bulletin of the Institute for Chemical Research, Kyoto University*, **60**, 260–268.

6 Kokubo, T. (1993) An Introduction to Bioceramics (eds L.L. Hench and J. Wilson), World Scientific,Singapore, pp. 75–88.

7 Hulbert, S.F. (1993) An Introduction to Bioceramics (eds L.L. Hench and J. Wilson), World Scientific, Singapore, pp. 25–40.

8 Park, J.B. and Lakes, R.S. (1992) Biomaterials, (2nd edn.) Plenum Press, New York, pp. 185–222.

9 Bonfield, W., Grynpas, M.D., Tully, A.E., Bowman, J. and Abram, J. (1981) Hydroxyapatite reinforced polyethylene – a mechanically compatible implant. *Biomaterials*, **2**, 185–186.

10 Bonfield, W. (1993) An Introduction to Bioceramics (eds L.L. Hench and J. Wilson), World Scientific,Singapore, pp. 299–303.

11 Wang, M., Hench, L.L. and Bonfield, W. (1998) Bioglass/high density polyethylene composite for soft tissue applications: Preparation and evaluation. *Journal of Biomedical Materials Research*, **42**, 577–586.

12 Juhasz, J.A., Best, S.M., Bonfield, W., Kawashita, M., Miyata, N., Kokubo, T. and Nakamura, T. (2003) Apatite-forming ability of glass-ceramic apatite-wollastonite – polyethylene composites: effect of filler content. *Journal of Materials Science-Materials in Medicine*, **14**, 489–95.

13 Neo, M., Kotani, S., Yamamuro, T., Ohtsuki, C., Kokubo, T. and Bando, Y. (1992) A comparative study of ultrastructures of the interfaces between four kinds of surface-active ceramic and bone. *Journal of Biomedical Materials Research*, **26**, 1419–1432.

14 Kokubo, T. (1991) Bioactive glass ceramics; Properties and applications. *Biomaterials*, **12**, 155–163.

15 Kokubo, T., Ito, S., Huang, Z.T., Hayashi, T., Sakka, S., Kitsugi, T. and Yamamuro, T. (1990) Ca, P-rich layer formed on high-strength bioactive glass-ceramic A-W. *Journal of Biomedical Materials Research*, **24**, 331–343.

16 Kokubo, T., Kushitani, H., Sakka, S., Kitsugi, T. and Yamamuro, T. (1990) Solution able to reproduce in vivo surface-structure changes in bioactive glass-ceramic A-W. *Journal of Biomedical Materials Research*, **24**, 721–734.

17 Cho, S.B., Nakanishi, K., Kokubo, T., Soga, N., Ohtsuki, C., Nakamura, T., Kitsugi, T. and Yamamuro, T. (1995) Dependence of apatite formation on silica gel on its structure: effect of heat treatment. *Journal of the American Ceramic Society*, **78**, 1769–1774.

18 Takadama, H., Hashimoto, M., Mizuno, M. and Kokubo, T. (2004) Round-robin test of SBF for in vitro measurement of apatite-forming ability of synthetic materials. *Phosphorus Research Bulletin*, **17**, 119–25.

19 Kokubo, T. and Takadama, H. (2006) How useful is SBF in predicting in vivo bone bioactivity? *Biomaterials*, **27**, 2907–2915.

20 Neuman, W.F. and Neuman, M.W. (1958) The Chemical Dynamics of Bone Mineral,The University of Chicago Press,Chicago, pp. 23–38.

21 Ohtsuki, C., Kokubo, T., Takatsuka, K. and Yamamuro, T. (1991) Compositional dependence of bioactivity of glasses in the system CaO-SiO_2-P_2O_5: its *in vitro* evaluation. *Journal of the Ceramic Society of Japan (Seramikkusu Ronbunshi)*, **99**, 1–6.

22 Ohtsuki, C., Kokubo, T. and Yamamuro, T. (1992) Mechanism of apatite formation on CaO-SiO_2-P_2O_5 glasses in a simulated body fluid. *Journal of Non-Crystalline Solids*, **143**, 84–92.

23 Li, P., Ohtsuki, C., Kokubo, T., Nakanishi, K., Soga, N., Nakamura, T. and Yamamuro, T. (1992) Apatite formation induced by silica gel in a simulated body fluid. *Journal of the American Ceramic Society*, **75**, 2094–2097.

24 Mackenzie, J.D., Chung, Y.J. and Hu, Y. (1992) Rubbery ormosils and their applications. *Journal of Non-Crystalline Solids*, **147&148**, 271–279.

25 Tsuru, K., Ohtsuki, C., Osaka, A., Iwamoto, T. and Mackenzie, J.D. (1997) Bioactivity of sol-gel derived organically modified silicates, Part I: *In vitro* examination. *Journal of Materials Science-Materials in Medicine*, **8**, 157–161.

26 Chen, Q., Miyata, N., Kokubo, T. and Nakamura, T. (2000) Bioactivity and mechanical properties of PDMS-modified CaO-SiO_2-TiO_2 hybrids prepared by sol-gel process. *Journal of Biomedical Materials Research*, **51**, 605–611.

27 Kamitakahara, M., Kawashita, M., Miyata, N., Kokubo, T. and Nakamura, T. (2001) Bioactivity and mechanical properties of polydimethylsiloxane (PDMS)-CaO-SiO_2 hybrids with different PDMS contents. *Journal of Sol-Gel Science and Technology*, **21**, 75–81.

28 Kamitakahara, M., Kawashita, M., Miyata, N., Kokubo, T. and Nakamura, T. (2002) Bioactivity and mechanical properties of polydimethylsiloxane (PDMS)-CaO-SiO_2 hybrids with different calcium contents. *Journal of Materials Science-Materials in Medicine*, **13**, 1015–1020.

29 Hench, L.L. and Wilson, J. (1993) An Introduction to Bioceramics (eds L.L. Hench and J. Wilson), World Scientific, Singapore, pp. 12.

30 Rhee, S.H., Choi, J.Y. and Kim, H.-M. (2002) Preparation of a bioactive and degradable poly(epsilon-caprolactone)/ silica hybrid through a sol-gel method. *Biomaterials*, **23**, 4915–4921.

31 Ohtsuki, C., Miyazaki, T. and Tanihara, M. (2002) Development of bioactive organic-inorganic hybrid for bone substitutes. *Materials Science and Engineering C*, **22**, 27–34.

32 Miyazaki, T., Ohtsuki, C. and Tanihara, M. (2003) Synthesis of bioactive organic-inorganic nano-hybrid for bone repair through sol-gel processing. *Journal of Nanoscience and Nanotechnology*, **3**, 511–515.

33 Ohtsuki, C., Miyazaki, T., Kyomoto, M., Tanihara, M. and Osaka, A. (2001) Development of bioactive PMMA-based cement by modification with alkoxysilane and calcium salt. *Journal of Materials Science-Materials in Medicine*, **12**, 895–899.

34 Miyazaki, T., Ohtsuki, C., Kyomoto, M., Tanihara, M., Mori, A. and Kuramoto, K. (2003) Bioactive PMMA bone cement prepared by modification with methacryloxypropyltrimethoxy silane and calcium chloride. *Journal of Biomedical Materials Research*, **67A**, 1417–1423.

35 Mori, A., Ohtsuki, C., Sugino, A., Kuramoto, K., Miyazaki, T., Tanihara, M. and Osaka, A. (2003)Bioactive PMMA-based bone cement modified with methacryloxypropyltrimethoxysilane and calcium salts – Effects of calcium salts on apatite-forming ability. *Journal of the Ceramic Society of Japan*, **111**, 738–742.

36 Shinzato, S., Kobayashi, M., Mousa, W.F., Kamimura, M., Neo, M., Kitamura, Y., Kokubo, T. and Nakamura, T. (2000) Bioactive polymethyl methacrylate-based bone cement: Comparison of glass beads, apatite- and wollastonite-containing glass-ceramic, and hydroxyapatite fillers on mechanical and biological properties. *Journal of Biomedical Materials Research*, **51**, 258–272.

37 Miyazaki, T., Akita, H., Ishida, E., Ashizuka, M. and Ohtsuki, C. (2006) Synthesis of bioactive organic-inorganic hybrids from tetraisopropyl titanate and hydroxyethylmethacrylate. *Journal of the Ceramic Society of Japan*, **114**, 87–91.

38 Kamitakahara, M., Kawashita, M., Miyata, N., Kokubo, T. and Nakamura, T. (2004) Degradation of bioactive polydimethylxiloxane- CaO-SiO_2-TiO_2 and poly(tetramethylene oxide)-CaO-TiO_2 hybrids in a simulated body fluid. *Journal of the American Ceramic Society*, **87**, 235–239.

39 Uchida, M., Kim, H.-M., Kokubo, T., Fujibayashi, S. and Nakamura, T. (2003) Structural dependence of apatite formation on titania gels in a simulated body fluid. *Journal of Biomedical Materials Research*, **64A**, 164–170.

40 Kamitakahara, M., Kawashita, M., Miyata, N., Kokubo, T. and Nakamura, T. (2003) Apatite formation on CaO-free polydimethylsiloxane(PDMS)-TiO_2 hybrids. *Journal of Materials Science-Materials in Medicine*, **14**, 1067–1072.

41 Wen, J. and Wilkes, G.L. (1996) Organic/inorganic hybrid network materials by the sol-gel approach. *Chemistry of Materials*, **8**, 1667–1681.

42 Kamitakahara, M., Kawashita, M., Miyata, N., Kokubo, T. and Nakamura, T. (2003) Apatite-forming ability and mechanical properties of CaO-free poly (tetramethylene oxide) (PTMO)-TiO_2 hybrids treated with hot water. *Biomaterials*, **24**, 1357–1363.

43 Miyazaki, T., Kim, H.-M., Kokubo, T., Kato, H. and Nakamura, T. (2001) Induction and acceleration of bonelike apatite formation on tantalum oxide

gel in simulated body fluid. *Journal of Sol-Gel Science and Technology*, **21**, 83–88.

44 Kamitakahara, M., Kawashita, M., Miyata, N., Kokubo, T. and Nakamura, T. (2007) Preparation of bioactive flexible poly(tetramethylene oxide) (PTMO)-CaO-Ta_2O_5 hybrids, *Journal of Materials Science-Materials in Medicine*, **18**, 1117–1124.

45 Uchino, T., Ohtsuki, C., Kamitakahara, M., Tanihara, M. and Miyazaki, T. (2006) Hydroxyapatite-forming ability and mechanical properties of organic–inorganic hybrids reinforced by calcium phosphates. *Journal of the Ceramic Society of Japan*, **114**, 692–696.

46 Abe, Y., Kokubo, T. and Yamamuro, T. (1990) Apatite coating on ceramics, metals and polymers utilizing a biological process. *Journal of Materials Science-Materials in Medicine*, **1**, 233–238.

47 Takadama, H., Kim, H.-M., Miyaji, F., Kokubo, T. and Nakamura, T. (2000) Mechanism of apatite formation induced by silanol groups – TEM observation. *Journal of the Ceramic Society of Japan*, **108**, 118–121.

48 Li, P., Ohtsuki, C., Kokubo, T., Nakanishi, K., Soga, N. and de Groot, K. (1994) The role of hydrated silica, titania and alumina in inducing apatite on implants. *Journal of Biomedical Materials Research*, **28**, 7–15.

49 Uchida, M., Kim, H.-M.Miyaji, F. Kokubo, T. and Nakamura, T. (2001) Bonelike apatite formation induced on zirconia gel in a simulated body fluid and its modified solutions. *Journal of the American Ceramic Society*, **84**, 2041–2044.

50 Miyazaki, T., Kim, H.-M., Kokubo, T., Ohtsuki, C. and Nakamura, T. (2001) Apatite-forming ability of niobium oxide gels in a simulated body fluid. *Journal of the Ceramic Society of Japan*, **109**, 929–933.

51 Tanahashi, M. and Matsuda, T. (1997) Surface functional group dependence on apatite formation on self-assembled monolayers in a simulated body fluid. *Journal of Biomedical Materials Research*, **34**, 305–315.

52 Kim, H.-M., Uenoyama, M., Kokubo, T., Minoda, M., Miyamoto, T. and Nakamura, T. (2001) Biomimetic apatite formation on polyethylene photografted with vinyltrimethoxysilane and hydrolyzed. *Biomaterials*, **22**, 2489–2494.

53 Oyane, A., Kawashita, M., Nakanishi, K., Kokubo, T., Minoda, M., Miyamoto, T. and Nakamura, T. (2003) Bonelike apatite formation on ethylene-vinyl alcohol copolymer modified with silane coupling agent and calcium silicate solutions. *Biomaterials*, **24**, 1729–1735.

54 Oyane, A., Kawashita, M., Kokubo, T., Minoda, M., Miyamoto, T. and Nakamura, T. (2002) Bonelike apatite formation on ethylene-vinyl alcohol copolymer modified with a silane coupling agent and titania solution. *Journal of the Ceramic Society of Japan*, **110**, 248–254.

55 Hosoya, K., Ohtsuki, C., Kawai, T., Kamitakahara, M., Ogata, S., Miyazaki, T. and Tanihara, M. (2004) A novel covalently crosslinked gel of alginate and silane with the ability to form bone-like apatite. *Journal of Biomedical Materials Research*, **71A**, 596–601.

56 Kamitakahara, M., Ohtsuki, C., Morihara, Y., Ogata, S. and Tanihara, M. (2005) Hydroxyapatite deposition on collagen-like polypeptide modified with silanol groups, in Archives of BioCeramics Research (eds F. Watari, T. Akazawa, M. Uo, T. Akasaka), **Vol.5**, pp. 210–213.

57 Kawashita, M., Nakao, M., Minoda, M., Kim, H.-M., Beppu, T., Miyamoto, T., Kokubo, T. and Nakamura, T. (2003) Apatite-forming ability of carboxyl group-containing polymer gels in a simulated body fluid. *Biomaterials*, **24**, 2477–2484.

58 Kokubo, T., Hanakawa, M., Kawashita, M., Minoda, M., Beppu, T., Miyamoto, T. and Nakamura, T. (2004) Apatite formation on non-woven fabric of carboxymethylated chitin in SBF. *Biomaterials*, **25**, 4485–4488.

59 Kokubo, T., Hanakawa, M., Kawashita, M., Minoda, M., Beppu, T., Miyamoto, T. and Nakamura, T. (2004) Apatite-forming ability of alginate fibers treated with calcium hydroxide solution. *Journal of Materials Science-Materials in Medicine*, **15**, 1007–1012.

60 Takeuchi, A., Ohtsuki, C., Miyazaki, T., Tanaka, H., Yamazaki, M. and Tanihara, M. (2003) Deposition of bone-like apatite on silk fiber in a solution that mimics extracellular fluid. *Journal of Biomedical Materials Research*, **65A**, 283–289.

61 Shimura, K. and Katagata, Y. (2000) Structure of silk yarn. Part B. Chemical structure and processing of silk yarn (ed N. Hojo), Science Publishers, Enfield, NH, pp. 3–20.

62 Komatsu, K. (2000) Structure of silk yarn. Part B. Chemical structure and processing of silk yarn (ed N. Hojo), Science Publishers, Enfield, NH, pp. 21–46.

63 Takeuchi, A., Ohtsuki, C., Miyazaki, T., Kamitakahara, M., Ogata, S., Yamazaki, M., Furutani, Y., Kinoshita, H. and Tanihara, M. (2005) Heterogeneous nucleation of apatite on protein: Structural effect of silk sericin. *Journal of the Royal Society Interface*, **2**, 373–378.

64 Miyazaki, T., Ohtsuki, C., Akioka, Y., Tanihara, M., Nakao, J., Sakaguchi, Y. and Konagaya, S. (2003)Apatite deposition on polyamide films containing carboxyl group in a biomimetic solution. *Journal of Materials Science-Materials in Medicine*, **14**, 569–574.

65 Konagaya, S. and Tokai, M. (2000) Synthesis of ternary copolyamides from aromatic diamine (*m*-phenylenediamine, diaminodiphenylsulfone), aromatic diamine with carboxyl or sulfonic group (3, 5-diaminobenzoic acid, 2, 4-diaminobenzenesulfonic acid), and iso- or terephthaloyl chloride. *Journal of Applied Polymer Science*, **76**, 913–920.

66 Kawai, T., Ohtsuki, C., Kamitakahara, M., Miyazaki, T., Tanihara, M., Sakaguchi, Y. and Konagaya, S. (2004) Coating of an apatite layer on polyamide films containing sulfonic groups by a biomimetic process. *Biomaterials*, **25**, 4529–4534.

67 Kawai, T., Ohtsuki, C., Kamitakahara, M., Tanihara, M., Miyazaki, T., Sakaguchi, Y. and Konagaya, S. (2005) A comparative study of apatite deposition on polyamide films containing different functional groups under a biomimetic condition. *Journal of the Ceramic Society of Japan*, **113**, 588–592.

68 Takadama, H., Kim, H.–M., Kokubo, T. and Nakamura, T. (2001)Mechanism of biomineralization of apatite on a sodium silicate glass: TEM-EDX study in vitro. *Chemistry of Materials*, **13**, 1108–1113.

69 Kim, H.-M., Kishimoto, K., Miyaji, F., Kokubo, T., Yao, T., Suetsugu, Y., Tanaka, J. and Nakamura, T. (1999) Composition and structure of the apatite formed on PETsubstrates in SBF modified with various ionic activity products. *Journal of Biomedical Materials Research*, **46**, 228–235.

70 Kim, H.-M., Kishimoto, K., Miyaji, F., Kokubo, T., Yao, T., Suetsugu, Y., Tanaka, J. and Nakamura, T. (2000) Composition and structure of apatite formed on organic polymer in simulated body fluid with a high content of carbonate ion. *Journal of Materials Science-Materials in Medicine*, **11**, 421–426.

71 Hata, K., Kokubo, T., Nakamura, T. and Yamamuro, T. (1995) Growth of a bonelike apatite layer on a substrate by a biomimetic process. *Journal of the American Ceramic Society*, **78**, 1049–1053.

72 Uchida, M., Oyane, A., Kim, H.-M., Kokubo, T. and Ito, A. (2004) Biomimetic coating of laminin-apatite composite on titanium metal and its excellent cell-adhesive properties. *Advanced Materials*, **16**, 1071–1074.

73 Oyane, A., Uchida, M. and Ito, A. (2005) Laminin-apatite composite coating to enhance cell adhesion to ethylene-vinyl alcohol copolymer. *Journal of Biomedical Materials Research*, **72A**, 168–174.

74 Leonor, I.B., Azevedo, H.S., Alves, C.M. and Reis, R.L. (2003) Effects of the incorporation of proteins and active enzymes on biomimetic calcium-phosphate coatings. *Key Engineering Materials*, **240–242**, 97–100.

12
Nanostructured Hybrid Materials for Bone Implants Fabrication

María Vallet-Regí, Daniel Arcos

12.1
Introduction

The role played by biomaterials science in the fields of orthopedic and dental surgery, is becoming more important as people's quality of life improves. The increase in the lifespan of the population has dramatically increased during the last four decades, and this scenario requires a larger number of materials aimed at regenerating and repairing the human body. The World Health Organization (WHO) has designated the 2000–2010 decade as the Bone and Joint Decade, responding to the higher incidence of bone diseases as a direct consequence of the population aging [1]. This organization estimates that several hundred million people already suffer from bone and joint diseases, with dramatic increases expected due to a doubling in the number of people over 50 years of age by 2020. The forecast concerning the female population is even more worrying: around 40 % of women older than 50 years will suffer an osteoporotic fracture [2–4]. "Bioinert" materials, such as metal alloy hip replacements tend to survive for 15–25 years, which is clearly insufficient for a patient having an implant at an age younger than 60. In order to overcome this situation the new trends in bone implants have shifted from tissue replacement to tissue regeneration [5], that is materials that can stimulate the body's own regenerative mechanisms, restoring diseased or damaged tissue to its original state and function.

Bone is one of the few tissues capable of self-regeneration during skeletal deficiency, but this regeneration is limited by the nature and size of the defect. In general, skeletal deficiency occurs as a result of trauma, tumor, bone disease, or abnormality. In the case of severe fracture, bone will not heal by itself. For this reason, artificial bone substitutes may be required to restore routine function without damaging living tissue, and the selection of the bone graft substitute is the most important factor for better performance *in vivo*.

Natural bone grafts would appear to provide the ideal material. However, autografts are necessarily limited in volume and xenografts or allografts should be considered with caution due to the potential risk for transmission of viruses or other non-conventional agents [6]. For these reasons, there has been a growing

Bio-inorganic Hybrid Nanomaterials. Edited by E. Ruiz-Hitzky, K. Ariga, Y. M. Lvov

ISBN: 978-3-527-31718-9

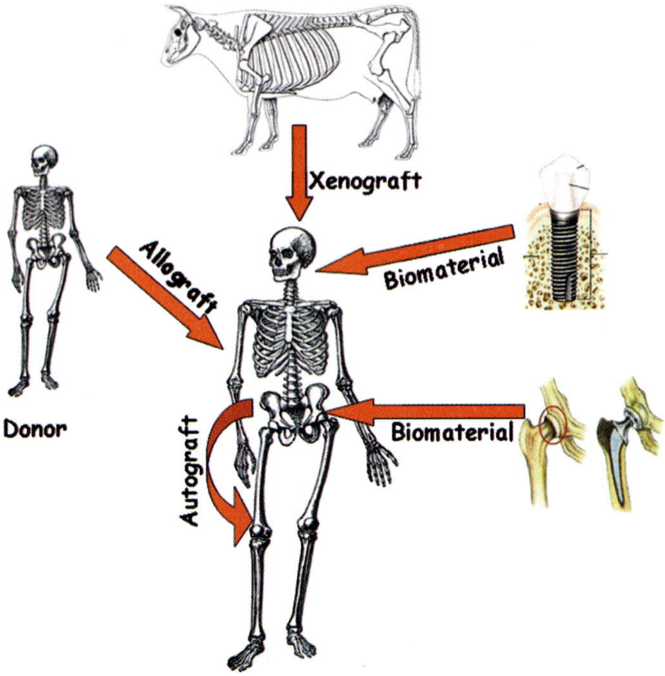

Fig. 12.1 Schematic representation of different bone grafts.

interest in developing innovative biomaterials for osseous and dental applications with improved functionality, which can be used for orthopedic, dental, and plastic surgery as well as tissue engineering. As *biomaterial* we mean "a synthetic or natural material used to replace part of a living system or to function in intimate contact with living tissue. Biomaterials are intended to interface with biological systems to emulate, treat, augment or replace any tissue, organ or function of the body" [7]. A *biomaterial* is different from a *biological material* such as bone that is produced by a biological system. Artificial hips, vascular stents, artificial pacemakers, and catheters are all made from different biomaterials and comprise different medical devices. Figure 12.1 is a schematic representation of the typical techniques for bone and dental grafting.

Currently, research on the ideal bone substitute is still in progress. The most direct approach toward the development of an artificial bone is to develop materials with similar composition and/or structure in nano-dimensional, physical, biochemical and biological response to natural bone. In this sense, the *organic–inorganic hybrid materials* provide a wide and very interesting alternative in the field of materials science [8–16]. These materials have the unique feature of combining the properties of traditional materials, such as ceramics and organic polymers, on the nanoscopic scale. As a rule, the general behavior of these organic–inorganic nanocomposites is dependent on the nature and relative content of the constitutive inorganic and organic components, although other parameters such as the synthesis conditions also determine the properties of the final materials. In this chapter we will consider the advances produced in the field of hybrid materials that are able to mimic some of the functional properties of bone. Special attention will be paid to the synthesis methodology that, as we will see, is closely related to the development of sol–gel science.

12.2
Bone: A Biological Hybrid Nanostructured Material

Bone is an extremely dense connective tissue that, in various shapes, constitutes the skeleton. Although it is one of the hardest structures in the body, bone maintains a degree of elasticity owing to its structure and composition. It possesses a hierarchical structure and, as most of the tissues, is nanostructured; in fact, it is a nanoscaled composite of collagen (organic extracellular matrix) and hydroxycarbonate apatite, (HCA, bone mineral). This nanostructure is in intimate contact with the bone cells (several microns in size), which result (at the macroscopic level) in the bone tissue. Figure 12.2 shows the bone hierarchical ordering from the bone to the crystalline structure of HCA.

There are two types of bone: (a) compact or cortical bone and (b) trabecular or cancellous bone. Cortical bone is found principally in the shafts (diaphyses) of long bones. It consists of a number of irregularly spaced overlapping cylindrical units termed Haversian systems. Each consists of a central Haversian canal surrounded by concentric lamellae of bony tissue. Trabecular bone is found principally at the ends of long bones and flat bones. It is composed of a meshwork of trabeculae within which

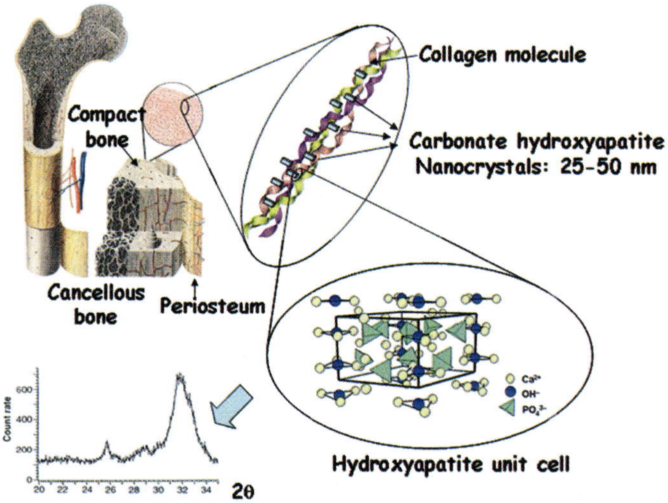

Fig. 12.2 Hierarchical organization of the organic and inorganic components of the bone tissue. Partially adapted from [30].

are intercommunicating spaces. It may appear that the trabeculae are arranged in a haphazard manner, but they are organized to provide maximum strength. The trabeculae of cancellous bone follow the lines of stress and can realign if the direction of stress changes.

The skeleton consists of approximately 80% cortical bone and 20% trabecular bone, mainly in the axial skeleton. These amounts vary according to site and relate to the need for mechanical support. While trabecular bone accounts for the minority of total skeletal tissue, it is the site of greater bone turnover because its total surface area is greater than that of cortical bone.

From a nanostructural point of view, bone is a natural composite material comprising an organic matrix with an array of inorganic apatite nanocrystals. This inorganic phase represents 45 % of the total mass of the bone, while the remainder is 30 % organic matter, and 25 % water. The major part of the organic matrix consists of type I collagen, a structural protein, which is associated with a smaller fraction of non-collageneous proteins. The collagen molecules are arranged to form fibrils, ordered in fibers and giving rise to the different macroscopic structures described above [17–19]. Collagen is the most abundant protein in mammals. A rare feature of collagen is its regular arrangement of amino acids in each of the alpha chains of the collagen sub-units. The sequence generally follows the pattern Gly-X-Y, where Gly is glycine, and X and Y are any amino acid residues. Most of the time, X is proline and Y is hydroxyproline. There are very few other proteins with such regularity. The inordinate number of Gly residues allows the tight coiling of each of the alpha chain subunits of tropocollagen, where there is a rise per turn of just 0.3 nm as opposed to the 0.36 nm of a regular alpha helical coil. Hydroxylysine and hydroxyproline play important roles in the stabilisation of the tropocollagen globular structure as well as the final fiber-shaped structure by forming covalent bonds. The resulting structure is called a collagen helix whose diameter varies from 100 to 2000 nm (see Figure 12.2).

The collagen fibers leave small compartments where apatite nanocrystals are deposited during a controlled biomineralization process [20]. The collagen acts as a structural framework in which plate-like nanocrystals of carbonated hydroxyapatite (CHA) are embedded to strengthen the bone. The chemical formula of biological CHA can be represented as follows:

$$Ca_{10-x}\, V^{(Ca)}\, [(PO_4)_{6-x-y}(HPO_4)_x(CO_3)_y]\, [(OH)_{2-x-y}(CO_3)_y V^{(OH)}{}_x]$$

where $V^{(Ca)}$ and $V^{(OH)}$ correspond to vacancies on the calcium and hydroxyl sites, respectively, and $2 - x - y = 0.4$. The CHA crystals are needle-like as a consequence of the unit cell hexagonal symmetry, which promotes anisotropic crystal growth along the *c* axis. These needle-like crystals are flat with sizes in the nanometer range and the long axis makes a small angle with the direction of the collagen fibrils. Biological apatites found in bones, dentine and enamel are always non-stoichiometric with a large compositional range in their three sublattices, always calcium deficient, with the presence of carbonates and with crystal sizes in the nanometer range. For example, in biological apatites $CO_3{}^{2-}$ and $HPO_4{}^{2-}$ by $PO_4{}^{3-}$ (type B) or OH^- (type A)

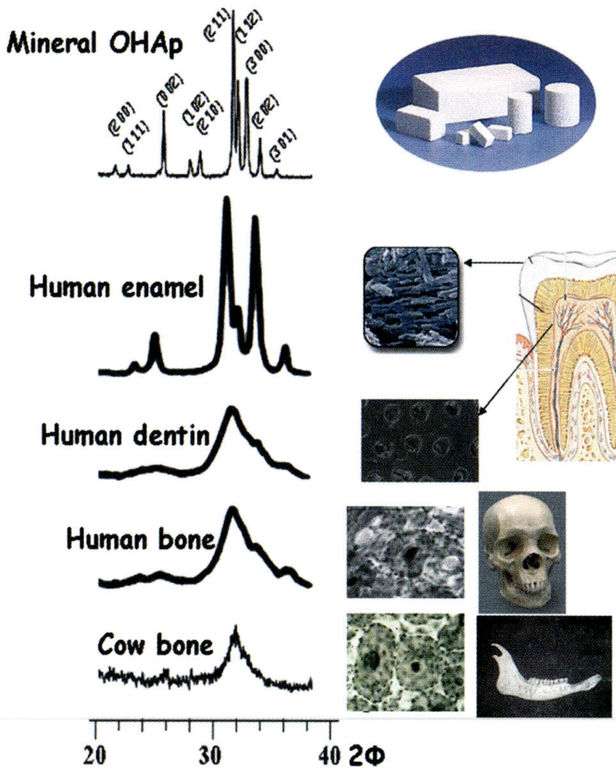

Fig. 12.3 Powder X-ray diffraction patterns of biological and synthetic hydroxyapatites.

are likely substitutions [21,22]. In the case of B-type carbonated apatites, neutrality is usually reached by the incorporation of single valence cations (Na^+ or K^+ or vacancies in the Ca^{2+} positions [23,24]. Moreover, the composition has been shown to vary as a function of species, age, and maturation stage. Figure 12.3 shows the X-ray diffraction (XRD) patterns obtained from cow and human bone as well as enamel and dentin obtained from human teeth after removing the organic fraction. The XRD pattern of synthetic HA obtained by the ceramic method (high temperature) is also shown for comparison. It can easily be observed that biological apatite samples appear different from those taken from synthetic hydroxyapatite. The profile broadening of biological samples indicates a smaller crystallite size compared with HA obtained at high temperature.

The principal cells in bone are the osteoclasts and osteoblasts. Osteoclasts, the cells responsible for resorption of bone, are derived from hematopoietic stem cells. Osteoblasts are derived from local mesenchymal cells. They are the pivotal bone cell, responsible for bone formation. Skeletal tissues are remodelled throughout a lifetime, alternating resorption phases by osteoclasts with periods of intense collagen synthesis. This balance is under the control of mechanical and hormonal stimuli, which ensure the appropriate performance of the bone. Skeletal tissues have three functions: (i) mechanical: supporting the body weight, (ii) protection of essential body organs and (iii) metabolic: a reservoir of mineral ions, mostly calcium and phosphate. The complex architecture of bone tissue determines not only the mechanical properties but also the physiological function of the bone. As in many biological materials, the hierarchical organisation at the nanometer-to-millimeter scale is essential to allow an appropriate physiological performance.

12.3 Biomimetic Materials for Bone Repair. The Hybrid Approach

Biomimetic materials are those materials not made by living organisms but having compositions and properties similar to those made by living organisms. For instance, some calcium phosphates and silica-based glasses used for bone and dental grafting, are able to develop a CHA phase in contact with the physiological fluids, in a similar way to that of vertebrate organisms. In the field of biomaterials science, these materials are commonly referred to as *bioactive materials* [25–33]. The concept of a bioactive synthetic material able to bond with tissues, started with the work of Hench *et al.* in the early 1970s [34]. It was discovered that certain glass compositions in the SiO_2–P_2O_5–Na_2O–CaO system (Bioglasses) could bond with bone when implanted. Later work by Wilson and Nolleti showed that certain compositions of the bioactive glasses also bond with soft tissues [35]. The bond-forming mechanism involves a series of controlled surface reactions in which a biologically active calcium phosphate layer is produced from species leached from the bulk glass and/or contained in the surrounding fluids. This calcium phosphate surface layer serves as the bonding site between the glass and bone, leading to the subsequent "bioactive bond".

The bonding mechanism between glass and bone has been described in detail [36]. The basis for bone bonding is the reaction of the glass with the surrounding solution. A sequence of interfacial reactions, which begin immediately after the bioactive material is implanted, leads to the formation of a CHA layer and the establishment of an interfacial bonding. The sequence of interfacial reactions can be summarized as follows:

1. Rapid exchange of Na^+ or Ca^{2+} with H^+ or H_3O^+ from solution and formation of silanols (Si–OH) at the glass surface.
2. Loss of soluble silica, in the form of $Si(OH)_4$ resulting from breaking of Si–O–Si bonds and formation of silanols.
3. Polycondensation of silanols to form a hydrated silica gel layer.
4. Migration of Ca^{2+} and ${PO_4}^{3-}$ groups to the surface through the silica layer.
5. Formation of a CaO–P_2O_5-rich film on the top of the silica-richer layer.
6. Growth of a SiO_2-rich layer by diffusion-controlled alkali ion exchange.
7. Growth of an amorphous calcium phosphate layer by incorporation of soluble calcium and phosphates from the solution.

8. Crystallization of the amorphous layer by incorporation of OH^-, CO_3^{2-} or F^- anions from solution to form a mixed hydroxyl carbonate fluorapatite layer (CHA).
9. Adsorption of biological moieties in the HCA layer.
10. Action of the macrophages.
11. Attachment of the stem cells.
12. Differentiation of the stem cells.
13. Generation of the collagen matrix.
14. Crystallization of the mineral matrix.

The first bioglasses were obtained by the conventional melting process and the Na_2O–CaO–SiO_2 diagram was chosen to design the first compositions (Figure 12.4). The glass composition of 45 % SiO_2–24.5 % Na_2O–24.5 % CaO–6 % P_2O_5 (45S5 Bioglass) was selected to provide a large amount of CaO with some P_2O_5 in a Na_2O–SiO_2 matrix. This composition is very close to a ternary eutectic, making it easy to melt. The *in vitro* tests showed that the 45S5 Bioglass composition developed a hydroxyapatite layer in test solutions. This rapid formation of HA *in vitro* was equivalent to the interfacial HA crystals of the bonded interface observed *in vivo* [37]. The HA crystals were bonded to layers of collagen fibrils produced at the interface by osteoblasts. The chemical bonding of the HA layer to collagen created the strongly bonded interface [37–40]. The rate of tissue bonding to bioactive glasses appears to depend on the rate of CHA formation, which depends on the glass composition [41,42]. As can be observed from Figure 12.4, these first *melt derived glasses* showed bioactive behavior for compositions containing silica in amounts lower than 60 % by weight.

In 1991, Li, Clark and Hench showed that a stable bioactive glass could be made by sol–gel processing [43]. By this procedure, bioactive glasses with up to 90 % SiO_2 content were obtained and, for the same silica content, the rate of CHA formation was higher in sol–gel-derived glasses than in melt-derived ones [28]. The higher bioactivity of the sol–gel glasses is attributed to the large surface area and a large concentration of silanol groups on the surface. These high surface area values are

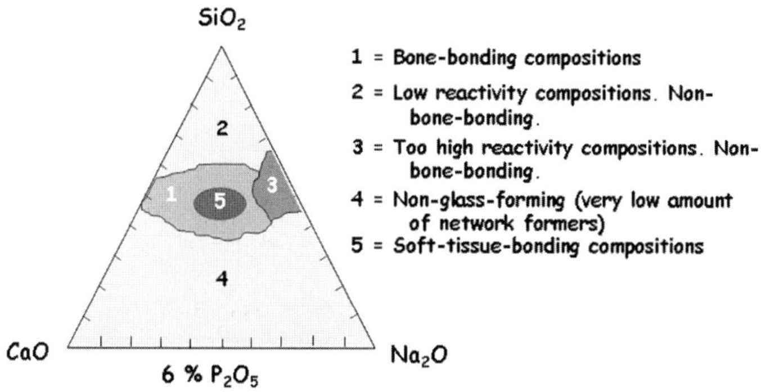

Fig. 12.4 Ternary phase diagram for SiO_2–CaO–Na_2O melt-derived glasses.

a consequence of the microporosity associated with the silica network. These features come from the sol–gel processing that allows the production of glasses and ceramics at much lower temperatures than with conventional methods. Basically, sol–gel processing of bioactive glasses is based on the hydrolysis and polycondensation of alkoxide precursors and/or inorganic salts, in order to obtain silica-based glasses with several components such as SiO_2–CaO–P_2O_5 or SiO_2–CaO systems [44–49].

Bioactive glasses are used primarily in non-loadbearing applications due to their poor mechanical properties. Glasses are rigid and brittle, whereas the bone shows excellent mechanical properties able to stand very strong impacts and strains. For this reason, obtaining materials with the bioactive properties of sol–gel glasses and acceptable mechanical properties is a very attractive aim, and the synthesis of organic–inorganic hybrid materials is one of the most promising methods to obtain artificial materials with these properties.

12.3.1 The Hybrid Approach

In recent years, research on organic–inorganic hybrid materials has become an important subject for biomedical materials and researchers. The concept of organic–inorganic hybrid materials appeared in the eighties with the expansion of soft inorganic chemistry processes [50–52]. These materials have the unique feature of combining the properties of traditional materials, such as ceramics and organic polymers, on the nanoscopic scale [53–55]. As a rule, the general behavior of hybrid materials depends on the nature and relative content of the constitutive inorganic and organic components, with a close dependence on the experimental conditions. The final product must be an intimate "mixture" where at least one of the domains (inorganic or organic) has a dimension ranging from a few angstroms to a few tens of nanometers.

The main goal when synthesising a silicate-containing hybrid material for any application, including biomedical ones, is to take advantage of both domains to improve the final properties. Table 12.1 collects some of the features that each domain can supply to the hybrid. The final properties are not only the addition of the

Tab. 12.1 Respective properties from the organic and inorganic domains, expected to be combined in hybrid materials.

Inorganic	Organic
hardness, brittleness	elasticity, plasticity
strength	low density
thermal stability	gas permeability
high density	hydrophobicity
high refractive index	selective complexation
mixed valence state (red-ox)	chemical reactivity
semiconduction	
bioactivity	
...	...

properties of the individual components but synergetic effects can be expected due to the high interfacial area.

Based on the nature of the interactions exchanged by both components, organic–inorganic hybrid materials can be classified as Class I or Class II [56]. Class I hybrid materials show weak interactions between both domains, such as van der Waals, hydrogen bonds and electrostatic interactions. No chemical links (covalent or iono-covalent) are present between the components. In these cases, silica is considered as an inorganic nanofiller incorporated to the organic component. In contrast, Class II organic–inorganic hybrid materials show chemical links between the components and, consequently strong interactions are produced. In this last case, the silicates are considered to be organically modified and they are usually referred to as *ormosils*.

In order to achieve a mechanical performance similar to that of natural bone as well as bone-bonding ability, many silica-containing hybrid materials have been developed. The high bioactivity of silicate-based glasses [57,58] suggests that the incorporation of silicate as the inorganic component would supply bioactivity to the organic component through the hybrid material synthesis.

12.4 Synthesis and Properties of Organic–Inorganic Hybrid Materials for Bone and Dental Applications

12.4.1 Class I Hybrid Materials

As mentioned above, Class I hybrid materials do not show strong chemical links (covalent bonds) between the organic and inorganic components. This fact provides a larger degree of freedom for the association of both domains. Consequently, the possibility to design new Class I hybrid materials associating biopolymers with mineral phases relies on the understanding and control of their mutual interaction. An interesting approach is synthesizing organic–inorganic hybrids based on bioactive gel glasses (BG) and a biocompatible hydrophilic organic polymer. In this section the reader will find the detailed synthesis for two systems: BG–poly(vinyl alcohol) (PVAL) and silica particles– poly(hydroxyethylmethacrylate) (pHEMA). By modifying one or both of the inorganic and organic components, these methodologies provide a wide family of Type I hybrid materials with potential applications for bone grafting.

12.4.1.1 BG–Poly(vinyl Alcohol)

The synthesis of BG–PVAL-based hybrid materials is aimed at obtaining a new family of compounds, which exhibits the bioactive behavior of sol–gel glasses together with the mechanical properties and bio-degradability of PVAL. The bioactive glass component can belong to the SiO_2–CaO–P_2O_5 or SiO_2–CaO systems. The presence of these kinds of components not only ensures the implant integration, but also stimulates the new bone formation due to the action of their degradation products

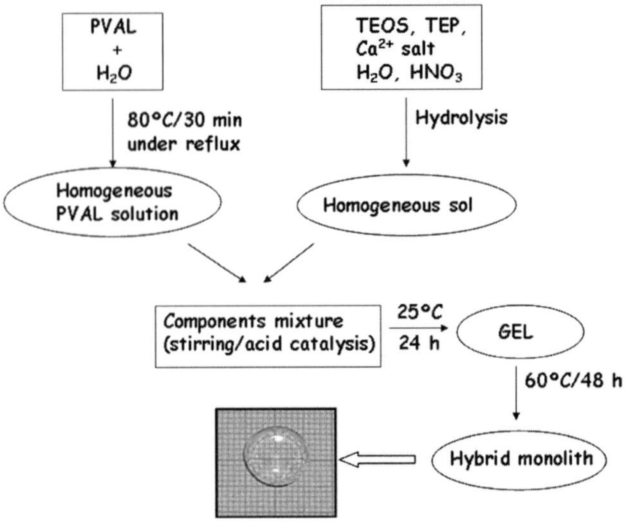

Fig. 12.5 Schematic representation of the process followed for the BG–PVAL synthesis.

(soluble silica, Ca^{2+} cations, etc.) on the gene expression of bone growth factors. Moreover, PVAL has been shown to be a very safe and biocompatible agent and years of experience have been gained with the use of this agent for embolization. Commonly, these materials exhibit infinite microstructure, can be transparent and easily shaped as films or monoliths. As a drawback, it must be said that they are generally polydisperse in size and locally heterogeneous in chemical composition.

BG–PVAL hybrid materials are formed because the organic component (in this case PVAL) is crosslinkable with the inorganic component through H-bonds or van der Waals interactions. The synthesis of these hybrids is carried out through the sol–gel synthesis under acidic conditions. For this purpose the hydrolysis of organically modified silicon and/or phosphorus alkoxides together with a calcium salt is a one-pot method, in the presence of dissolved PVAL. Figure 12.5 collects the most relevant steps of this kind of synthesis.

In a typical synthesis, the common reactants used for BG–PVAL hybrid materials synthesis can be tetraethyl orthosilicate (TEOS), triethyl phosphate (TEP) and calcium nitrate ($Ca(NO_3)_2{\cdot}4H_2O$), as precursors of SiO_2, P_2O_5 and CaO, respectively, and PVAL with average M_w between 89 000 and 98 000 g/mol. The alkoxide acid hydrolysis is commonly carried out in water and catalyzed with HNO_3. When a homogeneous sol is formed and before gelation, different amounts of PVAL are dissolved, thus obtaining Class I hybrids with different characteristics. Once gelation takes place (commonly, in 3 days at room temperature), the aging of the gel is carried out at 60 °C for 2 days and thereafter, the aged gel is dried at temperatures ranging between 60 and 80 °C, allowing the gases formed during the drying to leak out.

The final materials are Class I hybrid materials where PVAL tailors the hybrid degradation, that is the rate at which the hybrid material is dissolved by the physiological fluids, being replaced by biological newly formed bone.

PVAL has also been proposed for controlled release systems due to its biodegradability [59,60]. Pereira *et al.* have prepared multicomponent silicate–PVAL hybrids having inorganic phases with compositions containing silica, calcium and phosphorus, as transparent free-standing films [61]. Although these materials must be classified as Class I hybrid materials, the authors suggest that some of the C−OH bonds in PVAL are converted to C−O−Si bonds due to esterification reactions occurring during the synthesis of the hybrids. The results obtained from swelling experiments and infrared spectroscopy show that the crossover organic–inorganic crosslink density can be increased when the hybrids are prepared with larger concentrations of the inorganic component. Moreover, hybrids prepared at temperatures as high as 60 °C have, among other properties, greater crosslink densities and inorganic phases with larger amounts of Si−O−Si bridging bonds. The nanostructure of the materials must be closely controlled when designing silica–PVAL hybrids for bone and dental applications. Swelling experiments have shown that they vary in their reactivities ranging from fast dissolution to hydrogel properties, such as fluid uptake without material degradation.

Martín *et al.* have also reported on this system obtained as monoliths and characterized before and after being soaked in simulated body fluid (SBF) [62,63]. The biodegradability and bioactivity of the hybrids were studied as a function of the PVAL content. They observed that the addition of PVAL helped the synthesis of crack-free monoliths able to develop an apatite-like phase when soaked in SBF, Figure 12.6. However, higher amounts of P_2O_5 made the hybrid synthesis difficult and decreased the *in vitro* bioactivity, although it also contributed to the material degradability. Thus, hybrids with very high amounts of both PVAL and P_2O_5 showed such a fast

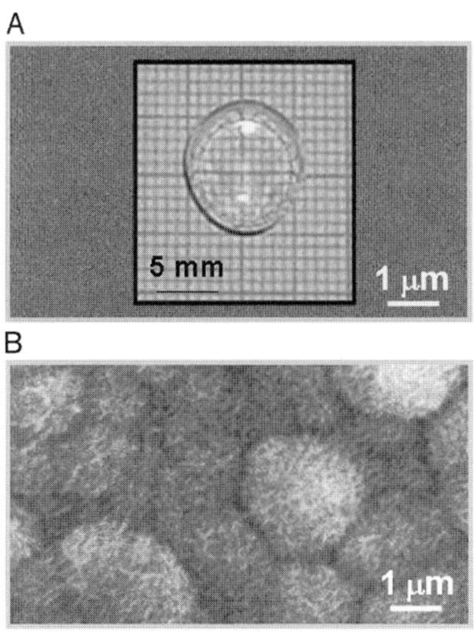

Fig. 12.6 Scanning electron micrographs obtained for the surface of a BG-PVAL hybrid material, before (A) and after (B) soaking in simulated body fluid.

degradation that apatite formation was impeded. The authors proposed these compounds as potential bone defects fillers in non-loadbearing applications or as matrices for controlled release systems.

12.4.1.2 Silica Particles–pHEMA

The synthesis of this kind of Class I hybrid material is based on the dispersion of well-defined nanobuilding blocks, which consist of preformed objects that keep their integrity in the final material. In this case, the inorganic component consists of clusters of silica nanoparticles, although nano-core shells, nanocrystalline ceramics and lamellar phosphates could also be applied, since they are easily intercalated with the organic component. The final hybrid is formed by nanoparticles connected through organic spacers, such as polymers or functional dendrimers. The main advantages of this synthesis route are that the hybrids obtained exhibit lower reactivity towards hydrolysis or attack by nucleophilic species and also that the inorganic blocks are nanometric, monodispersed and with well defined structures, which facilitates the characterization of the final materials.

Schiraldi *et al.* [64] have developed this kind of material by combining silica particles and pHEMA. pHEMA is a biocompatible hydrogel that has been widely studied in the past decades due to its chemical-physical structure and mechanical properties. It has been widely used in ophthalmic prostheses (contact or intraocular lenses), vascular prostheses, drug delivery systems and soft-tissue replacement [65]. These authors have shown that by incorporating silica nanoparticles, the resulting hybrid material is highly biocompatible and promotes bone cell adhesion and proliferation of bone cells seeded on it.[1)]

In the specific case of silica nanoparticles–pHEMA hybrid materials, the synthesis relies on obtaining a fine dispersion of silica nanoparticles (with a mean diameter of 7 nm) in HEMA monomers (liquid phase). When a homogeneous solution is obtained, a free radical initiator is added at a concentration based on the weight of the monomer mixture. After the initiator dissolution, the solution can be poured into molds or between two glass plates to obtain monoliths or uniform films, respectively, after being cured at temperatures around 60–85 °C for several hours.

12.4.2 Class II Hybrid Materials

On the basis of the sol–gel process, much more attention has been paid to silicate-containing Class II hybrids, also referred to as organically modified silicas, ormosils. In this section, the synthesis of some of the most characteristic ormosils for bone and dental applications are described.

1) Incorporation of silica nanoparticles as colloidal silica in ormosils, also contributes to improvement in the mechanical properties of Class II hybrid bioactive materials, as demonstrated by Aburatani *et al.* [66].

The strategy to synthesise Class II hybrid materials consists of making intentionally strong bonds (covalent or iono-covalent) between the organic and inorganic components. Organically modified metal alkoxides are hybrid molecular precursors that can be used for this purpose [67], but the chemistry of hybrid organic–inorganic networks is mainly developed around silicon-containing materials. Currently, the most common way to introduce an organic group into an inorganic silica network is to use organo-alkoxysilane molecular precursors or oligomers of general formula $R'_nSi(OR)_{4-n}$ or $(OR)_{4-n}Si–R''–Si(OR)_{4-n}$ with $n = 1, 2, 3$.

In most sol–gel conditions the Si–C bond remains stable during hydrolysis and the R′ group introduces new properties to the inorganic network (flexibility, hydrophobicity, biodegradability, mechanical endurance, etc.). Organic groups R′ can be introduced into an inorganic network in two different ways: as network modifiers or network formers, depending mainly on the kind of organofunctional groups contained in R′. If R′ is a simple nonhydrolyzable organic group ($Si–CH_3$, Si–phenyl, etc.) it will have a network modifying effect. However, if R′ can react with itself (R′ contains a vinyl, a methacryl or an epoxy group for example) or additional polymerizable monomers, it acts as a network former (see Figure 12.7).

The sol–gel synthesis of siloxane-based hybrid organic–inorganic implants usually involves di- or trifunctional organosilanes co-condensed with metal alkoxides, mainly $Si(OR)_4$ and $Ti(OR)_4$. As we will see in this section, the incorporation of Ca salts is a common strategy to provide bioactivity at the systems. Each of these components has specific roles that will be reviewed and discussed.

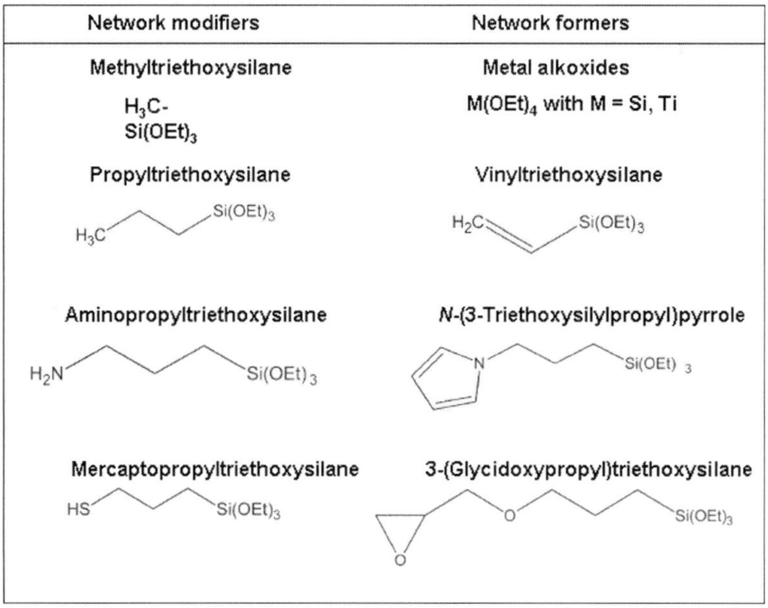

Fig. 12.7 Some examples of network formers and network modifiers used for the synthesis of ormosils.

12.4.2.1 PMMA–SiO_2 Ormosils

PMMA–silica hybrid composites have been prepared for dental restorative and bone replacement applications [68–70]. In this case, trimethoxysilyl functional groups, which are reactive in the sol–gel process, are introduced to polymethylmethacrylate through the copolymerization of 3-(trimethoxysilyl)propyl methacrylate (MSMA) with methyl methacrylate (MMA), using benzoyl peroxide (BPO) as initiator. A copolymer is obtained, hydrolyzed, and subsequently co-condensed with TEOS under acidic conditions. PMMA–SiO_2 hybrid materials exhibit growth of a low crystalline CHA layer on the surface when soaked in SBF, indicating the bioactive behavior of this hybrid. Biocompatibility tests have been carried out with this kind of material [70]. Mouse calvarial osteblast cell cultures showed better biological response when seeded on PMMA–SiO_2 hybrid materials than on PMMA in terms of cell attachment, proliferation and differentiation. The enhanced biocompatibility of the PMMA–SiO_2 hybrid was explained by two possible interrelated mechanisms: (i) the capability of inducing a calcium phosphate layer formation on the surface of the PMMA–SiO_2 in cell culture media and (ii) the capability to release silica (as silicic acid), which induces osteoblast early mineralization.

Due to the high bioactivity of PMMA–SiO_2 hybrids, this material can be used to improve the performance of acrylic bone cements. Yang *et al.* used PMMA–silica hybrid composites as filler to reinforce PMMA bone cement, changing the hybrid content from 0 to 60 % [69,71]. The Young's modulus and compressive strength are improved by the addition of the PMMA–SiO_2 hybrid material, although the tensile strength decreases due to the different particle size of PMMA–SiO_2 and PMMA.

12.4.2.2 PEG–SiO_2 Ormosils

Poly(oxyethylene)–SiO_2 ormosils have been prepared as an approach to the preparation of biologically active polymer–apatite composites. For this purpose, Yamamoto *et al.* [72] obtained these Class II hybrids from triethoxysilyl-terminated poly(oxyethylene) (PEG) and TEOS by using the *in situ* sol–gel process. After being subjected to the biomimetic process to form the bone-like apatite layer, it was found that a dense apatite layer could be prepared on the hybrid materials, indicating that the silanol groups provide effective sites for CHA nucleation and growth.

12.4.2.3 PDMS–CaO–SiO_2–TiO_2 Ormosils

One of the more thoroughly studied organic–inorganic hybrid systems for bone and dental repair is that including poly(dimethylsiloxane) (PDMS) as precursor. PDMS ($HO{-}[Si(CH_3)_2{-}O{-}]_nH$ has two silanol (Si–OH) end groups and can be incorporated into the hybrid network through condensation, together with metal or silicon alkoxides such as tetraethylorthotitanate (TEOT) or TEOS, respectively. An interesting discovery was that hybrids made from PDMS and TEOS could be rubbery even when the inorganic component weight was in excess of 70 %. These rubbery hybrid materials [73–75] showed properties comparable to those of organic rubbers, exhibiting more rubbery behavior when high acid concentration ($H+/Si = 0.3$) was used. The strong acidity of the medium increases the hydrolysis rates of Si–OR species, rapidly providing silanol groups, and slows down condensation reactions. The resulting material consists of a matrix of medium chain length and small silica particles with a porous structure. This porous texture produces the free volume needed for chain motion. As a consequence, the material exhibits rubbery elasticity.

Mackenzie *et al.* synthesised PDMS–SiO_2 ormosils from PDMS and TEOS [76]. The mechanical properties of these compounds could be controlled through the PDMS/TEOS ratio. The next step in the hybrid materials synthesis for biomedical application was in 1994, when Jones prepared similar composites of SiO_2–poly (methyl methacrylate) with Ca^{2+} ions that exhibited bioactivity although no mechanical properties were reported for these compounds [77]. Kamitakahara *et al.* obtained a highly deformable PDMS–TiO_2 hybrid with apatite-forming ability, by hydrolysis and polycondensation of PDMS and titanium ethoxide [78]. For this purpose, these authors synthesised several hybrids by changing the Si/Ti atomic ratios, observing that the stability of the material was a function of the Si/Ti atomic ratio. The hybrid material with Si/Ti atomic ratio of 2 formed many cracks. This was attributed to the insufficient content of the organic component, which gives flexibility and relaxes the stress during gelation and heat treatment. Hybrids with Si/Ti ratios of 5 and 10 were crack free, but they did not show bioactive behavior after soaking in SBF. The authors describe how to give bioactivity to the hybrids by treating with hot water at 80 °C for 7 days. Hybrid materials with an Si/Ti ratio of 5 exhibit many cracks. This might be attributed to large structural rearrangement of the hybrid to allow precipitation of anatase during the hot-water treatment, due to the large amount of titania contained in this material. The hybrid material with Si/Ti ratio of 10 was crack-free after the hot-water treatment. This hybrid material was observed to homogeneously precipitate anatase particles of 10–20 nm in size. Moreover, it formed an apatite-like phase on its surface when soaked in SBF. The amount of the newly formed apatite phase was closely related to the anatase content in the hybrid.

Chen *et al.* have worked extensively on the PDMS-modified CaO–SiO_2–TiO_2 system [79], obtaining dense and homogeneous monoliths composed of a silica and titania network incorporated with PDMS and the calcium ion ionically bonded to the network. The hybrids show relatively large amounts of calcium in their surfaces and an apatite-like phase is developed within 12 to 24 h in SBF [80,81]. The bioactivity and mechanical properties of these materials have been studied as a function of the PDMS/(TEOS + TiTP) weight ratio [82–84], observing that the apatite formation increases for lower PDMS/(TEOS + TiTP) ratios. Among the different compositions, the hybrids with PDMS/(TEOS + TiTP) of about 30 : 70 in weight showed fairly high apatite-forming ability, high extensibilities, and a Young's modulus almost equal to that of the human cancellous bone.

Both the mechanical and bioactive behavior are not only affected by chemical composition, but also by the synthesis parameters such as the thermal treatment [85]. When PDMS-modified CaO–SiO_2–TiO_2 hybrids are dried at 60 °C, an apatite-like phase grows on the surface in SBF within only one day, indicating high bioactivity. The bioactivity decreases slightly with heat treatment below 250 °C. The bending strength of the product was about 11 MPa, within the range of that of human cancellous bone. The materials strain failure decreased when the temperature of the thermal treatment increased until, after treatment in the temperature range 150 to 250 °C, it exhibited the same value as human cancellous bone.

By removing the titania component, a simpler $CaO–SiO_2$–PDMS system is obtained. With this system, the role of each component can be easily studied whereas it still combines in a single material the excellent bioactivity of the inorganic component, $CaO–SiO_2$, and the rubber-like mechanical properties induced by the organic constituent, PDMS. This system was first studied by Kamitakahra *et al.*, who focused on the bioactivity as a function of CaO content [86]. They obtained PDMS–$CaO–SiO_2$ hybrids with starting compositions containing PDMS/(TEOS + PDMS) in the weight ratio 0.30 and $Ca(NO_3)_2$/TEOS molar ratios ranging between 0 and 0.2. The apatite-forming ability of the hybrids increased with the calcium content in the $Ca(NO_3)_2$/TEOS molar ratio range 0–0.1. The hybrids with a $Ca(NO_3)_2$/TEOS molar ratio between 0.1 and 0.2 formed apatite on their surfaces in SBF within 12 h. The hybrid with a $Ca(NO_3)_2$/TEOS molar ratio of 0.10 showed an excellent apatite-forming ability in SBF with a low silicon release. These ormosils also showed mechanical properties analogous to those of human cancellous bones.

Our research group has further studied this system by using high-resolution transmission electron microscopy (HRTEM) [87]. The hybrid material showed the characteristic contrast distribution observed for amorphous materials, suggesting similar structural features to those of glasses, Figure 12.8. Energy dispersive spectroscopy (EDS) microanalysis results clearly showed the incorporation of Ca atoms randomly distributed in the SiO_2 cluster network. The nanostructural analysis revealed distances of 0.53 nm between the [SiO_4^{4-}] units. Non-bioactive $CaO–SiO_2$–PDMS materials were also synthesised. For this synthesis, the same amounts of reactants and catalyst as for the bioactive one were used, but in this case twice the amount of H_2O was used. The corresponding Fourier-filtered HRTEM image showed an average distance of 0.39 nm between [SiO_4^{4-}] units. This distance is clearly lower than the 0.53 nm measured for the bioactive hybrid, suggesting that Ca is not incorporated in the non-bioactive material. Since both hybrids exhibit different

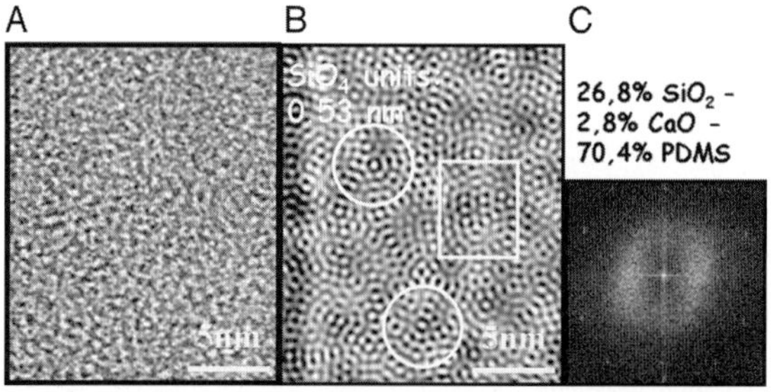

Fig. 12.8 Electron microscopy study of a PDMS–SiO_2–CaO ormosil: (A) Original HRTEM image of the amorphous matrix, (B) filtered HRTEM image and (C) Fourier transform pattern. Distances up to 0.53 nm for $(SiO_4)^{4-}$ can be observed in the filtered image, indicating the presence of Ca^{2+} between the tetrahedra.

kinetics of bioactive response, this behavior can be explained in terms of both nanostructure and chemical composition.

Ormosils have been also proposed as metallic coatings of $CaO–SiO_2$–PDMS organic–inorganic hybrids on Ti6Al4V substrates [88]. The coatings are applied by the dip-coating method. This technique is based on the sol–gel process and allows the deposition of organic–inorganic hybrids at low temperatures onto metallic substrates. These substrates are immersed in the aqueous multicomponent sols and then thermally treated. The coating method allows (i) avoidance of the corrosion of the metallic substrate due to the formation of a barrier to ion release from the implant to the body, (ii) the formation of an apatite layer on the surface improving the implant fixation, and (iii) the formation of a material intermediate between bone and the implant, which buffers their different mechanical properties. The viscosity of the sol together with the withdrawal speed of the substrate plays a key role in the final properties of the films. In general, a nanocrystalline CHA layer was formed over the film surface when soaked in SBF for 7 days.

12.4.2.4 PTMO–CaO–SiO_2–TiO_2 Hybrid Materials

Although PDMS-derived ormosils show high ductility, the mechanical strength and Young's modulus are much lower than those of human bones. In an attempt to improve these features, similar organic–inorganic hybrid materials have been synthesised by replacing PDMS with poly(tetramethylene oxide) (PTMO), expecting that a ductile bioactive material with higher mechanical strength could be obtained [83]. Kamitakahara *et al.* have obtained PTMO–TiO_2 hybrids materials that, as the analogous PDMS–TiO_2, showed bioactivity after being treated with hot water [89]. Even the hydrothermally treated hybrids still show mechanical properties analogous to those of human cancellous bones in bending strength and Young's modulus. The presence of TiO_2 in this kind of system seems to enhance the bioactivity of the hybrids. The PTMO-modified CaO–SiO_2–TiO_2 systems studied by Miyata *et al.* demonstrated that incorporation of the TiO_2 component in a PTMO–CaO–SiO_2 hybrid results in an increase in the apatite-forming ability in SBF [90]. Moreover, the TiO_2 does not significantly affect the overall mechanical properties such as Young's modulus, tensile strength and strain at failure, within the range of compositions studied by these authors.

12.4.2.5 MPS–HEMA Ormosils

Sometimes the resulting ormosils are not hydrophilic enough to promote ionic exchange with the surrounding physiological fluids. This problem can be overcome by including another organic component. In this way, bioactive organic–inorganic hybrids from 3-methacryloxypropyl-trimethoxysilane (MPS) and HEMA have been synthesised by Ohtsuki *et al.* [91,92]. MPS has alkoxysilane groups that provide silanol groups after hydrolysis, whereas HEMA may provide a hydrophilic polymer matrix in the hybrid. Once the polymer solution is obtained, $CaCl_2$ is added in order to provide better bioactivity. *In vitro* bioactivity is achieved for hybrids containing 10 mol% of MPS and with calcium ions incorporated into the hydrophilic organic polymer (pHEMA). These authors have also carried out modification of PMMA bone cement with alkoxysilanes and calcium salts. The modified cements were expected to bond tightly to living bone when implanted. The results indicated that the difference in the apatite formation is attributed to the solubility of the calcium salts. The acrylic cement (PMMA) becomes bioactive when it is modified with $CaCl_2$, $Ca(CH_3COO)_2$ or $Ca(OH)_2$ in addition to alkoxysilanes, whereas modification with $CaCO_3$ or β-TCP does not provide apatite-forming ability, even when the cement contains the alkoxysilanes.

12.4.2.6 Gelatine–SiO_2 Systems

The incorporation of gelatine is based on the assumption that the use of natural polymers, like proteins or polysaccharides, as the organic components should lead to both bioactivity and biodegradability of the resultant hybrids. Ren *et al.* have synthesized gelatine siloxane hybrids, where 3-(glycidoxypropyl) trimethoxysilane (GPSM) is grafted to the gelatine chains at the epoxy end and the grafted GPSM molecules form cross-links through hydrolysis and condensation of the methoxy silane groups at the other end [93]. The gelation of the hybrids is favoured when moderate incorporation of GPSM into the gelatine solution is carried out (mass fraction between 0.33 and 0.75) and the morphology of the gels was also dependent of the GPSM amount. As can be expected, the methoxy silane groups (Si–OCH_3) of GPSM are hydrolyzed to give silanol (Si–OH) groups and result in –Si–O–Si– bridging bonds. As indicated by ^{29}SiCP-MAS-NMR spectra, all of the Si atoms of GPSM are condensed to yield two or three bridging bonds, although the number of –Si–O–Si– bridging bonds decreases with gelatine content. Amino acid analysis demonstrated that the GPSM molecules were grafted to the gelatine chains due to the reactions between the epoxy groups of GPSM and the nucleophilic active groups of gelatine. Thus this process provides inorganic–organic hybrids in which the inorganic component of silicon chemically bonds to the natural polymer of gelatine with –Si–O–Si– bonds. The gel forming, the macro- and micro-structure, the phase separation and the crosslinking density of these hybrids depend on their molecular composition and can be tailored by varying the GPMS fraction.

Bioactive properties of gelatin–siloxane hybrids can be improved by adding a calcium source to the gelatine solution during the synthesis [94]. Post-gelation soaking in 1 M NH_4OH, Tris buffer (pH 10 and 7.0) or acetic acid-sodium acetate buffer (pH 5.0 and 3.0) solutions, followed by freeze drying leads to a macroporosity that can be controlled with the freezing temperature and the pH value of the soaking solutions. The pore size distributions range from 5 to 500 μm and the calcium content causes these hybrids to develop an apatite-like phase in contact with physiological fluids. For these reasons it has been proposed that these materials could find application as novel bioactive and biodegradable scaffolds in bone tissue engineering [95]. Further studies have demonstrated the biocompatibility of these hybrids with MC3T3-E1 cell cultures [96], demonstrating that the appropriate incorporation of Ca^{2+} ions stimulates *in vitro* osteoblast proliferation and differentiation.

12.4.2.7 Poly(ε-Caprolactone)–Silica Ormosils

Poly(ε-caprolactone) (PCL) is a well-known polymer that shows a unique set of properties, that is, biocompatibility, permeability and biodegradability. Depending on the polymerization mechanism, PCL can be end-capped by hydroxyl or vinyl groups, which are reactive with alkoxysilane. On this basis, biodegradable and biocompatible inorganic–organic hybrid materials have been prepared by Rhee *et al.* from tetraethoxysilane and end-reactive poly(ε-caprolactone) (TEOS-PCL) [97]. Nevertheless, even deprived of reactive end-groups, PCL can be successfully incorporated into the silica network, as a result of strong hydrogen bonding of the carbonyl ester group with the Si−OH groups of the silica network, resulting in Class I hybrid materials. Ormosils have been prepared by reacting α-ω-hydroxyl poly(ε-caprolactone) as organic reagent, followed by conversion of the hydroxyl end-groups into triethoxysilane end-groups [98]. Since the hydroxyl end-groups of PCL are less reactive than their triethoxysilane counterparts, the preliminary reaction of the hydroxyl end-groups with 3-isocyanatopropyltriethoxysilane is recommended in order to decrease the gelation time and the amount of PCL extractable from the final hybrid material. The amount of PCL incorporated into the network is limited by the early vitrification of the reactive system. However, thermal post-treatment can overcome these diffusional and/or kinetic limitations. TEM observations and TEM image analysis are in agreement with a co-continuous interpenetrating network of the inorganic component (silica) and the organic polymer (PCL), at least when the weight composition is close to 50 % (SiO_2/PCL). An average distance of 5 nm between the constituents in the case of a 50/50 weight composition has been calculated. However, the extent of PCL incorporation into the silica network also depends on other factors, such as the PCL molecular weight, number, and reactivity/number per chain of the PCL functional groups.

Although the biocompatibility and biodegradability of these materials were rapidly determined, the bioactivity of SiO_2–PCL hybrid materials was not studied until recently [99]. In order to provide bioactivity to SiO_2–PCL hybrid materials, Rhee prepared triethoxysilane end-capped poly(ε-caprolactone) which was then co-condensed with tetraethyl orthosilicate and calcium nitrate via the sol–gel method. The Ca-containing PCL/silica hybrid so obtained showed *in vitro* bioactivity and biodegradability. The hybridization procedure between the α,ω-hydroxyl PCL and silica phases was proposed to be as follows:

1. End-capping of both ends of α,ω-hydroxyl PCL with 3-isocyanatopropyl triethoxysilane (IPTS).
2. Two hydroxyl groups in the PCL structure react with the isocyanate group in the IPTS and form a urethane linkage.
3. Subsequent hydrolysis of triethoxysilane end-capped PCL (Si-PCL) and TEOS.
4. Condensation reaction between the hydrolyzed Si-PCL and TEOS.
5. Interpenetrating PCL–silica network is formed as in the scheme shown in Figure 12.9.

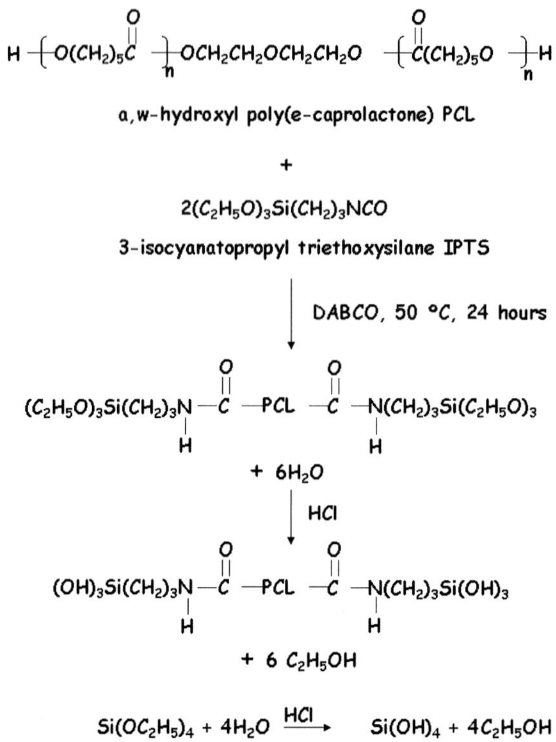

Fig. 12.9 Scheme of the formation of a hybrid between α,ω-hydroxyl PCL and silica phases.

When the PCL/silica hybrid containing calcium salt is soaked in SBF, the calcium ions are released, starting the reactions that result in nucleation and growth of a nanocrystalline apatite-like layer. Further studies with PCL/silica ormosils determined the effect of the calcium salt content on the nucleation and growth of the apatite-like layer [100]. For this purpose, the hybrids were prepared with low and high concentrations of calcium nitrate, following the same sol–gel method. Significant differences were found in the Ca/P ratio as well as in the apatite layer morphology, depending on the calcium salt concentration. Rhee explains these differences in terms of the degree of supersaturation of apatite in the SBF, as determined by the concentrations of the constituent ions of apatite and pH. He suggests that a dense bone-like apatite layer on the hybrids could be achieved *in vitro* and *in vivo* by including an appropriate calcium salt content. However, not only the calcium content determines the bioactivity, but the PCL content also modifies the bioactivity and the mechanical properties of the hybrids. The higher the PCL content in the hybrid, the lower the apatite-forming rate, showing polymer-like ductile-tough fracture behavior. On the contrary, lowering the PCL content in the hybrid leads to higher apatite-forming rates, showing ceramic-like hard-brittle fracture behavior.

Regardless of the composition, the synthesis conditions also play an important role in the bioactivity and mechanical properties [101]. When PCL/silica hybrid materials are treated at temperatures between 100 and 200 °C, the tensile strength and Young's modulus of the composites increase for high temperatures. However, the thermal treatment results in a decrease in the initial rate of apatite formation as a consequence of the condensation reaction that takes place among the silanol groups.

12.4.2.8 Bioactive Star Gels

Bioactive glasses are currently used as granulate for bone and dental grafting in small defects, or as powder incorporated into toothpaste. Although silica-based bioactive glasses meant an extraordinary advance in the field of bone tissue regeneration, their application as pieces for medium and large defects is not possible due to their very poor mechanical properties.

In 1995 DuPont Corp. developed the *star gels* materials [102–104]. Star gels are a type of organic–inorganic hybrid that presents a singular structure of an organic core surrounded by flexible arms, which are terminated in alkoxysilane groups. At the macroscopic level, star gels exhibit behavior in terms of mechanical properties between that of conventional glasses and highly cross-linked rubbers. Currently, star gels are one of the most interesting subjects in the field of hybrid materials due to their mechanical properties [105].

Very recently, the synthesis of *bioactive star gels* (BSG), that is star gels capable of integrating with bone tissue, has been developed [106]. These materials can be excellent candidates for bone tissue regeneration if several conditions are satisfied: (i) star gels must be obtained as monoliths of different shapes in order to fit any kind of medium or large bone defect; (ii) star gels must be structurally homogeneous to predict their biological and mechanical response when implanted; (c) star gels must be able to develop an apatite-like phase in contact with physiological fluids, that is they must be bioactive and (iv) star gels must exhibit mechanical properties significantly better than those exhibited by conventional bioactive glasses.

Like many other Class II hybrid materials, BSG are obtained by hydrolysis and condensation of alkoxysilanes containing precursors. In fact, star gels are formulated as single-component molecular precursors with flexibility built in at the molecular level. The starting materials comprise an organic core with multiple flexible arms that terminate in network-forming trialkoxysilane groups. The core can be a single silicon atom, a linear disiloxane segment, or a ring system, as can be seen in Figure 12.10.

The development of bioactive star gels is still in progress. Only the precursors marked A and B in Figure 12.10 have been used so far for the design of bioactive implants [106]. The basis of the star gels bioactivity consists in incorporating Ca^{2+} cations into the inorganic component of the hybrid structure, thus exhibiting similar properties to conventional SiO_2–CaO sol–gel glasses but having the flexibility supplied by the organic chains. Conventional sol–gel glasses commonly contain Ca^{2+} as CaO, which is often formed after calcination of an inorganic salt (calcium nitrate, etc.) at temperatures between 600 and 800 °C. Obviously, this thermal treatment is not compatible with the hybrid structure of the star gels and therefore an alternative synthesis route was devised by incorporating the Ca^{2+} as calcium alkoxide. The next section describes the synthesis of the bioactive star gels obtained to date.

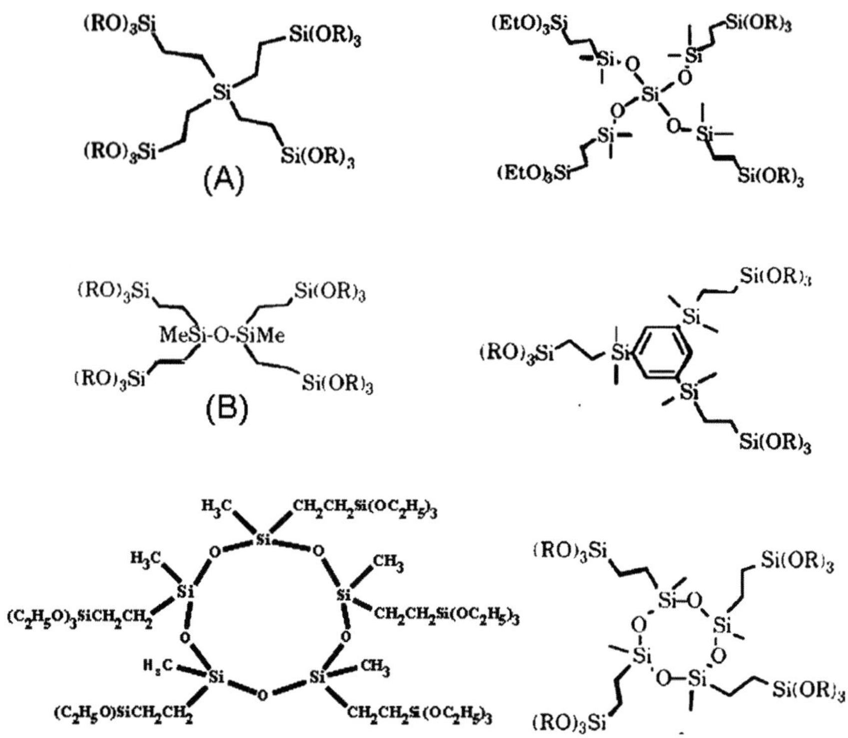

Fig. 12.10 Star gels precursors.

12.4.2.9 **The Synthesis of Bioactive Star Gels**

The synthesis of precursors A and B (see Figure 12.10) has been described by Michalczyk *et al.* [102]. These compounds can be synthesised from platinum-catalysed hydrosilylation reactions, that is addition reactions between Si–H and C=C groups in the presence of a catalyst. Once the pure precursors are obtained, BSG can be synthesised by incorporation of calcium alkoxide during polycondensation of the precursors.

In a typical synthesis, the corresponding precursor and calcium methoxyethoxide are dissolved in ethanol in an acidic medium (commonly HCl, 1 N) under an argon atmosphere. The critical factors in obtaining crack-free star gels monoliths are the temperature and processing times, as indicated in Table 12.2.

Tab. 12.2 Temperatures and processing times recommended for bioactive star gels fabrication.

Stage	Temperature/°C	Time/weeks
gelling	25	1
gel aging	60	4
gel drying	90	1

Fig. 12.11 Transparent star gel monoliths prepared with calcium alkoxide (right) and without calcium (left).

This process allows one to obtain homogeneous transparent monoliths, as shown in Figure 12.11, which have copied the shape of the reaction vessel. Obviously, this procedure allows one to tailor the shape and size of the monoliths by using vessels of different shapes and/or changing the amounts of reactants.

12.4.2.10 **How to Characterize Bioactive Star Gels?**

As mentioned before, star gels are organic–inorganic hybrid materials resulting from the hydrolysis and polycondensation of the alkoxysilane groups located at the end of the precursor arms. Ca^{2+} is incorporated into the inorganic component of the hybrid network, which allows us to understand the bioactive behavior of these materials. ^{25}Si CP-MAS NMR spectroscopy is an excellent tool to measure the degree of condensation (extent of reaction) in the gels by the relative abundance of trifunctional silicon nuclei from terminal R–Si sites. The study of the network connectivity evolution, as a function of the Ca content, is a good approximation to knowing the Ca^{2+} cations position in the network. Figure 12.12 shows the ^{29}Si CP-MAS NMR spectra of star gels type A and B with different amount of Ca^{2+}. The increase in the Ca content leads to lower T^3 populations whilst increasing those of T^2 and T^1. This is true for the two series and it in agreement with the role of Ca^{2+} as network modifier in bioactive sol–gel glasses.

12.4.2.11 **The Bioactivity of the Star Gels**

Not all the Ca^{2+}-containing star gels are bioactive. The relative amount of network formers (alkoxysilanes) and network modifiers (Ca^{2+} cations) determines the bioactivity. More specifically, the Si/Ca ratio provides a good approximation to predict whether a star gel will be bioactive or not. All compositions with Si/Ca ratios higher than 9 are not bioactive, due to the high stability of these star gels at physiological pH. The chemical composition and structure of the precursors must be known, since the number of Si atoms per unit formula must be determined. All the Si atoms must be taken into account, and not only those with hydrolyzable groups such as –Si–O–R. In this way, precursors A and B of Figure 12.10 contribute with their 5 and 9 Si atoms, respectively, to the Si/Ca ratio. Figure 12.13 is an example of the surface evolution for a star gel obtained from precursor A and with a Si/Ca ratio of 5. This figure shows the

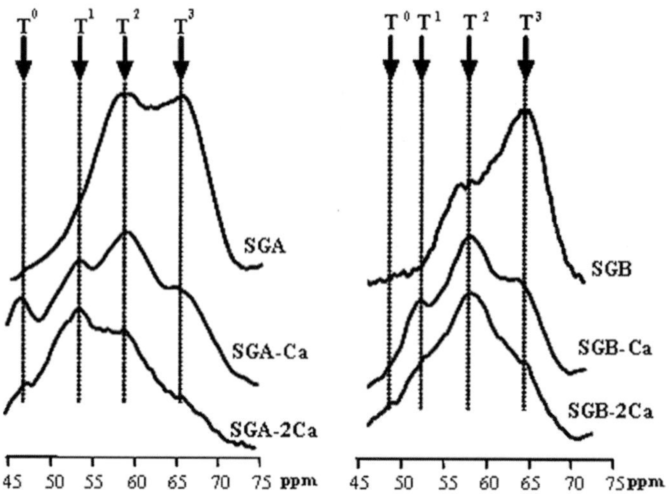

Fig. 12.12 ^{29}Si CP-MAS NMR spectra of star gels obtained from Star A and Star B precursors. Trifunctional silicon centers were named with the conventional T^n notation, where T refers to $(SiO)_nR'Si(OR)_{3-n}$ units and n to the number of bridging oxygen atoms surrounding the central silicon atom. In both BSG materials, the T^2, T^1 and T^0 relative contribution was observed to increase as the amount of Ca^{2+} introduced into the hybrid network increased. The percentage of T^3 species decreased when Ca^{2+} was added due to the role of Ca^{2+} in the network, that is reducing the cross-linking density and, therefore, reducing the amount of fully condensed Si–O–Si structures.

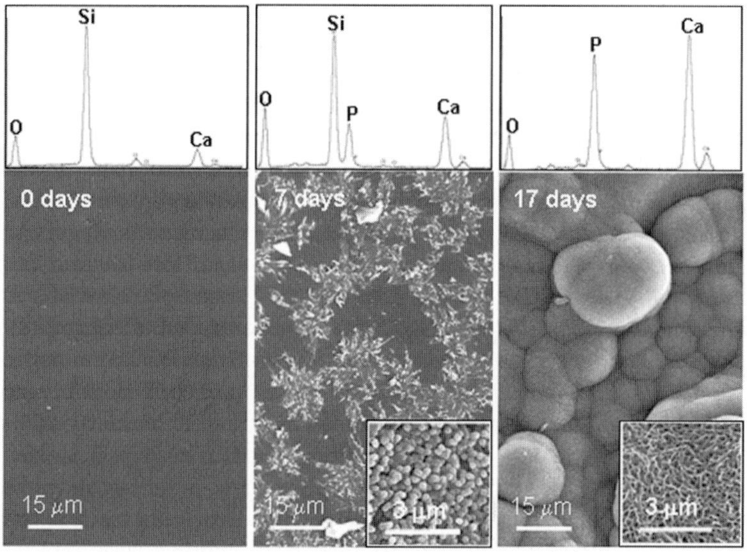

Fig. 12.13 SEM micrographs and EDX spectra of the SGA–Ca surface before and after soaking in SBF. The EDX spectra clearly show the evolution of the inorganic chemical composition on the surface: from silicon and calcium oxide to a calcium phosphate.

scanning electron micrographs for sample SGA-Ca before and after soaking in SBF for 7 and 17 days. Before soaking, the micrograph shows a smooth surface, characteristic of a non-porous and homogeneous gel. Moreover, EDX spectroscopy confirms the presence of Si and Ca as the only components of the inorganic phase. After 7 days in SFB, a new phase partially covers the star gel surface. This phase is formed by rounded submicron particles composed of Ca and P, as indicated by EDX spectroscopy. After 17 days in SBF, the monolith surface is fully covered by a layer constituted of spherical particles formed by numerous needle-shaped crystallites (characteristic of the apatite phase growth over a bioactive material's surface). At this point, the EDX spectrum indicates that the surface is fully covered by a calcium phosphate with a Ca/P ratio of 1.6, that is, corresponding to a calcium-deficient apatite. This surface evolution clearly indicates that star gels will undergo a high osteointegration process when implanted.

12.4.2.12 **The Mechanical Properties of Bioactive Star Gels**

The bioactive star gel monoliths exhibit substantially better mechanical behavior than conventional glasses. Figure 12.14 shows the Young's modulii and fracture toughness of star gels obtained from precursors A and B (denoted as Star gel (A) and Star gel (B), respectively) as well as for a conventional sol–gel glass of analogous Si and Ca content. In general, the values of the mechanical properties showed very low standard deviations, which allows one to affirm that the bioactive star gels are chemically homogeneous.

Young's modulus is a material constant that mainly depends on the chemical bond elastic forces. In the case of conventional sol–gel glasses, chemical bonds undergo very little deformation under a mechanical load, mainly due to their high bond energy that leads to rigidity and, consequently, very small deformations. The addition of an organic component results in a decreased average bond stiffness (due to this new more flexible geometrical configuration). In this case, the elastic deformation is higher under the same mechanical load. At this point, it is clear that SGB shows a very low Young's modulus and, therefore, it is not an appropriate material for structural purposes. However, the SGA material shows a Young's modulus very similar to

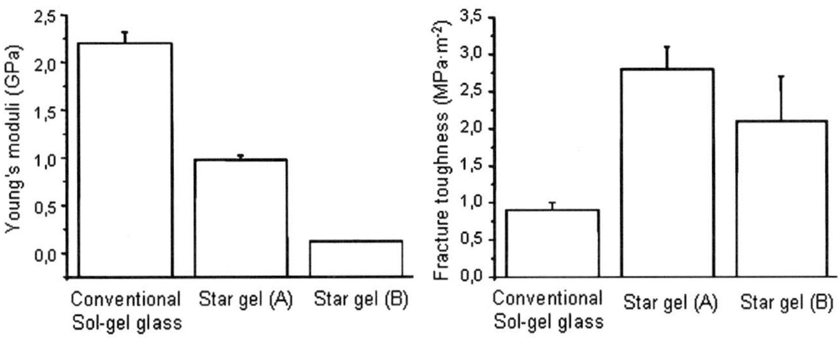

Fig. 12.14 Mechanical parameters calculated by nano-indentation tests for a conventional sol–gel glass and two different bioactive star gels with analogous Ca/Si molar ratio.

Tab. 12.3 Fracture toughness (K_{Ic}) values of a conventional sol–gel glass, star gels and those obtained for human cortical bone by different authors.

Specimen	Direction	K_{Ic} (MPa m$^{1/2}$)	Reference
sol–gel glass	N/A	0.8	—
SGA	N/A	2.8	—
SGB	N/A	2.1	—
human tibia	longitudinal	2.4–5.3	[108]
human tibia	longitudinal	3.7	[109]
human femur	transversal	6.4	[110]

cancellous bone (0.5–1.5 GPa) and lower than that of cortical bone (4–10 GPa) [107]. Conventional sol–gel glass evidences the highest Young's modulus value, followed by SGA and SGB (Figure 12.14a).

Fracture mechanics describes the fracture process in a material by relating the stress field close to the crack tip to the resultant extension of the crack length. This allows one to describe the resistance of a material to rapid crack propagation in terms of a basic parameter: the critical stress intensity factor (K_{Ic}) also known as *fracture toughness*. Fracture toughness values of star gels are significantly higher than conventional bioactive sol–gel glasses (Figure 12.14). This means that star gels are able to absorb more energy before breaking. Fracture toughness is a very important value when considering materials for clinical applications.

Conventional sol–gel glass is the material with the lowest K_{Ic} value, whereas the organic–inorganic hybrid materials show a significant increment of this value. The more flexible structure of star gels allows rotation and deformations that result in materials with better toughness than conventional sol–gel glasses. In fact, K_{Ic} values for star gels are of the same order of magnitude as those for cortical bone, determined by several authors as is shown in Table 12.3.

Currently, bioactive star gels have opened a new research field providing a range of novel bone degeneration treatment [111].

12.5 Conclusion

The fabrication of implants with controlled nanostructures has very important advantages from the point of view of clinical applications. An appropriate design at the nanostructural level results in better implant–body interactions, improved mechanical properties, and control of the material biodegradation.

Bioactive materials can be used as powders for filling small defects and as coatings that enhance metallic prosthesis fixation. However, when considering bioactive materials for bone regeneration in medium and large defects, bioactive pieces with appropriate mechanical properties are required. At this point, the key is to keep the properties provided by the nanostructure when processing a piece at the macroscopic level, that is the piece implanted by the surgeon. As we have seen in this chapter, bone is a hierarchically organized tissue, where the organic and inorganic components interact, leading to the different structures that constitute the skeleton. Traditionally, the fabrication of ceramic pieces is carried out through the so-called "ceramic process," which is not compatible with hierarchical structures. This is because the high temperatures used in this process result in crystal growth and nanostructure collapse of the as-synthesised powders. In order to avoid this inconvenience, other methods such as gel casting techniques have been used for the fabrication of ceramic pieces with medical applications [112,113]. However, the development of organic–inorganic hybrid materials, which is closely related to sol–gel chemistry, allows the synthesis of monoliths of any shape using a low-temperature process. Through this set of methodologies, we can tailor pieces with designed nanostructure and with the properties derived from this.

References

1 World Health Organization. http://www.boneandjointdecade.org/.

2 NIH Consensus Development Panel on Osteoporosis Prevention, Diagnosis, and Therapy (2001) Osteoporosis prevention, diagnosis, and therapy. *The Journal of the American Medical Association*, **285**, 785.

3 Melton, L.J., 3rd, Thamer, M., Ray, N. F., Chan, J.K., Chesnut, C.H., 3rd and Einhorn, T.A. (1997) Fractures attributable to osteoporosis: report from the National Osteoporosis Foundation. *Journal of Bone and Mineral Research*, **12**, 16–23.

4 Melton, L.J., Atkinson, E.J. and Khosla, S. (1999) Secondary osteoporosis and the risk of vertebral deformities in women. *Bone*, **24**, 49–55.

5 Hench, L.L. and Polak J.M. (2002) Third-generation biomedical materials. *Science*, **295**, 1014.

6 Murugan, R. and Ramakrishna S. (2005) Handbook of Nanostructured Biomaterials and Their Applications in Nanobiotechnology (ed. Hari Singh Nalwa), American Scientific Publishers,Stevenson Ranch, Vol. 2, pp. 141–148.

7 Society for Biomaterials Consensus Conference II. http://www.biomaterials.org/.

8 Schmidt, H. (1985) *Journal of Non-Crystalline Solids*, **73**, 681.

9 Huang, H.H., Orler, B. and Wilkes, G. L. (1985) Ceramers – hybrid materials incorporating polymeric oligomeric species with inorganic glasses by a sol-gel process. 2. Effect of acid content on the final properties. *Polymer Bulletin*, **14**, 557–564.

10 Mackenzie, J.D., Chung, Y.J. and Hu, Y. (1992) Rubbery ormosils and their applications. *Journal of Non-Crystalline Solids*, **147/148**, 271–279.

11 Mackenzie, J.D. (1994) Structures and properties of Ormosils. *Journal of Sol-Gel Science and Technology*, **2**, 81–86.

12 Motakef, S., Suratwala, T., Poncone, R. L., Boulton, J.M., Teowee, G. and Uhlmann, D.R. (1994) Bioactivity and mechanical properties of polydimethylsiloxane (PDMS)-CaO-SiO_2 hybrids with different PDMS contents. *Journal of Non-Crystalline Solids*, **178**, 37.

13 Tsuru, K., Ohtsuki, C., Osaka, A., Iwamoto, T. and Mackenzie, J.D. (1997)

Bioactivity of sol-gel derived organically modified silicates. 1. *In vitro* examination. *Journal of Materials Science–Materials in Medicine*, **8**, 157–161.

14 Chen, Q., Miyaji, F., Kokubo, T. and Nakamura, T. (1999) Apatite formation on PDMS-modified CaO-SiO_2-TiO_2 hybrids prepared by sol-gel process. *Biomaterials*, **20**, 1127–1132.

15 Sanchez, C., Lebeau, B., Chaput, F. and Boilot, J.P. (2003) Optical properties of functional hybrid organic-inorganic nanocomposites. *Advanced Materials*, **15**, 1969–1994.

16 Sanchez, C., Julián, B., Belleville, P. and Popall, M. (2005) Application of hybrid organic-inorganic nanocomposites. *Journal of Materials Chemistry*, **15**, 3559–3592.

17 Mann, S. (1989) Biomineralization; Chemical and Biochemical Perspectives (eds S. Mann, J. Webb, R.J.P. Williams), VCH, Weinheim, Chapter 2.

18 Nancollas, G.H. (1989) Biomineralization; Chemical and Biochemical Perspectives (eds S. Mann, J. Webb, R.J.P. Williams), VCH, Weinheim, Chapter 6.

19 Giraud-Guille, M.M., Besseau, L. and Martin, R. (2003) Liquid crystalline assemblies of collagen in bone and in vitro systems. *Journal of Biomechanics*, **36**, 1571–1579.

20 Glimcher, M.J. (1998) Metabolic Bone Disease and Related Disorders (eds Avioli and Krane), Academic Press, London, pp. 23–50.

21 Young, R.A. and Mackie P.E.(1980) Crystallography of human tooth enamel – initial structure refinement. *Materials Research Bulletin*, **15**, 17–29.

22 Wilson, R.M., Elliott, J.C. and Dowker, S.E.P. (1999) Rietveld refinement of the crystallographic structure of human dental enamel apatites. *American Mineralogist*, **84**, 1406–1414.

23 DeMaeyer, E.A.P., Verbeeck, R.M.H. and Naessens, D.E. (1993) Stoichiometry of Na^+ and CO_3^{2-} containing apatites obtained by hydrolysis of monetite. *Inorganic Chemistry*, **32**, 5709–5714.

24 Verbeeck, R.M.H., De Maeyer, E.A.P. and Driessens, F.C.M. (1995) Stoichiometry of potassium-containing and carbonate-containing apatites synthesized by solid-state reactions. *Inorganic Chemistry*, **34**, 2084–2088.

25 Vallet-Regí, M. (2001) Ceramics for medical applications. *Journal of the Chemical Society-Dalton Transactions*, 97–108.

26 Jones, J.R. and Hench L.L. (2003) Regeneration of trabecular bone using porous ceramics. *Current Opinion in Solid State & Materials Science*, **7**, 301–307.

27 Vallet-Regí, M. and Arcos D. (2005) Silicon substituted hydroxyapatites. A method to upgrade calcium phosphate based implants. *Journal of Materials Chemistry*, **15**, 1509–1516.

28 Vallet-Regí, M., Ragel, C.V. and Salinas, A.J. (2003) Glasses with medical applications. *European Journal of Inorganic Chemistry*, 1029–1042.

29 Hench, L.L. (1998) Biomaterials: a forecast for the future. *Biomaterials*, **19**, 1419–1423.

30 Vallet-Regí, M. and González-Calbet, J.M. (2004) Calcium phosphates as substitutes for bone tissues. *Progress in Solid State Chemistry*, **32**, 1–31.

31 Hench, L.L. and Wilson, J. (1984) Surface-active biomaterials. *Science*. **226**, 630–636.

32 Kokubo, T., Kushitani, H., Sakka, S., Kitsugi, T. and Yamamuro, T. (1990) Solutions able to reproduce in vivo surface-structure changes in bioactive glass-ceramic A-W. *Journal of Biomedical Materials Research*, **24**, 721–734.

33 Arcos, D., Greenspan, D.C. and Vallet-Regí, M. (2003) A new quantitative method to evaluate the in vitro bioactivity of melt and sol-gel-derived silicate glasses. *Journal of Biomedical Materials Research*, **65A**, 344–351.

34 Hench, L.L., Splinter, R.J., Greenlee, T. K. and Allen, W.C. (1971) Bonding mechanism at the interface of ceramic prosthetic materials. *Journal of Biomedical Materials Research*, **2**, 117.

35 Wilson, J. and Nolleti D. (1990) Handbook of Bioactive Ceramics (eds Yamamuro, Hench, Wilson), Boca Raton, Florida, pp. 283–302.

36 Hench, L.L. and Andersson O. (1993) bioactive Glasses. An Introduction to Bioceramics (eds L.L. Hench and J. Wilson), World Scientific Publishing, Singapore, pp. 41.

37 Hench, L.L. (2006) The story of Bioglass. *Journal of Materials Science–Materials in Medicine*, **17**, 967–978.

38 Hench, L.L., Clark, A.E. and Schaake, H.F. (1972) *Journal of Non-Crystalline Solids*, **8-10**, 837.

39 Hench, L.L. and Paschall H.A. (1973) Direct chemical bond of bioactive glass-ceramic materials to bone and muscle. *Journal of Biomedical Materials Research Symposium*, **4**, 25.

40 Hench, L.L. and Paschall H.A. (1974) Histochemical responses at a biomaterials interface. *Journal of Biomedical Materials Research Symposium*, **5**, 49.

41 Kim, C.Y., Clark, A.E. and Hench, L.L. (1992) Compositional dependence of calcium-phosphate layer formation in fluoride bioglasses. *Journal of Materials Science*, **26**, 1147–1161.

42 Ogino, M., Ohuchi, F. and Hench, L.L. (1980) Compositional dependence of the formation of calcium-phosphate films on bioglass. *Journal of Biomedical Materials Research*, **14**, 55–64.

43 Li, R., Clark, A.E. and Hench, L.L. (1991) An investigation of bioactive glass powders by sol-gel processing. *Journal of Applied Biomaterials*, **2**, 231–239.

44 Hench, L.L. and West J.K. (1990) The sol-gel process. *Chemical Reviews*, **90**, 33–72.

45 Vallet-Regí, M., Romero, A.M., Ragel, C.V. and LeGeros, R.Z. (1999) XRD, SEM-EDS, and FTIR studies of *in vitro* growth of an apatite-like layer on sol-gel glasses. *Journal of Biomedical Materials Research*, **44**, 416–421.

46 Peltola, M., Jokinen, M., Rahiala, H., Levanen, E., Rosenholm, J.B., Kangasniemi, I. and Yli-Urpo, A. (1999) Calcium phosphate formation on porous sol-gel-derived SiO_2 and CaO-P_2O_5-SiO_2 substrates *in vitro. Journal of Biomedical Materials Research*, **44**, 12–21.

47 Vallet-Regí, M., Arcos, D. and Pérez-Pariente, J. (2000) Evolution of porosity during *in vitro* hydroxycarbonate apatite growth in sol-gel glasses. *Journal of Biomedical Materials Research*, **51**, 23–28.

48 Izquierdo-Barba, I., Salinas, A.J. and Vallet-Regí, M. (1999) In vitro calcium phosphate layer formation on sol-gel glasses of the CaO-SiO_2 system. *Journal of Biomedical Materials Research*, **47**, 243–250.

49 Arcos, D., Greenspan, D.C. and Vallet-Regí, M. (2002) Influence of the stabilization temperature on textural and structural features and ion release in SiO_2-CaO-P_2O_5 sol-gel glasses. *Chemistry of Materials*, **14**, 1515–1522.

50 Schmidt, H., Kaiser, A., Patzelt, H. and Sholze, H. (1982) Mechanical and physical-properties of amorphous solids based on $(CH_3)_2SiO$-SiO_2 gels. *Journal of Physics*, **12**, 275–278.

51 Livage, J., Henry, M. and Sanchez, C. (1988) Sol-gel chemistry of transition-metal oxides. *Progress in Solid State Chemistry*, **18**, 259–341.

52 Avnir, D., Levy, D. and Reisfeld, R. (1984) The nature of the silica cage as reflected by spectral changes and enhanced photostability of trapped rhodamine-6G. *Journal of Physical Chemistry*, **88**, 5956–5959.

53 Novak, B.M. (1993) Hybrid nanocomposite materials – between inorganic glasses and organic polymers. *Advanced Materials*, **5**, 422–433.

54 Chujo, Y. and Saegusa T. (1992) Organic polymer hybrids with silica-gel formed by means of the sol-gel method. *Advances in Polymer Science*, **100**, 11–29.

55 Schubert, U., Hüsing, N. and Lorenz, A. (1995) Hybrid inorganic-organic materials by sol-gel processing of organofunctional metal alkoxides. *Chemistry of Materials*, **7**, 2010–2027.

56 Sanchez, C. and Ribot F. (1994) Design of hybrid organic-inorganic materials synthesized via sol-gel chemistry. *New Journal of Chemistry*, **18**, 1007–1047.

57 Hench, L.L. and LaTorre G.P. (1993) Bioceramics (eds Yamamuro, Kokubo, Nakamura, Kobunshi), Kankokai, Inc., Kyoto, **5**, pp. 67–74.

58 Ohtsuki, C., Kokubo, T. and Yamamuro, T. (1992) Mechanism of apatite formation on CaO-SiO_2-P_2O_5 glasses in a simulated body fluid. *Journal of Non-Crystalline Solids*, **143**, 84–92.

59 Sánchez-Chávez, M., Arranz, F. and Cortázar, M. (1998) Poly(vinyl alcohol) functionalized by monosuccinate groups. Coupling of bioactive amino compounds. *Polymer*, **39**, 2751–2757.

60 Soppimath, K.S., Kulkarni, A.R. and Aminabhave, M. (2000) Controlled release of antihypertensive drug from the interpenetrating network poly (vinyl alcohol)-guar gum hydrogel microspheres. *Journal of Biomaterials Science-Polymer Edition*, **11**, 27–43.

61 Pereira, A.P.V., Vasconcelos, W.L. and Oréfice, R.K. (2000) Novel multicomponent silicate-poly(vinyl alcohol) hybrids with controlled reactivity. *Journal of Non-Crystalline Solids*, **273**, 180–185.

62 Martín, A.I., Salinas, A.J. and Vallet-Regí, M. (2005) Bioactive and degradable organic-inorganic hybrids. *Journal of the European Ceramic Society*, **25**, 3533–3538.

63 Salinas, A.J., Merino, J.M., Hijón, N., Martín, A.I. and Vallet-Regí, M. (2004) Bioactive organic-inorganic hybrids based on CaO-SiO_2 sol-gel glasses. *Key Engineering Materials*, **254–256**, 481–484.

64 Schiraldi, C., DAgostino, A., Oliva, A., Flamma, F., De Rosa, A., Apicella, A., Aversa, R. and De Rosa, M. (2004) Development of hybrid materials based on hydroxyethylmethacrylate as supports for improving cell adhesion and proliferation. *Biomaterials*, **23**, 3645–3653.

65 Monthéard, J.P., Chatzopoulos, M. and Chappard, D. (1992) 2-hydroxyethyl methacrylate (HEMA) – chemical properties and applications in biomedical fields. *Journal of Macromolecular Science – Macromolecular Review*, **32**, 1–34.

66 Aburatani, Y., Tsuru, K., Hayakawa, S. and Osaka, A. (2002) Mechanical properties and microstructure of bioactive ORMOSILs containing silica particles. *Materials Science and Engineering C*, **20**, 195–198.

67 Schmidt, H. and Seiferling B. (1986) Chemistry and applications of inorganic-organic polymers. *Materials Research Society Symposium Proceedings*, **73**, 739.

68 Wei, Y. and Jin D.Sep 7 (1997) "A new class of organic-inorganic hybrid dental materials" Abstracts of papers of the American Chemical Society 214: 145-POLY Part 2.

69 Yang, J.M., Lu, C.S., Hsu, Y.G. and Shih, C.H. (1997) Mechanical properties of acrylic bone cement containing PMMA-SiO_2 hybrid sol-gel material. *Journal of Biomedical*

Materials Research Applied Biomaterials, **38**, 143–154.

70 Rhee, S. and Choi J. (2002) Preparation of a bioactive poly(methyl methacrylate)/silica nanocomposite. *Journal of the American Ceramic Society*, **85**, 1318–1320.

71 Yang, J.M., Shih, C.H., Chang, C.N., Lin, F.H., Jiang, J.M., Hsu, Y.G., Su, W. Y. and See, L.C. (2003) Preparation of epoxy-SiO_2 hybrid sol-gel material for bone cement. *Journal of Biomedical Materials Research*, **64A**, 138–146.

72 Yamamoto, S., Miyamoto, T., Kokubo, T. and Nakamura, T. (1998) Preparation of polymer-silicate hybrid materials bearing silanol groups and the apatite formation on/in the hybrid materials. *Polymer Bulletin*, **40**, 243–250.

73 Hu, Y. and Mackenzie J.D. (1992) Rubber-like elasticity of organically modified silicates. *Journal of Materials Science*, **27**, 4415–4420.

74 Chung, Y.J., Ting, S. and Mackenzie, J.D. (1990) Better Ceramics through Chemistry IV (eds B.J.J. Zelinski, C.J. Brinker, D.E. Clark, D.R. Ulrich), Materials Research Society, Pittsburg, PA, Vol. **180**, pp. 981.

75 Mackenzie, J.D., Chung, Y.J. and Hu, Y. (1992) Rubbery ormosils and their applications. *Journal of Non-Crystalline Solids*, **147**, 271–279.

76 Mackenzie, J.D., Huang, Q.H. and Iwamoto, T. (1996) Mechanical properties of ormosils. *Journal of Sol-Gel Science and Technology*, **7**, 151–161.

77 Jones, S.M., Friberg, S.E. and Sjoblom, J. (1994) A bioactive composite-material produced by the sol-gel method. *Journal of Materials Science*, **29**, 4075–4080.

78 Kamitakahara, M., Kawashita, M., Miyata, N., Kokubo, T. and Nakamura, T. (2003) Apatite formation on CaO-free polydimethylsiloxane (PDMS)-TiO_2 hybrids. *Journal of Materials Science–Materials in Medicine*, **14**, 1067–1072.

79 Chen, Q., Kamitakahara, M., Miyata, N., Kokubo, T. and Nakamura, T. (2000) Preparation of bioactive PDMS-modified CaO-SiO_2-TiO_2 hybrids by the sol-gel method. *Journal of Sol-Gel Science and Technology*, **19**, 101–105.

80 Chen, Q., Miyaji, F., Kokubo, T. and Nakamura, T. (1999) Apatite formation on PDMS-modified CaO-SiO_2-TiO_2 hybrids prepared by sol-gel process. *Biomaterials*, **20**, 1127–1132.

81 Miyata, N., Fuke, K., Chen, Q., Kawashita, M., Kokubo, T. and Nakamura, T. (2003) Preparation of PTMO-modified CaO-TiO_2 hybrids via sol-gel processing: Their apatite-forming ability and mechanical properties. *Journal of the Ceramic Society of Japan*, **111**, 555–559.

82 Chen, Q., Miyata, N. and Kokubo, T. (2003) Bioactivity and mechanical properties of poly (dimethylsiloxane) modified calcia-silica hybrids with added titania. *Journal of the American Ceramic Society*, **86**, 806–810.

83 Miyata, N., Fuke, K., Chen, Q., Kawashita, M., Kokubo, T. and Nakamura, T. (2002) Apatite-forming ability and mechanical properties of PTMO-modified CaO-SiO_2 hybrids prepared by sol-gel processing: Effect of CaO and PTMO contents. *Biomaterials*, **23**, 3033–3040.

84 Chen, Q., Miyata, N., Kokubo, T. and Nakamura, T. (2000) Bioactivity and mechanical properties of PDMS-modified CaO-SiO_2-TiO_2 hybrids prepared by sol-gel process. *Journal of Biomedical Materials Research*, **51**, 605–611.

85 Chen, Q., Miyata, N., Kokubo, T. and Nakamura, T. (2001) Effect of heat treatment on bioactivity and mechanical properties of PDMS-modified CaO-SiO_2-TiO_2 hybrids via sol-gel process. *Journal of Materials*

Science–Materials in Medicine, **12**, 515–522.

86 Kamitakahara, M., Kawashita, M., Miyata, N., Kokubo, T. and Nakamura, T. (2002) Bioactivity and mechanical properties of polydimethylsiloxane (PDMS)-CaO-SiO_2 hybrids with different calcium contents. *Journal of Materials Science–Materials in Medicine*, **13**, 1015–1020.

87 Vallet-Regí, M., Salinas, A.J., Ramírez-Castellanos, J. and González-Calbet, J. M. (2005) Nanostructure of bioactive sol-gel glasses and organic inorganic hybrids. *Chemistry of Materials*, **17**, 1874–1879.

88 Hijón, N., Manzano, M., Salinas, A.J. and Vallet-Regí, M. (2005) Bioactive CaO-SiO_2-PDMS coatings on Ti6Al4V substrates, *Chemistry of Materials*, **17**, 1591–1596.

89 Kamitakahara, M., Kawashita, M., Miyata, N., Kokubo, T. and Nakamura, T. (2003) Apatite-forming ability and mechanical properties of CaO-free poly (tetramethylene oxide) (PTMO)-TiO_2 hybrids treated with hot water. *Biomaterials*, **24**, 1357–1363.

90 Miyata, N., Fuke, K., Chen, Q., Masakazu, K., Kokubo, T. and Nakamura, T. (2004) Apatite-forming ability and mechanical properties of PTMO-modified CaO-SiO_2-TiO_2 hybrids derived from sol-gel processing. *Biomaterials*, **25**, 1–7.

91 Ohtsuki, C., Miyazaki, T. and Tanihara, M. (2002) Development of bioactive organic-inorganic hybrid for bone substitutes. *Materials Science and Engineering C*, **22**, 27–34.

92 Miyazaki, T., Ohtsuki, C. and Tanihara, M. (2003) Synthesis of bioactive organic-inorganic nanohybrid for bone repair through sol-gel processing. *Journal of Nanoscience and Nanotechnology*, **3**, 511–515.

93 Ren, L., Tsuru, K., Hayakawa, S. and Osaka, A. (2001) Synthesis and characterization of gelatin-siloxane hybrids derived through sol-gel procedure. *Journal of Sol-Gel Science and Technology*, **21**, 115–121.

94 Ren, L., Tsuru, K., Hayakawa, S. and Osaka, A. (2001) Sol-gel preparation and *in vitro* deposition of apatite on porous gelatin-siloxane hybrids. *Journal of Non-Crystalline Solids*, **285**, 116–122.

95 Ren, L., Tsuru, K., Hayakawa, S. and Osaka, A. (2002) Novel approach to fabricate porous gelatin-siloxane hybrids for bone tissue engineering. *Biomaterials*, **23**, 4765–4773.

96 Ren, L., Tsuru, K., Hayakawa, S. and Osaka, A. (2003) In vitro evaluation of osteoblast response to sol-gel derived gelatin-siloxane hybrids. *Journal of Sol-Gel Science and Technology*, **26**, 1137–1140.

97 Rhee, S.H., Choi, J.Y. and Kim, H.M. (2002) Preparation of a bioactive and degradable poly(epsilon-caprolactone)/silica hybrid through a sol-gel method. *Biomaterials*, **23**, 4915–4921.

98 Tian, D., Blacher, S., Ph Dubois and Jérome, R. (1998) Biodegradable and biocompatible inorganic-organic hybrid materials – 2. Dynamic mechanical properties, structure and morphology. *Polymer*, **39**, 855–864.

99 Rhee, S.H. (2004) Bone-like apatite-forming ability and mechanical properties of poly(ε-caprolactone)/silica hybrid as a function of poly (ε-caprolactone) content. *Biomaterials*, **25**, 1167–1175.

100 Rhee, S.H. (2003) Effect of calcium salt content in the poly(ε-caprolactone)/ silica nanocomposite on the nucleation and growth behavior of apatite layer. *Journal of Biomedical Materials Research*, **67A**, 1131–1138.

101 Yoo, J.J. and Rhee S.H. (2004) Evaluations of bioactivity and mechanical properties of poly (ε-caprolactone) silica nanocomposite following heat treatment. *Journal of*

Biomedical Materials Research, **68A**, 401–410.

102 Michalczyk, M.J. and Sharp K.G. US Patent 5,378,790 (1995).

103 Sharp, K.G. and Michalczyk M.J. (1997) Star gels: New hybrid network materials from polyfunctional single component precursors. *Journal of Sol-Gel Science and Technology*, **8**, 541–546.

104 Sharp, K.G. (1998) Inorganic/organic hybrid materials. *Advanced Materials*, **10**, 1243.

105 Sharp, K.G. (2005) Star alkoxysilane molecules, gels and appreciably tough glasses. *Journal of Materials Chemistry*, **15**, 3812–3820.

106 Manzano, M., Arcos, D., Rodríguez Delgado, M., Ruiz, E., Gil, F.J. and Vallet-Regí, M. (2006) Bioactive star gels. *Chemistry of Materials*, **18**, 5696–5703.

107 Currey, J.D. (1998) The effect of porosity and mineral-content on the Youngs modulus of elasticity of compact-bone. *Journal of Biomechanics*, **21**, 131–139.

108 Bonfield, W., Behiri, J.C. and Charambilides, B. (1984) Biomechanics: Current Interdisciplinary Research (eds S.M. Perrin and E. Scheider), Martinus Nijhoff, Dordrecht, pp. 36.

109 Norman, T.L., Vashischth, D. and Burr, D.B. (1995) Fracture-toughness of human bone under tension. *Journal of Biomechanics*, **28**, 309–320.

110 Zioupos, P. and Currey J.D. (1998) Changes in the stiffness, strength, and toughness of human cortical bone with age. *Bone*, **22**, 57–66.

111 Greene, M.E. (2007) Bone up on star gels – Research news. *Materials Today*, **10**, 15.

112 Padilla, S., Vallet-Regí, M., Ginebra, M. P. and Gil, F.J. (2005) Processing and mechanical properties of hydroxyapatite pieces obtained by the gelcasting method. *Journal of the European Ceramic Society*, **25**, 375–383.

113 Cabañas, M.V., Peña, J., Román, J. and Vallet-Regí, M. (2006) Room temperature synthesis of agarose/sol glass pieces with tailored interconnected porosity. *Journal of Biomedical Materials Research*, **78A**, 508–514.

13
Bio-inorganic Conjugates for Drug and Gene Delivery

Jin-Ho Choy, Jae-Min Oh, Soo-Jin Choi

13.1
Introduction

In the last century, many attempts were made to understand the structure–property relationships of various inorganic, organic, and biomolecules on the nanometer scale. However, the major research initiative in the twenty first century is the design of new structures and the creation of novel properties through the hybridization of these materials with one another leading to the convergence of related technologies, such as NT-BT (nanotechnology-biotechnology) [1,2], NT-IT (nanotechnology-information technology) [3], and BT-IT [4,5]. This new paradigm in science and technology could be considered as a revolutionary solution to solving current problems of overconsumption of energy, disease therapy, environmental pollution, and so on. Among various multidisciplinary hybrid materials, the conjugation between inorganic nanomaterials and biomolecules is one of the most attractive topics, since the bio-inorganic conjugates on the nanometer scale like those used in drug and gene delivery systems have been successfully demonstrated to have low toxicity and to provide target-oriented delivery of biomolecules [6–14].

The term "bio-inorganic conjugate" stands for a chemical nanohybrid in which biomolecules or bio-functionalized molecules are chemically conjugated with inorganic host materials on the nanoscale. Historically, research on bio-inorganic conjugates originated from the idea of sol–gel encapsulation of labile biological molecules which have catalytic or transduction functions [15–20]. Recently, rapid development of both inorganic nanoscience and biology has led to bio-inorganic conjugates with various functionalities for catalyst, reservoir, biosensor, drug delivery carrier, and so on. For the past decade, we have worked to create new bio-inorganic conjugates based on layered inorganic nanomaterials such as natural layered alluminosilicates or layered metal hydroxides [7–11,21–26].

Layered metal hydroxides can be categorized into several classes according to their structure (Figure 13.1). Layered double hydroxides (LDHs), which have a hydrotalcite-like structure [27–34], can be expressed as $[M^{2+}{}_{1-x}M^{3+}{}_{x}(OH)_2]^{-}(A^{m-})_{x/m}, nH_2O$,

Bio-inorganic Hybrid Nanomaterials. Edited by E. Ruiz-Hitzky, K. Ariga, Y. M. Lvov

ISBN: 978-3-527-31718-9

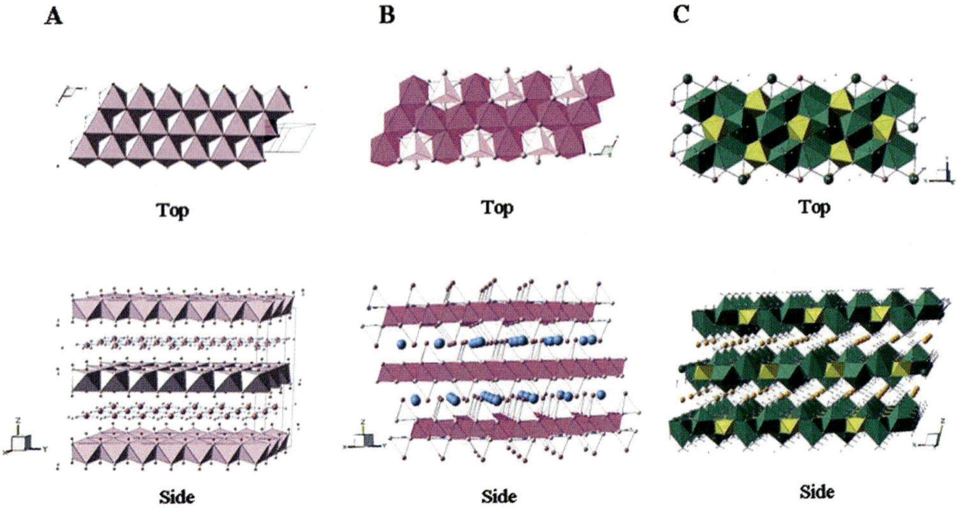

Fig. 13.1 Structures of various layered metal hydroxides: (A) Layered double hydroxide (LDH), (B) hydroxy double salt (HDS) and (C) hydrocalumite.

where the M^{n+} are metal cations (M^{2+}=Mg^{2+}, Zn^{2+}, Ni^{2+}, Cu^{2+}, ..., M^{3+}=Al^{3+}, Fe^{3+}, ...) and A^{m-} are interlayer anions (A^{m-}=CO_3^{2-}, NO_3^-, SO_4^{2-}, and other anionic species). The isomorphous substitution of M^{2+} with M^{3+} gives rise to a positive charge in a brucite-like layer and thus LDHs can accommodate charge-compensating anions (Figure 13.1(A)). Another form of layered metal hydroxides is hydrocalumite with general formula $[Ca_2M(OH)_6]^+ A^- \cdot mH_2O$ in which M^{3+} is a trivalent metal ion like Fe^{3+} or Al^{3+} [35–37]. As shown in Figure 13.1(C), hydrocalumite has corrugated brucite-like layers and this unique layer framework results in the characteristic XRD pattern distinguished it from LDHs, usually leading to better crystallinity. In this layered structure, Ca atoms are hepta-coordinated with six hydroxides and one interlayer water, which are edge-shared with octahedral trivalent metal cations. The other layered metal hydroxide type is, hydroxy double salts (HDSs) which have the general formula $[(M^{2+}{}_{1-x}Me^{2+}{}_{1+x})(OH)_{3(1-y)}]X^{n-}{}_{(1+3y)/n} \cdot zH_2O$ in which M^{2+} and Me^{2+} correspond to divalent metals such as Cu, Co, Ni, Mn, or Zn and X^{n-} is the exchangeable anion [38–44] (Figure 13.1(B)). Like LDHs and hydrocalumites, divalent metal ions stabilized in the HDS lattice form a brucite-like layer with some octahedral vacancies, however, some of the metal ions are positioned above and below the vacant sites outside the hydroxide layer to form tetrahedral coordination. An example of HDS is zinc basic salt, which has the formula $Zn_5(OH)_8(NO_3)_2 \cdot nH_2O$, in which one quarter of the Oh sites are vacant. It is worth noting here that the interlayered anions are generally exchangeable, regardless of the structural type. Among them, LDHs exhibit very attractive properties for both stabilization and reorganization of anionic biomolecules due to their high anion exchange capacity (2–5 meq/g), great flexibility of chemical composition, swelling properties, extreme affinity to carbonate ion, pH-dependent solubility, and high biocompatibility.

LDHs can encase many important small or macro-molecules with negative charge, including oligomers and polymers like single or double stranded DNA as well as simple molecules like nucleotides and vitamins [7,9,10,14,22,45–54]. These biomolecules are widely used as the ingredients of foods and medicines. Among them, pharmaceutical substances like antisense oligonucleotides or anticancer chemotherapeutic agents are attractive, since their therapeutic application can be expanded infinitely if safe protection, target-specific transport, and high delivery efficiency are achievable.

LDHs are good candidates for solution of the above problems by formation of bio-inorganic conjugates. Because of their positively charged inorganic framework and swellable interlayer space they could be hybridized with a variety of anionic biomolecules through an anion exchange reaction. Hybridization with LDHs can offer important features such as stabilization against physicochemical and biological degradation, controlled release, enhanced cellular uptake, and uniform two-dimensional rearrangement (controllable size and charge density).

In this chapter, we describe the synthetic strategy for bio-inorganic conjugates which can be used as drug or gene delivery systems and their efficient cellular uptake mechanism.

13.2 Synthesis of Bio-inorganic Conjugates

In this section, we demonstrate the synthetic strategies for bio-inorganic conjugates using 2-D layered inorganic nanoparticles. Among the various 2-D layered materials, LDH is an excellent candidate for bio-inorganic conjugates with biocompatibility and large anion exchange capacity. Due to the relatively high charge density ($\sim$25 $Å^-/e^-$) and flexible interlayer space, various anionic biomolecules can be tightly encapsulated in the gallery space of LDHs via co-precipitation and ion-exchange reaction.

The anticancer agent Methotrexate (MTX) was conjugated with LDH by a coprecipitation method. To synthesize the MTX-LDH conjugate, powdered MTX was dissolved in decarbonated water, and titrated with NaOH (0.5 M) solution to give a 0.043 M solution of MTX at pH 7. The mixed metal solution of $Mg(NO_3)_2{\cdot}6H_2O$ and $Al(NO_3)_3{\cdot}9H_2O$ with molar ratio Mg/Al = 2/1 was added to the MTX solution and the solution was titrated with NaOH solution until pH $\sim$ 9.5 to produce yellowish precipitates [7,26].

In order to synthesize the bio-inorganic conjugate for gene therapy, antisense *myc* oligonucleotide (As-*myc*: 5′ d (AACGTTGAGGGGCAT) 3′) was encapsulated by LDH through an ion-exchange reaction. First, pristine $Mg_2Al(OH)_6(NO_3)$-LDH was synthesized through a conventional coprecipitation method. The mixed solution was titrated with 0.5 M NaOH solution to pH $\sim$ 9.5 under a N_2 atmosphere and aged for 24 h with vigorous stirring. The resulting white precipitate was collected by centrifugation and washed thoroughly with decarbonated water [8,55]. The As-*myc* solution

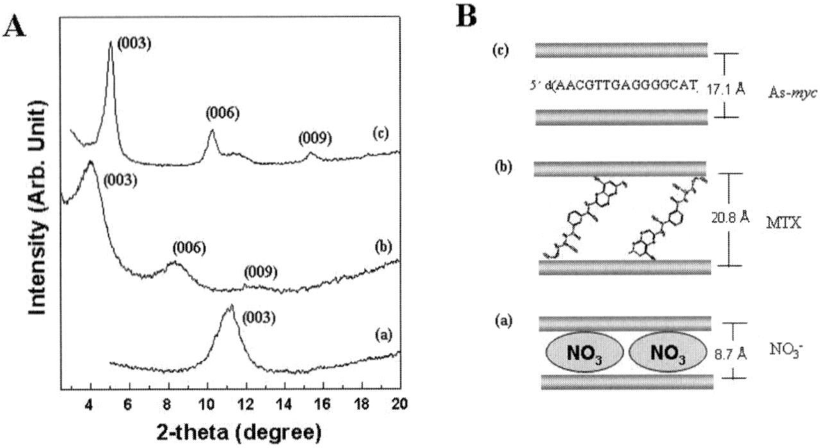

Fig. 13.2 Powder X-ray diffraction patterns (A) and corresponding schematic structural illustrations (B). (a) $Mg_2Al(OH)_6(NO_3)$-LDH (b) $Mg_2Al(OH)_6$(MTX)-LDH and (c) $Mg_2Al(OH)_6$(As-*myc*)-LDH.

was then added to a pristine LDH suspension (1 mg/1 mL) and mixed in a shaking incubator for 2 days at 65 °C. The resulting As-*myc*-LDH hybrids were collected by centrifugation and washed with decarbonated water.

The structure of the synthesized bio-inorganic conjugates is easily confirmed with X-ray diffraction (Figure 13.2). Since the LDHs are 2-dimensional layered materials, the interlayer distance along the *c*-axis changes upon biomolecule conjugation. It is calculated from the 2-theta values of the (00l) peaks that the interlayer distance of LDH increases from 0.87 nm (for NO_3^-) to 2.08 nm (MTX) and 1.71 nm (As-*myc*), respectively, upon encapsulating the biomolecules into the hydroxide layers (Figure 13.2). In the case of MTX-LDH conjugates, it was found that the MTX molecules of length (~2.30 nm) are larger than the gallery height (1.60 nm: (interlayer distance 2.08 nm) – (LDH layer thickness 0.48 nm)) and hence are packed with a tilting angle of ~44°.

In order to utilize these bio-inorganic conjugates as drug or gene delivery systems, the physicochemical properties of the encapsulant LDH, such as size and surface charge, should be controlled. Especially, the size must be confined to the range 50 ~ 500 nm, because too small or too large particles are easily eliminated via renal clearance or the reticuloendothelial system [56,57]. Fortunately, the size of LDH particles can be manipulated by adjusting the synthetic conditions such as concentration, temperature, and reaction time.

Typically, various sized LDH particles are synthesized under hydrothermal conditions by altering the aging time and reaction temperature. A clear metal solution was prepared with concentration 0.1 M, and the ratio of Mg: Al fixed at 2 : 1. The clear solution was titrated up to pH ≈ 9.5 with 0.5 M of NaOH solution containing Na_2CO_3, and samples were then aged in an autoclave at various temperatures for various reaction times. The resulting precipitates were centrifuged, repeatedly washed with de-ionized water and then freeze-dried for about 24 hours.

LDHs with exceptionally large particle size can be achieved through the urea hydrolysis method. Homogeneous hydrolysis of urea in the solution raises the solution pH homogeneously and thus the crystal growth can be maximized. Synthesis was carried out by varying parameters such as aging time and total metal concentration of $Mg(NO_3)_2{\cdot}6H_2O$ and $Al(NO_3)_3{\cdot}9H_2O$. In this reaction, the total metal concentration can be varied from 0.06 to 0.9 M, and the aging time can be varied from a few hours to several weeks. It is worth noting here that, the ratio of Mg: Al: urea was 2 : 1 : 9.9, and the temperature was controlled at 90 °C to induce effective hydrolysis of urea. After the reaction, the resulting precipitates were centrifuged, repeatedly washed with de-ionized water and then freeze-dried for about 24 hours.

As shown in Figure 13.3, the particle size of LDH particles can be controlled from a hundred nanometers to a few micrometers. The particle sizes obtained in various synthetic conditions are shown in Table 13.1.

It is clear that low concentration, high temperature and long reaction time resulted in large particles. The formation of LDHs occurs by precipitation from supersaturated metal solution upon adjusting the pH. The crystal growth of the precipitation

Fig. 13.3 Scanning electron microscopic images of LDH particles with various size (A) 100, (B) 200, (C) 1500, and (D) 4500 nm. LDH particles (A) and (B) were synthesized under hydrothermal conditions and (C) and (D) were prepared using hydrolysis of urea (see Table 13.1).

Tab. 13.1 Effect of synthesis parameters, aging time, reaction temperature, and concentration of metal ions, upon the particle size of the LDHs.

Synthetic method	Aging time (h)	Reaction Temperature (°C)	Total metal concentration (M)	Size (nm)	Standard deviation (nm)
hydrothermal synthesis	12	100	0.133	80	14
	24	100	0.133	100	15
	48	100	0.133	110	13
	72	100	0.133	120	23
	48	100	0.133	110	13
	48	125	0.133	200	28
	48	150	0.133	290	56
	48	180	0.133	350	29
urea hydrolysis	6	90	0.6	900	490
	30	90	0.6	1500	330
	45	90	0.6	2100	240
	69	90	0.6	2200	300
	672	65	0.87	1200	220
	672	65	0.65	2100	330
	672	65	0.44	2800	790
	672	65	0.065	4500	420

reaction has two steps; fast nucleation (formation of crystal nuclei) and subsequent growth of nuclei [58–60]. The synthesis parameters such as concentration, temperature, and reaction time can affect the two steps either thermodynamically or kinetically, thus determining the size of the product crystals. In the nucleation step, the higher the supersaturation ratio ($S = c/c_{eq}$, where S is the supersaturation ratio, c the actual concentration of the supersaturated solution, and c_{eq} the equilibrium concentration of the saturated solution) [58], the faster the nucleation rate, and thus the production of more nuclei. The relation between the final particle size and the synthesis parameters can be expressed by $L \propto C(1/N)^{1/3}$ where L is the final crystal size, C is a constant, and N is the number of nuclei. A lower supersaturation ratio results in the formation of fewer nuclei and thus the final particle size becomes larger.

In addition, the aging time and reaction temperature can influence the crystal growth kinetically or thermodynamically. The crystal growth rate after nucleation can be expressed as $\mathrm{d}L/\mathrm{d}t = C\exp(-1/RT)$ (C: constant), where L is the particle size, and T the reaction temperature [59]. According to the above formula the crystal growth rate is proportional to the reaction temperature and reaction time if other conditions are fixed. Therefore it is expected that higher reaction temperature and longer reaction time will result in larger particles, which is in good agreement with the experimental results shown in Table 13.1.

It is worth noting that the particle sizes of samples prepared by the urea hydrolysis method are larger than other LDHs. Such a finding can be rationalized by considering the slow hydrolysis of urea [61], since it can be classified as a very weak Brønsted base ($pK_b = 13.8$). Its hydrolysis mechanism consists of the formation of ammonium cyanate as the rate-determining step, followed by fast hydrolysis of the cyanate to ammonium carbonate:

$$CO(NH_2)_2 \rightarrow NH_4CNO \ : \text{slow}$$

$$NH_4CNO + 2H_2O \rightarrow (NH_4)_2CO_3$$

The hydrolysis gives a pH of about $7 \sim 9$, depending on the temperature. This pH is comparatively suitable for precipitating LDH. This hydrolysis progresses very slowly and homogeneously, which leads to a low degree of supersaturation during precipitation and, consequently, to large particle size [24,62,63].

13.3 Bio-inorganic Conjugate for Efficient Gene Delivery

13.3.1 Cellular Uptake Kinetics of LDH–FITC Into Cells

To be used as delivery carriers for biomolecules, first it is essential to check whether the conjugates can effectively enter cells. Cellular uptake experiments were performed by using fluorescein isothiocyanate (FITC) conjugated LDH as a probe. Cells (5×10^5/ 1 ml) were incubated with LDH-FITC and its uptake was measured by flow cytometry. As shown in Figure 13.4, the cellular uptake was time and concentration dependent and

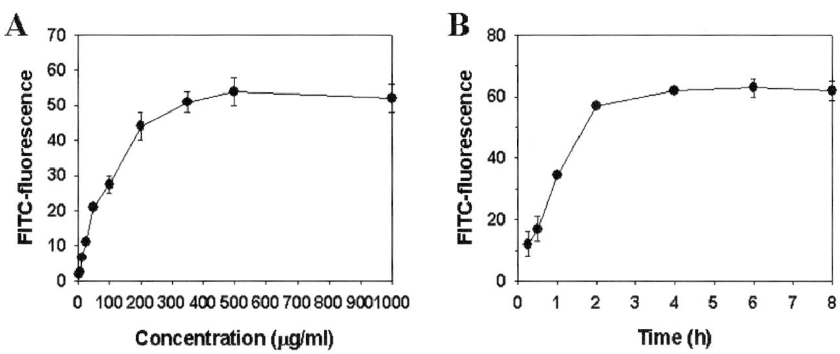

Fig. 13.4 Cellular uptake kinetics of LDH-FITC as a function of concentration (A) and incubation time (B) in HOS cells. Cells were incubated with LDH-FITC for 2 h (A), concentration of LDH-FITC was 200 μg/ml (B). The cellular uptake was quantified by flow cytometry.

reached a plateau after 2 hours incubation at 200 μg/ml. Especially remarkable is that the uptake of LDH-FITC was observed to be as rapid as within 1 hour. This implies that LDH can be effectively internalized into cells, suggesting its high potential as a delivery vehicle.

13.3.2
Effect of As-*myc*–LDH Hybrid on the Suppression of Cancer Cells

Over the last few years, antisense therapy has emerged as a promising strategy for cancer treatment [64–70]. Antisense technology works at the genetic level and its principle is the sequence-specific binding of an antisense oligonucleotide to target messenger RNA, thus selectively preventing the synthesis of the proteins related to cancer development. For example, high levels of expression of the *myc* gene family are known to have an important role in human cancers. However, poor delivery of small oligonucleotides into cells and its instability are critical problems whose solution can be suggested by using a delivery carrier. In this purpose, As- *myc*-LDH conjugate was synthesized and its efficacy was evaluated by MTT (3-(4,5- dimethylthiazol-2-yl)-2,5-diphenyl tetrazolium bromide) assay, a calorimetric method used for measuring cell proliferation. The yellow MTT reagent enters the cells and passes into the mitochondria where it is cleaved to an insoluble purple formazan product by metabolic active reduction systems. Leukemia cells, HL-60 were cultured at an initial concentration of 5×10^4/ml and exposed to As-*myc* or As-*myc*-LDH hybrid at a final concentration of 5, 10, 20 μM for 4 days. Control experiments were performed without As-*myc* or As-*myc*-LDH hybrid under the same conditions. As shown in Figure 13.5, As-*myc*-LDH hybrid showed considerable time and concentration-dependent inhibition of cell proliferation, about 65 % cell growth inhibition compared to untreated cells after 4 days at 20 μM. It is worth noting here that only 12 % growth inhibition was obtained by As-*myc* only, and no discernible cytotoxicity of LDH itself was determined. This result suggests that As-*myc* in As-*myc*-LDH hybrid is efficiently internalized into cells and eventually suppressed cell proliferation, probably due to LDH.

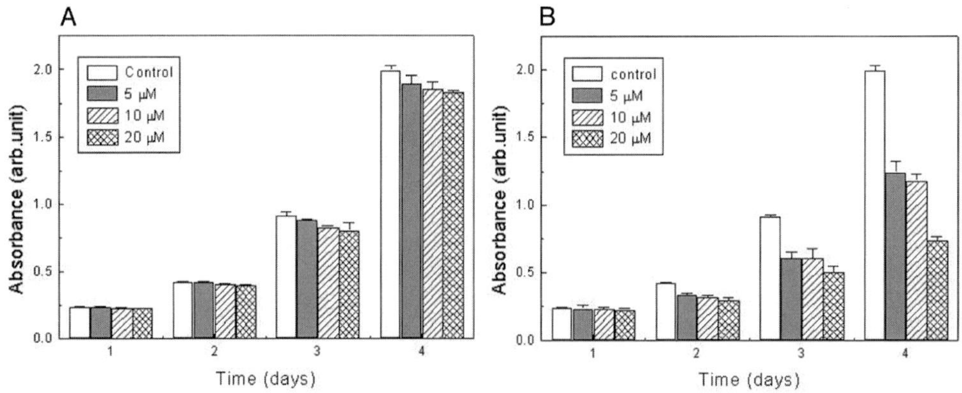

Fig. 13.5 Effect of As-*myc* (A) and As-*myc*-LDH (B) on the growth of HL-60.

13.4 Bio-inorganic Conjugate for Efficient Drug Delivery

13.4.1 Cellular Uptake of MTX–LDH Hybrid

Drug efficacy is directly related to its intracellular concentration level, so it is necessary to evaluate the MTX concentration in cells. In particular, MTX is a folate antagonist, thus it binds to dihydrofolate reductase in competition with folate [71–77]. A low intracellular level of MTX caused by high efflux and low uptake in resistant cells is also the main disadvantage of MTX medication [78,79]. This leads to a high dosage of MTX for cancer treatment, which is also directly associated with adverse effects.

Cellular uptake of MTX in an MTX–LDH hybrid was quantified by flow cytometry with FITC-conjugated MTX (MTX-FTIC). Human osteosarcoma (HOS) cells (5×10^5/1 ml) were treated with 50 μg/mL of MTX-FITC and 112 μg/mL of MTX-FITC–LDH, where the concentration of MTX-FITC in MTX-FITC–LDH was adjusted to maintain an equivalent amount of free MTX-FITC treated. After 1, 2, and 5 h, cellular uptake was about 55-fold higher for MTX-FITC–LDH than for free MTX-FITC, as determined by the remarkably high fluorescence intensity of MTX-FITC–LDH compared to free MTX-FITC (Figure 13.6). This is evidence that MTX delivery in cells is greatly enhanced by using the MTX–LDH hybrid.

13.4.2 Effect of MTX–LDH on Cell Proliferation and Viability

MTX interferes with the growth of cancer cells by inhibiting the metabolism of folic acid. Drug efficacy was evaluated *in vitro* by MTT assay, as described above, and by Trypan Blue exclusion. Trypan Blue is a non-vital dye excluded by viable cells, but selectively staining dead cells. According to Figure 13.7, higher suppression of cell

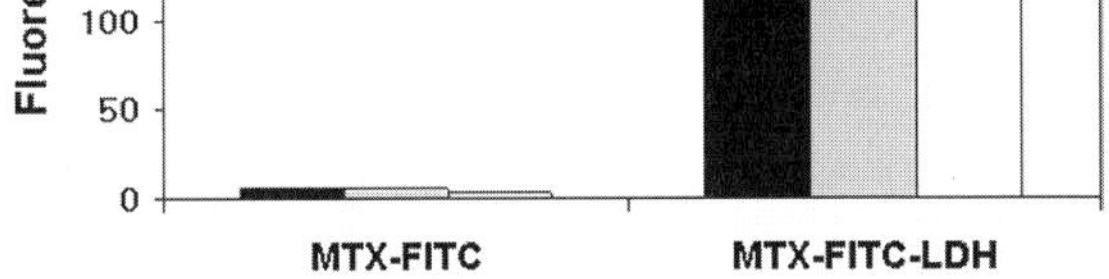

Fig. 13.6 Comparison of MTX-FITC uptake in HOS cells treated with free MTX-FITC or MTX-FITC-LDH hybrid. The concentration of the hybrid was adjusted to maintain equivalent molar amounts of MTX-FITC to that for the free MTX-FITC treatment.

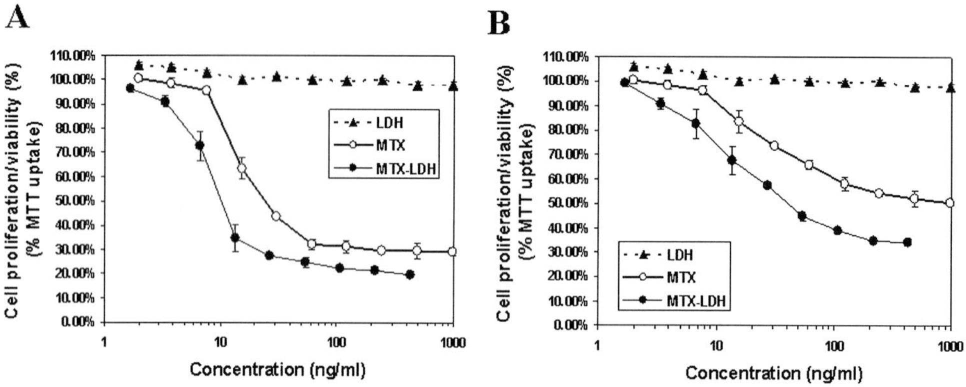

Fig. 13.7 Effect of MTX-LDH hybrid on cell proliferation (A) and viability (B) as determined by MTT assay and Trypan Blue exclusion, respectively. HOS cells were incubated with free MTX, MTX-LDH hybrid or LDH for 72 h.

proliferation was observed in cells treated with MTX–LDH than in those treated with free MTX, and the same result was obtained by trypan blue exclusion (Figure 13.7(B)), indicating enhanced drug efficacy in the MTX–LDH hybrid system. Efficient cellular uptake of LDH, as demonstrated in Figure 13.4 surely contributed to drug efficacy.

13.4.3 Effect of MTX–LDH Hybrid on the Cell Cycle

Another factor to be examined as an indicator of drug efficacy is its effect on cell cycle distribution. The cell cycle is the series of events resulting from cell growth, DNA replication, and cell division in eukaryotic cells and consists of four distinct phases: the G1, S, G2 phases (known as interphases) and the M phase. During the G1 phase, increase in cell size, production of RNA, and protein synthesis occur and an important cell cycle control mechanism is activated (G1 checkpoint). The S phase is the period of DNA replication to produce daughter cells. During the G2 phase, another control checkpoint (G2 checkpoint), the cell continues to grow and produces new proteins. Nuclear and cytoplasmic division occurs in the M phase. Therefore, cell cycle arrest by anticancer drugs most frequently occurs at the G1/S or the G2/M boundary [80].

For cell cycle analysis, HOS cells (1×10^6) were treated with the indicated concentrations (see Table 13.2) of MTX or MTX–LDH for 20 h, harvested by trypsin treatment, washed with PBS, and then fixed with cold 70 % ethanol on ice overnight. The fixed cells were incubated with propidium iodide and flow cytometric measurement was carried out. Over 20 h, an increase in the number of cells in the G1 phase resulted from MTX and MTX–LDH treatment compared to untreated cells, indicating arrest at the G1/S boundary. It is worth noting that the inhibition of the G1/S transition was more evident in the cells treated with MTX–LDH than in those treated with free MTX (85.59 % versus 66.62 % at 320 μM/ml). This is consistent with

Tab. 13.2 Cell cycle of MNNG/HOS cells treated with either MTX or MTX–LDH.

	G1 (%)	S (%)	G2 (%)
Control	50.72	25.00	24.41
LDH (384 μg/ml)	55.17	24.25	20.49
MTX (320 μg/ml)	66.62	17.73	15.73
MTX-LDH (704 μg/ml)	85.59	9.15	5.02

Note that the MTX contents in the two samples are the same.

the results obtained by MTT assay and Trypan Blue exclusion suggesting that LDH as a drug delivery carrier enhances drug efficacy.

13.4.4
Potential of Bio-inorganic Conjugates for Gene and Drug Delivery

Bio-inorganic conjugates also have great potential for biological purposes *in vivo*, since targeted delivery of pharmaceutical substances to tumor tissue can be obtained by nano-sized particles themselves due to the so-called EPR (enhanced permeability and retention) effect. The EPR effect is based on the phenomenon that tumor cells stimulate the production of blood vessels for fast cell growth, but these new blood vessels are leakier than regular blood vessels so macromolecules (~100 nm) are able to penetrate deeply into the leaky vasculature surrounding tumor cells [81,82]. In the case of cancer therapy, non-selectivity of anti-cancer agents is one of the most important problems that cause severe side effects and decrease in drug efficacy. Therefore, bio-inorganic conjugates with well controlled size will provide fascinating perspectives for cancer therapy.

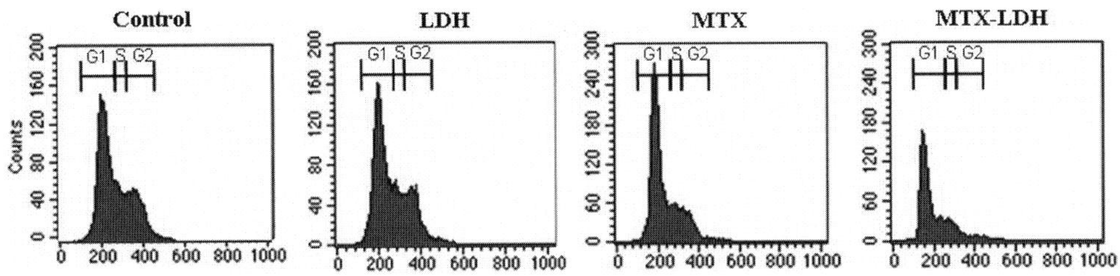

Fig. 13.8 G1 phase arrest of HOS cells exposed to either MTX or MTX-LDH for 20 h. The cell cycle was studied by FACS analysis of PI-stained cells. DNA histograms are shown, with the *x*-axis representing the DNA content and the *y*-axis representing the cell number. Note that 320 μg/ml MTX are equivalent to 704 μg/ml of MTX-LHD, based on MTX content in MTX-LDH.

13.5
Cellular Uptake Mechanism of LDH

13.5.1
Endocytosis of LDH

Because of the unusual high cellular uptake of MTX–LDH compared to MTX itself, we were able to suggest a different cellular uptake pathway of MTX–LDH conjugates. MTX is thought to enter into a cell through facilitated transport mediated by reduced folate carrier (RFC), which has been reported as a transporter protein for folate or folate derivatives [83,84]. Facilitated transport is a process of diffusion, a form of passive transport, where molecules diffuse across membranes with the assistance of transport proteins, and large molecules like MTX diffuse through the assistance of carrier proteins that change shape as the molecules are carried through. Therefore, the uptake of MTX might be limited by concentration or RFC activity. Furthermore, the negatively charged carboxylate functional groups in the MTX molecule are repelled by the anionic plasma membrane, consequently there is less chance for MTX to interact with RFC in the cell membrane. On the other hand, the surface of MTX–LDH conjugates is covered with positively charged LDHs, leading to stronger interaction with the membrane. After the enhanced interaction between the bio-inorganic conjugate and the membrane, MTX-LDH conjugates are thought to enter the cell massively via a different means than passive transport. Taking into account the massive uptake of MTX-LDH conjugates, endocytosis can be suggested as a possible method of cellular uptake.

In order to confirm the hypothesis, the cellular behavior of inorganic encapsulant LDHs was measured by transmission electron microscopy (TEM). LDH particles of a uniform size ~200 nm (Figure 13.9(C)) were synthesized according to the method mentioned in Section 13.2. For the cellular uptake, human osteosracoma, Saos-2 cells were treated with colloidal LDH and observed with microscopy. Typically, 75 cm^2 flasks containing 10 ml of medium, 2×10^5 Saos-2 cells were incubated for 4 days at 37 °C under CO_2 atmosphere, and then 50 μg/mL of LDHs and MTX–LDH nanohybrids were added to the flask. Every hour, the cells from each flask were washed with 1 × PBS buffer three times and collected by centrifugation. Every flask was washed with 1 × PBS buffer every 2 h in order to remove LDHs adsorbed on the cell surface. The cells were fixed with paraformaldehyde and glutaraldehyde mixed solution, and stained with OsO_4^-, then embedded in Epon-812 resin. The prepared blocks were trimmed and cross-sectioned with an ultramicrotome for TEM measurement.

Figure 13.9 represents the TEM image of LDH particles and their cellular internalization. As expected, LDH particles are internalized by endocytosis. Figure 13.9(A) shows the cellular uptake process of LDHs after 3 h of treatment, and demonstrates a successive entry of LDH by endocytosis; first the LDH particles were located around the cell membrane due to their positive charge (①), then they migrate to the membrane ruffles which are considered as endocytic bodies (②), finally the coated intracellular vesicles were formed as early endosomes (③). Figure 13.9(B)

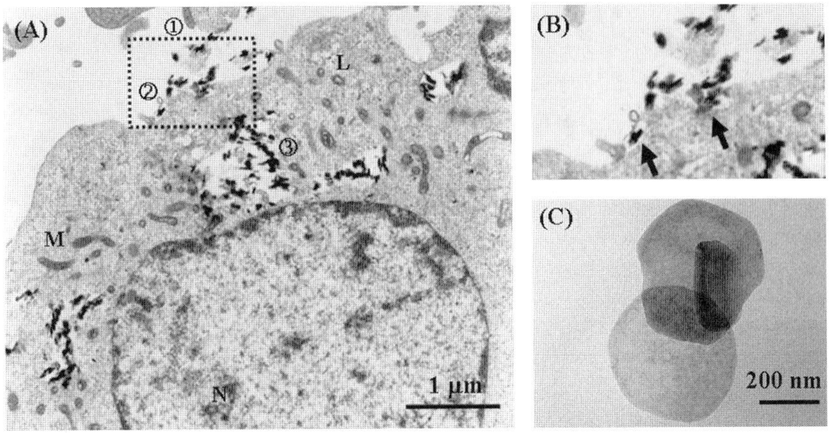

Fig. 13.9 Transmission electron microscopic data of endocytosis. Saos-2 cell was treated with LDH and after 1 h the cell was fixed and subjected to TEM measurement. (A) successive internalization of LDH via endocytosis: ① access of LDH around plasma membrane, ② formation of membrane ruffles and interaction between LDH and plasma membrane, ③ internalization of LDH into early endosome. (B) Magnified image of square in (A), arrow indicates the point of endocytosis (C) Transmission electron microscopic image of LDH (L: lysosome, M: mitochondria, N: nucleus).

is the magnified image of the membrane periphery and the flask-shaped membrane ruffles indicated by arrows show the moment of endocytosis.

13.5.2 Endocytic Pathway of LDH

Quite recently, much research has been focused on the cellular interaction of nanomaterials at molecular levels to understand their entry and subsequent intracellular movement. This can provide new strategies for the design of efficient drug delivery carriers. As is well documented, small molecules, such as amino acids, sugars and ions, can pass through the plasma membrane via ion pumps or channels. However, macromolecules or nanomaterials should be transported into cells through a process called endocytosis. Endocytosis is categorized into two broad classes, phagocytosis and pinocytosis. The former represents the uptake of large particles, and the latter is the uptake of fluid and solutes. In particular, pinocytosis occurs in all cell types and is mediated by at least four basic mechanisms: macropinocytosis, clathrin-mediated endocytosis, caveolae-mediated endocytosis, and clathrin-caveolae and dynamin-independent endocytosis [85].

Since endocytosis of LDH was confirmed by TEM images (Figure 13.9), for the next step, its specific endocytic pathway for membrane entry was determined by immunofluorescence and confocal microscopy. Cells were incubated with LDH-FITC, fixed with 3.7 % freshly made formaldehyde, and then stained with either anti-clathrin antibody or anti-caveolin-1 antibody both conjugated to the red fluorescent dye Texas Red (TR). The confocal microscopic images showed that green fluorescent

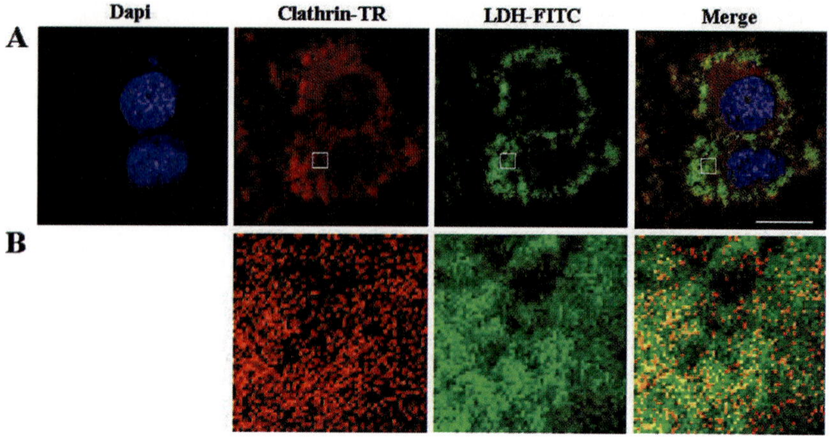

Fig. 13.10 Confocal microscopic images showing colocalization of LDH with clathrin in HOS cells. (A) Cells were incubated with LDH-FITC for 2 h, treated with anti-clathrin antibody, and then stained by TR. Nucleus was stained by Dapi. Scale bar represents 10 μm. (B) Magnified images of the white box in (A).

LDH-FITC is well overlapped with red fluorescent clathrin-TR, but not with caveolin-1-TR (Figure 13.10). This is clear evidence that clathrin-mediated endocytosis is the principal mechanism for the cellular internalization of LDH particles. Caveolae-mediated endocytosis, if any, seems not to be responsible for LDH uptake.

Membrane entry of LDH via clathrin-coated vesicles was further confirmed biochemically by modulating cellular uptake with endocytic inhibitors. Chlorpromazine (CPZ) is a cationic drug that prevents clathrin-mediated endocytosis by disrupting the assembly and disassembly of the clathrin lattice, as well as decreasing clathrin recycling [86,87]. Clathrin-coated vesicle formation is arrested due to potassium depletion after treating the cells with a hypotonic buffer [88]. Thus, cells were pre-incubated with CPZ or in a potassium-free buffer followed by further incubation with LDH-FITC, and then cellular uptake was quantified by flow cytometry. As expected, clathrin inhibition gave rise to a remarkable reduction in the uptake of LDH-FITC particles, as shown by the decreased intensity of the green fluorescence (Figure 13.11). Considering both experiments, it is concluded that LDH particles are primarily internalized into cells via clathrin-mediated endocytosis.

The fact that LDH selects clathrin-mediated endocytosis for cellular entry is very fascinating, because clathrin-mediated endocytosis is the most common mechanism of uptake in all mammalian cells and is of fundamental importance for many cellular activities, therefore, cellular uptake of LDH can be facilitated. The targeted entry of drug carrier into cells is very important in nanomedicine as it can also reduce drug toxicity and increase drug efficacy. It is now clear that selective entry of LDH via clathrin-mediated endocytosis can considerably enhance drug uptake as well as the therapeutic effects, suggesting it has promising potential as a delivery carrier.

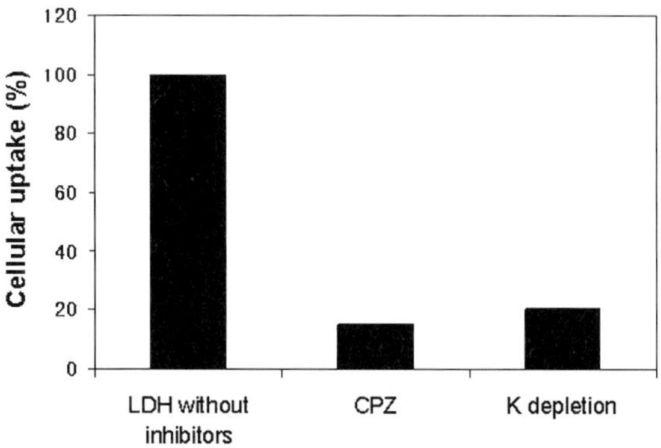

Fig. 13.11 Effects of clathrin-mediated endocytosis inhibitors on the internalization of LDH-FITC. Cellular uptake (%) of LDH was calculated by comparison with that in the absence of inhibitor (100%).

13.6 Conclusion

Hitherto, several bio-inorganic conjugates for gene and drug delivery have been investigated. Various biomolecules could be conjugated with the inorganic nanocarrier LDH through co-precipitation or ion-exchange reaction, and so on. The synthesized bio-inorganic conjugate showed a high cellular uptake rate and drug delivery efficacy. It was also verified that the efficiency of such bio-inorganic conjugates in drug delivery is due to the clathrin-mediated endocytic pathway of the inorganic nanocarrier LDH. These results will allow us to propose new directions in designing target-specific nanomedicine. In particular, the better understanding of their cellular uptake pathways and intracellular delivery mechanism may provide a new strategy in designing drug delivery nanovehicles based on metal hydroxide nanoparticles.

References

1 Gudiksen, M.S., Lauhon, L.J., Wang, J., Smith, D.C. and Lieber, C.M. (2002) *Nature*, **415**, 617.

2 Patolsky, F., Weizmann, Y. and Willner, I. (2004) *Nature Materials*, **3**, 692.

3 Krusin-Elbaum, L., Shibauchi, T., Argyle, B., Gignac, L. and Weller, D. (2001) *Nature*, **410**, 444.

4 Hultquist, S.J., Harrison, R. and Yang, Y.Z. (2002) *Nature Biotechnology*, **20**, 743.

5 Hultquist, S.J., Harrison, R. and Yang, Y.Z. (2002) *Nature Biotechnology*, **20**, 517.

6 Chowdhury, E.H. and Akaike, T. (2005) *Current Gene Therapy*, **5**, 669.

7 Choy, J.H., Jung, J.S., Oh, J.M., Park, M., Jeong, J., Kang, Y.K. and Han, O.J. (2004) *Biomaterials*, **25**, 3059.

8 Choy, J.H., Kwak, S.Y., Jeong, Y.J. and Park, J.S. (2000) *Angewandte Chemie-International Edition*, **39**, 4042.

9 Choy, J.H., Kwak, S.Y., Park, J.S. and Jeong, Y.J. (2001) *Journal of Materials Chemistry*, **11**, 1671.

10 Choy, J.H., Kwak, S.Y., Park, J.S., Jeong, Y.J. and Portier, J. (1999) *Journal of the American Chemical Society*, **121**, 1399.

11 Choy, J.H., Oh, J.M., Park, M., Sohn, K. M. and Kim, J.W. (2004) *Advanced Materials*, **16**, 1181.

12 Kutsuzawa, K., Chowdhury, E.H., Nagaoka, M., Maruyama, K., Akiyama, Y. and Akaike, T. (2006) *Biochemical and Biophysical Research Communications*, **350**, 514.

13 Ruiz, A.I., Darder, M., Aranda, P., Jimenez, R., VanDamme, H. and Ruiz-Hitzky, E. (2006) *Journal of Nanoscience and Nanotechnology*, **6**, 1602.

14 Zhi, P.X., Qing, H.Z., Gao, Q.L. and Ai, B.Y. (2006) *Chemical Engineering Science*, **61**, 1027.

15 Avnir, D., Braun, S., Lev, O. and Ottolenghi, M. (1994) *Chemistry of Materials*, **6**, 1605.

16 Braun, S., Rappoport, S., Zusman, R., Avnir, D. and Ottolenghi, M. (1990) *Materials Letters*, **10**, 1.

17 Ellerby, L.M., Nishida, C.R., Nishida, F., Yamanaka, S.A., Dunn, B., Valentine, J. S. and Zink, J.I. (1992) *Science*, **255**, 1113.

18 Gill, I. (2001) *Chemistry of Materials*, **13**, 3404.

19 Gill, I. and Ballesteros, A. (2000) *Trends in Biotechnology*, **18**, 282.

20 Glad, M., Norrlow, O., Sellergren, B., Siegbahn, N. and Mosbach, K. (1985) *Journal of Chromatography*, **347**, 11.

21 Choy, J.H., Jung, E.Y., Son, Y.H. and Park, M. (2004) *Journal of Physics and Chemistry of Solids*, **65**, 509.

22 Choy, J.H. and Son, Y.H. (2004) *Bulletin of the Korean Chemical Society*, **25**, 122.

23 Oh, J.M., Choi, S.J., Kim, S.T. and Choy, J.H. (2006) *Bioconjugate Chemistry*, **17**, 1411.

24 Oh, J.M., Hwang, S.H. and Choy, J.H. (2002) *Solid State Ionics*, **151**, 285.

25 Oh, J.M., Kwak, S.Y. and Choy, J.H. (2006) *Journal of Physics and Chemistry of Solids*, **67**, 1028.

26 Oh, J.-M., Park, M., Kim, S.-T., Jung, J.-Y., Kang, Y.-G. and Choy, J.-H. (2006) *Journal of Physics and Chemistry of Solids*, **67**, 1024.

27 Brindley, G.W. and Kikkawa, S. (1980) *Clays and Clay Minerals*, **28**, 87.

28 Cavani, F., Trifiro, F. and Vaccari, A. (1991) *Catalysis Today*, **11**, 173.

29 Clause, O., Rebours, B., Merlen, E., Trifiro, F. and Vaccari, A. (1992) *Journal of Catalysis*, **133**, 231.

30 Hernandezmoreno, M.J., Ulibarri, M. A., Rendon, J.L. and Serna, C.J. (1985) *Physics and Chemistry of Minerals*, **12**, 34.

31 Itaya, K., Chang, H.C. and Uchida, I. (1987) *Inorganic Chemistry*, **26**, 624.

32 Meyn, M., Beneke, K. and Lagaly, G. (1990) *Inorganic Chemistry*, **29**, 5201.

33 Miyata, S. (1983) *Clays and Clay Minerals*, **31**, 305.

34 Reichle, W.T. (1986) *Solid State Ionics*, **22**, 135.

35 Kalinichev, A.G., Kirkpatrick, R.J., and Cygan, R.T. (2000) *American Mineralogist*, **85**, 1046.

36 Kirkpatrick, R.J., Yu, P., Hou, X.Q. and Kim, Y. (1999) *American Mineralogist*, **84**, 1186.

37 Rousselot, I., Taviot-Gueho, C., Leroux, F., Leone, P., Palvadeau, P. and Besse, J. P. (2002) *Journal of Solid State Chemistry*, **167**, 137.

38 Meyn, M., Beneke, K. and Lagaly, G. (1993) *Inorganic Chemistry*, **32**, 1209.

39 Morioka, H., Tagaya, H., Karasu, M., Kadokawa, J. and Chiba, K. (1998) *Journal of Materials Research*, **13**, 848.

40 Morioka, H., Tagaya, H., Ogawa, A., Karasu, M., Kadokawa, J. and Chiba, K. (1998) *Abstracts of Papers of the American Chemical Society*, **215**, U445.

41 Newman, S.P. and Jones, W. (1999) *Journal of Solid State Chemistry*, **148**, 26.

42 Nishizawa, H. and Yuasa, K. (1998) *Journal of Solid State Chemistry*, **141**, 229.

43 Rajamathi, J.T., Britto, S. and Rajamathi, M. (2005) *Journal of Chemical Sciences*, **117**, 629.

44 Tagaya, H., Sasaki, N., Morioka, H. and Kadokawa, J. (2000) *Molecular Crystals and Liquid Crystals*, **341**, 1217.

45 Aisawa, S., Kudo, H., Hoshi, T., Takahashi, S., Hirahara, H., Umetsu, Y. and Narita, E. (2004) *Journal of Solid State Chemistry*, **177**, 3987.

46 Aisawa, S., Sasaki, S., Takahashi, S., Hirahara, H., Nakayama, H. and Narita, E. (2006) *Journal of Physics and Chemistry of Solids*, **67**, 920.

47 Burzlaff, A., Brethauer, S., Kasper, C., Jackisch, B.O. and Scheper, T. (2004) *Cytometry Part A*, **62A**, 65.

48 Darder, M., Ruiz, A.I., Aranda, P., VanDamme, H. and Ruiz-Hitzky, E. (2006) *Current Nanoscience*, **2**, 231.

49 Desigaux, L., Ben Belkacem, M., Richard, P., Cellier, J., Leone, P., Cario, L., Leroux, F., Taviot-Gueho, C. and Pitard, B. (2006) *Nano Letters*, **6**, 199.

50 Gerstel, P., Lipowsky, P., Durupthy, O., Hoffmann, R.C., Bellina, P., Bill, J. and Aldinger, F. (2006) *Journal of the Ceramic Society of Japan*, **114**, 911.

51 Hwang, S.H., Han, Y.S. and Choy, J.H. (2001) *Bulletin of the Korean Chemical Society*, **22**, 1019.

52 Li, S.P. (2006) *Colloids and Surfaces a-Physicochemical and Engineering Aspects*, **290**, 56.

53 Reinholdt, M.X. and Kirkpatrick, R.J. (2006) *Chemistry of Materials*, **18**, 2567.

54 Yuan, Q., Wei, M., Evans, D.G. and Duan, X. (2004) *Journal of Physical Chemistry B*, **108**, 12381.

55 Kwak, S.Y., Jeong, Y.J., Park, J.S. and Choy, J.H. (2002) *Solid State Ionics*, **151**, 229.

56 Fang, J., Sawa, T., Akaike, T., Akuta, T., Sahoo, S.K., Khaled, G., Hamada, A. and Maeda, H. (2003) *Cancer Research*, **63**, 3567.

57 Moghimi, S.M., Hunter, A.C. and Murray, J.C. (2001) *Pharmacological Reviews*, **53**, 283.

58 Gilman, J.J. (1963) *The art and science of growing crystals*, John Wiley Sons, Inc., New York.

59 Nyvlt, J., Sohnel, O., Matuchova, M., and Broul, M. (1985) *The kinetic of industrial crystallization, chemical engineering monographs*, Elsevier, Amsterdam, Vol. 19.

60 Randolph, A.D., Larson, M.A. (1988) *Theory of particulate processes, analysis and techniques of continuous crystallization*, Academic Press, 2nd edn.

61 Shaw, W.H.R. and Bordeaux, J.J. (1955) *Journal of the American Chemical Society*, **77**, 4729.

62 Costantino, U., Marmottini, F., Nocchetti, M., and Vivani, R. (1998) *European Journal of Inorganic Chemistry*, 1439.

63 Ogawa, M. andKaiho, H. (2002) *Langmuir*, **18**, 4240.

64 Agrawal, S., Temsamani, J. and Tang, J.Y. (1991) *Proceedings of the National Academy of Sciences of the United States of America*, **88**, 7595.

65 Burfeind, P., Chernicky, C.L., Rininsland, F., Ilan, J. and Ilan, J. (1996) *Proceedings of the National Academy of Sciences of the United States of America*, **93**, 7263.

66 Calabretta, B., Sims, R.B., Valtieri, M., Caracciolo, D., Szczylik, C., Venturelli, D., Ratajczak, M., Beran, M. and Gewirtz, A.M. (1991) *Proceedings of the National Academy of Sciences of the United States of America*, **88**, 2351.

67 Chu, P., Deforce, D., Pedersen, I.M., Kim, Y., Kitada, S., Reed, J.C. and Kipps, T.J. (2002) *Proceedings of the National Academy of Sciences of the United States of America*, **99**, 3854.

68 Wacheck, V. and Zangemeister-Wittke, U. (2006) *Critical Reviews in Oncology Hematology*, **59**, 65.

69 Yoshimura, K., Rosenfeld, M.A., Nakamura, H., Scherer, E.M., Pavirani, A., Lecocq, J.P. and Crystal, R.G. (1992) *Nucleic Acids Research*, **20**, 3233.

70 Yuen, A.R., Halsey, J., Fisher, G.A., Holmlund, J.T., Geary, R.S., Kwoh, T.J., Dorr, A. and Sikic, B.I. (1999) *Clinical Cancer Research*, **5**, 3357.

71 Ames, M.M., Powin, G., and Kovach, J. S. (1983) *Pharmaco kinetics of anticancer agents in humans*, Elsevier, Amsterdam, New York, Oxford.

72 Bailey, L.B. (1995) *Folate in health and disease*, Marcel Dekker.

73 Bertino, J.R. (1993) *Journal of Clinical Oncology*, **11**, 5.

74 Eksborg, S., Albertioni, F., Rask, C., Beck, O., Palm, C., Schroeder, H. and Peterson, C. (1996) *Cancer Letters*, **108**, 163.

75 Georgopapadakou, N.H. (1995) *Drug transport in antimicrobial and anticancer chemotherapy*, Marcel Dekker.

76 Gilman, A.G., Goodman, L.S., and Gilman, A. (1980) *The Pharmacological basis of therapeutics*, 6th edn. Macmillan Publishing Co. Inc., New York.

77 Hitchings, G.H. (1980) *Trends in Pharmacological Sciences*, **1**, 167.

78 Henderson, G.B. and Tsuji, J.M. (1987) *Journal of Biological Chemistry*, **262**, 13571.

79 Sirotnak, F.M. (1985) *Cancer Research*, **45**, 3992.

80 Shapiro, G.I. and Harper, J.W. (1999) *Journal of Clinical Investigation*, **104**, 1645.

81 Maeda, H. (2001) *Advances in Enzyme Regulation*, **41**, 189.

82 Maeda, H. (2001) *Advanced Drug Delivery Reviews*, **46**, 169.

83 Jansen, G., Westerhof, G.R., Jarmuszewski, M.J.A., Kathmann, I., Rijksen, G. and Schornagel, J.H. (1990) *Journal of Biological Chemistry*, **265**, 18272.

84 Moscow, J.A. (1998) *Leukemia and Lymphoma*, **30**, 215.

85 Conner, S.D. and Schmid, S.L. (2003) *Nature*, **422**, 37.

86 Kirchhausen, T. (2000) *Annual Review of Biochemistry*, **69**, 699.

87 Marsh, M. and Pelchen-Matthews, A. (2000) *Traffic*, **1**, 525.

88 Carpentier, J.L., Sawano, F., Geiger, D., Gorden, P., Perrelet, A. and Orci, L. (1989) *Journal of Cellular Physiology*, **138**, 519.

14
Halloysite Nanotubules, a Novel Substrate for the Controlled Delivery of Bioactive Molecules

Yuri M. Lvov, Ronald R. Price

14.1
Halloysite Structural Characterization

Halloysite ($Al_2Si_2O_5(OH)_4nH_2O$) is a two-layered (1 : 1) aluminosilicate which exhibits a range of morphologies. However, one predominant form is a hollow tubular structure in the submicron range and is chemically similar to kaolin (spherical halloysite was also found but we are not considering it in this chapter) [1–4]. When $n=2$ the clay is in the form of hydrated halloysite-10 Å with one layer of water molecules between the multilayers. The 10 Å designation indicates the spacing in the multilayer walls as the kaolinite plates roll up into a multiwalled cylinder. And at $n=0$, the structure is dehydrated halloysite-7 Å which may be obtained through an irreversible phase transition with loss of adsorbed water as the halloysite is heated to between 60 and 150 °C (halloysite has an endothermic peak at 500–600 °C due to structural dehydroxylation and then it remains stable till approximately 900 °C). It was recommended by the AIPEA Nomenclature Committee to rename metahalloysite (the dehydrated form) as halloysite-7 Å, but for simplicity we will call it halloysite.

The neighboring alumina and silica layers, and their water of hydration, create a packing disorder causing them to curve and roll up forming multilayer tubes. It is interesting, that even in the recent mineralogical review on halloysite clay [2], it is written that the reason why flat kaolinite rolls into halloysite tubules is unclear. It was assumed that this was due to an irregular replacement of Al^{3+} by Fe^{3+} in the octahedral sheet positions which is in accordance with the increased amount of Fe_2O_3 in halloysite as compared with kaolinite. The schematic representation of halloysite is shown in Figure 14.1.

Halloysite is an economically viable material that can be mined from deposits as a raw mineral. The size of halloysite tubules varies within 0.5–10 microns in length and 15–200 nm in inner diameter, depending on the deposit. There are between 15 and 20 aluminosilicate layers rolled in the multilayer tubule walls with a layer spacing of

Bio-inorganic Hybrid Nanomaterials. Edited by E. Ruiz-Hitzky, K. Ariga, Y. M. Lvov

ISBN: 978-3-527-31718-9

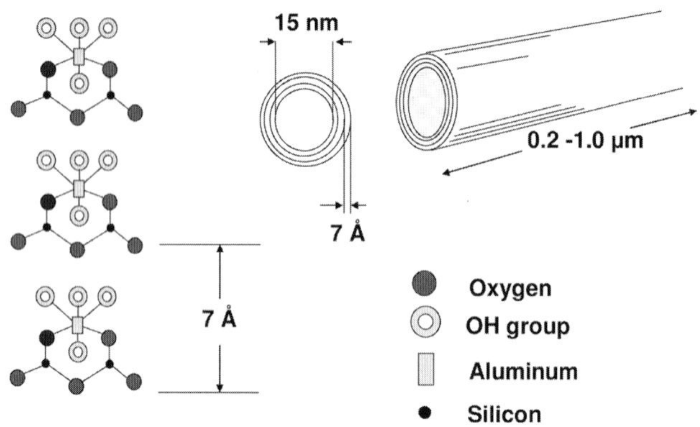

Fig. 14.1 Schematic representation of halloysite-7 Å nanotubules.

0.72 nm for the dehydrated halloysite. The most interesting, from the technical perspective, are cylinders with a very small inner diameter, and for this loading/release study we used halloysite nanotubules with a diameter of 50 nm and lumen of 15 nm which were obtained from Nanoclay and Technology Inc. and the Dragon Mine in Eureka, Utah. The specific surface area of this halloysite is 65 m^2/g; pore volume 1.25–1.34 mL/g; refractive index of 1.54; specific gravity 2.53 g/cm^3, and chemical composition Al_2O_3 – 36.9 %; SiO_2 – 48.0 %; Fe_2O_3 – 0.14 %, CaO – 0.20 %; TiO_2 – 0.04 %; LOI– 13.8 %.

The morphology of the clay varies by geographic location. Figure 14.2 shows TEM images of halloysite from the Dragon Mine. Halloysite is also available from Imerys Corporation, Auckland, New Zealand. Both halloysite samples have a tubule shape with external diameter of about 50 ± 10 nm for completely rolled clay nanotubes.

The outermost surface of the halloysite may be compared with silica and the inner lumen surface with alumina. In Figure 14.3 we have provided the zeta potential (electrokinetic potential) for halloysite and, for comparison, zeta potentials for silica and alumina nanoparticles which may represent the behavior of the outermost and innermost of the halloysite tubules. The surface potential of the halloysite falls between the curves of silica and alumina, indicating the predominance of silica properties on its exterior surface [4]. Below pH 8.5 the tubule lumen has a positive surface endorsing loading of negative macromolecules and preventing their adsorption on the negatively charged outer surface.

It is interesting to compare halloysite with carbon nanotubes (Table 14.1). One can see that halloysite has some advantages in applications which require a biocompatible nano-container, in addition the halloysite clay is far less expensive, the world supply is in excess of 50 000 tons per year. Carbon nanotubes have smaller diameter; they may be conductive and they are very strong.

Halloysite was found to be a viable and inexpensive nanoscale container for the encapsulation of biologically active molecules and drugs as was first demonstrated by Price *et al.* [5–8]. Its physicochemical characterization as a novel drug delivery system was also reported by Levis and Deasy *et al.* [9,10], who further demonstrated the controlled release of diltiazem hydrochloride and propanol hydrochloride

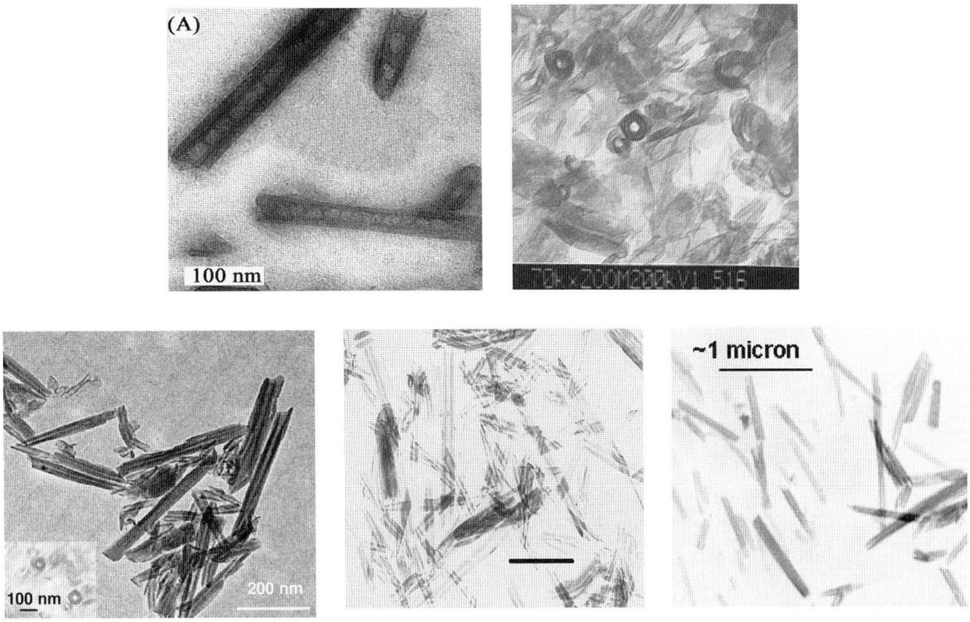

Fig. 14.2 Transmission electron microscopy images of halloysite from Nanoclay and Technologies Inc.: longitudinal and cross-sectional views (A, B) cross-section, and three different samples from supplies 2006–2007 (C–F).

[11,12]. A strong surface charge on the halloysite tubules has been exploited for designing nano-organized multilayers using the layer-by-layer (LbL) method of Lvov *et al.* [8,13,14]. The lumen of the halloysite has been used as an enzymatic nano-reactor by Shchukin *et al.* [15] The biocompatible nature of the halloysite was

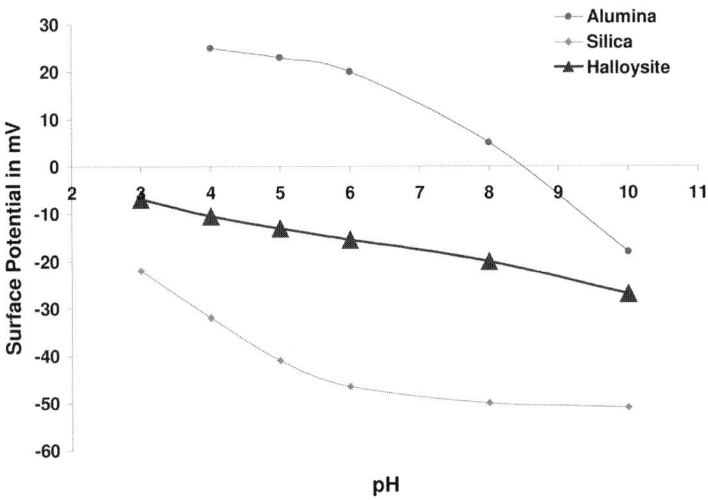

Fig. 14.3 Zeta-potential of halloysite, and silica and alumina nanoparticles (for comparison).

Tab. 14.1 Comparison of halloysite and carbon nanotubes.

	Halloysite nanotubules	Carbon nanotubes
Inner diameter/length	15 × 1000 nm	2 × 1000 nm
Biocompatibility	biocompatible	toxic
Price/availability	4$ per kg/tons	$200 000 per kg/grams

delineated when Kommireddy *et al.* [13] used halloysite as a substrate for stem cell attachment and proliferation. Other applications include cosmetics, transdermal drug delivery systems, bio-implants, the delivery of herbicides, insecticides, fungicides, and in advanced ceramic materials and synthetic catalysts. It is evident that halloysite is a promising biomaterial for pharmacology, dentistry, and for tissue scaffolds. Further, we will concentrate on the description of halloysite as a tubular substrate for the entrapment and controlled release of macromolecules.

14.2 Macromolecule Loading and Sustained Release

14.2.1 Nanotubule Loading Procedure

To entrap hydrophilic molecules, the halloysite was mixed as a dry powder with a saturated solution of the chosen active agent in water, ethanol or another solvent. A beaker containing the active agent and halloysite suspension was transferred to a vacuum jar which was then evacuated using a mechanical vacuum pump. Slight fizzing of the suspension indicated that air was being removed from the core of the halloysite tubules. When the fizzing was observed to stop, the sample was cycled back to atmospheric pressure, rinsed, filtered and dried. This procedure may be repeated 2–3 times to ensure that the halloysite is filled with the maximum amount of the material. Finally, the clay is rinsed rapidly in water, filtered and dried. For highly soluble materials, one can slow the release rate by blending them with viscose polymeric materials compatible with the active agent (like polyvinylpyrrolidone). Materials with low melting point may be loaded in their liquid form (like khellin [6]).

Determination of loading efficiency: In most cases, the concentration was determined by use of a UV/VIS spectrophotometer to measure the absorbence in solution in the UV range. The amount of active agent entrapped was defined as that released into solution following 30 min of sonication in a solvent solution that was not over 10 % saturation, the sonication being repeated until the amount of active agent in solution was below the limit of detection. This procedure was carried out after every release experiment. Entrapment efficiency was determined by first weighing the nano-cylinders and active agent to determine the total mass. The amount of released active agent was determined after the sonication procedure and a mass balance was used to determine the achievement of 100 % release.

14.2.2
Drugs and Biocides

We describe below the use of halloysite nanotubules for entrapment and the subsequent release of three drugs: Nifedipine (anti-anginal), Furosemide (anti-hypertension and diuretic), and Dexamethasone (synthetic corticosteroid). To increase the loading of these poorly soluble drugs in the tubules, we loaded them into a water–alcohol co-solvent solution (0, 10, and 50 % ethanol).

Loading was accomplished with saturated solutions of the drugs in 0, 10 and 50 % ethanol–water solvent by pulling a vacuum in two steps. In the first step the pre-weighed halloysite is soaked in saturated drug solution with the vacuum applied and released twice. Unentrapped drug was removed by centrifuging the halloysite and decanting the supernatant. In the second step, the drug containing halloysite is further soaked in a saturated drug solution and the same procedure was repeated, finally the tubules were washed and vacuum dried (all processing took 30 min).

We performed loading from alcohol-containing solvents to provide a higher solubility of the chosen drugs, and the release rate determination was conducted in physiological conditions at pH 7. For monitoring Furosemide, Dexamethasone, and Nifedipine concentrations, the absorbance study was performed at 277, 240, and 235 nm, respectively, at 22 °C. The release profiles of drugs from halloysite in comparison to that of the drug microcrystals are shown in Figure 14.4. It is evident that the drug release from the halloysite tubules lasted 50–100 times longer than that from the microcrystals. For Nifedipine it was 30 times longer than that from the drug

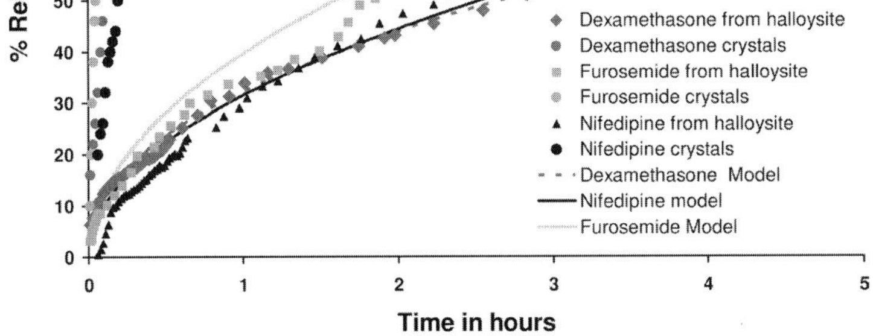

Fig. 14.4 Release profile of hydrophobic drugs from halloysite, pH 7 (loading in 10 % alcohol).

crystals, and in the case of Furosemide and Dexamethasone it was 100 times longer than that from the crystals. In all the three release curves the initial burst is observed within 10 min and is followed by a prolonged 6–10 h release. The initial burst is thought to be due to the release of exogenously bound drug.

The next release experiments were performed with loading using saturated drug solution in 0, 10 and 50 % ethanol solvents and continuous body sink conditions were maintained to simulate physiological conditions. Loading of the drugs was higher at higher alcohol ratios, at 10 % alcohol – 3 wt% of drugs, for 10 % alcohol – 4 wt% and for 50 % alcohol – 5 wt%. The release also appeared to become more linear for loading with higher alcohol ratio solvents. Earlier, we were able to reach linear 4 hours release of Nicotinamide Adenine Dinucleotide (NAD) from halloysite by increasing the viscosity of the loading solvent. Linear release means that the rate of release remains constant over time, which is a very desirable attribute for medical applications.

The release profile from the halloysite fits the Peppas model ($M_t/M_\infty = kt^n$, where M_t is the amount of drug released at time t, M_∞ is the amount of drug released at infinite time, n is the exponent characteristic of the release mechanism, and k is a constant). In the case of loading and release from 10 % alcohol as compared to the loading and release in water the value of n is around 0.5 for all the three drugs, indicating that the release mechanism is Fickian diffusion. In the case of loading using 50 % ethanol and release in water (Figure 14.5) the value of n is around 0.75 for all the drugs, indicating that the release mechanism is approaching zero-order kinetics. We consider this linear release may be due to the higher loading of the drug inside the halloysite tubule giving us a nano-pore controlled release. The absence of this mechanism found in the results from the loading and release from 10 % alcohol

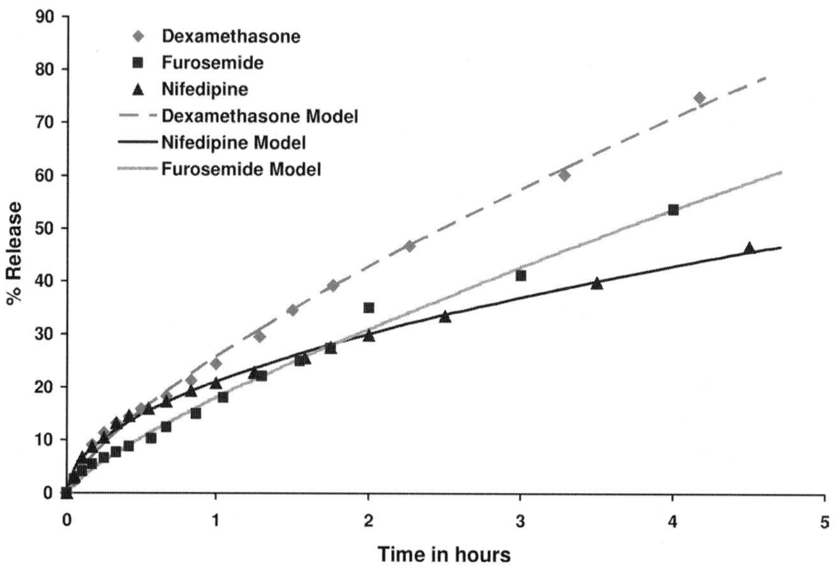

Fig. 14.5 Release curves from halloysite in water, pH 7, with loading using 50 % ethanol/water solvent.

loading may be due to the smaller amount of the drug spread along the lumen walls and not occluding the tube opening. Therefore, an extended 10–20 hours ofdrug release from 15 nm lumen diameter tubular halloysite has been demonstrated. Using alcohol drug solutions allowed higher concentrations and linear release approaching a zero-order rate.

If we assume that no loading occurs between the rolled layers then, from simple geometric consideration, one gets that maximal loading in the lumen equals $r^2/(R-r)^2$ which is of the order of 15 vol% (where $R = 25$ nm is the outer diameter, and $r = 7.5$ nm is the lumen radius). Our experimental estimation gave approximately this volume ratio of the loaded drugs to halloysite (approximately 12 vol%, as it will be demonstrated below). No drug was inserted between the halloysite clay layers in the roll, and this was confirmed by preserved 7.2 Å packing X-ray reflection. In review [2], it is indicated that it is possible to intercalate only ethylene glycol or small ions by drying hydrated halloysite (halloysite-10 Å) in the presence of NH_4-, K-, Cs- or Rb-salts, but in our experiments we worked only with thermodynamically stable dehydrated halloysite-7 Å.

The amount of exogenous drug is difficult to determine. Before conducting release experiments, loaded halloysite was always washed with a large amount of water during a 5 minute rinse to ensure removal of material from the surface. However, we cannot exclude some of the active agent being attached to the outer surface of the halloysite, especially in the natural gap-defects on the cylinder surface at the end of the rolled clay sheet (it is a natural pocket on the tubule surface). The typical 5–10 % initial release burst we observed may be related to dissolution of this material.

As we described above, in the pH range of 4–7 surface charge in the tubule lumen is positive and the outermost surface carries a negative charge. This allows more efficient loading of halloysite with negatively charged molecules at lower pH (e.g., for Dexamethasone as is demonstrated in Figure 14.6(a)). At low pH we have 5–6 wt% loading and at higher pH 2–3 wt% loading. Thus, low pH allowed us to reach loading close to that obtained from a 50 % alcohol solvent. Dexamethasone release was checked at pH 7.4 (Figure 14.6(B)). One can see a similar release rate for different loadings but the initial burst associated with rapid dissolution of the exogenous drugs was larger at lower pH. This may indicate that at low pH some part of the loaded material was associated with the external tubule surface.

A maximum loading of 5.5 wt% at pH 1 when both lumen and outermost surface are positive is close to full loading (about 12 vol% taking into account that the density of the clay is approximately twice that of the drug). Minimum 2 % loading was at pH 9 when both halloysite lumen and Dexamethasone have the same negative charge.

For well soluble compounds like the important co-factor NAD, the release was rather fast (within 1 hour) and to slow it we increased the viscosity inside the tubules, using as solvent water with 5 wt% of PVP (polyvinylpyrrolidone). In Figure 14.7 one can see 7 hours linear NAD release from halloysite [6]. An initial burst of 20 % in the release curve is typical for release from halloysite.

Another highly soluble bioactive compound, oxytetracycline HCl, loaded in halloysite gave 80 % release into water during 10 hours. The release rate was slowed

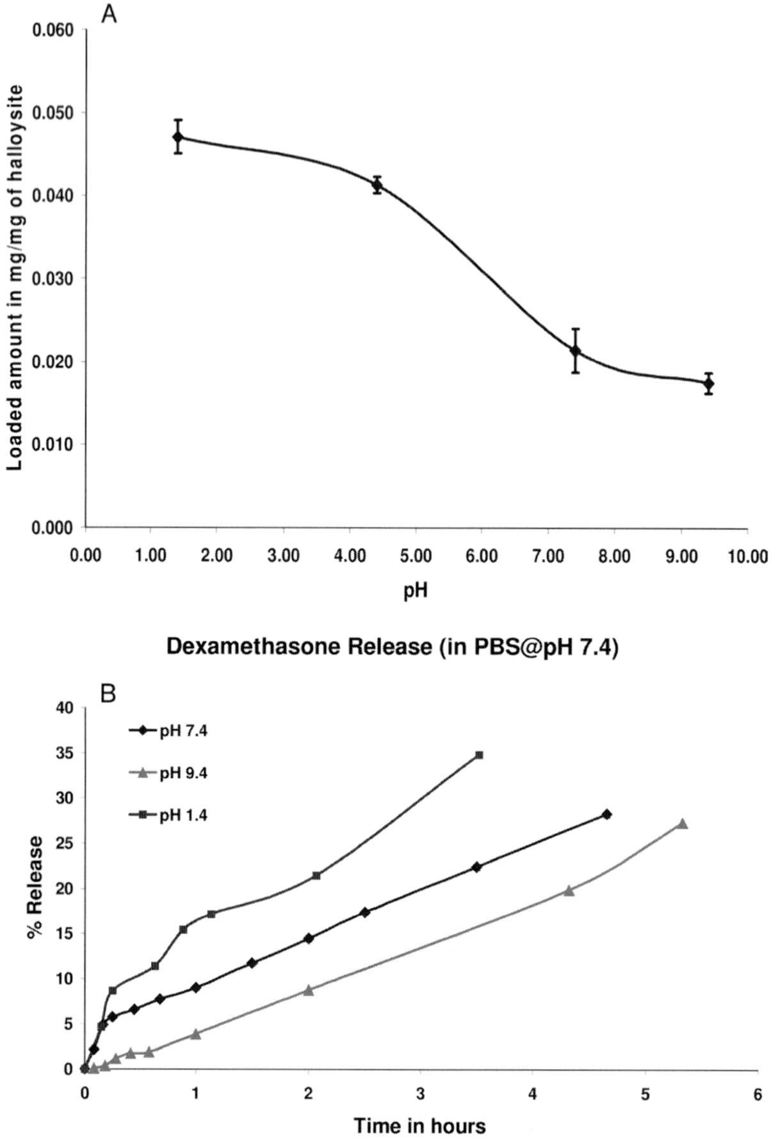

Fig. 14.6 (A) Amount of Dexamethasone loading depending on pH, and (B) its release at pH 7.4.

to 30 hours when loading was performed from water with the addition of 5 % of the water soluble epoxy Quitol 651. Therefore, one can adjust the release rate by using mixed solvent of a proper viscosity.

In conclusion, one can load drug molecules into halloysite nanotubules and get their slow release, typically during 5–30 hours. To achieve slower release kinetics, one needs a range of materials to act as stoppers on the ends of the tubules.

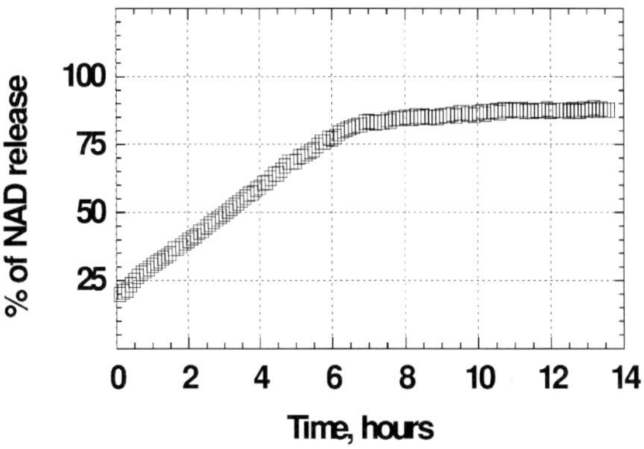

Fig. 14.7 Release curve of NAD from halloysite into water, pH 6.5.

Halloysite is a biocompatible material but its biodegradability is unclear. Therefore, its usage in medicine may be restricted for dermatological and dental applications or those associated with medical implants [11–14].

14.2.3
Globular Proteins

The loading of globular proteins is of special interest because their typical globule diameter is 2–9 nm which is close to the 15-nm lumen opening. Thus, one can fit 2–7 protein globules into the lumen diameter and in the total cross-section area only 4–20 proteins may be located. The behavior of large molecules in such narrow channels has not been described previously. A probable irreversible adsorption of protein monolayers on the lumen surface can make analysis especially complicated. We studied the entrapment and release properties of seven globular proteins with different diameters and surface charge: (1) insulin, diameter $D = 2$ nm, isoelectric point pK_a 5.9; (2) pepsin, $D = 3$ nm, pK_a 1; (3) urease, $D = 4.5$ nm, pK_a 6; (4) hemoglobin, D = 5.5 nm, pKa 6.8; (5) bovine serum albumin, elongated globule with axes of $11 \times 3 \times 3$ nm^3, p$K_a = 4.9$; (6) glucose oxidase, $D = 8$ nm, pK_a 4.1; and (7) catalyse, $D = 9$ nm, pK_a 5.5. Release experiments were carried out in water at pH 6.5, where protein #4 (hemoglobin) was positive, and all other proteins were negative.

As expected, the rate of protein release from halloysite was of longer duration, and reached 140 hours for insulin to achieve a 70% release (Figure 14.8). Diffusion coefficients for protein globules are larger than for lower molecular weight drugs but this provides only a partial explanation. This 6-day release can make halloysite a good medical carrier for insulin. The only problem is that even though halloysite is biocompatible it is not biodegradable and cannot be used for injection. Another small protein, pepsin, was released during 10 hours. Currently, transdermal or pulmanological applications of such systems are under consideration.

For other, larger proteins, the release was very slow – in 100 hours no more than 5% of loaded catalyse, urease, glucose oxidase or hemoglobin were released.

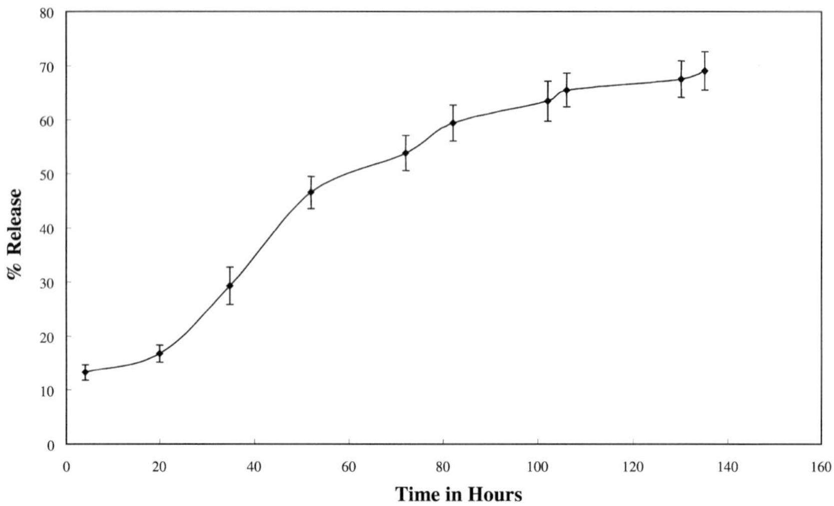

Fig. 14.8 Insulin release curve at pH 6.5 (average data for three experiments) [15].

Theoretical and model analysis based on a nanofluidic approach is needed for this situation. One may ask, is it possible to release proteins loaded in nanotubules? We have found that the addition of the polycation PEI in the release solvent resulted in much quicker protein release, as demonstrated in Figure 14.9. In this case, most of the insulin was released in 1 hour instead of 100 hours. 10–40 % of glucose oxidase, catalyse, and hemoglobin were released within 4 hours through complexation with PEI. It is unclear, whether the proteins were replaced by the polycation or released in a complex with PEI.

14.3 Nanoassembly on Tubules and at the Lumen Opening

A LbL self-assembly was used for formation of organized nanoshells on halloysite tubules. An ultrathin film assembly by means of alternate adsorption of oppositely charged linear polyelectrolytes was introduced in the 1990s by Decher *et al.* [16]. The basis of this method involves a re-saturation of the surface charge created by polyelectrolyte adsorption, resulting in the reversal of the terminal surface charge of the film after deposition of each layer. Thus, alternate positive and negative layer interaction allows organized multilayer film formation. The method provides the possibility of designing ultrathin multilayer films with a repeatability of better than one nanometer of defined molecular composition. The assembly process elaborated for planar solid supports was adapted for nano- and micro-templates (colloidal particles with sizes of 100 nm to microns, e.g., latex spheres, lipid tubules, and other colloids) [17]. In this process, a polycation solution is added to the suspension of

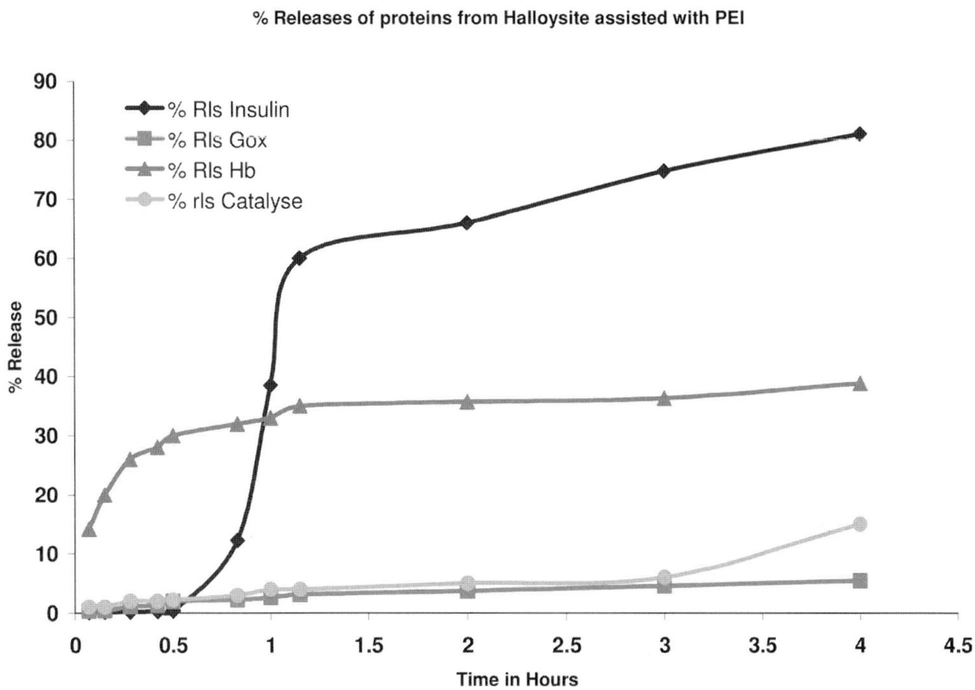

Fig. 14.9 Protein release initiated by addition of 1 mg/mL cationic PEI, pH 8.

colloidal particles and, after adsorption saturation, the particles are separated from free polycations in solution. A polyanion layer is then deposited. In the same manner, one can deposit any number of polyelectrolyte layers in the shell. An assembly of a polyelectrolyte nanoshell on halloysite may change its surface properties, and form stoppers at the lumen openings.

In addition the permeability of the polyelectrolyte shell can be varied by varying its composition and number of layers. As shown in Ref. [18] the wall permeability for macromolecules depends on the pH: at low pH the LbL walls were open and at a pH higher than 8 they were closed. By varying the pH, the shell permeability can be changed to increase or decrease the release rate. Experiments with opening and closing polyelectrolyte multilayer walls composed of polyion multilayers are based on the finding [19] that varying solution pH can induce charge imbalance in polycation–polyanion complexation in the multilayer, resulting in opening of the nanopores. This approach may be used for a triggered release via opening/closing LbL stoppers at the halloysite ends.

During the formation of polycation–polyanion multilayer coatings on halloysite, we monitored the surface potential (electrokinetic zeta potential). Initially negative halloysite (−40 mV) was converted to a positive surface with polycation layer adsorption in the first step of the LbLassembly (Figure 14.10). Adsorption of polyanions in the second step re-established the negative charge which was reversed

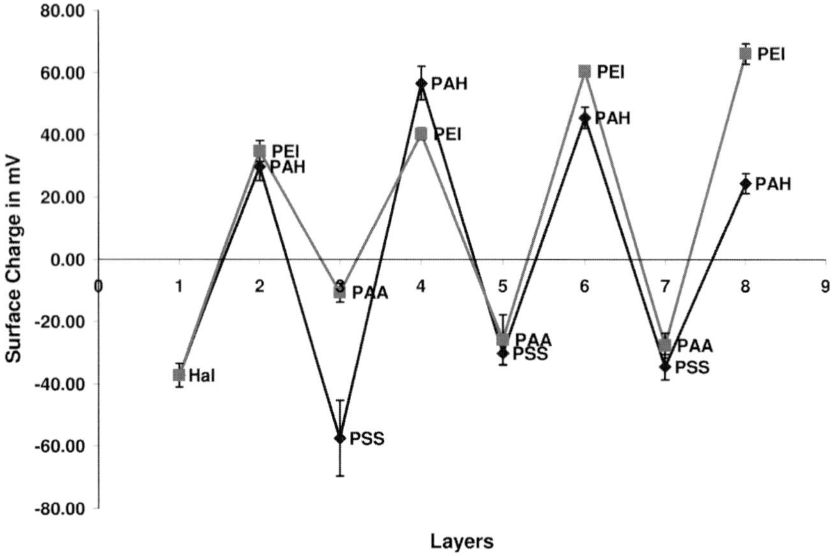

Fig. 14.10 Zeta-potential data for sequential deposition of polycations (PAH, PEI) and polyanions (PAA, PSS) on halloysite tubules, pH 6.5.

with each subsequent addition of polyion. For the LbL shell formation, we used polyallylamine hydrochloride (PAH) or polyethylenimine (PEI) as the polycations and sodium polystyrene sulfonate (PSS) or polyacrylic acid (PAA) as polyanions. Typical bilayer thickness for PAH/PSS was 4 nm [16]. Nano-organized shells containing 2, 3, and 4 polyelectrolyte bilayers on halloysite were prepared. A regular alternation of surface charge shown in Figure 14.10 demonstrates the success of the shelling procedure.

In a follow-on experiment, we formed a LbL shell on halloysite consisting of multilayers of the polycation PEI and anionic silica of 7 and 10 nm diameters (realizing by this positive/negative component alternation). Figure 14.11 shows the

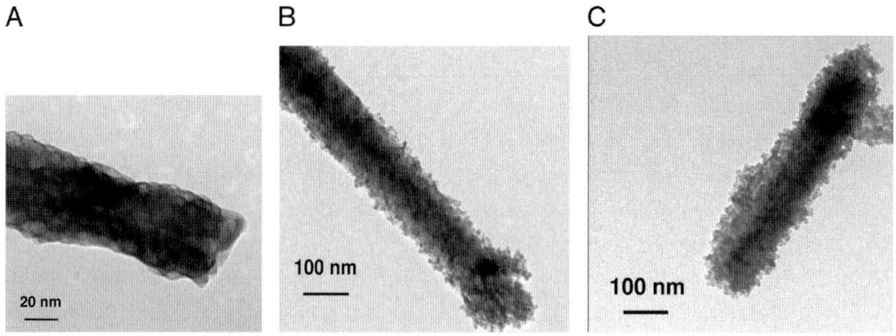

Fig. 14.11 LbL multilayer shells on halloysite: (A) PEI/7 nm silica/PEI coating (three layers), (B) (PEI/7 nm silica)$_2$ coating (four layers), (C) (PEI/10 nm silica)$_3$ coating, (six layers).

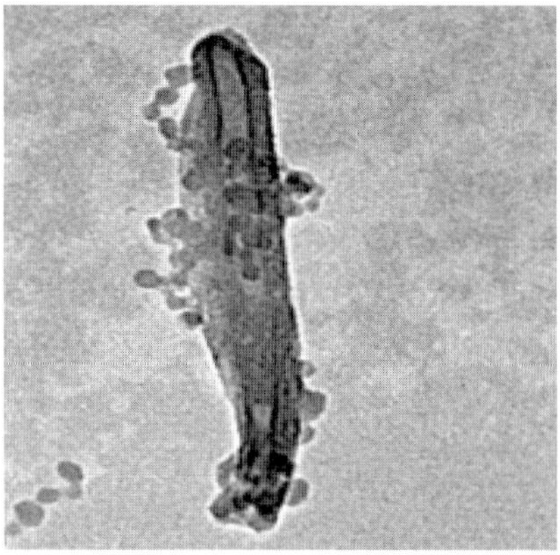

Fig. 14.12 Nanoparticle co-loading: 10-nm silica inside halloysite lumen.

formation of organized nanoparticle/polymer shells on the tubules. Therefore, it is possible to form nanoshells of thickness 10–50 nm on halloysite. Such a shell is not permeable to molecules with a molecular weight above 3000 and also decreases the release of smaller molecules from LbL-covered tubule openings. The next target of this research will be to examine the control that may be achieved from halloysite by optimization of the shell composition depending on pH or temperature. We are also considering the possibility of completely stopping the release of active agent by use of a nanoparticle stopper plugging the openings.

In this set of experiments, we avoided the formation of halloysite aggregates by working with diluted dispersions but microparticle formation through polycation bridging may also be useful for halloysite applications.

Nanoparticle co-loading. Another way to control release through decreasing tubule opening area is co-loading with admixed nanoparticles with a diameter slightly less than the lumen diameter (e.g., with 5 nm gold or 7–12 nm silica). We made such a co-loading (Figure 14.12) and demonstrated a fivefold decrease in the release rate of Dexamethasone (from 20 to 100 hours). Ideally, 2–3 nanoparticles for one halloysite tubule would be an optimal quantity.

14.4
Catalysis in a Nanoconstrained Volume of the Tubule Lumen

The biomimetic synthesis of inorganic composites is a rapidly developing research area lying at the nexus of chemistry, biology, medicine, and materials science. Biomimetic processes parallel the ways biological systems produce inorganic solids. Moreover, biomineralization is one of the most promising methods for forming hybrid inorganic–organic nanomaterials, which often display unique and desirable morphological, structural, and mechanical properties. These hybrid materials represent informative models for the design of complex functional structures. This process usually occurs at ambient temperatures and under mild conditions but requires the presence of organic templates (micro- or nano-confined reactors) like Langmuir monolayers, self-assembled monolayers, polyelectrolyte capsules, micelles, and emulsions; or a wide range of additives, such as organic molecules, peptides, and polymers. Depending on the structure of the organic substance and the reactor type, desired crystal modifications and unique properties can be imparted to the resulting inorganic material.

Among other templates employed for biomineralization, halloysite, a naturally occurring aluminosilicate nanotube, has been undeservedly forgotten. We employed halloysite tubules as hollow enzymatic nanoreactors [20]. The reaction of urease-catalyzed deposition of $CaCO_3$ from water solution containing $CaCl_2$ and urea was taken as a model reaction to demonstrate the possibility of conducting a biomineralization process inside the halloysite nanotube. Entrapment of the enzyme urease (MW ~480 000 and diameter 5 nm) inside the inner volume of the halloysite nanotube was performed according to an adaptation of the procedure described by Price *et al.* [6]. The urease isoelectric point is at pH 6, and in pure water it is negatively charged. This may enhance urease adsorption inside the positively charged halloysite lumen. Activity of the entrapped urease, measured by pH-sensitive dye, was about 20 % of the activity of initial urease dissolved in solution.

On the synthetic stage, urease-loaded halloysite particles were immersed in the mixture of 0.5 M urea and 1 M $CaCl_2$ for 30 min (Figure 14.13). The formation of $CaCO_3$ precipitate inside halloysite particles starts immediately after urea decomposition into ammonia and CO_3^{2-}, catalyzed by entrapped urease. Urea decomposition and CO_3^{2-} formation occur in the halloysite lumen as calcium cations diffuse from the surrounding solution. To prevent diffusion of CO_3^{2-} ions from the halloysite interior accompanied by the undesirable formation of $CaCO_3$ in solution, high concentrations of Ca^{2+}, that will consume all the CO_3^{2-}, are required. Reducing the Ca^{2+} concentration below 0.1 M leads to the formation of $CaCO_3$ particles within the solution phase. The hollow lumen of the urease-loaded halloysite nanotubes before

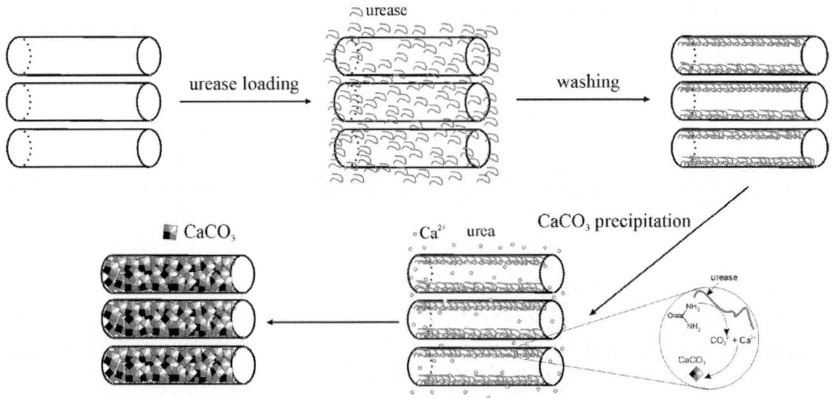

Fig. 14.13 Schematic illustration of the urease-catalyzed synthesis of $CaCO_3$ inside halloysite nanotubes.

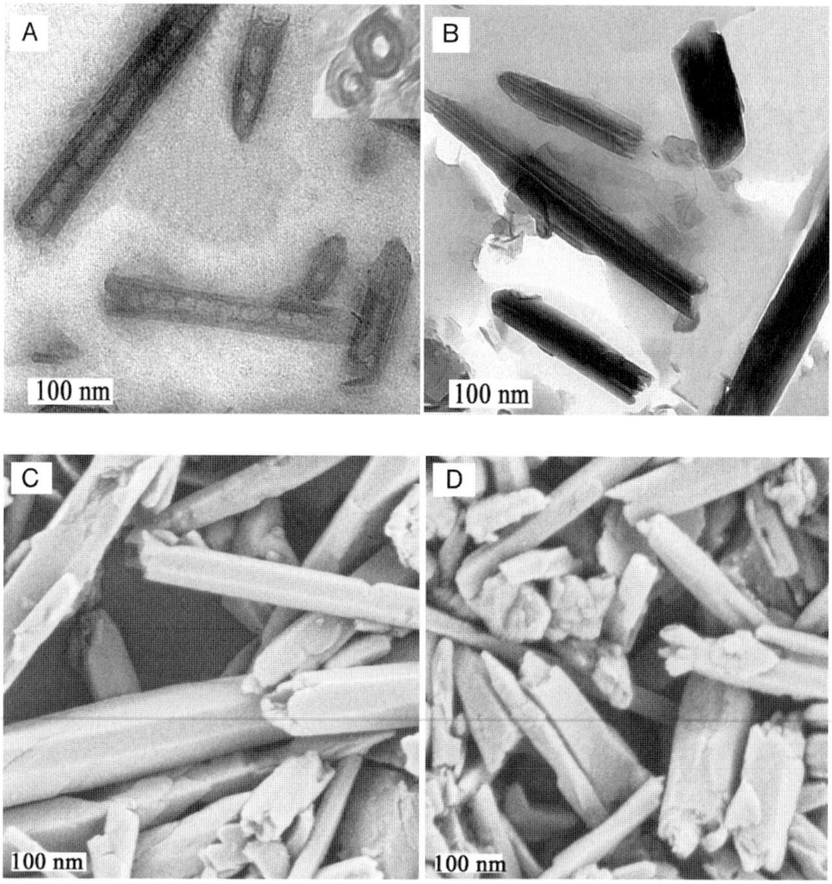

Fig. 14.14 Transmission electron microscopy images of ultra-microtomed halloysite G nanotubes before, longitudinal and, in the inset, perpendicular cross-section (A), and image after $CaCO_3$ formation (B). Scanning electron microscopy images of halloysite G nanotubes before (C) and after (D) $CaCO_3$ formation.

$CaCO_3$ precipitation is clearly visible in Figure 14.14, which is a TEM image of halloysite cross-sections. Halloysite has >90 % of the tubular form with outer diameter 50 ± 5 nm and inner diameter of the lumen 15 nm. The length of the initial halloysite is less than 1 micron, which results in a substantially short diffusion path length. The calculated value of the free inner space indicates the ability to load a maximum 14 ± 3 % of the total volume of the halloysite. The entrapment efficiency of 5 % by volume was estimated.

Complete $CaCO_3$ filling of the inner halloysite space is evident in the TEM image of a halloysite cross-section after the synthesis was carried out (Figure 14.14B). No empty space is seen inside the halloysite particle, indicating uniform calcium carbonate filling of the tubular lumens. This uniformity of the halloysite filling could be explained by the higher diffusion rate of Ca^{2+} ions as compared to the rate of urea decomposition, which is decreased due to the formation of the intermediate urea–Si_2O_5 complex. Ca^{2+} ions may diffuse inside the halloysite lumen not only from the open end of the clay nanotubes but also through the aluminosilicate layers composing the tube shell. The increase in the halloysite weight after $CaCO_3$ precipitation, measured by QCM, is 24 ± 3 %, which coincides with complete filling of the tubular space (21 % by the whole halloysite volume) with $CaCO_3$ deposit.

The urease catalyzed reaction of $CaCO_3$ deposition occurs exclusively inside halloysite nanotubes and no $CaCO_3$ precipitate was found on the outer surface of the nanotubes or in solution after complete filling of the halloysite interior. $CaCO_3$ formation on the nanotube end is also seen on SEM images. The initial, pseudo-cylindrical shape of the halloysite shell (Figure 14.14) undergoes a partial deformation, resulting in a more angular shell with less regular morphology. This may be caused by crystallization forces that appear inside the nanotube during the reaction and lead to an increased inner pressure accompanied by a "swelling" of the halloysite shell. Wide-angle X-ray analysis of the halloysite tubes after $CaCO_3$ synthesis revealed the appearance of additional peaks at the scattering angles $2\Theta = 27.07, 32.80, 43.880, 50.140, 55.850°$, which correspond to the metastable vaterite modification of $CaCO_3$, while, surprisingly, a more stable calcite phase was not found (Figure 14.15). The formation of a metastable vaterite phase was also observed during $CaCO_3$ biomineralization in the presence of different block-copolymers and inside micron-diameter polyelectrolyte capsules, which proves the strong influence of the confined reaction

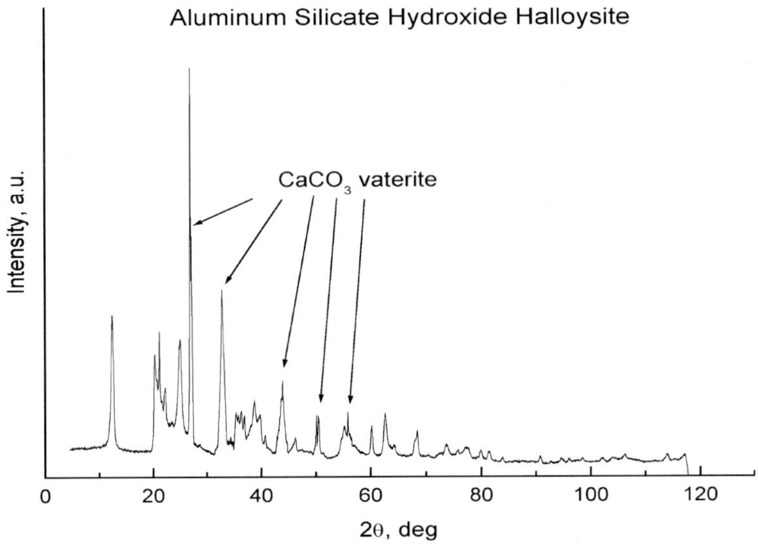

Fig. 14.15 X-ray diffraction patterns of the $CaCO_3$/halloysite composite after $CaCO_3$ formation. The first peak at 12.3° corresponds to 7.2 Å (dehydrated halloysite packing reflection). Arrows indicate peaks corresponding to the vaterite modification of $CaCO_3$. Calcite modification of $CaCO_3$ was not found. Peak assignment was made using the Powder Diffraction Database.

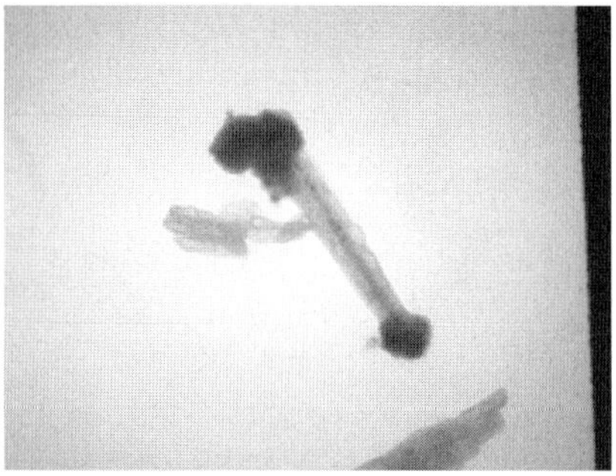

Fig. 14.16 "Nanobell" formation through catalytic reduction of copper from loaded copper salt.

micro-(nano-) volume on the polymorph modification of the resulting inorganic substances.

Thus the application of halloysite as a new biomineralization nanoreactor for carrying out enzyme-catalyzed inorganic synthesis exclusively inside hollow tubular lumen has been demonstrated. In addition it has been elucidated in this example of the urease-catalyzed hydrolysis of urea resulting in the formation of precipitated CO_3^{2-} ions. Synthesized $CaCO_3$ completely fills the inner halloysite lumens and exhibits the metastable vaterite phase. $CaCO_3$ formation was not observed to occur on the outer surface of the halloysite nanotube nor in solution. The "enzymatic" approach may be utilized for loading the interior of halloysite with different compounds or with bioactive material. The idea of using halloysite nano-volume lumens as biomimetic nanoreactors offers promising possibilities for studying crystal engineering and fundamental aspects of the biomineralization process.

Another example of synthesis inside halloysite is copper reduction in the lumen with formation of nanobells (Figure 14.16). Metal deposition within, on the ends, or on the exterior surface of the halloysite tubule, may be governed by the charge of the palladium catalyst used. Commercial stannous palladium catalysts carry a net negative charge on the stannous shell. By employing this catalyst it is possible to sorb the catalyst at the ends of the tubes or on the interior walls of the lumen. When commercial electroless metal deposition baths are used it is possible to deposit a range of metal coatings on the halloysite. Utilization of a polyion tie coat to reverse the surface charge of the halloysite will also render the outer surface active and allow for exterior coating as well. By utilizing dye titration it is possible to determine the correct amount of catalyst needed to ensure full binding and to reduce exogenous metal formation in the plating bath.

A range of metals including iron, copper, cobalt, permalloy and nickel were used to form conductive and paramagnetic coatings on the halloysite clay. These conductive small particles may then be used in polymeric systems to form composites with interesting electromagnetic properties.

20-nm Ni and 5-nm Pd nanoparticles were synthesized through electroless plating both on the outermost surface and in the lumen of halloysite [21–23]. Adsorption of uranium from aqueous solutions in halloysite clay was demonstrated in Ref. [24]. Using halloysite as hydrocarbon conversion catalysts was also proposed in Ref. [26].

14.5
Multilayer Halloysite Assembly for Organized Nanofilms. Forming Low Density Tubule Nanoporous Materials

14.5.1
Tubule–Polycation Multilayer

In this section, we give an example of the assembly of an anionic halloysite in organized multilayers through the alternate adsorption with linear polycations via a LbL technique [8]. As polycations, we used poly(ethyleneimine) at pH 8, 1 mg/mL. The film thickness calculated by frequency decrease utilizing a quartz crystal microbalance (QCM) gave for the halloysite/PEI bilayer a thickness of 55 ± 5 nm. This value is in good agreement with the mean diameter of the halloysite tubule, which is 50 nm. At the conclusion of every adsorption cycle the film thickness was found to increase by a value corresponding to the tubule diameter indicating that we have a regular assembly. Tubules in the monolayers are loosely packed, forming a network with considerable open space (Figure 14.17). The top view of the film examined by SEM confirms this conclusion. We chose a 15 minute adsorption time because it gave a film growth rate close to the tubule diameter (a longer adsorption time gives an insignificant increase of the layer). The tubules in the layer are not parallel, but randomly oriented, and the layer density is rather low. We obtained the density of the halloysite/PEI multilayer as $\rho = (0.48 \pm 0.8)$ g/cm^3. This is less than the density of water and demonstrates that there is no water in the voids between the halloysite tubules in the dry film. Taking the halloysite density as 2.53 g/cm^3, one may calculate that 20 vol% of the sample is occupied by halloysite and 80 % of the sample is air. Therefore, the halloysite coating is of rather low density.

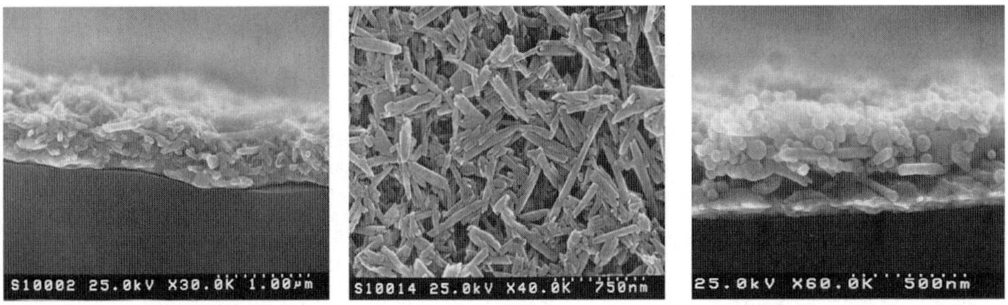

A B C

Fig. 14.17 SEM image of (halloysite/PEI)$_7$ multilayers: (A) cross-section view, (B) top view, and (C) cross-section view of the (halloysite/PEI)$_2$ + (PEI/silica/PEI/halloysite)$_3$ multilayer. The sample is coated with 2 nm Pt, SEM Hitachi S-900, Japan, voltage 25 kV.

From analysis of the film SEM images, the total thickness of the (halloysite/PEI)$_7$ film was found to be 360 nm and the thickness of (halloysite/PEI)$_4$ was 240 nm. Dividing these values by the number of layers we have the average thickness for the layer pair 56 ± 5 nm, which coincides with previous estimations. The branched polycation (PEI) connects and bridges tubules and recharges their surface, so that the next layer of halloysite may be adsorbed and fixed. Using a linear polycation, poly (dimethyldiallyl ammonium chloride), (PDDA) or poly(PAH) did not change the assembly process. In the described assembly we produced halloysite tubule multilayers with a controllable number of layers. We are developing this approach in two directions: (i) Assembly of tubule and spherical nanoparticles in the composite sandwich-like film. (ii) Assembly of proteins with co-enzyme loaded tubules to construct an enzymatic reactor.

14.5.2 Assembly of Tubule/Sphere Multilayer Nanocomposites

In this approach, we constructed a multilayer "sandwich" with an alternation of negatively charged halloysite tubules and spheres of similar diameter, which are kept together by the polycation connectors. Figure 14.17C gives the cross-section image of the (PEI/halloysite)$_2$ + (PEI/50-nm silica/PEI/halloysite)$_3$ film. One can see a closely packed film consisting of tubule layers alternated with sphere layers according to the above given architecture. The density of this film is higher than the density of the loose halloysite/PEI multilayer. The density of the film is (1.0 ± 0.08) g/cm^3. This is less than the density of 50-nm silica/PEI multilayer (1.2 g/cm^3), but more than the density of halloysite/PEI multilayer (0.48 g/cm^3).

Multilayer deposition of halloysite is possible not only on large solid surfaces but also on soft biological surfaces such as wood or cotton cellulose microfibers (Figure 14.18). This coating allows a drastic increase in the porosity of the fibers and materials made from them (paper and textile).

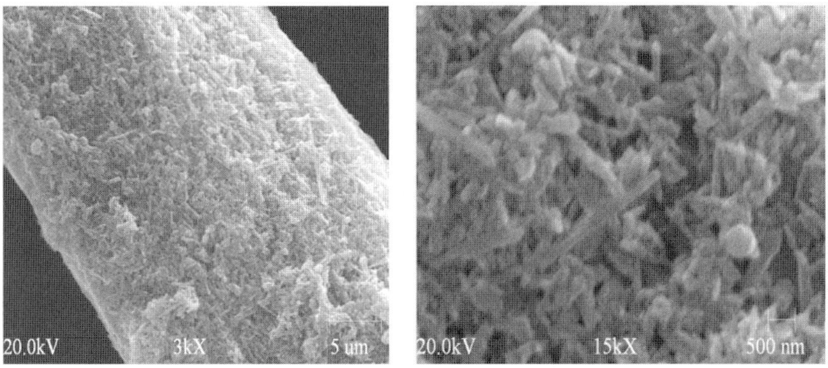

Fig. 14.18 Halloysite/PAH alternate multilayer on lignocellulose microfibers from soft wood at different magnifications (SEM images, AMRAY).

14.6
Applications: Current and Potential

One of the first applications of halloysite as a nanocontainer was in paint with antifouling properties where marine biocide was loaded onto halloysite which was then mixed with paint (Figure 14.19A). Mildewcides are also capable of sustained activity (Figure 14.19B), food additives, fertilizers, oil-well anti-clogging treatment, drug sustained release, plastic fillers, radio wave adsorbing coating with metallized halloysite, specific ion adsorbent, and even hydrogen storage are the list of possible applications based on loading bioactive molecules into the halloysite lumen and their slow release. Other applications include templating nanoparticle surfaces, cosmetics,

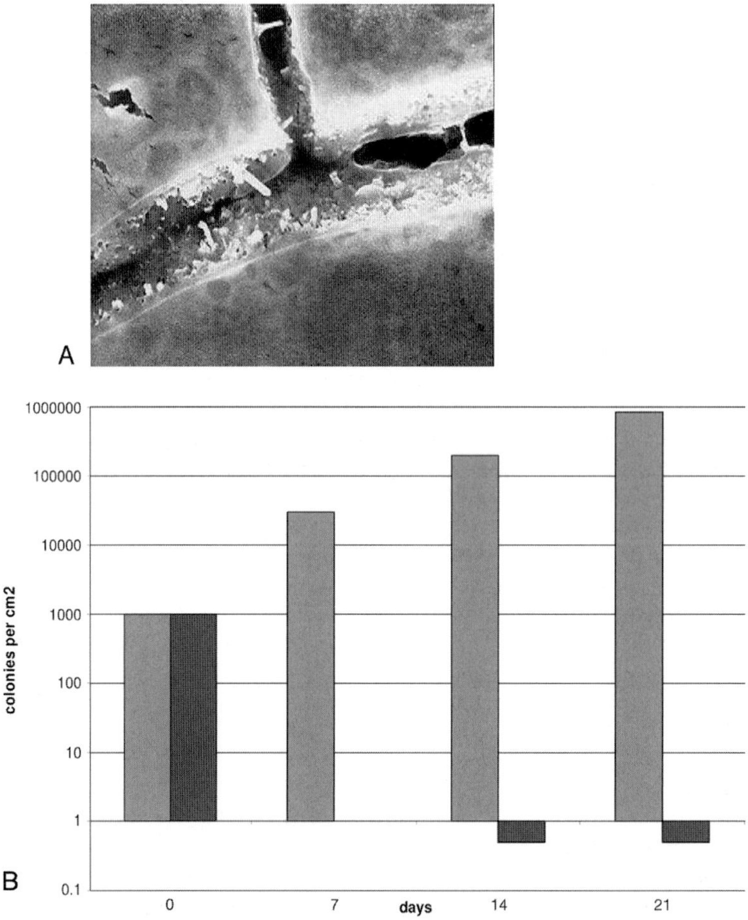

Fig. 14.19 (A) SEM image of cracks in a dry layer of paint with inclusion of halloysite nanotubules containing antifouling agent; (B) mold growth inhibition: number of colonies versus days (light shading – untreated and dark shading – treated growth media).

the delivery of herbicides, insecticides, fungicides and anti-microbials, use in advanced ceramic materials, bio-implants, transdermal drug delivery systems, synthetic catalytic materials and molecular sieves.

Our group is working on halloysite loading with anticorrosion agents. Mixing of halloysite tubules with polymers (like polypropylene) to increase the strength and thermal stability of the resulting composites is also under consideration [26], but more research on halloysite surface modification for better adhesion with plastics is needed. Inclusion of the halloysite as a nanoclay in polymers may be aided by the SiO_2 surface chemistry allowing a range of surface treatments. Common quaternary surfactants, as well as functional silanes, may render the clay easily wet by common polymers. Unlike platy clays, halloysite does not need exfoliation prior to use. In addition, unlike common nanoclays, the functionality of the halloysite as a controlled delivery system or as a conductive filler expands the range of polymeric applications, thus permitting greater creativity in product design.

Acknowledgements

We are grateful to N. Veerabadran and S. Balkundi who are working on the halloysite project at Louisiana Tech University. National Science Foundation NIRT-0210298 grant support is acknowledged. Any opinions and conclusions expressed are these of the authors and do not necessarily reflect the view of the NSF.

References

1 Bates, T., Hilderbrand, F. and Swineford, A. (1950) Morphology and structure of endellite and halloysite. *American Mineralogists*, 35, 463–485.

2 Joussein, E., Pitit, S., Churchman, J., Theng, B., Righi, D. and Delvaux, B. (2005) Halloysite clay minerals – A review. *Clay Minerals*, 40, 383–426.

3 Baral, S., Brandshow, S. and Gaber, B. (1993) Electrolysis metallization of halloysite, a hollow cylindrical 1:1 alumosilicate of submicron diameter. *Chemistry of Materials*, 5, 1227–1233.

4 Tari, G., Bobos, I., Gomes, C. and Ferreira, J. (1999) Modification of surface charge properties during kaolinate to halloysite-7A transformation. *Journal of Colloid and Interface Science*, 210, 360.

5 Price, R. and Gaber, B. "Controlled Release of Active Agents Using Inorganic Tubules," US Patent 5,651,976. Price, R. "Sustained Delivery of Active Compounds from Tubules with Rational Control," US Patent 5,705,191. Price, R. "Method of Controlled Release and Controlled Release Microstructures," US Patent 6,280,759. Price, R. "Efficient Method for Subsurface Treatment, Including Squeeze Treatment," US Patent6, 401,816.

6 Price, R., Gaber, B. and Lvov, Y. (2001) *In-vitro* release characteristics of tetracycline, khellin and nicotinamide adenine dinucleotide from halloysite; a cylindrical mineral for delivery of biologically active agents. *Journal of Microencapsulation*, 18, 713–723.

7 Gaber, B., Price, R., Santos, J.P. and Vittal, V. (2001) Entrapment and controlled release of active agents from halloysite. *Rocks & Minerals*, 76, 211–215.

8 Lvov, Y., Price, R., Gaber, B. and Ichinose, I. (2002) Thin film nanofabrication via layer-by-layer adsorption of tubule halloysite, spherical silica, proteins and polycations. *Colloids and Surfaces: Engineering*, 198–200, 375–382.

9 Levis, S. and Deasy, P. (2002) Characterization of halloysite for use as a microtubular drug delivery system. *Internal Journal of Pharmaceutics*, 243, 125–134.

10 Levis, S. and Deasy, P. (2003) Use of coated microtubular halloysite for the sustained release of diltiazem hydrochloride and propranolol hydrochloride. *International Journal of Pharmaceutics*, 253, 145–157.

11 Kelly, H., Deasy, P., Ziaka, E. and Claffey, N. (2004) Formulation and preliminary *in vivo* dog studies of a novel drug delivery system for the treatment of periodontitis. *Internal Journal of Pharmaceutics*, 274, 167–183.

12 Wagner, A., Cooper, S. and Riedlinger, M. (2005) Natural nanotubes enhance biodegradable and biocompatible nanocomposites. *Industrial Biotechnology*, 1, 190–193.

13 Kommireddy, D., Sriram, S., Lvov, Y. and Mill, D. (2006) Layer-by-layer assembled nanoparticle thin films – a new surface modification approach for stem cell attachment. *Biomaterials*, 27, 4296–4303.

14 Veerabadran, N., Price, R. and Lvov, Y. (2007) "Clay Nanotubules for Drug Encapsulation and Sustained Release." *Nano Journal*, 2, 215–222.

15 Balkundi, S., Veerabadran, N., Lvov, Y. and Price, R. (2007) "Protein Controlled Release from Halloysite Nanotubules." *Journal of Controlled Release* 90, 766–771.

16 (a) Decher, G. (1997) Fuzzy nanoassemblies: Toward layered polymeric multicomposites. *Science*, 227, 1232–1237. (b) Lvov, Y., Ariga, K., Ichinose, I. and Kunitake, T. (1995) Assembly of multicomponent protein films by means of electrostatic layer-by-layer adsorption. *Journal of the American Chemical Society*, 117, 6117–6123.

17 (a) Sukhorukov, G., Donath, E., Lichtehfeld, H., Knippel, E., Knippel, M., Badde, A. and Möhwald, H. (1998) Layer-by-layer self-assembly of polyelectrolytes on colloidal particles. *Colloids and Surfaces A-Physicochemical and Engineering Aspects*, 137, 253–261. (b) Caruso, F., Caruso, R. and Möhwald, H. Fabrication of hollow, spherical silica and composite shells via electrostatic self-assembly of nanocomposite multilayers on decomposable colloidal templates. *Science*, **282**, 1111–1114.

18 Caruso, F. and Sukhorukov, G. (2003) Coated Colloids: Preparation, Characterization, Assembly and Utilization, in Multilayer Thin Films: Sequential Assembly of Nanocomposite Materials (eds G. Decher and J. Schlenoff), Wiley-VCH, Weinheim, pp. 331–362.

19 Mendelson, J., Barrett, C., Chan, V., Pal, A., Mayes, A. and Rubner, M. (2000) Fabrication of microporous thin films from polyelectrolyte multilayers. *Langmuir*, 16, 5017–5023.

20 Shchukin, D., Price, R. and Lvov, Y. (2005) Biomimetic synthesis of vaterite in the interior of clay nanotubules. *Small (Nano, Micro)*, 1, 510–513.

21 Fu, Y., Zhang, L. and Zheng, J. (2005) *In-situ* deposition of Pd nanoparticles on tubular halloysite template for initial metallization. *Journal of Nanoscience and Nanotechnology*, 5, 558–564.

22 Fu, Y. and Zhang, L. (2005) Deposition feature of Ni nanoparticles on halloysite template and magnetic properties of the composite. *Journal of Nanoscience and Nanotechnology*, 5, 1113–1119.

23 Fu, Y. and Zang, L. (2005) Simultaneous deposition of Ni nanoparticles and wires on a tubular halloysite template: A novel metallized ceramic microstructure. *Journal of Solid State Chemistry*, 178, 3595–3600.

24 Kilisliglu, A. and Bilgin, B. (2002) Adsorption of uranium (U-YI) from aqueous solutions on halloysite clay. *Radiochimica Acta*, 90, 155–160.

25 Robson, A. "Synthetic Halloysite as Hydrocarbon Conversion Catalysts," US Patent 4,098,676.

26 Du, M.Gao and Jia, D. (2006) Thermal stability and flame retardant effect of halloysite nanotubules on Poly(propylene). *European Polymer Journal*, 42, 1362–1369.

15
Enzyme-based Bioinorganic Materials

Claude Forano, Vanessa Prévot

15.1
Introduction

Trends in research materials currently focus on innovation in the field of intelligent or smart materials [1]. Smart materials are materials that display functionalities able to detect a change in the near environment, answer to a stimulus, modify a property, store, transfer or convert a signal, in a sense behave like living cells. The design of smart materials focuses on the concept of biomimetic materials [2–4] in order to reproduce specific and selective properties not deliverable by standard chemical or physical processes. Hence, there has been much effort concentrated on the development of enzyme-based biohybrid materials. Enzyme-based biohybrid materials are under much investigation for the operation of physical or chemical functions. Enzymes display active and selective properties such as molecular recognition, complexation or carrier functions, and a wide range of catalytic reactions (hydrolysis, condensation, redox reaction, isomerization, reaction transfer). Enzymes are thus very attractive as an alternative to traditional catalysts when chemical routes are difficult to apply. Enzymes are able to catalyze regio- and stereo-selective reactions with rate accelerations up to 10–17 fold [17]. They are interesting materials for realization of sensing devices, recognition systems and nanoreactors with many potential industrial developments in biocatalysis [4] for the food industry or effluent treatment [5,6], in biomedical domains for medical implants or enzyme delivery [7], in bioanalysis and affinity chromatography [8,9], and in biosensor technology [10–12].

Engineering the development of biohybrid enzyme-based materials requires solution of the limitations of the use of enzymes due to their intrinsic fragility against the various environmental stresses and their narrow range of working conditions (e.g. short pH and temperature range). Embedding enzymes in protective matrices in order to solve these problems has been envisaged for many years [13,14]. Immobilization of enzymes in solid supports is seen as a strategy to preserve the active site, to improve the performance of the enzyme, that is, its activity, reaction rate and long-term stability, and to allow enzyme reuse. Physical and chemical properties of the solid supports such as particle size, surface area, pore diameter, mechanical

Bio-inorganic Hybrid Nanomaterials. Edited by E. Ruiz-Hitzky, K. Ariga, Y. M. Lvov

ISBN: 978-3-527-31718-9

strength, microbial resistance, thermal stability, chemical durability, hydrophobic/hydrophilic character, ease of regeneration, loading capacity and cost will have a great impact on the practical and technical implementation of a biocatalyst [1]. The use of inorganic matrices for enzyme immobilization offers many benefits compared to organic or bioorganic polymers, such as rigid frameworks, chemical biocompatibility, high chemical and thermal stabilities, tunable surface and porosity, ease of preparation and low cost.

Various inorganic host structures have been envisaged for the entrapment of enzymes and the elaboration of novel nanohybrid materials, depending on the desired property. The inorganic host structure may be selected according to the various properties of their framework: structural dimensionality, surface and porosity properties, acid/base character, redox and conducting behavior, surface chemical reactivity and hydrophilicity. Encapsulation of an enzyme in a non-porous matrix with a 3-D framework will be optimized providing a high surface area is promoted. Immobilization of enzymes at the surface of inorganic nanoparticles [15] will then be preferred. Mesoporous materials offer a tridimensional network of interconnected cages and channels beneficial for enzyme encapsulation and substrate/product diffusion. Sol–gel materials with mesoporous frameworks may be easily formed around the targetted biomolecules and have been intensively investigated for designing novel biohybrid materials. Inorganic materials with low dimensional structures (0-D, 1-D, 2-D) display a specific topochemical reactivities. Indeed, their intercalation, ion exchange or delamination properties are very favorable for large biomacromolecule immobilization and the elaboration of nanostructured materials. Their low-dimensional character offers open frameworks useful for the realization of self-assembled and nanostructured enzyme and inorganic building units. Moreover, one can envisage the synergic coupling of specific properties of both the host structure and the enzyme guest for the elaboration of a novel generation of biohybrid nanomaterials with multifunctional properties. Association of an electron conducting matrix with an enzyme able to perform redox processes (oxidase or reductase enzyme) will couple molecular and electronic recognition in one material and be useful for the design of a biosensor component. A host structure with ion exchange properties may facilitate diffusion of the substrate and/or reaction products, create selective recognition and improve the catalytic activity of the embedded enzyme.

Enzymes are versatile biological materials which display either electron or radical transfer, splitting or forming bonds, hydrolysis reaction, molecular isomerization or molecular recognition and may thus be used for many analytical purposes. One of the most interesting fields of application is the development of enzyme biosensors. Biosensors are under strong investigation due to their large potentiality of applications in term of molecular recognition, and more particularly for medical diagnosis and continuous monitoring devices for the environment [2]. For example, biosensors based on urease have been developed for the determination of urea in human bodies in order to diagnose diabetes or uremia pathology and dysfunction of the liver or kidneys [4,5].

Enzymes may be used in the treatment of some diseases but when they are unstable under physiological conditions they must be immobilized in injectable delivery matrices. Calcium titanium phosphate (CTP) is a bioactive ceramic with ion exchange and chemical adsorption properties [14]. It has been used for the immobilization of several enzymes [16] and has been successfully applied in the biomedical field. Glucocerebrosidase (GCR), used in the treatment of Gaucher disease (type I), has been embedded in CTP-alginate and hydroxy-apatitealginate (HAp-alginate) microspheres [17].

15.2 Enzymes versus Inorganic Host Properties

15.2.1 Enzyme Properties

Enzymes are biological materials with a high level of structural hierarchy. They are biopolymers wich result from sequential condensation of amino acid units. Segments of the polymeric chain display either helix (mainly α type), sheet (mainly β type) or unfolded conformations leading to a characteristic secondary structure. Hydrogen bonding between chains and sheets forces the folding of the overall polymer in a tertiary structure. The active site of the enzyme is protected inside the proteic cavity, accessible to the substrate. The properties of the amino acid side chains vary from hydrophobic to hydrophilic, polar to nonpolar, charged to neutral, small to large, and reactive to inert. The protein surfaces are thus capable of interacting with a variety of environments. The binding of proteins to various surfaces is mainly due to electrostatic interaction between the protein side chains and the host. However, covalent grafting of enzymes using a condensation reaction with coupling agents such as the biotine–avidine system has been implemented for the functionalization of enzymes or the realization of biohybrid materials.

Each enzyme has a working name, a specific name in relation to the enzyme action and a code of four numbers: the first indicates the type of catalysed reaction; the second and third, the sub- and sub-subclass of reaction and the fourth indentifies the enzyme [18]. In all relevant studies, it is necessary to state the source of the enzyme, the physical state of drying (lyophilized or air-dried), the purity and the catalytic activity. The main parameter, from an analytical viewpoint is the catalytic activity which is expressed in the enzyme Unit (U) or in katal. One U corresponds to the amount of enzyme that catalyzes the conversion of one micromole of substrate per minute whereas one katal (SI unit) is the amount of enzyme that converts 1 mole of substrate per second. The activity of the enzyme toward a specific reaction is evaluated by the rate of the catalytic reaction using the Michaelis–Menten equation: $V_0 = V_{max}[S]/([S] + K_M)$ where V_0 is the initial rate of the reaction, defined as the activity; V_{max} is the maximum rate, [S] the concentration of substrate and K_M the Michaelis constant which give the relative enzyme–substrate affinity.

The specific properties of a series of enzymes of interest for various applications and intensively investigated for immobilization are listed in Table 15.1.

Tab. 15.1 Specific characteristic of some enzymes or proteins.

Enzyme/Protein	Source	Activity
Urease	Jack bean	urea hydrolysis
Tyrosinase (Tyr)	mushroom	cresol and catechol oxidation
Laccase (Lac)	*Rhus vernicifera*	Cu-oxidoreductase, diphenols
Lipase (Lip)	*Candida rugosa*	esterification, transesterification
Glucose oxidase (GOD)	*Aspergillus niger VII*	redox glucose
Acid Phosphatase (AcP)	potato	Zn-phosphohydrolase
Alkaline Phosphatase (AlP)	*Escherichia coli*	Zn-phosphohydrolase
α-Chymotrypsin	bovine pancreas	hydrolase
Lysozyme (Lys)	chicken egg	hydrolase
Peroxidase (HRP)	horseradish	redox/H_2O_2 amine, phenol, organics oxidation
Hemoglobin (Hb)	bovine	redox/H_2O_2
Myoglobin (Mb)	horse muscle	redox/H_2O_2
Cytochrome c (Cyt-c)	horse heart	redox/H_2O_2

15.2.2 Inorganic Host Structures

In order to achieve a high level of enzyme uptake with a minimum of enzyme degradation or inactivation, different inorganic support matrices and techniques of immobilization have been studied. In general, the support must have a high capacity to bind enzyme. The available surface area, chemical inertia and mechanical stability appears as decisive factors. In addition, the charge of the surface and the hydrophilicity as well as the diffusion ability in the bulk must also be considered for enzyme activity. All these properties will affect the stability of the enzyme and the efficiency of its immobilization. Considering all these specific requirements, different inorganic materials have been used which will be further detailed in the following sections. The most commonly used inorganic solids for immobilization of enzymes are given in Table 15.2 with some specific properties. This series represents a large variety of inorganic matrices with a broad range of chemical, structural and conductive properties, ranging from conductor to insulator, hydrophilic to hydrophobic, neutral to charged framework or ion exchanger, acidic to basic, nonporous to micro- or mesoporous. Most supports have been used for the realization of enzyme biosensors while for biocatalytic purposes silicates were mainly chosen.

15.3 Immobilization Strategy

Preparation of biohybrid enzyme–inorganic materials imposes strict chemical requirements in order that the chemical conditions of the biohybrid preparation respect the preservation of the biomolecule integrity. Knowing that enzymes are very

Tab. 15.2 Various host structure for enzyme immobilization.

Material class	Materials	Chemical formula	Framework charge	Electronic prop	Ionic Exch. prop	p.z.c	Ref
2-D	magadiite	$Na_2Si_{14}O_{29} \cdot nH_2O$	–	insulating	cation	—	[19,20]
2-D	kaolinite	$Al_2SiO_5(OH)_4$	0	insulating	cation	3.6	[21]
2-D	montmorillonite	$A_x(Al_{2-x}Mg_x)(Si_4O_{10})\ (OH)_2 \cdot nH_2O$	–	insulating	cation	variable	[22]
2-D	layered double hydroxides	$[M^{II}{}_{1-x}M^{III}{}_x(OH)_2][X^{q-}_{x/q} \cdot nH_2O]$	+	insulating	anion	variable	[23]
2-D	Zr phosphate	α-$Zr(HPO_4)_2 \cdot H_2O$	–	insulating	cation	1.3	[24]
2-D	Zr phosphonate	α-$Zr(HPO_4)_{1-x}\ (RPO_4)_x \cdot H_2O$	0	insulating	f(R)	1.3	[25]
2-D	birnessite	MnO_2	0	conductive	0	2.7	[26]
2-D	vanadium pentoxide	V_2O_5	0	conductive	0	3.0	[27]
3-D	silica	SiO_2	0	insulating	0	1.8	[28]
3-D	mesoporous silica	SiO_2	0	insulating	0	1.8	[29]
3-D		TiO_2	0		0	6.7	[30]
3-D	alumina	Al_2O_3	0	insulating	0	8.5	[31]
3-D	zeolite NaY	$Na_xSi_{192-x}Al_xO_{384}$	–	insulating	cation	4.2–4.5	[32]

fragile materials whose bioactivity may deteriorate under various physical and chemical stresses, such as temperature or pH changes and chemical agression, immobilization or encapsulation of enzymes or proteins in inorganic host structures imposes the use of "Soft Chemistry" processes [33]. Adsorption onto the surface of particles, intercalation in porous materials, *in situ* encapsulation by sol–gel processes or precipitation reactions have been investigated for various inorganic matrices, mainly silica, silicate and aluminosilicate, metal phosphates, metal oxides and hydroxides and carbon nanotubes. Post-modification treatments such as thermal treatment or chemical anchoring have also been widely developed to solve the problem of immobilization reversibility and enzyme leaching. In any case, the immobilization method must be designed to minimize enzyme desorption, to maximize the stability of the enzyme on the supports and the access of the substrate to the active site of the enzyme.

15.3.1
Adsorption Process

Extracellular enzymes play a key role in the nutrient aquisition and degradation of organic matter in soils [34,35]. Most are adsorbed in mild conditions on soil minerals as enzyme–clay complexes preserving their biological activities. Such a natural soft chemistry process was developed for *in vitro* immobilization of enzymes in inorganic solids for the preparation of biohybrid materials. Adsorption is a phenomenon which occurs mainly at the surface of the crystallites resulting from a chemical equilibrium at the solid/liquid interface. Physical adsorption involves weak bonding (electrostatic interactions, hydrogen and van der Waals bondings) and so was originally the preferred process in terms of activity preservation. Adsorption experiments have been widely used to quantify the physi- and chemi- sorbed enzymes. They are performed in batch reactors. This method is simple, involving mixing enzyme and solid solutions under mild conditions in various molar ratios of adsorbate and sorbent. Nevertheless, some disavantages must be underlined, such as significant desorption of the enzyme and further non-specific absorption of species in the medium, resulting in decrease in the enzyme activity. To further stabilize the adsorption and limit enzyme leakage, cross-linking of the enzymes to the support may be achieved by glutaraldehyde treatment.

The adsorption capacities of the adsorbents are usually determined from modeling of the adsorption isotherms according to the Giles's classification [36] (Figure 15.1).

Most of the isotherms reported for the adsorption of enzymes on solids are of the H or L types and are best fitted either by the Langmuir or the Freundlich model (Table 15.3).

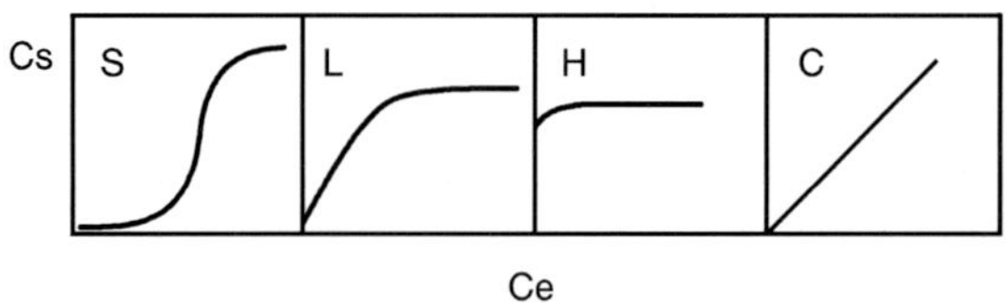

Fig. 15.1 Classification of adsorption isotherms according to Giles *et al.* [36].

Tab. 15.3 Adsorption models.

Langmuir model	Freundlich model
$C_e/C_s = C_e/C_m + 1/(C_m * L)$ C_m = maximum amount for the surface saturation by a monolayer of adsorbate L = adsorbate/adsorbent affinity	$C_s = K_f C_e\ 1/n_f$ K_f = adsorption capacity n_f = adsorption intensity

The amount of adsorbed enzymes depends on various intrinsic properties of the sorbent, including the charge density of the surface, ion exchange properties, structure, surface area and porosity. Electrostatic interactions between the oppositely charged of guest and host surfaces will enhance the amount of adsorbed proteins. Ionic strength, pH, solid/liquid ratio, temperature, buffering conditions, competing molecules, may also affect the adsorption. Enzymes with a low isoelectric point value (i.e.p.) will exhibit a negatively charged surface in neutral pH conditions and adsorption will be favored on positive or neutral surfaces, while for enzymes with high isoelectric point, a negative or basic suppport will adsorb higher amounts. Ion exchange matrices are very suitable host structures with selective adsorption properties for proteins with opposite surface charge. However, enzymes are macromolecules with a large cross-sectional area (Table 15.1) compared to the adsorption site surface of the support and display a heterogeneous distribution of charges and hydrophilic/hydrophobic domains on their surface. Adsorbed enzyme will interact with a high number of adsorption or exchange sites and the adsorption phenomenon will appear as non-specific. The determination of adsorption capacity will enable one to quantify the yield of enzyme immobilization in the biohybrid materials. Most of the adsorption phenomena of enzymes occur in one step with surface saturation reached at high equilibrium concentration, leading to a Langmuir or Freundlich type isotherm. A few examples have reported a multistep adsorption, explained by differentiation of interactions [37]. Adsorption will be favored for solid supports with high surface area and mesoporosity properties. 2-D and 1-D materials with high surface areas appear among the best materials. A colloidal solution of nanoparticles offers a high surface area for adsorption. Enzyme entrapment may also be obtained by reaggregation of colloids in a solution containing the biomolecules. Chevalier *et al.* [38] studied the entrapment of polyphenol oxidase in a hydrogel of the clay material, Laponite, for the construction of a biosensor. Colloidal solutions of layered $\alpha\text{-}Zr(HPO_4)_2 \cdot H_2O$ [39] and layered double hydroxide $Zn_2Al(OH)_6 \cdot Cl \cdot nH_2O$ [40] have also been used for the entrapment of enzymes.

15.3.2
Encapsulation Processes

In order to overcome some limitations of the adsorption process due to surface accessibility or diffusional hindering, immobilization of enzymes by direct *in situ* encapsulation has been investigated. When inorganic supports can be prepared in mild conditions compatible with the enzyme stability, then such processes allow the embedding of large amounts of enzymes. The sol–gel encapsulation method has attracted much attention due to its potential for creating a high concentration of encapsulated enzyme and soft immobilization conditions. Sol–gel processes have been used mainly for immobilization of enzymes in a silica matrix. Entrapment is accomplished through simple matrix polymerization performed in an aqueous solution containing the enzyme. This method affords a high uptake of enzyme without effect on the enzyme molecules. Sol–gel [41,42] silica and its hybrid derivatives are the most widely used inorganic matrices for encapsulation of biological entities [43,44] (among which is an impressive array of enzymes) since the sol–gel process has been described as suitable to trap biologicals, preserving the activities of the immobilized species and allowing diffusion of the substrate. In comparison with the organic polymers generally used for encapsulation (alginate, polyacrylamides), sol–gel matrices exhibit simple preparation, versatile composition and physico-chemical properties, higher chemical and thermal stability and better mechanical strength. The sol–gel process displays also the flexibility to shape the materials in various configurations (monoliths, thick or thin films. . .) according to the intended applications. Nevertheless, no generic conditions can be fixed and they must be optimized for each biohybrid material development.

However, since the enzyme is weakly bound, leakage of entrapped enzyme is often observed by diffusion out of the organic polymer matrix, dependent on the morphology and pore size. Another hurdle of the method could be that the matrix structure acts as a mass transfer barrier and limits or even hampers the diffusion of large substrate molecules to the embedded enzyme. Moreover, the fragility of the final gel structure may often lead to shrinkage of the gel, pore collapse and poor adhesion to the substrate. Some specific synthesis conditions (low pH values, organic solvent) may be too aggressive for the preservation of the structural integrity of the proteins. An alternative encapsulation process may be obtained by precipitation of metal oxides or hydroxides at a pH not far from the working pH of the enzymes and in aqueous solution. When a 3-D network is formed around the biomolecules, in many cases porosity is lost and accessibility to the enzyme is highly limited. To overcome this limitation, low-dimensional materials, and particularly layered materials, have interesting potential for biomolecule encapsulation [45]. The pronounced anisotropy of their structures and crystal shape (platelets, chains, nanotubes, balls or nanoparticles) allows one to build nanocomposite biohybrid materials with a high interface area and tunable self-assembling or aggregation. Porosity may be generated. Figure 15.2 displays the schematic representation of the encapsulation of enzymes in 0-D, 1-D, 2-D and 3-D porous matrices.

Many reviews describe in detail the current state of sol–gel bioencapsulation [46–48] and we will only describe some representative examples in this chapter.

15.3.3
Nanostructuring of Enzyme-based Films

Formation of multicomponent protein films by Layer-By-Layer (LBL) or Langmuir–Blodgett adsorption have been developed for a broad range of applications

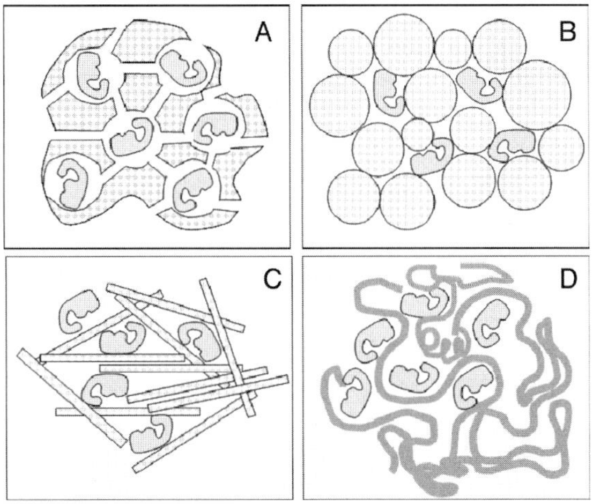

Fig. 15.2 Enzyme encapsulation in (A) 3-D, (B) 0-D, (C) 2-D and (D) 1-D matrices.

such as biological research, medical diagnostics and monitoring, biocatalysis, immunosensing, drug encapsulation, nanoreactors and biofuel cells [49]. Immobilization of enzymes in nanostructured films is a soft chemistry process which provides a unique way to build bio-nanomaterials with tunable multicomponent composition and multi-level architectures. Such nanomanufacturing has been mainly developed, so far, for embedding enzymes in polymers [50,51], amphiphilic monolayers [52] and polyions and clay platelet LBL films [53,54]. More recently, ultrathin multilayers of hemoglobin (Hb), myoglobin (Mb) and horseradish peroxidase (HRP) have been assembled with Montmorillonite nanoparticles [55,56]. Electrodes modified with these $\{clay/protein\}_n$ films are used for electrocatalytic applications.

Porous nanocrystalline metal oxide films are also attracting considerable attention with technological applications that include dye-sensitized solar cells [57,58], electrochromic windows [59] and ion-insertion batteries [60]. Such films are interesting for their electrochemical activity at potentials above their conduction band edge. Variety of proteins (Cyt-c, Hb) may be immobilized in nanoporous TiO_2 electrodes [61–64] for the development of optical and electrochemical biosensors. The nanoporous structure of the film enhances the active surface available for the binding of proteins and TiO_2 favors the electron transfer between protein molecules and the electrodes. With a higher isoelectric point (i.e.p. = 9.5 against 6.5 for TiO_2) and a lower conduction band edge than TiO_2, ZnO is a suitable alternative candidate to adsorb low i.e.p. proteins and display significant conductivities for potentials starting near 0 V [56]. Cytochrome-c (Cyt-c, i.e.p. = 10.5) and Green Fluorescent Protein (GFP, i.e.p. = 4.5) were selectively adsorbed on nanoporous films of TiO_2 and ZnO due to surface charge compatibility. 100 % surface coverage (10.6 nmol/cm^2) was reached for Cyt-c adsorption on TiO_2 against 66 % for ZnO, while the ZnO surface was saturated by GFP (26 nmol/cm^2) (40 % on TiO_2). The films loaded with proteins show the characteristic absorption spectra of the enzymes allowing one not only to monitor and to quantify the protein content in the films but also to confirm whether any detectable denaturation occurred. Spectroelectrochemical studies indicate that more than 90% of the Cyt-c immobilized in both films is electrochemically reducible. Interestingly, the use of nanoporous ZnO allows reduction of Cyt-c at less negative potential.

15.3.4
Covalent Grafting

Covalent binding of the enzymes to a solid support appears to be the most appropriate treatment when enzyme must be used in drastic environments where high stability is required or leakage must be prevented. This is the case, for example, in textile bleaching processes when a high concentration of hydrogen peroxide must be extensively washed at the end of the treatment. Catalase is used for H_2O_2 decomposition. In such destructive conditions, the enzyme must be embedded in a protective stable matrix [65]. Moreover, the selectivity of the grafting reactions may force the orientation of the enzyme at the surface of the material, leading to a narrow distribution of orientations and an enhancement of the catalytic activity.

Proteins display side chains of amino acids with reactive groups ($-OH$, $-SH$, $-NH_2$, $-NH_3^+$, $-COO^-$) able to undergo condensation reactions with various organic functions or complexation reactions with metallic cations. One can take advantage of this property to fix the biomolecule irreversibly onto the inorganic supports hence avoiding its release under working conditions. In many cases, grafting implies pre-existing inorganic support modification to introduce functionalities ($-NH_2$, $-COOH$, etc.) allowing the attachment, by means of a cross-linking agent or a condensing agent, of free functional groups to the enzyme. Often the solid surface must be activated by strong acid treatment to lead to more reactive surfaces. Various molecular groups in enzymes, such as amino, thiol or phenol functions, may be involved in covalent bonds, but due to their widespread surface exposure and reactivity, the lysine residues are the most reactive and useful groups [66]. During the binding reactions, it is important to preserve the active site of the enzyme which must be unaltered by the reagents and the conditions used. Then the covalent immobilization process must involve only functional groups of the enzyme that are not essential for catalytic action. In order to increase the yield of the immobilized enzyme, it is possible to increase the number of reactive functions of the enzyme residues, providing alternative reaction sites to those essential for enzymatic activity. To minimize side-reactions (non-specific reaction, enzyme denaturation), a serial processing method should be carried out.

Various organic modifiers, cross-linking agents and inorganic supports have been used to study enzyme covalent grafting. Glutaraldhyde is one of the most commonly used coupling agents because it may combine with several amino acids and proteins. The two aldehyde groups of the glutaraldehyde molecule can react with amino groups of two enzyme molecules, resulting in a cross-linking enzyme network. Glutaraldehyde is then widely used for the reticulation of enzymes or proteins and for the improvement of the immobilization when they are supported on various matrices [67]. For example, Papain was immobilized on layered double hydroxides activated by glutaraldehyde [68]. A high amount of glutaraldehyde is usually necessary for efficient stabilization but this leads to a partial blockage of the substrate diffusion in the membrane and to a decrease in accessibility to the active site. It has been shown that biosensor performance may be decreased significantly when the reticulation time is too long [69,70]. Furthermore, this cross-linking process does not anchor the biomolecules onto the inorganic surface.

Organosiloxane- and organophosphate-based molecules are other interesting reactive molecules for the covalent grafting of organic molecules onto inorganic surfaces [71–74]. These coupling agents have been used for the functionnalization of silicates, aluminosilicates, metal phosphate and metal oxide surfaces by organophilic molecules, metal complexing functions and catalyst species and may be used for the grafting of enzymes. This process involves a previous step of surface modification followed by adsorption of the enzyme and the grafting reaction. Control of the functionalization yield and enzyme diffusion are both limiting steps in the immobilization reaction. Reaction of silica gel or other hydroxylated supports with an aminosilane derivative (3-aminopropyltriethoxysilane...) is a general way to modify the surface that has been used for a number of years [75,76]. Compounds containing an alcohol or a phenol group (monoethanolamine, *p*-aminophenol) [77] or polyaldehyde [78] can also be used as coupling agents with inorganic supports having surface hydroxy groups. The surface-modified support can then be further coupled to the enzyme, either directly or after activation of the graft in the presence of a bifunctional reagent such as glutaraldhehyde. In all cases, the immobilization occurs mostly on the outer surface. Indeed the diffusion of the different species into the matrix induces clogging by occlusion of pore spaces thus preventing high enzyme loadings.

Nishimura *et al.* [79] explain that enzyme immobilization (glucoamylase, β-amylase, α-amylase, glutamate decarboxylase, glucose isomerase and inulase) is improved by using glutaraldehyde in the presence of a phenolic carboxylic acid having one or more hydroxy groups such as tannic acid or chitosan. According to the authors, these latter substances having specific adsorbability for enzymes, stabilize the enzymes and prevent their deactivation by the cross-linking agent during the immobilization process.

Covalent bonding can also be achieved using carbodiimides to modify enzymes on their carboxylate groups. Yang and coworkers [80] used this reaction to convert glucose oxidase into reactive intermediates which can further react with an amine functional hybrid sol–gel support obtained from the amine-based sol–gel monomer (3-aminopropyltriethoxysilane (APTS)). In a similar approach, enzymes have been bonded to ordered mesoporous silicas [81,82]. By cross-linking, α-chimotrypsin and lipases were successfully retained within hierarchically-ordered mesocellular mesoporous silica pores, providing high loading capacity [83].

In a related approach, alginate beads were modified with aminopropyl-silicate membrane (derived from APTS and tetramethoxysilane). The amino groups remaining on the surface are favorable for the anchoring of biomolecules [84].

In comparison to the other immobilization methods, this process is disadvantageous since generally the support must be modified which is time consuming and involves the use of expensive or hazardous reagents. Furthermore, the determination of optimal conditions is difficult, and production of an immobilized enzyme having high activity is also generally difficult.

15.4 Bioinorganic Nanohybrids

15.4.1 Immobilization of Enzymes in 2-D Inorganic Hosts

15.4.1.1 Immobilization in Clay Minerals and Related Materials

Clay minerals or phyllosilicates are lamellar natural and synthetic materials with high surface area, cation exchange and swelling properties, exfoliation ability, variable surface charge density and hydrophobic/hydrophilic character [85]. They are good host structures for intercalation or adsorption of organic molecules and macromolecules, particularly proteins. On the basis of the natural adsorption of proteins by clay minerals and various clay complexes that occurs in soils, many authors have investigated the use of clay and clay-derived materials as matrices for the immobilization of enzymes, either for environmental chemistry purpose or in the chemical and material industries.

The immobilization of α-chymotrypsin [86,87], peroxidase [88,89], invertase [90,91], lipase [92,93], and tyrosinase [94,95] on clay has been intensively studied. Layered phyllosilicates display a wide range of physico-chemical properties depending on their chemical composition and the type of isomorphous metal cation substitution in the tetraedral and octaedral layers and interlayers. Various silicate supports were tested for the immobilization of Rhizomucor Miehei (RM) and *Candida cylindracae* (CC) lipases and the evaluation of their enzymatic activity in a confined environment [92]. The amount of adsorbed lipases was strongly dependent on the type of support, varying from 44 to 88.2 and 78.7 mg of RM lipase per gram of sepiolite, palygorskite, montmorillonite, respectively, at 105 mg/g initial loading. For CC lipase, the imobilization yield on palygorskite and sepiolite was reduced to about 55 % compared to RM lipase amounts, showing that the enzyme source may affect significantly the immobilization performance. Even larger amount of lipase (*Candida rugosa*) was immobilized on Kaolinite (1.02 g/g) by adsorption experiments [93]. De Fuentes *et al.* [92] observed a reduction in the activity at high enzyme loading for palygorskite and montmorillonite supports due to a lower surface area available for adsorption compared to the sepiolite. With increased loading on the former supports, multilayer adsorption of lipase occurs, leading to the hindering of substrate diffusion with consequent reduction in activity. The mechanism of adsorption of lipases onto the various silicate supports is supposed to occur through a cation exchange reaction with protonated lysines of the enzyme. However, hydrophobic guest–host interactions may also be taken into account, as evidenced by the irreversible adsorption in an aqueous medium. Lipase immobilized in the fibrous silicates (sepiolite, palygorskite) resulted in a better catalytic behavior compared to the layered silicate (montmorillonite) due to a higher flexibility of the mineral crystallite and a greater accessibility.

The interlayered ions of the cationic clays may also have a great impact on their sorption properties [4]. In the case of horseradish peroxidase adsorption on montmorillonite, there was more adsorption on Na-Mont than on Ca-Mont, ~4 μmol/g against ~1 μmol/g at the highest 50 μmol/L equilibrium concentration. Na-Mont is known to give a better dispersion of crystallite, enhancing the surface at the solid/liquid interface and the adsorption amount of macromolecules. However, the enzymatic activity of HRP was almost completely lost when bound on Na-Mont, whereas much of the initial activity was retained with the Ca-Mont support. A multistep adsorption mechanism was shown on adsorption isotherm plots for both solid supports, involving successive surface adsorption, surface rearrangement of the macromolecules and interlayer penetration. The structural stabilization of the HRP active site and the maintenance of its activity involves the complexation of two Ca^{2+} ions [94]. Under adsorption of HRP on montmorillonite a cation exchange reaction may occur. The cation exchange between Ca^{2+}-HRP and the Na^{+}-Mont surface will then modify the secondary structure of the enzyme and reduce strongly its activity while for Ca^{2+}-Mont, Ca^{2+} exchange will not occur.

Layered polysilicate magadiite ($Na_2Si_{14}O_{29} \cdot nH_2O$) was used by Peng *et al.* [89,96] for the immobilization of myoglobin (Mb) and hemoglobin (Hb). Magadiite is a material made from a stacking of negative Si-containing tetrahedral layers with exchangeable sodium cation, ensuring charge compensation between the layers. It displays wide exchange and intercalation properties, with a highly reactive framework containing silanol groups. Intercalation of myoglobin and hemoglobin was performed with a previously exchanged large organic cation (tetrabutylammonium) in order to expand the interlayer galleries and to weaken the guest cations–inorganic layers interaction. High resolution transmission electron microscopy images are in good aggrement with the XRD data, showing expansion of basal spacing to 4.4–6.8 nm for Mb–magadiite and Hb–magadiite, respectively. The narrow XRD peaks for both biohybrid materials are attributed to the formation of layered biohybrid materials where most of the enzyme is entrapped between the inorganic layers with a large distribution of orientation. While the secondary structures of the protein appear undisturbed (infrared data), the appearance of a new band at 356 nm in the UV–Visible absorption spectrum indicates that the structure or the environment of the heme site was affected by the protein–magadiite interactions. The high values of binding constant determined from the adsorption isotherms for both Mb and Hb proteins show their high affinity to magadiite. The immobilized proteins retained most of their activity allowing their use as biological nanosensors.

The entrapment of various enzymes and proteins by clay minerals proceeds by weak interactions including electrostatic interactions, hydrogen and van der Waals bonding. Additivity of these various attractive forces renders the adsorption irreversible in some cases, but usually a leaching of enzyme is observed under working conditions. In order to fix the enzyme irreversibly at the surface of the clay layers different processes have been tried. In order to fix invertase on bentonite, Monsan and Durand [90] previously treated the clay mineral with a coupling agent, namely cyanuryl chloride, but the amount of invertase immobilized was poor even though a significant percentage (13.4%) of catalytic activity was retained after anchoring. However, as for other host structures, increase in stability against denaturing factors (pH, temperature) was observed.

The effects of organic molecules and phosphate on the adsorption of acid phosphatase on various minerals, and kaolinite in particular, have been investigated by Huang *et al.* [97]. The Langmuir affinity constant for AcP adsorption by kaolinite follows the series: tartrate ($K = 97.8$) > phosphate ($K = 48.6$) > oxalate ($K = 35.6$) > acetate ($K = 13.4$). At low concentration, acetate even promoted the adsorption of acid phosphatase. It was considered that competitive interactions between anionic adsorbates can occur directly through competition for surface sites and indirectly through effects of anion adsorption on the surface charge and protonation.

Jackbean urease was immobilized on kaolinite and montmorillonite [98]. The amounts of urease required for maximum immobilization were 70 and 90 mg g^{-1} of kaolinite and montmorillonite, respectively. The K_m values of immobilized urease (25.1–60.8 mM) were of the same order of magnitude as that of free urease (29.4 mM) but one order of magnitude higher than those of soil urease (1.77–2.90 mM). Immobilization of urease on clay surfaces leads to increases in the kinetic constants.

Montmorillonite (Mont) and other clay minerals coated or intercalated with aluminum or iron hydroxide clusters or nanoparticles are natural occuring nanocomposites [95,99–101]. However such nanocomposites can be interesting models for the development of surface modified or pillared clay materials. Aluminum hydroxide clusters or nanoparticles may be used for the expansion of the interlayer galleries and as reactive species for enzyme grafting. Naidja *et al.* [102,103] reported the adsorption of tyrosinase onto Ca-Mont and $Al(OH)_x$-Mont complexes. Upon loading aluminum hydroxide into Montmorillonite the basal spacing increases, replacement of Ca^{2+} occurs reducing the corresponding cation exchange capacity, simultaneously with the enhancement of the BET specific surface area of the modified clay. The tyrosinase adsorption on Ca-Mont leads to a shift in the *d* spacing from 1.49 to 1.95 nm ($\Delta d = 0.99$ nm) only at high loading (>255 mg/g), indicating that some intercalation of the enzyme has occurred. Adsorption by $Al(OH)_x$-Mont nanocomposites corresponds to isotherms fitting the Langmuir model. The maximum adsorption capacity was estimated to range from 354 to 429 mg/g of clay, depending on the level of $Al(OH)_x$ coating. At high concentration, the amount of adsorbed tyrosinase was double the amount adsorbed by unpillared Ca-Mont. Under pH working conditions (6.5), tyrosinase displays a net negative charge because of its lower isoelectric point (6.1) and is immobilized on Ca-Mont, mainly through hydrogen bonding and attractive van der Waals interactions. Coating of clay with $Al(OH)_x$ prevents any intercalation and forces the protein to be adsorbed on the external surface of the crystallites. However, the adsorption on $Al(OH)_x$-coated Mont was found to be irreversible because tyrosinase binds through a ligand exchange reaction, carboxylate groups displacing OH groups from Al hydroxide polymers. A similar approach was developped by Rao *et al.* [104], for the immobilization of acid phosphatase (AcP) on montmorillonite and $Al(OH)_{18}$-Montmorillonite. Adsorption of the enzyme on both supports leads to a reduction in enzymatic activity of 80 and

42 % respectively. The pH activity profile is strongly modified after immobilization. While free phosphatase displays an optimum pH at 4.5–5.0, the activities of Mont-AcP and $Al(OH)_{18}$-Mont-AcP increase continuously from the alkaline pH range to pH 4.0. The presence of AlOH species here also benefits the maintenance of the activity. The reduction of the negative charge of the clay layers by AlOH surface modification favors an increase in the OH^- ions concentration in the surroundings of the adsorbed enzyme, leading to a shift in the optimum activity towards more acid pH. However, changes in the activity due to structural modifications or orientational effects of the protein constrained in the clay matrix is not excluded, as suggested by Quiquampoix *et al.* [105,106]. The unusual lower Michaelis–Menten constant (K_m) values measured for the embedded enzyme evidences an unusually high enzyme–substrate affinity and confirms the maintenance of the native structure of the proteins.

Organoalkoxysilane has been used extensively for the functionnalization of silicate surfaces by organic molecules. Using the same approach, invertase was immobilized on Mont-K10 [88] by adsorption and covalent binding using glutaraldehyde or 3-aminopropyltriethoxysilane (APTES) treated montmorillonite. Functionalization is monitored by X-ray diffraction, showing the expansion of the *d*-spacing by 1.53 nm upon APTES intercalation and grafting. Immobilization of invertase leads to further expansion (2.25 nm), the consequence of some enzyme intercalation. This *d*-spacing value cannot account for the whole protein intercalation which is probably due to partial insertion of side-chain amino acid residues. The activity of the immobilized invertase for sucrose hydrolysis was evaluated. The maximum rate (V_m) of immobilized enzymes is much lower than that for free enzyme. The abnormal increase in the K_m value is due to the restriction of substrate diffusion. However, a increase in chemical stability for adsorbed and immobilized enzyme leads to an interesting re-usability of invertase even though some leaching is observed for non-cross-linked invertase.

Walcarius *et al.* [107] demonstrated recently that organically functionalized clays display high capacities of glucose oxidase (GOD) and polyphenol oxidase (PPO) anchoring. Smectite (Smec) was reacted with organosilane reagent APTS and *N*-trimethoxysilyl propyl-*N,N,N*-trimethylammonium (TMPA) for surface modification. Grafting of GOD with Smec-APTS was performed using glutaraldehyde (Glu) as a coupling agent. Covalent linking was then obtained for Smec–APTS–Glu–GOD. PPO was entrapped via electrostatic interaction in Smec–TMPA clay. The grafting process modified strongly the acid–base properties of the clay from 7.2 for the pristine materials to 10.1 or 4.5 for Smec–APTS and Smec–TMPA, respectively. Up to 60 mg of GOD per g of organoclay was covalently fixed, leading to the preparation of a glucose amperometric biosensor with higher sensitivity than ungrafted GOD/clay analogous material.

15.4.1.2 **Immobilization in Layered Double Hydroxides**

Layered double hydroxides (LDH), also referred to as anionic clays, are very useful materials due to their anion exchange properties. LDH display a layered structure built on a stacking of positive layers ($[M^{II}{}_{1-x}M^{III}{}_{x}(OH)_2]^{x+}$), separated by interlamellar domains constituted of anions and water molecules ($[X_{x/q} \cdot nH_2O]^{x-}$) [117].

The increasing interest in LDH as host structures arises from their versatile properties in terms of chemical composition of both layers and interlayers, their high and tunable layer charge density and anion exchange capacities, their bidimensional structure which can accommodate large anionic and/or polar molecules, their adjustable textural properties and ability to delaminate in solution. These intrinsic LDH properties make them very favorable host structures for the preparation of bio-LDH hybrid materials [108,111] and particularly for the confinement of negatively charged enzymes with isoelectric point varying over a large pH domain [112–114]. LDHs can be prepared in by the coprecipitation method [115,116]. Formation of LDHs is possible over a large range of pH values allowing the preparation method to be adapted to a large variety of enzymes with different isoelectric points. Immobilization of active proteins can then be realized in one step by addition of the biomolecules in the precipitation medium. The amount of immobilized enzymes is adjusted by tuning the charge density of the layer and the enzyme/LDH weight per weight ratio. Interestingly, there are a large number of preparation methods of hydrotalcite-like compounds allowing a great tuning of morphology, porosity and surface area properties [115].

Rhaman and coworkers [112,113] studied the adsorption of lipase on [MgAl] LDH and its biocatalytic activity for butyl oleate synthesis. They demonstrated that up to 277 and 531 $mg\,g^{-1}$ of lipase were adsorbed on [MgAl-NO_3] and [MgAl-Dodecylsulfate] LDH, respectively, showing the highest adsorption capacity of the anionic clays compared to smectite or inorganic phosphate. Recently, we reported the adsorption isotherms of urease on [Zn_RAl] LDH under various experimental conditions (pH, buffer) [117]. The kinetic study showed the fast adsorption process (less than 60 min) (Figure 15.3).

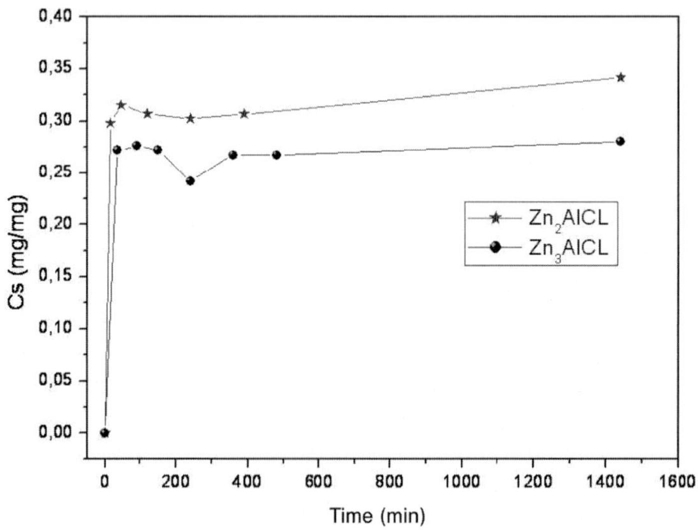

Fig. 15.3 Kinetic study of urease adsorption on Zn_2Al and Zn_3Al LDH.

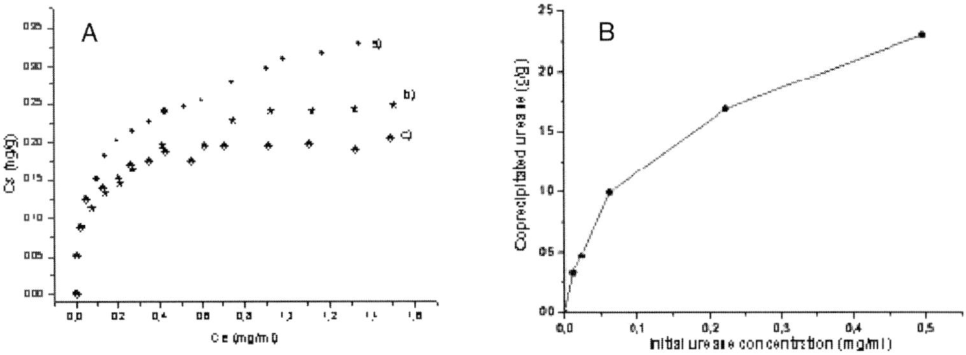

Fig. 15.4 (A) Urease adsorption isotherms for (a) Zn_2Al-Cl, (b) Zn_3Al-Cl, (c) Zn_4Al-Cl; (B) amount of immobilized urease in Zn_3-Al by the coprecipitation method.

Figure 15.4(A) shows the effect of the $R = Zn^{2+}/Al^{3+}$ ratio, which determines the charge density of the LDH layer, on the Freundlich adsorption isotherms. K_f values are far higher than those measured for smectite or other inorganic matrices. The increase in K_f with the charge density ($K_f = 215$, 228, 325 mg/g, respectively, for $R = 4$, 3 and 2) is supported by a mechanism of adsorption based on an anion exchange reaction. The desorption isotherms confirm that urease is chemically adsorbed by the LDH surface. The aggregation of the LDH platelets can affect noticeably their adsorption capacity for enzymes and the preparation of LDH adsorbant appears to be a determinant step for the immobilization efficiency. [Zn_RAl]-urease hybrid LDH was also prepared by coprecipitation with $R = 2$, 3 and 4 and $Q =$ urease/Zn_RAl from 1/3 up to 2.5. For $Q \leq 1.0$, 100 % of the urease is retained by the LDH matrix whatever the R value while for higher Q values an increase in the enzyme/LDH weight ratio leads to a decrease in the percentage of the immobilized amount.

The amount of urease immobilized by *in-situ* coprecipitation has been plotted versus the initial urease concentration (Figure 15.4(B)). An L-type adsorption isotherm profile is obtained, with an increase in the coprecipitated amount up to $Q = 1.0$ and a surface saturation behavior similar to that found in the adsorption experiment (Figure 15.4(A)). The amount of coprecipitated urease was always greater than 90 % of the initial value, regardless of the R and pH conditions. A comparison of Figure 15.4(A) and (B) clearly shows the best efficiency of the coprecipitation process for the immobilization of greater amounts of enzyme. Excess washing of the coprecipitated phases did not induce any urease release in this case, showing that the enzyme is irreversibly immobilized between the LDH layers and platelets. While, for the adsorption reaction, the desorption isotherm confirms that urease can be released by successive washings with water and even more efficiently with phosphate buffer often used for enzymatic tests.

A second process of enzyme entrapment in LDH was also performed using colloidal solutions of LDH. Such colloidal solutions can be prepared by delamination of organically modified LDH in an organic medium [118–120]. Exfoliation leads to a total deaggregation of secondary particles, the individual platelets retaining few units of stacked layers. LDH colloidal solutions can be mixed with aqueous solutions of urease, resulting in a restacking of the clay layers with entrapment of the enzyme [121]. As shown by Rhaman *et al.* [112] in the adsorption study of lipase, the use of LDH modified by anionic hydrophobic surfactant allows one to increase the amount of adsorbed protein. The modified clays display both hydrophobic and hydophilic domains well adapted to the similar ambivalent behavior of the enzyme. Both processes, coprecipitation and exfoliation, display the advantage for adsorption that once the biomolecules are immobilized no leaching is observed in aqueous solution.

Calcined [MgAl] LDH was also used to adsorb penicillin G acylase [121]. The calcined LDH phases have porous structures, large specific surface areas and abundant basic sites to bind the enzymes. The effect of varying the composition of the LDH precursor and calcination temperature on the activity of the immobilized enzyme has been reported. In this case, the percentage of immobilized proteins increases up to 88%.

A more sophisticated method was developed by Ren *et al.* [123] for the immobilization of penicillin G acylase by covalent grafting in the interlayer galleries of a LDH by a three-step procedure using a glutamate pillared LDH as the starting material.

15.4.1.3 **Immobilization in Layered Metal Oxides**

Birnessite γ-MnO_2 is a strong adsorbent which occurs as a natural mineral [124] and is widely studied as a material for solid state electrochemistry [125]. γ-MnO_2 has a layered structure whose chemical composition is more properly represented by the formula $A_xMnO_{2y} \cdot zH_2O$. As a layered host structure, it displays a high potentiality to immobilize or intercalate a large variety of chemical species such as enzymes [126–128]. The adsorption isotherm of tyrosinase on birnessite is well described by an H type, as classified by Giles *et al.* [129], modeled by the Langmuir equation with an adsorption capacity of 9.62 g of tyrosinase per g of solid and a binding affinity of $K = 3.57$, much higher than for adsorption of tyrosinase by Ca-montmorillonite [130]. Proteins were bonded to the external surface as shown by the unchanged X-ray diffraction patterns (Figure 15.5).

The shift of the amide I mode (FTIR spectra) from 1657 to 1646 cm^{-1} was attributed to a change in the α-helix native structure to β-sheets, secondary structure conformations. Atomic Force Microscopy (AFM) images display the coating of the manganese oxide surface as well as the unfolding in a ellipsoidal chain of the protein molecules after adsorption and immobilization on the surface.

A colloidal suspension of conductive vanadium pentoxide [130] can be used to perform intercalation, adsorption or encapsulation of electroactive molecules or biomolecules for electrodes or biosensor realization [131]. Encapsulation of glucose oxidase in nanocomposite films made with polyvinyl alcohol and V_2O_5 sol–gel matrix or in ferrocene intercalated V_2O_5 sol–gel [132] were envisaged to prepare glucose biosensors.

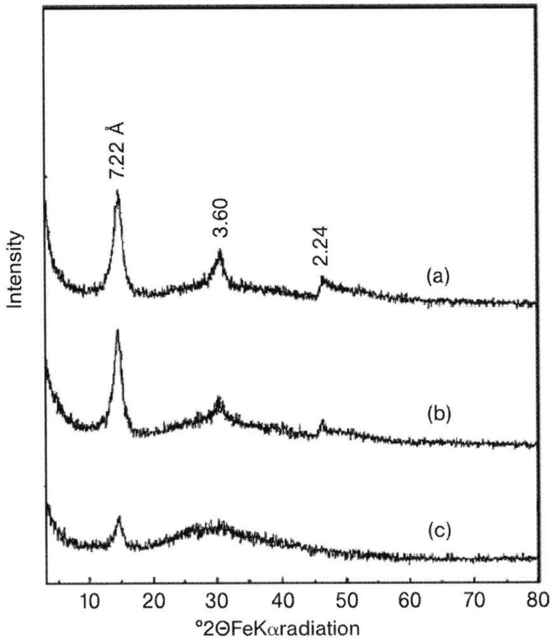

Fig. 15.5 PXRD patterns of (a) birnessite and the tyrosinase – birnessite complexes at (b) 0.21 g/g and (c) 2.74 g/g tyrosinase loading from Ref. [126].

15.4.1.4 **Immobilization in Layered Zirconium Phosphate and Phosphonate**

Layered inorganic phosphates have been used as host materials for nano- and micro-encapsulation of two redox active heme proteins, cytochrome c and horseradish peroxidase (HRP) [39]. Proteins having a number of positive charged residues are expected to show a high affinity for the anionic surfaces of the α-$Zr(HPO_4)_2 \cdot H_2O$ (α-ZrP). The excess of surface charge of the matrix may be tuned by adding appropriate electrolyte/buffer ions or by controlling the pH of the solution. As other layered compounds such as clay minerals, layered double hydroxide or layered sulfide, α-ZrP may be exfoliated to give a colloidal solution, hence enhancing its surface property. Preparation of biohybrid materials was performed by equilibration of buffered protein solutions and a colloidal suspension of exfoliated material (1 g L^{-1}) during 12 h. The binding stoichiometry of Cytc per α-ZrP was evaluated to be 95.6 mg/g. The binding constants were determined by the Scatchard model [133]. Electrostatic interactions between the positive surface of the enzyme and the negative surface of the matrix are responsible for the strong binding affinity constant. However, the binding affinity was much lower for Cyt c (42 μM^{-1}) than for HRP(1.5 μM^{-1}). UV-Visible and infrared spectroscopy data indicate that immobilization of Cyt c and HRP proceeds via intercalation in the α-ZrP galleries without disturbing the secondary structure of the proteins. The soret bands (UV-visible spectra) retain the spectroscopic feature of the free proteins and the vibration bands of the amide groups (FTIR spectra) are located at

Tab. 15.4 Adsorption and structural data for α-ZrP-protein biohybrid material from Kumar *et al.* [136].

Biohybrid	Binding stoichiometry (μM)	K_b (M^{-1})	*d* spacing (Å)
myoglobin	12	2×10^5	54
lysozyme	40	1.33×10^6	47
hemoglobin	14	5.4×10^6	66
chymotrypsin	3	2.5×10^6	62
glucose oxidase	1.1	5.6×10^4	112

the same positions, leading to superimposable spectra. Intercalation of protein in the galleries of α-ZrP is observed, as shown by the expansion of the interlamellar distance from 7.6 to 32 Å. The broadness of the diffraction lines is attributed to a large distribution of randomly orientated proteins in the interlayer spacing. Total reversibility of the adsorption phenomenon can be demonstrated when increasing the ionic strength of the buffered solution. The immobilized enzymes retain their redox and peroxidase activities after nanoencapsulation.

A series of enzyme and proteins (met-myoglobin, lysozyme, met-hemoglobin, glucose oxidase, α-chymotrypsin) was also immobilized in α-ZrP by Kumar *et al.* [134]. Binding constant values clearly confirm the high affinity of the various proteins with the host structure (Table 15.4).

Here again, protein immobilization proceeds through an intercalation mechanism which leads to the expansion of the basal spacing, proportional to the size of the enzyme (Table 15.4) without any structural change of the proteins, as shown by the infrared and circular dichroism data. *d*-spacing values correspond to the intercalation of a monolayer of protein except for glucose oxidase which is probably intercalated as a double layer with a *d* value nearly twice the size of the protein. The average area occupied by the protein thus depends on the binding constants as well as on their average size. The activity rates for lysozyme, chymotrypsin and glucose oxidase are even slightly higher than for the free proteins while the reduction in V_{max} for immobilized Mb is attributed to conformational restrictions imposed in the rigid matrix. Studies of the thermal stability of immobilized proteins can be monitored by changes in the X-ray diffraction pattern at low Bragg angles, an infrared feature in the amide groups region and emission fluorescence spectra under excitation in the 280–310 nm domain.

Layered phosphate/phosphonate and phosphonate materials, obtained by substitution of the phosphate moiety by phosphonate groups, display interesting tunable hydrophilic/organophilic properties for adsorption processes. When Candida rugosa lipase (CRL) is simply equilibrated with zirconium phosphate and phosphonate [135,136], immobilization was demonstrated to take place at the surface of the microcrystals. However, because lipase exhibits a strong hydrophobic character, its uptake by zirconium phosphate and phosphonate was much more related to the hydrophobic/hydrophilic character of the supports than to the surface area properties. A higher uptake is observed for zirconium-phenylphosphonate (78 %)

and zirconiumphosphate-phenylphosphonate (74%) than for α-ZrP, due to a co-operative positive effect of both protogenic and hydrophobic groups on the surface of the support. Immobilization was stable in various desorption conditions. The authors observed that increasing the loading of the proteins on the supports induces a loss of activity, explained by a decrease in the accessibility of the substrate.

Chemical modifications of the α-ZrP layers by substitution of phosphate groups by phosphonate ($O_3P–CH_2–COOH$: CMP, $O_3P–CH_2–CH_2–COOH$: CEP) leads to a tuning of the electrostatic attraction of proteins by the surface of the host matrix [137]. Among the three solids, the binding capacities for Mb, Hb, and GOD are in the order α-ZrP > α-ZrCMP > α-ZrCEP following the decrease in the acidic behavior and confirming the dominating character of the electrostatic interactions in the binding except for immobilization of lyzosyme. XRD patterns evidence the intercalation of protein within the α-ZrP galleries (Figure 15.6). The confinement leads to an increase in thermal stability (up to 90 °C for Cyt c for example) [138].

When Hb intercalated α-ZrP is heated over the denaturation temperature of the free enzyme a recovery of the enzyme activity is observed after an initial activity loss [139]. This unusual behavior for free protein is even enhanced for α-ZrCMP and α-ZrCEP. Hb activity was twice higher after thermal treatment than before heating for α-ZrCEP. Simultaneously a recovery of the structure was observed, showing that structural changes (dehydration, unfolding of the peptide chains, formation of random coils) may be reversible. The authors [139] evidenced that recovery of the activity after thermal treatment of Hb immobilized in the three solids leads to a new conformation of the protein resulting in an improved bioactivity. Using ITC analysis, the authors demonstrated that immobilization of Hemoglobin in α-ZrP was exothermic and proceeds via a single site binding model. The binding constant ($K = 0.9 \times 10^6\,M^{-1}$), the reaction enthalpy ($\Delta H = -31.0\,kcal\,mol^{-1}$) and entropy ($\Delta S = -77.0$ e.u.) and the number of binding sites ($n = 60$) were determined. K and n values are in good agreement with equilibrium binding data (Table 15.4).

Bellezza *et al.* [140] recently performed the adsorption of myoglobin onto amorphous porous α-ZrP and ZrBP prepared by direct template synthesis, where microcrystals of zirconium phosphate or phosphonate are precipitated at the interface of AOT micelles. Adsorption isotherms were fitted using the Langmuir equation, showing a strong affinity between the protein and the supports. The maximum uptake was reached at a pH between 5 and 6 where both enzyme and support display opposite charges. The surface coverage by myoglobin calculated from structural data was always lower than the specific surface area measured by the BET method. As for lipase immobilized non-porous zirconium phosphate/phosphonate, adsorption on porous ZrBP ($Q_{max} = 134$ mg/g, pH 5) was nearly twice that for porous α-ZrP ($Q_{max} = 75$ mg/g, pH = 5). The effect of temperature above 60 °C induces a conformational change in the protein with an increase in absorption caused by an enhancement of the hydrophobic guest–host interactions.

LiNa Geng *et al.* realized similar experiments for the adsorption of trypsin [141] and hemoglobin [142] on γ-zirconium phosphate (γ-ZrP) and organo γ-ZrP intercalated with butylammonium (BA) and tetrabutylammonium (TBA). Hb adsorbed in the galleries of BA-γ-ZrP and TBA-γ-ZrP mainly by hydrophobic

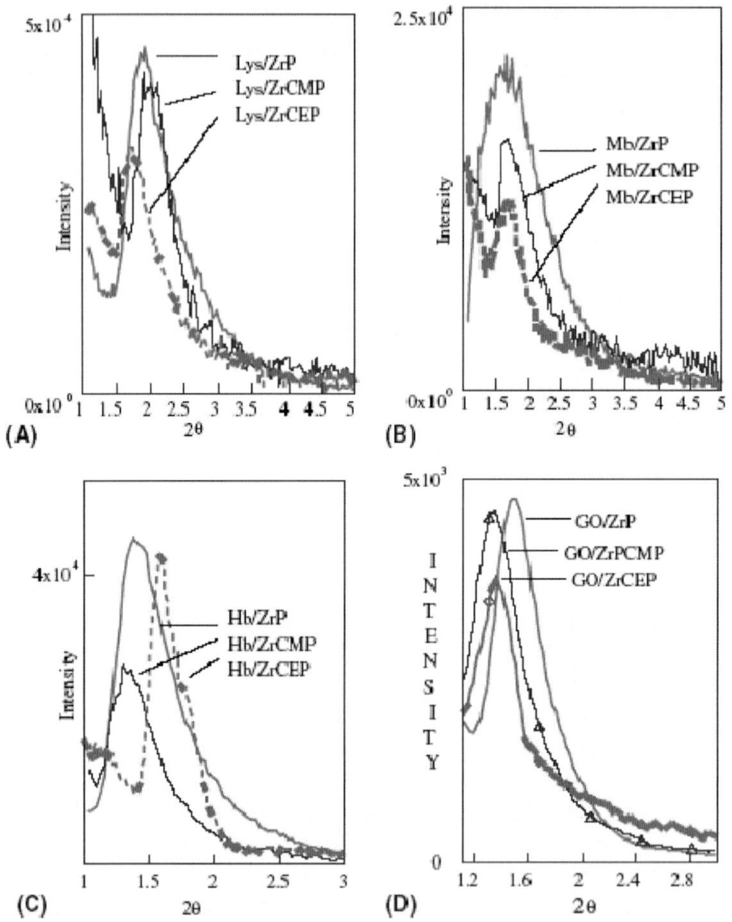

Fig. 15.6 Powder X-ray diffraction patterns of Lyz (A), Mb (B), Hb (C), and GO (D) bound to α-ZrP, α-ZrCMP and α-ZrCEP, respectively, from Ref. [137].

interactions with preintercalated alkyl chains then the helical content decreases because of structural deformation.

15.4.2 Immobilization of Enzymes in 3-D Inorganic Hosts

15.4.2.1 Immobilization in SiO_2

Since it is the major constituent of the Earth's crust, silica is very abundant in nature in amorphous or crystalline forms. Silica can be manufactured, for instance, by high-temperature processing (glasses) or by the sol–gel process. This latter occurs through a two-step reaction: hydrolysis followed by condensation. Silica gels are good candidates for use in enzyme immobilization. They display good optical and mechanical properties as well as high thermal and chemical stabilities. Moreover, they are resistant to microbial degradation. Their fabrication is simple and the surface can easily be modified to improve the immobilized enzyme yields. Note that for pure silica zero point charge (z.p.c.) is around 2. Thus for values above this, the silica surface will be negatively charged. Another advantage of silica gels is that they provide interesting tunable suface area and porosity. Template synthesis in the presence of organic molecules (surfactant, block copolymers, ...) allows the formation of highly ordered mesoporous silica with uniform pores or a cage-like structure interconnected via restrictions. Nanostructured silicas with pore sizes from 2 to 10 nm have been described (MCM-41, SBA-15, FDU) [143–145]. Emulsions appear also as an efficient template to generate larger pores (15–40 nm), leading to mesocellular foams [146].

Generally, an enzyme is immobilized by building a porous silica gel network by chemical condensation around the biological species. One of the main advantages of the sol–gel encapsulation is that, in varying the nature of the sol gel precursors (silicates, alkoxides, alkoxisilanes ...) and synthetic condtions [147–149], physical and chemical properties of the resulting sol–gel matrix, such as porosity [150], hydrophobicity/hydrophilicity [151,152] wettability [153] and so on can be tuned, leading to a vast number of inorganic and inorganic–organic hybrid sol–gel matrices. The sol–gel process has the ability to be adapted in regard to enzyme preservation, avoiding a too high level of alcohol [154] or too acidic conditions, which have, in some cases, a deactivating effect. One approach consists of first performing the acid-catalyzed hydrolysis step without the enzymes, which are introduced in the second condensation step [155,156]. A two-step strategy is also reported as an appropriate way to evaporate the alcohol produced by the hydrolysis before the introduction of enzymes. Ballesteros *et al.* describe a method to surmount the problem of alcohol release during the alkoxide hydrolysis by using more biocompatible precursors such as poly(glycerylsilicate) which liberate bioprotective alcohols (i.e., glycerol) after auto-hydrolysis [157]. Other silicas derived from more biocompatible precursors [158,159] (i.e., diglyceryl silane, gluconamidylsilane) have been proposed by Brennan *et al.* Such sol–gel precursors lead to highly porous bio-hybrid materials with activities approaching those of the free enzyme.

As an alternative to the sol–gel routes using alkoxides, other approaches using an aqueous process and aqueous precusors such as silicates or colloidal silica have been investigated [160,161]. A biomimetic approach is described by Wright and coworkers, employing a water-soluble template (PAMAM dendrimers) to drive the encapsulation of glucose oxidase in silica spheres starting from $Si(OH)_4$, with high loading and excellent long term stability [162]. In parallel, a modified sol–gel process in a water/oil microemulsion is proposed by Cellesi and Tirelli to minimize the unfavorable conditions and produce horseradish peroxidase encapsulated in silica gel nanoparticules [163]. In other studies, the ability of peptides [164] or proteins (such as lysozyme, lipases ...) [165] to catalyze the formation of silica has been used as a benign encapsulation method. Indeed the biosilification reaction can then serve as a more biologically compatible method for the entrapment of enzymes. Silaffin polypeptides from diatoms catalyze the precipitation of biosilica from silicic acid

[166,167]. Stone *et al.* use this method to simultaneously synthesize the silica and entrap the butyrylcholinesterase which retains all its activity after the process of encapsulation, a high enzyme loading (90 %) is reached and the stability is increased [168]. The method has been further developed to simultaneously entrap catalase and horseradish peroxidase with inorganic magnetic nanoparticles [169] which will facilitate the separation [170,171].

Since the nature of the precursors used during the sol–gel process influence strongly the hydrophilicity/hydrophobicity and porosity of the resulting sol–gel materials, various functional alkoxysilanes have been used. With regard to the structure and mode of action of lipases, Reetz and coworkers have shown that a hydrophobic sol–gel material obtained from alkyl and aryl siloxanes precursors (R–Si $(OR')_3$), is more suitable than a pure silica matrix improving the lipase activity to 400–8800 % [172,173]. It appears that in these hydrophobic organically modified gels (so-called ambigels) the unfavorable [174] contraction observed during the drying and conversion from hydrogel to xerogel is strongly decreased or even suppressed, the xerogel being generally preferred in regard to mechanical properties and chemical resistance. Note that to avoid the drawback of drying, aerogel obtained after supercritical drying can be used to entrap enzymes [175–177] and even self-organized protein superstructures [178].

A wide range of additives can also be introduced into the sol–gel matrices in order to modulate the hydrophobicity of the materials and to improve enzyme stability, activity and accessibility, leading to hybrid or even composite sol–gel matrices. Polymers [157,179,180] such as polyethyleneglycol, polyvinylpyrrolidone, polyvinylalcohol, polyglycidol, polyethyleneimine, polyacrylate have been simultaneously entrapped with enzymes in a siloxane matrix, as well as organic additives (sugar, amino acid) [181,182]. Inclusion of inorganic compounds such as graphite, quartz fiber felt, clays and metal oxides is another way to modify the porosity of the sol–gel matrices [157,183]. More recently, ionic liquids have also been used as additives. Lee and coworkers explain that ionic liquids may act as a template during the sol–gel process reducing the shrinkage of the matrix by pore filling and behave as a stabilizer to protect *Candida rugosa* lipases [184].

With regard to biosensor applications, a wide variety of electrochemically active species (ferrocene, ruthenium complexes, or carbon and metal (Pt, Pd, Au . . .) [185,186] were also introduced into the sol–gel matrices or adsorbed to improve the electron transfer from the biomolecules to the conductive support [187,188]. For instance, glucose oxidase has been trapped in organically modified sol–gel chitosan composite with adsorbed ferrocene to construct a low-cost biosensor exhibiting high sensitivity and good stability [189].

Intermediate strategies between organic polymer (chitosan, dextran, alginate, gelatin) and sol–gel encapsulation have been developed to improve the immobilization of enzymes or, more widely, biological entities, through the formation of composites or nanocomposites [190–193]. Enhancement of the chemical and mechanical stability of alginate microcapsules [194,195] can be achieved by silica coating whereas b- galactolsidase entrapped inside is not denatured. To perform the coating, the charged surface of the biopolymers must be compatible with the silica charge.

Since both are negatively charged, a first coating of the capsules' surface with L-polylysine is needed.

Nanostructured silica and ordered mesoporous silicas have been envisaged as small enzyme immobilization supports [196]. The encapsulation approach is required either to further immobilize enzymes adsorbed in the channels by reducing the pore opening by further silylation or by encapsulating the enzyme directly [197].

Encapsulation via the layer-by-layer assembly of multilayered polyelectrolyte (PE) or PE/nanoparticle nanocomposite thin shells of catalase in bimodal mesoporous silica spheres is also described by Wang and Caruso [198]. The use of a bimodal mesoporous structure allows faster immobilization rates and greater enzyme immobilization capacity (20–40 wt%) in comparison with a monomodal structure. The activity of the encapsulated catalase was retained (70 % after 25 successive batch reactions) and its stability enhanced.

Lower and coworkers [199] have investigated the adsorption of T4 lysozyme on colloidal silica. It was observed that the enzymatic activity decreased upon adsorption due to the differences in adsorbed enzyme structure and orientation as well as the electrostatic effects.

Diaz and Balkus [200] were the first to evaluate the adsorption of biomolecules on mesoporous silica (MCM-41) evidencing the influence of the enzyme molecular dimensions on the adsorption capacity, the pore size limiting the enzyme diffusion according to their molecular weights. The larger the pores, the greater the adsorption of enzyme with high molecular weight. This feature is of particular interest for application in enzyme separation. A study on the adsorption of lysozyme on mesoporous MCM-41 and SBA-15 [201] has shown that the adsorption fitted the Langmuir model displaying a type L isotherm at pH 6.5 to 10.5 while an S type is recorded at pH 12, indicating that the interaction between the lysozyme molecule and the silica surface is weak compared to the interaction between enzymes. The introduction of aluminum into the sol–gel matrix increased the lysozyme amount adsorbed (580 mg/g). The introduction of functional groups on silica appears as an alternative to modulate the surface interaction with the enzyme and then the adsorption [202].

15.4.2.2 **Immobilization on Alumina**

The alumina and titania sol–gel synthesis methods may be used for encapsulation of proteins, leading to gel structures less fragile than silica [203–207]. Encapsulation of enzymes in alumina sol–gel membranes has been widely developed for the realization of amperometric biosensors [208–215]. Heule *et al.* [203]reported the encapsulation of horseradish peroxidase in a ceramic microstructure of porous Al_2O_3. Wu *et al.* [216] extended this method for the realization of microfluidics enzyme-reactors where alumina and titania are loaded with bovine serum albumin (BSA) or trypsin. Immobilization of catalase on alumina [208,209,217] was performed following a cross-linking process. In the working pH conditions (7.0), catalase, due to its low i.e.p. (5.4), is negatively charged and hence favorably adsorbed by the positive surface of alumina (z.p.c. = 8.0). However, while the electrostatic interactions are favorable, the amount of catalase immobilized is very low (3.4 mg/g) and a chemical treatment of the alumina surface is necessary [210]. Treatment with $FeCl_2$ allows the creation of anchoring sites at the surface to bind the biomolecules and to increase the immobilization yield to 17.3 mg/g. It is reported that, while catalase is inactive when immobilized on alumina, the activity is promoted if adsorbed on $FeCl_2$-treated Al_2O_3. Cavaco-Paulo *et al.* [65,208] investigated the grafting of catalase (Bacillus SF) on alkoxylsilane-treated alumina. Alumina pellets were functionalized by silanization with γ-aminopropyltriethoxysilane and subsequently treated with glutaraldehyde. Immobilization of catalase on a thus modified alumina leads to a high immobilization yield (36.9 % from initial enzyme loading) and a maximum adsorption of 16 mg/g, a result similar to that obtained with a polymer host structure. The pH of optimum activity (pH = 8.0) was retained after immobilization. However, at pH over 10.0 the entrapped enzyme maintained more than 60 % of its optimal activity while it was lost for free enzyme. The thermal stability was not much improved by immobilization.

As already mentioned, the sol–gel process is a very useful technique for the co-immobilization of two enzymes. Chen *et al.* [210] demonstrated that when BSA was co-immobilized with glucose oxidase (GOD) in a sol–gel host the sensitivity to glucose was nearly double, allowing the realization of highly sensitive Al sol–gel/GOD/BSA electrodes. This good performance was explained by the large surface area and nanoporous character of the boehmite derived γ-alumina and the protective behavior of the BSA to GOD activity in the sol–gel environment. Such immobilization allows the realization of composite membranes which display a very good response to glucose (90 % of sensitivity) and an excellent linearity up to 35 mM.

An alumina matrix may be prepared with high pore density (more than 60 %) and pore diameters ranging from 5 to 250 nm. Ruiz-Hitzky *et al.* [214] immobilized GOD in nanoporous alumina membranes with regular hexagonal arrays of highly ordered cylindrical pores aligned perpendicularly to the membrane surface. GOD was anchored in the membrane by the highly hydrophilic chitosan biopolymer. Full activity was maintained for at least 50 hours.

The enzymatic hydrolysis of starch to low-molecular weight products (carbohydrates) is of great interest in the food, paper, and textile industries. α-Amylase is under investigation for such reactions. Reshmi *et al.* [215] immobilized α-amylase onto alumina by a direct adsorption procedure in phosphate buffer. α-Amylase was only adsorbed (10 mg/g) on the external surface of the support, as shown by the surface area decrease after immobilization. The disappearance of OH stretching bands evidences that α-amylase adsorbed via the OH groups. The high binding constant values (K_m) exhibited by the adsorbed enzyme were explained by some conformational changes or diffusional resistances. V_{max} values for the immobilized systems were lower than that of the free enzyme, indicating a lowering of activity of the enzyme on account of immobilization. At pH greater than 8.0 the lower loading of α-amylase and the probable change in conformation reduce the activity by nearly 50 %. The stability of the confined amylase was increased at higher pH.

15.4.2.3 Immobilization in Zeolite

Zeolites possess several features useful for the immobilization of enzymes [32]. They are insoluble host structures with a high surface area. Zeolites are microporous aluminosilicate materials with a 3-D oxygen framework carrying an excess negative charge compensated by exchangeable cations and possessing central cages with a free diameter up to 13 Å. They are cation exchangers with tunable acid–base properties. They display properties ranging from hydrophobic to highly hydrophilic, useful for enzymatic catalytic reactions in which diffusion of substrates and products involves molecular transfers from hydrophobic to hydrophilic domains. They have a high capability to adsorb free radicals which are known to inactivate some enzymes such as tyrosinase. The effects of the zeolite properties on the immobilization of enzymes have already been investigated [217].

Zeolites have been investigated as solid supports for the immobilization of enzymes used for biocatalysis reactions in nonaqueous media (organic solvents, supercritical fluids). Immobilization of enzymes onto zeolites is seen as a strategy to retain an optimal water amount necessary for the enzyme activity but not enough to promote hydrolytic activity. Indeed the water content of the support is also a paramount parameter when a catalytic reaction must be carried out in a nonaqueous media. Zeolites may be used as a water sorbent to prevent the accumulation of water in enzyme-catalyzed esterifications performed in organic media. However, in addition to higher synthetic yields, improved enzyme performance has also often been observed in the presence of zeolites. This may often be due to a fortunate coincidence of events: zeolites act as bases and commonly used enzymes (lipase, esterases, proteases) usually impair catalysts in acidic conditions. Zeolite may be used to tune both the water activity a_w and the ionization state of the enzyme. Zeolites with increasing basicity and with larger pores facilitating the inner ion-exchange process would probably be extremely efficient in promoting enhanced catalytic activities of this type of enzyme. The enzyme catalytic ability can be modified due to deviations from the native form, concerned with possible substitutions of amino acid residues, variations in the protonation state, in the partially but stable enzyme states, or dissociation. Moreover, immobilization on zeolite molecular sieves or nanocomposite membranes made with zeolites can favor the ionization state of the enzyme in a more basic form and then increase the catalytic activity [218].

The concept of zeolite action was tested in a particular reaction where the enzyme is exposed from the beginning to an acidic environment: the esterification of geraniol with acetic acid catalyzed by *Candida antarctica* lipase B immobilized on zeolite NaA [219]. Lipases have been used for the hydrolysis of triglycerides and due to their ambivalent hydrophobic/hydrophilic properties they are effective biocatalysts for the hydrolysis of hydrophobic substrates [220]. When water-soluble lipases are used in organic media they have to be immobilized on solid supports in order to exhibit significant catalytic activity.

Hydrolysis of tricaprylin in trimethylpentane by Fusarium solani pisi recombinant cutinase immobilized on various zeolites (NaA, NaX, NaY, LZY-82, dealuminated Y) was investigated in order to assess the effect of chemical composition (Si/Al ratio), hydrophilic character and acidity on the catalytic activity [221]. The adsorption of

5–50 mg of enzyme per g of zeolite was performed by simple contact for 1 min with stirring. Coverage of the external surface of the zeolite never reached the surface saturation (<78 %), even at high loading. With regard to the high values of the immobilization content (even for high acid support), cutinase shows a very high affinity towards the zeolite's surface. The chemical composition of the support did not affect significantly the activity but strongly affected the selectivity. For NaY, NaX, NaA, CaA, KA, CsA, LiA, $HNaZSM5_1$, $NaZSM5_1$, silica gel and alumina supports, adsorption ranged from 42 to 72 % [222]. The amount of cutinase immobilized on the various supports was found to be similar before and after reaction so that no significant desorption occurs. The global activities may be substantially increased for a high yield of enzyme immobilization (up to 25 mg/g). Measurements of the specific activity versus water activity show that a certain amount of water is necessary to hydrate the enzyme and allow it to function properly but as more water is made available, the hydrolytic transformation by water begins to compete with the reaction under study, resulting in a decrease in the alcoholysis reaction rate. Then the optimal water content for the alcoholysis reaction in the organic medium should be a compromise between the required hydration of the enzyme and a low hydrolytic activity. It was found equal to $a_w = 0.97$, approximately the same whatever the zeolite. The a_w value increases in the following order: $KA \approx CsA \approx NaA < LiA < CaA$: the higher the charge density of the cation ($Cs < K < Na < Li < Ca$) the higher the water content for optimal activity will be. On the other hand, the increase in the acidity of the support drastically reduced the catalytic activity. This behavior was explained by the direct interaction between acid sites of the zeolite with the enzyme or ion exchange reaction with the buffer solution. The negative effect of the acidity was clearly demonstrated by comparison of $NaZSM5_1$ with the strong Bro?nsted acid $HNaZSM5_1$ form wich displays nearly no activity. Cutinase immobilized on zeolite NaYpresents the highest activities for the alcoholysis reaction and the best stability [223]. Cutinase immobilized on zeolite NaYmaintains a very high activity even after a 150 °C thermal treatment.

Fontes *et al.* [224,225] addressed the acid—base effects of the zeolites on enzymes in nonaqueous media by looking at how these materials affected the catalytic activity of cross-linked subtilisin microcrystals in supercritical fluids (CO_2, ethane) and in polar and nonpolar organic solvents (acetonitrile, hexane) at controlled water activity (a_w). They were interested in how immobilization of subtilisin on zeolite could affected its ionization state and hence their catalytic performances. Transesterification activity of substilisin supported on NaA zeolite is improved up to 10-fold and 100-fold when performed under low a_w values in supercritical-CO_2 and supercritical-ethane respectively. The increase is also observed when increasing the amount of zeolite due not only to a dehydrating effect but also to a cation exchange process between the surface proton of the enzyme and the sodium ions of the zeolite. The resulting basic form of the enzyme enhances the catalytic activity. In organic solvent the activity was even more enhanced than in sc-hexane, 10-fold and 20-fold for acetonitrile and hexane, respectively, probably due to a difference in the solubility of the acid byproduct.

Production of drugs from enzyme activity is seen as an interesting and selective alternative to chemical production. Production of L-DOPA by tyrosinase was particularly investigated [226]. Before immobilization of the enzyme, the support was

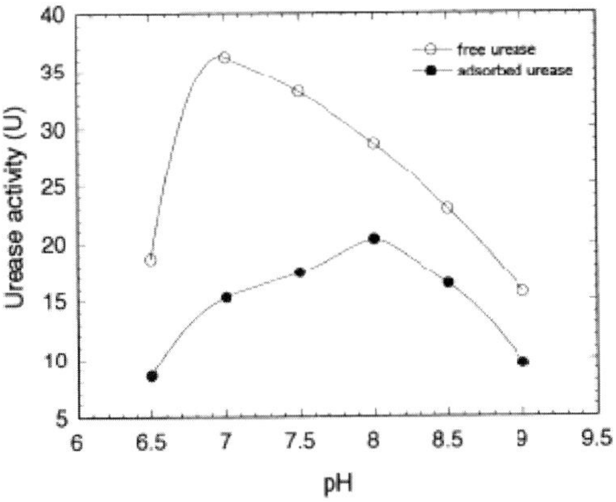

Fig. 15.7 Activity of urease free and adsorbed on hydroxyapatite versus pH from Ref. [228].

impregnated with a glutaraldehyde solution buffered at pH 7.0, washed with water and dried overnight at room temperature. Immobilization of tyrosinase on Na or Ca aluminosilicate was performed by incubation of the modified zeolite with the enzyme solution for 24 h. The enzyme uptake was similar on zeolite NaA (82 %) and CaA (85 %) and desorption was negligible. An average of $121 \pm 9\,mg\,L^{-1}\,h^{-1}$ of L-DOPA was produced. While soluble tyrosinase has an extremely short half-life ($t_{0.5} = 50$ min with ascorbate), NaA and CaA are excellent supports for stabilization leading to a stable activity after 48 and 40 h, respectively.

15.4.2.4 **Immobilization in Hydroxyapatite and Tricalciumphosphate**

For 3-D inorganic solids the higher dimensionality of the networks often limits the adsorption because of lower potential surface area. For example, the amount of urease adsorbed on a 3-D hydroxyapatite biocompatible framework does not exceed $55\,mg\,g^{-1}$ of solid [227,228]. The decrease in adsorbed capacity with increasing pH evidences that urease and hydroxyapatite with oppositely charged surfaces display pH-dependent interactions. Negative charges appear on both surfaces with increasing pH, leading to repulsive interactions. The optimal pH for maximum activity was shifted about one unit higher for adsorbed urease (8.0) compared to free enzyme (7.0) (Figure 15.7). This change has been attributed to the charge of the surface combined with the hydration level.

Tricalcium phosphate was also used as an enzyme embedding matrix. Das and coworkers [229] demonstrated that acid phosphatase and amylase immobilized on $Ca_3(PO_4)_2$ retained their activities with increased thermal stability.

15.5
Enzyme–Host Structure Interactions

Immobilization of the enzyme in an inorganic matrix may lead to loss of activity due to denaturation of the protein. Such denaturation may arise when changes in the enzyme conformation, protonation, charge distribution or ion exchange reaction occur. Weak interactions between the enzyme and the host structure are often a guarantee of enzyme structure preservation. Various spectroscopic and electrochemical investigations have been useful in elucidating the enzyme micorenvironements and the physico-chemical properties of biomolecules embedded in inorganic matrices. For example, the conformation and dynamics of biomolecules embedded in sol–gel matrices have been described in various papers and review [230–233]. It appears that, in most cases, the encapsulation restricts the enzyme diffusional movements but allows sufficient local conformational transitions needed by the enzymatic reactions (binding and release of substrates). Nevertheless, the aging of the sol–gel matrices must be considered, it induces an increase in the cross-linking and the solvent expulsion which is accompanied by important shrinkage which may limit the diffusion and alter the enzyme properties. The forces binding enzymes to support surfaces include electrostatic, hydrogen bonding and Van des Waals interactions. Electrostatic interactions play an important role, being the strongest. The surface adsorption capacities of the enzymes depend on pH, the maximum occurring at the p*I* of the protein, the enzyme adsorption being strongly conditioned by the surronding solution pH.

Modeling of enzyme adsorption isotherms on an inorganic support using the Scatchard equation Equation (1) allow one to determine the binding constant.

$$1/C_f = (K_b.n)\times 1/r - K_b \qquad (1)$$

with C_f being the free protein concentration, K_b the binding constant, *n* the binding size and *r* the binding density.

The binding density is the concentration of bound enzyme over the concentration of binding sites. The binding constant K_b is related to the affinity between the enzyme and the matrix, the greater K_b the higher the affinity. Figure 15.8 displays the Scatchard plot of Mb adsorbed on (A) α-ZrP and (B) magadiite, from Refs. [86] and [133], respectively.

The calculated binding constants are almost the same, $2 \times 10^5\ M^{-1}$ for Mb/α-ZrP and $4.6 \times 10^5\ M^{-1}$, for Mb/magadiite as that of chymotrypsin adsorbed on α-ZrP. In the case of Hb, higher values were measured 5.4×10^6 and $1.6 \times 10^6\ M^{-1}$ for Hb/α-ZrP [86] and Mb/magadiite [133], respectively. These differences may be related to

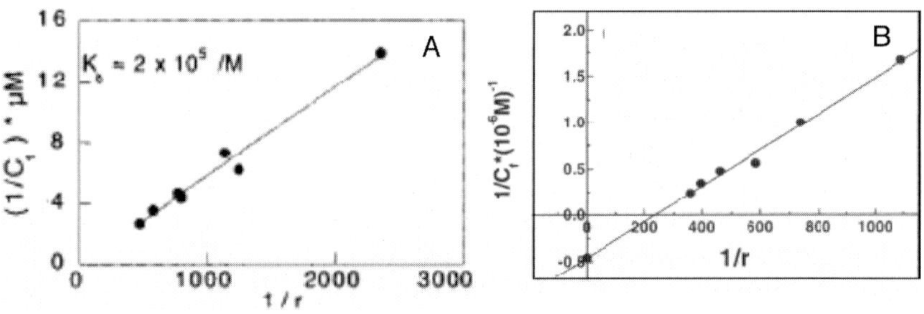

Fig. 15.8 Scatchard plots of myoglobin adsorbed on (A) α-ZrP and (B) magadiite, from Refs. [86] and [133]. respectively.

the difference in size between Mb ($3 \times 4 \times 4\,nm^3$) and Hb ($5.3 \times 5.4 \times 6.5\,nm^3$). Isothermal titration calorimetry (ITC) [234] is also a convenient method to quantify the thermodynamics of protein binding to the solid support and then evaluate the contribution of enthalpic/entropic terms to protein binding. The heat released during the titration is related to the binding constant, the number of binding sites available, and the thermodynamic binding parameters.

Infrared spectroscopy is one of the most frequently used characterization techniques to monitor the entrapment of enzyme in mineral matrices. Enzymes display a characteristic feature in the frequency range of amide groups. The amide (I, II, III) vibrational bands in the region 1000–1700 cm^{-1} are responsive to the hydrogen bond network in the environment of the amide groups. Indeed, protein denaturation modifies the interactions to a significant extent, and the amide vibrational frequencies shift. Accordingly, the amide I (assigned to the stretching of the carbonyl coupled to the (C–N) and amide II vibrational bands are used as sensitive markers of protein structural changes. Upon denaturation of the α-helix, for example, the amide band I shifts from 1650 to 1640 cm^{-1}, while the amide band II shifts to higher frequencies from 1550 cm^{-1}. Chaudhari *et al.* [136] showed that the infrared spectra of Lyz, Hb, Mb and GOD intercalated on α-ZrP are nearly superimposed on those of the native proteins, indicating that the secondary structure of the proteins is retained on encapsulation. α-Chymotrypsin was used as an enzyme model by Baron *et al.* [83,84] to quantify its adsorption on Montmorillonite and investigate the structural effect of the immobilization by FTIR spectroscopy. They pointed out, for pH around the i.e.p. (4.8), the strong effect of the hydrophilic and electronegative surface of montmorillonite on the protonation rate of Asp and Glu side chains which induces the unfolding of both external and internal α-helix domains of the BSA protein. On a hydrophobic talc surface, the pH dependent unfolding involves only external chains, adsorption on talc occurs via peripheral hydrophobic regions of the BSA which maintains its internal conformation.

Very few papers have been devoted to the investigation of the structural changes of an enzyme under immobilization in inorganic host structures by solid-state NMR. Probably due to the high level of technical challenge to get high-resolution spectra for non-liquid systems. Sanjay *et al.* [235,236] evidence by solid state ^{27}Al NMR that the adsorption of α-amylase on K-10 montmorillonite affects the Al tetrahedral sites while covalent binding occurs exclusively on the octahedral sites. More recently an NMR study [237] demonstrated for the first time that it is possible to obtain high resolution for proteins (Cyt-c, Cyt-b_5) entrapped in a sol–gel matrix, showing that the embedded proteins show little conformational change.

Enzyme structure may be studied by fluorescence spectroscopy [238–244]. Excitation in the 280–310 nm absorption bands of proteins, usually results in fluorescence from tryptophan (Trp) residues in the 310–390 nm region. The fluorescence from the Trp residues is a convenient marker for protein denaturation and large decreases or red-shifts in fluorescence are observed when proteins are denatured. These changes are most often due to the exposure of the Trp residues that are buried in the protein and may be due to the changes in the proximities of specific residues that may act as fluorescence quenchers. Fluorescence emission characterization of the immobilized cutinase has confirmed that quite different conformations can be induced by different supports, with important consequences for the enzyme activities.

For proteins, a common method of characterization is far-ultraviolet (far-UV, 170–250 nm) circular dichroism spectroscopy. The chiral nature of the protein building blocks and their arrangement in the secondary structure result in characteristic circular dichroism (CD) spectra [244]. All α-proteins show two strong negative ellipticities at 222 and 208–210 nm and a strong maximum at 191–193 nm, which are characteristic of α-helix. The intensity of the 3 CD bands reflects the amount of helicity of the protein. B-sheets exhibit a single broad negative peak at 212 nm and random coils shaow a strong sharp negative peak at 195 nm. Transition of secondary structure can be evaluated by α-helix content from CD results. However, the CD method is not able to provide information on the three-dimensional structure of the proteins.

In order to get a better insight into the local structure in contact with the enzyme, X-ray absorption spectroscopy (XAS) is a good tool. XAS was used to study the encapsulation of urease during the coprecipitation $Zn_RAl(OH)_6$–urease ($R=2$, 3) [117]. FT curves for Zn_3Al–urease phases with $Q=1/5$, 2 and 3 are displayed in Figure 15.9. The occurrence of the characteristic peaks (P1 to P4) confirms the formation of the LDH phases. However, the presence of the enzyme implies some modifications. Indeed, the urease-encapsulated phases prepared by coprecipitation with an amount $Q=1/5$ present an extracted signal superimposable on the reference material $Zn_3Al(OH)_6Cl \cdot nH_2O$. For a larger amount of urease $Q=2$, and 3, the main characteristic features are maintained but additional noise is recorded at high wave vector k over 10 Å^{-1}, that contributes to enlarging and shifting the peak on the FT curves. Qualitatively, Figure 15.9 shows that the oscillations of LDH–urease phases

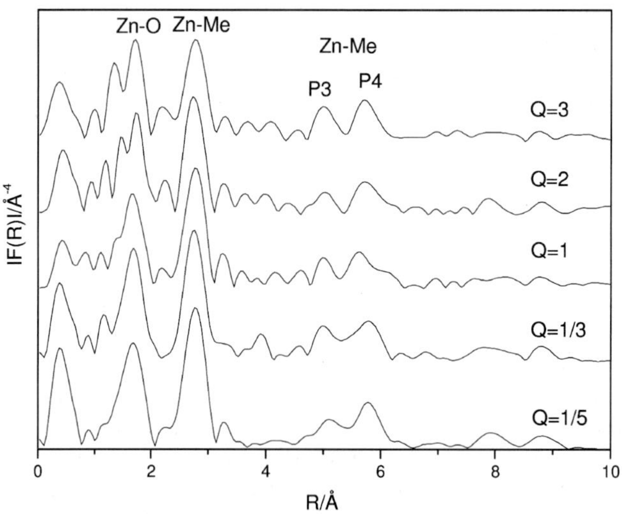

Fig. 15.9 Fourier transform spectra (Zn K-edge): Zn_3Al-urease with $Q=1/5$ to 3. Distances are given without phase shift corrections, from Ref. [117].

are quite superimposable on $Zn_3Al(OH)_6Cl \cdot nH_2O$ and suggests, from a local aspect, that the LDH framework is still the major component obtained by coprecipitation in the presence of the enzyme.

From the point of view of the morphology of enzyme-based biohydrid materials, it was observed that adsorption of enzyme at the surface of crystallites has little effect on the overall textural properties. However, when using the direct encapsulation process, either sol–gel or precipitation routes, the resulting morphology evidences the strong impact that the presence of enzyme in the reaction medium has on the gelation or nucleation/crystal growth. When biohybrid host–enzyme materials are directly prepared, the interactions between both partners develop at the nanoscale interface, leading to a homogeneous dispersion of the enzyme in the inorganic solids. For example, hybrid $Zn_RAl(OH)_6$–urease LDH particles prepared by coprecipitation [117] have a great tendency to preferential orientation, and the platelets aggregate face to face stuck together by urease enzymes (Figure 15.10).

Thick films are formed by a stacking of parallel particles. Individual platelets are still observed with a preferential parallel orientation in the film and a random orientation at the surface of the aggregates (Figure 15.10). These morphologies favor a strong association between urease and the ZnAl layers. Lyophilized biohybrid materials display a more porous morphology. The biomembrane may be flexible, as observed for the immobilization of α-chymotrypsin in the $Zn_3Al(OH)_6$ LDH matrix Figure 15.10 [245].

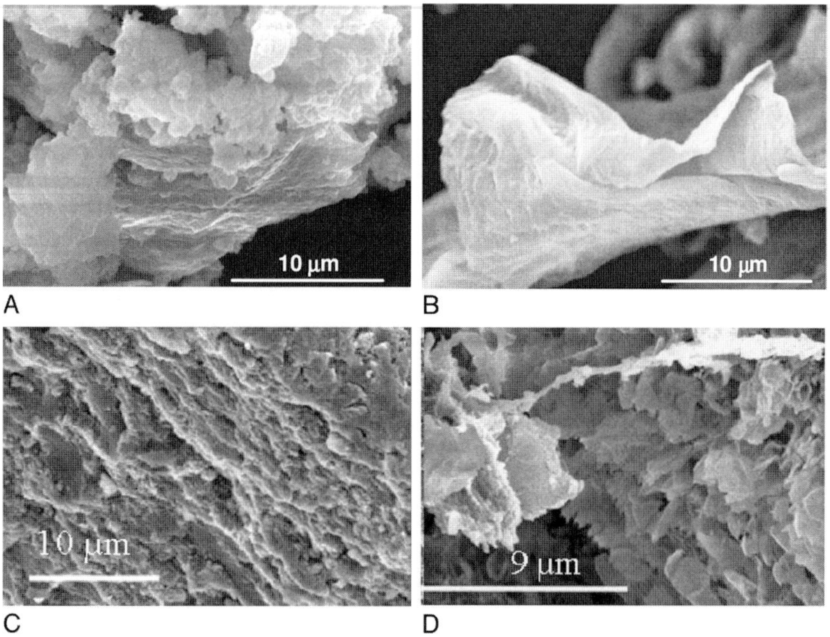

Fig. 15.10 SEM images of [Zn_3Al]-α-chymotrypsin prepared by (A) adsorption and (B) delamination and Zn_3Al-Cl-urease ($Q = 1$) prepared by coprecitation (C) air-dried and (D) lyophilized.

References

1 Schwartz, Mel (ed)(2002) Encyclopedia of Smart Materials, Volumes 1–2, John Wiley & Sons, New York.

2 Crowson, Andrew (ed)(1996) Smart Structures and Materials 1996: Smart Materials Technologies and Biomimetics, Proceedings of SPIE, Vol. 2716.

3 Liming, Dai (ed)(2004) Intelligent Macromolecules for Smart Devices: From Materials Synthesis to Device Applications, Springer-Verlag, Berlin, pp. 496.

4 Buchholz, K., Kasche, V. and Bornscheuer, U.T. (eds)(2005) Biocatalysts and Enzyme Technology, Wiley-VCH, Weinheim.

5 Cavaco-Paulo, A. and Gubitz G. (eds) (2003) Textile Processing with Enzymes, Woodhead Publishing, Cambridge.

6 Durán, N., Esposito, E., Innocentini-Mei, L.H. and Canhos, V.P. (1994) *Biodegradation*, **5**, 13–19.

7 Ribeiro, C.C., Barriasa, C.C. and Barbosa, M.A. (2004) *Biomaterials*, **25**, 4363–4373.

8 Duan, C. and Meyerhoff M.E. (1995) *Microchimica Acta*, **117**, 195–206.

9 Kandimalla, Vivek Babu, Tripathi, Vijay Shyam and Ju, Huangxian (2006) *Analytical Chemistry*, **36**, 73–106.

10 Ramsey, G. (1998) Commercial Biosensors: Applications to Clinical, Bioprocess, and Environmental Samples, Wiley-Interscience, New York.

11 Soldatkin, A.P., Burbriak, O.A., Starodub, N.F., El'skaya, A.E., Sandrovskii, A.K., Shul'ga, A.A. and Strikha, V.I. (1993) *Russian Journal of Electrochemistry*, **29**, 408–414.

12 Jdanova, A.S., Poyard, S., Soldatkin, A.P., Jaffrezic-Renault, N. and Martelet, C. (1996) *Analytica Chimica Acta*, **321**, 35–40.

13 Worsfold, P.J. (1995) *Pure and Applied Chemistry*, **67**, 597–600.

14 Torrey, S. (ed)(1983) Enzyme Technology, Preparation, Purification, Stabilization, Immobilization: Recent Advances, Noyes Publications, Bracknell.

15 Mandal, Saikat, Phadtare, Sumant and Sastry, Murali (2005) *Current Applied Physics*, **5**, 118–127.

16 Alamo, J. (1993) *Solid State Ionics*, **63**, 547–61.

17 Takahiro, S., Motohiro, T. and Hosono, H. (1991) *Journal of Fermentation and Bioengineering*, **72**, 384–391.

18 Commission of enzyme nomenclature, (1984) Enzyme nomenclature, Academic Press, Orlando.

19 Thomas, J. Pinnavaia, Iny, D. Johnson and Lipsicas, Max (1986) *Journal of Solid State Chemistry*, **63**, 118–121.

20 Wang, Yuh-Ruey, Wang, Sea-Fue and Chang, Li-Chung (2006) *Applied Clay Science*, **33**, 73–77.

21 Jackson, M.L. and Abdel-Kader F.H. (1978) *Clays and Clay Minerals*, **26**, 81–87.

22 Garwood, G.A., Mortland, M.M. and Pinnavaia, T.J. (1983) *Journal of Molecular Catalysis*, **22**, 153–163.

23 Rives, V. (2001) Layered Double Hydroxides: Present and Future, Nova Science Publishers, New York.

24 Clearfield, A. and Costantino U. (1996) Layered Metal Phosphates and Their Intercalation Chemistry, in Comprehensive Supramolecular Chemistry (eds G. Alberti and T. Bein) Pergamon Elsevier Ltd Press, Vol. **7**, pp. 107.

25 Clearfield, A. (1997) Metal Phosphonate Chemistry, inProgress in Inorganic Chemistry (ed K.D. Karlin), Wiley & Sons, NewYork, Vol. **47**, pp. 371–510.

26 Ammundsen, B., Wortham, E., Jones, D.J. and Roziere, J. (1998) *Molecular*

Crystals and Liquid Crystals, **311**, 735–740.

27 Inagaki, M., Nakamura, T. and Shimizu, A. (1998) *Journal of Materials Research*, **13**, 896–900.

28 Vakros, J., Kordulis, C. and Lycourghiotis, A. (2002) *Chemical Communications*, 1980–1981.

29 Zhao, D., Feng, J., Huo, Q., Melosh, N., Fredrickson, G.H., Chmelka, B.F. and Stucky G.D. (1998) *Science*, **279**, 548–552.

30 Scott, M.J. and Jones M.N. (2002) *Colloids and Surfaces, A: Physicochemical and Engineering Aspects*, **207**, 69–79.

31 Klein, C.P.A.T., De Groot, K., Vermeiden, J.P.W. and Van Kamp, G.J. (1980) *Biomaterial Research*, **14**, 705–12.

32 Breck, D.W. (1974) Zeolite Molecular Sieves, Wiley, New York.

33 Rouxel, J., Tournoux, M. and Brec, R. (eds)(1994) Soft Chemistry Routes to New Materials: Chimie Douce, Trans Tech Publications, pp. 422.

34 Leprince, F. and Quiquampoix H. (1996) *European Journal of Soil Science*, **47**, 511–522.

35 Burns, R.G., Dick, R.P. and Templeton, D.M. (eds)(2002) Enzymes in the Environment: Activity, Ecology, and Applications, Marcel Dekker, New York, pp. 640.

36 Giles, C.H., Mac Ewan, T.H., Nakhwa, S.N. and Smith, D. (1960) *Journal of the Chemical Society, London*, 3973–3999.

37 Lozzi, I., Calamai, L., Fusi, P., Bosetto, M. and Stotzky, G. (2001) *Soil Biology and Biochemistry*, **33**, 1021–1028.

38 Chevalier, Y., Coche-Guérente, L. and Labbé, P. (2002) *Materials Science and Engineering: C*, 81–89.

39 Kumar, C.V. and Mclendon G.L. (1997) *Chemistry of Materials*, **9**, 863–870.

40 Barhoumi, H., Maaref, A., Rammah, M., Martelet, C., Jaffrezic, N., Mousty, C., Vial, S. and Forano, C. (2006) *Materials Science and Engineering, C: Biomimetic and Supramolecular Systems*, **26**, 328–333.

41 Livage, J., Henry, M. and Sanchez, C. (1988) *Progress in Solid State Chemistry*, **18**, 259–341.

42 Hench, L.L. and West J.K. (1990) *Chemical Reviews*, **90**, 33–72.

43 Livage, J., Coradin, T. and Roux, C. (2001) *Journal of Physics: Condensed Matter*, **13**, R673–R691.

44 Avnir, D., Coradin, T., Lev, O. and Livage, J. (2006) *Journal of Materials Chemistry*, **16**, 1013–1030.

45 Carrado, K.A., Auerbach, A. and Auerbach S.M. (eds) (2004) Handbook of Layered Materials, CRC Press, Boca Raton, FL, pp. 646.

46 Gill, I. and Ballesteros A. (2000) *Trends in Biotechnology*, **18**, 282–296.

47 Jin, W. and Brennan J.D. (2002) *Analytica Chimica Acta*, **461**, 1–36.

48 Pierre, A. (2004) *Biocatalysis and Biotransformation*, **22**, 145–170.

49 Campas, M. and Katakis I. (2004) Trends in Electrochemistry and Corrosion at the Beginning of the 21st Century, (eds E. Brillas and P.-L. Cabot) Universitat de Barcelona, Barcelona, pp. 535.

50 Suxia, Z., Weiwei, Y., Yaming, N. and Changqing, S. (2004) *Sensors and Actuators B*, **101**, 387.

51 Haitao, Z., Hideo, O., Syu, N., Shin-ichiro, S. and Teruo, H. (2004) *Science and Technology of Advanced Materials*, **5**, 371–367.

52 Hou, Y., Tlili, C., Jaffrezic-Renault, N., Zhang, A., Martelet, C., Ponsonnet, L., Errachid, A., Samitier, J. and Bausells, J. (2004) *Biosensors and Bioelectronics*, **20**, 1126–1133.

53 Lvov, Y., Ariga, K., Ichinose, I. and Kunitake, T. (1995) *Journal of the American Chemical Society*, **117**, 6117.

54 Lvov, Y., Ariga, K., Ichinose, I. and Kunitake, T. (1996) *Langmuir*, **12**, 3038–3044.

55 Zhou, Y. and Li, Z. (2002) *Langmuir*, **18**, 8573–8579.

56 Li, Z. and Hu N. (2003) *Journal of Electroanalytical Chemistry*, **558**, 155.

57 O'Regan, B. and Grätzel M. (1991) *Nature*, **353**, 737–740.

58 Rensmo, H., Keis, K., Lindstrom, H., Sodergren, S., Solbrand, A., Hagfeldt, A., Lindquist, S.E. and Muhammed, L. N.M. (1997) *Journal of Physical Chemistry B*, **101**, 2598–2601.

59 Hagfeldt, A., Vlachopoulos, N. and Grätzel, M. (1994) *Journal of the Electrochemical Society*, **141**, 82–84.

60 Huang, S., Kavan, L., Exnar, I. and Grätzel, M. (1995) *Journal of the Electrochemical Society*, **142**, 182.

61 Topoglidis, E., Cass, A.E.G., Gilardi, G., Sadeghi, S. and Durrant, J.R. (1998) *Analytical Chemistry*, **70**, 5111–5113.

62 Topoglidis, E., Lutz, T., Willis, R.L., Barnett, C.J., Cass, A.E.G., Gilardi, G., Sadeghi, S., Beaumont, N. and Durrant, J.R. (2000) *Faraday Discussions*, **116**, 35–46.

63 Topoglidis, E., Campbell, C.J., Cass, A.E.G. and Durrant, J.R. (2001) *Langmuir*, **17**, 7899–7906.

64 Topoglidis, E., Cass, A.E.G., O'Regan, B. and Durrant, J.R. (2001) *Journal of Electroanalytical Chemistry*, **517**, 20–27.

65 Costa, S.A., Tzanov, T., Paar, A., Gudelj, M., Gubitz, G.M. and Cavaco-Paulo, A. (2001) *Enzyme and Microbial Technology*, **28**, 815–819.

66 Tischer Wedekind W. (1999) Biocatalysis-from Discovery to Application, Springer Verlag, Berlin, pp. 95–126.

67 Mousty, C. (2004) *Applied Clay Science*, **27**, 159–177.

68 Ji, H., Wang, Y.-h. and Ma, R.-y. (2004) Beijing Huagong Daxue Xuebao, Ziran Kexueban, 31, 26.

69 de-Melo, J.V., Cosnier, S., Mousty, C., Martelet, C. and Jaffrezic-Renault, N. (2002) *Analytical Chemistry*, **74**, 4037–4043.

70 Shan, D., Cosnier, S. and Mousty, C. (2004) *Biosensors and Bioelectronics*, **20**, 390–396.

71 Park, Man and Komarneni Sridhar (1998) *Microporous and Mesoporous Materials*, **25**, 75–80.

72 Kurusu, Yasuhiko (1995) *Reactive Polymers*, **25**, 63–68.

73 Jin, Ren-Hua and Kurusu Yasuhiko (1992) *Journal of Molecular Catalysis*, **73**, 215–224.

74 Mehdi, A. and Mutin P.H. (2006) *Journal of Materials Chemistry*, **16**, 1606–1607.

75 Weetall, H.H. (1974) *Advances in Lactoferrin Research*, **42**, 191–212.

76 Weetall, H.H. (1993) *Applied Biochemistry and Biotechnology*, **41**, 157–188.

77 Mirabel, US patent 4,258,133 (1981).

78 Le Fèvre, G.N. and Saville B.A. (1999) US patent 5998183.

79 Yuusaku, N., Masahiko, I. and Ryoichi, H. (1989) US patent 4,888,285.

80 Yang, X., Hua, L., Gong, H. and Tan, S.N. (2003) *Analytica Chimica acta*, **478**, 67–75.

81 Yiu, H.H.P., Wright, P.A. and Botting, N.P. (2001) *Journal of Molecular Catalysis B-Enzymatic*, **15**, 81–92.

82 Wang, P., Dai, S., Waezsada, S.D., Tsao, A.Y. and Davison, B.H. (2001) *Biotechnology and Bioengineering*, **74**, 249–255.

83 Kim, M.I., Kim, J., Lee, J., Jia, H., Na, H.B., Youn, J.K., Kwak, J.H., Dohnalkova, A., Grate, J.W., Wang, P., Hyeon, T., Park, H.G. and Chang, H.N. (2007) *Biotechnology and Bioengineering*, **96**, 210–218.

84 Sakai, S., Ono, T., Ijima, H. and Kawakami, K. (2001) *Biomaterials*, **22**, 2827–2834.

85 Bergaya, F., Theng, B.K.G. and Lagaly, G. (eds) (2006) Handbook of Clay Science, Elsevier, Amsterdam.

86 Baron, M.H., Revault, M., Servagent-Noinville, S., Abadie, J. and

Quiquampoix, H. (1999) *Journal of Colloid and Interface Science*, **214**, 319–332.

87 Servagent-Noinville, S., Revault, M., Quiquampoix, H. and Baron, M.H. (2000) *Journal of Colloid and Interface Science*, **221**, 273–283.

88 Lozzi, I., Calamai, L., Fusi, P., Bosetto, M. and Stotzky, G. (2001) *Soil Biology and Biochemistry*, **33**, 1021–1028.

89 Peng, S., Gao, Q., Wang, Q. and Shi, J. (2004) *Chemistry of Materials*, **16**, 2675–2684.

90 Monsan, P. and Durand G. (1971) *FEBS Letters*, **16**, 39–42.

91 Sanjay, G. and Sugunan S. (2005) *Catalysis Communications*, **6**, 81–86.

92 de Fuentes, I.E., Viseras, C.A., Ubiali, D., Terreni, M. and Alcantara, A.R. (2001) *Journal of Molecular Catalysis B-Enzymatic*, **11**, 657–663.

93 Abdul Rahman, M.B., Tajudin, S.Md. Hussein, M.Z., Abdul Rahman, R.N.Z.R., Sallhe, A.B. and Basri, M. (2005) *Applied Clay Science*, **29**, 111–116.

94 Ogawa, S., Shiro, Y. and Morishima, I., (1979) *Biochemical and Biophysical Research Communications*, **90**, 674–678.

95 Gianfreda, L., Rao, M.A. and Violante, A. (1991) *Soil Biology and Biochemistry*, **23**, 581–587.

96 Peng, S., Gao, Q., Du, Z. and Shi, J. (2006) *Applied Clay Science*, **31**, 229–237.

97 Huang, Q., Zhao, Z. and Chen, W. (2003) *Chemosphere*, **52**, 571–579.

98 Lai, C.M. and Tabatabai M.A. (1992) *Soil Biology and Biochemistry*, **24**, 225–228.

99 Gianfreda, L., Rao, M.A. and Violante, A. (1992) *Soil Biology and Biochemistry*, **24**, 51–58.

100 Gianfreda, L., Rao, M.A. and Violante, A. (1993) *Soil Biology and Biochemistry*, **25**, 671–677.

101 Fusi, P., Ristori, L., Calamai, L. and Stotzky, G. (1989) *Soil Biology and Biochemistry*, **21**, 911–920.

102 Naidja, A., Violante, A. and Huang, P.M. (1995) *Clays and Clay Minerals*, **43**, 647–655.

103 Naidja, A., Huang, P.M. and Bollag, J.M. (1997) *Journal of Molecular Catalysis A-Chemical*, **115**, 305–316.

104 Rao, M.A., Violante, A. and Gianfreda, L. (2000) *Soil Biology and Biochemistry*, **32**, 1007–1014.

105 Quiquampoix, H. (1987) *Biochimie*, **69**, 753–763.

106 Stauton, S. and Quiquampoix H. (1994) *Journal of Colloid and Interface Science*, **166**, 89–94.

107 Mbouguen, J.K., Ngameni, E. and Walcarius, A. (2006) *Analytica Chimica Acta*, **578**, 145–155.

108 Choy, J.H. (2004) *Journal of Physics and Chemistry of Solids*, **65**, 373–383.

109 Aisawa, S., Takahashi, S., Ogasawara, W., Umetsu, Y. and Narita, E. (2001) *Journal of Solid State Chemistry*, **162**, 52–62.

110 Kwak, S.Y., Jeong, Y.J., Park, J.S. and Choy, J.H. (2002) *Solid State Ionics*, 229–234.

111 Choy, J.H., Kwak, S.Y., Park, J.S., Jeong, Y.J. and Portier, J. (1999) *Journal of the American Chemical Society*, **121**, 1399–1400.

112 Rahman, M.B.A., Basri, M., Hussein, M.Z., Idris, M.N.H., Rahman, R.N.Z. R.A. and Salleh, A.B. (2004) *Catalysis Today*, **93–95**, 405–410.

113 Rahman, M.B.A., Basri, M., Hussein, M.Z., Rahman, R.N.Z.A., Zainol, D.H. and Salleh, A.B. (2004) *Applied Biochemistry and Biotechnology*, **118**, 313–320.

114 Adachi-Pagano, M., Forano, C. and Besse, J.-P. (2003) *Journal of Materials Chemistry*, **13**, 1988–1993.

115 de Roy, A., Forano, C., El Malki, M. and Besse, J.-P. (1992) Synthesis of Microporous Materials, inExpanded Clays and Other Microporous Solids

(eds M.L. Occelli and H.E. Robson) Van Nostrand Reinhold, New York, pp. 108.
116 Cavani, F., Trifirò, F. and Vaccari A. (1991) *Catalysis Today*, **11**, 173–301.
117 Vial, S., Prevot, V., Leroux, F. and Forano, C. Microporous and Mesoporous Materials, in press.
118 Leroux, F., Adachi-Pagano, M., Intissar, M., Chauviere, S., Forano, C. and Besse, J.-P. (2001) *Journal of Materials Chemistry*, **11**, 105–112.
119 Hibino, T. and Jones W. (2001) *Journal of Materials Chemistry*, **11**, 1321–1326.
120 O'Leary, S., O'Hare, D. and Seeley, G. (2002) *Chemical Communications*, **14**, 1506–1507.
121 Vial, S., Forano, C., Shan, D., Mousty, C., Barhoumi, H., Martelet, C. and Jaffrezic, N. (2006) *Materials Science and Engineering: C*, **26**, 387–393.
122 Ren, J.L., He, J., Evans, D.G., Xue, D.A. and Ma, R.Y. (2001) *Journal of Molecular Catalysis B-Enzymatic*, **16**, 65–71.
123 Ren, L., He, J., Zhang, S., Evans, D.G. and Duan, X. (2002) *Journal of Molecular Catalysis B-Enzymatic*, **18**, 3–11.
124 Mc Kenzie, R.M. (1989) Minerals in Soil Environments (eds J.B. Dixon and S.B. Weed) Soil Sciences Society of America, Madison, WI, pp. 439.
125 Abou-El-Sherbini, K.S., Askar, M.H. and Schollhorn, R. (2002) *Solid State Ionics*, **150**, 407–415.
126 Naidja, A., Liu, C. and Huang, P.M. (2002) *Journal of Colloid and Interface Science*, **251**, 46–56.
127 Ahn, M.Y., Martnez, C.E., Archibald, D.D., Zimmerman, A.R., Bollag, J.M. and Dec, J. (2006) *Soil Biology and Biochemistry*, **38**, 1015–1020.
128 Pal, S., Bollag, J.M. and Huang, P.M. (1994) *Soil Biology and Biochemistry*, **26**, 813–820.
129 Giles, C.H., Smith, D. and Huitson, A. J. (1974) *Journal of Colloid and Interface Science*, **47**, 755.
130 Bullot, J., Cordier, P., Gauthier, M. and Livage, J. (1984) *Journal of Non-Crystalline Solids*, **68**, 123–134.
131 Glezer, V. and Lev O. (1993) *Journal of the American Chemical Society*, **115**, 2533–2534.
132 Tsiafoulis, C.G., Florou, A.B., Trikalitis, P.N., Bakas, T. and Prodomidis, M.I. (2005) *Electrochemistry Communications*, **7**, 781–788.
133 Marshall, A.G. (1978) Biophysical Chemistry, Principles, Techniques and Applications, Wiley & Sons, New York, pp. 70.
134 Kumar, C.V. and Chaudhari A. (2000) *Journal of the American Chemical Society*, **122**, 830–837.
135 Belleza, F., Cipiciani, A., Costantino, U. and Negozio, M.E. (2002) *Langmuir*, **18**, 8737–8742.
136 Belleza, F., Cipiciani, A. and Costantino, U. (2003) *Journal of Molecular Catalysis B-Enzymatic*, **26**, 47–56.
137 Chaudhari, A. and Kumar C.V. (2005) *Microporous and Mesoporous Materials*, **77**, 175–187.
138 Kumar, C.V. and Chaudhari A. (2003) *Microporous and Mesoporous Materials*, **57**, 181–190.
139 Jagannadham, V., Bhambhani, A. and Kumar, C.V. (2006) *Microporous and Mesoporous Materials*, **88**, 275–282.
140 Belleza, F., Cipiciani, A., Costantino, U. and Marmottini, F. (2006) *Langmuir*, **22**, 5064–5069.
141 Geng, LiNa, Li, Na, Xiang, MingHui, Wen, XiaoFan, Xu, Da, Zhao, FengLin and Li, KeAn (2003) *Colloids and Surfaces B: Biointerfaces*, **30**, 99–109.
142 Geng, LiNaWang, Xu, Li, Na, Xiang, MingHui and Li, KeAn (2004) *Colloids and Surfaces B: Biointerfaces*, **34**, 231–238.
143 Beck, J.S., Vartuli, J.C., Roth, W.J., Leonowicz, M.E., kresge, C.T., Schmitt, K.D., Chu, C.T.W., Olson, D.H., Sheppard, E.W., McCullen, S.B.,

Higgins, J.B. and Schlenker, J.L. (1992) *Journal of the American Chemical Society*, **114**, 10834–10843.

144 Kresge, C.T., Leonowicz, M.E., Roth, W.J., Vartuli, J.C. and Beck, J.S. (1992) *Nature*, **359**, 710–712.

145 Zhoa, D.Y., Huo, Q.S., Feng, J.L., Chmelka, B.F. and Stucky, G.D. (1998) *Journal of the American Chemical Society*, **120**, 6024–6036.

146 Schmidt-Winkel, P., Lukens, P.P., Zhao, D., Yang, P., Chmelka, B.F. and Stucky, G.D. (1999) *Journal of the American Chemical Society*, **121**, 254–255.

147 Rao, M.S. and Dave B.C. (1998) *Journal of the American Chemical Society*, **120**, 13270–13271.

148 Chaudhury, N.K., Bhardwaj, R. and Murari, B.M. (2003) *Journal of Current Applied Physics*, **3**, 177–184.

149 Gupta, R., Mozumdar, S. and Chaudhury, N.K. (2005) *Biosensors and Bioelectronics*, **21**, 549–556.

150 Goring, G.L.G. and Brennan J.D. (2002) *Journal of Materials Chemistry*, **12**, 3400–3406.

151 Rassy, H. El., Buisson, S., Bouali, P., Perrard, A. and Pierre, A.C. (2003) *Langmuir*, **19**, 358–363.

152 Hüsing, N., Schubert, U., Misof, K. and Fratzi, P. (1998) *Chemistry of Materials*, **10**, 3024–3032.

153 Tripathi, V.S., Kandimalla, V.B. and Ju, H. (2006) *Sensors Actuators B*, **114**, 1071–1082.

154 Avnir, D. and Kaufman V.R. (1987) *Journal of Non-Crystalline Solids*, **192**, 180–182.

155 Ellerby, L., Nishida, C.R., Nishida, F., Yamanaka, S.A., Dunn, B., Valentile, J. S. and Zink, J.I. (1992) *Science*, **255** (5048), 1113–1115.

156 Avnir, D., Braun, S., Vev, O. and Ottolenghi, M. (1994) *Chemistry of Materials*, **6**, 1605–1614.

157 Gill, I. and Ballesteros A. (1998) *Journal of the American Chemical Society*, **120**, 8587–8598.

158 Besanger, T.R., Chen, Y., Deisingh, A.K., Hodgson, R.J., Jin, W., Mayer, S., Brook, M.A. and Brennan, J.D. (2003) *Analytical Chemistry*, **75**, 20382–2391.

159 Sui, X., Cruz-Aguado, J.A., Chen, Y., Zhang, Z., Brook, M. and Brennan, J. D. (2005) *Chemistry of Materials*, **17**, 1174–1182.

160 Liu D, M. and Chen J.W. (1999) *Acta Materialia*, **47**, 4535–4544.

161 Bhatia, R.B., Brinker, C.J., Gupta, A.K. and Singh, A.K. (2000) *Chemistry of Materials*, **12**, 2434–2441.

162 Miller, S.A., Hong, E.D. and Wright, D. (2006) *Macromolecular Bioscience*, **6**, 839–845.

163 Cellesi, F. and Tirelli N. (2006) *Colloids and Surfaces A-Physicochemical and Engineering Aspects*, **288**, 52–61.

164 Coradin, T., Durupthy, O. and Livage, J. (2002) *Langmuir*, **18**, 2331–2336.

165 Luckarift, H.R., Balasubramanian, S., Paliwal, S., Johnson, G.R. and Simonian, A.L. (2007) Colloids and Surfaces B-Biointerfaces, in press **58**, 28–33.

166 Kroger, N., Deutzmann, R. and Sumper, M. (1999) *Science*, **286**, 1129.

167 Naik, R.R., Whitlock, P.W., Rodriguez, F., Brott, L.L., Glawe, D., Clarson, S.J. and Stone, M.O. (2002) *Journal of Nanoscience and Nanotechnology*, **2**, 95–100.

168 Luckarift, H.R., Spain, J.C., Naik, R.R. and Stone, M.O. (2004) *Nature Biotechnology*, **22**, 211–213.

169 Naik, R.R., Tomczak, M.T., Luckarift, H.R., Spain, J.C. and Stone, M.O. (2004) *Chemical Communications*, 1684–1685.

170 Kim, J., Lee, J., Na, H.B., Kim, B.C., Youn, J.K., Kwak, J.H., Moon, K., Lee, E., Kim, J., Park, J., Bohnalkova, A., Park, H.G., Gu, M.B., Chang, H.N., Grate, J.W. and Hyeon, T. (2005) *Small*, **1**, 1203–1207.

171 Reetz, M.T., Zonta, A., Vijayakrishnan, V. and Schimossek, K. (1998) *Journal of Molecular Catalysis B*, **134**, 251–258.

172 Reetz, M.T. (1997) *Advanced Materials*, **9**, 943–954.

173 Reetz, M.T., Zonta, A. and Simpelkamp, J. (1995) *Angewandte Chemie-International Edition*, **34**, 301.

174 Kawakami, K. and Yoshida S. (1994) *Biotechnology Techniques*, **8**, 441–446.

175 Basso, A., DeMartin, L., Ebert, C., Gardossi, L., Tomat, A., Casarci, M. and LiRosi, O. (2000) *Tetrahedron Letters*, **41**, 8627–8630.

176 Orcaire, O., Buisson, P. and Pierre, A. C. (2006) *Journal of Molecular Catalysis B-Enzymatic*, **42**, 106–113.

177 Pierre, M., Buisson, P., Fache, F. and Pierre, A.C. (2000) *Biocatalysis and Biotransformation*, **18**, 237–251.

178 Wallace, J.M., Rice, J.K., Pietron, J.J., Stroud, R.M., Long, J.W. and Rolison, D.R. (2003) *Nano Letters*, **3**, 1463–1467.

179 Glucev, M.D., Goring, G.L.G., Rakic, M. and Brennan, J.D. (2002) *Analytica Chimica Acta*, **457**, 47–59.

180 Shtelzer, S., Rappoport, S., Avnir, D., Ottolenghi, M. and Braun, S. (1992) *Biotechnology and Applied Biochemistry*, **15**, 227–235.

181 Wei, Y., Xu, J., Feng, Q., Lin, M., Dong, H., Zhang, W.J. and Wang, C. (2001) *Journal of Nanoscience and Nanotechnology*, **1**, (1), 83–93.

182 Brennan, J.D., Benjamin, D., DiBattista, E. and Glucev, M.D. (2003) *Chemistry of Materials*, **15**, 737–745.

183 Buisson, P. and Pierre A.C. (2006) *Journal of Molecular Catalysis B-Enzymatic*, **39**, 77–82.

184 Lee, S.H., Doan, T.T.N., Ha, S.H. and Koo, Y.-M. (2007) *Journal of Molecular Catalysis B-Enzymatic*, **45**, 57–61.

185 Wang, J., Park, D.S. and Pamidi, P.V.A. (1997) *Journal of Electroanalytical Chemistry*, **434**, 185–189.

186 Nagata, R., Yokoyama, K., Durliat, H., Comtat, M., Clark, S.A. and Karube, I. (1995) *Electroanalysis*, **7**, 1027–1031.

187 Willner, I. and Katz E. (2000) *Angewandte Chemie-International Edition*, **39**, 1181–1218.

188 Dave, B.C., Dunn, B., Valentine, J.S. and Zink, J.I. (1994) *Analytical Chemistry*, **66**, 11204A–1127A.

189 Chen, X., Jia, J. and Dong, S. (2003) *Electroanalysis*, **15**, 608–612.

190 Heichal-Segal, O., Rappoport, S. and Braun, S. (1995) *Bio/Technology*, **13**, 798–800.

191 Miao, Y. and Tan S.N. (2001) *Analytica Chimica Acta*, **437**, 87–93.

192 Sakai, S., Ono, T., Ijima, H. and Kawakami, K. (2002) *Biotechnology Progress*, **18**, 401–403.

193 Allouche, J., Boissière, M., Hélary, C., Livage, J. and Coradin, T. (2006) *Journal of Materials Chemistry*, **16**, 3121–3126.

194 Coradin, T., Mercey, E., Lisnard, L. and Livage, J. (2001) *Chemical Communications*, 2496–2497.

195 Boissière, M., Meadows, P.J., Brayner, R., Hélary, C., Livage, J. and Coradin, T. (2006) *Journal of Materials Chemistry*, **16**, 1178–1182.

196 Yiu, H.H.P. and Wright P. (2005) *Journal of Materials Chemistry*, **15**, 3690–3700.

197 Mureseanu, M., Galarneau, A., Renard, G. and Fajula, F. (2005) *Langmuir*, **21**, 4648–4655.

198 Wang, Y. and Caruso F. (2005) *Chemistry of Materials*, **17**, 953–961.

199 Lower, C.K., Sananikone, S., Pothwell, M.K. and Mc Guire, J. (1999) *Biotechnology and Bioengineering*, **64**, 373–376.

200 Diaz, J.F. and Balkus K.J. (1996) *Journal of Molecular Catalysis B-Enzymatic*, **2**, 115–126.

201 Vinu, A., Murugesan, V. and Hartmann, M. (2004) *Journal of Physical Chemistry B*, **108**, 7323–7330.

202 Yiu, H.H.P., Botting, C.H., Botting, N. P. and Wright, P.A. (2001) *Physical Chemistry Chemical Physics*, **3**, 2983–2885.

203 Heule, M., Rezwan, K., Cavalli, L. and Gauckler, L. (2003) *Advanced Materials*, **15**, 1191–1194.

204 Zhang, T., Tian, B.Z., Kong, J.L., Yang, P.Y. and Liu, B.H. (2003) *Analytica Chimica Acta*, **489**, 199–206.

205 Jiang, D.C., Tang, J., Liu, B.H., Yang, P. Y. and Kong, J.L. (2003) *Analytical Chemistry*, **75**, 4578–4584.

206 Liu, Z., Deng, J. and Li, D. (2000) *Analytica Chimica Acta*, **407**, 87–96.

207 Liu, Z., Liu, B., Zhang, M., Kong, J. and Deng, J. (1999) *Analytica Chimica Acta*, **392**, 135.

208 Fruhwirth, G.O., Paar, A., Gudelj, M., Cavaco-Paulo, A., Robra, K.H. and Gubitz, G.M. (2002) *Applied Microbiology and Biotechnology*, **60**, 313–319.

209 Eberhardt, A.M., Pedroni, V., Volpe, M. and Ferreira, M.L. (2004) *Applied Catalysis B: Environmental*, **47**, 153–163.

210 Chen, X., Hu, Y. and Wilson, G.S. (2002) *Biosensors and Bioelectronics*, **17**, 1005–1013.

211 Ballarin, B., Brumlik, C.J., Lawson, D. R., Liang, W., Van Dycke, L.S. and Martin, C.R. (1992) *Analytical Chemistry*, **64**, 2647–2651.

212 Myler, S., Collyer, S.D., Bridge, K.A. and Higson, S.P.J. (2002) *Biosensors and Bioelectronics*, **17**, 35–43.

213 Heilman, A., Teuscher, N., Kiesow, A., Janasek, D. and Spohn, U. (2003) *Journal of Nanoscience and Nanotechnology*, **3**, 375–379.

214 Darder, M., Aranda, P., Hernandez-Velez, M., Manova, E. and Ruiz-Hitzky, E. (2006) *Thin Solid Films*, **495**, 321–326.

215 Reshmi, R., Sanjay, G. and Sugunan, S. (2006) *Catalysis Communications*, **7**, 460–465.

216 Wu, H., Tian, Y., Liu, B.H., Lu, H., Wang, X., Zhai, J., Jin, H., Yang, P., Xu, Y. and Wang, H. (2004) *Journal of Proteome Research*, **3**, 1201–1209.

217 Lima, F.V., Pyle, D.L. and Asenjo, J.A. (1995) *Biotechnology and Bioengineering*, **46**, 69.

218 Hudson, E.P., Eppler, R.K. and Clark, D.S. (2005) *Current Opinion in Biotechnology*, **16**, 637–643.

219 Peres, C., Harper, N., Gomes da Silva, M.D.R. and Barreiros, S. (2005) *Enzyme and Microbial Technology*, **37**, 145–149.

220 Rogalska, E., Ransac, S. and Verger, R. (1990) *Journal of Biological Chemistry*, **265**, 20271–20276.

221 Goncalves, A.P.V., Lopes, J.M., Lemos, F., Ramôa Ribeiro, F., Prazeres, D.M.F., Cabral, J.M.S. and Aires-Barros, M.R. (1996) *Journal of Molecular Catalysis B-Enzymatic*, **1**, 53–60.

222 Serralha, F.N., Lopes, J.M., Lemos, F., Prazeres, D.M.F., Aires-Barros, M.R., Cabral, J.M.S. and Ramôa Ribeiro, F. (1998) *Journal of Molecular Catalysis B*, **4**, 303–311.

223 Serralha, F.N., Lopes, J.M., Aires-Barros, M.R., Prazeres, D.M.F., Cabral, J.M.S., Lemos, F. and Ramôa Ribeiro, F. (2002) *Enzyme and Microbiological Technology*, **31**, 29–34.

224 Fontes, N., Almeida, M.C., Garcia, S., Peres, C., Partridge, J. and Halling, P.J. (2001) *Biotechnology Progress*, **17**, 355–358.

225 Fontes, N., Partridge, J., Halling, P.J. and Barreiros, S. (2002) *Biotechnology and Bioengineering*, **77**, 296–305.

226 Seetharam, G. and Saville B.A. (2002) *Enzyme and Microbial Technology*, **31**, 747–753.

227 Ciurli, S., Marzadori, C., Benini, S., Deiana, S. and Gessa, C. (1996) *Soil Biology and Biochemistry*, **28**, 811–817.

228 Marzadori, C., Miletti, S., Gessa, C. and Ciurli, S. (1998) *Soil Biology and Biochemistry*, **30**, 1485–1490.

229 Das, G. and Prabhu K.A. (1990) *Enzyme and Microbial Technology*, **12**, 625.

230 Dunn, B. and Zink J.I. (1997) *Chemistry of Materials*, **9**, 2280–2291.

231 Audebert, P., Demaille, C. and Sanchez, C. (1993) *Chemistry of Materials*, **5**, 911.

232 Keeling-Tucker, T. and Brennan J.D. (2001) *Chemistry of Materials*, **13**, 3331–3350.

233 Gupta and Chaudhury N. K. (2007) Biosensors and Bioelectronics, in press.

234 Blandamer, M.J., Cullis, P.M. and Engberts, J.B.F.N. (1998) *Journal of the Chemical Society-Faraday Transactions*, **94**, 2261.

235 Sanjay, G. and Sugunan S. (2005) *Clay Minerals*, **40**, 499–510.

236 Sanjay, G. and Sugunan S. (2006) *Catalysis Communications*, **7**, 1005–1011.

237 Wheeler, K.E., Nocek, J.M. and Hoffman, B.M. (2006) *Journal of the American Chemical Society*, **128**, 14782–14783.

238 Lakowitz, J.R. and Weber G. (1973) *Biochemistry*, **12**, 4171.

239 Eftink, M.R. and Ghiron C.A. (1975) *Proceedings of the National Academy of Sciences of the United States of America*, **72**, 3290.

240 Eftink, M.R. and Ghiron C.A. (1981) *Analytical Biochemistry*, **114**, 199.

241 Cantor, C. and Schimmel P.R. (1980) Biophysical Chemistry,W.H. Freeman Company, San Francisco, pp. 11.

242 Haouz, A., Twist, C., Zentz, C., de Kersabiac, A.M., Pin, S. and Alpert, B. (1998) *Chemical Physics Letters*, **294**, 19.

243 Serralha, F.N., Lopes, J.M., Vieira Ferreira, L.F., Lemos, F., Prazeres, D. M.F. and Aires-Barros, M.R. (2001) *Catalysis Letters*, **73**, 63–66.

244 Woody, R.W. (1996) Circular dichroism and conformational analysis of biomolecules (ed G.D. Fasman), Plenum Press, New York, pp. 25–27.

245 Forano, C., Vial, S. and Mousty, C. (2006) *Current Nanoscience, CNANO, Trends in Biohybrid Nanostructured Materials*, **2**, 283–294.

Index

Bio-inorganic Hybrid Nanomaterials. Edited by E. Ruiz-Hitzky, K. Ariga, Y. M. Lvov

ISBN: 978-3-527-31718-9

i

p